Seafloor Observatories

A new vision of the Earth from the Abyss

Paolo Favali, Laura Beranzoli and Angelo De Santis (Editors)

Seafloor Observatories

A new vision of the Earth from the Abyss

Springer

Published in association with
Praxis Publishing
Chichester, UK

PRAXIS

Editors

Paolo Favali
Istituto Nazionale di
Geofisica e Vulcanologia
Rome
Italy

Laura Beranzoli
Istituto Nazionale di
Geofisica e Vulcanologia
Rome
Italy

Angelo De Santis
Istituto Nazionale di
Geofisica e Vulcanologia
Rome
Italy

SPRINGER-PRAXIS BOOKS IN GEOPHYSICAL SCIENCES
SUBJECT *ADVISORY EDITOR*: Philippe Blondel, C.Geol., FGS., Ph.D., M.Sc.,
Senior Scientist, Department of Physics, University of Bath, UK

ISBN 978-3-642-11373-4 ISBN 978-3-642-11374-1 (eBook)
DOI 10.1007/978-3-642-11374-1

Library of Congress Control Number: 2014942448

Springer Heidelberg New York Dordrecht London

Cover design: Jim Wilkie
Project copy editor: Rachael Wilkie
Typesetting: David Peduzzi

Cover images
Front cover clockwise from the top: Mooring deployment; Tempo-mini multi-instrument platform developed by IFREMER; MODUS underwater vehicle for GEOSTAR-class observatory management developed by Beuth University of Applied Sciences, Berlin.
Back cover: Engineer at work on the MODUS vehicle.

Printed on acid-free paper

Springer-Verlag GmbH Berlin Heidelberg is part of Springer Science+Business Media (www.springer.com)

Contents

List of figures

List of tables

To the memory of Giuseppe (Pino) Smriglio,
brilliant scientist,
visionary ahead of his time,
great and unforgettable friend

1 Introduction

P. Favali, L. Beranzoli, and A. De Santis[1]

How can continuous and reliable monitoring at the seafloor and along the water column by means of seafloor observatories extend the exploration and improve knowledge of our planet?

The ocean covers 70% of the planetary surface, and yet we call our planet "Earth" and not – maybe more appropriately – "Ocean". The ocean exerts a pervasive influence on the Earth's environment but its nature is poorly understood. Developing a deep understanding of the ocean's role and elucidating the complex physical, biological, chemical, and geological processes operating therein represent a major challenge in the opening decades of the 21st century. Major questions face us concerning the quality of human life and the destiny of biodiversity in such a rapidly changing environment that makes the ocean the last frontier of our planetary understanding.

The establishment of a global network of ocean observatories to monitor the sub-seafloor, seafloor, and water column has been providing powerful insights into complex processes such as the geo-, bio- and hydrosphere interactions and their evolution and variability through time. However, there is still a long way to go before a comprehensive vision of processes in the oceans is reached. The importance of developing a global underwater network was initially recognized within the geophysical community, in particular by seismologists and geomagneticians, in the mid-1980s when innovative methodologies for studies of the Earth's interior came up against the impossibility of having an even coverage of planetary observations. Many international projects and initiatives have been carried out since the early 1990s to match this new need expressed by the scientific community. These projects have comprised experiments to validate long-term monitoring systems in very different environmental conditions: from shallow water to the deep sea, from coastal areas to open oceans, from temperate to polar latitudes. Since 1978, Japan has been the pioneer nation in managing cabled, and thus permanent, seafloor observatories driven by the need of limiting the damage of earthquakes and tsunamis to the social-economic fabric of the country.

1 Istituto Nazionale di Geofisica e Vulcanologia

Sub-seafloor, seafloor, and water column Observatory Science is now a new branch of Earth (and Ocean) Sciences. It is indeed interdisciplinary in its nature and has the potential to greatly advance important scientific themes: the ocean's influence on climate; the dynamics of the oceanic lithosphere and the imaging of the Earth's interior; the relationship between fluids and life in the ocean crust; man-induced and natural changes in the coastal environment; turbulent mixing and biophysical interactions; ecosystem dynamics and biodiversity. Such a new science offers scientists new opportunities for studying multiple, interrelated natural phenomena over time scales ranging from seconds to decades, from episodic to global and long-term processes. Episodic processes include volcanic eruptions, deep-ocean convection at high latitudes, earthquakes, tsunamis, seafloor instabilities, and biological, chemical and physical impacts of storm events. The deep underwater environment also provides a place to perform astroparticle physics experiments to detect neutrinos and, in turn, potentially disclose the secrets of the deep universe.

The experience gained in the last few decades has paved the way for large-scale international programmes and the permanent installation of systems capable of sustained and long-term observations. Many large-scale projects are currently being undertaken to establish permanent seafloor networks at the international level. Japan, the USA, Canada and Europe are the major present players, but other actors, like China, are now joining in the game. These efforts constitute the sub-sea segment of GEOSS – Global Earth Observation System of Systems (http://www.earthobservations.org/geoss.html). The purpose of GEOSS is to achieve comprehensive, coordinated and sustained observations of the Earth's system in order to improve monitoring of the state of the Earth, increase understanding of Earth processes, and enhance prediction of the behaviour of the Earth system.

This book will open a window on the state-of-the-art of a new science, "Seafloor Observatory Science", with articles from authors belonging to the most active world-wide groups and the most relevant scientists and technologists in the field. It addresses the important and apparently simple question: "How can continuous and reliable monitoring below and at the seafloor and through the water column by means of seafloor observatories extend exploration and improve knowledge of our planet?"

The book will lead the readers through:

1. the present scientific challenges to be addressed with seafloor observatories;
2. the technical solutions for their architecture;
3. an excursion through ongoing projects and programmes world-wide; and finally
4. some relevant scientific multidisciplinary results.

A presentation of long-term vision and perspectives for the upcoming years will complete the volume.

The exploration of the abyss is an exciting adventure closely comparable with the space endeavour but, in some aspects, facing a more severe and extreme constraining environment. Seafloor Observatory scientific results will have a strong impact not only on our present knowledge but also on human life during the coming century and beyond, leaving a legacy of awareness to future generations.

Part I

Present scientific challenges to be addressed using seafloor observatories

2 Integrating continuous observatory data from the coast to the abyss: Assembling a multidisciplinary view of the ocean in four dimensions

M.M.R. Best[1,2], C.R. Barnes[1,3], B.D. Bornhold[1,4] and S.K. Juniper[1,3,5]

2.1 Introduction

The world ocean covers over 70% of our planet and comprises over 95% of the living space. It is the dynamic engine of Earth, driving energy transport and elemental cycles of the globe. It is also the integrator between what are (on average) the fast processes of the atmosphere and the slower processes of the earth's crust. A key hurdle in coming to grips with understanding this earth–ocean–atmosphere system is the range of spatial and temporal scales at which processes occur, their complex interconnectedness and, in many cases, their catastrophic episodicity.

The development of our understanding of the ocean has been hampered by our terrestrial existence as a species. Through the millennia we have viewed the ocean as a shore or a surface, to travel on or dip blindly into for food. Even with the development of satellites and improvement of our seagoing vessels, much of our enquiry has been defined by this perspective of an opaque surface. As we have started to delve within the ocean volume, through

1 Previously NEPTUNE Canada and Ocean Networks Canada, University of Victoria, Victoria, BC, Canada

2 Consultant, and Adjunct in Department of Earth Sciences, Laurentian University, Sudbury, ON, Canada

3 School of Earth and Ocean Sciences, University of Victoria, Victoria, BC, Canada

4 Brian D. Bornhold Inc., Ladysmith, BC, Canada

5 Ocean Networks Canada, Univeristy of Victoria, Victoria, BC, Canada

SCUBA, ROVs, and manned submersibles and submarines, the significance of the resulting observations is often weighted by the 2-D areal coverage (surficial perspective), rather than the elucidation of the four dimensions of the planet's biosphere. More importantly, these forays have given glimpses of the fundamental and powerful dynamism of the ocean system – sufficient to underline the supreme importance of more adequate sampling through time, particularly in the face of a changing world (e.g., Juniper et al., 2007; Halpern et al., 2008; Bernard and Robinson, 2009; Glover et al., 2010; Keeling et al., 2010).

Through significant investment in ship-based programs, in some cases including battery powered autonomous instruments, precious time-series have been collected over a few variables. However, a continuous interactive presence in the ocean, more analogous to how our knowledge and intuition about terrestrial environments has built up, has been elusive. However, as we enter the 21st century, this is changing as a series of technological advances enables continuous virtual presence in the ocean, with sufficient power and bandwidth to operate a broad range of instruments and open the door for development of sensors that do not yet exist (e.g., Tunnicliffe et al., 2003; NRC 2003; Priede et al., 2005; Massion and Raybould, 2006; Favali and Beranzoli 2006; Barnes et al., 2008; Manuel-Lazaro et al., 2010; Favali et al., 2010; Ruhl et al., 2011; Pulliat et al., 2012; Matabos et al., 2012; Best et al. 2014). The range of these initiatives is evidenced by the chapters within this volume.

One category of this advance is the use of telecommunications cable technology to link the shore to the seafloor, extending our power and Internet systems into the ocean. NEPTUNE Canada (NC – now part of Ocean Networks Canada oceannetworks.ca) is among the pioneers of this approach, having taken the hard-won experience of early coastal observatories and extending it out into the deep sea (Barnes and Tunnicliffe 2008). This chapter draws from that experience to provide an example of the challenges and successes of marine sensor systems.

2.2 Spatial (environmental) scope

In order to frame earth–ocean processes, it is necessary to capture information from the major regions and transition zones. The underlying bones of the Earth, the tectonic structures, define much of the overlying regional distribution of environments and transition zones. With this in mind, NEPTUNE Canada, for example, was designed to extend west from the continental North American plate across the entire Juan de Fuca plate and onto the Pacific plate. As a result, the observatory spans both continental and oceanic crust, extending from the coast across the shelf, down the continental slope (including a submarine canyon), across the abyssal plain, to include the mid-ocean ridge (Figure 2.1).

By covering this range of tectonic and physiographic settings, observatories can also span major zones of the ocean, including a range of photic to aphotic environments, and photosynthetic- to chemosynthetic-energy-based ecosystems. Energy sources include both sunlight and geothermal heat, both of which drive ocean currents and geochemical cycles, as well as primary production. Sources of nutrients range from terrestrial outflow, to coastal upwelling, to cold seeps and hydrothermal vents; transport of resultant primary productivi-

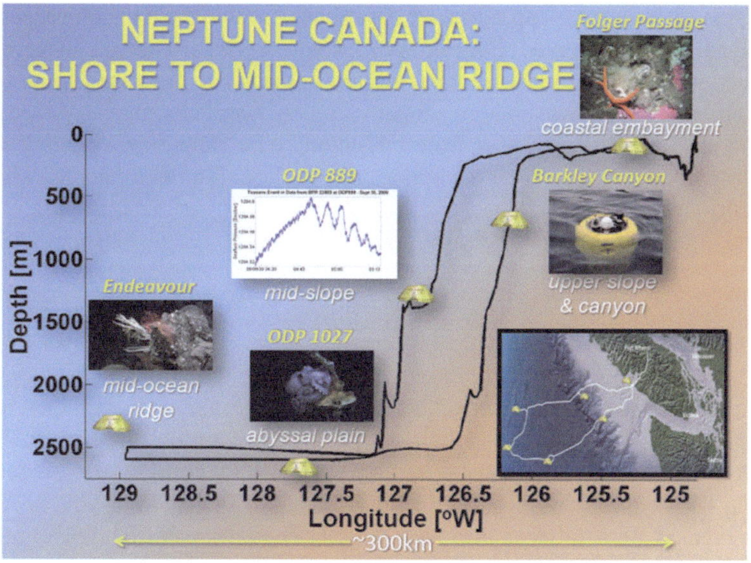

Figure 2.1 Depth profile of NEPTUNE Canada infrastructure starting in the east at Port Alberni, showing the 800km main cable, 5 nodes (expandable to 10 primary nodes), and an example of data from each area. The breadth of ocean environments is indicated below the node names. Note that sites ODP 889 and 1027 have recently been renamed as Clayoquot Slope and Cascadia Basin, respectively.

ty includes free fall, surface, mid-water and bottom currents (including through submarine canyons and ridge valleys), and tidal pumping. Benthic habitats include soft, firm and hard bottom environments in all of the shelf, slope, abyssal plain, and mid-ocean ridge regions, including unusual settings such as hydrothermal vents, hydrate seeps with carbonate crusts, submarine canyons, nepheloid layers, and wave-washed rocky pinnacles. Scales of life range across microbial, meiofaunal, mesofaunal, macrofaunal, and megafaunal; modes of life include planktic, pelagic, nektic, benthic (motile and sessile)-epifaunal, semi-infaunal, infaunal. It is key that, with continuous observations, it is possible to quantify the inherent temporal patchiness of biological occurrences.

2.3 Temporal scope

Processes within the earth–ocean system vary on scales of microseconds to millennia. Rapid processes require high-frequency sampling that may not be practical using autonomous battery-powered instruments, or shipboard sampling. Assembly of more than one data stream into variable space also requires time-coordination. The interpretation of longer-term processes often requires an adequate assessment of the short-term processes so that the latter signal can be removed. For example, climate trends are hard to discern if you do not have adequate sampling of the annual variability, and can show that the trend is more than a random sampling of seasonal or annual variation. Similarly, analysis of ocean waves on various scales needs to proceed from "de-tided" data. Therefore, continuous data at high sampling frequencies provide the opportunity for backing out trends at various scales (annual, multi-year, decadal).

2.4 Catastrophic episodicity

Many processes are also highly episodic and these, combined with rapid events at long intervals, make it difficult to plan sampling expeditions to capture the events. If the rapid processes are heterogeneously distributed through time (random, episodic), for example earthquakes, then continuous high-frequency time-coordinated sampling is required to capture the event at all. A solution is to have remote sensing capability in place all the time, waiting for the events to occur.

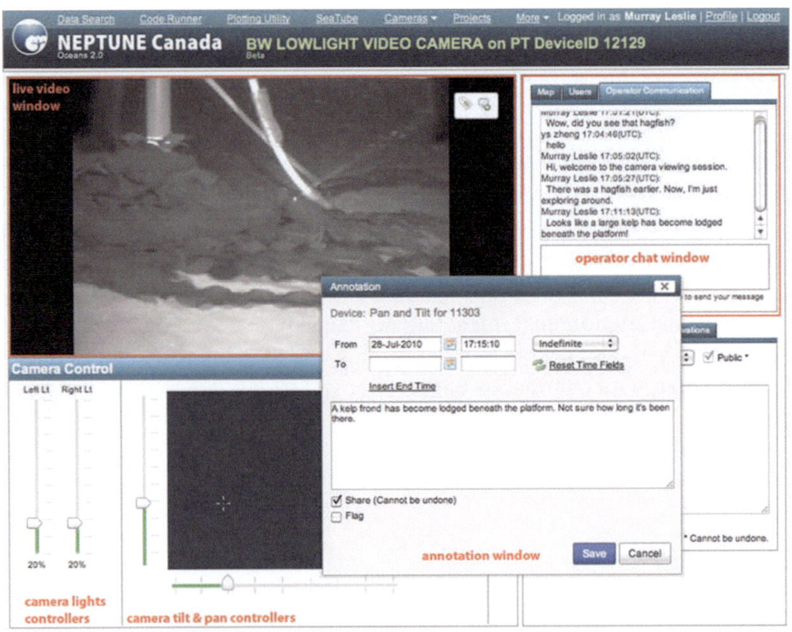

Figure 2.2 Camera control interface in NEPTUNE Canada's Oceans 2.0, including annotation and collaboration tools.

2.5 Complex interconnectedness

Perhaps even more powerful than the environmental and temporal scope above, however, is the realtime, coincident, coordinated sampling of hundreds of data streams across disciplines and the interactive remote control of instruments (Figure 2.2). This allows responsive realtime sampling of events, and analysis of relationships among the disparate data to test relationships and drivers across the earth–ocean system.

This not only requires high-resolution coincident sampling, but also extensive documentation of and access to meta-data in order to allow the scientific community the ability to judge the appropriateness of various data combinations for different analyses. This not only benefits from, but requires, an open data e-science approach, in order to allow for development of research initiatives across disciplinary lines, and across international experience. An example of the potential breadth of the scientific involvement is outlined below.

2.6 Challenges of multidisciplinarity

From the outset, NEPTUNE Canada was conceived as a multidisciplinary observatory, precisely providing the capacity to address interwoven earth–ocean processes. This also means that, from the beginning, it has required courage from the scientific community and supporters to proceed into this challenging approach to science. Not only are the technical aspects challenging in themselves, but perhaps more important is the difficulty of communication across disciplines and cultures; this is not only among scientific disciplines, but even more so among academic, government, engineering, finance and communications cultures, across public and private sectors. Success of such a large-scale interdisciplinary initiative requires contributions from all these sectors, and all involved need to learn new languages, and to accept that there is no one "right" way to proceed. While to a certain extent this faces all big science infrastructures in their initial instances (e.g., neutrino observatories, telescopes), the breadth of science disciplines/cultures involved is greater in the case of ocean observatories. That the concept of NC was multidisciplinary from the beginning helps navigate these challenges, as the initial "builder" community self-selected to face them, often bringing pioneering experience from other interdisciplinary ventures. An example is illustrated in Figure 2.3, where physical, chemical and biological oceanographers, benthic ecologists, sedimentary geologists and geomicrobiologists are all collaborating to understand carbon cycling in the slope/canyon region of Barkley Canyon.

2.7 Integrated networks

Building requires integration of the people networks discussed above with those of hardware and software. In the case of NEPTUNE Canada, hardware developed and installed includes: 800km powered fiber-optic backbone cable; nodes and junction boxes; mobile

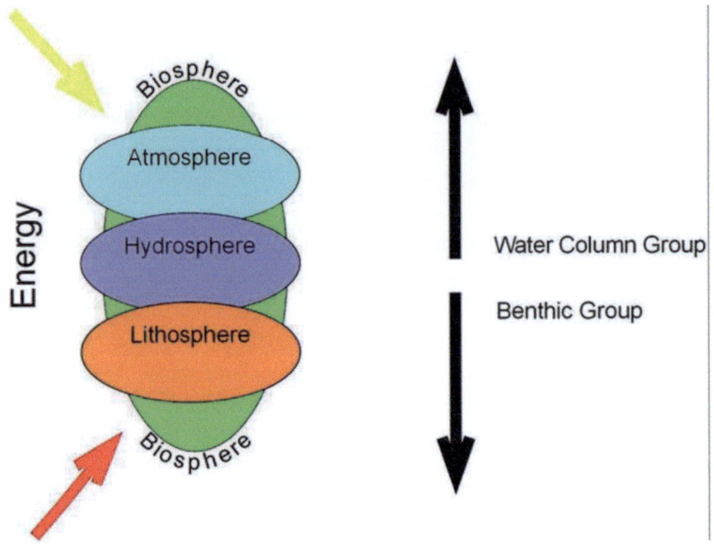

Figure 2.3 Diagram depicting the integrated study of water column and benthic carbon cycling by physical, chemical, biological and geological processes. Two interdisciplinary teams are involved in this study in the Barkley Canyon area.

platforms including a vertical profiler and a seabed crawler, and scores of instruments. The vast majority of this complex hardware network of over 1000 pieces is operating successfully following installation, a coup in these extreme environments (Barnes et al., 2011, 2013, 2015 this volume). A significant part of this success is due to the development of complex testing and integration procedures (Bornhold et al., 2010) and the logistical support of the ROPOS ROV team (ropos.com). In parallel, software and hardware systems are acquiring, archiving and delivering continuous real-time data (see Pirenne, this volume). As described below, this is delivered through the evolving Oceans 2.0 environment (Owens et al., 2010). A network of scientists, engineers and technicians continues to contribute to the process in every phase (see oceannetworks.ca).

2.8 Scientific initiatives

Here we provide greater detail on the initial proof-of-concept experiments that were installed at 5 network nodes during the installation and maintenance cruises 2009–2012. At inshore Folger Passage (Barkley Sound, west Vancouver Island; 2 study areas at 20–100m water depth), controls on biological productivity are used to evaluate the effects of marine processes on invertebrates, fish and marine mammals (e.g., Pawlowicz and McClure, 2010).

Experiments around Barkley Canyon (4 study areas at 400–1000m water depth) quantify changes in biological and chemical activity associated with nutrients and cross-shelf sediment transport at the shelf/slope break and through the canyon (e.g., Robert and Juniper, 2012; Purser et al., 2013). Along the mid-continental slope, exposed and shallowly buried hydrates (Barkley Canyon and ODP 889 – now Clayoquot Slope; 3 study areas, ~900–1200m water depth) allow monitoring of changes in their distribution, structure and venting, and relationships to earthquakes, slope failures and plate motions, as well as regional oceanography (e.g., Thomsen et al., 2012). Circulation Obviation Retrofit Kits (CORKs) at mid-plate ODP 1026-7 (node now called Cascadia Basin, ~2700m water depth) boreholes monitor real-time changes in crustal temperature and pressure, in response to earthquakes, hydrothermal convection or plate strain (e.g., NRC 2011). At Endeavour Ridge (7 study areas at 2000–2200m water depth), complex interactions among volcanic, tectonic, hydrothermal and biological processes are being quantified at the western edge of Juan de Fuca plate (e.g., Rona and Light, 2011). Across the network, high-resolution seismic information elucidates tectonic processes and earthquakes, and a tsunami system determines open ocean tsunami amplitude, propagation direction, and speed (e.g., Thomson et al., 2011). Again, this is a summary of initial proof-of-concept projects; The infrastructure has capacity to expand, thereby allowing growing participation in experiments, data analysis and technology development (http://www.oceannetworks.ca).

2.9 Scientific development

Involvement in ocean observatories can encompass a number of roles. Members of the NEPTUNE Canada scientific community can be described, for example, as builders, bridge-builders, data analysts, and knowledge beneficiaries. This breadth of interaction, along with a range of scientific cultures across disciplines, makes for powerful cross-fertilization at its best, though it can be a communications and management challenge when trying to design the observatory to meet scientific needs. In the following sections, the nature of these roles is developed based on the NEPTUNE Canada experience.

2.9.1 Builders

In 2004, approximately 250 scientists from Canada and internationally were participants in a series of NC workshops, and were fundamental designers of the system. Out of this process, the broad research themes and node locations outlined in Barnes et al. (2008, 2012, and this volume) were defined, and 17 proof-of-concept multidisciplinary proposals arose. These went out to an international review panel, from which nine multidisciplinary science teams were funded in 2005 (Best et al., 2007). Table 2.1 lists these projects, and further information is available at http://www.neptunecanada.ca/research/research-projects/.

It is important to note that the number of projects funded during this initial phase was not dependant on the number that were rated highly, rather the number that could be accommodated within the funds available for "proof-of-concept" within the initial infrastructure funding of NC. This means that, in addition to future initiatives, there remains

Project title	**Node areas**	**Initial instruments**	**Co-investigators**
A seafloor observatory for long-term integrated study of hydrates of thermogenic and microbial origins on the Cascadia Margin	1–2	12	11
Biophysical linkages underlying variability in coastal ocean processes off the west coast of Vancouver Island	1–2	13	9
Geophysical imaging of gas hydrate at the ODP889 Node-Bullseye Vent Site	1	3	7
The role of disturbance in deep-sea benthic ecosystems	2	13	8
Monitoring the Endeavour–Middle Valley Hydrothermal Systems: Integrated heat and mass fluxes and the response of biogeochemical and physical processes to perturbations	2	33	6
Ocean crustal hydrogeology	2–3	5	10
NEPTUNE Canada Seismograph Network	5	16	7
Ecological and biophysical determinants of primary and secondary production in a coastal marine ecosystem and implications for fish and whales	1	13	5
Towards real-time observations and modeling of tsunamis and other infra-gravity eaves off the west coast of North America	5	14	10

Table 2.1 Summary of initial interdisciplinary projects using NEPTUNE Canada.

a number of internationally considered highly-ranking initiatives and components of the existing initiatives that can be (and are being) pursued from this initial pool (see future initiatives discussions from NC's workshops, http://www.oceannetworks.ca/science/getting-involved/workshops).

These initial science teams consisted of ~80 scientists, technicians and students who were intimately involved in defining and developing NC through the planning and building phase, from 16 institutions in Canada, US and Europe. Their contribution first and foremost is conceptual, but extends to the very hands-on technical tasks involved in getting such a complex system running, including configuration and commissioning.

2.9.2 Future builders

The NC observatory network is expandable at every scale, with a large capacity available for new instruments which users can support through international science bodies, instrument manufacturer testing, and private sector testing/data collection.

All additions to the observatory or proposed experiments involving changes to existing instruments must go through a proposal and review process that requires approval by both ONC staff and the ONC Science Planning Committee. Funding must cover the instrument costs as well as connectivity, ship-time including retrieval, driver development, and yearly data costs. Costs to non-academic users are negotiated on a separate basis. Early examples of scientific instrumentation include: TEMPO-mini and Piezometer supported by Ifremer; COVIS acoustic plume imaging supported by NSF; CORKs and instruments supported by IODP – all of which are multimillion dollar outside investments in instrumentation to deploy on the NEPTUNE Canada infrastructure. Details about participating in NEPTUNE Canada are available through the Ocean Networks Canada website at http://www.ocean-networks.ca/science/getting-involved.

2.9.3 Bridge-builders

Cabled instruments provide the opportunity to temporally calibrate spatially broader shelf to deep ocean observations. In turn, broader spatial observations added to the temporal resolution such as NEPTUNE Canada data will expand spatial resolution and understanding of regional processes. This cross-calibration of observing methodologies started with the definition of the node locations – based on areas that had been identified as key pivot points during decades of previous ship-based research. In turn, NC is now starting to provide the temporal resolution never before available across a breadth of variables to help in

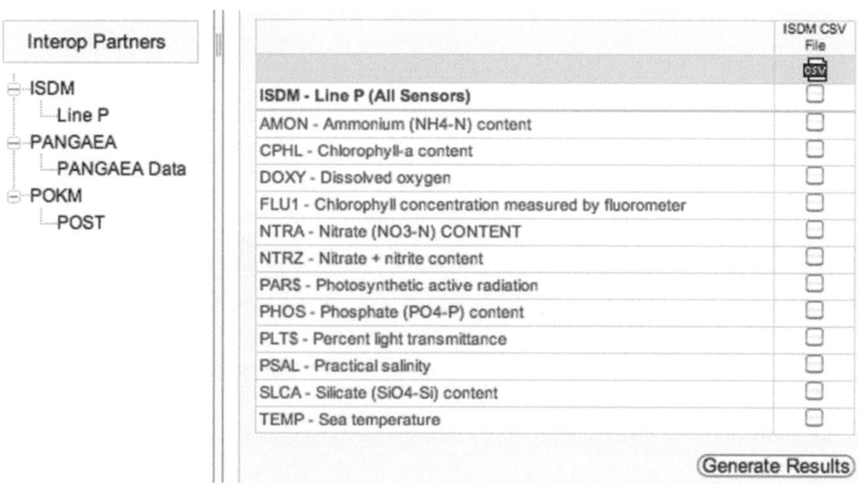

Figure 2.4 Data search interface including interoperable organizations housing NE Pacific data in addition to access to NEPTUNE Canada data. Further interoperability is also developed with IRIS (seismic) and PTWC (tsunami) data archives.

the temporal ground-truthing of archived data. This is facilitated by interoperability efforts between NC and other data archives (Figure 2.4).

Cross-calibration occurs with a range of other sampling methodologies, such as ship-based expeditions including sampling (e.g., DFO and other Canadian and international research initiatives), IODP CORK installations (e.g., ODP 889, 1027), and AUV and glider surveys (e.g., MBARI). As part of a collaboration with VENUS, gliders are being deployed among NC nodes, to cross-calibrate and expand the spatial dimension of NC cabled readings.

This cross-calibration and development of models is applicable not only to the North East Pacific, but also for any processes that sensibly have broader impact. The development of general models in many fields has often been based on data from a particular region that was well studied. Similarly, cabled observatories do not need to cover the whole seafloor in order to inform broader models of ocean processes. In particular, the coverage of "type" areas of the ocean allows others to use the data as a predictive tool for what might be occurring off other coasts. Continuous data can help constrain and target data parameters required to characterize a system elsewhere.

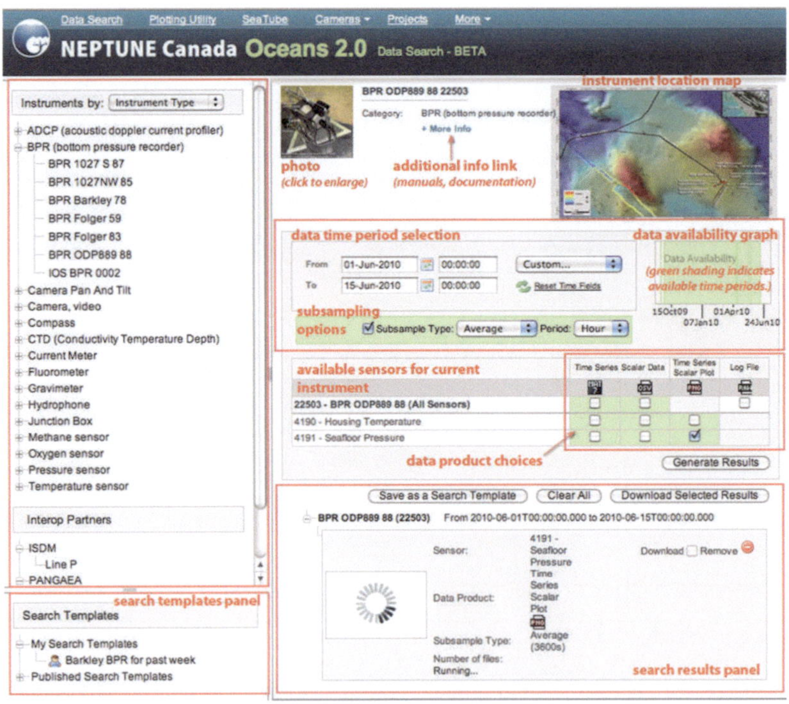

Figure 2.5 NEPTUNE Canada data search screen; data are freely and openly available.

2.9.4 Data analysts

NC has what remains a revolutionary data policy in ocean sciences (though inspired by other disciplines, e.g., the seismic community) – data are freely and openly available real-time, to anyone who simply registers for a user account (http://www.oceannetworks.ca/data-tools, see search interface in Figure 2.5).

This data policy arose out of Canadian and US workshops in the early part of the 21st century (e.g., NRC 2003) and the experience of other disciplines, with a growing appreciation that collection of data was a key part of the scientific endeavour, that there should be credit for this activity and that, in turn, the data should be available as openly and quickly as possible. In many cases, specific data streams are used in standard analysis, providing a real-time temporal resolution not previously available. Beyond this, a key development will continue to be Ocean Informatics' multivariate, interdisciplinary analysis of coordinated ocean–earth data, including biological, chemical, physical and geological parameters. It is expected that NC will act as a catalyst in this area, given the fact that it provides interdisciplinary, multivariate, coordinated ocean–earth data. We have seen this impact in other fields, when data have been made accessible in a common form and to basic standards, such as the growth of bioinformatics with the advent of data repositories such as GenBank. In addition to providing the data, NC is facilitating the development of tools for this analysis through the development of the Oceans 2.0 collaborative and analytical environment – a science-driven, web-based system of scientific and IT development to aid in the realization of the potential. This includes components to aid in collaboration such as the project spaces shown in Figure 2.6.

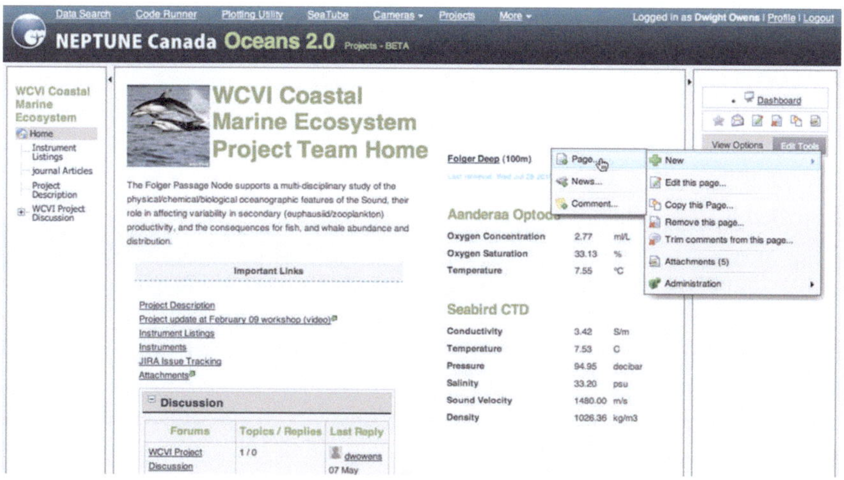

Figure 2.6 An example project space in Oceans 2.0, to foster collaboration in this virtual, distributed research community.

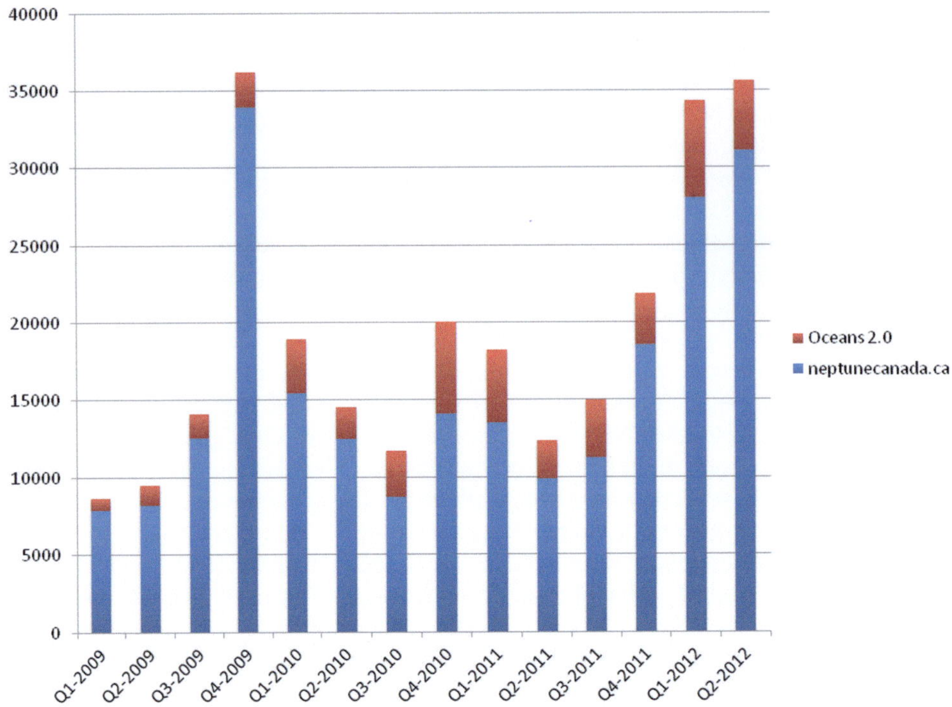

Figure 2.7 Quarterly traffic in different components of the NEPTUNE Canada web environment. www.neptunecanada.ca = general website; Oceans 2.0 = interactive working space with tools for collaboration, data access and analysis. This is aside from outside sites, such as YouTube and Flickr, where popular content is hosted.

2.9.5 Knowledge beneficiaries

This group of users includes those working in the realms of education, policy, the media, and the public. While initial NC funding did not provide for targeted focus on these groups, some data products have been developed for different user groups: K-12, post-secondary, policy-makers, science centers/aquaria, and television programing. For example, a number of school and university courses are using NC data in their programs. Oceans 2.0 is also accessible to these groups, as it evolves into an "Ocean lab" concept of collaborative user-designed web space. This realm of social computing brings the virtual observatory not only to the scientific community but to a broader reach, using approaches such as crowdsourcing and citizen science to enable not only passive but active involvement in the scientific endeavour by the community at large (Owens et al., 2010). Such engagement also cultivates the next generation of marine scientists (see Gunn and Thomsen, 2009).

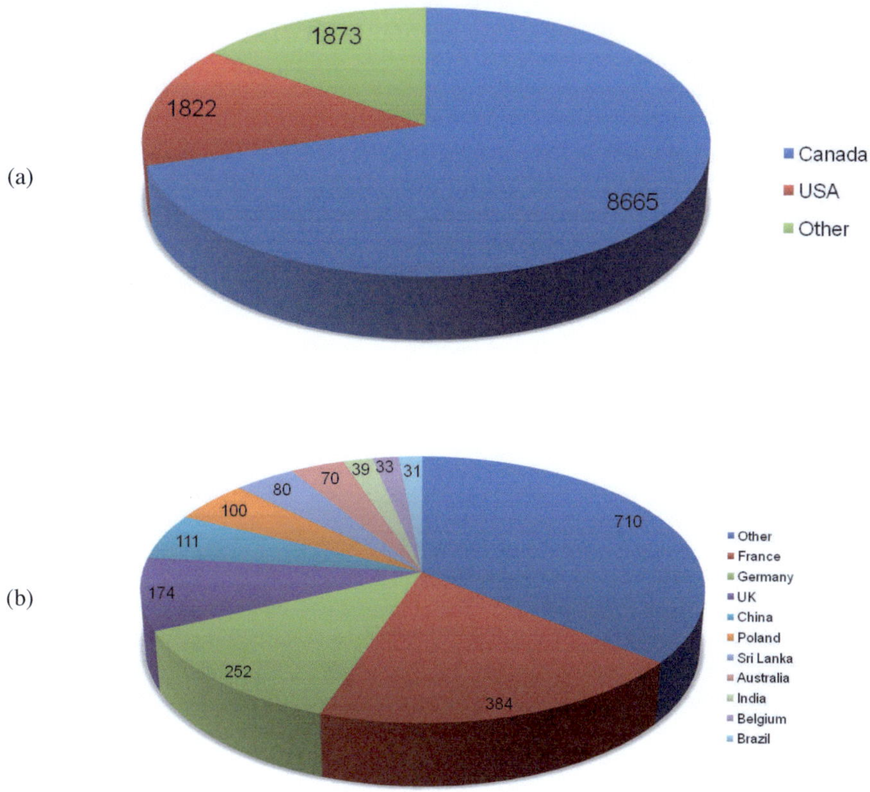

Figure 2.8 NEPTUNE Canada registered users by country (2009–2012), which shows the broad distribution of data users even at this early stage; (a) significant share of users from Canada and US; (b) distribution of users other than from Canada and the US.

2.10 Participants and data access

Figure 2.7 provides a summary of all website visits across different components of the NEPTUNE Canada web environment (2009–2012). The web environment is hugely important for a facility that is almost completely underwater; it is its only face to the world.

This captures those interested in general information about NEPTUNE Canada as well as those actively searching through data. It also captures activity around virtual installation cruises when scientists participate from around the world using live video streams, near-live annotated and searchable video, Twitter updates, blogs, Skype, email, and news instalments to follow and direct placement of instruments.

Registrations of NC database users prior to January 1, 2009 amounted to 220 individuals, predominantly drawn from those participants in the "builder" category described above. The period of January 2009 to March 2010 saw this number swell to 5781 registered users. This huge increase largely occurred in conjunction with the data launch event on December 8, 2009. This activity reflects both a commissioning phase by the first science users as well as a large initial set of registrants from the public. By early 2012, the number of registered users grew to over 14,000 (Figure 2.8a). If we consider the distribution of users by country, there is a very strong bias towards Canada and the USA; however, there are also hundreds of registered users across 141 other countries (Figure 2.8b). In addition, access of imagery has passed over a million views on YouTube and 150,000 Flickr searches. The number of users is expected to grow significantly as more instrumentation is installed and awareness of the opportunity provided by the facility increases.

2.11 Summary

Ocean observatories such as NEPTUNE Canada will transform our understanding of biological, chemical, physical and geological plate scale processes from the beach to the abyss. Real-time continuous monitoring, archiving, and long time series allow scientists to capture the temporal nature, characteristics and linkages of these natural processes in ways never before possible. As we attempt to trace linkages through the earth–ocean–atmosphere system, such continuous information on oceans will need to play a key role.

References

Barnes C.R. and Tunnicliffe V. 2008. Building the world's first multi-node cabled ocean observatories (NEPTUNE Canada and VENUS, Canada): Science, realities, challenges and opportunities. In: Proceedings of OCEANS 2008 - MTS/IEEE Kobe Techno-Ocean.

Barnes C.R., Best M.M.R., Johnson F., Phibbs P. and Pirenne B. 2008. Building the world's first regional cabled observatory: NEPTUNE Canada, Journal of Ocean Technology 3, 13–18.

Barnes C.R., Best M.M.R., Johnson F.R., Pautet L. and Pirenne B. 2011. Understanding Earth/ocean processes using real-time data from NEPTUNE Canada's widely distributed sensor networks, North-East Pacific. Geoscience Canada, March 2011.

Barnes C.R., Best M.M.R., Johnson F.R., Phibbs P. and Pirenne B. 2015. The NEPTUNE Canada Project: Installing the world's first regional cabled ocean observatory. Chapter 12 in P. Favali, L. Beranzoli and A. de Santis (Eds) Seafloor Observatories: A New Vision of the Earth from the Abyss. Chichester, Heidelberg: Springer Praxis (this volume).

Barnes C.R., Best, M.M.R., Johnson, F.R., Pautet, L. and Pirenne, B. 2013. Challenges, benefits and opportunities in installing and operating cabled ocean observatories: perspectives from NEPTUNE Canada. IEEE Journal of Oceanic Engineering, v. 28. no. 1, p. 144-157..

Best, M., P. Favali, L. Beranzoli, M. Cannat, N. Cagatay, J.J. Dañobeitia, E. Delory, H. de Stigter,B. Ferré, M. Gillooly, F. Grant, P.O.J. Hall, V. Lykousis, J. Mienert, J.M.A. de Miranda, G. Oaie,V. Radulescu, J.-F. Rolin, H. Ruhl, and C. Waldmann. 2014. EMSO: A distributed infrastructure for addressing geohazards and global ocean change. Oceanography 27(2):167–169,http://dx.doi.org/10.5670/oceanog.2014.52.

Best M.M.R., Bornhold B.D., Juniper S.K. and Barnes C.R. 2007. NEPTUNE Canada Regional Cabled Observatory: Science Plan. In; Proceedings of OCEANS 2007 - MTS/IEEE Vancouver.

Bernard E.N. and Robinson A.R. 2009. Introduction: Emergent findings and new directions in tsunami science. In: E.N. Bernard and A.R. Robinson (Eds) The Sea, Vol. 15, Cambridge, MA: Harvard University Press, pp. 1–22.

Bornhold B., Best M., Hasanen R., Hofmann M. and Leslie M., Shepherd K., Wallace K. and Williams J. 2010. Testing and integration of a cabled ocean observatory: Complexity is greater than the sum of the off-the-shelf parts. In: Proceedings of OCEANS 2010, MTS/IEEE Seattle.

Favali P. and Beranzoli L. 2006. Seafloor observatory science: A review. Annals of Geophysics 49, 515–567.

Favali P., Person R., Barnes C.R., Kaneda Y., Delaney J.R. and Hsu Shu-Kun. 2010. Seafloor Observatory Science. In: J. Hall et al. (Eds), Proceedings of the OceanObs'09: Sustained Ocean. Observations and Information for Society Conference 2, Venice, Italy, 21–25 September 2009. ESA Publication WPP-306.

Glover A.G., Gooday A., Bailey D.M., Billett D.S.M., Chevaldonne P., Colaco A., Cuvelier D., Desbruyeres D., Kalogeropoulou V., Klages M., Lampadiriou N., Lejeusne C., Mestre N.C., Paterson G.L., Perz T., Ruhl H., Sarrazin J., Soltwedel T., Soto E.H., Thatje S., Tselepides A., Van Gaever S., Vanreusel A. 2010. Temporal change in deep-sea benthic ecosystems: A review of the evidence from recent time-series studies. Advances in Marine Biology 58, 1–95.

Gunn V. and Thomsen L. (2009) The next generation: Providing inspiration and training for future marine scientists. Oceanography 22, 166–176.

Halpern, B.S., Walbridge S., Selkoe K.A., Kappel C.V., Micheli F., D'Agrosa C., Bruno J.F., Casey K.S., Ebert C., Fox H.E., Fujita R., Heinemann D., Lenihan H.S., Madin E.M.P., Perry M.T., Selig E.R., Spalding M., Steneck R. and Watson R. 2008. A global map of human impact on marine ecosystems. Science 319(5865) 948–952.

Juniper S.K., Escartin J. and Cannat M. 2007. Multidisciplinary, time-series observations at mid-ocean ridges. Oceanography 20, 102–111.

Keeling R.F., Körtzinger A. and Gruber N. 2010. Ocean deoxygenation in a warming world. In: C.A. Carlson and S.J. Giovannoni (Eds) Annual Review of Marine Science, Vol. 2. Palo Alto, CA: Annual Reviews, pp. 199–229.

Manuel-Lazaro A. and Nogueras M. and del Rio J. 2010. OBSEA: An expandable seafloor observatory. Sea Technology 51, 37–39.

Massion G. and Raybould K. 2006. MARS: The Monterey Accelerated Research System. Sea Technology 47, 39–42

Matabos M., Best M., Blandin J., Hoeberechts M., Juniper S.K., Pirenne B., Robert K., Ruhl H., Sarrazin J. and Vardaro M. 2012. Seafloor observatories. Chapter 14 in M. Clark and M. Consalvey (Eds) Biological Sampling in the Deep-Sea: An (Illustrated) Manual of Tools and Techniques. New York and Oxford: Wiley-Blackwell.

NRC (U.S. National Research Council). 2003. Enabling Ocean Research in the 21st Century: Implementation of a Network of Ocean Oobservatories. Washington, DC: The National Academies Press.

NRC (U.S. National Research Council). 2011. Scientific Ocean Drilling: Accomplishments and Challenges. Washington, DC: The National Academies Press.

Owens D., Best M., Guillemot E., Jenkyns R. and Pirenne B. 2010. Ocean observatories and social computing: Potential and progress. In: Proceedings of OCEANS 2010, MTS/IEEE Seattle.

Pawlowicz R. and McClure B. 2010. Inverted echosounder for continuous high-resolution water column profiling from the NEPTUNE (Canada) ocean observatory. In: Proceedings of OCEANS 2010, MTS/IEEE Seattle.

Pirenne B. 2015. The role of Information Communication Technologies (ICT) for seafloor observatories: Acquisition, archival, analysis, interoperability. Chapter 6 in P. Favali, L. Beranzoli and A. de Santis (Eds) Seafloor Observatories: A New Vision of the Earth from the Abyss. Chichester, Heidelberg: Springer Praxis Books (this volume).

Priede I.G., Person R. and Favali P. 2005. European seafloor observatory network. Sea Technology 46, 45–49.

Puillat I., Lanteri N., Drogou J.F., Blandin J., Géli L., Sarrazin J., Sarradin P.M., Auffret Y., Rolin J.F. and Léon P. 2012. Open-sea observatories: A new technology to bring the pulse of the sea to human awareness. In: Marco Marcelli (Ed.) Oceanography. InTech. Available from: http://www.intechopen.com/books/oceanography/open-sea-observatories-a-new-technology-to-take-the-pulse-of-the-sea-with-internet-in-the-ocean

Purser, A., Thomsen L., Barnes, C., Best M., Chapman, R., Hofbauer, M., Menzel, M., and Wagner, H. 2013. Temporal and spatial benthic data collection via Internet operated Deep Sea Crawler. Methods in Oceanography, v. 5, p. 1-18, doi:10.1016/j.mio.2013.07.001

Robert K. and Juniper S.K. 2012. Surface-sediment bioturbation quantified with cameras on the NEPTUNE Canada cabled observatory. Mar Ecol Prog Ser 453, 137–149.

Rona P. and Light R. 2011. Sonar images hydrothermal vents in seafloor observatory. Eos, Transactions, American Geophysical Union 92(20), 169–170.

Ruhl A.R., André M., Beranzoli L., Çağatay M.N., Cannat M., Danobeitia J.J., Favali P., Géli L., Gillooly M., Greinert J., Hall P.O.J., Huber R., Karstensen J., Lampitt R.S., Larkin K.E., Lykousis V., Mienert J., Miranda J.M., Person R., Priede I.G., Puillat I., Thomsen L. and Waldmann C. 2011. Societal need for improved understanding of climate change, anthropogenic impacts, and geo-hazard warning drive development of ocean observatories in European Seas. Progress in Oceanography 91, 1–33.

Thompson R., Fine I., Rabinovich A., Mihály S., Davis E., Heesemann M. and Krassovski M. 2011. Observation of the 2009 Samoa tsunami by the NEPTUNE Canada cabled observatory: Test data for an operational regional tsunami forecast model. Geophysical Research Letters 38(L11701), 5.

Thomsen, L., Barnes C., Best M.M.R., Chapman R., Pirenne B., Thomson R. and Vogt J. 2012. Ocean circulation promotes methane release from gas hydrate outcrops at the NEPTUNE Canada Barkley Canyon node. Geophysical Research Letters. doi:10.1029/2012GL052462.

Tunnicliffe V., Dewey R. and Smith D. 2003. Research plans for a mid-depth cabled seafloor observatory in Western Canada. Oceanography 16, 53–59.

3 Underwater neutrino telescopes: Detectors for astro-particle physics and a gateway for deep-sea laboratories

E. Migneco, P. Piattelli and G. Riccobene[1]

3.1 Introduction

The simultaneous observation of cosmic hadrons (protons and nuclei), photons and neutrinos (i.e., the so-called multimessenger astronomy) is providing a deeper understanding of the high-energy Universe and on the cosmic sources of high-energy particles. Our comprehension of the high-energy Universe is indeed rapidly increasing thanks to novel experiments that allow for precise measurements of ultra-high energy cosmic ray fluxes and the identification of several tens of astrophysical sources of gamma rays. Although photons are copiously produced in cosmic sources, the horizon for very high-energy (10^{12} eV<E<10^{15} eV) photons is limited by their interaction with the diffuse cosmic microwave and infrared background. The gamma-ray horizon is, indeed, some ten millions parsecs (1 pc ≈ 3.16 light years ≈ 10^{16}m) at 10^{12} eV and about 100 kpc at 10^{15} eV. On the other hand, charged particles are deflected by magnetic fields. Their directions are therefore scrambled and do not allow for source pointing. Only protons with E>10^{20} eV, that are only slightly deflected by magnetic field, could be used as astronomical probes, but they are absorbed in the Universe within 100 Mpc, due to their interaction with the Cosmic Microwave Background Radiation, known as GZK effect (Greisen et al., 1966, Zatsepin and Kuzmin, 1966). Neutrinos, light and neutral particles, are more difficult to detect but, contrary to protons and gamma rays, can reach the Earth by travelling through dense astrophysical environments and the whole Universe without being deflected or absorbed. Cosmic neutrino detection could therefore provide a fundamental piece of information in high-energy astrophysics (Figure 3.1).

1 Laboratori Nazionali del Sud – Istituto Nazionale di Fisica Nucleare. Catania, Italy

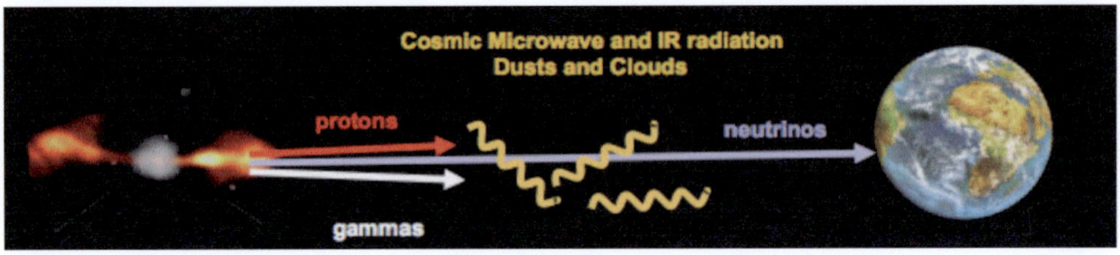

Figure 3.1 Unlike charged particles and photons, neutrinos are neither deflected nor absorbed in their journey to the Earth, and allow for high energy astronomy.

Astrophysical environments where explosive phenomena generate shock waves are suggested as candidate sources of high-energy cosmic rays (protons and heavier nuclei), gamma rays and neutrinos. In particular, supernova remnants in our galaxy and active galactic nuclei and gamma ray bursters in the far Universe seem to play major roles. Galactic supernova remnants are suggested as the major sources of the cosmic rays detected at Earth with energies up to about $Z \times 10^{14}$ eV (where Z is the charge number of the nucleus), while powerful extragalactic sources are supposed to be responsible for the cosmic ray flux up to energies of 10^{21} eV. In all these sources, E>TeV (10^{12} eV $\approx$ 1 erg = 10^{-7}) gamma rays and neutrino fluxes can be also produced by the interaction of accelerated protons and heavier ions with the ambient medium or with the low-energy ambient photons. Inelastic scattering originates neutral and charged pions, whose decay results in gamma rays and neutrinos fluxes, respectively.

Theoretical expectations on high-energy neutrino fluxes from astrophysical sources vary a lot according to different source models. A large part of these uncertainties is due to the incomplete knowledge of the astrophysical source environment. More robust estimates on neutrino fluxes can be obtained by extrapolating available experimental data on cosmic ray and gamma ray fluxes. At $E<10^{15}$ eV novel gamma-ray detectors are providing strong constraints on sources luminosity, particle acceleration and interaction mechanisms. At extreme energies ($E>10^{18}$ eV), based on the observed cosmic ray flux (Figure 3.2), a "guaranteed" flux of so-called cosmogenic neutrinos, produced from the interaction of cosmic protons on the cosmic microwave background radiation, is also expected.

Present estimates indicate that masses of target media of the order of several gigatons are needed to detect astrophysical neutrinos. The use of natural media as neutrino target, such as the deep polar ice-cap or deep-sea water (1 gigaton of water corresponds to a volume of 1 km^3), is therefore mandatory to build cosmic neutrino detectors within an affordable budget. These detectors are also named *km^3-scale underwater* (or under-ice) *neutrino telescopes*, since they are capable of measuring the energy and the astrophysical neutrino source direction, thus allowing astronomy to be studied (Katz and Spiering, 2012).

Such neutrino detectors will be also unique infrastructures to host sensors devoted to Earth and sea sciences in a hostile and unexplored environment such as the deep sea and the polar ice. In the last decade, indeed, the novel technologies for mechanics, power supply and data transmission systems in the deep sea, developed to match the astroparticle physics community's requirements, have permitted the installation of several cabled abyssal laboratories, used in collaboration with the Earth and sea science community.

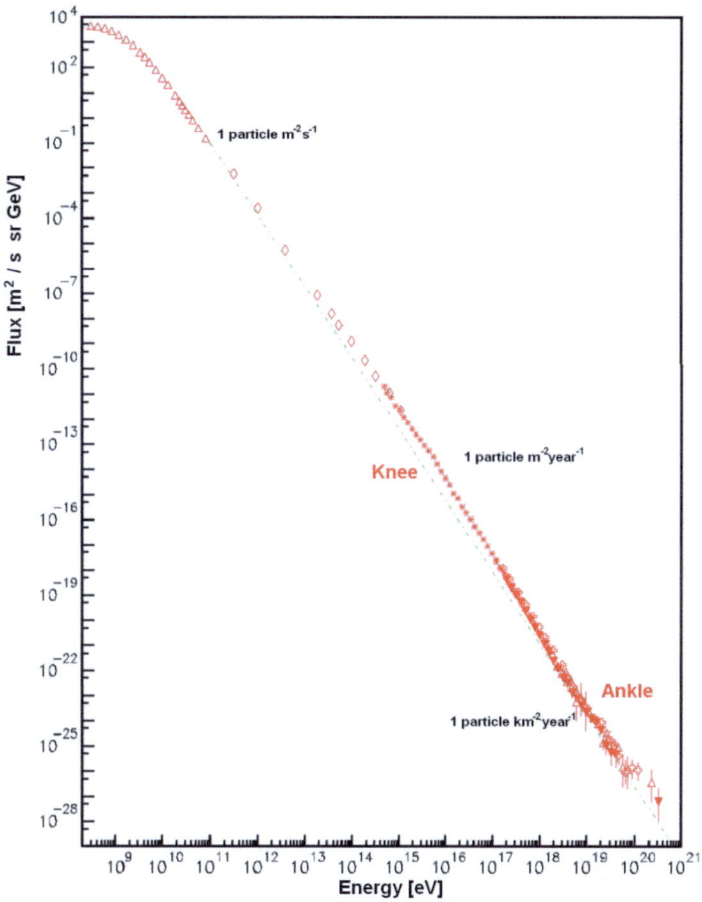

Figure 3.2 The cosmic ray spectrum (Cronin et al., 1997).

3.2 High-energy neutrino astronomy

The observed cosmic ray spectrum shown in Figure 3.2 is often explained in terms of charged particles accelerated in astrophysical sources through the so-called Fermi-Bell acceleration mechanism (Fermi, 1949; Bell, 1978). This mechanism takes place in sources where a plasma of charged particles is contained by magnetic fields and is driven by shock waves. In these sources, charged particles gain energy statistically, crossing shock-fronts generated during powerful explosions. The astrophysical sources proposed as sites of cosmic ray acceleration and production of gamma ray and neutrino fluxes are both galactic and extra-galactic. The first ones are less luminous (about 6 orders of magnitude than extra-galactic ones) and less powerful (the maximum achievable energy is about 10^{16} eV) but, thanks to their proximity to Earth, their high-energy particle flux can still be observed at Earth as a point-like source. Extra-galactic sources are extremely luminous (typically $10^{42\div46}$ erg/s – $10^{33\div39}$ J/s – for active galactic nuclei and 10^{53} erg –10^{46} J– for gamma ray bursters) and powerful sources, but their observation is so far limited to <10 TeV energies due to proton and gamma ray deflection and/or absorption in the Universe.

Neutrino observation is thus expected to open a new window on the high-energy Universe: the arrival direction of neutrinos will allow for tracing them back to the source and thus to establish a connection between the observed high-energy cosmic ray flux and the astrophysical sources.

A number of arguments indicate that galactic supernova remnants (SNR) are the most probable sources of the cosmic ray flux below 10^{18} eV (Berezinsky, 2008; Aharonian et al., 2004a, b). In this energy region the arrival directions of cosmic rays on Earth are randomized by the galactic magnetic field and connection between cosmic ray arrival direction and their sources cannot be achieved. In recent years, high-energy gamma-ray observations from air Cherenkov telescopes (MAGIC, Albert et al., 2005; HESS, Aharonian, 2007; and VERITAS, Holder et al., 2006) allowed a detailed survey of the accessible TeV gamma-ray sky and led to the discovery of about 100 TeV gamma sources (Aharonian, 2007; De Angelis et al., 2008)), most of which are harbored in the Milky Way or in close galaxies. Most of these sources show gamma-ray emission features confirming the "standard electromagnetic" astrophysical scenario, in which high-energy gamma ray fluxes are likely to originate from relativistic electrons/positrons through synchrotron and inverse Compton scattering. Nevertheless, some sources – namely galactic supernova remnants close to molecular clouds – show gamma-ray emission compatible with a "hadronic acceleration" scenario. In a simplified picture, charged particles (protons and heavier nuclei) accelerated in shock waves at the source (Fermi-Bell acceleration), can interact with close molecular clouds that provide dense gas targets. The outcome is a flux of high-energy gamma rays and neutrinos generated respectively by neutral and charged pions, produced in the beam dump. Since TeV gamma ray detection cannot definitively rule out purely electromagnetic processes, the observation of astrophysical neutrinos will be the "smoking gun" to demonstrate that supernova remnants are the sources of the detected cosmic ray spectrum. Recent papers (Villante and Vissani, 2008; Morlino et al., 2009) suggest that some galactic supernova remnants can be within the reach of a km^3-scale neutrino detector.

Another class of candidate galactic neutrino sources contains microquasars, binary stellar systems formed by a compact object called an "accretor" (a neutron star or black hole) and a "donor" star orbiting around it. Microquasars exhibit relativistic radio jets (Mirabel and Rodriguez, 1998). The detection of very high-energy gamma rays from microquasars (Khangulyan et al., 2008) clearly demonstrates that these objects are sites of effective acceleration of charged particles to multi-TeV energies.

A detectable high-energy neutrino flux is expected from the Galactic Centre region, a complex environment extending over about 500 pc composed by the candidate super-massive black hole Sagittarius A*, a population of supernova remnants and other compact and extended objects and molecular clouds, studied in multi-wavelength surveys. The Galactic Centre region has already been identified as an intense source of TeV gamma rays (Aharonian et al., 2006) and is a major candidate source for cosmic neutrino detection.

Unlike galactic gamma-ray sources, for which the knowledge of the very high-energy gamma-ray spectrum provides a rather good estimate of the expected neutrino fluxes, the situation is more complex in the case of extragalactic sources such as active galactic nuclei (AGN) and gamma ray bursters (GRB).

As already mentioned, gamma-ray absorption on the infrared and cosmic microwave background does not allow for the detection of very high-energy gamma-ray sources at distances larger than about 100 Mpc. On the other hand, protons having $E > 10^{19.5}$ eV, likely accelerated in these extragalactic sources, are considered good astrophysical probes being only slightly bent by cosmic magnetic fields. The interaction of these protons with cosmic microwave background, however, reduces the region of the Universe accessible to extremely high-energy proton detectors, and thus the number of candidate sources suitable for proton astronomy. Present experimental results concerning correlation between extremely high-energy proton arrival directions and active galactic nuclei are, indeed, still inconclusive (The Pierre Auger Collaboration, 2010). Active galactic nuclei and gamma ray bursters are also expected to produce neutrino fluxes extending up to 10^{19} eV, well above the other known neutrino sources, but the neutrino flux estimates are very uncertain. Waxman and Bahcall (1999) set an upper bound (the so-called Waxman & Bahcall limit) for the high-energy neutrino diffuse flux, that can be detected at the Earth. Although the Waxman & Bahcall limit ($10^{-7.5}$ GeV cm^{-2} s^{-1} sr^{-1}) relies on several uncertain parameters, it is considered as a "reference" for the expected neutrino fluxes and sets the dimensions of high-energy neutrino telescopes to km^3-scale.

Active galactic nuclei (AGN), the most luminous persistent object observed in the sky, are also candidate point-like neutrino sources. The standard AGN scenario assumes the presence of a very massive central black hole ($10^{6\div8}$ solar masses), swallowing huge quantities of surrounding matter from an accretion disk, and two relativistic jets where particles are accelerated up to the highest energies. Hadronic acceleration mechanisms were proposed to describe the observed gamma-ray emissions (Mannheim, 1995). Neutrinos can be produced by high-energy proton beam dump in several regions: (a) close to the AGN core (a "thick" environment that does not allow escape of both cosmic rays and gamma rays); (b) inside the AGN jet from protons accelerated by internal shocks; and (c) close to the radio lobes, at the end point of the jets (Torres and Anchorddoqui, 2004). However, the most optimistic neutrino production models (Stecker et al., 1991) were already excluded by observations and the question about the origin of the high-energy gamma emission observed in AGNs is still open.

Gamma ray bursters are astrophysical objects originating extremely luminous flashes of gamma rays: the gamma ray bursts (GRB). The total energy release of a typical GRB is huge (order of 10^{51} erg – 10^{44} J). Their emission in hard X-ray and MeV gamma rays lasts from a millisecond to several hundreds of seconds, with a late afterglow in the infrared, radio and optical bands (for a review, see Meszaros, 2006). The leading model for the gamma-ray production in gamma ray bursters MacFadyen and Woosley, 1999) assumes that in the catastrophic event due to collapse or merger into a black hole, surrounded by rapidly accreting masses, a fireball is created which expands at a highly relativistic velocity. The nonthermal features that dominate the gamma-ray spectrum are therefore due to charged particle interaction with the shock waves created as a consequence of the fireball expansion. On the basis of energetics, Waxman (1995) proposed a scenario where extremely high-energy cosmic rays are produced by gamma ray bursters via a Fermi mechanism occurring in internal shocks. Gamma ray bursters are thus expected to emit neutrinos during several stages of their evolution. First of all, although never observed, quite a large fraction of the whole energy released is expected to be carried out by low-energy neutrinos (at about 10 MeV) and gravitational waves in the first stages following the collapse. A "precursor" TeV neutrino flux, without gamma counterpart, due to the large optical depth of the medium, is expected about 100s before the gamma-ray flash, that is seen only when the jet escapes the external progenitor shells (Guetta and Granot, 2003). Then, hadron acceleration in both internal shocks (jet) and external shocks (afterglow) would lead to high-energy neutrino production due to proton–proton or proton–photon interaction. GRB neutrino detection is extremely appealing. In fact, this detection is almost background free, triggered by satellite alerts that allow tuning the search for neutrino signals from the burst direction in a time window around the GRB. However, first operations of neutrino detectors have not yet led to GRB neutrino observations (Abbasi et al., 2003).

A "guaranteed" flux of high-energy neutrinos is expected to be produced, through the GZK mechanism, by the interaction of $E>10^{19}$ eV protons with the cosmic microwave background (Berezinsky and Zatsepin, 1969). The energy spectrum of cosmogenic neutrinos is expected to span the range between approximately 10^{16} eV$<E<10^{20}$ eV, while the shape and absolute flux depend on several factors such as the primary cosmic ray flux, spectral index and composition. This leads to theoretical predictions showing remarkable discrepancies in the expected cosmogenic neutrino fluxes (Protheroe and Johnson, 1996; Kalashev et al., 2002; Ahlers et al., 2010). The recent results by the Pierre Auger Observatory (The Pierre Auger Collaboration, 2010) seem to show that composition of cosmic ray flux becomes heavier with increasing energy. A possible interpretation is that the observed cosmic ray flux reduction is only due to limited, intrinsic, acceleration power of the astrophysical sources. In this latter case the net effect will be a very limited flux of cosmogenic neutrinos.

3.3 High-energy neutrino detection

At energies above a few tens of GeV, neutrinos are detected through deep-inelastic scattering of the neutrino with a target nucleon. In neutrino–nucleon interaction at high energy ($E_\nu>1$ TeV) about 80% of the initial neutrino energy goes, on average, to a lepton (Quigg and Reno, 1996), which preserves the neutrino "flavour" (i.e., electron e, muon μ or tau τ). In the so-called neutral current interactions, the outgoing lepton is a neutrino, thus only a "hadronic" cascade is detectable; in charged-current interactions the neutrino interaction produces both a charged lepton and a hadronic cascade.

Since the neutrino cross-section (thus its probability of interaction with matter) increases as a function of energy, at $E > 100$ TeV neutrinos strongly absorbed crossing the Earth diameter. This effect plays an important role on the energy range accessible to different experimental techniques.

Due to the low neutrino cross-section and to the faint expected astrophysical fluxes, cosmic neutrino detectors must have a huge interaction target mass: of the order of several

GTons for neutrinos in the 10^{12} eV<E<10^{17} eV energy range and even larger at higher energies. Moreover, in order to act as telescopes, astrophysical high-energy neutrino detectors must be designed to reconstruct both the neutrino energy and direction. Among different experimental techniques suggested to detect high-energy neutrinos, the underwater/ice Cherenkov technique is, at present, the most promising and advanced.

3.3.1 The Cherenkov detection technique in transparent natural media

This technique, first suggested by Markov and Zheleznykh (1961), is based on instrumenting a large volume of a natural transparent medium (sea or lake water or polar ice) in order to detect the charged leptons (in particular muons, as we will discuss) emerging from a charged current (CC) neutrino interaction. Neutrino detectors of this type are comprised of a large array of optical sensors, typically photomultiplier tubes (PMTs), which allow the tracking of the charged leptons by detecting the Cherenkov light wave-front radiated by these particles.

In water and ice, relativistic particles radiate Cherenkov light mainly in the UV-blue wavelengths. In both media, the refractive index in this spectral region is $n \approx 1.35$, and photons are emitted along the particle track at an angle $\vartheta \approx 42°$ and symmetrically in azimuth. The time sequence of Cherenkov photons hits on the sensors constituting the array is thus correlated by a space-time causality relation used to reconstruct the Cherenkov wave front, and therefore the particle track, from the experimental data.

The reconstructed track direction is, however, affected by some experimental uncertainties: the uncertainty on the knowledge of the PMTs' positions and the uncertainty on the absolute photon hit time, which in turn is due to a combination of several effects (photon scattering in the medium, PMT transit time spread and detector time calibration). Photon scattering in the medium is an important issue, since scattering deflects Cherenkov photons, affecting the track direction reconstruction. In ice the scattering length of blue light is only few tens of cm – though most of the light is forward scattered at narrow angles – while in water it is about 100m (Mobley, 1994).

The outgoing muon energy converted into Cherenkov radiation is only a tiny fraction of the total one, and the number of Cherenkov photons (UV-blue) is only 300 per cm of the muon track. Given this small amount of light, photons can be detected only if produced at a distance from the sensor not much larger than the light absorption length in the medium. Ultimately, the medium optical properties determine the detector granularity (i.e., the optical sensor density) and its size. Water is transparent only to a narrow range of wavelengths (350 nm ÷ 550 nm). In particular, for deep polar ice the absorption length of blue light is larger than 100m (Ackermann et al., 2006), and it is about 70m for clear ocean waters (The KM3NeT Consortium, 2008). Assuming an average spacing between sensors of 100m, at least 5000 optical sensors are needed to fill up a sensitive volume of about 1km^3 (Katz and Spiering, 2012).

The "golden channel" for astrophysical neutrino detection is the muon one. High-energy muons (E>TeV), that are produced by charged current neutrino interactions, have a range in water of the order of kilometres. Therefore, the neutrino interaction can take place either within the detector or far outside of it, providing a flux of high-energy muons, either contained or crossing the detector. The muon direction is reconstructed from the Cherenkov wave front, radiated along the muon track, within the detector "instrumented" volume.

Since the average angle between the direction of the secondary muon and that of the interacting neutrino is, for energies above 1 TeV less than 1° ($< \vartheta_{\mu\nu} > \approx 0.7\ E_\nu[\mathrm{TeV}]^{0.6}$), the tracking of the muon allows to point back at the cosmic n source, thus allowing for *neutrino astronomy*. Photon yield is also effectively used to estimate the neutrino energy, since the muon generates high-energy showers along its path, strongly increasing the total number of photons. The charge dynamic range of the optical sensors and their readout electronics is therefore an important parameter. For the muon neutrino detection, up-going or horizontal muon tracks are preferred. In fact, when an upward-going muon is reconstructed this is a unique signature of a neutrino event, since the up-going atmospheric muon background is completely filtered out within few tens of km of water. The suppression of the intense down-going atmospheric muon flux is achieved by installing the detector at large water (-ice) depth: the muon stopping power of 3000m of water is equivalent to the one of 1km of rock. Water and ice have, therefore, a threefold function: huge (and inexpensive) neutrino target, Cherenkov light radiator and shield against cosmic muon background and daylight.

Neutrino telescopes are also expected to distinguish between neutrino flavours by reconstructing the Cherenkov wave-front shape of the event, which depends on the different propagation of e, μ and τ in water (and ice). In the case of ν_e charged current interactions, the final state involves high-energy electrons that provide a high-energy electromagnetic shower superimposed on the hadronic one. Both showers extend for a few tens of metres from the neutrino interaction point; thus, only interactions that are fully contained into the detector instrumented volume, or very close to it, can be identified. At a distance of few hundred meters from the shower, the shape of the light wave front is similar to an expanding sphere, and the neutrino direction can be reconstructed with an angular resolution poorer compared to the muon channel one. On the other hand, showers involve a large number of charged particles radiating Cherenkov light and, in this case, the lepton energy can be well estimated from the shower light yield. The scenario depicted for ν_e is similar to the case of tau neutrino detection: up to 1 PeV the unstable tau particle decays after a very short path in water and it is almost impossible to disentangle the shower produced by the tau decay from the hadronic shower originated at neutrino interaction vertex. When the tau path is about 100m long, the two cascades can be separated and the event topology shows the typical signature of a "double light bang". In the case of $\tau \rightarrow \mu\nu_\mu\nu_\tau$ decay, the event will look like a ν_μ charged current interaction.

A parameter usually quoted to describe detector performance for muon neutrino detection is its effective area for neutrino or muon fluxes, i.e., the surface intersecting the neutrino or neutrino-induced muon flux multiplied by the apparatus detection efficiency for muons. Assuming the neutrino flux given by Waxman and Bahcall one gets a rate of about 10^2 events per year for a $1\mathrm{km}^2$ effective area detector with $E \approx 1$ TeV threshold. This number sets the scale of dimension for astrophysical neutrino detectors. Due to photon detector and installation costs, the affordable size of this apparatus is of the order of few km^3. This size is optimal for the exploration of the $10^{11}\ \mathrm{eV} < E < 10^{17}$ eV energy range. The detailed study of detector performance requires Monte Carlo simulations that have to take into account the detector layout, the characteristics of the Cherenkov radiator which surrounds the detector (light refraction index, light absorption and scattering lengths) and the sources of background (The KM3NeT Consortium, 2010).

The major sources of background for astrophysical discoveries with underwater Cherenkov neutrino telescopes are the atmospheric neutrino and muon fluxes. Atmospheric muons and neutrinos are produced by the interaction of primary cosmic rays with the atmosphere. The cosmic muon flux decreases below sea surface as a function of depth and as a function of zenith angle: it falls to zero near the horizon and below where the large slant of water and the Earth shield all the muons. This is the reason why astrophysical neutrino signals are mainly searched among upward-going muons, which can only originate from interactions of neutrinos crossing the Earth. At 3000m depth, an underwater neutrino telescope "observes" a cosmic muon flux still about 10^6 times higher than the up-going atmospheric neutrino signal (that is, neutrinos that undergo interaction and produce muons), thus accurate reconstruction procedures and quality cuts are needed to get rid of the atmospheric muon tracks mis-reconstructed as "fake" up-going. The atmospheric neutrinos are an unavoidable source of background and only energy cuts and statistical arguments on direction allow discrimination of these events from astrophysical ones during data analysis. In fact, the atmospheric neutrino flux is expected to produce diffuse events with a spectrum that falls steeply with energy (Honda et al., 2007) while astrophysical neutrino fluxes are expected to have a harder spectrum. Diffuse neutrino fluxes searches look for an excess of signals above the atmospheric neutrino flux for a given energy threshold of about 10^{15} eV. Searches for point-like sources look for accumulations of neutrino events in a narrow region of the sky, whose dimension is essentially given by the detector angular resolution.

Both atmospheric muons and neutrinos are, besides being sources of background as discussed above, very useful sources of calibration for the detector. In particular, atmospheric muons can be used to verify both the energy calibration of the detector and the pointing accuracy of the neutrino telescope and its absolute positioning by detecting the Moon shadow. The Moon acts, indeed, as a filter for primary cosmic rays, thus a suppression of the atmospheric muon flux can be measured in the direction of the Moon, allowing also calibration of the absolute and relative pointing accuracy of the detector (The KM3NeT Consortium, 2010). The effectiveness of the "Moon shadow" reconstruction technique has been recently demonstrated by detailed analysis of the IceCube detector data (Boersma et al., 2011).

3.3.2 Underwater Cherenkov neutrino telescopes

The construction of km^3-scale neutrino Cherenkov detectors both under ice and in the deep sea has required long R&D activities in order to overcome the huge technological challenges arising from the hostile operational environment, such as high pressure, low temperature and corrosion.

After the pioneering experience made by the DUMAND Collaboration offshore the Hawaii Island (Roberts et al., 1992), the Baikal collaboration was the first to operate a neutrino telescope underwater. The detector is deployed in Lake Baikal (Siberia) at depths between 1000 and 1100m (Wischnewski, 2005). In 1998, Baikal NT-200 was completed and has taken data since then. Baikal NT-200 is an array suspended in lake water from an "umbrella-like" mechanical structure, with a 72m height and a diameter of 43m. It is composed of eight strings, each with 24 pairs of down-looking optical modules (OM), forming a high granularity detector (Ayutdinov, 2009). A rather low light transmission length in water, about 20m, high sedimentation, bio-fouling and a high optical background rate due to bioluminescence limit the NT-200 detector's performance as a high-energy astrophysical neutrino telescope. In order to improve the reconstruction of high-energy cascade events, an upgrade of the NT-200 telescope, named NT-200+, was undertaken in 2005 by adding three outer strings. Each string of NT-200+ is made of 12 OM each with a larger spacing. NT-200+ has an enclosed detector volume of about 5 MTons and it is expected to increase the sensitivity to diffuse fluxes by almost a factor of four with respect to NT-200.

The largest effort to date in this field was made by the IceCube collaboration. Since the end of the 1990s, the collaboration carried out the construction and operation of AMANDA (Antarctic Muon And Neutrino Detector Array) (Andres et al., 2000) and is now operating the first km^3-scale neutrino telescope: IceCube. Both detectors were built at the Amundsen-Scott South Pole Station. AMANDA was a first-generation instrument that served as the test bench for technologies and as the prototype for IceCube. The operation of AMANDA (in its final configuration consisting of 677 OMs arranged in 19 strings (Ahrens et al., 2004)) started in 2000 and ran until its decommissioning in 2009, accumulating a total exposure of about four years. The high-energy atmospheric neutrino flux has been measured with unprecedented statistics. A search for astrophysical neutrino point sources was performed, albeit with a rather poor angular resolution of about 2 degrees. No excess with high statistical significance was found.

The installation of the IceCube km^3 detector started in 2004 and was completed at the beginning of 2011. In its complete configuration, IceCube consists of 86 vertical strings, each equipped with 60 digital optical modules (DOM), deployed between 1450m and 2450m below the surface, as shown in Figure 3.3. The DOM spacing along the string is 17m and the strings are placed on a hexagonal layout with 125m spacing, giving $1km^3$ of instrumented volume. An improvement of the telescope sensitivity in the low-energy range was also achieved by the addition of six more densely instrumented strings that are deployed in the bottom centre of the telescope to form the so-called DeepCore detector that will lower the IceCube threshold for muons to about 10 GeV, thus allowing IceCube to address more effectively low-energy physics issues and especially to increase the sensitivity for the indirect search of Dark Matter (Resconi, 2009). In addition, on the ice surface above the IceCube strings, the IceTop air shower array is installed. IceTop consists of 180 ice-filled tanks of about $1m^3$ volume, each equipped with two DOMs (Gaisser, 2011). IceTop allows study of high-energy cosmic ray showers.

Thanks to much larger volume and better angular resolution (less than 1°), the sensitivity for neutrino point-like sources with IceCube is strongly improved compared to AMANDA. The full neutrino sky-map seen by IceCube including both up-going and down-going neutrinos is reported in Figure 3.4. Down-going events are selected using an energy threshold ($E > 10^{14}$ eV) sufficiently high to get rid of atmospheric muons. No significant excess from any sky direction has been observed up to now (Abbasi et al., 2011).

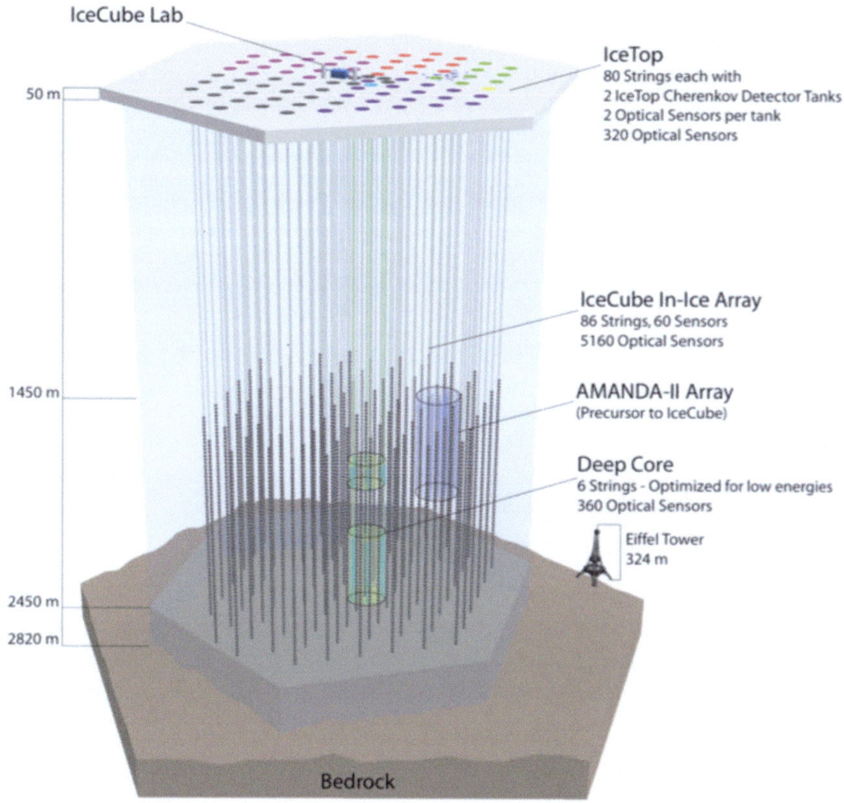

Figure 3.3 Schematic view of the IceCube detector at the South Pole (see: http://icecube.wisc.edu/).

3.4 Towards a deep-sea infrastructure for neutrino astronomy and Earth and sea science in the Mediterranean Sea

The Mediterranean Sea offers optimal conditions, on a worldwide scale, for locating an underwater neutrino telescope: clear sea waters, long periods of good weather conditions for sea operations, proximity to large industrial, commercial and scientific infrastructures. For more than a decade, the feasibility of neutrino astronomy with a detector in the depths of the Mediterranean Sea has been investigated in three pilot projects. In each of these, different configurations and techniques have been explored. These three projects (ANTARES,

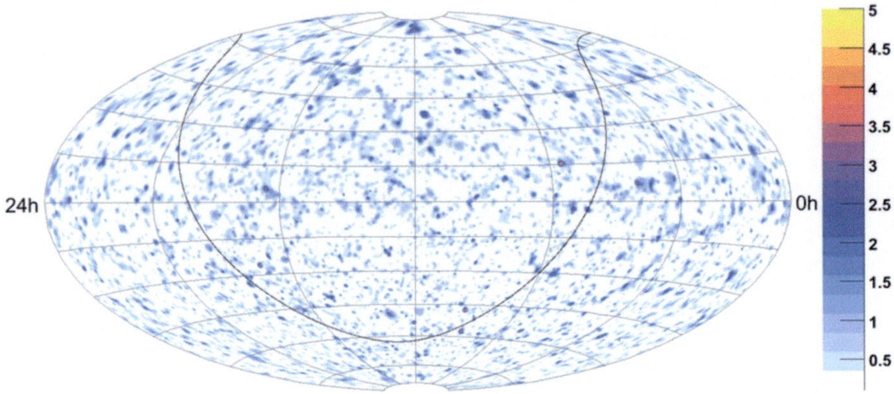

Figure 3.4 The neutrino sky seen by IceCube (see: http://icecube.wisc.edu/).

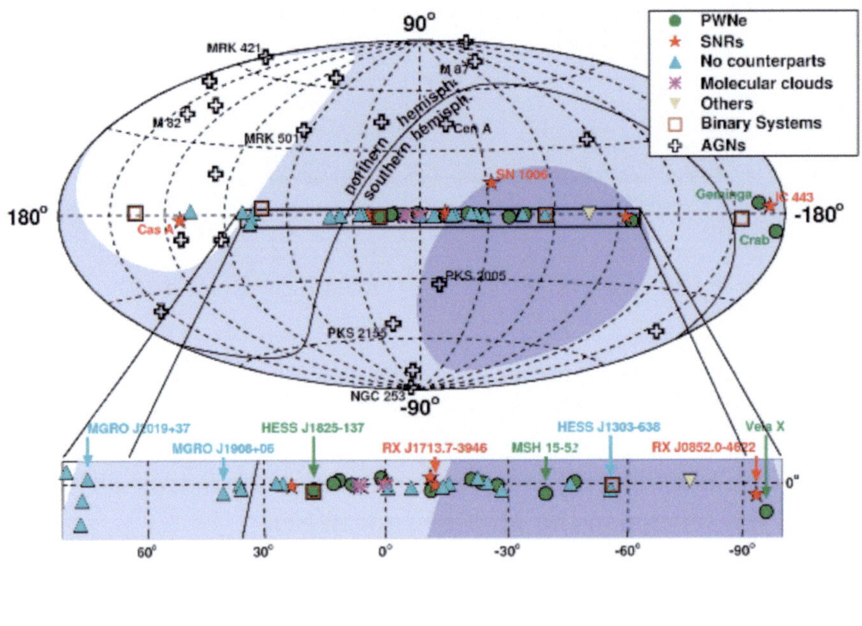

Figure 3.5 Sky coverage in Galactic coordinates for detectors located in the Mediterranean Sea and at the South Pole, where only the northern hemisphere can be observed.The shading indicates the visibility for a detector in the Mediterranean with 2π downward coverage; dark (light) blue areas are visible at least 75% (25%) of the time. The locations of observed sources of high-energy γ rays are also indicated (The KM3NeT Consortium, 2010).

NEMO and NESTOR, described below) have reached maturity and have demonstrated the potential of the detection technique by reconstructing the tracks of muons, the possible reaction products of the sought-after neutrinos. These projects have provided a wealth of information on the technologies required for a large deep-sea neutrino telescope.

More recently, the scientists of the three collaborations have joined the KM3NeT consortium (see http://www.km3net.org), funded by the EU under the sixth and seventh framework programmes, with the aim of designing and building a deep-sea infrastructure hosting a multi-km^3 neutrino detector and nodes for Earth and sea sciences. In recognition of its unique, multidisciplinary scientific potential and its advanced technical status, KM3NeT was selected as one of the European priority research infrastructures by the ESFRI panel (see http://ec.europa.eu/research/esfri/).

The telescope's location in the Mediterranean Sea allows the search for up-going neutrinos from a larger fraction of the Universe compared to IceCube. Moreover, KM3NeT will survey a large part of the Galactic Plane, including the Galactic Centre, which is not in the field of view of IceCube (Figure 3.5). The most intense gamma-ray emitters supernova remnants known up to date, RXJ1713.7-3946 and Vela-Jr (RXJ0852.0-4622), are both clearly in the field of view of KM3NeT.

Thanks to larger dimension and total photocathode area (more than 6 times than IceCube) and thanks to better properties of seawater as a blue light radiator, KM3NeT is expected to reach an angular resolution better than 0.1° above 30 TeV, increasing the signal (astrophysical ν) to background (atmospheric ν) ratio for point-like source searches, thus allowing for a discovery potential significantly exceeding that of IceCube (Figure 3.6) (Mobley, 1994).

3.4.1 NESTOR: Neutrino Extended Submarine Telescope with Oceanographic Research

NESTOR was the first collaboration that suggested the deployment of a neutrino telescope in the Mediterranean Sea, in a site offshore Pylos (Greece) (Trasatti, 1999), where seabed depth ranges between 3800m and 5000m. A semi-rigid structure (the NESTOR *tower*), 360m high and 32m in diameter, equipped with 168 PMTs was designed as "detection unit" for the neutrino telescope. The basic element of the NESTOR tower is a hexagonal floor or star with two optical modules, one upward-looking and the other downward-looking, at the apex of each arm. The optical modules consist of a 15-inch diameter photomultiplier tube enclosed in a spherical glass housing which can withstand the hydrostatic pressure up to 630 bar. The electronics is placed at the center of the floor, housed in a 1m-diameter titanium sphere.

In March 2003, such a floor, with a reduced size of 12m and equipped with 12 PMTs, data acquisition and transmission electronics, and associated environmental sensors, was deployed. The cable for connection to shore, previously deployed at 3850m and connected to the offshore station, was brought to the surface so that the floor was attached, cabled and redeployed to a depth of 3800m. This array was operational for about one month and the 745 down-going muon reconstructed events allowed to measure cosmic ray muon flux at the installation depth (Aggouras et al., 2005a, b)

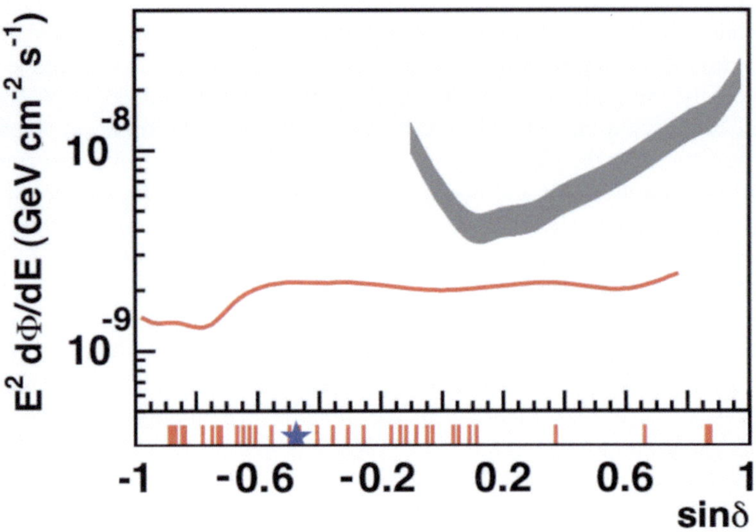

Figure 3.6 The discovery potential of KM3NeT for neutrino point sources with an E^{-2} spectrum as a function of source declination (full line) compared to the corresponding IceCube values (shaded band). Both curves are for one year of data taking. The red ticks at the bottom of the horizontal axis show the positions of Galactic gamma ray sources; the position of the Galactic Centre is indicated by a blue star (The KM3NeT Consortium, 2010).

3.4.2 ANTARES: Astronomy with a Neutrino Telescope and Abyss environmental RESearch

ANTARES is at present the largest neutrino telescope operating in the Northern Hemisphere (Bertin et al., 2014, this volume). The R&D activities towards the construction of ANTARES started in 1996, including site evaluation campaigns and the construction of some prototype lines that allow testing of critical components and technologies.

The ANTARES telescope (Figure 3.7) is made of 12 strings, about 70m apart, each with 25 storeys vertically spaced by 14.5m. Each storey is equipped with three pressure-resistant OMs, each one containing a 10in PMT oriented at 45° down-looking. The covered footprint is about 0.03km^2. In addition, a so-called instrumentation line (IL07) equipped with several oceanographic probes is part of the telescope (see Section 3.6). The strings are moored at a depth of 2475m and interlinked to a junction box connected with a 45km long electro-optical cable to the shore station at La Seyne sur Mer (close to Toulon, France).

The site shows underwater currents which change in intensity and direction as a function of time; this effect is correlated with variation of optical background due to bioluminescence activity that can reach values of several hundred kHz (Vallage, 2006). The data transmission is based on the "all-data-to-shore" concept, i.e., all PMT signals above a 0.3

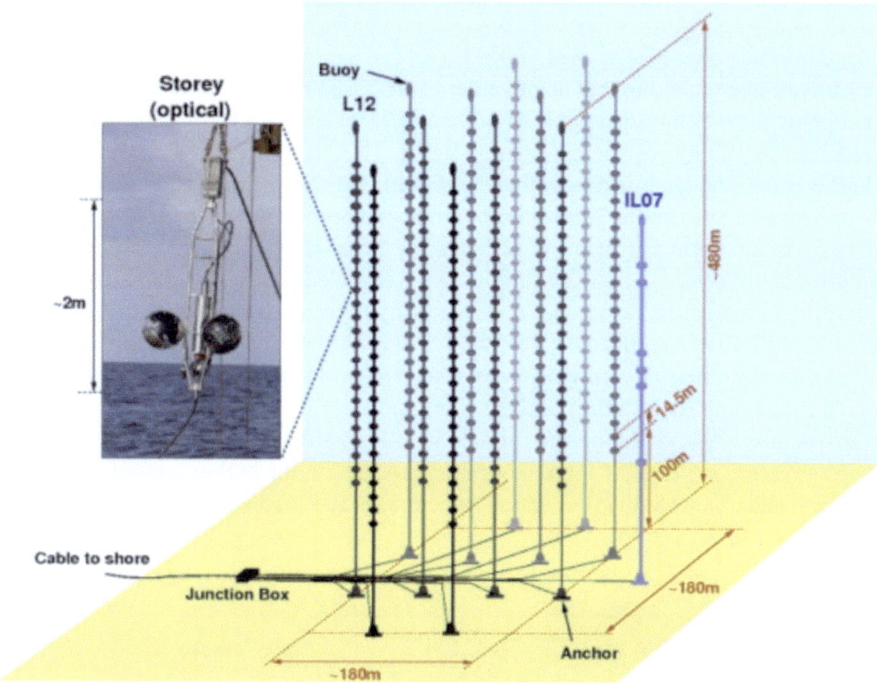

Figure 3.7 Scheme of ANTARES (Abassi et al., 2011).

photo-electron threshold are sent to shore where a computer farm performs the filtering. Several different triggers looking for specific neutrino signals are implemented. Thanks to its high PMT granularity, the ANTARES telescope has an energy neutrino threshold of about some tens GeV for reconstructed muon events and detector performance dramatically improves with increasing energy reaching at neutrino energy of 10 TeV an angular resolution of 0.3 degrees. Data taking started in March 2006 after the deployment of the first detector line and the installation of all the 12 lines was completed in May 2008. Recent results of the ANTARES detector are reported elsewhere in this publication.

3.4.3 NEMO: NEutrino Mediterranean Observatory

Since 1999, the NEMO Collaboration has carried out an R&D programme aiming at the construction of a km^3-scale neutrino telescope at depths of over 3000m (Migneco et al., 2008). After an intense site exploration and monitoring activity, a deep-sea site, located at a depth of 3500m and 80km off the Sicilian coast, was selected. This site was extensively studied to characterize it for the installation of the km^3 detector. Optical properties, such as light adsorption, attenuation length and optical background, have been studied. A light absorption length of 70m (close to optically pure water) in the blue region of the spectrum and a optical background of 30 kHz on 10 inch PMTs (at 0.5 s.p.e. threshold) have been measured. Oceanographic properties were also studied by measuring deep-sea currents and the sedimentation rate (Riccobene et al., 2007).

In order to validate technological achievements proposed for the km^3 detector, the Collaboration operated the demonstrator NEMO Phase-1 (Figure 3.8, left) in a test site located 2100m undersea and 25km offshore Catania (Italy). The NEMO Phase-1 project started in 2002 and it was completed in December 2006 with the deployment and connection of a submarine junction box and a 300m high 4-floor prototype tower structure hosting 16 OMs. All the key components of an underwater neutrino detector were included: optical and environmental sensors, power supply, front-end electronics and data acquisition, time and PMT position calibration systems, slow control systems, on-shore data processing (Ameli et al., 2008). Five-months data were analyzed providing information on the correct behaviour of the apparatus and allowing the measurement of the vertical muon intensity as a function of the angular distribution of the muon (Aiello et al., 2010).

An infrastructure at the candidate site for the km^3 installation has recently been completed. This infrastructure includes: a shore laboratory (located in the harbor area of Portopalo di Capo Passero and connected via dedicated optical fiber link to the high speed internet); a

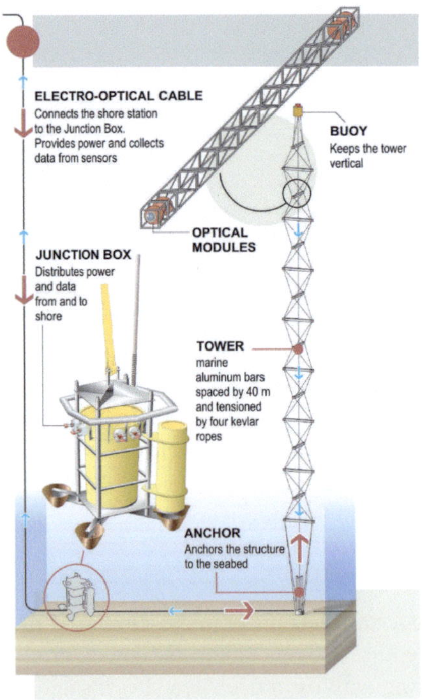

Figure 3.8 (Left) Artist's view of the NEMO Tower. (Right) the NEMO Phase II tower ready for deployment.

100km long cable with 20 optical fibers and single copper conductor, which links the shore station to the underwater infrastructure; a 10kW deep-sea DC/DC power converter. The shore lab is also connected via dedicated optical fiber link (up to 10 Gbps) to the internet, permitting data distribution in real time to collaborating institutes. The site will soon host the NEMO Phase-2 detector, a prototype tower detection unit equipped with 32 optical modules, 18 hydrophones and several oceanographic probes that will be the deepest neutrino detector prototype in the Mediterranean Sea (Figure 3.8, right).

3.4.4 KM3NeT

In 2010, the KM3NeT consortium reached an important milestone with the publication of the Technical Design Report, in which the investigation of different technical solutions concerning all aspects of the design, construction, installation and maintenance of the telescope and their impact on physics performance are reported (The KM3NeT Consortium, 2010). In parallel, the consortium is addressing, within the framework of the preparatory phase, the political, funding, governance and strategic issues that need to be settled before the start of construction. The full KM3NeT detector will consist of 300 vertical structures, called detection units (DUs) that are anchored on the seafloor and kept vertical by subtended buoys. Each detection unit supports OMs to form a 3D-array of photosensors. The detection units are connected to shore via a sea-bottom network of electro-optical cables and junction boxes for power distribution and data transmission. Sea-bottom connections between the DUs and the cable and junction box network are carried out through the use of deep-sea remotely operated vehicles (ROVs).

The full telescope of 300 detection units will consist of several smaller building blocks: it has been noticed that the sensitivity improves slightly when the detection units are arranged in this way (The KM3NeT Consortium, 2010). The reason is that the angular resolution does not improve when the tracklength exceeds about a kilometre, whereas the effective volume of the telescope is merely the product of the range of the produced muon and the surface subtended perpendicular to the direction being investigated. The latter is larger when the same number of towers is distributed over smaller blocks. Technically, the separation of the telescope into a network of more and smaller building blocks is also advantageous as power and data distribution becomes simpler. The impact of possible failures is reduced and contained within these smaller blocks.

For the installation of the detector three candidate sites, shown in Figure 3.9, are proposed. The site choice as well as the possibility of a multisite option is one of the strategic issues that will be addressed in the close future.

Using the ANTARES facilities, the Consortium is presently undergoing the installation of demonstrator digital readout optical modules (DOMs) equipped with 31 small area PMTs and acoustic sensors. A pre-production model detection unit, equipped with 3 DOMs, will be deployed in mid 2013. A major step of the project will be the installation of KM3NeT Italia that foresees the deployment of about 20 detection units in the Capo Passero site, before the end of 2015.

KM3NeT will also provide dedicated science nodes for Earth and sea science. These nodes will benefit from the infrastructure power feeding and data-transmission system and can provide information for detector environmental monitoring. Data from calibration

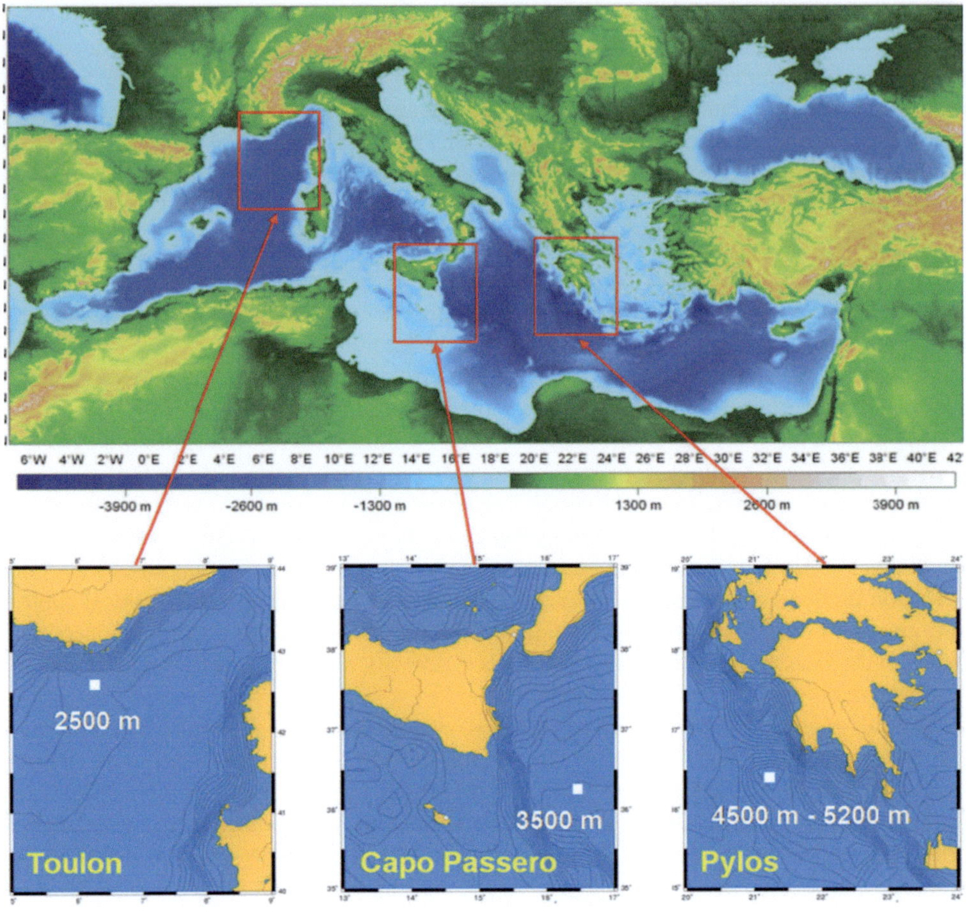

Figure 3.9 The candidate deep-sea sites for the KM3NeT detector.

systems, of multidisciplinary interest, could also be shared among the two communities, making of KM3NeT the largest underwater multidisciplinary laboratory.

In the following section, the design concepts for the final design and the most recent developments on prototyping activities are described.

3.4.4.1 Optical modules

The optical module is a glass vessel that houses the photomultiplier tubes (PMTs) and associated equipment, protecting them against hydrostatic pressure and seawater. Borosilicate glass spheres with diameters of 13 or 17 inches (about 33cm and 43cm, respectively) are commercially available and have been extensively used in all the deep-sea and ice experiments.

The main requirements for the optical modules are a large photocatode area and a good optical background rejection that is based on the multiphotons versus single photon discrimination. The major active optical component of the KM3NeT optical module is the photomultiplier. Unlike the standard optical modules used by ANTARES and NEMO, that use OMs equipped with a single large area PMT, the KM3NeT DOM is composed of a glass sphere that houses 31 PMTs (3 inch diameter – about 7.6cm – 19 in the lower hemisphere and 12 in the upper). Following the IceCube example, all digitizing and readout electronics is also housed in the digital optical module (Figure 3.10). Finally, the DOM includes instrumentation that allows for the reconstruction of its position (acoustic sensors), determination of the orientation (compass and tilt-meter) and calibration of the timing (LED nano-beacon). The electronics necessary for the readout of the DOM is based on an FPGA that also controls the photomultiplier high voltages and the additional sensors in the DOM. The DOM "multi-PMT" configuration provides high efficiency, photon counting, and purity of the signal and some information on the photon arrival direction that can be used to improve track reconstruction.

3.4.4.2 Detection unit mechanical structure

The sensors must be distributed over a large volume of seawater and, to that end, are supported on narrow vertical structures that are kept on the seabed by a deadweight and held vertical with the aid of a buoy at the top of the structure. For the performance of the detector, it turns out to be advantageous to have a structure horizontally extended. This allows for more efficient triggering and an increase of efficiency for low-energy neutrinos and a reduction of ambiguities in the tracking. The size of the horizontal extent is a compromise in performance and ease of handling of the structure during assembly and

Figure 3.10 The KM3NeT Digital Optical Module (DOM) prototype (The KM3NeT Consortium, 2010).

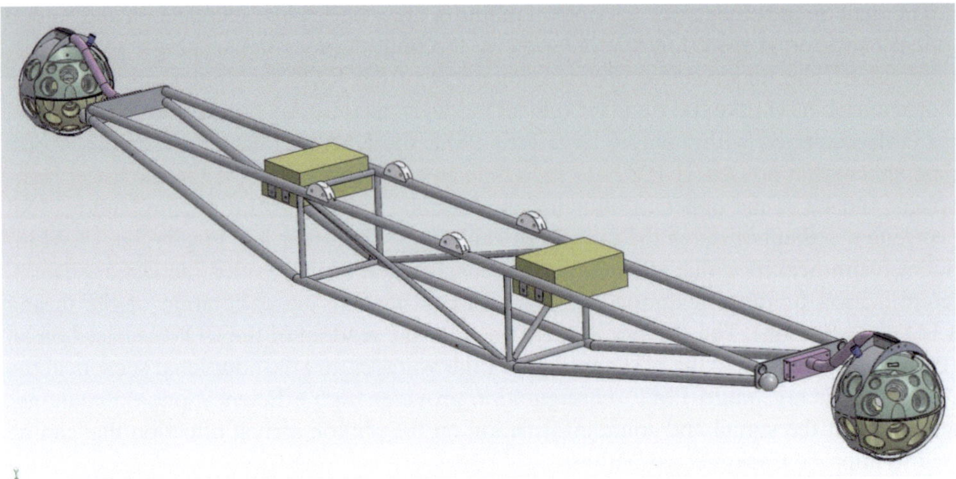

Figure 3.11 A storey of the KM3NeT detection unit (The KM3NeT Consortium, 2010).

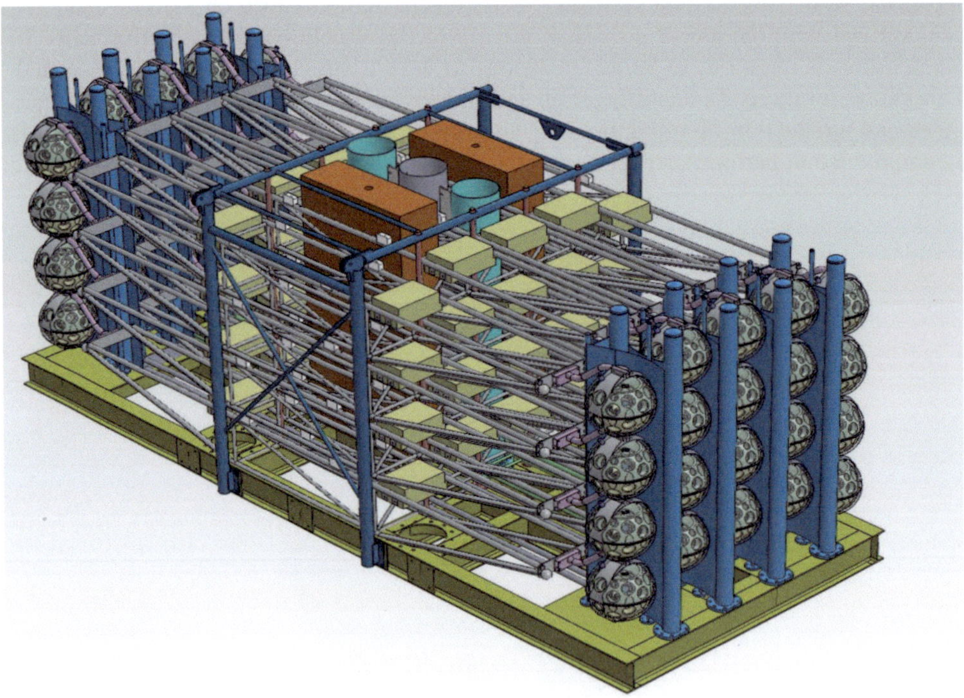

Figure 3.12 The KM3NeT detection unit in its packed configuration (The KM3NeT Consortium, 2010).

deployment on the seabed. These considerations have led to a structure of 6m length with a DOM positioned at each end (Figure 3.11). Twenty such storeys are then suspended, alternately perpendicular to each other, from four 5mm Dyneema® ropes with an inter-storey distance of 40m. The active storeys start at 100m from the seabed, so that the total height of the structure (a tower) is 860m. Two cables, carrying optical fibers for the readout and conductors for the power, run the length of the structure. For the deployment a full tower is brought as a single package to the seabed (Figure 3.12). The buoy is released and the storeys are pulled from the package one-by-one. During this process the ropes are unwound from braked synchronized drums, thus being kept taut. The read-out cables are stored, spiralled around two of the ropes. The structure includes just sufficient buoyancy to stabilize the structure during unfurling, while still allowing for a controlled release of each storey. Hydrodynamic calculations have shown that the structure can be held vertical within 150m (top displacement) in sea currents of up to 30cm per second.

Another mechanical option under study is the use of ANTARES-like stings, equipped with DOMs; despite an expected reduction of track reconstruction performances of the single detection unit, the string solution provides a well-tested alternative to 3D detection units.

In both cases, regular calibration on positioning and orientation is necessary to account for the movement of the DUs in the sea current. As in ANTARES, acoustic triangulation methods will be applied, together with orientation measurements by compasses and tilt-meters.

3.4.4.3 Readout technology

To allow for the data selection to take place in a computer farm onshore, where the data from the full detector are brought together, the connection from the telescope to the shore must have a high bandwidth. Taking into account the counting rate induced by the ^{40}K background and the overall size of the detector, this leads to a transfer rate of several hundred gigabits per second. Therefore the readout technology uses optical fibers. The scheme uses dense wavelength division multiplexing (DWDM) to provide up to 80 read-out channels (wavelengths) per optical fiber running from telescope to shore. Each DOM will have its own optical channel directly to shore. All communications lasers will be housed on shore and are of the continuous wave type. These carrier waves are amplified in the deep sea and split such that individual wavelengths are sent, on one downlink fiber, to each tower. There the wavelengths are split into separate fibers. each one of which runs to a DOM. At the DOM, the carrier wave is modulated with the data from the photomultipliers and reflected back using reflective electro-absorption modulators (REAM). The data from all DOM fibers are collected and multiplexed then transferred via a down-link fiber to shore. Because of the reflection at the DOM it is possible to incorporate a system to measure the time delay incurred by the signals in the transport to shore. The system has been lab-tested and industry standard bit-error rates have been achieved for transfers over 100km of fiber. The timing accuracy has been measured to be better than 50ps.

A system using partial modulation of the downlink carrier signal provides a method for the control of the DOM high voltages and instrumentation.

3.4.4.4 Sea-floor network

KM3NeT is a deep-sea multidisciplinary observatory that will provide innovative science opportunities not only for neutrino astronomy, but also for Earth and sea science. This will be possible through the synergy created by the use of a common infrastructure allowing for long-term continuous operation of a neutrino telescope and marine instrumentation. The technical specifications for the deep-sea infrastructure as well as installation, deployment and maintenance procedures, have been addressed, taking also into account the requirements for marine and environmental research activities.

The deep-sea cable network consists of one or a few main electro-optical cables from shore to primary junction boxes, from where it branches via secondary junction boxes to the DUs. Since the footprint of the detector is not yet decided, the exact configuration of the network is still open. The functionalities of cables, connectors and junction boxes in terms of electrical power distribution and data transmission have been studied in detail and are well defined. They will be implemented based on the existing experience from ANTARES and NEMO. The overall power consumption will be about 125kW and in the all-data-to-shore scenario the data rate sent to shore will be of the order of 25 GByte/s.

3.5 Beyond the km^3: New techniques for ultra-high-energy neutrino detection

At energies higher than 10^{17} eV the expected neutrino flux is so low that km^3-scale detectors are too small to detect such neutrino events. The distance between structures hosting optical sensors in Cherenkov telescopes is, indeed, of the order of 100m, due to light absorption length in water and ice. Therefore the cost of sensors, hardware, deployment and installation limits the affordable detector size to some km^2 effective area for neutrino-induced muons, which is not enough for the detection of the expected neutrino fluxes at higher energies. For this reason, different but complementary techniques have been investigated with the aim of observing extremely high-energy neutrino events, e.g., cosmogenic neutrinos.

These techniques rely on the identification of a neutrino interaction through the detection of the coherent radiation, which is produced by neutrino-induced cascades, that propagates in dense media for very large distances. Hadronic showers, produced at the vertex of the neutrino interaction, or electromagnetic showers, produced by charged leptons outgoing a neutrino charged current interaction, radiate coherent radio and acoustic emissions.

Radio waves have typical attenuation lengths of a few km in the glacial ice and the attenuation length for acoustic waves in the sea is also of the order of several km. Therefore, a sparse array of acoustic or radio sensors can be used to reconstruct the neutrino interaction vertex. Several experiments are now pursuing the measurement of the cosmogenic neutrino fluxes through the Askaryan radio technique (Askaryan, 1962).

Meanwhile studies on acoustic technique are still in an early stage, but the potential use of this technique is appealing to build very large acoustic neutrino detectors, thanks to the optimal properties of water as sound propagator (Urik, 1982). In the last decade, the possibility of using hydrophones installed on military arrays, and the infrastructures of new underwater and under-ice Cherenkov telescopes, has permitted several experimental groups to start R&D activities on acoustic detection.

In the framework of the activities of the ANTARES neutrino telescope, the AMADEUS (ANTARES Modules for Acoustic Detection Under the Sea) group has deployed tens of hydrophones aboard two strings. Hydrophones are both commercial piezo-ceramic hydrophones, self-made piezo-ceramic hydrophones and self-made hydrophones hosted in and acoustically coupled with 17in pressure-resistant glass spheres. The system is operating continuously and automatically, requiring little human intervention. AMADEUS (Migneco, 2008) allows for extensive studies of both transient signals and ambient noise in the deep sea, as well as signal correlations on several length scales and localization of acoustic point sources. The array has shown to be able to identify and locate signals emitted by the beacon used for the ANTARES acoustic positioning system and from biological sources (Aguilar et al., 2010). The system is configured to measure the bipolar pulses expected to originate from neutrino interaction.

The NEMO Collaboration is also conducting the first studies for acoustic neutrino detection. In 2005, the Collaboration deployed OνDE (Ocean noise Detection Experiment) at the NEMO test site, 2100m depth, 25km off the coast of Sicily (Riccobene et al., 2009). The detector comprised four hydrophones arranged on a pyramidal-shaped titanium frame. Acoustic noise was studied as a function of time, weather conditions, presence of ships and biological sources, with important drawbacks in bioacoustics (Nosengo, 2009). A larger array composed of 18 acoustic sensors (see SMO, described in Section 3.6.5) is installed aboard the NEMO Phase-2 prototype that will be deployed in Capo Passero in Spring 2013. The system will test different acoustic sensor technologies: piezo-electric acoustics sensors will be mounted inside glass spheres hosting PMTs, thus not sustaining high pressure; free flooded ring hydrophones, that could be used also as acoustic transmitters; novel large-bandwidth pressure-calibrated hydrophones. The detector will be able to work both as positioning system for the tower, and as an independent acoustic detector to study environmental noise and to develop techniques for search of neutrino acoustic signals.

Acoustic sensors (hydrophones and piezo in the DOM) will be also installed on the KM3NeT prototype detection unit, in order to test and validate the optimal technology for acoustic positioning and neutrino detection.

3.6 Deep-sea science with neutrino telescopes

The construction of underwater infrastructures for neutrino telescopes has led to improvements and novel solutions in deep-sea technology and science. A mandatory request for this apparatus is, indeed, the permanent connection between a shore laboratory and the abyssal detector to power-up the sensors and electronics and allow high data-rate transmission to shore. The neutrino telescopes infrastructures have thus become optimal sites for installation of Earth and sea science nodes, allowing long-term and real-time access for oceanographic, geophysical and biological instrumentation already deployed in deep-sea.

The ANTARES and NEMO deep-sea installations already provide access for the Earth and sea science community through direct connection to the detectors and/or through dedicated instrumentation connected to the junction boxes. Such instrumentation for environmental monitoring provides an important tool for neutrino telescope calibration: measurement of water optical and oceanographic properties, behaviour of bioluminescent organisms, measurement of sea currents, and identification of acoustic noise sources provide information – in the domain of the Earth and sea science community – required to calibrate and monitor the response of underwater neutrino detectors.

3.6.1 Bioluminescence

A main source of background for underwater neutrino telescopes is, indeed, "optical noise" – i.e., light bursts or flashes in sea water – that superimposes random photon hits to the (time-correlated) Cherenkov photons radiated along the charged lepton tracks. This background is due to the presence of radioactive isotopes and bioluminescent organisms. Radioactive elements in seawater (mainly ^{40}K contained in salt) emit electrons above the Cherenkov threshold. The uncorrelated background produced by ^{40}K decay is about few tens of kHz (for 10" PMT at 0.5 single photoelectron threshold). This background is almost site independent since variations in the salt concentration in deep-sea are small. Optical noise is also due to organisms living in deep water and producing their light in a chemical process called bioluminescence. Bioluminescent species are extremely diverse and occur in a wide taxonomic range, from bacteria to squids, crustaceans and fish. These organisms produce long lasting (from ms to s) bursts of light that may saturate close PMTs for the period of emission. Results obtained in the very deep waters of the Mediterranean Sea show that bioluminescence bursts are rare (few per hour) and do not affect the average optical noise rate on PMTs, as measured at 3000m depth in the Ionian Sea, at the Capo Passero site (The KM3NeT Consortium, 2008). On the contrary, in biologically active waters, bioluminescence signals may produce an intense background noise up to several hundred kHz (on 10" PMTs, 0.5 s.p.e). These background hits have to be monitored and, eventually, filtered off-line from the data by event triggers and reconstruction algorithms. Extremely high rates of bioluminescence can, however, deteriorate the track reconstruction quality and, in some cases, veto the detector operation for significant fractions of time.

In order to better study the behaviour of bioluminescent organisms and their effect on telescopes, further studies on bioluminescence have been carried out in strong collaboration with biologists. Thanks to a multiyear operation, ANTARES has collected an unprecedented set of data on bioluminescence in the deep sea. Preliminary results (Al Ali et al., 2010) indicate that a time-dependent continuous bioluminescent background in sea water is due to bacteria, as well as frequent and diverse light flashes attributed to larger organisms. These results show that the bioluminescence activity at great depths in the Mediterranean Sea is higher than anticipated. Onboard the ANTARES instrumentation line, two sensitive CCD cameras have been deployed in order to observe the occurrence of macroscopic bioluminescence events (Figure 3.13). Despite the sensitivity of the used cameras not being enough to precisely recognize bioluminescent organism, obtained results show the great potential of this technique and open the way to future improvements.

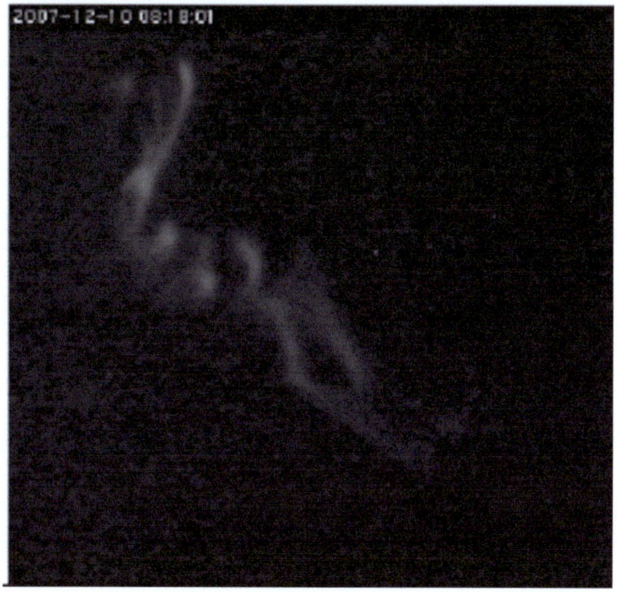

Figure 3.13 Bioluminescence event (2 seconds of exposure) recorded by the ANTARES deep-sea bio-camera (AXIS221). The camera has a field of view (max 90°) and exposure time settable from shore.

Data have been also acquired using stand-alone mooring stations or profilers. An ISIT (Intensified Silicon Intensifier Target) camera profiler has been used to study bioluminescence in the proposed KM3NeT installation sites. An exponential relationship between the density of bioluminescence near seabed and depth was found, and the decreasing trend of bioluminescence from the Western to Eastern Mediterranean Sea was confirmed (Craig et al., 2009). Other measurements based on water sampling at different depths and direct count of bioluminescent organism were also performed, confirming the scarce contribution of bioluminescence at >2500m depth in the Eastern and Central Mediterranean Sea.

3.6.2 Seawater optical properties

The study of optical properties in the telescope installation site is very important. The effective detector area is not only directly determined by the extension of the instrumented volume but is strongly affected by the light transmission in the water. A muon track crossing the water at about 50m distance from the detector can be easily observed since the emitted photons have a non-vanishing probability to reach the optical sensors.

Two microscopic processes mainly affect the propagation of light in the water: absorption and scattering. Light absorption directly reduces the effective area of the detector; the scattering has a negative effect on tracks reconstruction (causing a dispersion of the photon arrival times on the photon detectors). In pure water, light absorption and scattering are

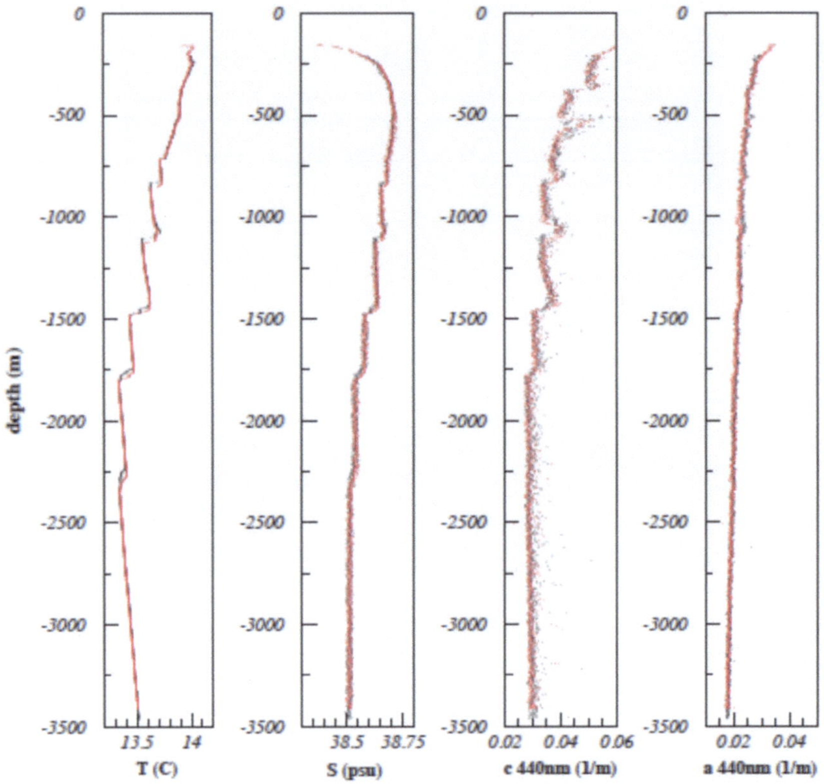

Figure 3.14 Temperature, salinity, light attenuation and absorption coefficients (at λ=440nm) as a function of depth, measured in two deployments (red and black, almost fully superimposed) in the Thyrrenian Sea, close to Alicudi Island.

strongly wavelength dependent. In particular, light transmission in pure water is extremely favoured in the 350–550 nm range, overlapping the region in which PMTs usually reach the highest quantum efficiency. In the visible region of the electromagnetic spectrum, light absorption steeply decreases as a function of wavelength and reaches its minimum at about 420 nm. Seawater optical properties depend also on water temperature, salinity and as a function of concentration, dimension and refraction index of the dissolved or suspended particulate. These parameters are different in different marine sites and change as a function of time. Since water temperature and salinity and particulate concentration may vary significantly in different marine sites, and also as a function of time, it is extremely important to measure optical parameters in situ.

Thanks to the activities of the KM3NeT groups, monitoring of water optical properties in the deep Mediterranean Sea has been extensively conducted with several experimental devices. Results confirm that deep-sea water of the Central and Eastern Mediterranean Sea shows optical properties close to optically pure water. In this area, optimal light transmission is observed at depths larger than 2500m; worsening of light transmission is observed in shallow water and close to the seabed (Aguilar et al., 2005; Capone et al., 2002; Anassontzis et al., 2010). Light transmission shows a seasonal effect only in shallow waters (down to about 1000m depth). It is worth mentioning that, thanks to these measurements, for the first time the well-studied "layers" observed by oceanographers in the Tyrrhenian Sea (i.e., waters layers with evident leaps of salinity and temperature) were also observed in correlation with changes of the light attenuation coefficient, as shown in Figure 3.14 (Capone et al., 2002).

3.6.3 Biofouling and sedimentation

The presence of sediments in the water affects the performances of the detector, increasing the light scattering and worsening the track reconstruction angular resolution. On the other hand, the deposit or growth of fouling (biofouling) on the sensitive part of photon detectors reduces, as a function of time, the global detector efficiency. Study of these phenomena is therefore important for the detector calibration.

Standard sedimentation measurements were conducted with recoverable mooring lines equipped with sedimentation traps in the three KM3NeT candidate sites. The Central and Eastern Mediterranean Sea was again confirmed as an oligotrophic environment with reduced biological growth and sedimentation. Also in this case, ANTARES provides an optimal infrastructure for real-time and very long-term observations of sedimentation and biological growth, exploited by deep-sea biologists (Amran et al., 2003). Biological growth was also studied at the NESTOR site with long-term deployment of dedicated mooring-lines equipped with different substrates (glass, plastics and metals), covering a wide range of depths (from 1500 to 4500m). Visual and microscope-aided inspections were complemented by molecular-biological approaches (Bellou et al., 2011), providing results also for the study of behavior of different materials in deep sea.

3.6.4 Underwater currents

Studies for neutrino telescopes require information on underwater currents to study the mechanical characteristics and behavior of the telescope detection units. Moreover, currents have the effect of stimulating bioluminescence and their measurement is required to better understand the increase of optical noise. Current-meters will therefore be deployed in the neutrino telescope installations.

Current-meter mooring lines are also commonly used by oceanographers. Long-term measurements carried out at the three KM3NeT sites have shown that underwater currents are very small (the average value measured at 3000m depth in Capo Passero is about 3cm/s) and permit the installation of long detection units. Water current data have interest also for oceanographers in finding new trends in Mediterranean deep-sea water circulations, and permitted the discovery of medium-scale structures, such as deep water eddies (Rubino et al., 2012).

An ADCP (Acoustic Doppler Current Profiler) was moored at the deep-sea site of the ANTARES to monitor seawater currents and their effect on the bending on the detector lines. The ADCP provided also a unique opportunity to compare high-resolution acoustic and optical observations in deep-sea. The ADCP measured downward vertical currents of magnitudes up to 3cm/s in late winter and early spring 2006. In the same period, observations were made of enhanced levels of acoustic reflection, interpreted as suspended particles including zooplankton, by a factor of about 10 and of horizontal currents reaching 35cm/s, in coincidence with high light levels detected by the telescope, interpreted as increased bioluminescence (van Haren et al., 2011).

3.6.5 Bioacoustics

A strong interplay between astroparticle physics and biology was experienced with the installation of hydrophones at the NEMO test site and at the ANTARES site, as described in Section 3.5.

The AMADEUS and OvDE detectors demonstrated the potential of this novel technique, providing one of the largest data archive of acoustic data from these depths to date. Recorded data had a groundbreaking influence on the marine biologists community. OvDE discovered, since first recordings, biological sounds revealing marine mammals passing or living in the area (Figure 3.15). The detection of sperm whales was an especially exciting finding. Biologists knew that whales travel through the whole Mediterranean Sea; the recordings provided evidence for record numbers of transiting whales and for a prolonged presence of them in the waters of Eastern Sicily. Both AMADEUS and OvDE provide data to the *Listen to the deep* project (http://www.listentothedeep.net) funded under the ESONeT (http://www.esonet-noe.org) FP6 Network of Excellence.

The activities of the two groups are today converging in a common stream finalized to the development of acoustic sensors to be installed aboard the KM3NeT detector. The joint efforts of AMADEUS and NEMO, conducted to the construction of a large acoustic array of 18 hydrophones, hosted aboard the NEMO Phase II detector that will be soon installed in Capo Passero and to the construction of a common acoustic system installed aboard the KM3NeT-Pre Production Module. In these contexts, the hydrophones' data will be used both for detector positioning and Earth and sea science. In particular, acoustic submarine data received from NEMO Phase II will be distributed, under the activities of the SMO (Submarine Multidisciplinary Observatory) projects to the bioacoustics science community, to search for cetacean's signatures (http://web.infn.it/smo).

In addition, two novel submarine acoustic antennas (NEMO-SN1 and OvDE-2), equipped with 4 large bandwidth hydrophones, were deployed in June 2012 in the Catania test site, that is East Sicily node of EMSO (http://www.emso-eu.org). The observatory is recording marine mammals' signals (dolphins and sperm-whales mainly) and is performing a continuous real-time analysis of acoustic background, correlating information on the transit of large ships in the area. Another scientific application of these antennas is the investigation of occurrence of acoustic events correlated with tsunami waves, as described below.

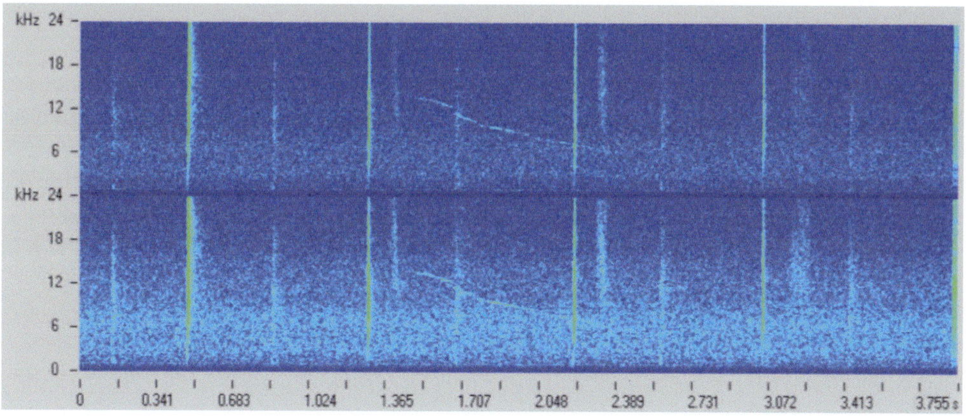

Figure 3.15 Spectrogram of a continuous recording taken with two of the four hydrophones of NEMO-OvDE. About 4s of data are shown. Dolphin whistles and sperm whale clicks are visible. The color scale indicates the relative sound amplitude (a.u.): impulsive clicks and tonal whistles (cyan) are clearly observable over background noise (blue). A software band-pass filter (1 kHz–24 kHz) is applied.

3.6.6 Geophysics

KM3NeT infrastructures have also offered open deep-sea interfaces to geophysics. The ANTARES telescope has hosted a GURALP seismometer for several years. Since 2002, the SN1 multidisciplinary seafloor observatory (Favali and Beranzoli, 2006) has been deployed off the coast of Sicily and connected to shore via the electro-optical cable of the Catania test site.

Mainly addressed to seismology and oceanography, SN1 was designed as a reduced-size version of GEOSTAR class stations. SN1 provided a real-time monitoring of seismic events in Eastern Sicily, a particularly active seismic area already covered by a dense network of land seismic monitors. Thanks to excellent signal-to-noise ratio of recorded data, SN1 permitted an improvement of regional seismic event location, extending – for the first time – the seismic sensor network to deep-sea. SN1 pursues also other scientific objectives such as the study of energy propagation within the Ionian Sea and monitoring the behavior of the Etna Volcano, from a "privileged" observation site, at about 25km off the volcano axis and, therefore, unaffected by the noise induced by the different phases of volcanic activity. The SN1 station has been recently refurbished and deployed in the EMSO East Sicily node. SN1 is now equipped with a prototypal tsunami early-warning alert system, based on correlation between seismic, magneto-metric and very sensitive pressure gauges data. The presence of a large bandwith acoustic antenna (see above) will be used to study the detection of acoustic waves produced by tsunami events, aiming at the improvement the tsunami-early warning system (Chierici et al., 2010).

3.7 Conclusions

Neutrino telescopes are expected to provide new information on the high-energy Universe and investigate occurrence of hadronic processes in astrophysical sources. A major goal of these detectors is the identification of the sources of cosmic rays.

In the last decade, huge technological progress has been achieved and the first neutrino telescopes have been in operation. At the South Pole, IceCube has reached the sensitivity to detect first astrophysical neutrino sources and results will be delivered in the coming years. After a long period of activity in prototyping and validating deep-sea technologies, the ANTARES underwater telescope is now fully operational. The KM3NeT Consortium, combining all the expertise gathered by European groups, has converged, has developed a full design for a multi-km^3 scale underwater telescope and is now ready for the start of the construction phase.

A new interest is also addressed in the detection of UHE neutrinos through radio and acoustic techniques that may in the future allow for the extension of the sensitivity of neutrino telescope by reaching much larger effective areas.

In collaboration among the astroparticle physics and Earth and sea science communities, also the installation of large-scale, deep-sea laboratories in Europe is anticipated. ANTARES and NEMO already provide permanent infrastructures for multidisciplinary activities. The joint operation of EMSO and KM3NeT will permit the operation of the largest deep-sea laboratory for marine ecosystem monitoring, marine mammals research, climate change and geo-hazard.

References

Abbasi R. et al. (2011) Constraints on the extremely-high energy cosmic neutrino flux with the IceCube 2008–2009 data. Physical Review D83, 092003

Abbasi R. et al. for the IceCube Collaboration (2012) An absence of neutrinos associated with cosmic-ray acceleration in gamma-ray bursts. Nature 484, 351–354.

Ackermann M. et al. (2006) Optical properties of deep glacial ice at the South Pole. Journal of Geophysical Research 111, D13203.

Aggouras G. et al. for the NESTOR Collaboration (2005a) A measurement of the cosmic-ray muon flux with a module of the NESTOR neutrino telescope. Astroparticle Physics 23, 377.

Aggouras G. et al. for the NESTOR Collaboration (2005b) Operation and performances of the NESTOR test detector. Nuclear Instruments and Methods A 552, 2005 420–439.

Aguilar J.A., et al. (2005) Transmission of light in deep sea water at the site of the Antares neutrino telescope. Astroparticle Physics 23(1), 131–155.

Aguilar J.A. et al. (2010) AMADEUS – The acoustic neutrino detection test system of the deep-sea ANTARES neutrino telescope. Nucl. Instr. and Meth. A626-627, 128–143.

Aharonian F. (2007) The very-high-energy gamma-ray sky. Science 315(5808), 70.

Aharonian F. et al. (2004a) High-energy particle acceleration in the shell of a supernova remnant. Nature 432, 75.

Aharonian F. et al. (2004b) Calibration of cameras of the H.E.S.S. detector. Astropart. Phys. 22, 109–125.

Aharonian F. et al. (2006) Discovery of very-high-energy γ-rays from the Galactic Centre ridge. Nature 439(1), 1038.

Ahlers M., Anchordoqui L.A., Gonzalez-Garcia M.C., Halzen F.and Sarkar Subir (2010) GZK neutrinos after the Fermi-LAT Diffuse Photon Flux Measurement. Astroparticle Physics 34, 106–115

Ahrens J. et al. (2004) Search for extraterrestrial point sources of neutrinos with AMANDA-II. Physical Review Letters 92, 071102.

Aiello S. et al. (2010) Measurement of the atmospheric muon flux with the NEMO Phase-1 detector. Astroparticle Phys. 33, 263.

Al Ali B., Garel M., Cuny P., Miquel J.-C., Toubal T., Robert A. and Tamburini C. (2010) Luminous bacteria in the deep-sea waters near the ANTARES underwater neutrino telescope (Mediterranean Sea). Chemistry and Ecology 26(1), 57–72.

Albert J. et al. (2005) Physics and astrophysics with a ground-based gamma-ray telescope of low energy threshold. Astroparticle Physics 23, 493.

Ameli F. et al. (2008) The data acquisition and transport design for NEMO Phase-1. IEEE Trans. Nucl. Sci. 55(1), 233.

Amram P. et al. (2003) Sedimentation and fouling of optical surfaces at the Antares site. Astroparticle Physics 19(2), 253–267.

Anassontzis E.G. et al. (2010) Water transparency measurements in the deep Ionian sea. Astroparticle Physics 34, 187.

Andres E. et al. (2000) The AMANDA neutrino telescope: Principle of operation and first results. Astroparticle Physics 13, 1.

Askaryan G.A. (1962) Excess negative charge of an electron-photon shower and its coherent radio emission. Soviet Physics JETP 14 (2), 441–443.

Ayutdinov V. for the Baikal Collaboration (2009) The prototype string for the km^3-scale Baikal neutrino telescope. Nuclear Instruments and Methods A 602, 14.

Bell A.R. (1978) The acceleration of cosmic rays in shock fronts. Monthly Notice Royal Astronomical Society 182, 147 and 443.

Bellou N., Colijn F. and Papathanassiou E. (2011) Experimental settlement study in the Eastern Mediterranean deep sea (Ionian Sea). Nuclear Instruments and Methods A 626-627, S102–S105.

Berezinsky V. (2008) Astroparticle physics: Puzzles and discoveries. Journal of Physics Conference Series, 120, 012001.

Berezinsky V.S. and Zatsepin G.T. (1969) Cosmic rays at ultra high energies (neutrino?). Physics Letters B, 28, 423.

Bertin V., Brunner J., Carr J., Curtil C., Destelle J.-J., Deschamps A., Escoffier S., Graf K., Höß, J., Lahmann R., Lefèvre D., Tamburini C., Schuller J.-P., and van Haren H. (2014) ANTARES neutrino telescope. In: P. Favali (Ed.) *Seafloor Observatories*. Chichester/ Heidelberg: Springer-Praxis (this volume).

Boersma D.J., Gladstone L.E. and Blumenthal J. for the IceCube Collaboration (2011) The shadow of the moon in cosmic rays measured with IceCube. e-print archive arXiv:1111.2741

Capone A. et al. (2002) Measurements of light transmission in deep sea with the AC9 trasmissometer. Nucl. Instr. and Meth. A 487, 423–434.

Chierici F., Pignaioli L. and Emnriaco D. (2010) Modeling of the hydroacoustic signal and tsunami wave generated by seafloor motion including a porous seabed. J. Geophys. Res. 115(3).

Craig J., Jamieson A.J., Heger A. and Priede I.G. (2009) Distribution of bioluminescent organisms in the Mediterranean Sea and predicted effects on a deep-sea neutrino telescope. Nuclear Instruments and Methods in Physics Research Section A 602, 224–226.

Cronin J.W., Gaisser T.K. and Swordy S.P. (1997) Cosmic rays at the energy frontier. Scientific American 276(7), 32–37.

De Angelis A. et al. (2008) Very-high energy gamma astrophysics. Rivista del Nuovo Cimento 31, 131.

Favali P. and Beranzoli L. (2006) Seafloor observatory science: A review. Annals of Geophysics 49(2/3), April/June 2006.

Fermi E. (1949) On the origin of the cosmic radiation. Physical Review 75, 1169.

Gaisser T.K. (2011) Cosmic-ray physics with IceCube. arxiv-1107.1690

Gandhi R., Quigg C., Reno M.H. and Sarcevic I. (1996) Ultra high-energy neutrino interactions. Astroparticle Physics 5, 81–110.

Greisen K. et al. (1966) End to the cosmic ray spectrum? Physical Review Letters 16, 748–650.

Guetta D. and Granot J. (2003) High-energy emission from the prompt gamma-ray burst. Astrophysics 585, 885.

Holder J. et al. (2006) The first VERITAS telescope. Astroparticle Physics 679, 1427.

Honda M. et al. (2007) Comparison of 3-dimensional and 1-dimensional schemes in the calculation of atmospheric neutrino. Physical Review D 75, 043006.

Kalashev O.E. et al. (2002) Ultrahigh-energy neutrino fluxes and their constraints. Physical Review D, 66, 063004.

Katz U.F. and Spiering Ch. (2012) High-energy neutrino astrophysics: Status and perspectives. Progress in Particle and Nuclear Physics 67, 651–704.

Khangulyan D., Aharonian F. and Bosch-Ramon V. (2008) On the formation of TeV radiation in LS 5039. Monthly Notice Royal Astronomical Society 383(2), 467–478.

KM3NeT Consortium (2008) Conceptual design report for a deep-sea research infrastructure incorporating a very large volume neutrino telescope in the Mediterranean Sea. http://www.km3net.org/CDR/CDR-KM3NeT.pdf.

KM3NeT Consortium (2010) Technical design report for a deep-sea research infrastructure in the Mediterranean Sea incorporating a very large volume neutrino telescope. http://www.km3net.org/TDR/KM3NeT-TDR.pdf.

MacFadyen A.I. and Woosley S. (1999) Collapsars: Gamma-ray bursts and explosions in "Failed Supernovae". Astrophysical Journal 524, 262–289.

Mannhein K. (1995) High-energy neutrinos from extragalactic jets. Astroparticle Physics 3, 295.

Markov M.A. and Zheleznykh I.M. (1961) Large-scale Cherenkov detectors in ocean, atmosphere and ice. Nuclear Physics 27, 385.

Meszaros P. (2006) Gamma-ray bursts. Reports on Progress in Physics 69, 2259.

Migneco E. et al. (2008) Recent achievements of the NEMO project. Nuclear Instruments and Methods A, 588, 111.

Mirabel I.F. and Rodriguez L.F. (1998) Microquasars in our Galaxy. Nature 392, 673.

Mobley C.D. (1994) Light and Water: Radiative transfer in natural waters. San Diego: Academic Press.

Morlino G., Blasi P. and Amato E. (2009) Gamma rays and neutrinos from SNR RX J1713.7-3946. Astroparticle Physics 31(5), 376.

Nosengo N. (2009) The neutrino and the whale. Nature 426, 560.

Pierre Auger Collaboration (2010a) Update on the correlation of the highest energy cosmic rays with nearby extragalactic matter. Astroparticle Physics 34(5), 314–326.

Pierre Auger Collaboration (2010b). Measurement of the energy spectrum of cosmic rays above 10**18** eV using the Pierre Auger Observatory. Physics Letters B 685(4–5), 239–246.

Protheroe R.J. and Johnson P.A. (1996) Are topological defects responsible for the 300 EeV cosmic rays? Astroparticle Physics 4, 253.

Resconi E. for the IceCube Collaboration (2009) Status and prospects of the IceCube neutrino telescope. Nucl. Instr. And Meth A 602(1), 7–13.

Riccobene G. et al. (2007) Deep seawater inherent optical properties in the Southern Ionian Sea. Astroparticle Physics 27, 1.

Riccobene G. et al. (2009) Long-term measurements of acoustic background noise in very deep sea. Nuclear Instruments and Methods in Physics Research Section A, 604(1–2), S149–S157.

Roberts A. et al. (1992) The birth of high-energy neutrino astronomy: A personal history of the DUMAND project. Review of Modern Physics 64, 259.

Rubino A., Falcini F., Zanchettin D., Bouche V., Salusti E., Bensi M., Riccobene G., De Bonis G., Masullo R., Simeone F. , Piattelli P., Sapienza P., Russo S. et al. (2012) Abyssal undular vortices in the Eastern Mediterranean basin. Nature Communications 3, 834.

Stecker F.W. et al. (1991) High-energy neutrinos from active galactic nuclei. Physical Review Letters 66 2697; 59 (1999) 023002.

Torres D.F. and Anchordoqui L.A. (2004) Astrophysical origins of ultrahigh energy cosmic rays. Reports on Progress in Physics 67, 1663.

Trasatti L. (1999) NESTOR: An underwater neutrino observatory in the Ionian Sea. Nuclear Physics B P.S. 70, 442–444.

Urik R.J. (1982) Sound Propagation in the Sea. Peninsula Publishing

Vallage B. for the ANTARES Collaboration (2006) Status report on the ANTARES project. Nucl. Phys. B. 151, 407.

van Haren H. et al. (2011) Acoustic abyssal undular vortices in the Eastern Mediterranean basin, and optical variations during rapid downward motion episodes in the deep north-western Mediterranean Sea. Deep-Sea Research I 58, 875–884.

Villante F.L. and Vissani F. (2008) How precisely can neutrino emission from supernova remnants be constrained by gamma ray observations? Physical Review D 78–10, 103007.

Waxman E. (1995) Cosmological gamma-ray bursts and the highest energy cosmic rays. Physical Review Letters 75, 386.

Waxman E. and Bachall J. (1999) High energy neutrinos from astrophysical sources: An upper bound. Phys. Rev. D. 59, 023002.

Wischnewski R. for the Baikal Collaboration (2005) The Baikal neutrino telescope – Results and plans. International Journal of Modern Physics A 20, 6932.

Zatsepin G.T. and Kuzmin V.A. (1996) Upper limit of the spectrum of cosmic rays. Journal of Experimental and Theoretical Physics Letters 4, 78–80.

Web resources

http://icecube.wisc.edu/

http://www.km3net.org

http://ec.europa.eu/research/esfri/

http://www.listentothedeep.net

http://www.esonet-noe.org

http://web.infn.it/smo

http://www.emso-eu.org

4 Seafloor observations and observatory activities in the Sea of Marmara

M.N. Çağatay[1], L. Geli[2], L. Gasperini[3], P. Henry[4], C. Gürbüz[5] and N. Görür[1]

4.1 Introduction

The Sea of Marmara (SOM) is located on the North Anatolian Fault (NAF) zone in NW Turkey (Figure 4.1; Barka, 1992; Straub and Kahle, 1997; Okay et al., 2000; İmren et al., 2001; Le Pichon et al., 2001, 2003; Armijo et al., 2002; Çağatay et al., 2002; Şengör et al., 2004). The NAF is a major transform-plate boundary that has produced devastating historical earthquakes along its 1600km length (Ambraseys, 2002). The most active northern branch of the NAF cuts across the Marmara in an east–west direction and continues westward into the NE Agean Sea (Figures 4.1 and 4.2). The earthquake events along the NAF have a westward progression with a 60-year sequence of rupturing toward Istanbul, in which one event promotes the next (Toksöz et al., 1979; Stein et al., 1997). After the 1999 Izmit and Düzce earthquakes, the next large (Mw>7) earthquake is expected in the SOM close to Istanbul, an important cultural centre and a mega-metropolis with 15 million inhabitants (Parsons, 2000; Parsons et al., 2000, 2004).

The SOM has three ~1250m-deep strike-slip basins that are separated by NE-trending transpressional highs. The slopes leading to the deep basins are steep (>18°) and carry the scars of many paleo-landslides and submarine canyons (Figure 4.2). They also have some unstable areas near the shelf break and in the upper slope that may slide during future

1 EMCOL, Faculty of Mines, İstanbul Technical University, Istanbul, Turkey
2 IFREMER, Brest, France
3 CNR-ISMAR, Bologna, Italy
4 CNRS and College de France, Aix-en Provence, France
5 KOERI, Boğaziçi University, Istanbul, Turkey

seismic events. Over the past two millennium the earthquake activity in the SOM has produced more than 30 tsunami events having about 6m high run-ups in the coastal areas (Yalçıner et al., 2002). Most of these tsunamis have probably been caused by submarine landslides. Any future earthquakes and tsunamis would have a devastating effect, not only in Istanbul but in the whole of the SOM coastal areas in which more than 20% of Turkey's population and 50% of its industry are located.

The SOM is also interesting in terms of its oceanographic setting. It is connected to the low salinity (S=18‰) Black Sea via the İstanbul (Bosphorus) Strait and to the normal marine (S=38.5‰) Aegean Sea via the Çanakkale (Dardanelles) Strait; the two straits have sill depths of –65 and –35m, respectively. There is a permanent two-layer flow in the straits and the SOM with the halocline located at –25m between the waters of Mediterranean and Black Sea origins (Ünlüata et al., 1990; Beşiktepe et al., 1994).

The active faults in the SOM basins are associated with cold fluid venting, carbonate crusts, sulphidic sediments and chemosynthetic benthic communities (Armijo et al., 2005; Zitter et al., 2008; Geli et al., 2008). Its high tectonic activity with geohazard risk, as well as its special oceanographic setting as a gateway between the Mediterrranean and Black Seas, makes the SOM a natural laboratory for multidisciplinary seafloor observations for geohazard and oceanographic monitoring. Seafloor observatories would therefore enable earth and ocean scientists to study multiple interrelated processes that occur over time scales ranging from seconds to decades.

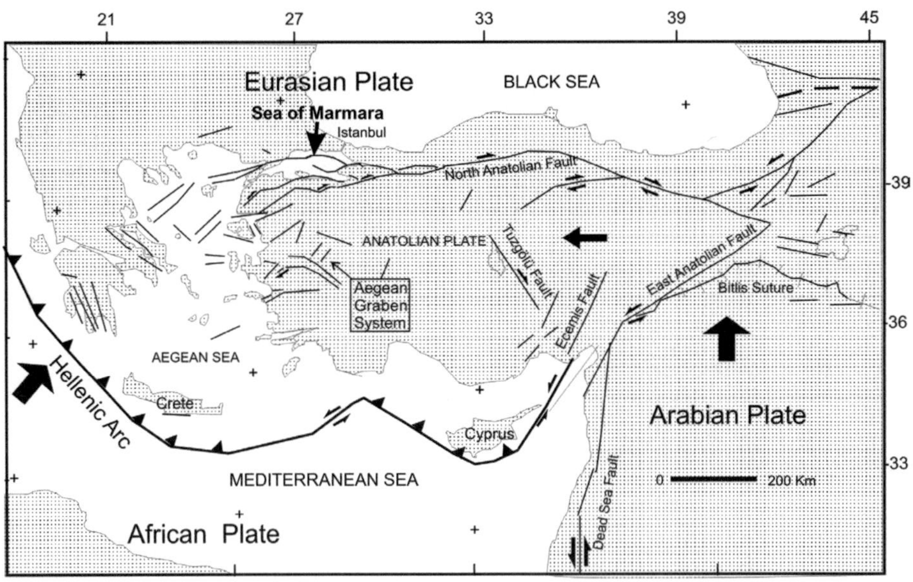

Figure 4.1 Active tectonics of the eastern Mediterranean showing the location of the Sea of Marmara on the North Anatolian Fault zone.

4.2 Geohazards in the Sea of Marmara

4.2.1 The Sea of Marmara seismic gap

The westward progression of major ruptures along the NAF (Barka, 1996), culminating with the destructive earthquake events of 1999, leaves a seismic gap along the SOM capable of generating large earthquakes (Reilinger et al., 2000; Hubert-Ferrari et al., 2000). The last destructive earthquakes occurred at the western and eastern edges of the Marmara basin (i.e., Ganos (1912) and the İzmit and Düzce earthquakes (1999)). It is likely that fault ruptures will fill this gap in the coming decades (Pınar and Lahn, 2001; Parsons et al., 2000; Atakan et al., 2002; Oglesby et al., 2008).

The hazard facing Istanbul and adjacent areas varies widely depending on where and how the predicted Marmara seismic rupture will take place. This region of the NAF is thus critical to our understanding of fault interactions, stress build-up during seismic cycle and seismic hazard in the Istanbul area. Estimates of the slip rate, distribution of motion, microseismicity along the strands of the NAF, detection of fluid outflow (Zitter et al., 2008; Géli et al., 2008) and the analysis of how it relates to the seismic activity can help reach a realistic assessment of seismic hazards for this densely populated area of Turkey.

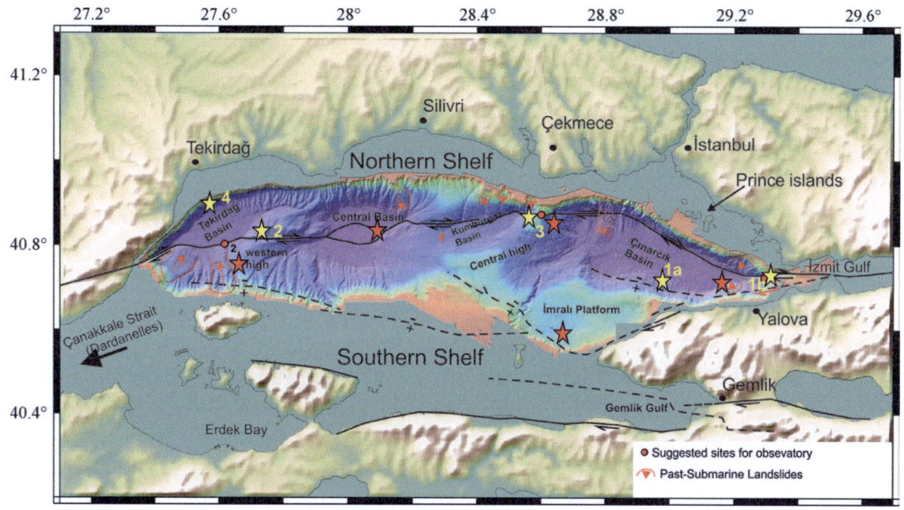

Figure 4.2 Morphotectonic map of the Sea of Marmara showing the active faults and submarine landslides (bathymetry after Le Pichon et al., 2000; faults modified after Armijo et al., 2005). Red stars indicate the five cabled seafloor observatories that were deployed in 2011 by KOERI as part of the Turkish national seismological network. Yellow stars indicate those sites selected for implementing multidisciplinary seafloor observatories to monitor fluids and seismicity within the ESONET Marmara-DM project.

4.2.2 Submarine landslides

The steep slopes of the SOM are marked by numerous scars and deposits of paleo-landslides (Figure 4.2). The shelf break and upper slope areas are characterized by a relatively high rate of sediment deposited during the Quaternary sea-level lowstands. Moreover, the northern margins of the deep basins are composed of clayey Palaeozoic rocks in the east and weakly cemented Tertiary siliciclastic rocks in the west (Görür and Çağatay, 2010). These sediments and sedimentary rocks on the shelf edge and slopes are prone to mass failures triggered mainly by earthquakes. The largest landslides are found in the eastern SOM, which involve the Palaeozoic shales. Most of the landslides probably occurred during lowstand periods, when fresh to brackish water lacustrine conditions prevailed. This conclusion is supported by the deep basin sedimentary records in the SOM in the form of submarine landslides, turbidites and homogenities. For example, the c. 15m thick homogenite deposition in the Central Basin (Beck et al., 2007) and the large landslide on the NE slope of the Çınarcık Basin south of Tuzla (Figure 4.2; Görür and Çağatay, 2010; Özeren et al., 2010) both occurred during late glacial period. The presence of gas hydrates (Bourry et al., 2009), their possible dissociation during lowstand periods in the SOM, might have also triggered mass-flows (e.g. Sultan et al., 2004; Görür and Çağatay, 2010). Gas venting and gas hydrate dissociation from the fault plain at the base of the slope may also occur during an earthquake cycle and destabilize the slope sediments.

4.2.3 Tsunamis

Considering the fact that the predominant faulting style along the NAF is strike-slip, it is more likely that most tsunamis in the SOM have been associated with landslides triggered by strong earthquake shaking. However, the normal faulting on the southern margin of the Çınarcık Basin (McHugh et al., 2006) might have also caused tsunamis. Modelling studies indicate that some of the large submarine landsides, such as the one south of Tuzla Peninsula (Figure 4.2) might have created tsunami waves with wave height up to 10m (Özeren et al., 2010).

The most recent tsunami event was associated with the Gölcük earthquake of August 1999, when a tsunami wave with up to 2.5m run-up inundated coastal areas in the central and eastern parts of the Gulf of İzmit (Altınok et al., 2001; Alpar, 1999; Yalçıner et al., 1999, 2002; Hebert et al., 2005). The source of the tsunami was on the south side of the Gulf where a coastal land mass slid into the sea.

Historical records reveal that more than 30 tsunami events occurred in the past two millennia, with heights up to about 6m in the coastal areas (Papazachos et al., 1986; Ambraseys and Finkel, 1990, 1991; Guidoboni et al., 1994; Altınok et al., 2003; Demirkent, 2001; Ozansoy 2001; Yalçıner et al., 2002). In the Byzantine period, the following tsunami events are mentioned in connection with major earthquakes: 358 Izmit, 447 Istanbul, 478 Istanbul and Izmit, 554 Istanbul and Izmit, 989 Izmit, 1332 Istanbul, 1343 Istanbul, 1344 Western Marmara and 1419 Izmit. The Ottoman record includes 1509 Istanbul, 1754 Gulf of Izmit, 1766 Istanbul, 1878 Izmit and 1894 Istanbul earthquakes and associated tsunamis after the Ottoman conquest of Istanbul.

4.3 Fluids and seismicity in the Sea of Marmara

A crucial aspect in geohazards research is the coupling between stress (and seismic stress release), pore pressure pulses, and other strain indicators. The question of coupling processes between fluids and crustal strain (seismic and aseismic) has been addressed for land environments over the past 20 years and more (e.g., Roeloffs, 1998; Muir-Wood and King, 1993; Trique et al., 1999). However, the importance of fluids in the dynamics of submarine faults has been realized relatively recently with the progress of deep seafloor exploration, in particular with the observation that fluid emission sites are often associated with active faults (e.g., Moore et al., 1990, Le Pichon et al., 1992; Henry et al., 2002; Brown et al., 2005).

In the deeper parts of the SOM, fluid outflow sites (cold seeps) manifested by carbonate crusts, black patches and bacterial mats are commonly observed along or near active faults (Figures 4.3 and 4.4; Armijo et al., 2005; Zitter et al., 2008; Géli et al., 2008). The cold seeps are associated with a rich chemosynthetic community of organisms consisting of bacterial mats, polychaetes, mytilids, cardiids and sea urchins (Figure 4.4bd,e; Ritt et al., 2010). The biodiversity and ecology of organisms living in the deep basins of the SOM, and especially the biologic activity associated with cold seeps along the active faults, are important areas of research.

Free gas emissions are common and appear to be influenced by earthquake occurrence. In the Gulf of Izmit, repeated surveys showed that the intensity of methane emissions increased after the earthquake of August 17, 1999 (Alpar, 1999; Kuşçu et al., 2005). In May to June 2007 during the MarNaut cruise and in November to December 2009 during the Marmesonet cruise, sounding with EK60 and EM302 and observations with Nautile manned submersible showed the presence of numerous spontaneous bubble emissions from faults in the SOM (Figures 4.3 and 4.4; Geli et al., 2008). The distribution of gas seeps in the main part of the SOM is uneven, with less activity on the linear fault segment crossing the Central High. This segment has not ruptured since at least 1766 and thus forms a seismic gap (Ambraseys and Jackson, 2000; Parsons, 2004; Pondard et al., 2007). In contrast, bubbling was observed above a buried transtensional fault zone (Carton et al., 2007; Okay et al., 2000) along the southern edge of the Çınarcık Basin (Figure 4.4; Geli et al., 2008), which displayed microseismic activity after the 1999 events (Özalaybey et al., 2002; Karabulut et al., 2002).

The origin of the gas in the SOM is diverse. While gas emitted from the Çınarcik basin is predominantly microbial methane of relatively shallow origin, the hydrocarbon gases expelled from faults cutting the highs are of deep thermal origin (Figure 4.4; Bourry et al., 2009). Furthermore, there is mantle helium (^{3}He) emission from a fault in the western Tekirdağ basin (Figure 4.4a; Burnard et al., 2012).

The observations listed above suggest two major questions of primary importance that could be addressed with multidisciplinary seafloor observatories:

1. Is there a hydrogeologic connection between some gas seeps and the seismogenic zone of the NAF?

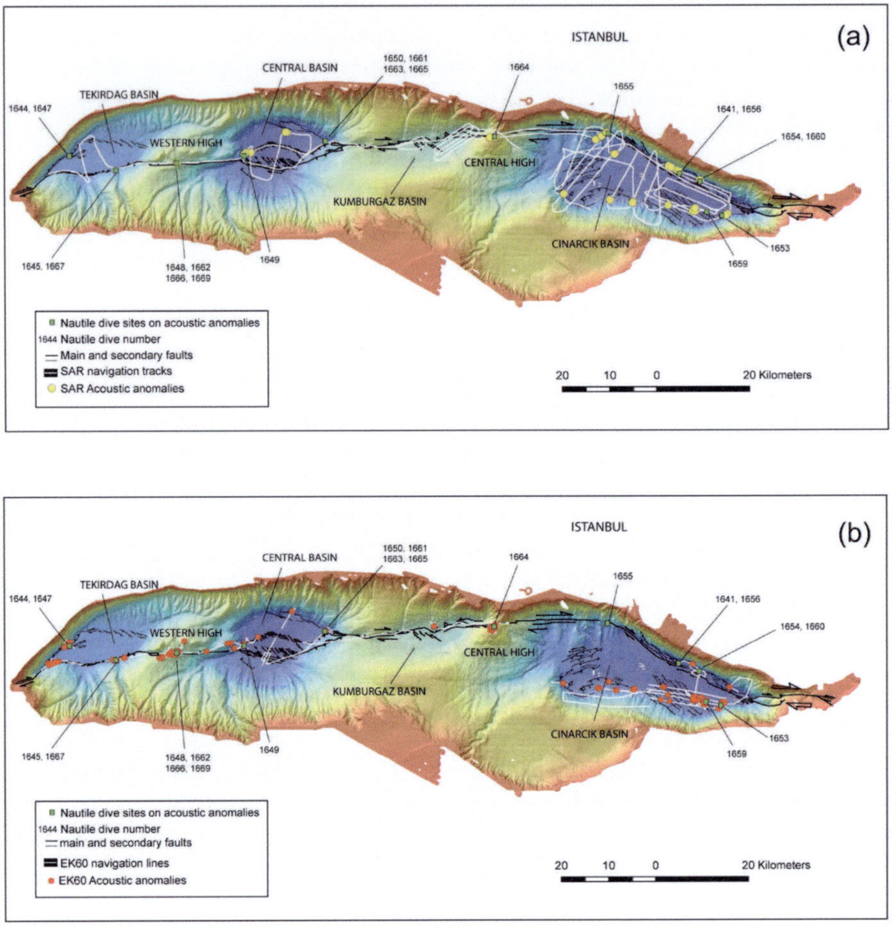

Figure 4.3 Distribution of acoustic anomalies, superimposed on the bathymetric map of Le Pichon et al. (2001) in the deeper parts of the Marmara Sea (after Géli et al., 2008). Active fault traces are also indicated based on Rangin et al. (2004). (a) White lines and yellow dots show routes followed by R/V Le Suroit in 2000 and location of acoustic anomalies detected in September 2000 using the deep tow SAR. (b) White lines and red dots respectively indicate routes followed by R/V L'Atalante in May–June 2007 and location of acoustic anomalies detected using the EK-60 sonar. Note that most gas emissions in the water column are found near the surface expression of known active faults. The linear fault segment, which crosses the Central High and forms a seismic gap, exhibits relatively less gas emission; this segment has not ruptured since 1766, based on historical seismicity. In the eastern Sea of Marmara, active gas emissions are also found above a buried transtensional fault zone, which displayed microseismic activity after the 1999 event. Remarkably, this zone of gas emission extends westward all along the southern edge of Çınarcık Basin, well beyond the zone where 1999 aftershocks were observed. Our findings suggest that, at least in some settings, distribution of gas seeps may provide indications of fault activity and even help identify buried structures.

2. Can the stress building and release processes (seismic and aseismic) during the seismic cycle generate detectable changes in fluid parameters (flow rate, fluid pressure, composition) at the seafloor?

These scientific questions can be addressed by seafloor observations using specific sensors described in Section 4.5 below.

4.4 Oceanographic and environmental sensitivity of the Sea of Marmara

Located on an important gateway between the Mediterranean and the Black Sea, the SOM is sensitive to climatic and environmental changes in the eastern Mediterranean region. The lower layer waters are of Mediterranean origin and their monitoring for salinity, temperature and oxygen is needed in climate (global warming) and environmental change studies. Monitoring of gas hydrate sites on active faults (i.e., the Western High) is also necessary for global warming as well as earthquake studies.

The SOM is also under a serious environmental risk. High population density, together with intense maritime traffic and large inputs of industrial and municipal pollutants from its drainage area and the Black Sea have caused serious environmental degradation, such as spoiled water and sediment quality and depleted oxygen levels in its deep waters. Because it is located on one of the busiest maritime transport routes, the SOM is often subjected to large ship accidents causing spillage of oil and other hazardous substances that lead to serious environmental pollution. Seafloor observatories are therefore needed in the SOM for environmental monitoring and security.

4.5 Sensors for seafloor observations ın the Sea of Marmara

4.5.1 Seismic motion

Broadband Ocean Bottom Seismometers ((BB-OBS) having bandwith of 0.03–30 Hz are recommended (Table 4.1). There are two leading manufacturers of BB sensors used in Ocean Bottom Seismology: Guralp (www.guralp.com) and Kinemetrics (www.kinemetrics.com). Cabled BB-OBSs may also be complemented by numerous, short period (>1 Hz), autonomous OBSs which are suited for local applications. Studying fluids and seismicity relations requires precise event location, within less than a few hundreds of meters. Therefore, the ideal design could consist of one broadband OBS at one given site and a cluster of portable, short-period OBSs, spaced by a few kilometres apart.

4.5.2 Flowmeters

Flowmeters are used to measure chemical fluxes across the sediment–water interface. Chemical and Aqueous Transfers (CAT) meters, such as those developed by the Scripps

Sensor	Manufacturer and model	Sampling rate
Seismometer	GURALP CMG-40T	100 Hz
Piezometer	Ifremer	
Current meter	NOBSKA MAVS-3	5Hz
CTD	SeaBird SBE-16plus	1 sample/10min
Turbidity meter	WET LABS Echo-BBRTD	1 sample/10min
CH_4#1	Franatech METS	1 Hz
CH_4#2	Contros HydroC	1 Hz
Oxygen	Aanderaa Oxygen Optode 3830	1 Hz

Table 4.1 List of sensors for the Sea of Marmara configuration.

Institution of Oceanography, have been very successful in obtaining time-series samples and measuring the fluid flow in both seep and non-seep environments with flow rates on the order of 0.01cm/yr to 100m/yr (Figure 4.5; Tryon et al., 2001).

4.5.3 Piezometers (pore-pressure sensors)

Many efforts have been made in recent years to monitor sediment pore pressure in submarine environments (Figure 4.6). Slip at active faults theoretically results in a change of mean stress and induces volumetric strain in the sediment. For instance, pore pressure monitoring in boreholes in Nankai subduction recorded secular variations of several kPa/yr for strain rates of the order of 10^{-6} yr^{-1} and short-term transients of more than 100kPa associated with aseismic slip events on the main subduction plane (Davis et al., 2006). It is thus hypothesized that strain events within active fault zones may be recorded as pore pressure variations on the order of a few kPa and can be used for monitoring background strain. Piezometers are also very useful for assessing the failure potential of submarine slopes (Sultan et al., 2006).

4.5.4 Gas-bubble monitoring

Standard and well-known acoustic technology, such as high directivity single beam or multibeam echo-sounders, could be very useful to map and quantify gas bubble emissions from the seafloor and monitor their temporal variability (Greinert, 2008; Greinert et al., 2006). These echo-sounders are ideally combined with 70–300 KHz ADCPs systems to determine the horizontal and vertical velocity of the bubbles. This helps in identifying different seeps in the data sets, and under specific circumstances might even allow determination of the changes in the velocity of the gas bubbles/the plume.

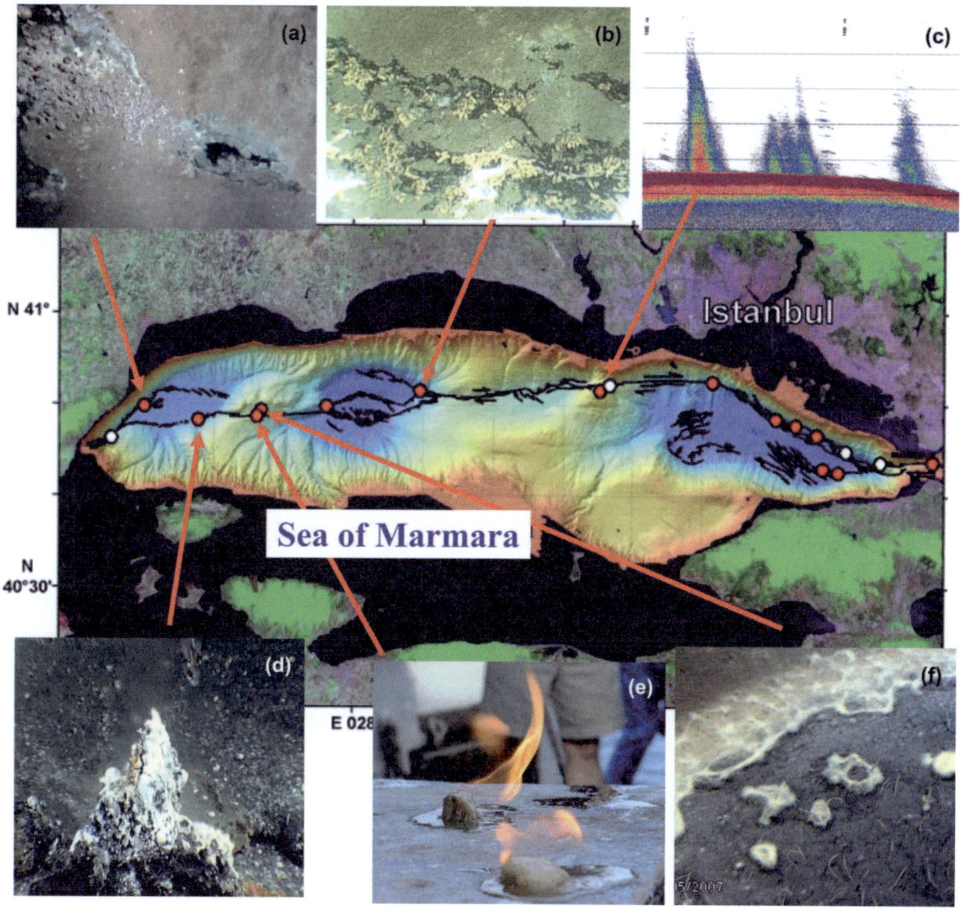

Figure 4.4 Location of Nautile dive sites during the MARNAUT Cruise (May 12–June 12, 2007). Red dots indicate sites where cold seeps were found; white dots are dive sites without cold seeps. (a) Gas bubbles of mantle ^{3}He from tensional faults cutting the north-western escarpment of the Tekirdağ basin. (b) Carbonate crust broken up by fault activity; its fractures occupy with black sulphidic sediment that is colonized by 1–2cm long mytilid bivalves. (c) Example of acoustic anomaly detected on the Central High using the 38 kHz, SIMRAD EK-60 echosounder. (d) A 40cm-high carbonate chimney emitting cold (14.5°C) brackish water in southeast of Tekirdağ Basin. (e) Gas hydrate recovered from the Central High. (f) Black patch of reduced (sulphidic) sediment with polychaete tubes (3cm length) and sulfide oxidizing bacterial colonies.

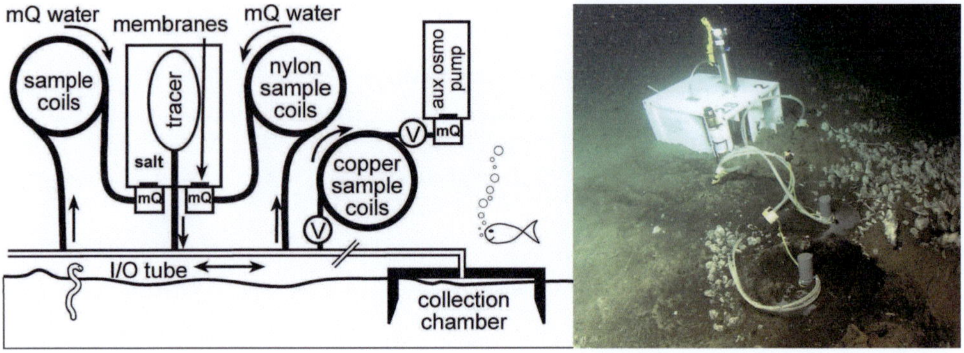

Figure 4.5 Chemical and Aqueous Transport (CAT) deployed on the seafloor (right) and schematic (left) (after Tryon et al., 2001, 2009). The CAT meter uses the dilution of a chemical tracer to measure flow through the outlet tubing exiting the top of a collection chamber. The pump contains two osmotic membranes that separate the chambers containing pure water from the saline side that is held at saturation levels by an excess of NaCl. Due to the constant gradient, distilled water is drawn from the fresh water chamber through the osmotic membrane into the saline chamber at a rate that is constant for a given temperature. The saline output side of the pump system is rigged to inject the tracer while the distilled input side of the two pumps are connected to separate sample coils into which they draw fluid from either side of the tracer injection point. Each sample coil is initially filled with de-ionized water. Having two sample coils allows both inflow and outflow to be measured. A unique pattern of chemical tracer distribution is recorded in the sample coils allowing a serial record of the flow rates to be determined. Upon recovery of the instruments the sample coils are subsampled at appropriate intervals and analyzed. Both tracer concentration and ion concentrations (e.g., Na, Ca, Mg, S, K, Sr, B, Li) are determined simultaneously. Scientific results of CAT meter samples in Sea of Marmara are given in Tryon et al. (2010).

4.5.5 Methane sensor

Methane is a useful parameter to monitor at the seep sites. It is expected that episodic methane release would occur during seismic activity (Alpar, 1999; Kuşçu et al., 2005) and have an important effect on bottom water and sediments biogeochemistry (Çağatay et al., 2012). However, to date there is no autonomous sensor available that meets the requirements in sensitivity, reliability and accuracy required for long-term methane measurements in the water column. Most particularly, none has yet the sensitivity to analyze the background methane concentrations of ~10^{-2} ηL/L. In addition, the commercial sensors do not allow continuous measurements for periods longer than a few weeks, due to drift, corrosion or bio-fouling problems that strongly affect the analytical accuracy. New perspectives are expected from infra-red (IR) detection techniques. The German CONTROS company, for instance, has developed a new sensor (HydroC™ /CH_4, Hydrocarbon & Methane Sensor) with substantially improved performances, but only for relatively short-term measurements. The solution might eventually come from autonomous, in-situ mass-spectrometers (ISMS) that may allow functioning autonomously over periods of a few months, or even longer. Important efforts have been made in the last few years to develop an "open-source"

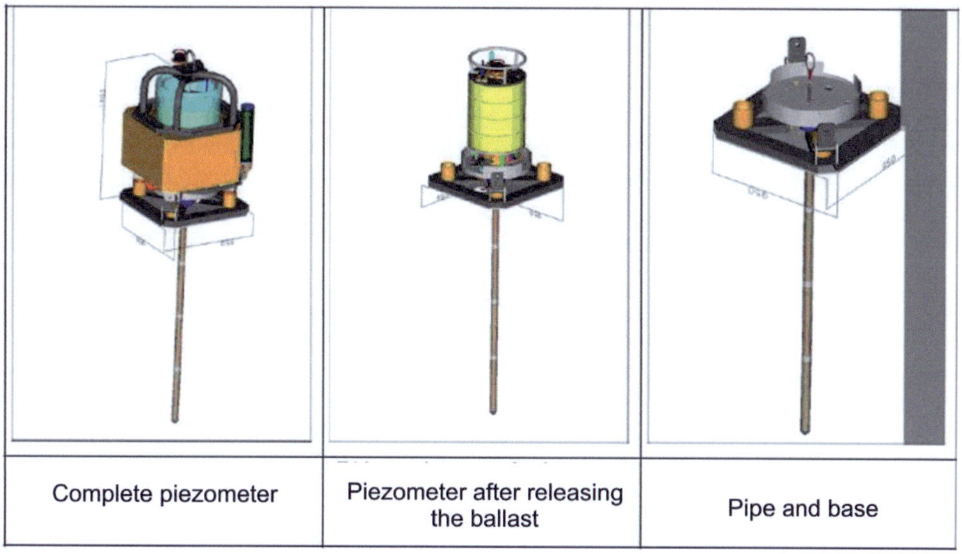

Figure 4.6 Ifremer piezometer: a free-fall device with a sediment-piercing lance attached to a recoverable instrumented part. It is ballasted with lead weights to penetrate a range of sediment types in water depths of up to 6000m. The length of the used lance depends on the stiffness of the sediment. For soft sediment, a 12m lance can be used successfully. Pore pressures are measured relative to hydrostatic pressure at maximum 10 ports on the 60mm diameter lance using specially adapted differential pressure transducers hydraulically communicating between the pressure ports at multiple depth levels and a sea water reference port at the top. The piezometer pore pressure sensors have a resolution of ± 0.4 kPa.

ISMS for use on deep submergence vehicles, on standalone moorings and cabled observatories, to measure many volatile compounds relevant to microbial and macrofaunal metabolic processes, including methane, oxygen, hydrogen sulphide, carbon monoxide, carbon dioxide, nitrogen, nitrous oxide, dimethyl sulphide, among others, to varying degrees of sensitivity from nM to μM depending on the compound (e.g., Tortell 2005; Short et al., 2006; Camilli and Hemond, 2004).

4.5.6 Oceanographic sensors

Oceanographic parameters, such as current velocity, temperature and salinity, would also be useful to measure in the SOM for both earthquake and oceanographic monitoring. Because the water column is stratified, bottom water oxygen concentrations are low (1–4mg/L; Ünlüata et al., 1990) and have a strong influence on the benthic communities (including microorganisms), both at fluid outflow sites and in the background. It is expected that oxygen (as well as other biogeochemical parameters) concentration will be modified by episodic methane release during seismic events.

4.6 Recommended observatory sites

Site 1 is located in the Çınarcık Basin (Figure 4.2). **Site 1a** is where active gas emission sites have been found during the 2000 and 2007 surveys; it is located above a buried trans-tensional zone extending in the prolongation of the fault that ruptured during the 1999 İzmit earthquake. **Site 1b** is located at the entrance of the Gulf of Izmit, where the main deformation zone of the NAF is less than some tens of meters wide (Figure 4.2). Site 1b is the location of the SN-4 experiment, explained in the next section. Sites 1a and 1b are particularly well suited to investigate the relationship between crustal strain and pore fluid pressure, because the Eastern Çınarcık area has persistent crustal microseismicity, including one cluster localized immediately beneath the zone where gas seeps are observed. Analysis of GPS data shows that the 1999 Izmit earthquake has an important post-seismic motion, corresponding to 50% of the plate motion six years after the earthquake (Ergintav et al., 2009). It is also suggested by this study that the Çınarcık basin is affected by aseismic slip on faults resulting in extension rates of the order of 10^{-7} yr^{-1} (if averaged over the basin width) not accounted for by interseismic loading models. This strain could cause detectable pore pressure drift in the sediment (in the 100–1000 Pa/yr range) and even larger perturbations may be produced if strain is localized and/or transient.

Site 2 is located on the Central High, 15km south-west of İstanbul (Figure 4.2). This site is situated on a fault segment constituting the "seismic gap" that has not ruptured since at least 1766. There is little evidence for fluid emission along the fault itself, but there is thermal hydrocarbon outflow from an anticlinal axis, a few kilometers south of the main fault trace (Bourry et al., 2009). It is therefore of critical importance to deploy instruments such as seismometers and piezometers on top of the Central High and within the fault valley, in order to establish comparisons between these two different environments. Site 2 is a good location also for oceanographic monitoring of the Mediterranean water on its way to the Istanbul Strait.

Site 3 is centered on the Western High, an area where gas hydrates were discovered during the 2007 MARNAUT cruise onboard R/V L'Atalante (Bourry et al., 2009; Figure 4.2). Hydrocarbon gases from this site have gas and isotopic compositions similar to those of the gases from the Thrace Basin. This suggests that the fault is acting as a conduit for the transfer of gases from the deep hydrocarbon reservoirs. Hence, this site is particularly suited to investigate the hydrogeologic connection between the gas seeps and the seismogenic zone of the NAF. Site 3 is a good place also for monitoring the Mediterranenan water, gas hydrates and cold seeps and their associated biological activity.

Site 4 is located on a secondary fault at the base of the Ganos slope in the Tekirdağ Basin, where strong bubbling of gas with 70% mantle helium (^3He) (Figure 4.4a) and deep crustal seismicity is observed (Figure 4.2). Combined geochemical sampling and monitoring of gas flux and microseismicity will be highly valuable at site 4.

4.7 Present initiatives for seafloor observatories in the Sea of Marmara

At present, there are two separate but complementary initiatives to implement seafloor observatories in the SOM: the Kandilli Marmara Sea Bottom Observatory (MSBO) project and the ESONET Marmara Demonstration Mission project.

4.7.1 Marmara Sea Bottom Observatory (MSBO) project

This project, funded by Turkish Telecom, deployed a total of five cabled seafloor observatories in 2009 (Figure 4.2). The observatories are operated by the Kandilli Observatory and Earthquake Research Institute (KOERI), as part of the Turkish national network for earthquake and tsunami hazards monitoring. The seafloor observatories consist of a three-component broadband velocity sensor with a natural period of 360s and digitizers, three component accelerometer, differential pressure meter, hydrophone, temperature meter, 3D current meter, camera, Flux Gate Compass and a tiltmeter sensor.

Deployment of the seafloor observatories and some new land stations within the framework of the MSBO project provides a uniform station distribution in the Marmara region, which is needed for effective monitoring of the earthquake activity along NAF zone. The network of seafloor and land seismological stations have the capability for: (a) early warning of tsunamis; (b) detailed earthquake source studies and tomography of the source; (c) determination of the crustal structure; (d) highly accurate determination of the earthquake magnitude and epicentre; (e) reduction in the earthquake location threshold to less than M=1.0; (f) determination of the source of seismic noise and nature of deformation and faulting; and (g) detection of fluid activity and gas-hydrate dissociation using temperature, current and pressure measurements.

Each observatory is connected to a land station by a fiber optic cable. Data flow is continuous and transmitted to KOERI by satellite transmission from the land station. Seismic data from the sea bottom observatory and land seismic network are integrated and used to locate earthquakes occurring in the SOM.

4.7.2 The ESONET Marmara-Demonstration Mission project

This project, funded by the ESONET-NoE, aims at collecting long-term multidisciplinary data in the SOM, with special emphasis on the fluid- and seismic-activity interactions. The ultimate objective of the project is to establish the most suitable sites and parameters for permanent multidisciplinary earthquake observatories in the SOM. In October 2009, six piezometers and 10 short-period (4.5 Hz) OBSs were deployed for six months (October 2009–March 2010) with the Italian vessel R/V Urania, just after the Marmesonet site survey cruises of R/V Le Suroit. A prototype instrument developed by IFREMER for monitoring gas bubble emissions (BOB) was deployed for a week with R/V Le Suroit in the Çınarcık Basin, where variations in the methane seepage was detected.

An important experiment was carried out at Site 1b (–200 m) in the entrance of the Gulf of Izmit (Figure 4.2). Starting from September 2009, the SN4 observatory was deployed from R/V Urania for a year (Figure 4.7). SN4, equipped with several sensors, including a

Figure 4.7 The SN4 observatory that was developed in the framework of ORION (Ocean Research by Integrated Observatory Networks) EC project ((Favali et al., 2006) was deployed under 200m water depth, at the entrance of the Gulf of Izmit with R/V Urania for one year starting September 2009. The new payload includes a broadband three-component seismometer, triaxial current meter, CTD, two methane sensors, oxygen sensor, and a turbidity sensor (Table 4.1). With all these sensors, SN4 detected some interesting methane anomalies that are presently being assessed to determine if there are any temporal relationships between geochemical signals related to fluid expulsion and seismic activity.

broadband, three-component seismometer, CONTROS-type methane sensors and oceanographic sensors (CTD, current meter) (Favali et al., 2006), recorded some interesting methane anomalies, the possible temporal relations of which to seismic activity or physical environmental variables (e.g., water temperature, current) are presently being evaluated. Site 1b is a key observation point to understand the behaviour of the NAF, because here the main displacement zone of the submerged strand of the NAF is extremely narrow and shows an almost pure strike-slip deformation pattern. In some of these areas, due to the presence of piercing points, we were able to estimate very accurately horizontal and vertical deformation rates along these fault strands during the Holocene (Polonia et al., 2004).

Moreover, recent analyses of geophysical imagery (multibeam bathymetry and side-scan sonar mosaics) and direct observations carried out using the Nautile submersible during MARNAUT cruise, suggest that the surface rupture of the 1999 terminated close to the Izmit Gulf entrance (Gasperini et al., 2011).

4.8 Conclusions

The SOM, with its high tectonic rates, high seismic risk and interesting oceanographic setting, is an important natural laboratory for testing different hypotheses using seafloor observatories. The correlation between fluid circulation and active faults in the SOM is now established, but the relationships between gas emissions and the seismic cycle are not well understood. Testing the hypothesis that physical and chemical changes of a precursory nature occur during an earthquake cycle is a major issue. To investigate this important issue, there is an urgent need to deploy multidisciplinary observatories incorporating a variety of geochemical, hydrological and seismic sensors in the SOM at seep sites associated with tectonic features. Provided that correlations between strain rate and fluid and seismic activity exist and may be determined over a limited time-scale experiment, we will have defined a tool that may be used to better constrain probabilistic estimates of earthquake occurrence in the marine environment. The observatories will also be useful for oceanographic and environmental monitoring of the Mediterranean water in the SOM as it moves over the deep basins and highs from the west towards the Istanbul Strait. The multidisciplinary observatory stations with flexible design to monitor different parameters in the SOM are of major importance for studies of fault activity, slope stability, tsunami, environmental and climatic changes, environmental pollution and safety, and oceanography.

References

Alpar B. 1999. Underwater signatures of the Kocaeli Earthquake (August 17, 1999). Turkish Journal of Marine Sciences 5, 111–129.

Altınok Y., Tinti S., Alpar B., Yalçıner A.C., Ersoy S., Bortolucci E. and Armigliato A. (2001) The tsunami of August 17, 1999 in Izmit Bay, Turkey. Natural Hazards 24, 133–146.

Altınok Y., Alpar B. and Yaltırak C. (2003) Sarköy-Murefte 1912 earthquake's tsunami: Extension of the associated faulting in the Marmara Sea, Turkey. J Seismol 7, 329–346.

Ambraseys N.N. (2002) The seismic activity of the Marmara Sea region over the last 2000 years. Bull Seismo Soc Am 92, 1–18.

Ambraseys N.N. and Finkel C.F. (1990) The Marmara Sea earthquake of 1509. Terra Nova 2, 167–174

Ambraseys N.N. and Finkel C.F. (1991) Long term seismicity of Istanbul and the Sea of Marmara region. Terra Nova 3, 527–529.

Ambraseys N.N. and Jackson, J.A. (2000) Seismicity of the Sea of Marmara (Turkey) since 1500. Geophys 141, F1–F6.

Armijo R., Meyer B., Navarro S., King G. and Barka A. (2002) Asymmetric slip partitioning in the Sea of Marmara pull-apart: A clue to propagation processes of the North Anatolian Fault? Terranova 14, 80–86.

Armijo R., Pondard N., Meyer B., Mercier de Lapinay B., Ucarkus G. and the MARMARASCARPS Cruise Party (2005) Submarine fault scarps in the Sea of Marmara pull-apart (North Anatolian Fault): Implications for seismic hazard in Istanbul. Geochem. Geophys. Geosyst. 6, 1–29.

Atakan K., Ojeda A. Meghraoui M. Barka A. Erdik M. and Bodare A. (2002) Seismic hazard in Istanbul following the 17 August 1999 İzmit and 12 November 1999 Düzce earthquakes. Bull. Seism. Soc. Am. 92, 466–482.

Barka A. (1992) The North Anatolian Fault Zone. Ann. Tecton. 6, 164–195.

Barka A. (1996) Slip distribution along the North Anatolian Fault associated with large earthquakes of the period 1939 to 1967. Bull. Seismol. Soc. Am. 86, 1238–1254.

Beck C., Mercier de Lapinay B., Schneider J.L., Cremer M., Çağatay N., Wendenbaum E., Boutareaoud S., Menot G., Schmidt S., Weber O., Eriş K., Armijo R., Meyer B., Pondard N., Gutcher M.A., Turon J.L., Labeyrie L., Cortijo E., Gallet Y., Bouquerel H., Görür N., Geravis A., Castera M.H., Londeix L., de Resseguier A. and Jaouen A. 2007 Late Quaternary co-seismic sedimentation in the Sea of Marmara's deep basins. Sedimentary Geology 199, 65–89.

Beşiktepe Ş.T., Sur İ.H., Özsoy E., Abdul Latif M., Oğuz, T. and Ünlüata Ü. (1994) The circulation and hydrography of the Sea of Marmara. Progress in Oceanography 34, 285–334.

Bourry C., Chazallon B., Charlou J-L, Donval J.P., Ruffine L., Henry P., Geli L., Çagatay M.N., Sedat İ, and Moreau M. (2009) Free gas and gas hydrates from the Sea of Marmara, Turkey: Chemical and structural characterization. Chemical Geology, doi:10.1016/j.chemgeo.2009.03.007

Brown K.M., Tryon M.D., DeShon H.R., Dorman L.M. and Schwartz S.Y. (2005) Correlated transient fluid pulsing and seismic tremor in the Costa Rica subduction zone. Earth Planet. Sci. Lett. 238, 189–203.

Burnard P., Bourlange S., Henry P., Geli L., Tryon M.D., Natal'in B., Şengör A.M.C., Özeren M.S. and Çağatay M.N. (2012) Constraints on fluid origins and migration velocities along the Marmara Main Fault (Sea of Marmara, Turkey) using helium isotopes. Earth Planet. Sci. Lett. 341–344, 68–78.

Çağatay M.N., Erel L., Belucci L.G., Polonia, A., Gasperini L., Eriş K., Sancar Ü., Biltekin

D., Uçarkuş G., Ülgen U.B. and Damcı E. (2012) Sedimentary earthquake records in the İzmit Gulf, Sea of Marmara, Turkey. Sedimentary Geology 282, 347–359.

Çağatay M.N., Görür N. and Alpar B. (2002) Western extension of the North Anatolian Fault and associated structures in the Gulf of Saros, NE Aegean Sea. In: N. Görür et al. (Eds) NATO Advanced Research Seminar "Integration of Earth Sciences Research on the 1999 Turkish and Greek Earthquakes". Kluwer Academic Publishers, pp. 60–71.

Camilli R., Hemond H. F. (2004) NEREUS/Kemonaut, a mobile autonomous underwater mass spectrometer. Trends in Analytical Chemistry 23(4), 307–313.

Carton H., Singh S.C., Hirn A., Bazin S., de Voogd B., Vigner A., Ricolleau A., Cetin S., Ocakoğlu N., Karakoc F. and Sevilgen V. (2007) Seismic imaging of the three-dimensional architecture of the Cinarcık Basin along the North Anatolian Fault. Jour. Geophys. Research-Solid Earth 112, B6, B06101.

Cormier M.-H., Seeber L. McHugh C.G., Polonia A., Çağatay N., Emre Ö., Gasperini L., Görür N., Bortoluzzi G., Bonatti E., Ryan W.B.F. and Newman R. (2006) Journal of Geophysical Research Vol.111, B04102, doi:10.1029/2005JB003633

Davis E.E., Becker K., Wang K., Obara K., Ito Y. and Kinoshita M. (2006) A discrete episode of seismic and aseismic deformation of the Nankai trough subduction zone accretionary prism and incoming Philippine Sea plate. Earth Plan. Sci. Let. 242, 73–84.

Demirkent I. (2001) Earthquakes in and around Istanbul during the IV–XI centuries according to the Byzantium records. In: I. Şahin I. (Ed.) Tarih boyunca Anadoluda doğal afetler ve deprem semineri, 22–23 May 2000 (in Turkish). Istanbul: Globus Dunya Publ. House, pp. 51–65.

Ergintav S., McClusky S., Hearn E., Reilinger R., Cakmak R., Herring T., Ozener H., Lenk O. and Tari, E. (2009) Seven years of postseismic feformation following the 1999, M=7.4, and M=7.2, Izmit-Duzce, Turkey Earthquake Sequence. J. Geophys. Res., doi:10.1029/2008JB006021

Favali P., Beranzoli L., D'Anna G., Gasparoni F., Marvaldi J., Clauss G., Gerber H.W., Nicot M., Marani M.P., Gamberi F., Millot C. and Flueh E.R. (2006) A fleet of multiparameter observatories for geophysical and environmental monitoring at seafloor. Annals of Geophysics 49, 2/3, 659–680.

Gasperini L., Polonia A., Çağatay M.N., Bortoluzzi G., Ferrante V. (2011) Geological slip rates along the North Anatolian Fault in the Marmara region. Tectonics, doi:10.1029/2011TC002906.

Géli L., Henry P., Zitter T., Dupré S., Tryon M., Çağatay M.N., de Lépinay B., Mercier Le Pichon X., Sengör A.M.C., Görür N., Natalin B., Uçarkus G., Özeren S., Volker D., Gasperini L. Burnard P., Bourlange S. and the Marnaut Scientific Party (2008) Gas emissions and active tectonics within the submerged section of the North Anatolian Fault zone in the Sea of Marmara. Earth Planet. Sci. Lett. 274(1–2), 34–39.

Görür N. and Çağatay M.N. (2010) Geohazards rooted from the northern margin of the Sea of Marmara since the late Pleistocene: A review of recent results. Natural Hazards 54(2), 583603.

Greinert, J. (2008) Monitoring temporal variability of bubble release at seeps: The hydroacoustic swath system GasQuant. Journal of Geophysical Research 113, C07048, doi:10.1029/2007JC004704

Greinert J., Egorov Y.A V., De Batist M. and McGinnis D. (2006). 1300m-high rising bubbles from mud volcanoes at 2080m in the Black Sea: Hydroacoustic characteristics and temporal variability. Earth and Planetary Science Letters 244, 1–15.

Guidoboni E., Comastri A. and Trania G. (1994) Catalogue of ancient earthquakes in the Mediterranean area up to the 10th century. Instituto, Nazionale di Geofisica, Rome.

Hebert H., Schindele F., Altinok Y., Alpar B. and Gazioglu C. (2005) Tsunami hazard in the Marmara Sea (Turkey): A numerical approach to discuss active faulting and impact on the Istanbul coastal areas. Marine Geology 215, 23–43.

Henry P., Lallemant S.J., Nakamura K., Tsunogai U., Mazzotti S. and Kobayashi K. (2002) Surface expression of fluid venting at the toe of the Nankai wedge and implications for flow paths. Marine Geology 187, 119–143.

Hubert-Ferrari A., Barka A., Jacques E., Nalbant S., Meyer B., Armijo R., Tapponier P. and King G.C.P. (2000) Seismic hazard in the Sea of Marmara following the 17 August 1999 Izmit earthquake. Nature 404, 269–272.

Gasperini L., Polonia L., Bortoluzzi G., Henry P., Le Pichon X., Tryon M., Çağatay N. and Louis Géli. (2011) How far did the surface rupture of the 1999 İzmit earthquake reach in Sea of Marmara? Tectonics Volume 30, TC1010, 11 PP., 2011 doi:10.1029/2010TC002726

Görür N., Çağatay M.N. (2009) Geohazards rooted from the northern margin of the Sea of Marmara since the late Pleistocene: A review of recent results. Natural Hazards 54(2), 583–603.

İmren C., Le Pichon X., Rangin C., Demirbağ E., Ecevitoglu B. and Gorur N. (2001) The North Anatolian Fault within the Sea of Marmara: A new interpretation based on multichannel seismic and multibeam bathymetry data. Earth and Planetary Sc.Letters.186, 143–158.

Karabulut H., Bouin M.-P., Bouchon M., Dietrich M., Cornou C. and Aktar M. (2002) The seismicity in the Eastern Marmara Sea after the 17 August 1999 Izmit earthquake. Bull. Seismol. Soc. Am. 92, 387–393.

Kuşçu I., Okamura M., Matsuoka H., Gokasan E., Awata Y., Tur H. and Şimşek M. (2005) Seafloor gas seeps and sediment failures triggered by the August 17, 1999 earthquake in the Eastern part of the Gulf of Izmit, Sea of Marmara, NW Turkey. Marine Geology 215, 193–214.

Le Pichon X, et al. (1992) Fluid venting activity within the Eastern Nankai Trough

Accretionary Wedge: A summary of the 1989 Kaiko-Nankai results. Earth Plan. Sci. Let. 109 (3–4), 303–318.

Le Pichon X., Şengör A.M.C. and Demirbağ E. (2000) Marine Atlas of the Sea of Marmara (Turkey). Geneva: Ifremer.

Le Pichon X., Şengör A.M.C., Demirbağ E., Rangin C., İmren C., Armijo R., Görür N., Çağatay N., Mercier De Lepinay B., Meyer B. Saatçılar R. and Tok B. (2001) The active Main Marmara Fault. Earth Plan. Sci. Let. 192, 595–616.

Le Pichon X., Chamot-Rooke N., Rangin C. and Şengör A.M.C. (2003) The North Anatolian fault in the Sea of Marmara. J. Geophys. Res. 108 (B4), 2179, doi: 10.1029/2002JB001862

McHugh C.M.G., Seeber L., Cormier M.-H., Dutton J., Çagatay M.N., Polonia A., Ryan W.B.F. and Görür N. (2006) Submarine earthquake geology along the North Anatolian Fault in the Marmara Sea, Turkey: A model for transform basin sedimentation. Earth and Planetary Science Letters 248, 661–684.

Moore J.C., Orange D. and Kulm L.D. (1990) Interrelationship of fluid venting and structural evolution - Alvin observations from the frontal accretionary prism, Oregon. J. Geophys. Res. 95 (B6), 8795–8808.

Muir-Wood R. and King G.C.P. (1993) Hydrological signature of earthquake strain. J. Geoph. Res. 98, B12, 22035–22068.

Oglesby D.D., Mai P.M., Atakan K. and Pucci, S. (2008) Dynamic models of earthquakes on the North Anatolian fault zone under the Sea of Marmara: Effect of hypocenter location. Geophysical Research Letters 35, L18302, doi:10.1029/2008GL035037, 2008

Okay A.I., Kaşlılar-Özcan A., Imren C., Boztepe-Güney A., Demirbağ E. and Kuşçu I. (2000) Active faults and evolving strike-slip basins in the Sea of Marmara, northwest Turkey: A multichannel seismic reflection study. Tectonophysics 321, 189–218.

Özalabey S., Ergin M., Aktar M., Tapirdamaz C., Biçmen F. and Yörük A. (2002) The 1999 Izmit earthquake sequence in Turkey: Seismological and tectonic aspects. Bull. Seism. Soc. Am. 92, 376–386.

Ozansoy E. (2001) 1200–1453 Istanbul earthquakes according to the Byzantium records. In: I. Şahin (Ed.) Tarih boyunca Anadoluda doğal afetler ve deprem semineri, 22–23 May 2000. Istanbul: Globus Dunya Publ. House, pp. 1–29 (in Turkish).

Özeren M.S., Çağatay M.N., Postacıoğlu N., Şengör A.M.C., Görür N. and Eriş K. (2010) Mathematical modelling of a potential tsunami associated with a late glacial submarine landslide in the Sea of Marmara. Geo-Marine Letters, doi 10.1007/s00367-010-0191-1

Papazachos C.B., Koutitas C.H., Hatzidimitriou M.P., Karacostas G.B. and Papaioannou A.Ch. (1986) Tsunami hazard in Greece and the surrounding area. Annales Geophysicae 4, 79–90.

Parsons T. (2004) Recalculated probability of M>7 earthquakes beneath the Sea of Marmara. Journal of Geophysical Research 109, doi:10.1029/2003JB002667

Parsons T., Toda S., Stein R.S., Barka A. and Dieterich J.H. (2000) Heightened odds of large earthquakes near Istanbul: An interaction based probability calculation. Science 288, 661–665.

Pınar N. and Lahn E. (2001) Explanatory catalogue of earthquakes of Turkey (in Turkish). Istanbul: Yıldız Technical University Foundation Publ.

Polonia A., Gasperini L., Amorosi A., Bonatti E., Çağatay N., Capotondi L., Cormier M.-H., Görür N., McHugh C. and Seeber L. (2004) Holocene slip rate of the North Anatolian Fault beneath the Sea of Marmara. Earth and Planet. Sci. Letters 227, 411–426.

Pondard N., Armijo R., King G.C.P., Meyer B. and Flerit F. (2007) Fault interactions in the Sea of Marmara pull-apart (North Anatolian Fault): Earthquake clustering and propagating earthquake sequences. Geophys. J. Int. 171, 1185–1197, doi: 10;1111/j.1365-246X.2007.03580.x

Reilinger R.E., Ergintav S., Burgmann R., McClusky S.,Lenk O., Barka A., Gurkan O., Hearn L., Feigl K.L., Cakmak R., Aktug B., Ozener H. and Toksoz, M.N. (2000) Coseismic and postseismic fault slip for the 17 August 1999, M=7.5, Izmit, Turkey earthquake, 2000. Science 289, 1519–1524.

Ritt B., Sarrazin J., Caprais J-C., Noel P., Gauthier O., Pierre C., Henry P. and Desbruyeres D. (2010) First insights into the stucture and environmental setting of vold-seep communities in the Marmara Sea. Deep-Sea Research I, doi: 10.1016/j.dsr.2010.05.011

Roeloffs E.A. (1998) Hydrologic precursors of earthquakes: A review. Pure. App. Geophys. 126, 177–209.

Şengör A.M.C., Tüysüz O., İmren C., Sakinç M., Eyidoğan H., Görür N., Le Pichon X. and Rangin C. (2004) The North Anatolian Fault: A new look. Annu. Rev. Earth Planet. Sci. 33, 1–75.

Short R.T., Toler S.K., Kibelka G.P.G., Rueda Roa D.T., Bell R.J. and Byrne R.H. (2006) Detection and quantification of chemical plumes using a portable underwater membrane introduction mass spectrometer. Trends Anal. Chem. 25 (7), 637–646.

Stein R.S., Barka A. and Dieterich J.H. (1997) Progressive failure on the North Anatolian fault since 1939 by earthquake stress triggering. Geophys. J. Int. 128, 594–604.

Straub C. and Kahle H.-G. (1997) Recent crustal deformation and strain accumulation in the Marmara Sea region, NW Anatolia, inferred from repeated GPS measurements in active tectonics of Northwest Anatolia. In: C. Schindler and M. Pfister (Eds) The Marmara Poly-project. Zurich: Hochschulverlag AG an der ETH, pp. 417–447.

Sultan N., Cochonat P., Canals M., Cattaneo A., Dennielou B., Haflidason H., Laberg J.S., Long D., Mienert J., Trincardi F., Urgeles R., Vorren T.O. and Wilson C. (2004) Triggering mechanisms of slope instability processes and sediment failures on continental margins: A geotechnical approach. Mar Geol 213, 291–321.

Sultan N., Voisset M., Marsset B., Marsset T., Cauquil E. and Colliat, J.-L. (2006) Potential role of compressional structures in generating submarine slope failures in the Niger delta. Marine Geology, doi: 10.1016/j.margeo.2006.11.002

Toksöz M.N., Shakal A.F. and Michael A.J. (1979) Space-time migration of earthquakes along the North Anatolian fault zone and seismic gaps. Pure Appl. Geophys. 117, 1258–1270.

Tortell P.D. (2005) Dissolved gas measurements in oceanic waters made by membrane inlet mass spectrometry. Limnol. Oceanogr. Methods 3, 24–37.

Trique M., Richon P. and Perrier F. (1999) Radon emanation and electric potential variations associated with transient deformation near reservoir lakes. Nature 399, 137–141.

Tryon M.D., Brown K.M., Dorman L.M. and Sauter A. (2001) A new benthic aqueous flux meter for very low to moderate discharge rates. Deep-Sea Research I 48(9), 2121–2146.

Tryon M.D., Çağatay M.N., Zitter T.A.C., Géli L., Gasperini L,. Burnard P., Bourlange, S. and Grall C. (2010) Pore fluid chemistry of the North Anatolian Fault Zone in the Sea of Marmara: A diversity of sources and processes. Geochemistry, Geophysics, Geosystems, Vol.11, dx.doi.org/10.1029/2010GC003177

Ünlüata Ü., Oğuz T., Latif M.A. and Özsoy E. (1990) On the physical oceanography of the Turkish Straits. In: L. J. Pratt (Ed.) The Physical Oceanography of Sea Straits. Dordrecht: NATO/ASI Series, Kluwer, pp. 25–60.

Yalçıner, A.C, Synolakis, C.E., Borrero, J., Altınok, Y., Watts, P., Imamura, F., Kuran, U., Ersoy, Ş., Kanoğlu, U., Tinti, S. (1999). Tsunami generation in Izmit Bay by the Izmit earthquake, Proceed. ITU-IAHS, Intern. Confer. on the Kocaeli Earthquake 17 August 1999, 217-212.

Yalçıner A.C., Alpar B., Altınok Y., Özbay I. and Imamura F. (2002) Tsunamis in the Sea of Marmara: Historical documents for the past, models for future. Mar. Geol. 190, 445–463.

Zitter T.A.C., Henry P., Aloisi G., Delaygue G., Çağatay M.N., Mercier de Lepinay B., Al-Samir M.F., Fornacciari M., Pekdeger A., Wallmann K. and Lericolais G. (2008) Cold seeps along the main Marmara fault in the Sea of Marmara (Turkey). Deep Sea Research, Part I, 55, 552–570.

5 The Hellenic deep sea observatory: Science objectives and implementation

V. Lykousis, K. Nittis, D. Ballas, L. Perivoliotis, D. Kassis, P. Pagonis and D. Sakellariou[1]

5.1 Introduction

The Hellenic observatory in the SE Ionian Sea is located on the steep marginal slopes and basins of the western segment of the Hellenic Arc and Trench System (HATS) (Figure 5.1). The Hellenic Arc and Trench System is a typical subduction zone that is the most tectonically active region in Europe, implying a complicated submarine morphology with numerous, deep, pull-apart sub-basins and a variety of geohazards like high seismicity, slope instabilities and basin wide tsunamis. Fluid emissions, methane gas seeps and mud volcanism in the Eastern Mediterranean accretionary prism provide the substrate for specific deep sea ecosystems. The deep waters of the sub-basins of the western sector of the Hellenic Trench are also sensitive to climate changes, since the deep sub-basins are the ultimate destination of dense (cold and salt) water masses formed in the Adriatic and Aegean Seas. Consequently, the temperature and salinity of bottom waters provides a year-to-year record of the (winter) low temperature and/or precipitation-aridity of the NE Mediterranean Sea. Being the most oligotrophic European sea the deep ecosystem of the Mediterranean Sea is sensitive to climate changes. It is also known that the Hellenic Trench region is a major pathway and important feeding ground for cetaceans in the Eastern Mediterranean (SE Sicily-SW Peloponnesus-S Crete-W Cyprus) (Reeves and Di Sciara, 2006).

1 Hellenic Centre for Marine Research, Anavyssos, Athens, Greece

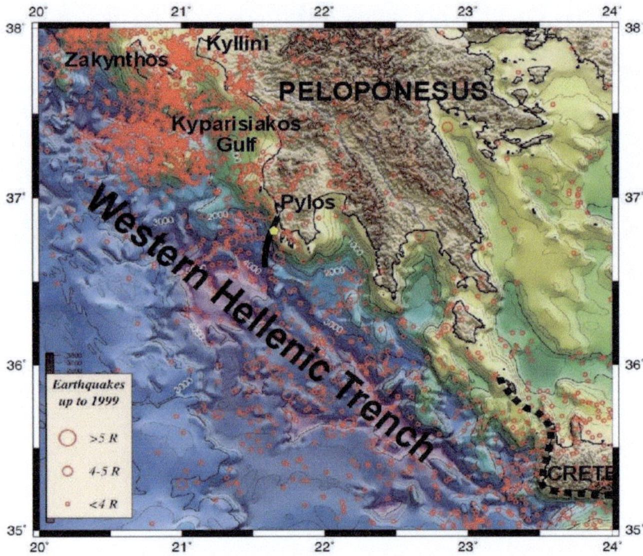

Figure 5.1 Regional map of the western segment of the Hellenic Arc and Trench System with the seismic epicentres of the last century. Dashed line denotes the uplifted coastal zones during the 365AD earthquake and solid line shows the existing electro-optical cable of the NESTOR Institute. The yellow pentagon indicates the deployment site of POSEIDON-Pylos observatory in the Hellenic site.

The scientific significance of the western sector of the Hellenic Arc is evidenced by numerous EU and international projects that are implemented in the region within the FP6-FP7: HERMES-IP, HERMIONE, EUROCEANS-NoE, CoralFish, SESAME-IP (marine ecosystem research), 3-HAZ, SEHELLARC, TRANSFER (geohazards) and Eurosites, Poseidon I, II; regarding water-column monitoring, marine forecast and modelling.

Within the framework of the European projects ESONET (European Seas Observatory NETwork) and the European Research Infrastructure EMSO (European Multidisciplinary Seafloor Observation), three Mediterranean sites have been identified as prospective European observatory nodes. They are characterized by having existing cable connections to the shore and are operated in synergy with the ANTARES (Ligurian Sea), NEMO (East Sicily) and NESTOR (Hellenic site) neutrino detectors (also candidate sites for the ESFRI KM3NET). These sites are of multidisciplinary scientific interest, with seismic activity near dense human habitation areas, slope instabilities with turbidity currents in canyons, past tsunamis and long-term studies of biological and physical oceanography.

The Hellenic node, located SW of the Peloponnesus, is a point in the seas around Europe with unique scientific characteristics for the monitoring of geohazards, climate variability and ecological parameters (4000–5000m). There is an operating stand-alone observatory with multiparametric water column and seabed sensors (Poseidon I-II-III), as well as the significant onshore infrastructure (buildings, specialized labs) and offshore installations (fiber optic cable, deployment platform, etc.) of the NESTOR Institute. Future prospects within KM3NET-ESFRI (deployment of new fiber optics cable), ESONET-NoE and EMSO-ESFRI implies that the Hellenic site is a promising offshore observatory within the frame of the European deep sea observation network. It should be noted that the first pilot for multidisciplinary cabled observatories was implemented for a short period by the NESTOR experiment during 2002 (Anassontzis and Koske, 2003).

5.2 Hellenic observatory: Science objectives

5.2.1 Geodynamics and seismicity

The greater region of the Hellenic observatory site is the western part of the Hellenic Arc and Trench System off the southwestern Peloponnesus, Eastern Mediterranean Sea. The Hellenic Arc marks the southern termination of the Aegean tectonic plate and has developed at the boundary with the subducting African–East Mediterranean plate. The Arc has migrated southward several hundreds of kilometers relative to Eurasia since Eocene times as a result of processes including south(west)ward rollback of the subducting African lithosphere and associated backarc extension, gravitational collapse, and westward extrusion of Anatolia (Dewey and Şengör, 1979; Le Pichon et al., 1982; Meulenkamp et al., 1988; Jacobshagen, 1994; Fassoulas et al., 1994; Meijer and Wortel, 1997; Cianetti et al., 2001; Jolivet, 2001; Armijo et al., 2004; Kreemer and Chamot-Rooke, 2004).

The Eastern Mediterranean Sea constitutes the last remnant of the Mesozoic-Cenozoic oceanic basin of Tethys, which is now almost totally consumed as a result of the long-term plate convergence between Eurasia and Africa. The morphology of the East Mediterranean seafloor relates both to the early history of the formation of the deep basins and the recent geodynamic interactions between interfering microplates. From the southwest Peloponnesus to south of Crete and Rhodes, the Eastern Mediterranean seafloor is characterized by a 1500km long and 200–250km wide, arc-shaped, sedimentary wedge/accretionary prism, known as the Mediterranean Ridge (Heezen and Ewing, 1963; Emery et al., 1966). The Mediterranean Ridge results from the relatively rapid (>3cm/year) subduction of the African plate beneath Eastern Europe and is formed in front of the overriding Aegean microplate (Le Pichon et al., 1995; Dewey and Şengör, 1979; Kreemer and Chamot-Rooke, 2004; McClusky et al., 2000).

The characteristics and morphostructural variability of the Mediterranean Ridge are related to: (1) incipient collision south of Crete, where the African and Aegean continental margins are nearly in contact; (2) unique regional kinematics, controlled by frontal convergence south of Crete (central Mediterranean Ridge) and oblique subduction with an opposing sense of lateral shear for the western (Ionian – dextral shear) and eastern (Levantine – sinistral shear) domains of the Mediterranean Ridge; and (3) peculiarities of its sedimentary cover, which includes massive salt layers within the outer Mediterranean Ridge and local salt deposits within the inner domains, both of them controlling the north–south morphostructural variability of the sedimentary wedge (Huguen et al., 2006).

The Western Hellenic Trench particularly displays a tectonically complex, largely compressive setting, related to (i) the northeastward underplating and subduction of the Ionian oceanic lithosphere below the overriding Aegean microplate and (ii) the "escape tectonics"regime accommodated by dextral strike slip along the trench, at the interface between the deformed sediments of the East Mediterranean Ridge and the rigid backstop of the Aegean microplate (McKenzie, 1978; Le Pichon et al., 1995; Kreemer and Chamot-Rooke, 2004). Chamot-Rooke et al. (2005) have shown that active shear is localized at the prism-backstop contact, which combines a major dextral flower structure and a site of massive mud expulsion.

The deformation pattern affecting the western sector of the Hellenic Trench during the Neogene and Quaternary produced a complex bottom physiography comprising basin plains, canyons, hummocky ("cobblestone") topography, ridges, steep slopes and perched basins on the continental slope. The structural and stratigraphic configurations of this region have been investigated in detail by means of several closely spaced, high-resolution seismic surveys (Mascle et al., 1982; Chaumillon et al., 1996; Chaumillon and Mascle, 1997; Chamot-Rooke et al., 2005). These studies reveal a structurally deformed, unconsolidated sediment sequence of highly variable thickness. This series, dated as probable Pliocene and Quaternary ages, lies above indurate strata locally comprising deposits of Cretaceous, Miocene, including probable Messinian evaporates. The influence of recent tectonics on the sediments in the trench, as revealed from the high-resolution sub-bottom surveys, shows deformation of unconsolidated sediments, ponding, and syndepositional tilting, folding and faulting of the uppermost deposits.

The seismicity of the western segment of the Hellenic Arc and Trench system could have two components: (i) predominantly small earthquakes associated with salt tectonics and (ii) high seismicity related mainly to thrusting and strike slip tectonics due to the subduction of the Ionian lithosphere, the lateral escape of the accreted sediments and the deformation of the overriding plate, as well as to normal faulting at localized, apparently extensional, structures behind the active arc. The seismic activity of the wider area around the Hellenic site during the last century includes hundreds of earthquakes of magnitude M>4, many others of magnitude M>5 and a considerable number of events with magnitude M>7. The expected peak ground accelerations due to seismic activity have been estimated to be 100–110cm s^{-2} (0.1–0.11g), with a 70% probability of not being exceeded in 50 years (Makropoulos and Burton, 1985). The maximum expected peak ground accelerations for a 25-year period in the greater region of the Hellenic site is 100–125cm s^{-2}, for a 50-year period it is 150–200cm s^{-2}, for a 75-year period it is 200–250cm s^{-2}, while for a 100-year period it is 250–300cm s^{-2} (Stavrakakis et al., 1989).

5.2.2 Seafloor instabilities

A variety of mass wasting on the continental shelf, marginal slopes and the deep subbasins of the trench has been identified on the tectonically active margins of the western segment of the Hellenic Arc.

The most predominant type of instabilities on the continental shelves are the small-scale rotational slumps that occur on modern (Holocene) prodelta deposits (i.e., Kyparisiakos-Kilini shelf; Ferentinos et al., 1985; Hasiotis et al., 2002; Lykousis et al., 2009). This type of slope instability is common to low-angle (0.5°–2°) prodelta slopes on the seismically active shelves of western Greece (the western Gulf of Corinth, the Patraikos and Kyparisiakos Gulfs). In these areas of normally consolidated prodelta deposits (mean sedimentation rates from 3–5m ka^{-1}), slope failures manifest themselves as shallow-seated, arcuate or curved slide plains (gravity faults) with small vertical displacement. These ruptures are slip planes associated with a sequence, or peripheral (rotational) block slumps restricted to the late Holocene (HST) foresets. The individual slump blocks are about 80–150m long and have slip planes extending to a mean depth of 10–15m. The deeper, transparent (early HST bottomsets), together with the late Pleistocene transgressive sequences (TST), are mostly unaffected. Gas charged sediment occurs between the early HST bottomsets and the lower section of the HST forests.

Similar failure processes have also been observed within Late Pleistocene prodelta sequences on the shelf-break, deposited during the late glacial maximum (sea level lowstand). Stability analysis and assessment of post-deformational processes (erosion, and the accumulation of undisturbed sediments in the non-eroded areas above the slip planes) indicate present inactivity and termination of the failure period between 18 and 10 kyears BP (Lykousis and Chronis, 1989).

Large-scale translational slides and slumps were detected on continental slopes of the west Hellenic Arc associated with major active fault zones, strong earthquake motions and relatively high sediment accumulation rates, e.g., Kyparisiakos and Killini slopes offshore of the W. Peloponnesus (Ferentinos et al., 1985; Hasiotis et al., 2002, 2005). In areas where the shelf is negligible, these failures are associated with significant upslope retrogradation that has caused coastal retreat with important human and economic impact (coastline, destruction of major coastal roads, beach submergence, etc.) during the last 200 years (i.e., western Gulf of Corinth and Patraikos Gulf, Lykousis et al., 2009). Potential failure planes in the western Hellenic Arc slopes are: (a) the thin muddy hemipelagic layers (high water content–low strength) that have been deposited during Pleistocene interglacial periods (sea-level high stand); (b) sandy layers that were deposited from maximum flooding during glacial stages (sea-level lowstand); and, probably (c) the relatively low shear strength of older (S_4–S_5) sapropelic layers (Lykousis, 1991).

Turbidites, homogenites and debris flow deposits are widespread in the deep basins of the Hellenic Trench. Coarser grained (sandy) turbidites are usually associated with strong earthquake motions (seismoturbidites) (Anastasakis and Piper, 1991), while muddy turbidites are present along sediment cores from most of the deep basins (unpublished data; Alves et al., 2007) (Figure 5.2). Debris flows are dispersed along the Zakynthos-Kefallinia deep valley and homogenites in most of the SW Ionian sub-basins (Stanley et al., 1978; Poulos et al., 1999).

Slope stability was calculated using the Normalised-Soil-Parameter (NSP) method under undrained conditions of the normally consolidated continental slope sediments (Lee and Edwards, 1986). This analysis indicates that for slopes of relatively rapidly accumulated sediments (i.e., Kyparisiakos-Killini slope) instabilities could be induced by critical earthquake ground accelerations of 0.12–0.14g (Lykousis et al., 2009; Lykousis et al., in preparation). Consequently, the sediments of the continental slopes of the western segment of the Hellenic Arc are highly unstable since the regionally expected peak ground accelerations (PGA) for the next 50 years range from 20 to 30% g (Makropoulos and Burton, 1985).

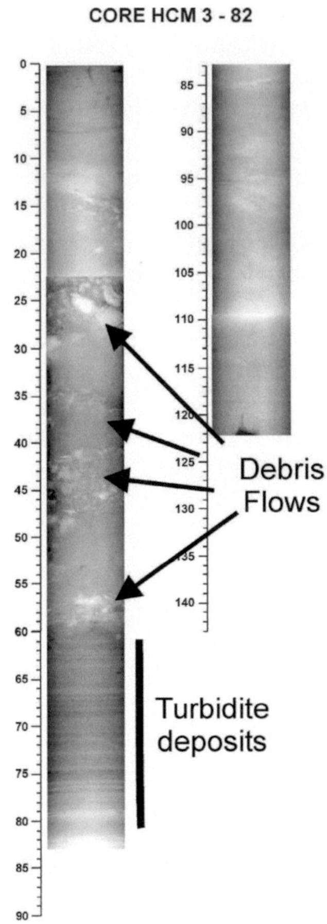

Figure 5.2 X-ray image of a typical sediment core from the Hellenic Trench indicating debris flow deposits in the upper part and a sequence of turbidites below.

5.2.3 Tsunamis

More than ten significant tsunamis have been reported in the western segment of the Hellenic Arc during the last 2000 years (Galanopoulos, 1960; Papadopoulos and Fokaefs, 2005).

Major active faults and high seismicity, complex seafloor relief and shallow submarine mass movements make the tsunami risk in the eastern Mediterranean one of the largest in the European seas. Statistical tsunami recurrence ranges from 8–11 years, regardless of tsunami intensities (Soloviev et al., 2000; Schielein et al., 2007). From a geographical point of view, the very strong tsunamis are associated either with highly seismogenic structures such as the Messina straits, South Italy, the Corinth Gulf, central Greece, and the

Hellenic Arc, or with the active volcanic complexes such as those in Thera, South Aegean Sea. Galanopoulos (1960) reported over ten significant tsunamis in the western segment of the Hellenic Arc during the last 2000 years, while Papadopoulos and Fokaefs (2005) list at least 17 tsunami events in the Ionian Sea since the 13th century AD. Papadopoulos (2005) has classified the western Hellenic Arc as one of the highest tsunamigenic potential in European seas, according to the frequency of occurrence and intensity.

The most devastating tsunami occurred on July 21, 365 AD, triggered by a large, M: 8–8.5 estimated magnitude, earthquake with the epicenter offshore of Western Crete. An extensive coastal zone of Antikythera island, western and south-western Crete was co-seismically uplifted by up to 7–9m. The basin-wide tsunami affected the entire Eastern Mediterranean, destroyed cities and drowned thousands of people in the coastal regions from the Nile Delta to Sicily and Croatia, including the coastal zone of Alexandria. Shaw et al. (2008) estimated about 5000 years for the repeat time of tsunamigenic events on the same fault in western Crete, but only 800 years if the same process takes place along the full length of the Hellenic subduction zone.

5.2.4 Fluid flow and mud volcanism

Extensive fluid flow and mud volcanism is evidenced by the presence of gas seeps and pockmarks on the continental margin and coastal zone of the NW Peloponnesus and mud volcanism with brine pools along the western segment of the Mediterranean ridge (Cita et al., 1980, 1981; Camerlenghi et al., 1992). The deformed sediments of the ridge are thrust northwestwards, with dextral shearing predominating (strike-slip faulting). Active faulting, salt tectonics and active shear strain are responsible for the migration of deeper fluids to the surface and for the formation of an extensive belt of mud volcanoes and brine lakes.

Fluid flow is evidenced by seafloor pockmarks that are spread wide on the continental shelf of the NW Peloponnesus and generally on the greater Greek-Ionian shelves (Papatheodorou et al., 1993; Hasiotis et al., 2002; Christodoulou et al., 2003). Offshore hydrocarbon leakage was observed in the continental margin south from Zakynthos Island. at depths of around 150m (HCMR, unpublished data). Shallow gas seepage is usually activated following strong earthquake motions (i.e., Gulf of Patras, Hasiotis et al., 1996).

Gas seepage along the coastal zone of the northwestern Peloponnesus (Killini, Katakolo and Kaiafas sites) reflects deep hydrocarbon-generation processes and upward gas escape (Etiope et al., 2006). Methane microseepage, gas concentration in offshore and onshore vents, and gas dissolved in water springs, including the isotopic analysis of methane, have shown that the seeps are caused by thermogenic methane accumulated in Mesozoic limestone that had migrated upward through faults, or zones of weakness, induced by salt diapirism. A link between local seismicity and salt tectonics is suggested by the analyses of hypocenter distribution. The gas seeps (10^1–10^5mg m^{-2} day^{-1}) are caused by salt tectonics producing zones of weakness and faults along which CH_4–H_2S migrate upward. The distribution and depth of the hypocenters in the Katakolo zone suggest that the thrusting faults in the pierced crust might be seismogenic. Accordingly, gas migration and seepage might respond mainly to diapirism-linked seismicity because the stress-strain field is closer to the gas accumulations and may easily perturb the fluid-pressure gradients and the gas-bearing property of the faults (Etiope et al., 2006).

The first indication of mud volcanism in the western segment of the Hellenic Arc and trench system ridge was in 1978. During the R/V Eastward cruise, the first sample of mud breccia was collected and studied in detail from a dome-shaped structure discovered west of Crete on the Mediterranean ridge ("Prometheus dome", Cita et al., 1980). Mud breccia was again recovered in the same region during the R/V Bannock cruise (1981), known as the Cobblestone 3 Area (Cita et al., 1982). After several years of investigation on the mud diapiric fluids of the central part of the Mediterranean Ridge, a new study of the western sector was initiated in 1994 in the framework of the UNESCO "Training-through-research (Floating University)" program with R/V Gelendzhik (TTR-4 cruise) (Limonov et al., 1995). Three new mud volcanoes were discovered and sampled during the cruise, the Novorossiysk Dome, Aros Dome, and Unnamed Dome. Mud breccia was recovered from all these mud volcanoes, containing numerous clasts from deeper strata in the Mediterranean Ridge (Limonov et al., 1995). More recent studies that included detailed multibeam and high-resolution sub-bottom profiling surveys provided indications of extensive mud volcanism associated with fluid flow and brine pools (Mascle et al., 1982; Chamot-Rooke et al., 2005). Most of the inferred mud volcanoes have not been sampled by sediment coring or visually observed by ROV or submersibles and consequently remain unexplored.

5.2.5 Thermohaline circulation and climate change

The Ionian Sea has a complex hydrology and can be considered as a transitional area where water masses formed in the Levantine, Aegean and Adriatic Seas meet and interact with the water masses of the Western Mediterranean that enter through the Sicily straits. The typical water mass-structure of this area includes: the Modified Atlantic Waters (MAW) characterized by a salinity minimum and occupying approximately the upper 25–100m of the water column; the Levantine Intermediate Waters (LIW) that occupy typically the 100–500m layer and are characterized by a salinity maximum; the Transitional Waters (TW) occupying a layer 500–1200m deep; and the Eastern Mediterranean Deep Waters (EMDW) filling the layers below 1200m (Nittis et al., 1993; Malanotte-Rizzoli et al., 1997). The Eastern Mediterranean Transient (EMT) of the early 1990s in association with massive production of dense water in the Aegean Sea further complicated this structure (Theocharis et al., 1999). The bottom waters of the Ionian Sea were occupied by the newly-formed warm and saline Aegean or Cretan Deep Water (CDW) which lifted the older EMDW of Adriatic origin into the transitional layers. In addition, the intermediate water masses were occupied by Cretan Intermediate Waters (CIW) with characteristics similar to LIW (Malanotte-Rizzoli et al., 1999). The origin of the EMT has been attributed to both regional meteorological anomalies (cold winters, reduced precipitation) and long-term salinity changes in the Eastern Mediterranean, related to climatic variability and anthropogenic activities. This major change of the thermohaline circulation in the Mediterranean Sea has demonstrated the vulnerability of the basin to external forcing, reaffirming the region as a perfect laboratory for studies of ocean dynamics, including climate variability effects.

The general circulation of the basin exhibits strong inter-annual variability. The large-scale anticyclonic circulation observed during the 1980s and early 1990s has been confined to the southern part and has been replaced by a cyclonic circulation in the northern Ionian during the mid-to-late 1990s (Larnicol et al., 2002). This large-scale circulation is in most cases superimposed by strong mesoscale eddies. Indeed, the south-eastern Ionian, where the Hellenic deep observatory is being developed, is in the vicinity of the strongest of these mesoscale features: the Pelops anticyclone gyre (Malanotte-Rizzoli et al., 1997). This gyre is intensified at intermediate and deep layers, having a strong temperature signal in the upper 500m and a salinity signal below (Nittis et al., 1993). Furthermore, the south-eastern Ionian is characterized by the northward flow of water masses (CIW, CDW) exiting from the western straits of the Aegean Sea, especially at intermediate and deeper layers, and flowing northwards towards the Otranto strait along the western Hellenic Arc. Therefore, the observatory is strategically located on the pathway of these waters, allowing the long-term monitoring of their properties and, eventually, the study of mixing processes with the deep waters of Adriatic origin. Overall, the observatory is situated in an ideal location to investigate the East Mediterranean Sea thermohaline circulation and its variability attributed to regional or to larger-scale forcing and climate change. For this reason, emphasis has been put in this initial phase on the equipment of the observatory: appropriate sensors to measure physical parameters in the whole water column, including the near-bottom layer.

5.3 Existing stand-alone observatory (Poseidon system – Pylos site)

The deep-sea observatory in the Hellenic region (SW Peloponnesus margin – Figures 5.1, 5.3) has been operating since 2007 in the SE Ionian Sea (Pylos slope at 1680m depth) and includes a mooring line hosting sensors on its upper 1000m, a surface buoy for real-time telemetry and air–sea interaction measurements, as well as an autonomous seabed platform. The observing system is complemented by a 4000m long mooring deployed in the deep basin (4500m depth) and operates as an offline standalone module for time-series measurements (sediment traps/current meters) at various depths (600m, 1100m, 2000m, 3000m and 4100m).

5.3.1 Surface buoy: Air–sea interaction monitoring

The buoy used at the POSEIDON-Pylos site is a Seawatch-Wavescan type, combining a multi-parametric instrumentation platform suitable for deployment in deep-offshore locations. The buoy includes meteorological sensors for wind speed and direction, air temperature and atmospheric pressure. Wave height, period and direction measurements are carried out by an Oceanor-Wavesense sensor. The surface (1m) oceanographic sensors hosted by the buoy include an SBE-SIP for temperature and salinity measurement, and a Nortek-Aquadopp for current speed and direction. The surface buoy also receives data from water column sensors and seabed modules through inductive and hydro-acoustic modems. All data are preprocessed and sent to the Operational Centre of HCMR every three hours (or longer in the case of tsunami detection) using cellular telephony (GPRS) and/or INMARSAT-C communication systems. The energy autonomy of the system relies on rechargeable batteries, solar panels and a wind power generator.

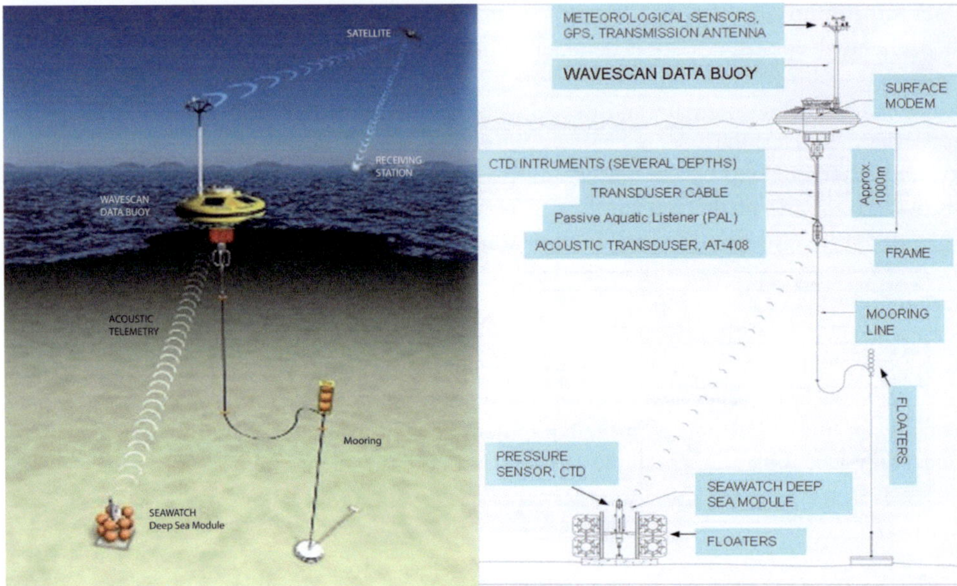

a) b)

Figure 5.3 Schematic presentation of the existing stand-alone POSEIDON-Pylos (Hellenic) observatory. (a) A representation of the acoustic telecommunication between the seabed platform and the buoy. (b) A more detailed representation of the system configuration.

5.3.2 Water column monitoring

The mooring line hosts CTD (SBE37 and SBE16) sensors at 1, 20, 50, 75, 100, 250, 400, 600 and 1000m. Thus, the observatory provides, for the first time, near real-time hydrological data from a key area where the Cretan Intermediate and Deep Waters (CIW, CDW) spread northwards towards the Adriatic and meet with the Eastern Mediterranean Deep Waters (EMDW) (Theocharis et al., 2002). The sampling interval is set to one hour while data are transferred to the surface buoy every three hours through the inductive mooring cable and appropriate modems. The mooring line is also hosting a Passive Aquatic Listener (PAL) at 500m depth for rainfall and wind estimates as well as for marine mammal detection. A PAL consists of a broadband, low noise omni-directional (zenith angle) hydrophone (Hi-Tech-92WB), a signal processing board, a low-power microprocessor (Tattletale-8) with a 100 kHz A/D digitizer, a 2 GB memory card and a 60 Amp-hour battery pack. The sampling strategy can be designed to allow autonomous operations for up to one year. These systems have been recently evaluated against X-band radar measurements in this area and were found to provide realistic estimates of precipitation (Anagnostou et al., 2008). The integration in the water column module of the observatory is done through an inductive modem that allows data transmission to the surface buoy every three hours.

5.3.3 Seabed platform

The platform is equipped with a high-accuracy pressure sensor (Paroscientific 43K-101) for tsunami detection, as part of a future early warning system (Table 5.1). The sampling rate for the pressure sensor is set to 15 sec but in standard mode data are transferred to the surface buoy every 3h. If a tsunami event is detected, the platform switches into an alarm mode sending data every 15 sec to the surface buoy and from there to the operational center. The software for on-board pressure data analysis and tsunami detection is based on the NOAA DART system of the Pacific Ocean (http://nctr.pmel.noaa.gov/tda_documentation.html). The platform is also equipped with a CTD probe (SBE 16) and a dissolved oxygen sensor (Aanderaa Optode) for near-bed physical hydrological measurements. The sampling interval for these sensors is set to 1h. These near-bed measurements allow monitoring of the evolution of deep-water properties (EMDW of either Adriatic of Aegean origin) and detection of possible dense-water formation events in the adjacent seas. The communication with the surface buoy is done through an acoustic modem (Benthos SR-100) at a band rate of 1200 bps. The battery pack of the central system (pressure sensors, processing and data transmission units) allows an autonomous operation of the platform for approximately 12 months, while the peripheral sensors (CTD, DO) have their own energy system of higher autonomy.

5.4 Ongoing operation management

5.4.1 Data flow, management and quality control procedures

The data transmission and control of POSEIDON-Pylos observatory utilizes a dual GSM/GPRS and INMARSAT-C satellite system. The collected data are transmitted by means of a GSM cellular network chosen carefully to provide sufficient signal strength at the station deployment site. This communication system is currently used in dial-up mode creating a link to the central receiving station at HCMR/Athens and works as a primary communication mode. One benefit of this system is that the operators get a bi-directional real-time communication link for low level debugging and possible reconfiguration of the buoy in case of a problem. Should the GSM data transmission fail, the secondary system (satellite INMARSAT-C) automatically tries to transmit data, the only difference being that the resulting data do not contain the frequency spectrum of the wave measurements. This is done to save on the size of the data transmitted, as the cost of using the satellite compared to the GSM is significantly higher. The station is programmed to collect the data every three hours starting on 00:00 UTC plus an offset of three minute increments for every remote station to avoid having two stations of the POSEIDON network transmitting at the same time. This would result in a delay or loss of data from a station since the GSM receiving modem can only accept one call at a time. This problem will be solved with the use of GPRS. The actual data collected by the station are saved locally inside a flash memory prior to transmission and remain there to be downloaded, usually on every maintenance survey. Additionally, the sensors used on this site are configured to store the measured data

Parameter	Depths measured (m)	Sensor(s) used	Accuracy
Air–Sea interface module (surface buoy)			
Wind speed/dir.	Surface	Young 04106	1m/s, 10 deg
Air pressure	Surface	Vaisala PTB 220A	0.10hPa
Air temperature	Surface	Omega	0.1 °C
Wave height, direction, period	Surface	Fugro OCEANOR Wavesense	0.1m, 0.5 deg, 0.5s
SST, SSS	Surface (1m)	Seabird SIP	0.1°C, 0.003mS/cm
Currents	Surface (1m)	Nortek Aquadopp current meter	Sp: 0.5cm/sec Dir: 2 deg
Water column module (mooring line)			
Temperature	20, 50, 75, 100 250, 400, 600, 1000m	Seabird 37-IM CT Seabird 16plus-IMP CT	0.005°C
Salinity	20, 50, 75, 100 250, 400, 600, 1000m	Seabird 37-IM CT Seabird 16plus-IMP CT	0.0005 S/m
Pressure	250m	Seabird 37-IM CTD	0.1% FS
Bottom module (seabed platform)			
Pressure	Seabed (1680m)	Paroscientific 43K-101	0.01% FS
Temperature	Seabed (1680m)	Seabird 16plus CTD	0.005°C
Salinity	Seabed (1680m)	Seabird 16plus CTD	0.0005 S/m
Pressure	Seabed (1680m)	Seabird 16plus CTD	0.1% FS
Dissolved 0xygen	Seabed (1680m)	Aanderaa Optode	<5% saturation

Table 5.1 Sensors used at the different modules of the POSEIDON-Pylos observatory.

in the internal memory so that if the acquisition system or the cable connecting the sensor to the acquisition system fail, the data can be recovered during the maintenance missions when the sensors are recovered. The data are stored in a binary encoded format to optimize for size.

The transmitted data are collected at the POSEIDON operational center, where automatic near real-time quality control checks are applied upon the time series on a daily basis. This analysis was established during the previous EU collaborative research projects (MFSTEP, MERSEA) and is performed before data are archived and released as standard quality- controlled daily products. Once the data are decoded, a date/hour confirmation check is applied to ensure no corruption has occurred. Then, several tests and specific flags are applied to data which fail or pass these numerical checks. Quality tests are based on some pre-assigned principles. Firstly, values must vary between certain bounds which are determined by the instrument measure range and the regional climatology. Furthermore, values may vary with a maximum rate of change (within a specific timeframe) which is related to a threshold value applied to each measured parameter. To check the correct functioning of sensors over time, data have to pass the "stationarity" test that shows whether values are stuck and recur in continuous measurements. It is obvious that, in order to apply the appropriate flags to each value, a combined knowledge is needed of the physical processes, technical details such as analog to digital conversions, transmission methods, etc.

After the first level of data-quality assurance, two types of daily files are produced, ASCII (Medatlas) and Binary (NetCDF), in order to comply with standard formats developed/used by international and European programs (OceanSITES, SEPRISE, MyOcean, etc.). Finally data are stored in the POSEIDON database to a normalized MYSQL database. Its design supports the demand for quick search and reliable results on the parameter values and their metadata. The table that contains the data information associates the values with metadata and a flag, which shows if the parameter has passed through a quality control process and serves quality checking purposes. In delayed mode, a visual inspection of the data is also performed on a regular basis, inherent in the maintenance of the sensors that takes place after every new deployment. The recovered sensors are checked in the laboratory of HCMR and raw data are extracted.

5.4.2 Data and information product dissemination

All data produced by the POSEIDON-Pylos observatory are integrated into the POSEIDON information system. The daily files in ASCII/MEDATLAS or NetCDF formats are made available to the scientific community through an ftp server. This type of dissemination is mainly for operational systems that use automated download procedures. Offline data are also made available through the web system of the POSEIDON system (www.poseidon.hcmr.gr)(i.e., Figure 5.4). The page offers several functionalities among which:

- A quick look at the latest data (a subsample of parameters) in a tabular form. This is an open service that allows users to have an overview of the latest conditions in their area of interest. It is mainly targeted at the wider user community (general public); thus, only meteorological and surface oceanographic (temperature, waves, currents) parameters are displayed.
- A viewing service (static graphical display) of the time-series of the last 15 days. This service is also open to all users and allows a detailed overview of the variability of the last two weeks. The service is targeted at both the wider public and scientific users of the systems, and thus includes all parameters monitored by the buoy network.
- A discovery and download service for all data from the system. The data that have been collected during the whole POSEIDON-network operation are easily accessible through a web interface to the system's database. The database is updated on a daily basis offering full access to both archived and recent data. The user also has access to multiple metadata information regarding the buoys' sensors, location, operational period and other combined information. This service is targeted at the scientific community and is restricted to registered users.

5.4.3 Operation of the POSEIDON-Pylos observatory, 2007–2010

The POSEIDON-Pylos observatory has recently completed more than three years of operation. The mooring line with the surface buoy was deployed on November 2007 and has operated since then, delivering near real-time data every three hours. The operation of the system was discontinued between October and November 2009 due to a major failure of the surface buoy, which could not be repaired on-site.

Data collected during the operation of the POSEIDON-Pylos observatory are presented in Figure 5.5. The collected wave data allow us to characterize, for the first time, the wave regime in the SE Ionian Sea, by deriving the appropriate statistics. The temperature

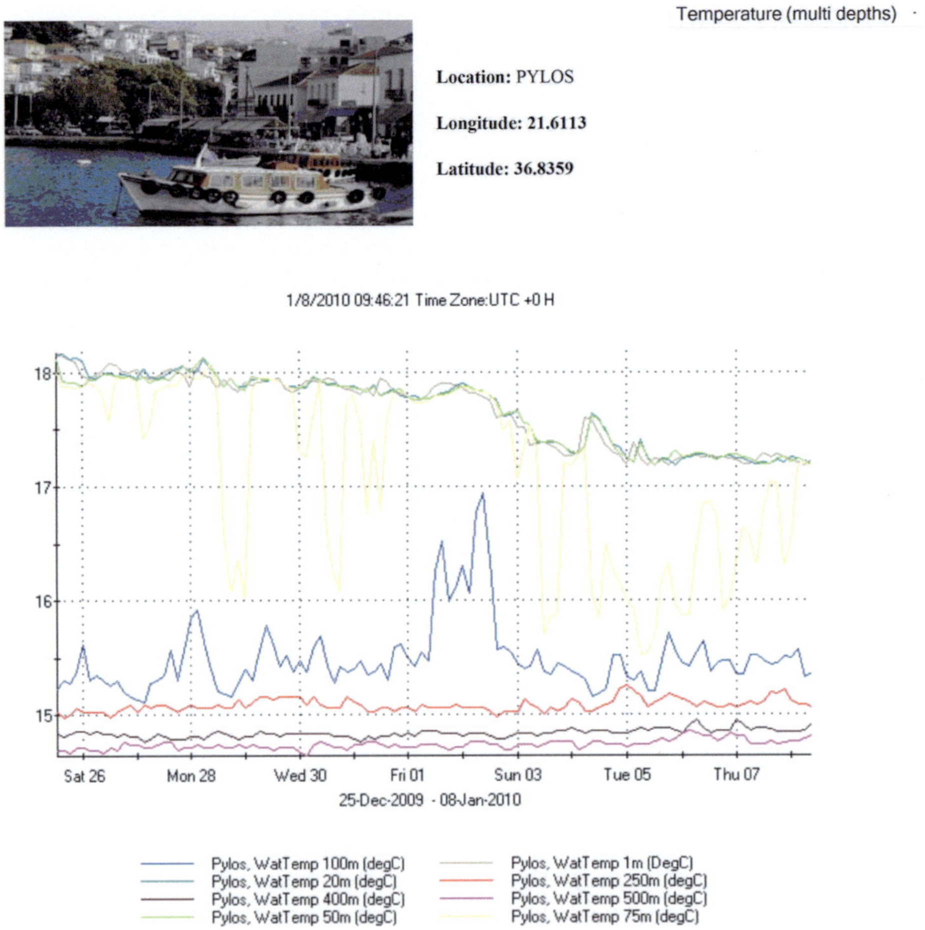

Figure 5.4 Example of the data viewing service offered by the POSEIDON web page (www.poseidon.hcmr.gr) displays the last 15 days' temperature time series at different depths from the POSEIDON-Pylos (Hellenic) observatory.

time-series demonstrate the synoptic and seasonal variability which are the dominant signals in this area. Although the time-series are relatively short, one can also observe inter-annual variability signals such as the stronger deep mixing during winter 2008–2009, the more abrupt mixing of the upper layers in summer 2009 and the slight warming of intermediate layers in 2009, compared to 2008.

The autonomous seabed platform was deployed in November 2008 for a pilot-operation period of four months. During this period, erroneous pressure data led to false tsunami alarms and unsuccessful communication events between the platform and the surface buoy (Figure 5.6).

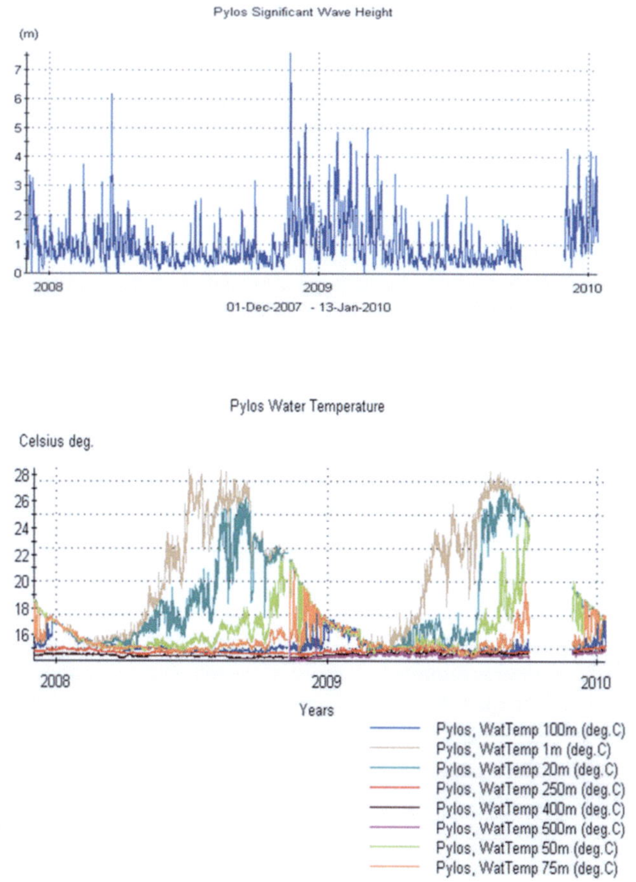

Figure 5.5 Time series of wave height (upper panel) and temperature at different depths (1–1000m) from the 2 years of operation of the POSEIDON-Pylos (Hellenic) observatory.

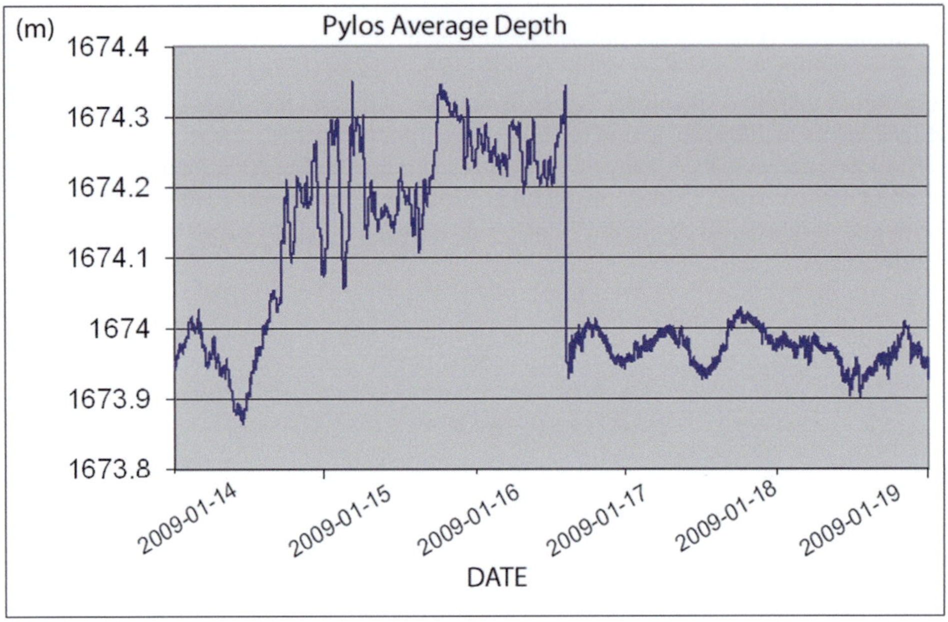

Figure 5.6 Time series of high resolution (15 sec) pressure data from the seabed platform. The spikes between 14 and 17 January 2009 indicate malfunction of the pressure (tsunami) sensor.

The seabed platform was recovered in March 2009 in order to continue laboratory tests and resolve these problems as well as to carry out other minor software upgrades (new bios, etc.). Following communication with the pressure sensor manufacturer, the erroneous data have been attributed to gas bubbles trapped in the instrument. The problem was solved by gas removal before the next deployment, following a standard maintenance procedure. The communication problems between the platform and the buoy are attributed to shadowing effects of seabed anomalies. A more appropriate area with smaller topographic anomalies was selected during the following deployment of the seabed platform, which took place in early December 2009.

The operation of the platform after the second deployment indicates that the outstanding problems were indeed solved. The pressure variability demonstrated in Figure 5.7 has a clearly distinguished high frequency component attributed to tides, which are known to have a weak signal in the Mediterranean Sea, as well as a lower frequency component related to synoptic variability of atmospheric pressure and wind effects.

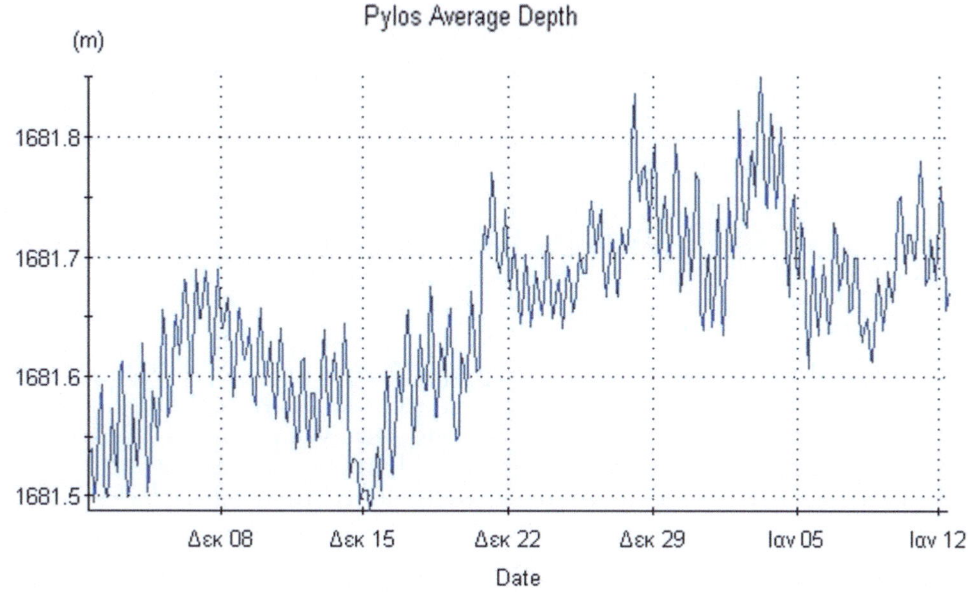

Figure 5.7 Time series of high resolution (15 sec) pressure data (tsunami sensor) from the seabed platform during the second deployment.

5.5 Concluding remarks

Over the three years from 2007 to 2010, the first elements of an integrated multidisciplinary deep sea observing system (Hellenic Observatory) have been developed in the south-eastern Ionian Sea. The selected site is an ideal area regarding research on:

(a) Climatic variability and impacts on the thermohaline circulation in the Mediterranean Sea. The complex hydrology of the area has been significantly impacted during the last two decades by the "Eastern Mediterranean Transient-EMT" that affected the thermohaline circulation of the whole basin, with its signal reaching the Atlantic. The vicinity of the site to the Aegean Sea, which is the source of the massive dense water formation that initiated the EMT, allows close monitoring of the relaxation process and makes early detection possible of future events in the region. The E. Mediterranean is the most oligotrophic environment of the European seas. The impact of climatic variability (Eastern Mediterranean Transient-EMT) on ecosystem functioning and biodiversity is a key scientific issue.

(b) Geohazards like seismicity, active faulting, slope failures and tsunamis, since the Western Hellenic Arc is one of the most active geodynamic systems with the highest tsunamigenic potential in European seas.

(c) Active sea bed fluid flow (methane and fluid seeps), deep sea geo-environments (mud volcanoes, brine lakes, etc.) and associated extreme ecosystems.

(d) Marine mammal listening (bioacoustics) and environmental (ambient) noise monitoring like rainfall, waves, ship noise, etc.

The first elements of the observatory were installed at the end of 2007 and include a deep mooring equipped with a surface buoy and sensors in the upper 1000m, as well as an autonomous seabed platform that delivers data in near real time using underwater acoustic technology (POSEIDON-Pylos). This system has been developed in the framework of the POSEIDON-II project which, among other projects, expanded and upgraded the operational buoy network of the Hellenic Seas and re-constructed the E1-M3A water column observatory of the Cretan Sea. This observatory is part of the OceanSITES global network of reference stations (www.oceansites.org), which has been operational since 2000 and has provided the necessary experience for the design and rapid implementation of the POSEIDON-Pylos site. Based on this experience the surface and water column components of the new observatory have operated since their first deployment in November 2007 without major problems (apart from accidents or vandalism). The seabed component of the observatory, i.e., the autonomous platform, operated in a pilot phase between November 2008 and March 2009. The identified problems (shadow effects of topography affecting acoustic communication, gas bubbles affecting the pressure sensor) have been addressed and no longer affect the operation of the system during its second deployment. Furthermore, the maintenance of the system has been integrated into the Greek national buoy program (POSEIDON) and its long-term operation, which is an important element for such investments, has been secured. Our plan includes the regular operation in the observatory of new sensors like CO2 and PAL (Passive Acoustic Listener) that have already been experimentally operated during 2008–2009 (Anagnostou et al., 2011).

The synergy and integration of the EU projects ESONET NoE and EMSO-KM3NET (ESFRI) with the EFTA funding POSEIDON projects in the Hellenic site provide the technological and scientific framework of a future cabled seabed and water column observatory. There are three feasible scenarios regarding the implementation of seabed observation in the Hellenic site (SW Peloponnesus-Pylos margin). The first (and simplest) scenario foresees a cable connection with the shore infrastructure of the POSEIDON-III sea bed platform. Within this new project (2010–2011), the existing sea bed platform is being upgraded in order to be ready for cable connection with ROV intervention. The second scenario incorporates the use of the cable connection of the deep sea (4000–5000m) installations of the KM3 neutrino telescope in order to deploy seabed observatories (scientific nodes) in the periphery of the base structure. The third solution could take advantage of the deployment of a new fiber optical cable that is designed as a 48km loop starting and ending in the NESTOR buildings in Pylos, in order to install deep sea observatories along the cable route.

Acknowledgments

This work has been supported by the POSEIDON-II and III projects co-funded by EEA Grants and the Hellenic Ministry of National Economy, the ESONET project co-funded by EU-FP6 and the Hellenic Ministry of Education (General Secretary for Research and Technology) and the EMSO-ESFRI project. The authors are gratefully acknowledge Dr T. Alves and the anonymous reviewer for their constructive comments.

References

Alves T.M., Lykousis V., Sakellariou D., Alexandri S. and P. Nomikou (2007) Constraining the origin and evolution of confined turbidite systems: Southern Cretan margin, Eastern Mediterranean Sea (34°30–36°N). GeoMarine Letters 27, 41–61.

Anagnostou M.N., Nystuen J.A., Anagnostou, E.N., Nikolopoulos E. and Amitai E. (2008) IEEE Trans Geosci. Rem. Sens. 46, 2936–2937.

Anagnostou M., Nystuen J., Anagnostou E, Papadopoulos A. and Lykousis V. (2011) Passive Aquatic Listener (PAL): An adoptive underwater acoustic recording system for the marine environment. Nuclear Instruments and Methods in Physics Research (A) 626–627, S94–S98

Anassontzis e. and Koske P. (2003) Deep sea station connected by cable to shore. Sea Technology 44(7), 10–14.

Anastasakis G. and Piper D.J.W. (1991), The character of seismoturbidites in the S-1 sapropel, Zakynthos and Strofadhes basins, Greece. Sedimentology 38, 717–733.

Armijo R., Flerit F., King G. and Meyer B. (2004) Linear elastic fracture mechanics explains the past and present evolution of the Aegean. Earth and Planetary Science Letters 217, 1–2, 85–95, doi: 10.1016/S0012-821X(03)00590-9.

Chamot-Rooke N., Rabaute A. and Kreemer C. (2005) Western Mediterranean ridge mud belt correlates with active shear strain at the prism-backstop geological contact. Geology 33, 861–864.

Camerlenghi A., Cita M.B., Hieke W. and Ricchiuto T. (1992) Geological evidence for mud diapirism on the Mediterranean Ridge accretionary complex. Earth Planet. Sci. Lett. 109, 493–504.

Chaumillon E., Mascle J. and Hoffmann H.J. (1996) Deformation of the western Mediterranean Ridge: Importance of Messinian evaporitic formations. Tectonophysics 263, 163–190.

Chaumillon E., and Mascle J. (1997) From foreland to forearc domains: New multichannel seismic reflection survey of the Mediterranean ridge accretionary complex (Eastern Mediterranean). Marine Geology 138, 237–259.

Cianetti S., Gasperini P., Giunchi C. and Boschi E. (2001) Numerical modelling of the Aegean–Anatolian region: Geodynamical constraints from observed rheological heterogeneities. Geophysical Journal Int. 146, 760–780, doi: 10.1046/j.1365-246X.2001.00492.

Cita M.B., Ryan W.B.F. and Paggi L. (1980) Prometheus mud breccia: An example of shale diapirism in the Western Mediterranean Ridge. Ann. Geol. Pays. Hellen. 30, 543–570.

Cita M.B., Bossio A., Colombo A., Gnaccolini M., Salvatorini G., Kastens K.A., McCoy F.W., Broglia C., Camerlenghi A., Catrullo D. et al. (1982) Stratigraphy and neotectonics in the Eastern Mediterranean Ridge. Cobblestone Area 3 re-visited. Boll. Soc. Geol. Ital. 24, 443–458.

Christodoulou D., Papatheodorou G., Ferentinos G. and Masson G. (2003) Active seepage in two contrasting pockmark field in the Patras and Corinth Gulfs, Greece. Geo-Mar. Lett. doi: 0.1007/500367-003-0151-0.

Dewey, J.F. and Şengör C.A.M. (1979) Aegean and surrounding regions: Complex multiplate and continuum tectonics in a convergent zone. Geol. Soc. Am. Bull. 90, 84–92.

Emery K.O., Heezen B. and Allan T.D. (1966) Bathymetry of the Eastern Mediterranean sea. Deep Sea Res. 13, 173–192.

Etiope G., Papatheodorou G., Christodoulou D., Ferentinos G., Sokos E. and Favali P. (2006) Methane and hydrogen sulfide seepage in the NW Peloponnesus petroliferous basin (Greece): Origin and geohazard. American Association of Petroleum Geologists Bulletin 90 (5), 701–713.

Fassoulas C., Kilias V. and Mountrakis D. (1994) Postnappe stacking extension and exhumation of highpressure/low-temperature rocks in the island of Crete, Greece. Tectonics 13, 127–138, doi: 10.1029/93TC01955.

Ferentinos G., Collins M., Pattiaratchi C. and Taylor P. (1985) Mechanisms of sediment transport and dispersion in a tectonically active submarine valley/canyon system: Zakynthos straits, NW Hellenic Trench. Marine Geology 65, 243–269.

Galanopoulos A.G. (1960) Tsunamis observed on the coasts of Greece from antiquity to present time. Annali di Geofisica 13, 369–386.

Hasiotis T., Papatheodorou G., Kastanos N. and Ferentinos G. (1996) A pockmark field in the Patras Gulf (Greece) and its activation during the 14/7/93 seismic event. Marine Geology 130, 333–344.

Hasiotis T., Papatheodorou G. and Ferentinos G. (2002) A string of large and deep gas-induced depressions (pockmarks) offshore of the Killini peninsula, Western Greece. Geo-Mar. Lett. 22 (3), 142–149.

Hasiotis T., Papatheodorou G. and Ferentinos G. (2005) A high resolution approach in the recent sedimentation processes at the head of Zakynthos Canyon, western Greece. Marine Geology 214, 49–73.

Heezen B.C. and Ewing M. (1963) The Mid Oceanic Ridge. In M.N. Hill (ed.) *The Seas*, Vol. 3. New York: Interscience, pp. 388–410.

Huguen C., Chamot-Rooke N., Loubrieu B. and Mascle J. (2006) Morphology of a pre-collisional, salt-bearing, accretionary complex: The Mediterranean Ridge (Eastern Mediterranean). Marine Geophysical Researches 27, 61–75.

Jacobshagen V. (1994) Orogenic evolution of the Hellenides: New aspects. Geologische Rundschau 83, 249–256.

Jolivet L. (2001) A comparison of geodetic and finite strain pattern in the Aegean, geodynamic implications. Earth and Planetary Science Letters 187, 95–104, doi: 10.1016/S0012-821X(01)00277-1.

Kreemer C. and Chamot-Rooke N. (2004) Contemporary kinematics of the southern Aegean and the Mediterranean Ridge. Geophysical Journal Int. 157, 1377–1392, doi: 10.1111/j.1365-246X.2004.02270.

Larnicol G., Ayoub N., Le Traon P.Y. (2002) Major changes in Mediterranean Sea level variability for 7 years of TOPEX/Poseidon and ERS-1/2 data. Journal of Marine Systems 33–34, 63–89.

Lee H.J. and Edwards B.D. (1986) Regional method to assess offshore slope stability. J. Geo. Eng. 112(5), 489–509.

Le Pichon X., Angelier J. and Sibuet J.C. (1982) Plate boundaries and extensional tectonics. Tectonophysics 81, 239–256, doi: 10.1016/0040-1951(82)90131-7.

Le Pichon X., Chamot-Rooke N., Lallemant S., Noomen R. and Veis G. (1995) Geodetic determination of the kinematics of central Greece with respect to Europe: Implications for Eastern Mediterranean tectonics. J. Geophys. Res. 100, 12675–12690.

Limonov A.F., Kenyon N.H., Ivanov M.K. and Woodside J.M. (Eds) (1995) Deep-sea depositional systems of the Western Mediterranean and mud volcanism on the Mediterranean Ridge. UNESCO Rep. Mar. Sci. **67**.

Lykousis V. (1991) Submarine slope instabilities in the Hellenic Arc region, Northeastern Mediterranean Sea. Marine Geotechnology 10, 83–96.

Lykousis V., Sakellariou D. and Rousakis G. (2009) Slope failures and stability analysis of shallow water prodeltas in the active margins of Western Greece, northeastern Mediterranean Sea. Journal of Earth Sciences 98, 807–822.

Lykousis V. and Chronis G. (1989) Mass movements, geotechnical properties and slope stability in the Outer Shelf Upper Slope, N.W. Aegean Sea. Marine Geotechnology 8(3), 231–247.

Makropoulos K. and Burton P. (1985) Seismic hazard in Greece, II: Ground acceleration. Tectonophysics 117, 259–294.

Malanotte-Rizzoli P., Manca B., Ribera D'Alcala M., Theocharis A., Bergamasco A., Bregant D., Budillon G., Civitarese G., Georgopoulos D. et al. (1997) A synthesis of the

Ionian Sea hydrography, circulation and water mass pathways during POEM-Phase I. Progress in Oceanography 39, 153–204.

Malanotte-Rizzoli P., Manca B., Ribera d'Alcala M., Theocharis A., Brenner S., Budillon G. and Ozsoy E. (1999) The Eastern Mediterranean in the 80s and in the 90s: The big transition in the intermediate and deep circulations. Dynamics of Atmospheres and Oceans 29, 365–395.

Mascle J., Le Quellec P., Leite O. and Jongsma D. (1982) Structural sketch of the Hellenic continental margin between the western Peloponnesus and eastern Crete. Geology 10, 113–116.

McClusky S., Balassanian S., Barka A., Demir C., Ergintav S., Georgiev I., Gurkan O., Hamburger M., Hurst K. and Kahle H. (2000) Global Positioning System constraints on plate kinematics and dynamics in the eastern Mediterranean and Caucasus. J. Geophys. Res. 105, 5695–5719.

McKenzie D.P. (1978) Active tectonics of the Alpine-Himalayan belt: The Aegean Sea and surrounding regions. Geophys. J. R. Astron. Soc. 55, 217–254.

Meijer P.Th. and Wortel M.J.R (1997) Present-day dynamics of the Aegean region: A model analysis of the horizontal pattern of stress and deformation. Tectonics 16, 879–895. doi: 10.1029/97TC02004.

Meulenkamp, J.E., Wortel M.J.R, Van Wamel W.A., Spakman W. and Hoogerduyn Strating E. (1988) On the Hellenic subduction zone and the geodynamical evolution of Crete since the late Middle Miocene. Tectonophysics 146, 203–215. doi: 10.1016/0040-1951(88)90091-1.

Nittis K., Pinardi N. and Lascaratos A. (1993) Characteristics of the summer 1987 flow field in the Ionian Sea. J. Geophys. Res. 98, 171–184.

Papadopoulos G.A. (2005) *Tsunamis in the Mediterranean Sea: The physical geography of the Mediterranean Sea*. Oxford: Oxford University Press.

Papadopoulos,G.A. and Fokaefs A. (2005) Strong tsunamis in the Mediterranean Sea: A re-evaluation. ISET Journal of Earthquake Technology 42 (4), 159–170.

Papatheodorou G., Hassiotis T. and Ferentinos G. (1993) Gas charged sediments in the Aegean and Ionian Seas, Greece. Marine Geology 112, 171–184.

Poulos S., Lykousis V., Collins M.B., Rolling E.J. and Pattiaratchi C.V. (1999) Sedimentation processes in a tectonically active environment: The Kerkyra-Kefalonia submarine valley system (NE Ionian Sea). Marine Geology 160, 25–44.

Reeves R. and Notarbartolo di Sciara G. (compilers and editors) (2006) *The Status and Distribution of Cetaceans in the Black Sea and Mediterranean Sea*. IUCN Centre for Mediterranean Cooperation, Malaga, Spain, 137 pp. http://www.cetaceanalliance.org/conservation.htm

Schielein P., Zschau J., Woith H. and Schellmann G. (2007) Tsunamigefaerdung im Mittelmeer – Eine Analyse Geomorphologisher and Historischer Zeugnisse. Bamberger Geograph. Schrift 22, 153–199.

Shaw B., Ambraseys N.N., England P.C., Floyd M.A., Gorman G.J., Higham T.F.G., Jackson J.A., Nocquet J.N., Pain C.C. and Piggott M.D. (2008) Eastern Mediterranean tectonics and tsunami hazard inferred from the AD 365 earthquake. Nature Geoscience. doi: 10.1038/ngeo151.

Soloviev S.L., Solovieva O.N., Go C.N., Kim K.S. and Shchetnikov N.A. (2000) *Tsunamis in the Mediterranean Sea 2000 BC–2000AD*. Dordrecht: Kluwer.

Stanley D.J., Knight J., Stuckenarth R. and Catani J.P. (1978) High sedimentation rates and variable dispersal patterns in the western Hellenic trench. Nature 273, 110–113.

Stavrakakis G., Papoulia J. and Drakopoulos J. (1989) A fault rupture model for seismic hazard analysis in the area of Messinia, SW. Peloponnesus. Bull. Geol. Soc. Greece XXIII/3, 365–371.

Theocharis A., Nittis K., Kontoyannis H., Papageorgiou E. and Balopoulos E. (1999) Climatic changes in the Aegean Sea influence the thermohaline circulation of the Eastern Mediterranean (1986–1997). Geophysical Research Letters 20(11), 1617–1620.

6 Marine seismogenic-tsunamigenic prone areas: The Gulf of Cadiz

J.M. Miranda[1], L. Matias[1], P. Terrinha[2], N. Zitellini[3], M.A. Baptista[1,4], F. Chierici[5], D. Embriaco[6], G. Marinaro[6], S. Monna[6] and L. Pignagnoli[2]

6.1 Introduction

It is well documented that, in the past, catastrophic earthquakes and tsunamis affected the area as described in the historical records from the countries bordering the Gulf of Cadiz: Portugal, Spain and Morocco (cf. Baptista and Miranda, 2009; Figure 6.1). The main mechanism behind geohazard generation in this area is the interplate deformation arising from the relative motion between the Eurasian and Nubian (African) plates. The plate boundary crosses the Gulf of Cadiz but the precise location of its trace is not yet well understood. In the Gulf of Cadiz the morphology is complex and characterized by a series of large ridges and seamounts, the Gorringe Bank, the Coral Patch and Ampère seamounts (cf. Figure 6.1 for locations), bounding major morphological depressions such as the Horseshoe and Tagus abyssal plains and discrete segments of plate boundary are hard to identify due to diffuse deformation over an area 200km wide (Sartori et al., 1994; Tortella et al., 1997). In Figure 6.1, we plot the epicenters of earthquakes with magnitude greater than 2, from the ISC catalog (International Seismological Centre), for the period 1970–2000. Focal mechanisms (Borges et al., 2001, Buforn et al., 2004) indicate right lateral and reverse faulting

1 University of Lisbon, Instituto Dom Luiz, Lisbon, Portugal
2 Laboratório Nacional de Energia e Geologia, I.P., Portugal
3 Istituto Science Marine, Sede Bologna, Bologna, Italy
4 Instituto Superior de Engenharia de Lisboa, IPL, Portugal
5 Istituto di Radioastronomia-Istituto Nazionale di Astrofisica, Bologna, Italy
6 Istituto Nazionale di Geofisica e Vulcanologia, Rome, Italy

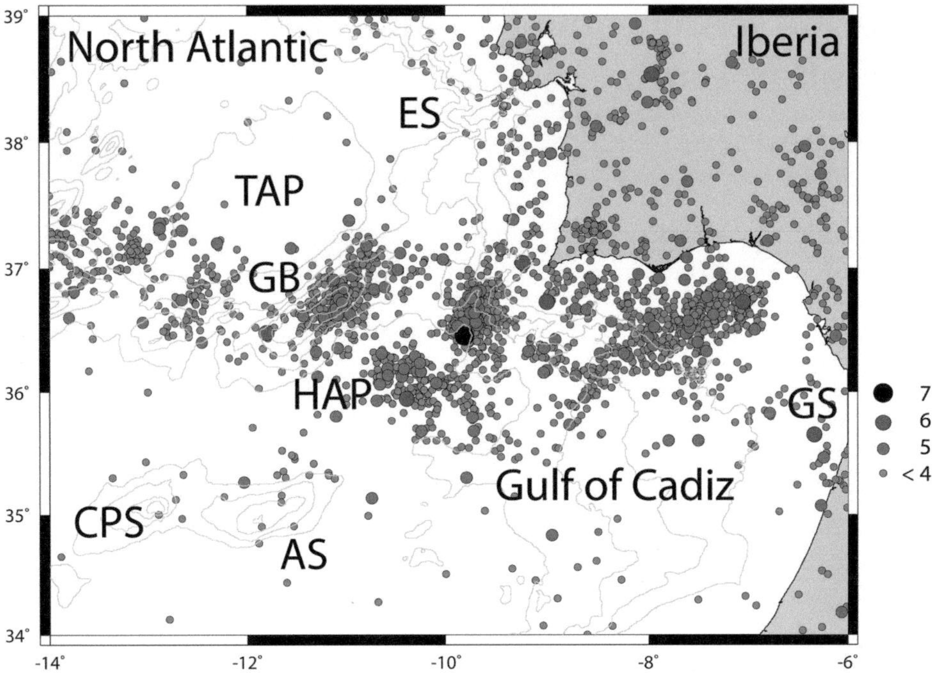

Figure 6.1 Seismicity of the Azores Gibraltar area. Data from ISC Catalogue for the period 1970–2010 and magnitudes larger than 2. HAP: Horseshoe Abyssal Plain; TAP: Tagus Abyssal Plain; AS: Ampère Seamount; GB: Gorringe Bank; CPS: Coral Patch Seamount; ES: Estremadura Spur; GS: Gibraltar Straight. Isobaths are plotted every 1000m.

on roughly NE–SW oriented structures. This is usually interpreted in kinematic plate models (e.g., DeMets et al., 1994, Sella et al., 2002; Fernandes et al., 2003, 2007) as the result of the relatively low, interplate, compressive motion (ca. 4mm/y). Tsunami events are in relation to this tectonic environment (Baptista et al., 1998a, 1998b; Zitellini et al., 2001; Gràcia et al., 2003; Gutscher et al., 2006) and to the transcurrent motion along the Gloria Fault and surrounding area (Kaabouben et al., 2008) or to the effect of faraway sources (e.g., Azores or Grand Banks). In Figure 6.2 we plot the assumed location of the major tsunami events in this area.

The definition of a submarine monitoring program for large earthquakes and tsunamis is strongly limited by the unknowns on the major seismogenic sources. Several attempts at using the available 1755 tsunami information on both sides of the Atlantic failed to constrain its source location (Baptista et al., 1998b; Baptista et al., 2003; Gutscher et al., 2006; Barkan et al., 2009). In addition, a number of studies tried to establish a relationship between the main tectonic structures and the large earthquake and tsunami events in the Gulf of Cadiz (e.g., Levret, 1991; Johnston, 1996; Baptista et al., 1998b; Gutscher et al., 2002; Zitellini et al., 2001, 2004, 2009). The conclusions drawn out from these studies reinforced our belief that, with the exception of a subduction slab hypothesized by Gutscher

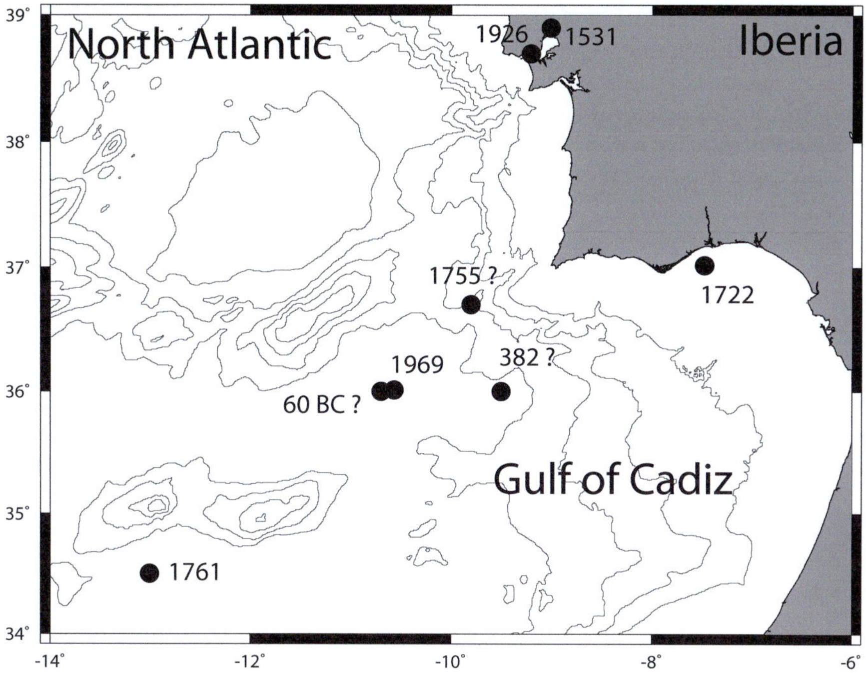

Figure 6.2 Tsunamis in SW Iberia. We present the approximate location of all events included in the tsunami catalogue of Baptista and Miranda (2009). Isobaths are plotted every 1000m.

et al. (2002), no single tectonic structure is large enough to generate the 1755 Lisbon Earthquake.

Although the precise localization of the tectonic source of the 1755 tsunami is still an open question, one clear result from the many geophysical surveys in the Gulf of Cadiz is the presence of many well constrained, potentially tsunamigenic sources, all of which are located near shore. This location implies a very short travel time from the source of the potential tsunamis to the shore, and thus the inadequacy of the present-day tsunami warning procedure. The warning procedure commonly used can be summarized as follows: after a shallow submarine earthquake exceeding a given magnitude, the seismic signals are used to issue warning messages which are successively confirmed or cleared by the sea level height measurements. More specifically, the most advanced tsunami warning systems are based on the evaluation of the earthquake magnitude and hypocenter depth (seismic events are considered potentially tsunamigenic when they have magnitude larger than 7.0 and depth shallower than 60km). Event magnitude and depth are estimated using on-land based seismic networks: this procedure takes advantage of the different propagation speed of the tsunami and earthquake waves. The epicenter location of the seismic event, its magnitude and the bathymetry of the area threatened by the possible tsunami are used as input parameters to very fast numerical modeling which computes expected tsunami wave heights and arrival times at the coast (Titov and Gonzalez, 1997). To minimize false alarms, sea level heights measured in open ocean or at distant tide gauges are used to confirm the actual generation of the tsunami. This technique, which requires in its first step enough time to estimate the earthquake parameters and more time to compute the tsunami heights and arrival times, has proven to be very effective for distant tsunamigenic sources (as, for instance, the 27 February 2010 Chilean earthquake, when the Pacific Tsunami Early Warning Center was able to release a warning after 12min from the event). On the other hand, this approach shows an intrinsic weakness in the case of nearshore sources and, in general, when the tsunami arrival times are very short, a situation which is unfortunately common to the Gulf of Cadiz, to all the Mediterranean Basin and to Japan, or in the case of tsunamis not triggered by seismic events.

The disadvantage coming from reduced warning times for near shore tsunamigenic sources in the Gulf of Cadiz can be partially compensated by their well constrained geometrical structures. In fact, identifying the tsunamigenic sources and their seismic regime allows for their continuous monitoring with dedicated instruments (at restrained costs), thus enabling the very quick evaluation of the source parameters needed in tsunami warning, and the use of ready-made scenarios to issue the watch and warning for the area. This strategy, which can considerably shorten the warning times, has been followed in the European Union (EU) project NEAREST (http://nearest.bo.ismar.cnr.it/), by deploying for one year the abyssal station GEOSTAR, Geophysical and Oceanographic Station for Abyssal Research (Favali et al., 2006), for seismic and oceanographic long-term monitoring. On board the abyssal station, a new tsunameter prototype was installed which was able to operate in the generation area.

The geological structures of this area were studied and mapped in the framework of the European Union (EU) projects: NEAREST and BIGSETS and the European Science Foundation project SWIM. In this work, we review and discuss the recent progress in the identification of the seismogenic zone which was considered for the deployment of the abyssal observatory GEOSTAR, hosting the prototype of the tsunameter, close to one of the most potential tsunamigenic structures of the area. We then describe the main strategy for seafloor monitoring in this localized tsunamigenic area and present some preliminary results for the first two years of the mission of the observatory in the Gulf of Cadiz.

6.2 Large earthquakes and tsunamis in the Gulf of Cadiz

The first compilations of earthquakes relevant for the Portuguese territory were made in the 18th century. The oldest Portuguese catalogue of earthquakes and tsunamis is called the "Universal History of Earthquakes which occurred around the world, since its creation up to the current century", and was published in 1758 by Joaquim Moreira de Mendonça. Most of this book is devoted to the 1755 event but previous sources are also quoted to describe older events. Perrey (1847) and Mallet (1852) also presented compilations of catastrophic events, integrating data from Mendonça (1758) with other sources.

The first seismic catalogue, in a form which is closer to our present understanding, was established for Iberia by Rodriguez (1932, 1940) encompassing the period 1032 BC to 1933 AD. After the 1970s, seismic catalogues which included large events in the Gulf of Cadiz were prepared by the Portuguese and Spanish authorities. Some of these were partially revised and updated (e.g., LNEC, 1986; Martins and Victor, 2001). Sousa et al. (1991) prepared a new seismic compilation for the Iberian region to be used in hazard studies, merging the regional or local compilations and the worldwide catalogues selected for the region.

Nearly 30 earthquakes with magnitudes larger than 7 originating in the SW Iberian margin are included in the regional catalogue. Among all the large historical events, the "Lisbon" earthquake of 1 November 1755 plays a decisive role. Its magnitude was greater than 8.7 (Johnston, 1996). An extensive set of documents concerning the 1755 Lisbon earthquake was published, including the documental compilation of Sousa (1919), isoseismal maps (Reid, 1914; Solares et al., 1979; Mezcua 1982; Solares and Lopez-Arroyo, 2004) and attenuation models (Oliveira, 1987; Levret, 1991; Sousa, 1996; Solares and Lopez-Arroyo, 2004).

In 1992, an MCS (multichannel seismic) study conducted by Zitellini et al. (see Zitellini et al., 2001), detected the seismic expression of a large compressive structure, the Marques de Pombal thrust, which is close to the location considered by Baptista et al. (1998b) as the most probable source area from the point of view of tsunami data. An alternative tectonic source for the main shock was proposed by Gutscher et al. (2002) relating to an eastward dipping lithospheric block beneath Gibraltar. A triggered earthquake in the Tagus Valley was also invoked by Vilanova et al. (2003). Even though the "tectonic puzzle" is far from being solved, the best geophysical and geological information suggests that the 1755 earthquake main shock source must lie closer to the SW Portuguese shore than to the Horseshoe Abyssal Plain. This event, together with historical evidence validating the occurrence of former large events, shows that relevant geohazards can originate in the Gulf of Cadiz.

The analysis of the Portuguese tsunami catalog (Baptista and Miranda, 2009) shows that the majority of the North East Atlantic (NEA) tsunami events are the result of large earthquakes. Since 1881, we have had access to tsunami instrumental data in the Portuguese coasts: the first Portuguese tide station was installed at Cascais in 1881 and the second one at Lagos in 1908; together with the Spanish tide-gauges, there are instrumental data for all 20th-century events, including: the 1929 Grand Banks tsunami (Fine et al., 2005); 25 November 1941 (Baptista and Miranda, 2009; Kaabouben et al, 2009); 28 February 1969 (Gjevik et al., 1997); 26 May 1975 (Kaabouben et al., 2008); and 1 January 1980.

Baptista and Miranda (2009) presented a revision of the Portuguese catalogue of tsunamis where instrumental data and historical documents were considered. The list of tsunamis in the Portuguese coasts includes 17 reliable events since 60 BC, 14 of these generated by earthquakes. A group of events can be considered regional or basin-wide tsunamis: 60 BC, 382, 1755, 1761 and 1929 AD. With the exception of the 1929 Grand Banks event, all were generated by M≥=8 earthquakes. They caused inundation of large areas, affected infrastructures close to shore and the draw-down at the harbors, damaging anchored ships. Morphological changes are claimed for three events: the 60 BC, 382 and 1755 AD. If we consider the period after 60 BC we can conclude that the earthquake events with magnitude larger than 7.5 and epicenters offshore Portugal generated the tsunamis.

6.3 Main hazard source zones in SW Iberia

The sources of the large events in SW Iberia correspond to different tectonic processes. We will consider in detail the most important ones in the following sections.

6.3.1 Gloria Fault

The Gloria Fault is located west of the Gulf of Cadiz (Figure 6.1). It is a large linear structure, of scarce seismicity characterized by right-lateral strike-slip motion (Buforn et al., 1988, 2004; Argus et al., 1989) where several instrumental tsunami events were generated, affecting Iberia and northwest Africa: 25 November 1941 M8.25 (Udias et al., 1976); 6 September 1969 M5.8 (Argus et al., 1989); 26 May 1975 M7.9 (Buforn, 1988; Argus et al., 1989), and 17 October 1983 (Argus et al., 1989). According to Baptista et al. (2006) it is also the probable source of the 31 March 1761 tsunami. The location of these events not only follows the Gloria Fault but also a second source zone striking almost EW, ca. 200km south (Lynnes and Ruff, 1985; Buforn et al., 1988). This can correspond to either the development of a tectonic microblock (Buforn et al., 1988) or the re-activation of a previous transform domain (Kaabouben et al., 2008). In any case, all known tsunami events correspond to large strike-slip faults with average slips that can reach up to 11m (Lynnes and Ruff, 1985), large enough to generate tsunamis.

6.3.2 SW Iberian transpressive domain

Zitellini et al. (2009) used a new swath bathymetry compilation of the area offshore SW Portugal to map a set of almost linear and sub-parallel dextral strike-slip faults, called SWIM (Southwestern Iberian Margin) Faults (Figure 6.1). These faults form a narrow band of deformation over a length of 600km, the SWIM Fault Zone, coincident with a small circle centered on the present-day pole of rotation of Nubia with respect to Eurasia (Sella et al., 2002; Fernandes et al., 2003). This deformation band connects the Gloria Fault to the Rif-Tell Fault Zone, two segments of the plate boundary between Africa and Eurasia. In this tectonic framework, the SWIM Fault Zone is interpreted as a precursor to the formation of a new transcurrent plate boundary controlling the present-day plate interaction between Iberia and Africa. Zitellini et al. (2009) showed that seismic strain is accommodated only by structures located north of the SWIM Fault Zone and maximum values coincide with the active morpho-tectonic features and faults already recognized and mapped by several authors (Zitellini et al., 2001, 2004; Gràcia et al., 2003; Terrinha et al., 2009). These include the Marques de Pombal Structure, the Gorringe Bank, the Ampere and Coral Patch Seamounts, the Portimao Bank fault, shown in Figure 6.1.

Most damaging earthquakes and tsunamis that affected the coasts of Portugal, Morocco and Spain were generated in this zone, here including the 1 November 1755 mega-tsunami. The largest instrumental event is the 28.02.1969 Horseshoe Ms 8.0 (cf. Figure 6.3 for locations). In this case, fault plane solutions and slip vectors reveal a NW–SE to NNW–SSE horizontal maximum compression (Buforn et al., 1988, 2004; Stich et al., 2003). Large thrust earthquakes show N–S to NW–SE orientation of P axes, compatible to the plate relative motion. This horizontal compression coexists with E–W to NE–SW horizontal tension in the Betics, Alboran Sea and northern Morocco (Negredo et al., 2002).

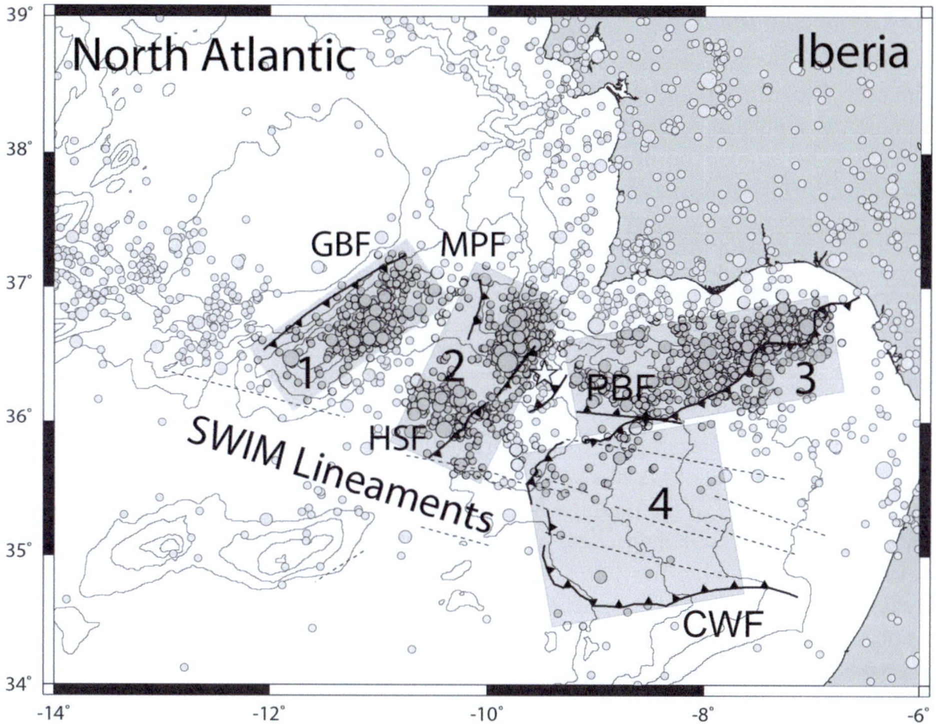

Figure 6.3 Main seismogenic faults and source zones. Thrusts in solid line correspond to the structures that are candidates for M~8 events that correspond in this environment to the worst yet credible case. The location of the GEOSTAR station is indicated with a star. Bathymetry is plotted every 1000m. GBF: Gorringe Bank Fault; MPF: Marques de Pombal Fault; HSF: Horseshoe Fault; PBF: Portimão Bank Fault; CWF: Cadiz Wedge Fault. The numbers 1–4 indicate the four main tsunami source areas in the Gulf of Cadiz. Isobaths are plotted every 1000m

As stated above, the southern limit of the SWIT Domain is marked by the series of long accidents which have been interpreted as dextral strike-slip faults trending WNWESE cutting across all tectonic structures and geological formations from the Mesozoic basement through the Holocene sediments, deeply rooted into Mesozoic basement rift faults and control the location of active mud volcanoes. The longest of these lineaments exceeds 600 km, and can be followed from the Hirondelle Seamount to the Morocco shelf (Zitellini et al., 2009). This interpretation is controversial and, in the absence of seismic activity (cf. Figure 6.1), we did not include it as a major tsunami source.

6.3.2.1 Gorringe Bank zone

The Gorringe Bank (GB, Figure 6.1) is a large uplifted block of oceanic lithosphere approximately 180km long and 60–70km wide, trending N55E. The bank itself is a huge morphological high that reaches 25m below sea level. This lithospheric block is usually interpreted as an almost continuous section of oceanic crust and upper mantle (Bergeron and Bonnin, 1991) and is associated with a large isostatic anomaly (ca. 300 mGal), which indicates the presence of a thick, high-density body close to the seafloor (Sborschikov et al., 1988; Bergeron and Bonnin, 1991; Sartori et al., 1994). For a long time, this anomaly has been interpreted as a sign of the lack of isostatic equilibrium (Bergeron and Bonnin, 1991) and, thus, a potential seismogenic feature. The existence of subduction below the GB and its interpretation as a piece of the plate boundary was suggested by Purdy (1975). After the earthquake of 28 February 1969, Fukao (1973) suggested that the Gorringe Bank could be considered as bounded by two thrust faults cutting through the crust at both its northern and southern flanks. Shortening (or extension) on either one of these faults would result in the uplift (or subsidence) of the block of crust. Bergeron and Bonnin (1991) made an evaluation of the magnitude of the regional field needed to support the GB out of equilibrium.

Sartori et al. (1994) published a deep-penetrating multichannel seismic (MCS) reflection line, AR92-03-08, crossing the Tagus Abyssal Plain, the Gorringe Ridge, the Horseshoe Abyssal Plain and the Coral Patch Seamount. The MCS line was acquired with a NW–SE direction, roughly parallel to the main NW–SE trending slip vectors of Nubia–Eurasia plate convergence. This line clearly showed the absence of subduction below GB and, instead, the presence of compressional deformation spread across an area about 200km wide, spanning from the GB to the Seine Plain. This deformation pattern was explained by Sartori et al. (1994) as being due to the similarity of the plates (EU and NU) in terms of age, density, rigidity and thickness, and because of the very low overall rate of convergence, ~4mm/yr. Under such circumstances, stresses are supposed to be released from the whole lithosphere across wide sectors of the converging plates without a well-defined plate margin (Hayward et al., 1999). These hypotheses were supported by the results of deep multichannel seismic profiling (Tortella et al., 1997). Sartori et al. (1994) proposed regional buckling as an explanation for the uplift of Gorringe Bank. Hayward et al. (1999) estimate that a maximum of 50km shortening has occurred at the Gorringe Bank, through thickening, folding and thrusting, since the mid Miocene. In spite of Sartori et al.'s (1994) arguments, the Gorringe Bank corresponds to a major, probably uncompensated, mass with its northern flank corresponding to an important stratigraphic discontinuity having widespread seismicity. This source zone is identified as "1" in Figure 6.3.

6.3.2.2 Horseshoe Marques-de-Pombal zone

Horseshoe Abyssal Plain (HAP; Figure 6.1) is an E–W elongated feature bounded by the Ampère and Coral Patch seamounts to the south, the Gorringe Bank to the north, the Madeira Tore rise to the west and the Iberian continental margin to the east. It is crossed by the SWIM Faults and, in its deeper part, the crust is around 15km thick. There is strong evidence for active compressional deformation in the HAP east of 12° W (Zitellini et al., 2004) while there is no evidence of major recent tectonic activity in the western sector (Purdy, 1975). In the eastern part of the plain, the oceanic basement is affected by large vertical offsets along both the north and south boundaries. Even though Sartori et al. (1994) argue that the overall pattern of deformation in the HAP is inconsistent with subduction in this area, there are significant seismic events located in this sector, in particular the M7.9 of 28 February 1969. The two major faults identified in this zone are the Horseshoe Fault (Gràcia et al., 2003; Zitellini et al., 2004) and the Marques de Pombal Fault (Zitellini et al., 2001).

The Horseshoe Fault is a reverse fault oriented perpendicular to the present-day kinematic displacement of Nubia with respect to Iberia. It is one of the most important tectonic elements identified in the area, where no large earthquakes have yet been located. The Marques de Pombal Fault is a large active compressive tectonic structure located 100km offshore SW Cape S. Vicente. It displays a pronounced drag fold on the fault hanging-wall and the height of the escarpment is taller in the north where it reaches 1.2km. The northernmost segment is offset by a WNW–ESE wrench fault. Besides the activity demonstrated by deformed sediments from the Holocene age on multichannel seismics and high resolution CHIRP and TOPAS records, the hanging-wall of this thrust displays a fair amount of slide scars and slumps at the seafloor surface. Mass wasting deposits associated with the uplift are also visible, such as debris flow deposits and ridges that correspond to turbidite levees (Gràcia et al.; 2003, Terrinha et al., 2003).

North of the Horseshoe source zone there is an extensional structure called Pereira de Sousa fault (Terrinha et al., 2003). However, this fault and its associated structures acted as extensional faults in the rifting of the western Iberian margin and do not show any evidence of compression (Terrinha et al., 2003, 2009). This source zone is identified as "2" in Figure 6.3.

6.3.2.3 The Algarve Margin

The Gulf of Cadiz developed as a result of interaction between the southern end of the Iberia paleo-margin, the westward migration of the Gibraltar arc, and the convergence between the African and Eurasian plates. The collision between the African and Eurasian plates started at the Upper Cretaceous with NE direction. After a decrease in the convergence in the Maastrichtian, the collision is renewed with N–NE direction in post-lower Eocene times, with the Iberian plate welded to Eurasia. The main collision stage occurred between the upper Eocene and the lower Oligocene, and was followed by a change in the direction of compression towards the N–NW, and this has been the present compression since the upper Miocene (Zülkhe et al., 2004). At the same time, the Alboran domain migrated towards the west during the formation of the Betic-Riffean orogenic belt, or Gibraltar Arc, which thrusted over the continental margins of Iberia and Africa. The emplacement of the orogenic front originated a series of overthrusted allochthonous wedges over the eastern part of the Gulf of Cadiz. These allochthonous units were later reactivated along with emplacement of olistostrome bodies (Torelli et al., 1997) in the central and western parts of the Gulf of Cadiz by a combined mechanism of mass gravity sliding and collapse along the slope, and latterly reactivated tectonically due to the N–NW convergence (Medialdea et al., 2004). These two tectonic mechanisms imply two different structural behaviours in the Gulf of Cadiz: in the eastern part, from the Horseshoe Abyssal Plain towards the Gibraltar Arc, a mixture of thin skinned and thick skinned tectonics comprising westward thrusting of the upper sedimentary units and the whole crust respectively; in the western part, towards the Gorringe Bank and Coral Patch, a thick skinned tectonics encompassing the whole crust, and possibly the upper mantle too (Stich et al., 2005), by means of high angle reverse faults (Medialdea et al., 2004).

Inspection of multichannel seismic lines shows two major faults that can be associated with neotectonic activity. The longer one, trending NE to SW (named the Cadiz Wedge Fault, CWF, in Figure 6.3) is interpreted as a transpressive strike-slip fault, and characterized by the lack of major vertical offsets or important escarpments. This accident can be related with the westwards translation of an Alboran crustal block, as suggested by the GPS kinematic velocity field (Fernandes et al., 2006). The other fault follows the southern flank of the Portimão Bank, and is named Portimão Bank Fault (PBF) in Figure 6.3. This source zone is identified as "3" in Figure 6.3.

6.3.2.4 East dipping subduction slab

East of the Gibraltar Straight lies the Alboran Sea, between the Betic Cordillera in southern Iberia and the Rif-Tell fold and thrust belt in northern Morocco. During the early and middle Miocene, the Alboran domain underwent extension and thinning of a previously thickened crust. The development of the Alboran Sea, the Betics, Rif and Tell cordilleras is interpreted as being the result of the collapse of the Alpine orogen (Platt and Vissers, 1989; Sanz de Galdeano, 1990), involving extension and westward movement of the Ibero-Maghrebian region, and simultaneous northwest directed thrusting on the Iberian margin and south to southwest thrusting on the Nubian margin (Sanz de Galdeano, 1990). This geodynamic environment led to the formation of an eastward-thickened wedge of deformed sediments west of Gibraltar (Torelli et al., 1997; Gutscher et al., 2002). The deformation of the sediments is associated to thrusts that sole out to an east-dipping décollement indicating thus the presence of an accretionary wedge. Several hypotheses have been proposed for this evolution: (1) radial extensional collapse caused by rapid convective removal (Platt & Vissers, 1989) or continental lithospheric delamination (Docherty and Banda, 1995; Seber et al., 1996; Calvert et al., 2000); (2) convective thinning (Houseman, 1996); (3) subduction and breaking of a slab of material (Zeck, 1996); (4) back-arc extension caused by subduction rollback (Lonergan and White, 1997; Michard et al., 2002). The fourth hypothesis is supported by Morales et al. (1999) and Gutscher et al. (2002) who claim to have found compelling evidence for present tectonic activity associated with this lithospheric block.

This slab was interpreted by Gutscher et al. (2002) as a potential seismogenic and tsunamigenic feature. This interpretation was mainly based on thermo-mechanical modeling, and on the assumption that megatsunamis are mainly produced in subduction environments. Gutscher et al. (2002) argue that the relative velocity of the accretionary wedge vis-à-vis the surrounding lithosphere can be considered twice the relative motion between Nubia and Eurasia, and so it is closer to the average velocity needed to generate ~M9 earthquakes every 2000 years as supposed by most geological studies. Nevertheless, Zitellini et al. (2009) argued against an active seismogenic role of the slab based on the fact that the accretionary wedge has no important seismicity and that it is sealed by an almost undeformed package of sediments deposited since the Early Pliocene at the Rharb Valley, in the Moroccan offshore. This source zone is identified as "4" in Figure 6.3.

6.4 The strategy for seafloor continuous monitoring

As a result of the research described above, we have an almost complete mapping of the seafloor, an extensive collection of geological and geophysical data, and increased knowledge of the main tectonic units. Firstly, it is now well established that the main tsunamigenic tectonic sources are located between the long ESE–WNW strike-slip SWIM Faults (Figure 2 of Zitellini et al. (2009)), and the Iberian coastline: the SWIT zone. Secondly, the main active tsunamigenic tectonic sources are few and geometrically well confined: the Marques de Pombal, the Horseshoe Fault, the Cadiz Fault and the Portimão Bank Fault.

On the other hand, despite the regional and global importance of earthquake and tsunami hazards in this area, the seismic activity is not properly monitored and even the location of the plate boundary is uncertain. The earthquake cycle, the processes associated with seismogenesis and tsunamigenesis or the distribution of the seismic energy release by small events, is not represented in our datasets, in spite of the effort being made to reinforce seismic observation in southern Portugal. To better understand the mechanisms behind geohazard generation in the Gulf of Cadiz we must extend the land observations offshore.

The fact that the active tectonic tsunamigenic structures are few and geometrically well defined allows for accurate on-site monitoring, and for the mitigation of the associated geohazard. These confined structures may be seen as a natural laboratory to understanding the process of tsunami generation. The Gulf of Cadiz is, hence, one of the more suitable areas to establish long-term monitoring with a geophysical and oceanographic network. This monitoring effort, together with the measurement and characterization of the seismic activity, should also include a set of oceanographic and geophysical instruments useful to mitigate tsunami hazards and to study oceanographic processes in this critical area between the Mediterranean and the North Atlantic.

The proximity of the possible tsunamigenic sources to the coast calls for the development of new methods of tsunami detection. The tsunami, due to its high propagation speed, can cover the distance between the generation area and the nearest populated sites in only a few tens of minutes; therefore, an early warning system is needed. We propose a novel early warning system based on a multiparameter approach, which takes into account seismological and other geophysical and oceanographic measurements near the source.

A first step to achieving this goal has been recently accomplished in the framework of the EU project NEAREST: a new Tsunami Early Warning System (TEWS), which can be used in tsunami generation areas and was developed and installed in the Gulf of Cadiz. The core of the system is a tsunami detector prototype (tsunameter) installed on-board GEOSTAR, and a dual acoustic link between GEOSTAR and a surface buoy (Figure 6.4). The tsunameter includes a set of sensors that are installed on the seafloor on-board GEOSTAR, and connected to a processing unit: a bottom pressure sensor, an accelerometer which is used to correct the pressure data, and a seismometer. The data processing unit allows the detection of the tsunami wave and local strong and potentially dangerous seismic events, thanks to a new tsunami detection algorithm able to detect tsunami signals down to few millimeters (Chierici et al., 2007). The tsunameter performs a crosscheck between seismic and pressure signal.

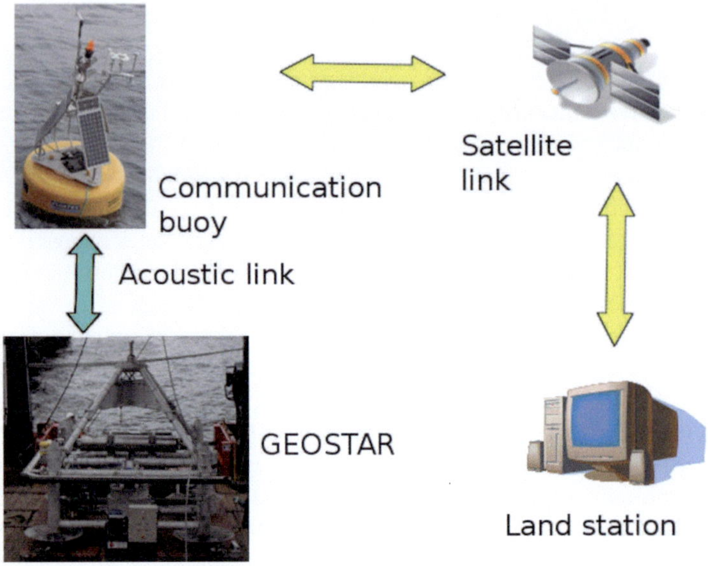

Figure 6.4 Two-way communication scheme of GEOSTAR observatory deployed during the second mission in November 2009.

The seafloor system communicates through a dual acoustic link with the surface buoy hosting the surface devices such as GPS, tilt meter, meteorological station, and the satellite dual link communication system connected with the control land station (Figure 6.4). In case of event detection (a tsunami wave and/or a relevant seismic event) an automated message (first notification) will be delivered to the land station. Afterwards, pressure data, sampled at 15 seconds, are delivered every 10 minutes to the shore station, for the whole event duration, by the dispatch of dedicated messages. The pressure data can be used to improve our warning assessment also by integrating them with other data from on land or on near-shore networks. Continuous data for the whole mission (seismic, oceanographic and geophysical) are finally obtained after the recovery of the abyssal station, and used for a posteriori analysis.

6.4.1 First results

The system was designed for near-field conditions and successfully operated from August 2007 to August 2008, 100km SW of Cabo de São Vicente (Portugal). The GEOSTAR station was deployed at 3207m depth at 36°21.875' N; 9°28.885' W at few hundreds of meters from the buoy anchorage system. Among the instruments available on the seafloor platform, the hydrophone, the three component broad-band Guralp seismometer CMG-40T and accelerometer CMG-5T are the ones which are most important to ensure high-quality seismic detection close to the seismic sources. On GEOSTAR there is also a set of

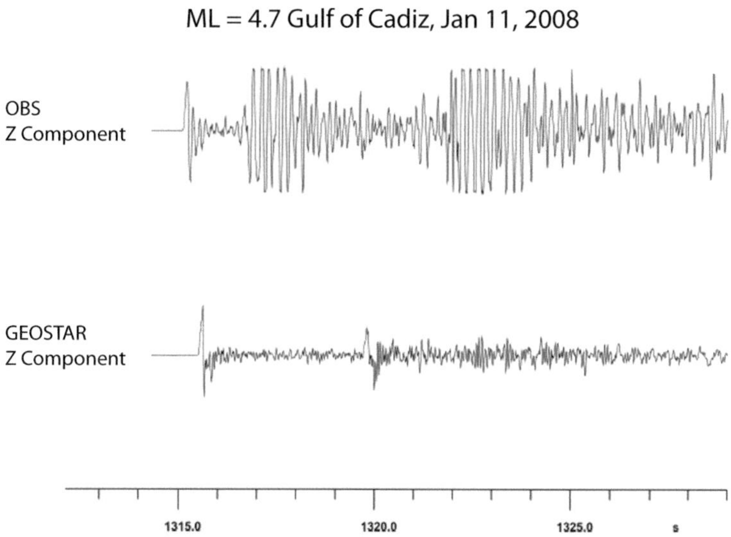

Figure 6.5 A M_L=4.7 event detected by GEOSTAR observatory (bottom trace) and also by an OBS deployed during NEAREST (top trace). Although the two sensors are separated by a distance of only 9km and are deployed on seabeds of similar lithology, the bottom trace shows very good sensor-ground coupling. The upper trace is also clipped due to the high gain chosen for the NEAREST OBS.

instruments for oceanographic and geophysical monitoring, i.e., punctual three axial current-meter, acoustic Doppler current profiler, CTD, turbidimeter and gravimeter. Within the NEAREST project an OBS survey was performed in the same area: 24 OBS spaced at about 50km were deployed for one year and used for both active and seismic studies (Geissler et al., 2010). During the first year of deployment a 7-month record of bottom pressure data, sampled every 15 seconds, and 11 months of seismic data, sampled at 100 Hz, has been acquired. The new real-time tsunami detection algorithm, able to automatically detect tsunami signals, has been successfully tested.

In Figure 6.5 we present a comparison between recordings by the GEOSTAR seismometer and a very close OBS, for a ML=4.7 event. One of the advantages of the GEOSTAR-type seismometer packaging and deployment is a good sensor–seafloor coupling, as shown by the high frequency content of the waveform (Monna et al., 2005).

In Figure 6.6 we present an example of one month of pressure data (September 2007) and a detailed zoom on two hours of data. The high quality of pressure data with a sensibility of less than a tenth of mbar and the stability over long periods needed for accurate tsunami detection is supplied by a Paroscientific model 8CB4000-I pressure sensor.

A new mission began on 10 November 2009, around the same location but on behalf of EU NEAREST and the EU ESONET (European Seas Observatory NETwork) NoE project: GEOSTAR station is now part of the ESONET seafloor monitoring network. In both missions, very critical sea conditions caused the failure of the mooring and the buoy was

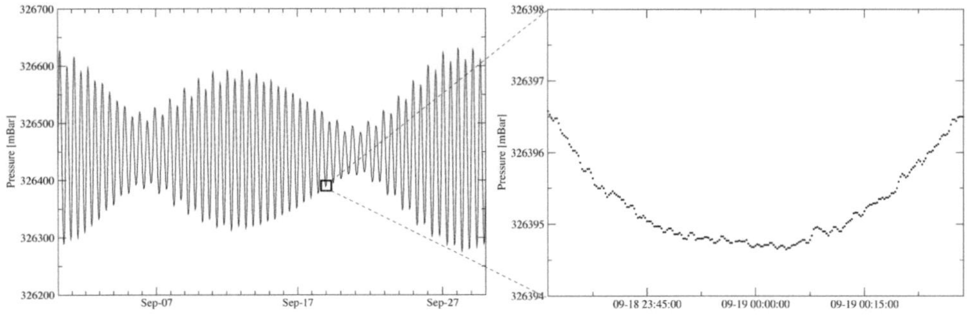

Figure 6.6 One month of bottom pressure data acquired during the mission 2007–2008 (http://www.moist.it/sites/iberian_margin/1/NEAREST) by GEOSTAR (left) and the detail of one hour data (right) (Marino, 2007).

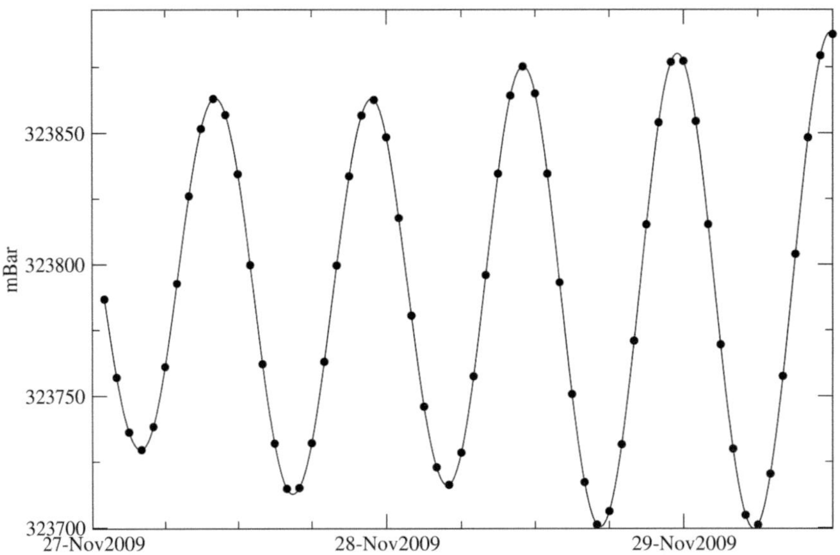

Figure 6.7 An example of decimated sea water pressure data acquired by the GEOSTAR observatory and dispatched to land station through the communication system during the second mission started in November 2009 (Embriaco, 2009).

operative only for a few months. There was also instability in communications due to incomplete satellite coverage. Nevertheless, during the second deployment the near real-time communication system has been successfully tested, with the automatic issuing of periodic messages containing a summary of the data acquired by the GEOSTAR observatory from the seafloor to the land station via satellite link. A sample of the bottom pressure data is shown in Figure 6.7, where the tides are the most powerful component of the signal. The pressure data, decimated at one hour, were automatically retrieved from the periodic messages sent by the abyssal station to land every six hours.

6.5 Conclusions

Over the past years a great effort has been made to have an accurate localization of the potentially tsunamigenic tectonic sources in the Gulf of Cadiz. This knowledge is necessary for the investigation of the tsunami generation processes and for the long-term monitoring of the sources. The deployment of a tsunameter near the potential tsunamigenic sources can significantly shorten the warning time.

Thanks to this experimental setting we can improve our understanding of the generation processes, achieve a fine tuning of the current tsunami generation model and possibly supply measured sets of initial conditions necessary for propagation models. Furthermore, continuous and long-term monitoring can help us to better understand oceanographic processes in the Gulf of Cadiz, located between the Mediterranean and the north Atlantic.

During the two missions, many geophysical and oceanographic data have been acquired by the GEOSTAR observatory; the greatest problem encountered in both missions was the instabilities in the acoustic communications. Important improvements were achieved in the second mission, although the buoy mooring remains a vulnerable part of the communication chain. A first step towards a fully operating tsunameter deployed in the generation area has been accomplished, with a test of all its constitutive parts assembled and hosted on board GEOSTAR. The future challenge is to bring the communication channel to the same level of reliability achieved in data acquisition and processing, possibly by a cable connection to SW Portugal.

We advocate the deployment of tsunameters of the type described in this chapter in at least three sites in the Gulf of Cadiz to have minimal safety coverage of the shore areas that could be possibly hit by a tsunami wave. The Gulf of Cadiz can be the appropriate site for the deployment of a permanent network of seafloor observatories for oceanographic, geophysical and warning purposes.

Acknowledgments

We acknowledge financial support from the EU Specific Programme "Integrating and Strengthening the European Research Area", Sub-Priority 1.1.6.3, "Global Change and Ecosystems", contract n. 037110 (NEAREST), ESONET-LIDO Demo Mission, contract n. 036851, ISMAR contribution n. 1709 and KINEMA project (PTDC/CTE-GIN/82681/2006, Portuguese Science Foundation).

References

Argus D.F., Gordon R.G., DeMets C. and Stein S. (1989) Closure of the Africa-Eurasia-North America plate motion circuit and tectonics of the Gloria fault. J. Geophys Res. 94, 5585–5602.

Baptista M.A. and Miranda J.M. (2009) Revision of the Portuguese Catalog of Tsunamis. Natural Hazards and Earth Systems Sciences 9, 25–42.

Baptista, M.A., Heitor S., Miranda J.M., Miranda P. and Mendes Victor L. (1998a) The 1755 Lisbon tsunami: Evaluation of the tsunami parameters. Journal of Geodynamics 25, 143–157.

Baptista M.A., Miranda P.M.A., Miranda, J.M. and Mendes Victor, L. (1998b) Constraints on the source of the 1755 Lisbon tsunami from numerical modelling of historical data. Journal of Geodynamics 25, 159–174.

Baptista M.A., Miranda J.M., Chierici F. and Zitellini N. (2003) New study of the 1755 earthquake source based on multi-channel seismic survey data and tsunami modeling. Natural Hazards and Earth System Sciences 3, 333–340.

Baptista M.A., Miranda J.M. and Luis J.F. (2006) In search of the 31st March 1761 earthquake and tsunami source. Bulletin of the Seismological Society of America 96, 713–721.

Barkan R., Ten-Brink U. and Lin J. (2009) Far field tsunami simulations of the 1755 Lisbon earthquake: Implications for tsunami hazard to the U.S. East Coast and the Caribbean. Mar. Geol. 264, 109–122.

Bergeron A. and Bonnin J. (1991) The deep structure of Gorringe Bank (NE Atlantic) and its surrounding area. Geophysical Journal International 105, 491–502.

Borges J.F., Fitas A.J.S., Bezzeghoud M. and Teves-Costa P. (2001) Seismotectonics of Portugal and its adjacent Atlantic area. Tectonophysics 337, 373–387.

Buforn E., Udias A. and Colombas M. (1988) Seismicity, source mechanisms and tectonics of the Azores–Gibraltar plate-boundary. Tectonophysics 152, 89–118.

Buforn E., Bezzegoud M., Udias A. and Pro C. (2004) Seismic sources on the Iberia-African plate boundary and their tectonic implications. Pure and Appl. Geophys. 161, 623–626.

Calvert A., Sandvol E., Seber D., Barazangi M., Roecker S., Mourabit T., Vidal F., Alguacil G. and Jabour N. (2000) Geodynamic evolution of the lithosphere and upper mantle beneath the Alboran region of the western Mediterranean: Constraints from travel time tomography. J. Geophys. Res. 105, 10871–10898.

Chierici F., Beranzoli L., Embriaco D., Favali P., Marinaro G., Monna S., Pignagnoli L., Zitellini N., Bruni F., Furlan F. and Gasparoni F. (2007) An innovative tsunami detector operating in tsunami generation environment. American Geophysical Union Fall Meeting 2007.

DeMets C., Gordon R., Argus D. and Stein S. (1994) Effect of recent revisions to the geomagnetic reversal time scale on estimates of current plate motions. Geophys. Res. Lett. 21, 2191–2194.

Docherty C. and Banda E. (1995) Evidence for the eastward migration of the Alboran Sea based on regional subsidence analysis: A case for basin formation by delamination of the subcrustal lithosphere? Tectonics 14, 804–818.

Embriaco D. (2009) LIDO 1 campaign using INGV/GEOSTAR seafloor platform during ESONET-NOE-LIDO project in Iberian Margin site (Gulf of Cadiz), part of EMSO network (http://www.moist.it/sites/iberian_margin/1/LIDO1).

Favali P., Beranzoli L., D'anna G., Gasparoni F., Marvaldi J., Clauss G., Gerber H.W., Nicot M., Marani M.P., Gamberi F., Millot C. and Flueh E.R. (2006) A fleet of multiparameter observatories for geophysical and environmental monitoring at seafloor. Annals of Geophysics 49 (2-3), 659–680.

Fernandes R.M.S, Ambrosius B.A.C., Noomen R., Bastos L., Wortel M.J.R., Spakman W. and Rovers G. (2003) The relative motion between Africa and Eurasia as derived from ITRF2000 and GPS data. Geophys. Res. Letters 30, 1828.

Fernandes R.M.S., Bastos L., Miranda J.M., Lourenço N., Ambrosius B.A.C., Noomen R. and Simons W. (2006) Defining the plate boundaries in the Azores region. Journal of Volcanology and Geothermal Research 156(1–2), 1–9.

Fernandes R.M.S., Miranda J.M., Meijninger B.M.L., Bos M.S., Noomen R., Bastos L., Ambrosius B.A.C. and Riva R.E.M. (2007) Surface velocity field of the Ibero-Maghrebian segment of the Eurasia-Nubia plate boundary. Geophysical Journal International 169(1), 315–324.

Fine I.V., Rabinovich A.B., Bornhold B.D., Thomson R.E. and Kulikov E.A. (2005) The Grand Banks landslide-generated tsunami of November 18, 1929: Preliminary analysis and numerical modeling. Marine Geology 215, 45–57.

Fukao Y. (1973) Thrust faulting at a lithospheric plate boundary: The Portugal earthquake of 1969. Earth Planet. Sci. Lett. 18, 205–216.

Geissler W.H. et al. (2010) Focal mechanisms for sub crustal earthquakes in the Gulf of Cadiz from a dense OBS deployment. Geophys. Res. Lett. 37, L18309, doi:10.1029/2010GL044289.

Gjevik, B.G., Pedersen G., Dybesland E., Harbitz C.B., Miranda P.M.A., Baptista M.A., Mendes-Victor L., Heinrich P., Roche R. and Guesmia M. (1997) Modeling tsunamis from earthquake sources near Gorringe Bank southwest of Portugal. J. Geophys. Res. 102, 27,931–27, 949.

Gràcia E., Dañobeita J., Vergés J., and the PARSIFAL team (2003) Mapping active faults offshore Portugal (36°N–38°N): Implications for seismic hazard assessment along the southwest Iberia margin. Geology 31, 83–86.

Gutscher M., Malod J., Rehault J.-P., Contruci I., Klingelhoefer F., Mendes-Victor L. and SpakmanW. (2002) Evidence for active subduction beneath Gibraltar. Geology 30 (12), 1071–1074.

Gutscher M.-A., Baptista M.A. and Miranda J.M. (2006) The Gibraltar Arc seismogenic zone (part 2): Constraints on a shallow east dipping fault plane source for the 1755 Lisbon earthquake provided by tsunami modelling and seismic intensity. Tectonophysics 426, 1–2.

Hayward N., Watts A.B., Westbrook G.K. and Collier J.S. (1999) A seismic reflection and GLORIA study of compressional deformation in the Gorringe Bank region, eastern North Atlantic. Geophysical Journal International 138, 831850.

Houseman G. (1996). Earth science, from mountains to basins. Nature 379, 771.

International Seismological Centre (2001) On-line bulletin: http://www.isc.ac.uk, Internatinal Seismic Centre, Thatcham, United Kingdom.

Johnston A. (1996) Seismic moment assessment of earthquakes in stable continental regions - III. New Madrid, 1811–1812, Charleston 1886 and Lisbon 1755. Geophys. J. Int. 126, 314–344.

Kaabouben F., Baptista M.A., Iben Brahim A., El Mouraouah A. and Toto A. (2009) On the Moroccan tsunami catalogue. Nat. Hazards Earth Syst. Sci. 9, 1227–1236.

Kaabouben F., Ibn Brahim A., Toto E.A., Baptista M.A., Miranda J.M., Soares P. and Luis J.F. (2008) On the focal mechanism of the 26.05.1975 North Atlantic Event; Contribution from tsunami modelling. Journal of Seismology 12(4), 575–583.

Levret A. (1991) The effects of the November 1, 1755 “Lisbon” earthquake in Morocco. Tectonophysics 193, 83–94.

LNEC (1986) A sismicidade histórica e a revisão do catálogo sísmico. In: Proc. 36/11/7368. Laboratório Nacional de Engenharia Civil. Ministério do Equipamento Social (in Portuguese).

Lonergan L. and White N. (1997) Origin of the Betic-Rif mountain belt. Tectonics 16, 504–522.

Lynnes C.S. and Ruff L.J. (1985) Source process and tectonic implications of the great 1975 North Atlantic earthquake. Geophys. J. R. Astr. Soc. 82, 497–510.

Mallet R. (1852) Third report of the facts of earthquake phenomena. Catalogue of recorded earthquakes from 1606 B.C. to A.D. 850 (1755–1784). Report of the Twenty-Third Meeting of the British Association of Advancement of Science.

Marinaro G. (2007) NEAREST 1 campaign using INGV/GEOSTAR seafloor platform during NEAREST project in Iberian Margin site (Gulf of Cadiz), part of EMSO network (http://www.moist.it/sites/iberian_margin/1/NEAREST).

Martins I. and Victor L.A.M. (2001) Contribution to the study of seismicity in the West Margin of Iberia. Univ Lisboa, IGIDL Pub 25, ISSN 0870-2748.

Medialdea T., Vegas R., Somoza L., Vázquez J.T., Maldonado A., Díaz-Del-Río V., Maestro A., Córdoba D., Fernández-Puga M.C. (2004) Structure and evolution of the "Olistostrome" complex of the Gibraltar Arc in the Gulf of Cadiz (Eastern Central Atlantic): Evidence from two long seismic cross-sections. Marine Geology 209, 173–198.

Mendonça J.J.M. (1758) *Historia Universal dos Terramostos que tem havido no Mundo de que há Notícia desde a sua Criação até ao Seculo Presente*. Biblioteca Nacional de Lisboa, Portugal, 272pp.

Mezcua J. (1982) *Catalogo General de Isossistas de a Península Iberica*. Instituto Geografico Nacional, Madrid, Spain.

Michard A., Chalouan A., Feinberg H., Goffé B. and Montigny R. (2002) How does the Alpine belt end between Spain and Morocco. Bulletin de la Société Géologique de France 173(1), 3–15.

Monna S., Frugoni F., Montuori C., Beranzoli L. and Favali P. (2005) High quality seismological recordings from the SN-1 deep seafloor observatory in the Mt. Etna region. Geophys. Res. Lett. 32, L07303, doi: 10.1029/2004GL021975.

Morales J.I., Serrano A., Jabaloy J., Galindo-Zaldivar D., Zhao F., Torcal F., Vidal F. and Gonzalez-Lodeiro F. (1999) Active continental subduction beneath the Betic Cordillera and Alboran Sea. Geology 27, 735–738.

Negredo A.M., Bird P., Galdeano S.C. and Buforn E. (2002) Neotectonic modeling of the Ibero-Maghrebian region. J. Geophys. Res. 107 (B11).

Oliveira C.S. (1987) Strong ground motion seismology. NATO ASI Series. Series C: Mathematical and Physical Sciences. In: Mustafa Özder Erdik and M. Nafi Toksöz (Eds) Vol. 204, pp. 405–460.

Perrey A. (1847) Sur les tremblements de terre de la Péninsule Ibérique. Annales des sciences physiques et naturelles, d'agriculture et d'industrie, X. Société Royale d'agriculture, d'histoire naturelle et des arts utiles, Lyon.

Platt J.P. and Vissers R.L.M. (1989) Extensional collapse of thickened continental lithosphere: A working hypothesis for the Alborán Sea and Gibraltar arc. Geology 17, 540–543.

Purdy G.M. (1975) The eastern end of the Azores-Gibraltar plate boundary. Geophysical Journal of the Royal Astronomical Society 43, 973–1000.

Reid H.F. (1914) The Lisbon earthquake of November 1, 1755. Bulletin of the Seismological Society of America 4, 53–80.

Rodriguez J.G. (1932) Catalogo sísmico de a zona compreendida entre los meridianos 5 e 20W de Greenwich e los paralelos 45 y 25 N., vol. I. Imprenta de Romano Velasco, Madrid, Spain.

Rodriguez J.G. (1940) Catalogo sísmico de a zona compreendida entre los meridianos 5 e 20W de Greenwich e los paralelos 45 y 25 N., vol. II. Imprenta de Romano Velasco, Madrid, Spain.

Sanz de Galdeano A. (1990) Geologic evolution of the Betic Cordilleras in the Western Mediterranean, Miocene to the present. Tectonophysics 172, 107–119.

Sartori R., Torelli L., Zitellini N., Peis D. and Lodolo E. (1994) Eastern segment of the Azores-Gibraltar line (central-eastern Atlantic): An oceanic plate boundary with diffuse compressional deformation. Geology 22, 555–558.

Sborschikov I.M., Shreyder A.A., Rimskiy-Korsakov N.A. and Yastrebov V.S. (1988) The Gorringe Bank and the tectonics of the Azores-Gibraltar fracture zone. Oceanology 28, 742–746.

Seber D., Barazangi M., Ibenbrahim A. and Demnati A. (1996) Geophysical evidence for lithospheric delamination beneath the Alborán Sea and Rif–Betic mountains. Nature 379, 785–790.

Sella G.F., Dixon T.H. and Mao A. (2002) REVEL: A model for recent plate velocities from space geodesy. J. Geophys. Res. 107, 2081.

Solares J.M. and Lopez-Arroyo A. (2004) The great historical 1755 earthquake: Effects and damage in Spain. J. Seismol. 8, 275 –294.

Solares J.M.M., Arroyo A.L. and Mezcua J. (1979) Isoseismal map of 1755 Lisbon earthquake obtained from Spanish data. Tectonophysics 53(3–4), 301–313.

Sousa F.L.P. (1919) O terremoto do 1° de Novembro de 1755 em Portugal e um estudo demográfico, Serviços Geológicos, 4 volumes (in Portuguese).

Sousa M.L. (1996) Modelos probabilistas para avaliação da casualidade sísmica em Portugal Continental. Msc Thesis, Universidade Técnica de Lisboa.

Sousa M.L., Martins A. and Oliveira C.S. (1991) Compilation of seismic catalogues of the Iberian region. L.N.E.C., Report 36/92-NDA, pp. 250.

Stich D., Ammon, C.J. and Morales J. (2003). Moment tensor solutions for small and moderate earthquakes in the Ibero-Maghreb region, J. Geophys. Res. 108(B3), 2148.

Stich D., Mancilla F.L. and Morales J. (2005) Crust-mantle coupling in the Gulf Of Cadiz (SW Iberia). Geophys. Res. Letts. 32, L13306.

Terrinha P., Matias L., Vicente J., Duarte J., Luís J., Pinheiro L., Lourenço N., Diez S., Rosas F., Magalhães V., Valadares V., Zitellini N., Mendes Víctor L. and MATESPRO Team (2009) Morphotectonics and strain partitioning at the Iberia-Africa plate boundary from multibeam and seismic reflection data. Marine Geology 267, 156–174.

Terrinha P., Pinheiro L.M., Henriet J.-P., Matias L., Ivanov M.K., Monteiro J.H., Akhmetzhanov A., Volkonskaya A., Cunha T., Shaskin P. and Rovere M. (2003) Tsunamigenic-seismogenic structures, neotectonics, sedimentary processes and slope instability on the southwest Portuguese margin. Marine Geology 195, 55–73.

Titov V.V. and Gonzalez F.I. (1997) Implementation and testing of the Method of Splitting Tsunami (MOST) model NOAA Technical Memorandum ERL PMEL-112.

Torelli L., Sartori R. and N. Zitellini (1997) The giant body in the Atlantic Ocean off Gibraltar: New results from a deepseismic reflection survey. Marine and Petroleum Geology 14(2), 125–138.

Tortella D., Torne M. and Pérez-Estaún A. (1997) Geodynamic evolution of the eastern segment of the Azores-Gibraltar zone: The Gorringe Bank and the Gulf of Cadiz region. Marine Geophysical Researches 19, 211–230.

Udias A., Lopez-Arroyo A. and Mezcua J. (1976) Seismotectonic of the Azores–Alboran region. Tectonophysics 31, 259–289.

Vilanova S.P., Nunes C.F. and Foncesca J.F.B.D. (2003) Lisbon 1755: A case of triggered onshore rupture? Bull. Seism. Soc. Am. 93, 2056–2068.

Zeck H.P. (1996) Betic-Rif orogeny: Subduction of Mesozoic Thetys under E-ward drifting Iberia, slab detachemant shortly before 22 Ma and subsequent uplift and extensional tectonics. Tectonophysics 254, 1–16.

Zitellini N., Gracia E., Matias L., Terrinha P., Abreu M.A., DeAlteriis G., Henriet J.P., Danobeitia J.J., Masson D.G., Ramella R., Somoza L. and Diez S. (2009) The quest for the Africa-Eurasia plate boundary west of the Strait of Gibraltar. Earth and Planetary Science Letters 280, 13–50.

Zitellini N., Mendes L.A., Cordoba D., Danobeitia J.J., Nicolich R., Pellis G., Ribeiro A., Sartori R. and Torelli L. (2001) Source of the 1755 Lisbon earthquake and tsunami investigated. Eos Trans. AGU 82(26), 285–291.

Zitellini N., Rovere M., Terrinha P., Chierici F., Matias L. and the BIGSETS team (2004) Neogene through Quaternary tectonic reactivation of SW Iberian passive margin. Pure Appl. Geophys. 161, 565–587.

Zühlke R., Bouaouda M.S., Ouajhain B., Bechstädt T. and Leinfelder R. (2004) Quantitative Meso-/Cenozoic development of the eastern Central Atlantic continental shelf, western High Atlas, Morocco. Marine and Petroleum Geology 21 (2), 225–276.

PART II

Technical solutions for seafloor observatory architecture

7 The role of Information Communication Technologies (ICT) for seafloor observatories: Acquisition, archival, analysis, interoperability

B. Pirenne[1]

On behalf of the Data Management and Archiving System Team at NEPTUNE Canada, past and present: Kimia Abhar, Darryl Bidulock, Tim Boesenkool, Ray Bon, Dennis Breckenridge, Yan Chen, Anna Cox, John Dorocicz, Karen Douglas, Dmytro Draga, Juan-Carlos Garcia, Ryan Graves, Eric Guilemot, Martin Hofmann, Shane Kerschtien, Eric Kolb, Tim Lavallee, Murray Leslie, Tony Lin, Khai Ong, Jeff Proctor, Daisy Qi, Yigal Rachmann, Kalpana Rawat, Reyna Jenkyns, Jason Rush, Ron Schouten, Nic Scott, Caleb Shortt, Karen Tang, Bernie Till, Mitozcelle Valenzuella, Aurora Walker, Liyan Zhao, Yingsong Zheng

7.1 Introduction

The revolution in seafloor observatories comes from their ability to: (a) allow for the coverage of wide swathes of ocean space in four dimensions at high temporal resolutions for decades; (b) provide quasi real-time data access as well as the ability to interact in very low latency (ms) with instruments hundreds of kilometers away from shore; and (c) offer new ways to perform science experiments where software plays a role in automating tasks as well as empowering interactive users.

This chapter addresses the role of cyber-infrastructures in the framework of online seafloor observatories that represents an extension of the Internet under the Ocean. Under this premise, this section explores the underwater infrastructure design from a network topology point of view, and considers how this design will enable new multidisciplinary science. Finally, the chapter addresses the social changes – or paradigm shifts – that users should expect to experience and how new tools under development will both facilitate the transition and improve scientific productivity.

1 Associate Director IT, NEPTUNE Canada, University of Victoria, Victoria, BC Canada

7.2 Different types of ocean observatories

Many systems can be called ocean observatories: from short cabled systems to buoy arrays to isolated tidal gauge systems. This chapter addresses systems that are online (via cable or satellite communication) and have sufficient power to provide uninterrupted data flow covering multiple environmental parameters, at high resolution, and in a four dimensional (4D) space. These observatory systems will provide a significant ability to remotely manage their assets. They will also have the mandate from their sponsors to have an archive that will support the long-term access to the 4D record and have a data policy that will open this archive to anyone.

There is, however, a significant distinction to be drawn between observatories with little power and limited data transmission capabilities and those that provide ample power and wired connectivity back to shore. While the former category has been around in one form or another for decades, the latter represents a new category that not only profoundly changes the access which users have to the seabed and water column but also challenges them by the amount of data produced. The new types of instruments it accommodates and the unprecedented time resolution implies significant data exploitation challenges.

7.3 Benefits of ICT for an ocean observatory

It has been suggested that central software systems fronting observatories are unnecessary and that one would be better off using the IT project money to buy more instruments. This is based on the point of view that, since the underwater infrastructure provides an Ethernet port to connect instruments, the observatory operators should therefore act like Internet Service Providers (ISPs) and let users deal with their instruments as they please. One analogy is that of a home or office Internet access service where the telecom company providing the service does not manage the user computers and peripherals on their network. In the case of an observatory, however, it is not quite the same as a wire going to a home or an office building. Space missions and other big science objectives have faced a similar dilemma that was resolved by calling certain experiments a "PI mission" and others an "Observatory mission". In the former case, a single, integrated team designs, manages and operates the observing facility or the experiment providing no explicit access to other non-members. There is no contention or access conflict. Such missions usually have a well-defined goal, a scientific question to be addressed. In the latter case, a (usually much more expensive) infrastructure is built by an implementing organization (IO) on behalf of entire communities of scientists. The IO will design, build and operate the facility in consultation with its stakeholders. Furthermore, they will not only implement the facility access and sharing rules, but typically will also attempt to guarantee a level of data quality and system availability to the users.

Arguments for a software system in a shared infrastructure are reviewed below:

1. Providing Internet access under the ocean is an expensive but worthwhile proposition. Typically the cost of the connection exceeds the cost of the instrument to be connected.

This is a Big Science argument (also applying to space experiments) which postulates that no single scientist or group of scientists should be given unrestricted, private access to research infrastructure of that scale, for obvious costs reasons. This research infrastructure has been funded by tax money **and its use should be shared, optimized and audited**.

2. The observatory should guarantee that all data are saved in an archive for the duration of the project and beyond. They should be accessible to the larger community and appropriately archived. The reasons are the same as in point 1 above where the cost of the initial infrastructure is significantly high and therefore its *use* (even "second hand" through the archive) *should be maximized*. Moreover, it is not clear that individual scientists would be or want to be in a position to guarantee an effective, secure and accessible data archive of their instruments for such a period of time. Making data accessible in a common format agreed upon by all users of the observatory would likely represent a difficult task, one that is not the first priority of a scientist, and would require funding to be allocated to different individuals or organizations to all do essentially the same thing. Economies of scale are therefore best realized with a **centralized archive**.
3. Significant obstacles must be overcome to connect new instruments to a complex underwater infrastructure. The cost of the connectivity is such that *there are limited connection points available for instruments as well as constraints on power and bandwidth*. Again, in the hope of maximizing the use of the facility, archive access or shared real-time access to data and underwater assets must be encouraged and enabled. A **software system will have to guarantee and implement** this policy to ensure fair share of the resources.
4. The tools provided by the software systems have complementary roles. They:
   - enable scientists to *cross-correlate real-time data with existing data from the same or similar sensors*
   - greatly *ease the access and comparison to data from various instruments simultaneously* – thanks to a single data access protocol and location
   - allow scientists to *monitor environmental parameters* for deviation from the norm *using an autonomous approach*. This latter point is very important as autonomous event detection in the face of large uninterrupted data flows is key to successful exploitation of the data
   - allow for a *uniform set of control systems* for similar instruments, rather than imposing the use of each instrument manufacturer's software on users
   - are used to create a web-based, *integrated interactive environment* where data can be *accessed, analyzed and visualized*; where instruments can be interacted with and where the scientific process that involves interaction between science team members will be greatly facilitated.
5. Finally, the software system only represents 5–10% of the total construction budget and about 10–20% of the operating budget. Experience in, e.g., large astronomy and space projects indicates that these ratios are true for most large-scale shared science (as

opposed to PI-based) projects. With this proportion of the cost and the potential for new science as well as a wider participation of more community members, this is certainly a good investment.

7.4 Mandate of a software infrastructure for ocean observatories

The role of software (or cyber) infrastructure can therefore be summarized in the following way. It will:

- Implement the vision of the observatory as an extension of the Internet under the Ocean
- Provide a long-term, accessible, organized archive for all the data captured by the successive generations of instruments that will be connected to the system
- Provide an integrated, secure environment to access the underwater systems
- Automate the data acquisition, data quality assessment, data exploitation as much as possible
- Support science, outreach and education activities by providing interfaces that address the needs of these various user categories.

Typically, "support science" could mean anything from providing simple data access and basic search capabilities to delivering quick-look/preview products to supporting science data analysis through tight integration with data processing systems, to enhancing scientific productivity by offering all matters of advanced analysis tools and collaboration environments.

The cyber infrastructure supporting ocean observatories should not attempt to provide data interpretation as products as such an activity fits best in the realm of science.

One mandate that will play a significant role in the system design is the possible need for the observatory to play a public policy role, such as one in which it provides data that is of critical importance in the forecast and mitigation of disasters or other natural phenomena that have an impact on human lives and personal property. In such a case, the organization as a whole has to implement tools to provide a highly reliable service to a "client" that is or represents a government agency. The software infrastructure has, in this case, to establish a Service Level Agreement (SLA) with the service recipient and give itself the means and resources to abide by its contractual obligation to provide the service.

7.5 Observatory system design

For the purpose of this chapter, the example of the NEPTUNE Canada and VENUS cabled observatories will be used to illustrate a recipe that can be further extended to any sensor network. We will not, however, consider directly specific requirements stemming from

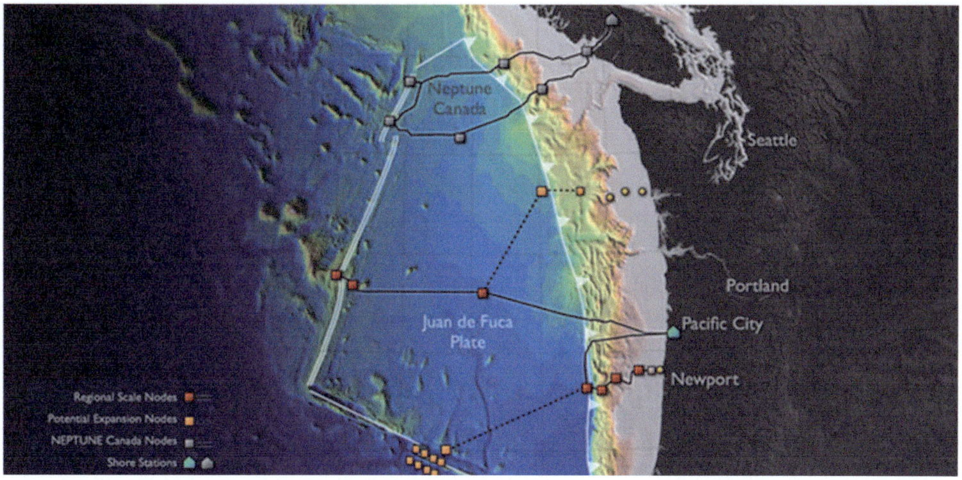

Figure 7.1 Example of network designs: NEPTUNE Canada and the Regional Nodes of the Ocean Observatory Initiative show either a double-ended loop or a number of linear cables respectively that provide power and communication capability from shore to instruments. Autonomous buoys with local cables to the seabed can also be considered.

the need to provide a critical public policy service, as such a mandate would introduce increased requirements on the robustness of the entire system.

The following section will consider the environment in which the system will operate; from the geographical point of view to the logical topology and selected network design.

7.5.1 Design decisions imposed on the ICT

The physical structure of a cabled observatory is normally that of a linear cable connecting a number of points of interest on the seabed. The cable provides power and data communication media and allows power to be carried from the shore to instruments on the seabed and, similarly, transport data from the seabed to the shore, as well as commands from shore to underwater devices. Cables can have one end at the surface (for power and signal feed) and the other in the water, or have a double termination above water, thereby creating a loop. Depending on the total system length, power will be provided by a single or multiple power conductors at a relatively high voltage. Communication presently relies on optical fibers. More details on undersea cables are provided in other sections of this book. Suffice it to say that at places of interest, power and communication media are available in adequate quantities to run and operate specific end-devices. Figure 7.1 shows examples of ocean cabled observatories that are capable of being managed by this software system design.

7.5.2 Network design considerations

The network design implements the vision of an Internet-based system, where every instrument and device is either a leaf on the tree structure or a junction point where multiple branches come together. The tree is of variable and arbitrary depth and does not impose conditions on its topology other than the fact that communication to other parts of the network will always propagate up the tree to the first common junction point between the two devices. This model is fairly straightforward and very common in today's electronic communication landscape.

To minimize the cost of the system and to re-use existing off-the-shelf technologies, the use of the Internet Protocol (IP) is preferred as a transport mechanism for data packets at the user/application level. Distances and fiber technology may require another transport mechanism at the lower level. So, in this instance, the Open Systems Interconnection (OSI) model[2] layer 1 can be implemented using fiber optics, lasers and repeaters, on which the Synchronous Optical Networking (SONET) protocol will be running. SONET generic criteria are described in Telcordia Technologies Generic Requirements document GR-253-CORE[3]. SONET packets will encapsulate layer 2 Ethernet (802.3) packets and deliver them to their end-point thanks to this standard's addressing system. At that level, a traditional network is available for implementing data communication, transport, routing, security, etc.

The details of the possibilities for implementing redundancy in a network design are not considered in this chapter as they have little impact on the high-level software managing the data from the underwater system.

As indicated in Figure 7.2, currently available oceanographic science instruments are of a legacy design, optimized for power consumption, internal recording and short stays in the water. Their typical data communication interface will be of the serial type (e.g., RS 232, RS 822 or RS 825). To implement the vision of the observatory representing the extension of the Internet underwater, it is necessary to convert the communication protocol of the instrument to IP as close as possible to the instrument. This can be done with simple devices, typically called "terminal servers" to be enclosed either in the original instrument, in a can on the cable linking it to a junction box or within the junction box itself, often only meters away from the instrument.

7.5.3 National security issues

An often unsuspected issue that faces planners in charge of designing a civilian underwater observing system is the concerns that the military (specifically the navies) might have with it, in particular due to the sensitivity of some of the passive acoustic instruments (e.g., hydrophones, seismometers). Accompanied by proper sophisticated data processing/pattern matching software, those instruments can be trained to detect specific ship propeller signatures. As is often the case, civilian scientific systems will have an open, public data

2 See http://en.wikipedia.org/wiki/OSI_model for a good overview of the various data communication layers identified in the OSI model

3 Telcordia GR-253-CORE, Synchronous Optical Network (SONET) Transport Systems: Common Generic Criteria (October 2009). Issue 5.

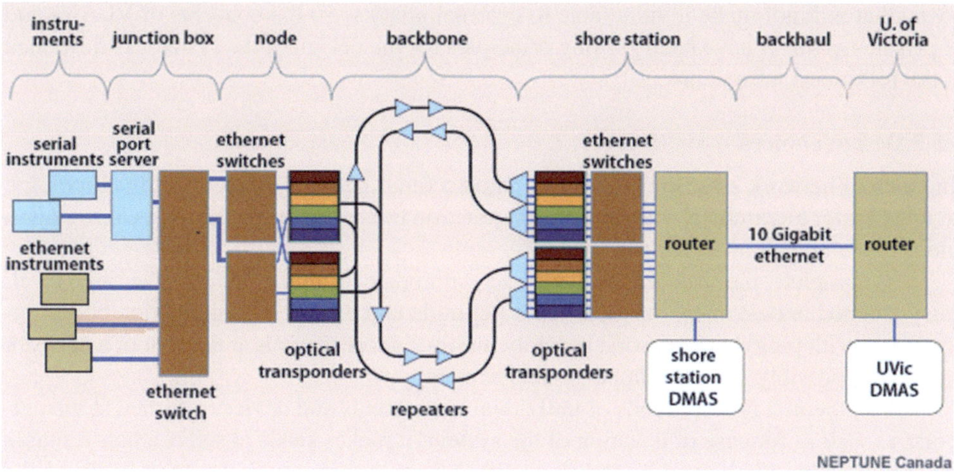

Figure 7.2 Example of the lower ISO Layers of the NEPTUNE Canada network design.

policy where data are immediately available to anyone. This is a clear concern for military powers whose assets regularly cruise in the area covered by the observatory. In the case of NEPTUNE Canada and VENUS (and this scheme presumably will be implemented for the US Ocean Observatory Initiative (OOI)), these issues can be addressed with a mitigation scheme that will (a) make use of the concept of Virtual Local Area Networks (VLAN), and (b) will implement data diversion of the streams from instruments deemed susceptible to detection of the presence of military assets nearby.

The concept of VLAN for the network makes sense, if only for the basic security of the system. For this particular navy issue, it was decided to gather all instruments of concern to the military in one specific VLAN and the others in a different one. Data Diversion can be implemented with a simple switch/router that captures all the data streams from all instruments in the reserved VLAN and redirects the flow of data to a military processing center until the vessel has sailed away from the sensor. To limit the loss of science data to a minimum, a preprocessing stage hosted on this switch computer will filter data in real time and send the nonsensitive part of the spectrum back to the regular science data flow. The current filtering mechanism still leaves frequency ranges lower than 1 Hz and higher than 5 kHz in the signal, so users can easily find out data that have been altered for reasons of national security. Moreover, the file names indicate whether the data in the file are complete or not.

The implementation of VLANs imposes a constraint on the system such that all network switches residing under water have to be manageable.

7.5.4 General network security threats mitigation

As described above, military concerns impose a minimum set of network security considerations for the entire system. Generally, it is always a good practice to set up a network in a way that will not make it vulnerable to external attacks. So the concepts of VLANs and of private networks have been guiding principles for the design of the system, as described in the following subsections.

7.5.5 Design choices

The overall network structure has to take into account the main elements of the topology: an underwater hierarchical tree with nodes, junction boxes and instruments, a shore station and a data and operations centre.

To be complete, the structure must also accommodate: multiple nodes at the same level; daisy-chained nodes; many junction boxes per node and daisy-chained junction boxes; instruments with piggyback sensors; possibly multiple shore stations at the root of a network and finally possibly several redundant data centers.

With a potential for thousands of individual instruments and devices attached to the network, as well as for ease of isolation of the system, it makes sense to select a non-routable set of addresses, as allowed by the IP protocol. In this case, given the complexity of the network, the familiar 10.0.0.0 address space (RFC 1918 – address allocation for private Internets, see http://tools.ietf.org/html/rfc1918) was selected. It allows system managers and security analysts to only worry about a few selected bridges between the outside world and the private network, while allowing complete freedom of address allocation and division into VLANs, etc. within the private domain.

IPv6 (Internet address composed of 16 bytes rather than 4 – see e.g., http://en.wikipedia.org/wiki/IPv6) is an option that can be seriously considered. However, its lack of general use, the lack of all-round experience with it and the fear that some of the hardware might not support it will likely limit the enthusiasm for it in many applications.

7.5.6 Private network and IP address range

The observatories are extensions of the Internet under the ocean. In that respect, a tree structure rooted at the data centre will be the best representation of the overall topology; main branches leading first to a number of shore stations, then branching off to nodes, to extension boxes and finally to instruments completing the picture.

The scheme described here allows each element of the network to be identified by an IP address and a port number that are unique across all observatories.

It could be countered that multiple sets of independent subnets, one for each observatory or even shore station, could be considered. However, to facilitate the operation of different infrastructures with the same staff, and perhaps even have instruments on one observatory eventually interacting with instruments on another one, a single address space covering present and future needs is preferred.

Given the "dumb" nature of most of today's instruments, the present model assumes that the addresses are assigned manually and remain static, as opposed to dynamically allocated using protocols like DHCP (see for example http://en.wikipedia.org/wiki/Dynamic_Host_Configuration_Protocol). Static addresses offer the security advantage of being predictable and can be pre-assigned. To ease the configuration of each individual device, *static DHCP allocation* is highly recommended. It assigns a predetermined IP address to each connected device based on its MAC address.

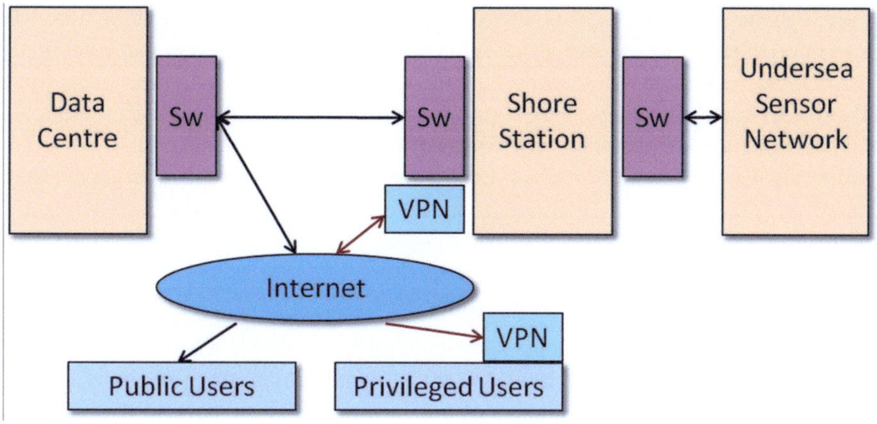

Figure 7.3 Options for user access to the infrastructure: through a controlled interface/proxy at the point of entry at the data center OR through VPN access to the shore station. “Sw” boxes represent switches.

7.5.7 Access only through VPN or through software proxies

A consequence of the choice of a private network for the observatory is the need to define bridges between the “external world” and the private system for basic, controlled access. Virtual Private Network (VPN) is a simple option that can be implemented to control external access for interactive use of the resources within the private network. VPN will create a virtual point-to-point connection between one’s computer anywhere on the Internet and another network point inside the private network. External users with the need for private, ad hoc access to a resource can use this option. A typical configuration would be one where a VPN server hosted at the shore station will identify each user individually and direct each user only to the device(s) that s/he is allowed to access.

Software proxies such as web interfaces are a good way to access shareable resources on a multi-user facility. Those software interfaces will allow a better control, log of the activities and will be less of a burden for users. The example of camera control where only one active user should be granted active control at any one time is typical. A web interface as proxy also allows multiple users to view currently ongoing activities with the resources and offers ubiquitous access to the system.

Figure 7.3 indicates two typical paths for users to access resources underwater.

7.5.8 Isolation of VLANs to isolate instrument categories from one another

Virtual Local Area Networks (VLAN – IEEE 802.1Q) offer service segmentation and will be the tool of choice if special categories of instruments need to be isolated from one another for security reasons. VLANs are an OSI layer 2 feature (see footnote 1), meaning they are defined at the level just above the physical medium and configured in network switches. There are multiple examples that can be considered where VLAN use would make considerable sense in the set up of an observatory. The example of a separate management VLAN comes to mind where all non-user accessible devices will be isolated in a special management VLAN. Such devices will not only include all network devices on the system (on land as well as underwater) such as switches, routers, media converters, serial-IP converters, but also the facility control computers, precision clocks, etc.

Another VLAN that should be considered is one that will host all instruments that are considered of "national security concern" and would need to be especially protected or have a different management policy.

7.5.9 User authentication and authorization

User access control and monitoring for such large infrastructure is a necessity. Control will help prevent accidental damage to the infrastructure and limit malevolent activities. Monitoring/auditing will allow the managers of the system and their stakeholders to see how much the system is being used and for what purpose.

The cost of the infrastructure and its public nature, and the need to provide service with minimum interruption (in, e.g., a response to a service level agreement) impose the set up of a controlled access policy and of its enforcement. Control will take the form of the determination of who is entitled to certain privileges on the system through the definition of roles.

Monitoring of the activities will not only serve the purpose of understanding changes that have occurred in the system configuration and their impact but also will help demonstrate to funding agencies and sponsors how much the facility is being used and for what purposes.

Monitoring will therefore require auditing changes to the infrastructure configuration and activity recording. (Who did what and when? How much was used? And so on.)

A typical implementation of the control and monitoring can be done through the definition of accounts, groups of users as well as privileges that can be granted or revoked.

7.5.10 Timing and time signals

An underwater observatory will likely represent a multidisciplinary resource, allowing many users to coordinate real-time or archived data from different sources at identical or separate locations. Correlation of data from different origins, often acquired at different sampling rates, will have to be done based on their time of acquisition. This requires that all clocks on the system be synchronized with a master clock to ensure that all data have the same time baseline. If data local to one observatory must be compared with those from outside the infrastructure, the system's master clock must itself be synchronized with a standard atomic time, as provided for example by the National Physics Lab in the UK, the National Research Council in Canada or the National Institute of Standards and Technology (NIST) in the USA.

While the above considerations make sense, in reality things can be somewhat more complicated: instruments to be deployed will most likely arrive with a design that assumes independent and autonomous deployment. Therefore, the emphasis is typically put on having a very stable clock in the instrument, but not necessarily on providing software tools or hardware connections as synchronization methods. In long-term deployments, no matter the quality of each clock, eventually they will drift apart making their individual measurements less than useful for users with highly precise time correlation requirements.

There are several approaches, not necessarily mutually exclusive, to solve this issue.

1. Convince instrument manufacturers to create "smart" instrument interfaces that will not only be capable of directly speaking IP, but also understand the Network Time Protocol (NTP) or the Precision Time Protocol (PTP). In this way, each instrument will individually keep itself synchronized with the rest of the infrastructure clocks. This is the ideal solution and will conservatively provide timing of measurements accurate to 10 ms or 10 μs with NTP and PTP respectively. Those values are estimates provided by Alcatel-Lucent when designing the NEPTUNE Canada cabled system, taking into account the fiber lengths, heavy traffic on the lines and a number of switches between the shore station and the instruments. The theoretical lab bench values would typically be an order of magnitude better for both protocols.

2. Implement in a shore-based software driver the ability to reset the clock of any instrument to the correct time. The procedure will be instrument dependent and the best accuracy one can reasonably hope to achieve likely cannot be better than 1 second for instruments using the serial protocol. Such a procedure will also possibly create an interruption of the regular data flow as the sampling will likely have to be suspended while the clock is being re-synchronized. The operation may need to be executed as often as daily.

3. Another method involves the time-tagging of each data record received from each instrument at the shore station. The method has the advantage of being generic and of offering a homogeneous time stamping across all instruments of the system. The accuracy of the timing will however be affected by anything between the instrument and the shore: (e.g., high traffic volume on the network, speed of switches and routers, efficacy of the software) and by physics (3 μs/km – determined by the speed of light). On a system with ample bandwidth and of a few hundred kilometres in size, this method can easily guarantee 100 ms accuracy for the measurements, albeit with non-negligible jitter.

A pragmatic approach is to implement option 3 above, while convincing instrument manufacturers and scientists that solution 1 is essential to guarantee the accuracy of experiments requiring either real-time co-ordination, or for subsequent exploitation of archived data.

7.6 Data acquisition

In this section, the emphasis is on capturing all data from all devices connected to the system. The word device here is to be understood as to mean any subsystem connected to the infrastructure that can produce data. We are therefore not only considering science instruments and their sensors: all other elements of the system, even if they do not participate in the collection of the science data are nevertheless important as they hold the clues to the health and safety of the entire observatory. In systems as complex as the ones examined here, elements at the shore station, in the nodes and junction boxes can produce large amounts of important data to help understand, or predict system failures. It also turns out that a very large part of the data volume produced by the observatory will emanate from those non-science data providers.

7.6.1 Data types in ocean sciences

A part of the complexity of ocean sciences stems from its plurality: an observatory as described in this chapter is serving many different communities with different goals and relying on different types of instruments to achieve their goals: physical oceanographers and chemists will have sensors measuring directly phenomena of interest while biologists will usually rely on proxies to derive populations, species and abundances. This is reflected in the instrumentation that has to be hosted on the system.

Typical instruments will therefore usually fall into one of three categories from a data management point of view, as shown in Table 7.1.

7.6.1.1 Data flow as streams – Data flow as an event management problem

For the purpose of designing a software system to manage the data flow coming from various devices connected to the infrastructure, a simple approach can be considered where all instruments are considered as sending a *stream* of data.

At the highest level of abstraction, given the individual duty cycles of each instrument, all categories will, from time to time, return their measurements as a string of bytes. A CTD may be returning the values of its sensors every second for months on end; a still camera may be programmed to take a picture every day, a video camera may be operated periodically and return a rapid succession of images.

At the same time that each instrument can be considered as a producer of a more-or-less continuous stream of bytes, another way to look at the problem is to see every new stream of bytes as an *event* that just occurred and for which some specific processing is required.

The design presented in this section assumes a combination of both approaches to deal with the data flow. Each instrument will be assumed to produce data in an ad hoc, but not necessarily predictable, fashion. The (a)synchronous occurrence of a new sequence of data will trigger the execution of a predetermined set of processing stages, the last of which will be the archival of said stream.

In Figure 7.4, data coming from different types of instruments have reached the shore-based systems. Each arrival of new data triggers a sequence of processing steps, from record parsing to QA/QC checks to archival. The use of the "publish and subscribe" model is useful to deal with the asynchronous nature of the streams. This is the model selected

Scalar	CTDs, chemical sensors, ...	Return lists of values at regular intervals
Complex	ADCP, still cameras, ...	Return n-dimensional matrices on a regular basis
Stream	Video cameras, hydrophones	Return uninterrupted streams of bytes

Table 7.1 Various data categories produced by the instrumentation to be supported by an ocean observatory.

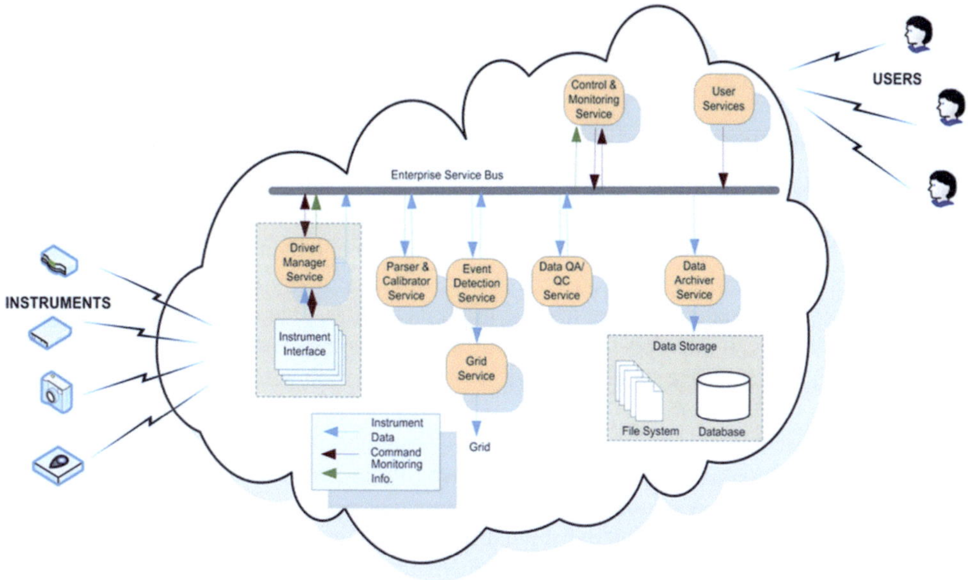

Figure 7.4 The stream processing event-driven Data Acquisition Framework adopted by the NEPTUNE Canada and VENUS Data Management and Archive System (DMAS).

for the Data Acquisition Framework of the NEPTUNE Canada and VENUS Data Management and Archiving System.

7.6.1.2 Interfaces to many different types of instruments

There is a clear need to adapt, translate and transform data from each instrument type to make them easily exchangeable with other users or other pieces of software. If one of the goals of the underwater observatory is to make possible cross-correlation between data from various sensors and therefore foster serendipitous discoveries, data must be not only accompanied by appropriate descriptions (metadata) but also organized in a way that will facilitate comparisons, matches and other correlations. Moreover, applying standard operators on the data (averages, decimation, identification of minima and maxima within a period of time) will be easier if data are organized in a systematic, standard way within a database or file system; hence the need to standardize data organization at the earliest possible stage.

The first stage in the stream/event model illustrated in Section 7.6.1.1 above is the access to instrumental data. Regardless of the selected design of the data management system, such an interface is necessary and has to be adapted to each type of instrument, each vendor, each model and each firmware revision. Therefore, significant resources have to be devoted to the understanding, coding, testing, debugging and the maintenance of the access to the various instruments. We estimate that expert software developers and testers will accomplish the various tasks involved with preparing instrument "drivers" (or interfaces) at a cost of about one person-month for each new instrument type. This value is based on several years' experience in developing and testing "driver" code for more than 30 different instrument types for the NEPTUNE Canada and VENUS systems.

7.6.1.3 Interoperability

Recognizing the challenges and the expenses incurred by the multiplicity of instrumentation interfaces and protocols, efforts are currently ongoing to propose data interface standards to be shared between manufacturers and implemented throughout their palette of models. This is, however, a long-term effort that will need many more years of meetings and lobbying to reach a ubiquitous state of implementation. One of the lead standards at present is SensorML. It attempts to accompany sensor readings with all relevant metadata. All values returned are surrounded by XML tags predefined in agreed-upon and published dictionaries (or ontologies)[4].

Reaching agreement at this level and having instruments return their measurement and accompanying metadata following SensorML conventions promises to achieve interoperability at the sensor level. A standard for data representation is, however, not sufficient. Two extra elements have to be considered for interoperability at the sensor level to truly take place: a transport protocol and a discovery mechanism. Sensor Web Enablement (SWE) is a proposed transport and discovery mechanism. The examples listed here are being proposed and supported by the Open Geospatial Consortium (OGC[5]) and represent one of the more dynamic initiatives in the area of interoperability at the sensor level. Under the auspices of the OGC, tests have been carried out with prototype instruments located at different places and making their data available to software that can subscribe to any one of the streams. For more information on SensorML, please consult the related OGC white paper[6].

There is another interoperability level that should be considered. It concerns *transparent data access at the level of data centers*. It addresses the ability to submit queries for archived data at any participating site and allows users to forgo prior knowledge of where data actually reside, how they are stored and the hoops to go through to access them. At the same time, this level of interoperability also solves the issue of data formats and transport as interoperable archives return requested data in compatible containers, typically using a variant of NetCDF[7].

7.6.1.4 Science data vs. engineering data

Science data collection is the primary goal of any ocean observatory; however, as described above, sensors and instruments are attached to an infrastructure that allows them to

4 An example of SensorML description for a CTD can be found at http://vast.uah.edu/downloads/sensorML/v1.0/examples/sensors/MBARI_CTD.xml

5 The Open Geospatial Consortium (OGC) can be found at http://www.ogcnetwork.net/

6 OGC White Paper: OGC® Sensor Web Enablement: Overview And High Level Architecture, downloadable at http://www.sensorml.net/

7 NetCDF (or Network Common Data Form) is a set of software libraries and machine-independent data formats that support the creation, access, and sharing of array-oriented scientific data. More description can be found at http://www.unidata.ucar.edu/software/netcdf/.

operate. The infrastructure typically provides power and communication media to instruments and their hosted sensors. Therefore, unless the infrastructure is "somebody else's problem" (such as is the case when all or part of the infrastructure, e.g., satellite data transmission is contracted out to an external organization), and regulated through a service level agreement (SLA), the organization operating the facility has to perform and support a potentially significant number of activities having to do with the overseeing of the entire system.

Overseeing the system is usually a 24/7 task that involves the monitoring of a large number of subsystems dealing with power and power distribution as well as with data transmission. All of those subsystems can themselves be considered as sensors that produce engineering data. The engineering data has to be acquired, converted, verified and checked against ceilings and thresholds on a regular basis. Any value identified as going beyond pre-set bounds will generate alerts to be dealt with by observatory personnel.

In the example of NEPTUNE Canada, nodes and junction boxes, distributing power and communication facilities to science instruments, are equipped with a large amount of electrical and environmental sensors. Such sensors typically return data at the rate of 1 Hz. As of 2010, just the nodes and junction boxes alone connected to the NEPTUNE Canada network produce about 8 TB of raw scalar data per year.

However, the data are essential to help predict trends, and offer the ability to conduct forensic analysis to understand the cause of element failure. An example where trending will help observatory managers extend the lifetime of the infrastructure and establish a priority list for maintenance and recovery, is the analysis of the stability of the various ground leak current sensors. A slowly increasing leak current (or reduced resistivity to ground) is an indication that something is amiss somewhere and could lead to accelerated corrosion of subsystems. Switching them off early will increase the lifetime of the remaining system.

Tools have to be provided to engineers and "wet plant" system managers to access, examine and react to events happening underwater such as the ground fault in Figure 7.5. The large number of individual sensors that have to be monitored requires systems that will automatically and constantly verify that all variables remain within their pre-set boundaries. A network management system (NMS) will collect alerts that come from any subsystem (power or communication) and draw the attention of system operators when they occur. Automating such tasks is essential to limit the operating costs of the infrastructure to a minimum and to avoid the need for a 24/7 coverage of the operation of the system, limiting the service requirement to having personnel on call.

7.6.2 Data archive and distribution management

7.6.2.1 The cost of a 25-year mandate

Big Science infrastructure, such as astronomical observatories, large vessels and nuclear reactors, is typically designed and built to last between 25 and 50 years after which it has to either be decommissioned or undergo significant refurbishment, upgrades and modernization. The case of an ocean observatory is no exception, but will likely have a life expectancy towards the lower end of the range, mostly due to the lack of experience with such system as well as the harsh and corrosive environmental conditions to which the various elements of the infrastructure are subjected.

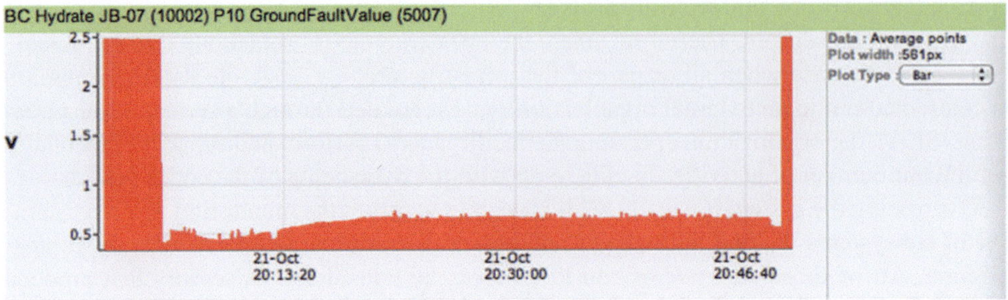

Figure 7.5 Example of the changes of ground fault (GF) value measured at a junction box port between the off and the on status of the faulty instrument connected to it. The diagram shows about 30 min of data during a test run following the deployment of an instrument platform at the Barkley Canyon node of NEPTUNE Canada. Ideally, the GF value should have remained at a value of around 2.5V (corresponding to an almost infinite resistance between the power lines and the seawater).

Consequently, with funding hopefully in place to support operations during the entire period, the software systems used to acquire and store the data, monitor and control the infrastructure should be sustained and provide access to the sum total of data, information and knowledge accumulated during the complete history of the facility. This is one of the fundamental requirements of the software system running the observatory and the reason why the underwater infrastructure does not merely extend the Internet under the ocean.

The design of a hardware/software system capable of supporting a large-scale experiment such as an underwater observatory has to take a number of considerations into account. Experience from other big science experiments is summarized in Table 7.2.

Based on the numbers in Table 7.2, it is clear that, at any one time, managers in charge of the entire system will essentially be involved in a constant state of change, migrating this or that subsystem to its next generation of hardware, software or design. This clearly demonstrates the need to keep funding in this area flowing at a reasonable level.

Failure to support this technology refresh cycle will only lead to early obsolescence and ultimately to a more costly observatory operation:

- old hardware will cost more and more to keep running (e.g., keeping lots of small disk drives in operation rather than a few large ones)
- old software implementation (legacy software) may make it more difficult to find suitable developers who know about the language, OS, etc.
- novel instrumentation design or radically different ways of using the underwater infrastructure might lead to the impossibility of continuing to operating with the assumptions that led to the elaboration of the system to that date.

Therefore, there is a need to evolve the system and maintain its capabilities aligned with the current state of the art software and hardware.

High-level design, topology, external environment	Lifetime
Software architecture	10–15 years
Programming language	10+ years
Operating system	10 years
Storage technology	8–10 years
Main software element design	7 years
Computers running the software	4–5 years
Storage system	3–5 years

Table 7.2 Longevity of the different elements constituting a software infrastructure, based of this author's previous experience in similar context of long-term big science projects (Albrecht and Pirenne, 1994).

Experience with the establishment of a cyber-infrastructure for an ocean observatory is in line with that of other large big science facilities and lies at about 5–10% of the total infrastructure costs, depending of the sort of experiment (whether it has a large or small engineering component, whether it is ground- or space-based, etc.). Yearly operational costs, however, must be maintained at the level of about 10–20% of overall O&M costs to keep the system's capabilities at a competitive level with other similar facilities around the world, thereby sustaining the large initial investment.

7.6.3 Data repository growth: Constant, linear or exponential?

7.6.3.1 Types of products

In the context of a scientific experiment destined to measure/survey a given environment over a long period of time (such as in astronomy, physics, biology, etc.), the host archive of the accumulated data will clearly grow as long as the infrastructure is in active operation. Predicting the costs of the system years ahead appears at first glance difficult, but a few simple assumptions can help determine the *growth regime* in which the archive will find itself for most of its lifetime.

Determining the growth regime depends on a number of factors:

- Raw data input:
  - Will the instruments return the same type of data and at the same rate for a long time?
  - Will the number of instruments change over time and, if ye,s at what rate?
  - Will new instruments be produced that will change the data rate dramatically?

- Products:
  - Will data products generated from the raw data be archived?
  - Will they undergo multiple successive reprocessing stages?

While the question of derived product storage can be managed and dealt with operationally, one can assume that complicated, unwieldy raw data will always yield more practical, science-ready products. They will typically be the result of a need to facilitate the work of scientists by performing standard data reduction. The extreme case of this would be in radio astronomy where detectors currently being developed will generate data at Tera-bits per second rates, clearly inappropriate for practical data analysis. The generation of products here will typically reduce the data rate by two or more orders of magnitude. But, by the same token, the storage of such derived products will result in only a small extraneous, fractional cost for the storage.

Derived products whose size is comparable to the original raw data impose on the archive the management of a potentially unnecessary load. In this case, an on-the-fly generation of products should be investigated.

7.6.3.2 Evolution of raw data rate

If derived products are either very small or otherwise manageable, what is to be considered is the original raw data rate. The evolution of raw data volume depends on a number of criteria that can be summarized in the following formula:

$\partial D/\partial t = F(I, N, s)$

where:

- $\partial D/\partial t$ is the evolution of the data rate over time
- I is the number of sensors
- N is the number of dimensions (N=0: scalar, N=1: vector, N>1: matrix)
- s is the sampling rate in Hz

The parameters of the function are illustrated with a few examples below:

I, the number of sensors, will certainly increase with time, potentially by one or two orders of magnitude, given both the evolution of technology that packs more and more sensors in small, low-power packages as well as the hierarchical structure described in Section 7.5.2 that allows for almost unlimited grafting of new instruments to the infrastructure.

The dimensionality of the structures returned by sensors (*N*) will also change. A simple example is that of current meters: they can periodically return a single number (current intensity in one direction – N=0) or a complicated matrix where N=5 with 3 directional vectors at multiple depths, integrated for a given amount of time. Another example that can only be imagined, but could possibly be true some day: going from a simple sensor measuring temperature at one point to a thermal infrared camera that will produce temperature gradient matrices of entire volume of water on a regular basis (e.g., across hot vent plumes).

The sampling rate *s* is clearly another parameter that could rapidly change over time. In battery-operated installations, length of deployment was favoured over high sampling rate, so hourly measurements were not uncommon to save battery power. With power constantly available to instruments on the seabed, sampling rate is only limited by physics and by the integration time of sensors, potentially leading to interesting ultra-short-term variation discoveries. So orders of magnitude increases are possible here as scientists venture in the very short time scale phenomena studies.

What this all means to archive growth in an ocean observatory context can be illustrated in a diagram such as the one in Figure 7.6. In this, several scenarios are considered.

- The constant scenario has no new instrument added/removed that would have an impact on the data received over a 10-year period. Clearly, this is not a likely scenario.
- The linear growth is projecting a 10% growth (in this example) per year. The approximation is more realistic, but does not take into account the fact that instruments will sometimes fail, will not necessarily be replaced; that due to funding lulls or different priorities, instruments will not be changed for a year or so and that, conversely, there will occasionally be a new instrument delivered that will create far more data than any other.
- An exponential growth is again more likely than the constant one, but suffers from the same issues discussed in the context of the constant growth.
- Finally a "likely" scenario is also shown. It is based on a simulation of: events plaguing the underwater system, technical issues with new instrument developments and lack of funding to renew and upgrade the instrumentation, but also considers a "reasonable" increase in the number of sensors and their longevity in the water.

7.6.3.3 Adapting the storage structure to expected use

One of the dominant characteristics of ocean observatories is the diversity of data types that their instruments produce (see Section 1.6.1). To further complicate the problem, users typically need different ways to deal with data, including selecting, browsing, retrieving and processing.

Browsing and searching are activities that have to be performed interactively and for which a rapid response time to key strokes or mouse clicks must provide a feeling of immediacy. Retrieving and processing can be allowed to take more time, particularly if the user is made aware of the size of the dataset to be transferred or the complexity of the processing to be performed.

To provide adequate response times to queries, typical practices involve the use of database management systems to store metadata and scalar data where an index on time and sensor is provided to speed up search. Scalar data will particularly benefit from being hosted in a database for the rapid access it will allow to quick plot tools and other easily generated preview products.

Even if technically possible, it does not make sense to store all data in traditional relational database management systems (RDBMS) as the issues of backup, space allocation constraints, etc. render the use of RDBMS' cumbersome as a mean for storage of large data "blobs". For datasets produced by complex instruments (n-dimensional matrices, streams) it is often preferable to store those on file systems and simply keep the registry of the files in a database. This is the model that the NEPTUNE Canada and VENUS Data Management

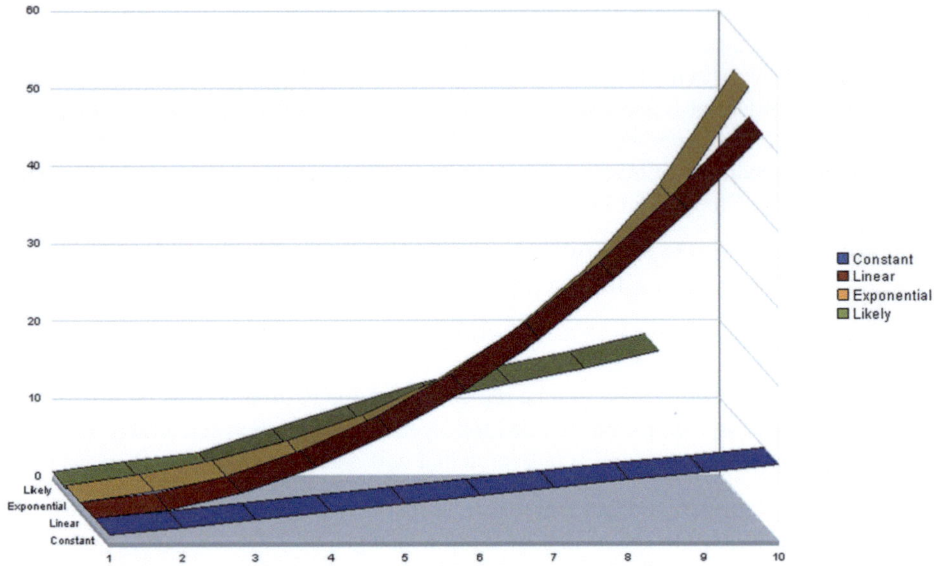

Figure 7.6 Expected growth of the archive given four different scenarios: constant input, linear with limited growth of 10%/year, exponential (with a slow start) and a more likely scenario that takes into account issues with instruments, the underwater environment, the capabilities of scientists to deal with the volume, etc. The "likely" scenario moves more in leaps and bounds, sometimes gets much more and sometimes gets a big drop of data rate with key instruments failing. The X-axis shows a 10-year simulation and the Y-axis represents an arbitrary data volume unit that starts at 1 in the first year.

and Archiving System has chosen to adopt. It is also the model that has been in place for many years in other scientific archive systems such as those developed for astronomy by this author and collaborators, in support of such large systems such as the Hubble Space Telescope, the European Southern Observatory telescope or the Canadian Astronomy Data Centre supported assets.

7.6.3.4 Observatory assets management and operation support

A large underwater observatory has many physical components. It also represents a facility that has to have a long lifetime and will therefore host several generations of caretakers. The complexity is so large that it is impossible for a single person or small group of people to remember everything about the system. Examples of essential information abound: installation date and position, date of recalibration of an instrument and the formulae that have to be used for each of its sensors; when the instrument was turned on and off and by whom, etc. This information is absolutely critical in order to understand the data that any instrument produces. Moreover, when dealing with a multiyear archive of data from instruments with a complicated history, understanding that history is necessary for data users to have some trust in the data quality.

Device Id: **19** Device Name: **SeaBird CTD 16 plus 4996**

General | Sensor | Ip | Electrical Rating | Data Rating | Nameplate | Port | Physical Characteristics | Device Action | Site | Procurement

Site Device Id	Site	Location	Region	Date From	Latitude	Longitude	Depth(m)
15	VIP-03	Central Node (SI)	Saanich Inlet	05-Aug-2006 00:00:00	48°39.0719'	-123°29.1605'	98.0
75	VIP-04	Central Node (SI)	Saanich Inlet	07-Nov-2006 00:00:00	48°39.0733'	-123°29.1632'	98.0
30018	On Ship or Shore	VENUS offices	VENUS Headquarters	07-Nov-2006 00:00:00			0.0
30019	On Ship or Shore	VENUS offices	VENUS Headquarters	01-Feb-2007 00:00:00			0.0
1123	VIP-05	Central Node (SI)	Saanich Inlet	03-Feb-2007 00:00:00	48°39.0722'	-123°29.1685'	96.0
22607	VENUS office/integration Lab	VENUS offices	VENUS Headquarters	11-Sep-2007 00:00:00	48°28.1154'	-123°18.5622'	-63.0
1552	VIP-08	Central Node (SI)	Saanich Inlet	27-Sep-2008 00:00:00	48°39.0707'	-123°29.1772'	97.4
30126	MTC NEPTUNE Integration Lab	NEPTUNE offices	NEPTUNE Headquarters	12-Feb-2009 02:00:00	48°38.9532'	-123°26.7723'	-20.0
30154	VIP-09	Central Node (SI)	Saanich Inlet	15-Feb-2009 17:00:00	48°39.0707'	-123°29.1772'	97.4
117461	On Ship or Shore	VENUS offices	VENUS Headquarters	26-Sep-2009 00:00:00			0.0
117488	BC Axis (POD 1)	Barkley Canyon	North East Pacific	30-Sep-2009 23:32:50	48°19.0030'	-126°03.0087'	984.4
117489	VENUS office/integration Lab	VENUS offices	VENUS Headquarters	30-Sep-2009 23:42:36	48°28.1154'	-123°18.5622'	-63.0

Figure 7.7 The importance of keeping track of observatory assets: the successive positions of a VENUS CTD over the course of about 3 years. To properly describe its data, all possible information on every instrument must be kept and made available.

The considerations above imply that the amount of information to be recorded, maintained and presented to users about any component of the observatory is tremendous and usually much more considerable than what casual observers would imagine. In Figure 7.7, we show a web screen with information about a single instrument on the VENUS observatory. The number of tabs showing the categories of information indicates the complexity of the structures involved. The example of the "Site" tab indicates the successive locations of that particular instrument throughout its lifetime at the observatory. Even though that instrument has been part of VENUS for just three years, the successive positions it has occupied over time indicate its complex history. Clearly, this information will be essential in explaining the data stemming from its sensors going forward.

7.6.3.5 Data access and analysis

An observatory is nothing without instruments and, likewise, raw data are useless without calibration and processing. Raw data coming from instruments almost always need to be processed before being presented to users so as to make sense even to non-field specialists.

Beyond visualization, there is a need to perform comparison and correlations, look for trends, and mine the noise for hidden features. Most often, this involves multiple sensors, long time series and large data volumes.

Moreover, observatories are bound to produce data volumes several orders of magnitude (perhaps >6) larger than previously obtained for the region they cover, and for this to happen every year. The data will not come and go, but will accumulate in archives. However, it is not anticipated that the number of scientists in a position to deal with that data volume will actually increase, and certainly not in proportions similar to the increase in data volume. This widening gap (Figure 7.8) is a real challenge that requires novel software approaches and new ways for users to approach the data.

The traditional model consists of providing search screens and a result download facility. Users end up with a number of files that they will have to individually process, usually in the isolation of their labs, with local resources and locally developed or installed software.

The model that is currently emerging involves a shift away from the search-download-process approach. The concept of Web 2.0 currently in vogue can be summarized with two keywords: collaboration and contribution. This participatory approach invites people to work together towards the solution of a problem and to contribute any information, data or interpretation they might have obtained to the existing archive. This approach has led the authors of this article to build something quite different, where users use web browsers to perform all activities related to the scientific process. Through this web interface, they can:

- collaborate online with remote colleagues and students. Data volumes are so large and so multidisciplinary that it will often be necessary to seek out the support and advice of colleagues in different disciplines to support a particular project execution. The new collaborators will not be co-located and may work at different times but a "work space" is available for all members of a work group to perform all tasks from data search and examination all the way to the redaction of the final paper
- search and sift through data, but this is done using other criteria and sources of information than previously available such as annotations provided by "crowd-sourcing" activities and data from other observatories using interoperability concepts

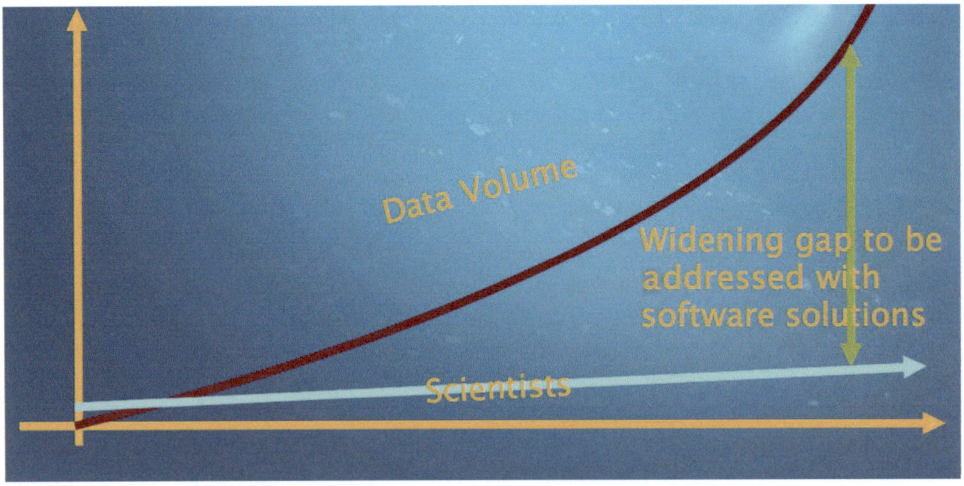

Figure 7.8 Graph indicating the widening gap between the amount of data accumulating about the oceans and the available scientific manpower to extract new knowledge from it. The text shows the approach that has been adopted at NEPTUNE Canada to address that issue.

- avoid downloading data: data processing facilities on the Grid or in a compute "Cloud" will be available through privileged links with the observatory. Data processing software libraries and templates are available to run against the data. Instead of downloading data, the new concept favours the upload of new code on the server. New code can first be tested, refined and later made available on the server for all to use.

With computing facilities becoming utilities and with storage capacity available on the network, there is no need to spend money and time maintaining one's own infrastructure. Shared infrastructures are always available at the other end of the high-capacity network. Shared cyber-infrastructure can also be located close to green power sources to avoid the need to supply them with "dirty" electricity, such as that produced by coal-fired plants.

The NEPTUNE Canada DMAS team is implementing the features described above under the name "Oceans 2.0" with the help of CANARIE Inc., the provider of research networking in Canada.

7.6.3.6 Remote use of underwater assets

Cabled observatories, or any system with direct access links to the rest of the world using, e.g., satellite communication, offers not only the ability to get real-time data from installed sensors, but also provides users with the potential for remote control of the facility. Interactive use of instruments such as pan and tilt cameras becomes a possibility provided there is a very low latency in the communication link.

Preprogrammed/scheduled routine surveys are also a very attractive possibility with the ability to run coordinated programs on multiple observatory locations from a central place on land.

The ability to install "smart" instrumentation allows for autonomous behaviour and reliable reaction to events in case network connectivity is not guaranteed or not permanent. In some cases, a lack of connectivity is unavoidable: e.g., gliders and autonomous underwater vehicles (AUV) will have to possess internal computation capabilities commensurate to their need to control their motion and find their way home.

The two concepts of cabled observatory and of smart instruments are not mutually exclusive. On the contrary, they complement each other and allow for novel experiments where the cable observatory hosts docking stations near places of interest. Observatory instruments (e.g., seismometers) can detect events in reaction to which a programme will be uploaded to an autonomous vehicle that will immediately embark on a survey mission of the area likely to be affected by the event. That can happen even if the scientist in charge of the experiment is not actively interacting with the infrastructure.

7.7 Summary

Ambitious ocean observatories are being established in many places around the world. Much of the emphasis, care, effort and budgets are going toward the underwater hardware. While it is a natural reaction to give most of the attention to these engineering aspects, one should not ignore or underestimate the importance of sound software systems to support all of its various aspects: data collection, archival and distribution but also all of the observatory management functions. If an observatory is to be defined as a facility (which it should if the investment is significant), a clear responsibility lies with the sponsors of the system to make sure that it is reliable, accessible to the largest possible audience and that it produces and keeps trustworthy data. Moreover, its management must be efficient and require a reasonable amount of effort, staff and financial resource. Software technologies represent a means to implement the above constraints.

This chapter makes the case for a sound system design and architecture and highlights most of the functions that have to be provided by the observatory – through its cyber-infrastructure – if it is to deserve the qualification of "research facility".

Name	Location	Operating/owner institution	Principal purpose	Size, instruments
Aloha	HI, USA	University of Hawaii	General purpose	100km, a few
Antares	Nice, France	European/French consortium, including Ifremer	(High-Energy Physics) Neutrino observatory	Oceanographic instruments piggy-back on neutrino observatory
Bonne-Bay	NL, Canada	Memorial University	Coastal science	Decommissioned
Donet	Nankai, Japan	JAMSTEC	Seismicity and tsunamis	90km, in deployment
LEO-15	NJ, USA	Rutgers University	Shallow waters	~12km, ~20
MARS	CA, USA	MBARI	Technological research	60 km, ~ 10 instruments
Martha's Vineyard	MA, USA	WHOI	Coastal sciences	~2km
NEPTUNE Canada	BC, Canada	University of Victoria	Coastal and deep ocean sciences	800km, 100+ instruments
VENUS	BC, Canada	University of Victoria	Coastal sciences	40 + 3km, 20+ instruments
NEMO-SN1 (EMSO operating node)	Off Eastern Sicily (Western Ionian Sea)	INGV-INFN	Geo-hazards, bioacoustics, oceanography	28km, 22 instruments

Table 7.3 A non-exhaustive list of cabled ocean observatories that have been or are still in existence at the time of writing. Several more observatories with short cables do exist. Many are being deployed or in the design phase.

7.8 Non-exhaustive list of ocean observatories

In this section, a non-exhaustive list of existing or out of commission ocean observatories is provided in Table 7.3. The table attempts to provide some of the salient characteristics of the system. The information provided here will possibly be incomplete or out-of-date. It relied on information available on these initiative's respective websites.

Reference

Albrecht R. and Pirenne B. (1994) Astrophysics and Space Science, Vol. 228, Proceedings of the 4th United Nations/European Space Agency Workshop, Cairo, Egypt, 27 June–1 July 1994, UN/ESA Workshops Vol. 6, p. 379–383.

Glossary of acronyms

ADCP	Acoustic Doppler Current Profiler
AUV	Autonomous Underwater Vehicle
CTD	Conductivity-Temperature-Depth sensor
DHCP	Dynamic Host Configuration Protocol
DMAS	Data Management and Archiving System (for NEPTUNE Canada and VENUS)
GF	Ground Fault or current leak back into seawater
ICT	Information and Computer Technologies
IEEE	Institute of Electrical and Electronics Engineers
IO	Implementing Organization
IP	Internet Protocol
ISO	International Standards Organization
ISP	Internet Service Provider
IT	Information Technology
MAC	Media Access Control (address)
NEPTUNE	North-East Pacific Time-series Underwater Network Experiment
NetCDF	Network Common Data Form
NMS	Network Management System
NTP	Network Time Protocol
O&M	Operations and Maintenance
OGC	Open Geospatial Consortium
OOI	Ocean Observatory Initiative
OS	Computer Operating System
PI	Principal Investigator

PTP	Precision Time Protocol
QA/QC	Quality Assessment/Quality Control
RDBMS	Relational Data Base Management System
RFC	Request for Change
SensorML	Sensor Markup Language
SLA	Service Level Agreement
SONet	Synchronous optical networking
SWE	Sensor Web Enablement
US	United States of America
VENUS	Victoria Experimental Network Under the Sea
VLAN	Virtual Local Area Network
VPN	Virtual Private Network
XML	Extended Markup Language

8 Long-term subsea observatories: Comparison of architectures and solutions for infrastructure design, interfaces, materials, sensor protection and deployment operations

Y. Auffret[1,3], J. Blandin[1], D. Choqueuse[1], C. Compère[1], L. Delauney[1], J.-F. Drogou[2], P. Jégou[1], C. Lévêque[2], J.-F. Rolin[1] and P. Valdy[2]

8.1 Introduction

These subsea observatory infrastructures may have different architectures; the ESONET and EMSO (see www.esonet-emso.org) projects have identified five main types of observatories:

1. Cabled observatory: distance from coast up to approx. 400km with several nodes and several junction boxes (Barnes et al., 2008; Rolin et al., 2007; Phibbs, 2009).
2. Simplified cabled observatory: distance from coast up to approx. 20km with one node or one junction box. This architecture is frequently implemented to extend an existing cabled infrastructure.
3. Stand-alone acoustic link observatory, very mobile observatory (Blandin et al., 2007).
4. Stand-alone observatory with an electro-optico-mechanical umbilical between seafloor and surface, mobile observatory.
5. Stand-alone winch observatory for measurements near polar ice caps (Arctic) or in other extreme environments.

This chapter focuses primarily on buoy-based (stand-alone) and cabled observatories.

1 IFREMER – French Research Institute for Exploitation of the Sea, Plouzané, France

2 IFREMER – French Research Institute for Exploitation of the Sea, La Seyne/Mer, France

3 ISEN-Brest – Engineering school, SEACom lab., Brest, France

8.2 Comparison between observatory architectures

The specific case of subsea observatories in deep seas around Europe shows a diversity of scientific and logistic requirements. One type of observatory is often better suited for one site than another. Four main criteria determine suitability:

- The duration of observation: does the scientific requirements for the implementation of a long-term observatory imply permanent monitoring (geohazard, climate change) or 5–10-year time series to establish scientific understanding of a poorly known site (biological and chemical processes, etc.)? For the latter case, cabled observatories are not indispensable.
- The need for real-time data: when near real-time data is sufficient, enabling technical verification, configuration modification and downloading, cabled observatories are not indispensable.
- The site of implementation: when the scientific demonstration of the interest of one site vs. another has not been agreed upon within the scientific community, the decision to implement a cabled observatory is not mature enough.
- The budget (Capital and Operational Expenditures): cabled observatories may incur lower expenses in the very long-term compared to stand-alone observatories, whose lifetime is estimated at 4–5 years. However, the initial cost is high for cabled observatories and must be justified by the other main criteria.

8.2.1 Vertically cabled architecture

In this architecture, the power and data transmission from/to the seabed units are provided by a surface buoy, through a cable. The cable ensures at least the mechanical function of keeping the buoy within its watch circle radius and the electrical function of bringing the surface-generated power to the seabed. The data transmission may be performed over the cable either electrically or optically, if a high data rate is required. Depending on the power level required by the seabed units, the power might be generated on the buoy by a set of solar panels, wind generator(s) or, in certain demanding applications, by diesel engines (see http://www.mcs.harris.com/oceannet/). It must be noted, however, that no reference of long-term operational implementation of this last solution has been found in the literature.

The real challenge of a such architecture lies in the long-term survival of the electro-(opto-) mechanical cable to the efforts, movements hence fatigue imposed by the current and surface wave conditions. In this regard, the most advanced studies and operational references to date are provided by the Monterey Bay Aquarium Research Institute (MBARI), which successfully conducted an 18-month experiment using such a cable in the Eastern Pacific (Han and Grosenbaugh, 2006; Frye et al., 2004).

8.2.2 Non-cabled architecture

Among the various types of subsea observatories, non-cabled infrastructures are worth considering, because of the following features that make them attractive for numerous scientific areas and situations:

Figure 8.1 Architecture of a non-cabled observatory.

3 Technical supervision infrastructure; 6 Land-sea communication segment – Satellite communication; 7 Buoy (node); 8 Acoustic telemetry - Branch extension of the network; 9 Junction box; 10 Link to instruments; 11 Individual instrument.

- They cost one to two orders of magnitude less than cabled infrastructures, making them more affordable for modest-sized multidisciplinary teams.
- They can successively be implemented at different sites within their lifetime, with an acquisition of time series ranging from months to several years during each deployment.

Non-cabled architectures represent a complementary alternative to very long-term cabled infrastructures. For instance, they can be very useful for investigating a site before deciding whether to build a heavier, cabled infrastructure. In non-cabled architectures, observation data are typically transmitted from the seabed to users by means of an acoustic modem, the receiver being placed either on a ship for occasional transmission or on a relay-buoy towards a permanent data server on shore (Figure 8.1) (Blandin et al., 2007).

The main limitation of this type of architecture lies in the modest amount of data it can transmit between servicing interventions on the seabed. This amount of data is the product of the quantity of embedded energy times the energy efficiency of the acoustic modem installed on the seabed. The energy efficiency is expressed in bits/J and is the ratio of the number of bits successfully transmitted to the surface and the corresponding number of Joules consumed on the seabed. Hence, the energy density (in mass and volume) of the embedded batteries and the choice of acoustic modem are of paramount importance.

IFREMER recently performed tests at 2200m water depth and showed that the latest state-of-the-art acoustic modems eventually reach a mean energy efficiency in the long run of around 90 bits/J, for a range of 3500m. Given a reasonable amount of embedded energy (i.e., 4000 Wh) devoted to data transmission on a seabed station, this efficiency allows the transmission of around 1.2 gigabytes of data before a servicing intervention takes place. This opens the door to retrieving subsets of data from ADCPs or still cameras, for instance, that have been so far accessible only through cabled infrastructures. In the case of seabed-to-shore data transmission, this improvement in the energy efficiency of acoustic modems tends to shift the bottlenecks in the transmission chain to the satellite data transmission segment.

8.2.3 Cabled architecture

In general, the choice of implementing a cabled network rather than a non-cabled network depends on bandwidth and power requirements and on the duration of operations.

If it is farther than the conventional wireless communication range (approx. 10km to 50km for 3G, WiMAX, long-range WiFi, etc.), the communication segment between a non-cabled network and the shore usually depends on a satellite link with a limited bandwidth compared to a cabled network. In addition, if the expected operational duration is more than 4–5 years, it is better to install a cabled network to reduce maintenance costs, avoiding buoys, batteries, etc. (Figure 8.2).

A cabled network can be considered as an Internet extension under the sea. Despite its cost, this type of infrastructure provides many interesting features when permanent monitoring is required:

- "virtually unlimited" power and bandwidth
- various data communication interfaces (Ethernet, serial, etc.)
- absolute clock synchronization (NTP or PTP IEEE1588 V2 for Ethernet instruments and Pulse Per Second (PPS)+NMEA (time code) for serial instruments)
- plug-and-play capabilities
- to improve reliability, the instrument driver can be located on shore if the infrastructure has a permanent link between the instrument and the shore station.

The technology of subsea telecommunication cables is well defined and mastered by industry. However, there are new constraints due the requirements of ocean research and particle astrophysics research disciplines to implement instruments underneath, above and on the seabed.

Figure 8.2 Cabled observatory architecture.
3 Land-based termination of sea infrastructure; 6 Land-sea communication segment; 7 Node from branching unit to node/extension; 8 Branch extension of the network; 9 Junction box; 10 Link to instruments; 11 Individual instrument

Power supply is an additional difficulty that pioneers in subsea observatories faced (Harris and Duennebier, 2002) due to critical components such as subsea wet-mateable connectors and critical operations such as plugging and unplugging. One example of this problem is set out below.

8.2.3.1 Architecture and mechanical design of a node and a junction box

Example: extending the ANTARES neutrino telescope to include an environmental observatory.

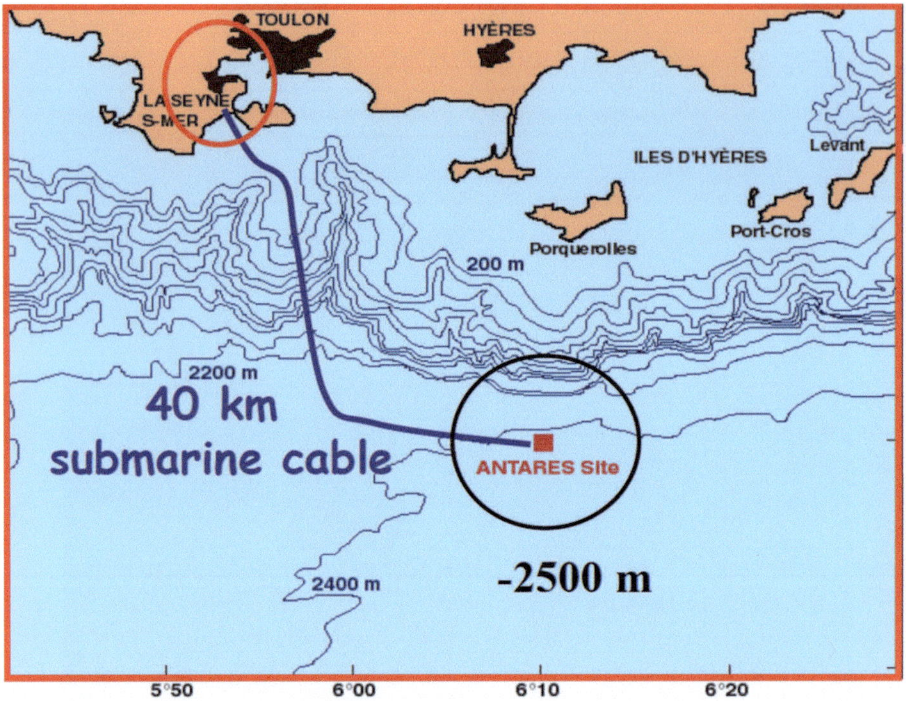

Figure 8.3 Location of the ANTARES site.

The international ANTARES collaboration (Astronomy with a Neutrino Telescope and Abyss environmental RESearch) aims to detect and study the production of high-energy neutrinos in the Universe. Located a few miles south of Toulon (France) (Figure 8.3), the ANTARES infrastructure is also a permanent marine observatory providing high-bandwidth, real-time data transmission from the deep sea, and thereby represents a unique opportunity for workers in the geosciences and marine environmental sciences. Extension of the initial detection line network by means of a secondary junction box (SJB) (Figures 8.4 and 8.5) offers a range of solutions to connect scientific equipment at 2500m depth, including:

- power and Ethernet ports for local experimentation (from 10m to 1000m from the SJB)
- Deep Sea Net (DSN) single-fiber optic link which allows data transmission of remote autonomous sensors connected to a DSN node.

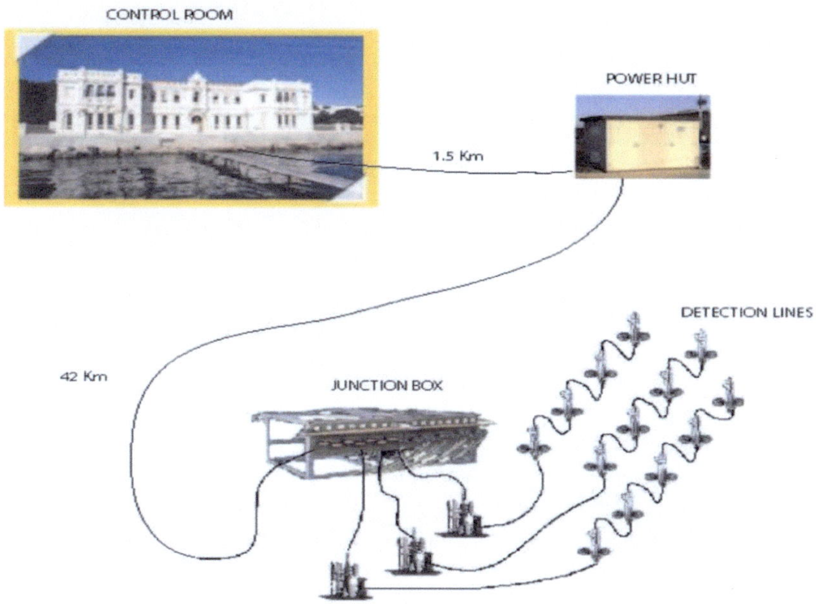

Figure 8.4 ANTARES telescope architecture.

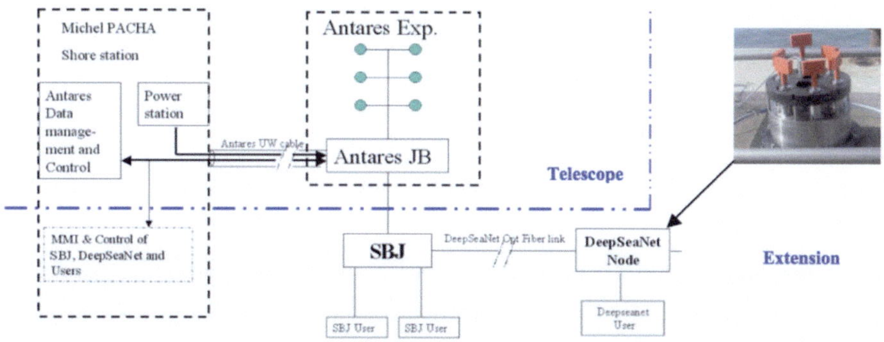

Note: "SJB users" can either connect instruments to the secondary junction box or connect several sensors to a scientific instrument interface module (SIIM) which is connected to the SJB, as done in the NEPTUNE Canada (Barnes et al., 2008; and see http://www.neptunecanada.ca) and VENUS Canada (http://www.venus.uvic.ca) observatories.

Figure 8.5 Architecture of the ANTARES telescope and extension from the "SJB" secondary junction box.

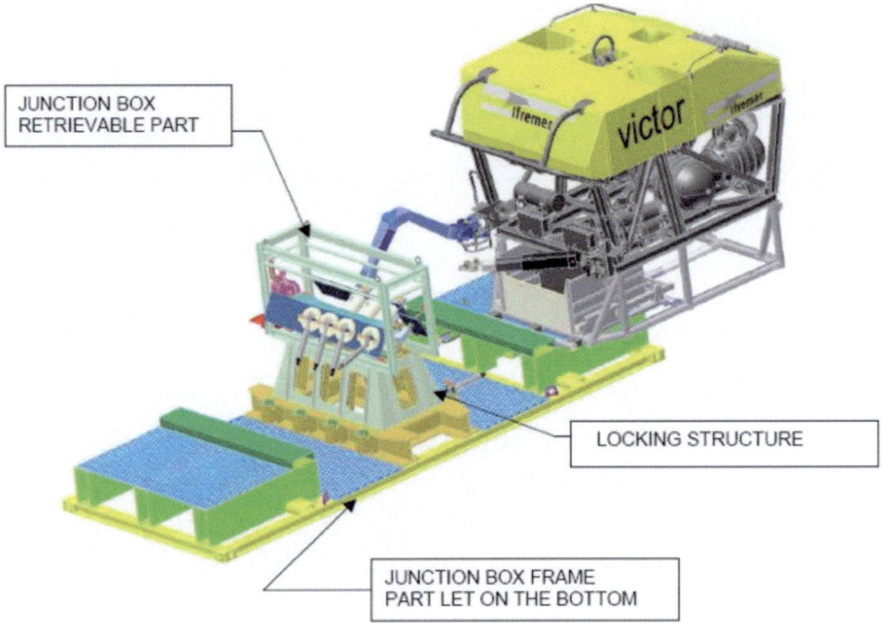

Figure 8.6 Overview of the ANTARES secondary junction box.

Mechanical design overview

The particular requirements of a secondary junction box design must:

- allow port connection and disconnection by an ROV
- ensure a permanent location for the node despite the recovery and re-installation operations required for its maintenance
- ensure a 20-year lifetime for any permanent bottom component and a 5-year lifetime for recoverable components
- ensure local jumper protection against damage from operations around the node
- offer receptacles at the sea bottom for unplugged jumpers during recovery of the node for its maintenance.

SJB design solutions

The node structure is split into two units in order to meet the above specifications:

- an anchor frame composed of a 6m × 2.5m platform, with a locking structure in its center to position the junction box and an ROV pad with a docking bar on each side of the locking structure

Figure 8.7 Plugging guide.

- a recoverable module with a receptacle form that is complementary to the locking structure on the anchor frame to hold the pressure-housing cylinder of the junction box electronics (Figure 8.6).

Several operational tests carried out with an ROV helped in defining the best port arrangement.

First, both input and output ports must be distributed along the two opposite generative lines of the housing cylinder. Second, for connector settings, to facilitate connection with an ROV, the best results were obtained when the connector axis was positioned with a downside inclination of about 30° (instead of a horizontal axis). This allowed for easier manipulator operation and provided a better view for operating from a camera. Third, a specific plugging guide (Figure 8.7) was defined to allow for the:

- easy localization using an entrance cone
- previous orientation of pins
- possibility of releasing the connector in a stable condition before inserting pins
- insertion of the pins only by pushing on the connector.

8.3 Recommendations for signals, protocols and connector pin-out between infrastructure and instrumentation

Several European countries as well as ESONET/EMSO academic and industrial partners already have various experiences with non-cabled and cabled observatories. These experiences and the requirements for ESONET and EMSO projects must be taken into account.

It is nearly impossible to meet the technological requirements and wishes of all partners involved, but the most interesting and the most challenging task is to define a generic interface and generic services between the undersea infrastructure and all the instruments attached to the infrastructure.

These generic interface and services are required to improve the compatibility and the interoperability between the different types of subsea observatories. These specifications will help partners to design new, specific instruments and will improve their existing design and/or technologies to be compliant with the upcoming ESONET/EMSO observatories.

The easiest way to ensure compatibility and interoperability between networks and instruments is to define the interface requirements and a generic connector pin-out between infrastructures and instruments. The specifications so defined for the interface are standards which must be met to obtain the ESONET/EMSO Label.

Tables 8.1–8.7 describe minimal recommendations in terms of:

- Table 8.1, voltage/power
- Table 8.2, data interfaces
- Table 8.3, clock synchronization
- Table 8.4, plug-and-play capabilities
- Tables 8.5, 8.6 and 8.7, connector pin-out.

Voltage/power	**Cabled observatory**	**Stand-alone acoustic link observatory**
375–400 VDC	Yes	Not a requirement
48 VDC	Yes	Yes
15 VDC	Yes	Yes
Remote power control (power up and power down)	Yes	Yes
Additional services: ground fault detection, power management, short circuit management...	Yes	Yes
600 W min. available (at least one port)	Yes	No
200 W min. available (at least two ports)	Yes	Not a requirement
20 W min. available (at least four ports)	Yes	Yes

Table 8.1 Recommendations for infrastructure interface (instrument side): voltage and power.

Data interfaces	Cabled observatory	Stand-alone acoustic link observatory
RS-232 (3 wires TX, RX, GND)	Yes	Yes
RS-422 (4 wires)	Yes	Yes
RS- 485 (2 wires)	Yes	Yes
Ethernet 100BaseT (copper)	Yes	Yes
Ethernet 1000BaseT (copper)	Yes	Not a requirement
Ethernet 1000LX or 1000ZX (fiber)	Optional - only for long-range extension	Not a requirement

Table 8.2 Recommendations for infrastructure interface (instrument side): data interfaces.

Clock	Cabled observatory	Stand-alone acoustic link observatory
NTP/Ethernet instrument: Network Time Protocol (2–10 ms)	Yes	Yes (generated with a local clock reference)
PTPv2/IEEE1588-2008/Ethernet instrument: Precision Time Protocol (better than 1 microsecond)	Yes	Yes (generated with a local clock reference)
Underwater GPS clock emulation PPS (pulse per second) + NMEA Time code/Serial instrument	Yes	Yes (generated with a local clock reference)
Local time stamping service	Yes	Yes

Table 8.3 Recommendations for infrastructure interface (instrument side): clock synchronization.

Plug-and-play capabilities	Cabled observatory	Stand-alone acoustic link observatory
1-wire tag for metadata information, compatible with any data interface (inside instrument or cable)	Optional - Not a requirement	Optional - Not a requirement
Zeroconf - industry-standard Zero-Configuration Networking solution for Ethernet instruments	Yes	Yes
Puck for serial instruments	Yes	Yes

Table 8.4 Recommendations for the infrastructure interface (instrument side): plug-and-play capabilities.

Junction box connector pin-out	Signal
1	GND_48VDC_15VDC
2	48VDC
3	15VDC
4	Data1
5	Data2
6	Data3
7	Data4
8	1-wire tag for metadata information (optional plug-and-play capabilities)
9	PPS1
10	PPS2
11	NMEA1
12	NMEA2

Table 8.5 Recommendations for the infrastructure interface (instrument side): connector pin-out (copper) for short-range connection.

GND_48VDC_15VDC: Power supply ground for the instrument
48VDC: 48 VDC power supply for the instrument
15VDC: 15 VDC power supply for the instrument
Data[1..4]: Serial (RS-232/RS-422/RS-485) or Ethernet 100BaseT
1-Wire tag: metadata information memory that contains Transducer Electronic Data Sheets (TEDS) such as identification, calibration, correction data, manufacturer-related information, technical parameters...
PPS[1..2]: Pulse Per Second output
NMEA[1..2]: NMEA serial interface that contains Time Code ASCII message.

Junction box connector pin-out	Signal
1	GND_375VDC
2	375VDC
3	VDSL2 - 1
4	VDSL2 - 2

Table 8.6 Recommendations for the infrastructure interface (instrument side): connector pin-out VDSL2 (copper) for long-range connection.

GND_375VDC: Power supply ground for the instrument
375VDC: 375 VDC power supply for the instrument
VDSL2 1 & 2: Twisted pair for VDSL2

Junction box connector pin-out	Signal
1	GND_375VDC
2	375VDC
Optical fiber 1	Optical fiber 1 - Ethernet 1000LX or Ethernet 1000ZX
Optical fiber 2	Optical fiber 2 - Ethernet 1000LX or Ethernet 1000ZX

Table 8.7 Recommendations for the infrastructure interface (instrument side): connector pin-out hybrid (copper and optical fiber) for long-range connection.

GND_375VDC: Power supply ground for the instrument
375VDC: 375 VDC power supply for the instrument
Optical fiber 1 & 2: Ethernet 1000LX or Ethernet 1000ZX

8.4 Long-term deployment: Materials for subsea observatories

In the definition of subsea observatories, from the scientific vision to cost estimates, long-term sustainable operation is a key issue and probably constitutes a limitation. Any improvement in ageing of materials and components merits further study. Based on the experience gained mainly in the offshore industry (Roche, 1978) where materials are exposed in deep sea (up to 2000m) for a long time (up to 25 years), in the framework of ESONET/ EMSO program guidelines for the choice and selection of materials for the long term, deep-sea exposure was proposed.

From these guidelines, based on existing literature (Hasson and Crowe, 1988), feedback from previous experiences and an initial reference paper (Rolin J.F., Bompais X., May 2011), the following points are addressed:

- Description of the deep-sea environment highlighting the parameters acting on the degradation process (pressure, oxygen concentration, fouling, etc.)
- Review of materials used in service deep-sea applications:
  - metallic material
  - non-metallic materials
- Associated protection
  - cathodic protection, including the choice and design of cathodic protection systems
  - paints and coatings
- Assembly
- Sub-components: moorings, landers, connectors, junction boxes, pressure housing, buoyancy, etc.
- Guide for performance evaluation.

It must be mentioned that experienced feedback is of primary importance for the use of most of the materials. Attention must be paid on the assembly of dissimilar material where galvanic corrosion could be initiated. Design of structure, choice of material and the associated protection method should be performed by skilled people in order to avoid reinventing solutions already evaluated.

In general, the long-term behavior of the material is not or is only weakly affected by pressure. While pressure loading must be taken into account in terms of mechanical loading on hulls for instance, the intrinsic properties of the material are generally not affected by pressure. Materials subject to creep, such as some thermoplastics, can be easily replaced with materials having more suitable characteristics.

In order to avoid most of the problems of corrosion the use of cathodic protection for metallic structure is strongly recommended and that will limit the use of "exotic" and "expensive" materials which could be proposed (e.g., high corrosion resistance alloys, titanium alloys, ...).

For polymer and composite materials (Choqueuse and Davies, 2002) good knowledge of behavior in water has to be considered in order to limit the risk of long-term detrimental degradation processes (hydrolysis, etc.). However, it must be noted that degradation of such material is generally thermally activated and, except in really specific areas (black smokers, etc.), temperature is low enough (around 4°C) to avoid initiation of degradation processes.

An approach based on accelerated test using time-temperature equivalence can be used to predict long-term performance of polymeric materials (Choqueuse and Davies, 2008; Davies and Evrard, 2007); however, good knowledge of degradation phenomena is needed in order to guarantee pertinence of the accelerated test.

For specific materials such as syntactic foam, synthetic fiber for mooring cable, knowledge of long-term behavior has already been addressed through specific programs related to the offshore industry (Grosjean et al., 2009; Desrombise et al., 2008).

8.5 Long-term deployment: Biofouling protection for marine environmental devices and sensors

Today, many autonomous marine environment monitoring networks have been set up throughout the world, and they commonly use sensors for dissolved oxygen, turbidity, conductivity, pH or fluorescence units, etc. and, in some specific situations, underwater video systems such as cameras, video equipment and lights.

For deep-sea research, down to 3000m, specialized autonomous stations, either cabled or storing their own energy, perform physico-chemical measurements and record pictures and movies. Some areas of interest include fumes at hydrothermal sites (Auffret et al., 2009). For these applications, the experienced autonomy is up to one year, while durations of decades are planned for subsea observatories. The compactness of the monitoring stations is also a crucial point because a remotely operating vehicle deploys the equipment.

These autonomous monitoring systems are affected by a well-known phenomenon in seawater conditions, called biofouling (Lehaitre et al., 2008). The major goal for this

Figure 8.8 Turbidimeter probe after 14 days in Mont St.-Michel Bay (France) in the spring.

Figure 8.9 Fluorometer after 90 days in the roadstead of Brest (France) in the summer.

equipment is to provide in reliable and real-time measurements without costly frequent maintenance. In deep-sea conditions, servicing is nearly impossible to carry out more than twice a year and most often servicing intervals run to one year. For unprotected coastal applications, a two-month servicing interval is a generally accepted minimum for economically viable in situ monitoring systems (Blain et al., 2004; Satpathy, 2006). Consequently, deep-sea systems without efficient biofouling protection are useless. Antifouling protection must be applied to sensors and underwater communications equipment based on acoustic technologies.

Marine biofouling species can grow very rapidly during their productive period and lead to poor data quality in less than two weeks. As shown in Figures 8.8 and 8.9, the biofouling species involved can vary greatly from one location to another.

This development of biofouling species very often gives rise to a continuous shift of measurements. Consequently, measurements can be out of tolerance and data may be unusable. Video systems such as cameras, video equipment and lights are also disrupted by biofouling. Images become blurred and lights lose efficiency because light intensity decreases due to the screen effect of biofilm and macro-fouling organisms.

8.5.1 Biofouling protection by "controlled" biocide generation: Localized seawater electro-chlorination system

This technique, widely used in seawater-cooling systems for industrial applications (Satpathy, 2006), has been adapted to protect in situ oceanographic sensors from biofouling. For deep-sea applications, only the sensor interface will be protected, hence the term "localized". The biocide is generated by seawater electrolysis. Using this technique, a powerful biocide, hypochloric acid, which can be concentrated as much as required, is used on the sensor interface. This technique has numerous advantages:

- Biocide generation is controlled. Consequently, the quantity of biocide can be adjusted and on/off periods can be arranged as needed. On/off periods are useful for setting up biocide-free periods to perform measurements in suitable environmental conditions. Moreover, the control of the biocide concentration makes it possible to adapt the biocide concentration as a function of the colonization of fouling organisms which might be encountered.
- The energy needed for such systems is fully compatible with autonomous coastal monitoring systems and deep-sea autonomous monitoring stations.
- The system is very robust and reliable since there are no moving parts.
- The system is easily adaptable to existing sensors, even at great depths.
- Manufacturers can directly integrate the system to the sensors.

As shown in Figure 8.10, the system is made of an electrode placed around the sensor interface or around the optical port. This electrode is connected to an electro-chlorination unit. The electro-chlorination unit can be in a separate electronic container or can be integrated inside the instrument.

This biofouling protection technique has been successfully used in many in situ coastal monitoring systems (Delaunay et al., 2002) – even immersed at shallow depths of 2–3m,

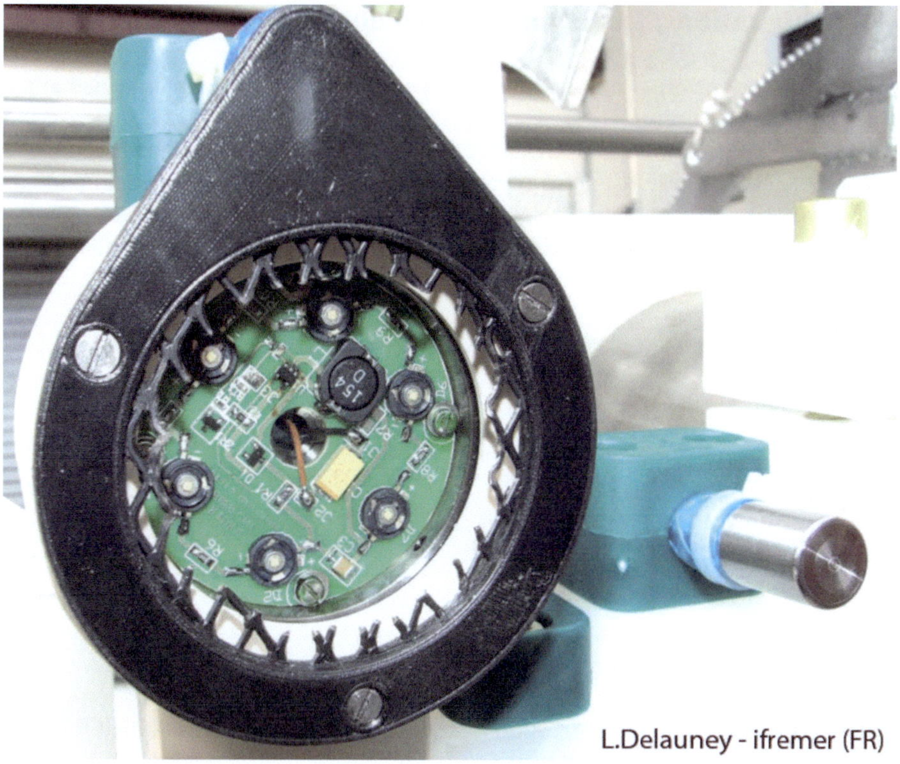

Figure 8.10 Biofouling protection of a Tempo-Mini (Grosjean et al., 2009) lighting device by localized seawater electro-chlorination (IFREMER).

where biofouling development is intense – as well as for medium depth stations (Marvaldi et al., 2006) (15–100m) or for stations down to 2000m where biofouling can appear near hydrothermal smokers (Sarrazin et al., 2007).

The local chlorination technique has been applied to various instrument technologies, optics (turbidity, fluorescence, oxygen, video and lights), electrodes (conductivity) and glass membranes (pH). To implement the system on instruments, it is necessary to check for a possible adverse effect on measurements.

The system can be easily adapted to many kinds of instruments. The energy needed is compatible with autonomous monitoring (one D cell for 3 months). The system is now used by IFREMER for autonomous monitoring, providing high-quality measurements at a reasonable servicing frequency.

8.6 ROV operations: Deployment and maintenance operations

How can a network of observatories around Europe be deployed and maintained at a reasonable cost?

Different concepts of observatories have been described in the previous sections. The main components of these different types are cables, nodes, junction boxes, permanent cabled instrumentation, short-term scientific packages, moored buoys with or without electro-optical-mechanical (EOM) cables. Whatever the type of observatory, the construction and maintenance phases of an underwater observatory have various steps, each one calling for specific skills:

- site surveys
- module lifting and lowering to seabed
- cable laying and underwater connections
- inspection and maintenance works.

These operations require, at different levels:

- naval means
- underwater intervention systems (ROVs, etc.)
- techniques and procedures.

Considering the cost of at-sea intervention, the ESONET project has identified four objectives to demonstrate that technologies and tools for construction, servicing and repair exist, and are reliable at an affordable cost.

- **Precise technical conditions to increase the number and flexibility of welcoming vessels (ships and submersibles).**

  Clearly, the intervention equipment (submarines, ROVs, AUVs) currently available to the scientific community will dictate the initial intervention procedures implemented on underwater observatories. It is foreseen that the involvement of the European initiative EUROFLEETS (http://www.eurofleets.eu) will guarantee that any European ROV can be easily launched from any of several European ships to avoid unnecessary transit times and/or transport costs.

- **Facilitate exchanges of sensors, equipment and payloads** among the different vehicles, by employing "standard" interfaces.

- **Provide scientists and operators with standard proven procedures or recommended practices** to operate equipment in a safe and productive way.

  Recommended practices for each specific construction and maintenance activity will provide useful guidelines for field-proven procedures, adequate construction equipment and support vessel, crew training, and methodology to conduct operations at sea.

  The recommended practices for inspection and maintenance should lead to the establishment of a comprehensive inspection and maintenance plan, for the duration of the plant to optimize availability and minimize downtime and repair work.

The offshore industry has gained extensive experience in designing, installing and maintaining underwater structures. The use or adaptation of some of these already developed standards would be of great benefit in some cases. However, a direct transfer of the offshore approach to the installation of underwater scientific modules may be inappropriate for the scientific community in terms of equipment availability and installation costs. In contrast to the offshore approach which demands uncompromising equipment performance, a more flexible approach should be adopted by the scientific community by evaluating performances of alternative methods, such as smart rigging and lower cost of support ships, versus sea state capability and global cost.

- **Provide recommendations for training (crew) and testing (procedures).**
  The successful underwater installation, maintenance and recovery of observatory platforms and infrastructures require a shared operations scheme, allowing scheduled maintenance periods to be carried out independently of a single operation team and hardware platform. This will require common understanding of both the process of dedicated intervention on a given structure as well as the underwater hardware interfacing, shared among different possible scientific ROVs and vehicle operators. To achieve this common understanding, training will be necessary aside from seagoing cruises and expensive "on-site" operations. The goal of training should be to ensure quality work that is comparable among all participating operators, so that the observatory missions and their associated scientific tasks can be successfully carried out over long-term deployment periods.

8.7 Conclusion and next steps

In 2010–2012, the experiments of the main partners of ESONET have supported the Preparatory Phase of EMSO by an acquisition of practical knowledge. It is centered around the continuation of the Esonet Demonstration Missions. As an example, IFREMER with its partners is deploying:

- a buoy-based network to monitor hydrothermal vents in mid-Atlantic ridge in real time as part of the ESONET demonstration mission (MOMARSat, MOMAR-D), a step forward in view of permanent monitoring at the EMSO Azores node
- an ANTARES Extension called Secondary Junction Box (SJB), a step towards establishing the EMSO Ligurian Sea node infrastructure
- monitoring equipment on the seismically active fault for the EMSO Marmara Sea node
- specific instrumentation on the NEPTUNE Canada cabled network: Tempo-Mini (Auffret et al., 2009) at the Endeavour site for hydrothermal monitoring and a pore pressure sensor at the ODP1027 site
- coastal monitoring network of a marine protected area near Brest, France for the Marine e-Data Observatory Network (MeDON) project (http://medon.info).

Marine biofouling is a natural process occurring in the ocean as soon as a material is moored underwater and biofouling effects on marine instrumentation are numerous.

The most convenient antifouling strategies must be determined during a rigorous qualification phase and should not interfere with the measurements produced by the instruments or with the environment. Particular attention must be paid to calibrations of the sensors with the antifouling system in place, before and after deployment.

For subsea observatories, biofouling protection on sensors must be involved since deployment duration is commonly long (one year). Active biofouling protection is very convenient since a free biocide production period can be managed in order not to disrupt the sensor and the measured environment. Local chlorination is a response to these requests.

In marine protected areas, where environmental impact must be minimized, subsea and sea-level networks based on mixed cabled and non-cabled architectures can take advantage of both systems. For this, low-power micro-cable networks with low-power meshed wireless networks will be combined. The power requirement is thus optimized to reduce the impact on marine life: high-sensibility cameras without lights in shallow water, passive acoustic monitoring, etc. This new concept of a coastal monitoring network for marine protected areas forms the basis of the MeDON project (http://medon.info).

Collaborations at European level on all these technologies and corresponding standards will be continued in the ESONET-Vi consortium that will follow the ESONET Network of Excellence. It will be applied by the EMSO large-scale infrastructure.

Acknowledgments

IFREMER extends its sincere thanks to the European Commission for funding ESONET (FP6 Network of Excellence), EMSO (Preparatory Phase -FP7-INFRASTRUCTURE 2007 Project 211816) projects and the MeDON project within the scope of INTERREG IV A France (Channel) – England cross-border project.

Special thanks to ANTARES, NEPTUNE and VENUS Canada and OBSEA networks for fruitful collaboration on subsea networks.

References

Auffret Y., Sarrazin J., Sarradin P.M. et al. (2009) TEMPO-Mini: A custom-designed instrument for real-time monitoring of hydrothermal vent ecosystems. MARTECH09, Conference Proceedings, Vilanova I La Geltru, Spain.

Barnes C.R., Best M.M.R. and Zielinski A. (2008) The NEPTUNE Canada regional cabled ocean observatory. Sea Technology 49(7), 10–14.

Blain S., Guillou J., Tréguer P., Woerther P. and Delauney L. (2004) High frequency monitoring of the coastal marine environment using the MAREL buoy. J. Environ. Monit. 6, 569–575.

Blandin J., Vangriesheim A., Legrand J., Coail J.Y. and Leildé B. (2007) Real time transmission of current and turbidity data from the near bottom Var canyon system.

Martech 2007 Second International Workshop on Marine Technology, 15–16 November 2007, Barcelona, Spain.

Choqueuse D. and Davies P. (2002) Durabilité des polymères et composites pour application sous marine, Revue des composites et des matériaux avancés.

Choqueuse D. and Davies P. (2008) Ageing of composites in underwater applications. In: R. Martin (Ed.) Ageing of Composites. Cambridge: Woodhead.

Davies P. and Evrard G. (2007) Accelerated ageing of polyurethanes for marine applications. Polymer Degradation and Stability 92(8), 1455–1464.

Delauney L., Lepage V. and Festy D. BRIMOM (Biofouling Resistant Infrastructure for Measuring, Observing and Monitoring), EC-Project number EVR1-CT-2002-40023.

Desrombise G., Van Scoors L., Davies P. and Dussud L. (2008) Durability of aramid ropes in a marine environment, OMAE 2008-57199

Frye D., Hamilton A., Grosenbaugh M., Paul W. and Chaffey M. (2004)Deepwater mooring designs for ocean observatory science. Marine Technology Society Journal 38(2), 2004.

Grosjean F., Bouchonneau N., Choqueuse D. and Sauvant V. (2009) Comprehensive analyses of syntactic foam behaviour in deepwater environment. Journal of Materials Science 44(6), 1462–1468.

Han S. and Grosenbaugh M. (2006) On the design of single-point cable-linked moorings for ocean observatories. IEEE Journal of Oceanic Engineering 31(3), 585–598.

Harris D.W. and Duennebier F.K. (2002) Powering ocean-bottom observatories. IEEE Journal of Oceanic Engineering 27(2), 202–211.

Hasson D.F. and Crowe C.R. (Eds) (1998) Materials for Marine Systems and Structures. New York: Academic Press.

Lehaitre M., Delauney L. and Compère C. Real-time coastal observing systems for marine ecosystem synamics and harmful algal blooms. Chapter 12 in: Biofouling and Underwater Measurements. Geneva: UNESCO.

Manov D., Chang G.C and Dickey T. (2004) Methods for reducing biofouling of moored optical sensors. Journal of Atmospheric and Oceanic Technology 21, 958–968.

Marvaldi J., Coail J.Y. and Legrand J. (2006) ROSE, CMM'06 Caractérisation du Milieu Marin, 16–19 Oct 2006.

Phibbs P.G. (2009) System engineering at the edge of a cabled ocean observatory, Oceans 09 MTS/IEEE conference in Bremen.

Roche M. (1978) Protection Contre la Corrosion des Ouvrages Maritime Pétroliers.

Rolin J.F., Bompais X. (May 2011) ESONET LABEL Definition http://wwz.ifremer.fr/esonet_emso/content/download/42247/574588/file/Deliverable_D68_esonet-label-definition_1.0.pdf

Rolin J.F., Waterworth G., Gillooly M., Priede I.G., Pfannkuche O. and Nolan G. (2007) Design of the scientific cabled network between Ireland and Porcupine Seabight and Abyssal Plain, study site of the European Seafloor Observatory Network Implementation Model (ESONIM)- Baltimore - SubOptic 07.

Satpathy K.K. (2006) Biofouling and its control in power plants cooling water system. International Conference on Recent Advances in Marine Antifouling Technology (RAMAT) 6–8 Nov. 2006, Chennai (Madras), India.

Sarrazin J., Blandin J. et al. (2007) TEMPO: A new ecological module for studying deep-sea community dynamics at hydrothermal vents. OCEANS'07 IEEE Aberdeen Conf. Proceedings (IEEE), 2007/0.

Web resources

ESONET and EMSO website: www.esonet-emso.org

Harris/OceanNet website: http://www.mcs.harris.com/oceannet/

NEPTUNE Canada website: http://www.neptunecanada.ca

VENUS Canada website: http://www.venus.uvic.ca

EUROFLEETS project website: http://www.eurofleets.eu

MeDON project website: http://medon.info

Glossary

3G	Third generation mobile telecommunications standards
ADCP	Acoustic Doppler Current Profiler
ANTARES	Astronomy with a Neutrino Telescope and Abyss environmental RESearch
DSN	Deep Sea Net
EMSO	European Multidisciplinary Seafloor Observatory http://www.emso-eu.org
ESONET	
EUROFLEETS	Alliance of European Research Fleets project
IFREMER	French Research Institute for Exploration of the Sea
MeDON	Marine e-Data Observatory Network project
MOMARSat & MOMAR-D	Monitoring the Mid-Atlantic Ridge projects, ESONET demo mission
NEPTUNE Canada	NorthEast Pacific Time-Series Undersea Networked Experiments project
NMEA	Electrical and data specification for data exchange between marine electronic devices

NTP	Network Time Protocol
PPS	Pulse Per Second
PTPv2 / IEEE 1588-2008	Precision Time Protocol
PUCK	Protocol from Marine Plug-and-Work Consortium (MBARI) http://www.mbari.org/pw/puck.htm
ROV	Remotely Operated Vehicles
SIIM	Scientific instrument interface module
SJB	Secondary Junction Box
TEDS	Transducer Electronic Data Sheets contain identification, calibration, correction data, manufacturer-related information, technical parameters...
VDSL2	Very-high-speed digital subscriber line 2
VENUS Canada	Victoria Experimental Network Under the Sea
WiFi	Wireless Fidelity - Wireless communication standards
WiMAX	Worldwide Interoperability for Microwave Access
Zeroconf protocol	Industry-standard Zero-Configuration Networking solution

9 Development and demonstration of a mobile response observatory prototype for subsea environmental monitoring: The case of ROSE

J. Marvaldi[1], J. Legrand[1], J.F. Masset[1], L. Delauney[1], M. Nicot[2], B. Darbot[2], Y. Degres[3], M. Jouannic[3], F. Cabioch[4], J. Guyomarch[4], C. Lietard[5], P. Billand[5], A.M. Caujan[5], S. Hibral[5] and C. Laot[6]

9.1 Introduction

The severe oil spill hazards that endangered environment and ecosystem equilibria in western European seas such as "Erika" (1999) and "Prestige" (2002), led the French government to open research funds under the RITMER (Réseau de recherche et d'Innovation Technologiques pollutions Marines accidentelles et conséquences Ecologiques suR le littoral) Program. The ROSE project (December 2003–May 2006), for Acoustic Network for Wreck Monitoring, was selected in this program. Its main objective was to define and study a new monitoring system adapted to such polluting wrecks, and based on deployment of genuine buoyant midwater stations using radio-electric link to communicate with the onshore control station.

The ROSE program development was carried out according to three axes: an Operation axis, a Sensor axis and a System axis.

1 Ifremer Centre de Brest, Plouzané, France

2 Underwater Acoustic Division, SERCEL Brest, France

3 NKE instrumentation, Hennebont, France

4 Cedre, Brest France

5 Altran-Atlantide, Brest, France

6 Telecom Bretagne, Technopôle Brest Iroise, Brest France

The scope of the "Operation axis" was to define the requirements to be satisfied and system functional specifications. The first step was to analyze documented cases of maritime accidents at the origin of effective or latent pollutions. The corresponding tasks led to the definition of methods and means necessary to operate the system and definition of the assistance to be provided to the system operators mainly as regards display of collected data.

The scope of the "Sensor axis" was to define the parameters to be measured and select existing sensors and qualify selected sensors.

The scope of the "System axis" was to draw up the technical specifications of sub-systems, implement the studies, build a prototype and test it in a sea water tank on shore before finally carrying out a demonstration at sea.

The ROSE project was conducted in partnership between Ifremer (French Institute for Ocean Research), as coordinator, Cedre (the French expert organization in pollution), companies Altran-Atlantide, NKE instrumentation, Sercel Underwater Acoustic Division and the telecommunications engineer school Telecom Bretagne.

9.2 System specifications

The ROSE system was initially conceived to be deployed around a sunken ship in order to detect pollution emissions, quantify them as much as possible and measure oceanographic parameters to forecast of pollution spreading and later interventions on the wreck. The ROSE system consists of a set of autonomous sea bottom stations which constitutes the nodes of an underwater acoustic network, a buoy moored within the deployment area of stations and communicating with them by acoustics, and an on-shore control station communicating with the buoy via a radio-electric link (Figure 9.1).

9.2.1 Functional specifications

ROSE system functions consist of "data acquisition and recording", "data transmission", and "system control and data processing".

Data acquisition and recording are performed by sea bottom stations. Pollution and environmental parameters, as well as monitoring of station operating status, are measured and recorded by the station electronic unit and can be unloaded after station recovery.

In order to enable data transmission, a bi-directional near-real-time link is implemented between the subsea stations and the on-shore control station. The link consists of an acoustic network between the sea bottom stations and the surface buoy, which is extended by a radio-electric link from the buoy to the on-shore control station. The acoustic network can be implemented either through a direct path station-buoy (Blandin et al., 2003), or a multi-segment path transiting through one or several other stations (Beranzoli et al. 2004). The choice of the radio-electric link between the buoy and the on-shore control station was ruled primarily by the range of available links as mobile phone ("short range"), VHF (Very High Frequency) radio ("mid range") or satellite ("long range"), and by operating costs. Information circulating through the communication system are either "periodical data", "alarms data" or "commands data". Additional data transmission can also be possible via

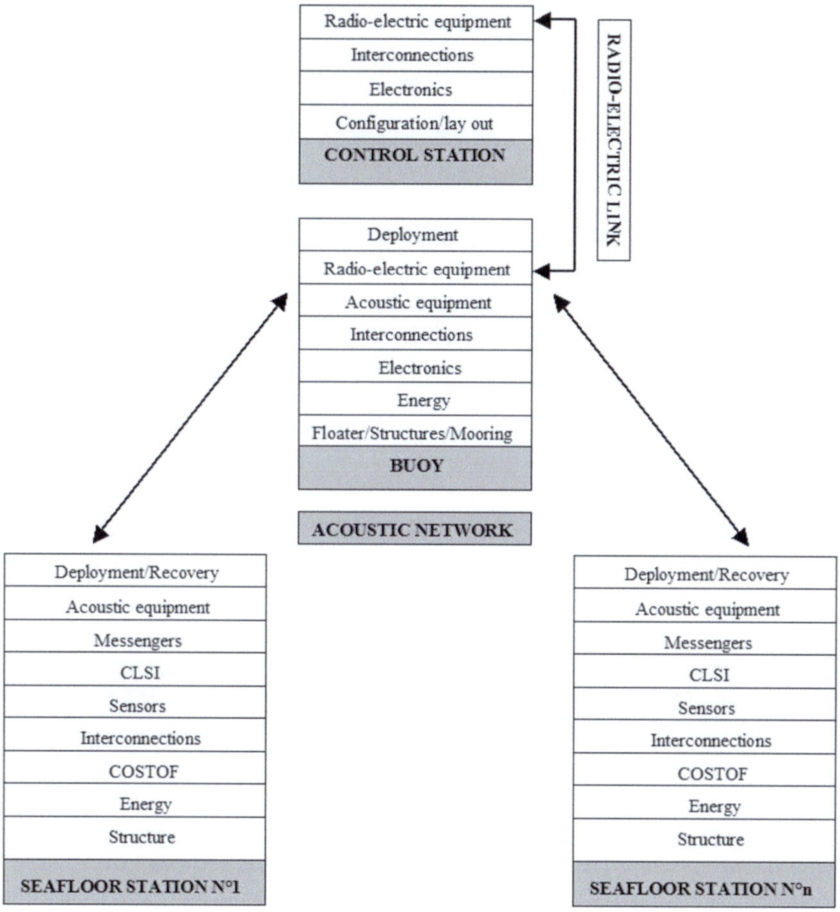

Figure 9.1 Concept of the ROSE system breakdown structure. Several seafloor stations (n) can be deployed at distances of a few km (determined by an acoustic transmission simulation). In the case of the Douarnenez ROSE demonstration (see text for details), only two stations (n=2) were deployed.

CLSI (Contact Less Serial Interface) and Messengers. CLSI is an inductive link enabling communication with the station electronic unit either in air or in immersion.

The system control, including the data processing, is implemented from an on-shore control station and can be packaged as an easily transportable set.

Station deployment conditions may differ as regards to the lay-out of stations around the wreck or within the perimeter to be monitored, and the lay-out of sensors. Three cases can be envisaged for the sensor lay-out. The first one corresponds to sensors placed on the station structure. The second case corresponds to sensors fixed on a floating line attached to the station, which is a necessary option when the sensor measurement field has to be positioned at a given height above sea bottom. The last case corresponds to sensors positioned at some distance from the station.

Two station deployment methods can be used: a cable lowering and free fall descent with or without assistance of a ROV (Remote Operated Vehicle) at the bottom. For the cable deployment and recovery, the method requires a precise positioning system on the ship, a convenient winch and cable, and an acoustic release at the cable end. Free ascent recovery requires the fitting of an acoustic release and expendable deadweight on the station. The usual procedure to deploy buoy and mooring is the so-called "free fall deadweight last" method.

Design orientations were worked according to several requirements. Components and subsystems have been qualified according to the French standard NF X 10800 (http://www.ifremer.fr/tsi/qualite/specif/31SE09.pdf) for oceanographic equipment. To cope with the wide range of operational situations, design efforts on sensor payload and embarked energy were directed in order to make it more adaptable and adjustable. One of the requirements is the integration of some previously-implemented and proved developments, for instance, the station electronics and message protocol developed for the ASSEM project (Blandin et al., 2003), and the acoustic and radio-electric communication system based on Sercel acoustic equipment MATS 200 Net including self-adaptive algorithms (Labat et al. 1998; Laot et al., 2001).

9.2.2 Technical specifications

Technical specifications provide a general description of constituent sub-systems and are the basis for development of the ROSE demonstration system.

9.2.2.1 Acoustic network

The acoustic network can be composed of up to N nodes. Depending on applications, entry point to the network can be the relay buoy or a portable equipment on a ship. Data transmission operates from a source node to a destination node and gets through one or several intermediate nodes when the direct path is impossible (e.g., nodes are too distant or an acoustic mask stands between them). Only two nodes are activated at a given time to prevent blockage situations. Each node is identified by its address and a routing table defines its possible connections to adjacent nodes. Routing tables are set up once the network physical configuration is known, and can be modified from the on-shore control station in case of modification of configuration.

Transmission principles in the acoustic network are based on data exchanges, which take place according to three phases: emission, reception and sleep (Lurton, 2002). The emission phase begins with a sequence of channel listening to ensure channel availability. If the channel is free, a sequence takes place to send service messages and characteristics of data to be transmitted. Then begins the proper phase of data transmission, which ends with a connection closure sequence.

The reception phase begins by receiving service messages and characteristics of data to be received. After interpreting the service messages and data characteristics, the phase of reception and interpretation of transmitted data begins, which ends with a connection closure sequence.

The sleeping phase enables the reduction of node energy consumption. Transmission parameters, emission power and data flow rate are adapted to environmental conditions and type of data to be transmitted to optimize energy consumption.

Acoustic transmissions generally use the 10–14 kHz frequency band, which offers a good trade-off between range and data flow rate, which can range from 20 to 200 bits/s according to the chosen modulation. Three types of robust modulations with respect to multiple acoustic paths and Doppler effect are implemented. An error correcting code enables the increase of acoustic transmission reliability. Three levels of emission power are available from 173 to 185 dB/micro Pa by a step of 6 dB. To increase range or economize on energy, one may use directional transducers in place of omnidirectional ones.

Constraints and limits of this system are, first, the maximum possible distance between adjacent nodes, which is imposed by the physics of acoustic propagation. It is of prime importance to conveniently position the acoustic transducers on stations so as to prevent any acoustic masking by sea bottom relief. A second constraint is inherent in the half-duplex type of acoustic transmission, which means that a node cannot emit and receive simultaneously. A third constraint is linked to channel selection procedure. Emission phase begins by a listening sequence during which the node performs a measure of ambient noise. If noise level is over a defined level, transmission will be delayed. If noise level maintains high, data will only be emitted after a defined number of noise measurement sequences. So a noisy acoustic channel will limit reactivity and reliability of network communications. Finally, the data flow rate limitation at some hundreds bits/s will necessarily put limits to transmissible data volume.

9.2.2.2 Radio-electric link

The architecture of the radio-electric link is composed of an electronic unit connected to an antenna on the relay buoy and of a PC connected to an antenna or to phone network (depending on the type of link used) set up on the on shore control station. The type of link used can be mobile phone, VHF radio or satellite.

Mobile phone transmission is chosen for coastal deployment, up to several kilometers from shore. Currently, available modems enable the choice between GSM (Global System for Mobile communications) or GPRS (General Packet Radio Service) systems. GPRS advantages are the direct access to the Internet, the possibility of having several control stations on shore, a higher data flow rate – 40 kbits/s against 9.6 kbits/s for GSM – and payment charge based on transmitted data volume. The TCP-IP layer in place in GPRS between the GSM layer and the Internet layer offers several interesting functions. For this type of link, the on-shore control station equipment is composed of a PC loaded with the control MMI (Man Machine Interface) and of an external Switched Phone Net modem. Buoy equipment consists of a GSM or GPRS modem connected to an application board, which may be included in the modem. Transmission uses either the 900 or 1800 MHz frequency bands, but most recent modems can use both and automatically shift from one to the other in case of connection difficulty.

The VHF radio transmission is a convenient solution for a continental shelf (up to some ten of kilometers). For this choice, the on-shore control station equipment is composed of a radio modem and a VHF antenna. A connection to the phone network enables another operator to communicate thanks to the MMI on PC. The on-shore control station must be in optical view with the buoy. Buoy equipment consists of a radio modem, an application board and a VHF antenna. Commonly authorized VHF frequencies are in the band of some hundreds of MHz. Data flow rate is typically around 9600 bits/s and emission power is generally limited to 5 Watts.

Satellite transmission is, however, the only solution for long-distance communications. In that case, the on-shore control station equipment is composed of a PC and an Internet connection or a connection to a dedicated server. Buoy equipment consists of a satellite modem, an application board and an antenna.

The operating principles are based on the activation of the radio-electric link by an operator from the on-shore control station for transmissions of commands or data requests to subsea stations. The link is activated by the buoy for periodical transmission of data from subsea stations or transmission of alarms. Communication takes place in two phases. The first phase enables the transmission parameters to be configured. Then a call phase enables the connection and transfer of the data. In the case of failure of connection after expiration of a time-out, a defined number of attempts are made. If all these attempts fail, the entire cycle is resumed, either up to implementation of the connection or a defined number of times.

Some constraints and limits have, however, been observed with this system. The first constraint relates to implementation of communication at the initiative of the operator, when the connection is done through the Internet and requires use of a TCP-IP layer. In case of a GPRS system, it is necessary to use the "ring indicator", a modem option in GSM mode, to establish the communication. The problem is not always easy to solve where communication is through the Internet by VHF radio or satellite links.

The second type of limitation is the telephone and VHF radio range. Satellite communication solves this limitation but is highly energy consuming (except for the Iridium system) and particularly expensive.

9.2.2.3 Information flows

Information exchanged through this system is classed into three groups: "Periodical", "Alarms" and "Commands" data.

Periodical data correspond to sensor and technical data messages emitted a regular intervals. Each sensor has its own Measurement Period (MP) (set by software between three and 65,535 minutes). At the end of each MP, the sensor emits a data package comprising raw data to be stored locally and reduced data to be acoustically transmitted and also stored. These data are time stamped by the station with a precision of one minute (sometimes up to one microsecond). Each station has also its own Acoustic Transmission Period (ATP) programmable from some minutes to infinity. At the end of each ATP, reduced data are emitted towards the buoy and thus forwarded to the on-shore control station. Subsea stations also elaborate periodical technical messages containing information on station operating status. When a station is equipped with Messengers (small floats released to reach surface and emit), periodical data transfer towards a Messenger is performed through a Messenger control unit at a predetermined frequency.

Alarms data are sent to the buoy when a sensor's data leave a defined mini or maxi range (or abnormal technical data). Alarms data are immediately forwarded to the on-shore control station.

Commands data can be emitted from the on-shore control station, a ship on site equipped with an acoustic unit and from a submersible or ROV equipped with a CLSI inductive coupling. Commands can request the emission of periodical data, to configure various parameters, and to activate sensor or technical measurements or various devices.

Data message generic structure is composed of a destination address (8 bits), a source address (8 bits), a message header (8 bits, 1 ASCII character) and a message content.

Transmission principles differ from one segment of the communication link to the other.

The Internet segment enables a group of users to have access to the data server from a PC connected to the Internet. Communication principles are those of the Internet.

The radio-electric segment allows communication between the server and the buoy. Whatever the type of radio-electric link, it uses functions provided by the TCP-IP layer. An encapsulation process is performed by the application at the point of origin of the message. An inverse process is performed at the destination point.

As regards the subsea acoustic network, this layer attributes an address to each station, and each sensor of a station must also have an address so that it can be identified by an 8-bit address. Within the acoustic network, messages are delimited in the same way as in the radio-electric segment. Data structure, length and content are adapted to data format transitting in the network layer. Each transmission is carried on as a unique packet, so that a data packet loss does not damage the integrity of the following one. Size of packets transmitted by the acoustic network is optimized in terms of energy and in all cases the size of messages will not exceed 2–4 kbytes.

9.2.2.4 Sea bottom stations

The concept of a sea bottom station is given in Figure 9.2. As utilization of 4000 meters capable equipment is a penalty for "coastal" or "shelf" stations in terms of mass and cost increase, two station depth classes have been defined, one less than 400m and one down to 4000m. Differences relate to the type of buoyancy and possibly the dimensioning of pressure-resistant containers.

The sea bottom station structure consists of a welded assembly of tubes in sea water-resistant aluminum alloy. Zinc anodes are bolted on to the structure to ensure protection against corrosion. The structure's configuration must comply with several requirements. Components have to be laid out according to their functionalities, to ensure the station's equilibrium and to facilitate mounting and dismounting operations. The station must also be compact to facilitate handling at sea, and robust while keeping mass as low as possible. It is powered by primary battery packs delivering both 12 and 24 Volts housed in one or two watertight containers of standard dimensions, each housing up to 120 LSH 20 batteries, and with a maximum service depth of 4000m.

Station operation is controlled by an electronic unit, called COSTOF, housed in a pressure-resistant container. COSTOF hardware is composed of the basket board, which can be connected up to eight bridgeboards, each of which be fitted with an extension 64 Mbytes memory board. Five bridgeboards are dedicated to sensors, the other three are dedicated to the MATS 200 Net acoustic modem, the Messenger control electronics and the CLSI con-

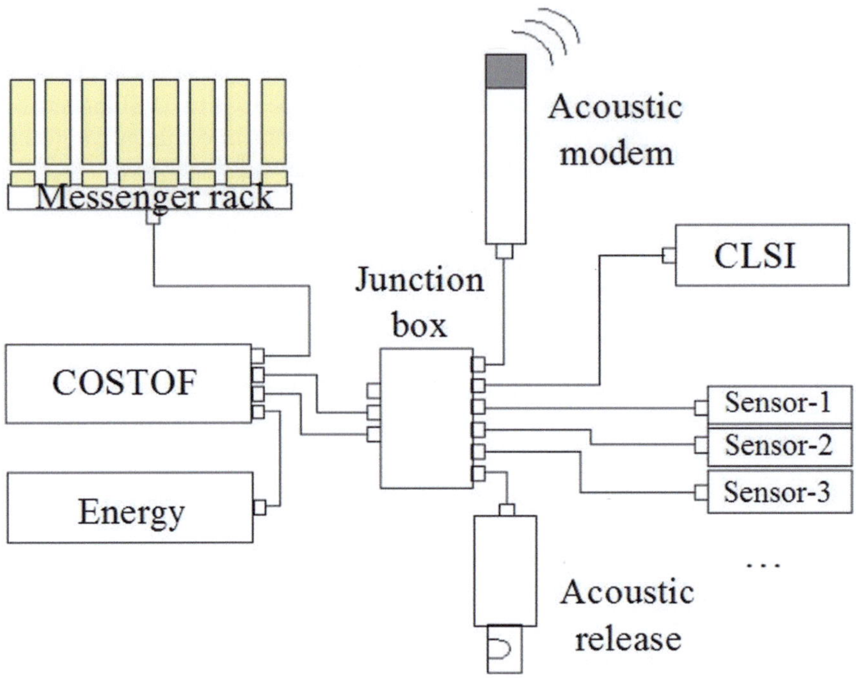

Figure 9.2 Concept of the sea bottom station. The standard energy container is constituted by hard anodized 7075 aluminum alloy; external diameter is 176mm, internal diameter is 145mm and length is 850mm. The COSTOF container is constituted by hardly anodized 7075 aluminum alloy; external diameter is 176mm, internal diameter is 145mm and length is 560mm.

nection. The CLSI bridgeboard is also coupled to an ancillary board ensuring monitoring of technical parameters. Communication between bridgeboards is ensured by the CAN field bus with CAN Open protocol.

Where COSTOF stores sensor data, it is also in charge of monitoring battery voltage (12 Volts for the COSTOF and sensors and 24 Volts for the acoustic modem) and the station remaining autonomy. It is also in charge of checking the absence of water entries (through internal nitrogen pressure controlled at a stable value), and the good operating status of each bridgeboard.

The communication equipment of the sea bottom station includes acoustical equipment, consisting of a Sercel MATS 200 Net modem with an omnidirectional transducer integrated into the modem container, a CLSI and a Messenger rack equipped with eight slots. Energy and data interconnections are implemented using pressure compensated cabling in oil and underwater connectors.

Fittings of a sea bottom station depend on station deployment and recovery mode. Fittings attached to the station comprise the hardware used to latch the station to handling cable or a ROV, additional buoyancy, the expendable deadweight and its release device. The additional buoyancy is used when it is necessary to modify the station weight in water during deployment operations. This buoyancy is brought about by hollow plastic floats down to 400m, or syntactic foam blocks down to 4000m.

9.2.2.5 The buoy

The buoy is composed of a foam float protected by a polyurethane (or equivalent) material cover, an aerial structure carrying the warning equipment and the antenna of the radio-electric link, and an immersed metallic structure weighted at the foot to ensure buoy stability.

The buoy can be linked to the sea bottom station by two types of mooring line. The "Small depth type" (50–100m water depth) is made up of a segment of large diameter steel cable, a short segment of nylon cable acting as tension damper, a segment of steel chain and the anchoring deadweight. The "Great depth type" (over 100m water depth) is made up of successive segments of different cables (large diameter steel cable, nylon cable and polypropylene cable of light positive buoyancy), a float supporting one to two acoustic releases in parallel and a segment of steel cable attached to the anchoring deadweight. Cable segments are terminated by splices on thimbles and are linked together by sets of shackles and rings. Anodes are bolted on shackles to ensure protection against corrosion.

The buoy is supplied by rechargeable acid batteries in a watertight container, and connected to solar panels if necessary.

As with the sea bottom station, the buoy's communication equipment is composed, of a Sercel MATS 200 Net modem connected to a separate omnidirectional transducer. The transducer can be attached to the steel cable segment of the mooring line at a depth of around 10m. A variant solution can be to fix the transducer on a tube some meters long, which can be lifted for cleaning or intervention on the transducer. In this case, the transducer, which is closer to the sea surface, must be protected from wave noise by a foam mask. The buoy is also equipped with radio-electric equipment, including the modem and the antenna fixed on to its aerial structure.

Electronic parts of the buoy are composed of an application board, which controls data exchange between the acoustic network and the radio-electric link. A common watertight container houses the application board and the acoustic and radio-electric modems.

As regards buoy deployment, the "Free fall deadweight last" procedure is used. The procedure consists of putting the buoy at sea first, paying out the mooring line and finally dropping the deadweight. In case of a "Great depth" mooring, the procedure must be numerically simulated to predict the point where the deadweight will reach the seabed and be implemented accordingly.

9.2.2.6 On-shore control station

The on-shore control station remotely monitors and controls the network of subsea stations via the relay buoy. It is generally positioned on shore but it must be portable aboard a ship. A PC houses the supervisory application of the acoustic network and the control application of the stations. Communication equipment depends on the type of radio-electric link in place.

9.2.2.7 Messengers

Messengers can transmit alarms and a selection of data messages through the ARGOS satellite system. Although Messengers are considered as expendable devices, all the stored data can be unloaded if the Messenger is recovered.

A Messenger is composed of a 4000m pressure-resistant tubular container, which houses the data storage and transmission electronics, the energy source, a float which ensures its ascent to the surface and buoyancy, a submersible ARGOS antenna and a releasable latching device to the Messenger rack. An interface electronic board ensures data transmission by inductive link between the Messenger bridgeboard in COSTOF and one of the eight Messengers in the rack.

The Messenger has several functions: data processing, release and ascent and data transmission by satellite or through the inductive link.

The data processing is realized through an inductive half-duplex link, which ensures a minimum data flow rate of 9600 bits/s to the Messenger. A secured protocol guarantees exchange integrity. The Messenger bridgeboard first sorts the data into two classes: current data and data to be emitted by satellite. Then it forwards the data to the Messenger through the interface board. Data storage in the Messenger is implemented in a nonvolatile memory of 4 or 8 Mbyte capacity. Data volume to be transmitted by the satellite ARGOS network can range up to some tenths of Kbytes, depending on the Messenger's available energy, emission period of ARGOS messages and number of repetitions of each ARGOS message to guarantee high enough reception probability.

Messenger release is activated by the Messenger bridgeboard after a predetermined time, emission of an alarm or reception of a command message through the acoustic modem or CLSI interface of the station. Release of the Messenger is based on the electrolysis of the Inconel wire, which latches Messenger to the rack. The wire's diameter is determined according to a trade-off between mechanical resistance and consumption duration, set at 15 minutes approximately. Messenger ascent is ensured by a syntactic foam float. To limit ascent duration and subsequent drift of Messenger with current, ascent speed is set between 0.6 and 1 m/s, that is less than one hour from 2000m.

Data transmission is carried out after a defined time lapse starting with the release command. The Messenger emits frames towards the ARGOS satellites. The time lapse is intended to guarantee Messenger surfacing before starting emission and can be adjusted to planned operations, for example 2 minutes for an ascent from 100m at 1 m/s or 66 minutes for an ascent from 4000m. Data are emitted in frames with a recurrence chosen to guarantee a fair probability level of reception. For example, emission every 40 seconds during 80 hours guarantees a recurrence of 53 for the emission of 134 frames each containing 31 bytes, which corresponds to the emission of about 4 Kbytes of data. A lower recurrence will permit the emission of a greater volume of data, for example 32 Kbytes with recurrence of 6, in return for a higher risk of non-received.

If the data transmission by satellite does not work, it is possible, after the Messenger has been recovered, to unload all the stored data through the Messenger CLSI link using a dedicated MMI on a PC. PC-implemented MMI enables each Messenger to be configured and tested individually through its inductive link. Messengers are accessible for system tests via the station CLSI link, when they are stored in the rack.

Energy capacity provided to a Messenger enables it to sustain the energy autonomy of the five following life cycles: a one-year storing after delivery; a one-year stay on the station at the sea bottom before release; at least two loadings of 8 Mbytes memory full capacity; release and emission towards the ARGOS satellite with prescribed redundancy; and at least two readings of the 8 Mbytes full memory capacity after recovery.

9.3 Study and construction of a prototype system

Technical specifications were aimed at meeting all possible cases of system utilization. On the other hand, the scope of prototype system was chosen to be limited to a demonstration in coastal area, with water depth less than 50m, for a duration of one or two months and using a coastal oceanographic ship of the Ifremer fleet for deployment and recovery. Moreover, the system is assembled from both specifically-built components and existing pieces of equipment. It is composed of two stations, one equipped with two Messengers, a simplified acoustic node and a relay buoy.

9.3.1 Seafloor stations

The initial idea was to adapt the concept of the ASSEM (Array of Sensors for long-term SeabEd Monitoring of geohazards) station (Figure 9.3) (Blandin et al. 2003).

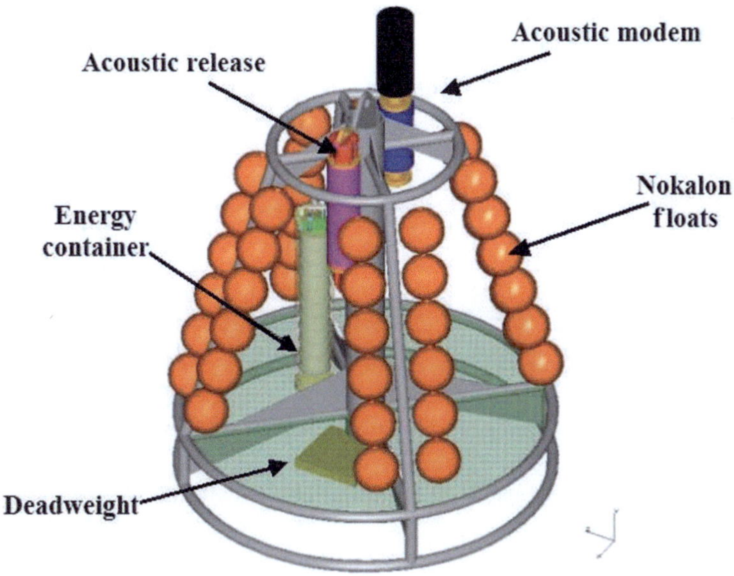

Figure 9.3 ASSEM bottom station.

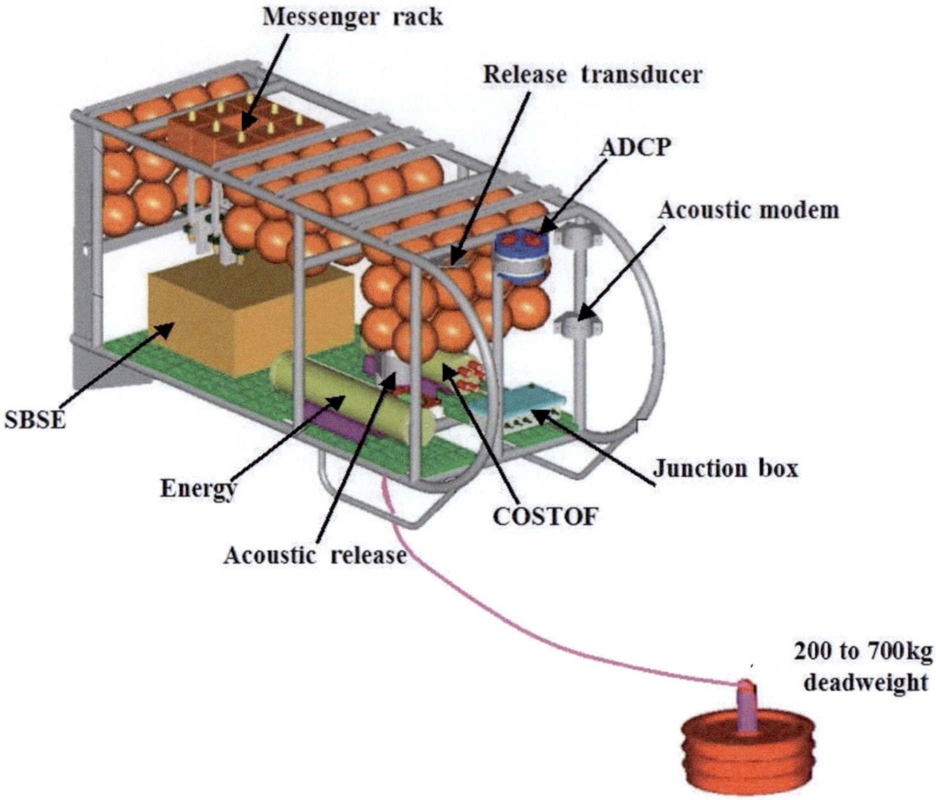

Figure 9.4 ROSE station: general view. Station structure is built of 5754 aluminum alloy tubes. Overall station dimensions are: length: 2.70m × width: 1.00m × height: 1.30m. Mass about 300kg and buoyancy around 30 daN. Maximum service depth: 6000m. Service load: 1250 daN. Release load: 500 daN. Tension in the 3m-long mooring line ranges from 60–150 daN in current from 2 to 3 knots, which is well under the release load. For the demonstration mission, a 200kg deadweight made of thick steel plates was used.

To meet the possible requirement of carrying out measurements some meters above the seafloor, it had been devised to set sensors on a buoyant line unfurled above the station. Further thinking led to the genuinely innovative concept of a buoyant station anchored above seafloor, the ROSE station (Figure 9.4).

The station consists of an original box-shaped structure made buoyant by the addition of floats in the upper part and anchored to the seafloor by a mooring line. The line attachment point is located in the first third of the structure. Components are laid out on the station so as to keep it in a horizontal attitude. Vertical plates on the rear enable the station to align itself with the direction of the current. A curved plate in the lower part generates lift, which balances the effect of current drag so that the station always remains horizontal. The station

Figure 9.5 Messenger rack on Station-1. Station-1 was equipped with an 8-slot Messenger rack. Main characteristics: height 480mm; tube diameter 38mm; buoyancy block 150 × 150 × 150mm; mass 32 kg; buoyancy 0.6 daN.

floor is composed of a perforated plank in glass-reinforced plastic, on which components are latched by collars. The Messenger rack is located at the upper rear. The rack structure is made of Ertacetal polyamide. Free space under the rack is available to house additional equipment on request (Figure 9.5).

Buoyancy modules are fitted in the upper part. For the "Coastal and Shelf" version buoyancy consists of plastic hollow floats with 1200m service depth capability.

The station is fitted with an autonomous acoustic release placed vertically above the attachment point of the mooring line. The separate release transducer is placed at the left fore part of station. Symmetrically, the acoustic transmission modem is placed on the right .

To ensure communication, stations were equipped with a Sercel MATS 200 Net modem with omnidirectional transducer fixed on the modem container.

The station is anchored to a deadweight by a 2m length of synthetic fiber cable. The line is secured to the hook of the autonomous acoustic release. To recover the station, a release command is sent by the acoustic release modem Ixsea AR861-CE-HD, with separate transducer.

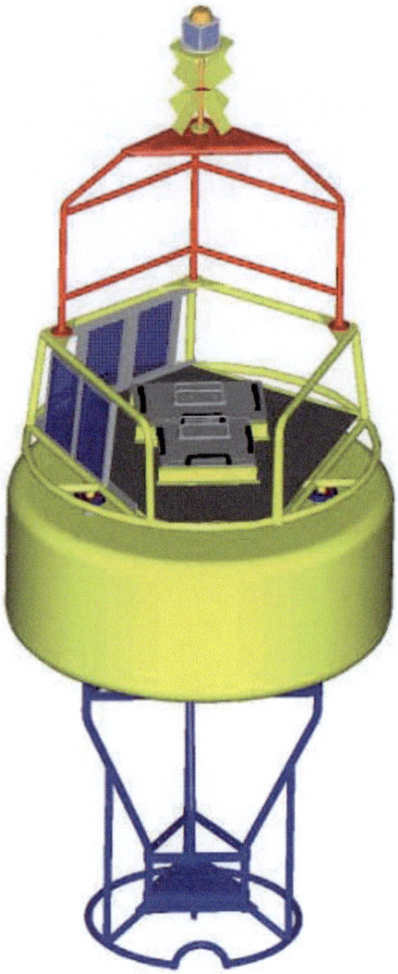

Figure 9.6 ROSE prototype system buoy. The float dimensions are: 2.2m outside diameter, 1.0m inside diameter and 0.9m height, for a volume of 2.7m^3. The overall dimensions of the assembled buoy (except mobile equipment) are 2.2m diameter and 5.3m height, for a 730kg mass. Unitary characteristics of solar panels are: surface 0.2m^2; crest power 20 Watts; maximum voltage 17 Volts.

Prototype stations were fitted with only one standard energy container, housing 6 blocks of 12 LSH 20 lithium primary batteries, providing a global 78 Ah energy capacity under 12 and 24 Volts. As COSTOF bridgeboards are programmed according to their use, some software was available from previous experiments whilst others, such as Messengers or hydrocarbon fluorometers, had to be programmed for systems or sensors used for the first time. Interconnections are implemented with pressure compensated cable in oil and SUB-CONN connectors.

9.3.2 Buoy

The buoy and "Small depth" mooring of the prototype system were built according to the technical specifications described Section 9.2.2.5. The buoy (Figure 9.6) is composed of a toroidal float, an aerial and a submarine structure.

The toroidal float is of rounded rectangular sections made of polyolefin foam protected by a reinforcing cover. The float is fitted with three steel tubes with flanges, on which aerial and submarine structures are bolted. The submarine structure is of galvanized steel ballasted at the foot with galvanized steel plates, to which the mooring line is attached.

On the aerial structure are fixed watertight containers for energy and electronics, solar panels, marking equipment, the GPRS phone antenna and a lightning conductor linked to a cable immersed in seawater.

The buoy is linked to the station by a mooring line, comprising 12ms of 28mm steel cable, 1m of 22mm polyamide cable acting as damper, 40m of chain of 14mm size. Mooring overall weight for 30m depth amounts to 230kg. Wind and current exert a 130 daN traction force on the buoy. In nominal conditions, float draft is 0.3m for a 950kg displacement, which leaves an ample margin of 1900kg buoyancy reserve.

The buoy electronics are supplied by two lead-acid 38 Ah batteries, charged by eight Solems-Kyocera KS20 solar panels. Batteries and charge regulator were placed in a watertight container in aluminum alloy. The container was fitted with a security valve against internal accidental overpressure.

Communication equipment is the same as those described in the technical specifications of Section 9.2.2.5.

9.3.3 Sensors

There are five sensors in all: two on station-1 (a backscatter meter and a dissolved oxygen optode); three on station-2 (a hydrocarbon fluorometer, a CTD probe and an ADCP profiler):

- The TRIOS enviroFlu-HC hydrocarbon fluorometer measures hydrocarbons by fluorometry, with illumination at 254nm and detection centered on 360nm. Two measuring ranges are available, one at 0–50 μg/l and one at 0–500 μg/l. The detection threshold is 0.1 μg/l. The sensors were factory calibrated with naphthalene solution in water, and checked by the Cedre Laboratory.
- The Sea-Bird MicroCAT SBE 37-SMP CTD probe with circulating pump has measure ranges of 0–0mS/cm for conductivity, –5–35°C for temperature and is operational between 0 and 7000m depth. The sensor was factory calibrated.
- The WET Labs BBRTD-226R backscatter meter has measure range of 0–7.5 NTU. The sensor was calibrated in the Ifremer Laboratory in a suspension of formazine.
- The Aanderaa 3080 dissolved oxygen optode incorporates a temperature sensor. The O_2 sensor was calibrated in the Ifremer Laboratory by the Winkler method.
- The ADCP RDI profiler 300 kHz was lent for the duration of the demonstration by an Ifremer laboratory.

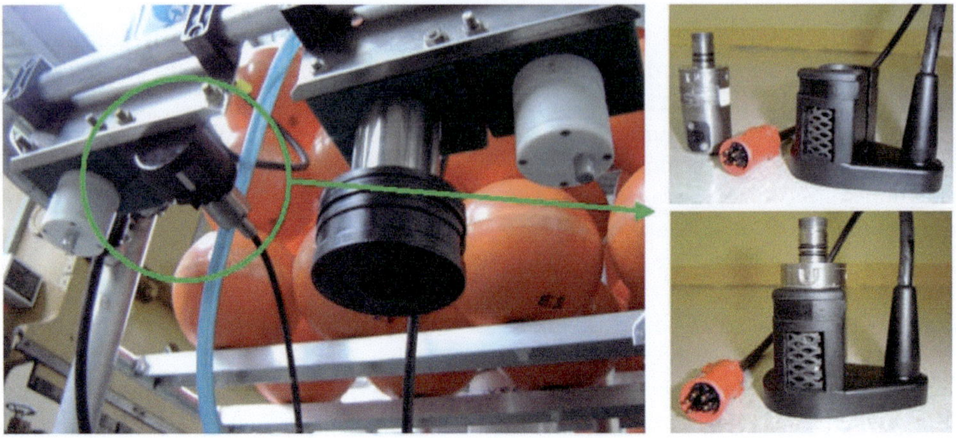

Figure 9.7 Aanderaa optode biofouling protection device.

Sensors are strapped to fixing pieces welded on structure tubes by polyamide collars, which ensure galvanic isolation. The three optical sensors – the fluorometer, the backscatter meter and the O_2 optode – were placed so that the glass windows were looking downwards to prevent deposit of sediment. Glass windows of optical sensors were protected against biofouling development by a local electrolytic chlorination (Figure 9.7) (Delauney and Compère, 2009; Delauney et al., 2010).

9.4 Prototype tests in Ifremer seawater tank

Tests were carried out during one week in the large seawater tank of Ifremer Center in Brest (http://www.ifremer.fr/dtmsi/moyens_essais/bassin.htm). Tank tests are a mandatory step before a sea deployment and enable the system to be operated in conditions similar to its future environment, although limited tank dimensions are not able to test all open sea conditions.

Tested equipments comprised station-1 and 2, fully equipped (including sensors and the Messenger rack with two Messengers on station-1), and the buoy communication equipment. This last one is composed of an acoustic modem linked to a submersed transducer, the application board and a GPRS modem, which enabled either local connection in air with the antenna of a local PC or transmission via GPRS network to Sercel premises. Seawater tank tests focused on the stations and on the Messenger.

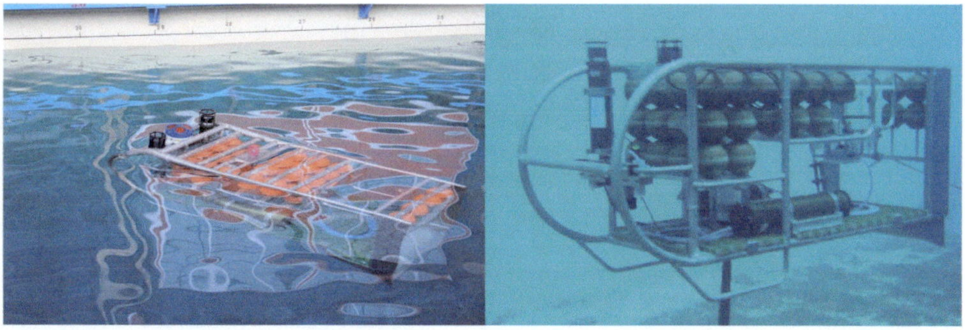

Figure 9.8 Station tests in Ifremer seawater tank.

9.4.1 Station tests

The hydro-mechanical behavior of the stations was tested. Tests on weight and equilibrium in air and water gave good results with measured values quite close to the computed ones. Release and ascent tests were performed on both stations anchored at a depth of 10m. The station ascended along the vertical with no perturbations at approximate 0.5m/s speed, which confirmed drag estimate. Five towing tests were performed at increasing speeds from 0.25 to 0.85m/s. Station kept a quasi-horizontal and stable attitude, without roll, heave or swaying (Figure 9.8).

The station had been programmed to operate continuously, first over two nights without acoustic emission, then over one night with emissions. Later, it operated continuously for about 2.5 days with periodical acoustic emissions and GPRS transmissions. Data recorded by sensors – except for the ADCP that could not be operated in the basin – have been in accordance with values measured by the tank's own monitoring equipment as regards temperature and conductivity. The acoustic link has been successfully tested with both stations, which enabled station status data and sensor data to be recovered. Of course, the tank's dimensions only allowed acoustic tests over short distances of some meters and along paths at small angles with respect to the vertical. During the 2.5-day period of continuous operation of the whole communication link, data were received at the simulated shore station in Sercel premises and stored in a database.

9.4.2 Messenger tests

The hydro-mechanical behavior of the Messenger was tested and gave weight check results (Messenger-1: 3.253kg and Messenger-2: 3.240kg), very close to computed ones (3.314kg). Messenger behavior was checked during ascent from a depth of 20m. The Messenger ascended along the vertical with small undulating movements, at an estimated speed of about 0.6m/s. Observation of Messenger's motions in waves, 20cm in height and for 1-second periods, demonstrated a satisfying stability.

The Messenger operation was also tested. Latching wire electrolysis was predicted to take around 20 minutes from laboratory tests. Five releases were performed in the seawater

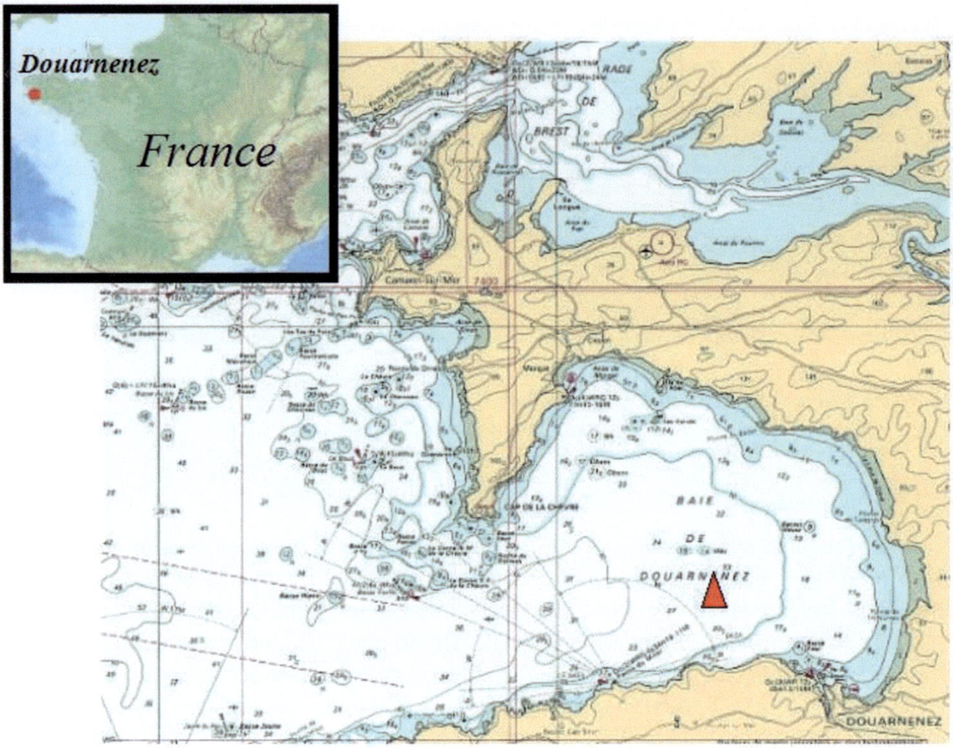

Figure 9.9 Location of the ROSE deployment in Douarnenez Bay.

tank with durations from 15 to 25 minutes. The whole set of data transfer and storage functions were tested in the tank. ARGOS data transfer was tested with a Messenger floating in a small tank in the open air.

9.5 Demonstration at sea

The system demonstration was deployed in the Douarnenez Bay (Brittany, France), at four nautical miles from the Douarnenez harbor (48°N 08' 04°W 25') (Figure 9.9). Demonstration tests were set between 14 June 2006 and 5 September 2006. The system was deployed in an area with a flat seafloor, and a water depth of 25–27m and where maximum tide amplitude reached 7m at 105 tide coefficient in mid August.

All sea operations were performed using an Ifremer team and the Ifremer coastal ship Thalia. Demonstration at sea consisted of two stations, one simplified acoustic Node-3 and buoys and attachments.

Figure 9.10 Deployment of the ROSE prototype.

9.5.1 Sea operations

9.5.1.1 System deployment

Operations were carried out over three days (Figure 9.10). On day 1, the two stations and the acoustic node (820kg) were loaded on the Ifremer ship Thalia in Brest harbor and set sail to Douarnenez harbor. Day 1 was also devoted to assembling the buoy and testing the GPRS link. During day 2, the buoy assembly was finished and programs and tests started on the stations. Materials were sent to the deployment site. As the buoy (1830kg) could not be loaded on board, it had to be towed from Douarnenez. Later, stations were deployed and put at sea using the ship's 3-ton working load tilting A-frame and winch. The configuration of the deployment was like a Y: the buoy was moored at the Y's branching point, the two ROSE stations at the ends of the short Y branches at about 750m from the buoy and the simplified Node-3 at the extremity of the foot of the Y, 1000m from the buoy. Day 3 was devoted to acoustic transmission tests. The on-shore control station was located in Sercel premises in Brest.

9.5.1.2 System operation from mid-June to early September

Stations and sensors were programmed with an acquisition frequency of 15 minutes and 1 hour respectively. Data acoustic transmissions were programmed every 4 hours at: 00, 04, 08, 12, 16 and 20 hours UT, with a 0.5 hour time lapse between stations. Consequently, sizes of acoustic messages were 430 bytes for station-1, 2500 bytes for station-2 (due to the volume of ADCP data) and 180 bytes for Node-3 (1080 bytes from the end of July for test purposes). Acoustic modems operating parameters had been set with an acoustic power of 179 dB/μPa at 1m and a data flow rate of 100 bits/s.

Thus the on-shore control station received data sequences of 4-hour duration. Sequences without any errors were only transmitted to end-users as time series. During the demonstration, a daily journal was completed, which noted for each data sequence the time segments received without error, the total duration of measurements without error and the ratio of that duration to 24 hours. Measurement durations were summed every month and related to hourly month duration.

Messengers were programmed to be released 15 and 30 days after deployment.

An inspection of stations by divers was planned and corrective actions were devised, which consisted mainly of positioning the transducer higher above the station and fitting it with a foam mask to prevent possible interferences between the acoustic beam and station structure. Diver intervention could only take place one month after deployment and modification could only be made on station-1.

9.5.1.3 System recovery

For the preparatory operations at sea and for the recovery, a team of divers with a rubber boat was necessary in addition of the technical team of Ifremer. Preparatory operations by divers consisted of marking precisely the positions of the stations by fitting a rope to the anchoring deadweight with a buoy at the surface. The recovery (Figure 9.11) was performed over two days. Day 1 was devoted to the successive recovery of the two stations, the Node-3 and their anchoring deadweights, the buoy and its mooring and deadweight. Later, teams and materials were returned to Douarnenez, with the buoy towed behind the ship. Day 2 was devoted to the disassembly of the buoy on quay and loading it and the stations on trucks for the return to Ifremer Brest.

9.5.2 Analyses of at-sea demonstration results

9.5.2.1 Communication system and station operation

The station electronics as well as acoustic and radio-electric equipment appeared to work properly. Acoustic emissions from Node-3 were received satisfactorily. Difficulties in acoustic communication with ROSE anchored floating stations were a somewhat disappointing issue of system operation. Several hypothetical reasons were put forwards, ranging from problems with deployment of stations, which might not be floating correctly, to acoustical reasons, e.g., interferences between acoustic transducer beam and station structure or differences in propagation conditions between bottom laid and floating stations.

Data messages received periodically at the on-shore control station were processed and delivered as time series, which enabled environmental parameter evolution to be monitored in spite of lacking data sequences. The rate of messages received without error depended on periods, stations and measurements. Globally, 90% of data messages were received without error from Node-3, but only between 45% and 65% from the ROSE stations. Node-3 reception rates were at the expected level with respect to former experience, but

Figure 9.11 Recovery of the ROSE prototype.

with a tendency to decline from June to August. Stations-1 and 2 reception rates were both around 20% in June; in July, they increased to a mean value of 61% for station-1 and 47% for station-2, but declined to about 20% in August.

Possible interference between the modem acoustic beam and station structure, especially the hollow floats, was deemed to be the most probable reason for difficulties encountered in communicating with the ROSE anchored floating stations. Modifications were implemented by divers on station-1 during the demonstration. They brought some improvement without, however, enhancing performances to the level attained with Node-3 laid on the bottom. Initial and modified modem configurations are shown in Figure 9.12.

Analyses of factors that might affect acoustic transmissions were carried out. Tide influence could arise from two factors: station heading variations with current and variation of water height above station. Graphs of good reception periods were superimposed with hours of high and low tide levels or tide coefficient curve, without revealing correlations.

Periods of stronger wind and rougher sea or rain events might have increased ambient noise and disturbed communications. But meteorological records of the days of low success transmission did not show more disturbed environmental conditions.

Analysis of successful transmission rates shows a performance decrease in August compared to July. Biofouling development, observed on transducers during an inspection dive (see Sections 9.5.2.2 and 9.5.2.3) certainly has a negative impact. Thus, transducer protection by a local chlorination device is worth considering.

a)

b)

Figure 9.12 (a) Biofouling; (b) Initial and modified configurations of the acoustic modem on station-1.

Directivity diagram measurements were carried out in the seawater tank on the two configurations of acoustic modem after demonstration at sea. Directivity diagrams of the initial configuration show an up to 10 dB loss in rear direction in both the vertical symmetry plane and horizontal plane. For the modified configuration, losses are reduced and generally do not exceed 3 dB. More important modifications have been devised for future implementations to prevent interferences between the acoustic beam and station structure, to place the modem about the mid-length of the station in a higher position, or to cover the buoyancy floats with a plank of absorbing foam. Of course, the proposed modifications should be proved in the tank before deployment at sea.

From the daily journal, cumulated daily reception duration free of errors for the various messages was derived, which were presented as bar graphs for each sensor and for technical data of each station (Figure 9.13). A 100% reception rate corresponds to 24 hours of sensor data or 24 messages of technical data. It shows in July principally periods of good reception interrupted by successions of 1- to 3-day periods without any reception free of errors.

During the station operations, technical parameters were also monitored. Amongst technical parameters transmitted to the on-shore control station were battery voltage, remaining energy under 12 and 24 Volts and internal pressure of the COSTOF container. On the two stations and Node-3, parameters were measured every hour, but the period was reduced to 10 minutes on Node-3 from the end of July for test purposes.

The analyses of the technical parameter results showed that the 12 Volts nominal voltage for the COSTOF and sensors was particularly stable. However, the 24 Volts nominal voltage for the acoustic system was stable for the two stations but not for the Node-3. Indeed, a decrease was observed on Node-3 following an increase of transmitted message size after 27 July.

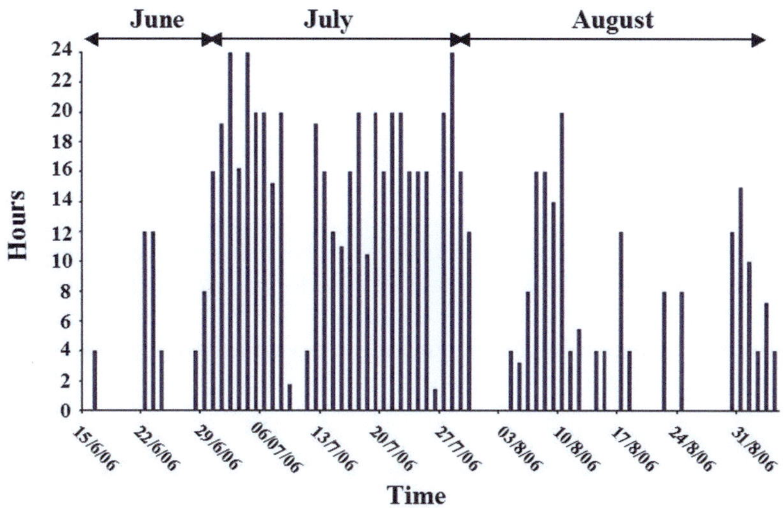

Figure 9.13 Example of error free measurement reception bar graph of the backscatter meter from station-1.

Figure 9.14 Hydrocarbon fluorometer glass window of station-2 after recovery.

Energy consumptions under 12 Volts depended on the station's sensor payload, as illustrated by the higher values for station-2 which is related to the ADCP operation.

9.5.2.2 Biofouling

Sensor antifouling protection devices had been programmed according to usual values of current intensity and alternate periods on/off. A day-to-day review of data received from stations-1 and 2 demonstrated that the devices functioned properly (Figure 9.14).

After the recovery, we observed that the stations' structures and components were largely covered with biofouling. The major part, except calcareous concretions, could be removed by high-pressure jet washing. Sensors were also largely covered with biofouling and some calcareous concretions, except on and around glass windows, which had been efficiently protected by the local chlorination device. ADCP transducers were quite clean. No corrosion traces were visible, except the start of cavernous corrosion on the stainless steel container of the fluorometer under fixing collars. Protection zinc anodes were moderately consumed. The buoy was recovered in excellent condition with only slight biofouling on the submersed part of the float. The same was true for mooring line components.

In conclusion, there were no problems originating from the design, but constant attention must be paid to careful implementation of best practice at each manufacturing and assembly step to guarantee benefits of good design.

9.5.2.3 Messenger

The first Messenger should have been released 15 days after deployment. However, during a diver inspection, an important development of biofouling on station structure and equipments was observed, and divers found the first Messenger in the rack (but not the second one). It was concluded that latching wire electrolysis had occurred on the due date but that the Messenger had remained stuck in the rack by fouling until divers had released it by moving the station. Messengers are normally intended for stations deployed far away from shore in great water depth, where fouling development is much less important than in coastal areas. As the release operation had been tested successfully several times in tank on shore, it was concluded that fouling in the rack clogged the Messengers, and their small 0.6 daN buoyancy was insufficient to pull them out. Modifications have been devised, consisting of improving the Messenger guiding device in the rack and possibly in supplementing its buoyancy by a spring push force.

Once released, the Messenger was recovered at the surface and put in emission; but the internal battery had been too much depleted during the jamming period for it to emit exploitable messages through ARGOS. However, all expected data were unloaded from the Messenger mass memory. The same problem was encountered with the second Messenger. A solution to help the Messenger release from the rack was devised.

9.5.2.4 Sensors

In spite of discontinuities in data series, it has been possible to monitor the evolution of environmental parameters thanks to COSTOF records. These have been processed to obtain full time series from mid June to early September. First conclusions were that sensors operated quite satisfactorily. On data curves, data received during the demonstration (onshore control station) are pointed in red while complete data series from COSTOF are pointed in blue.

Several results were extracted and interpreted from the different sensors. The fluorometer showed a signal fluctuating between 150 and 250 units (Figure 9.15), indicating the absence of hydrocarbon pollution. Should traces of pollution have occurred, the signal would have jumped to considerably higher values. As regards corrosion resistance, this experience has pointed out the necessity of replacing the stainless steel housing of the fluorometer with a titanium container. Moreover, this type of container would enable deployment down to a depth of 4000m.

The mean signal of the backscatter meter progressively increased from 0 to 1.2 NTU, with daily fluctuations up to 2.5 NTU (Figure 9.16). The slow signal increase along 2.5 months might be due to slight biofouling development on sensor glass window in spite of the chlorination.

The Aanderaa probe O_2 of station-1 showed evidence of under/over saturation (Figure 9.17). Measured temperature values of the Aanderaa probe on station-1 were in fair agreement with those measured by the CTD of station-2 (Figure 9.18). Pressure variations in Figure 9.19 correspond to variations in water height with tides, which are in fair agreement with values calculated from tide tables on some test dates.

As regards the ADCP profiler, rather important treatment of ADCP raw data is required to get physical values. Only a rapid overview of data was performed, which demonstrated proper functioning of the sensor.

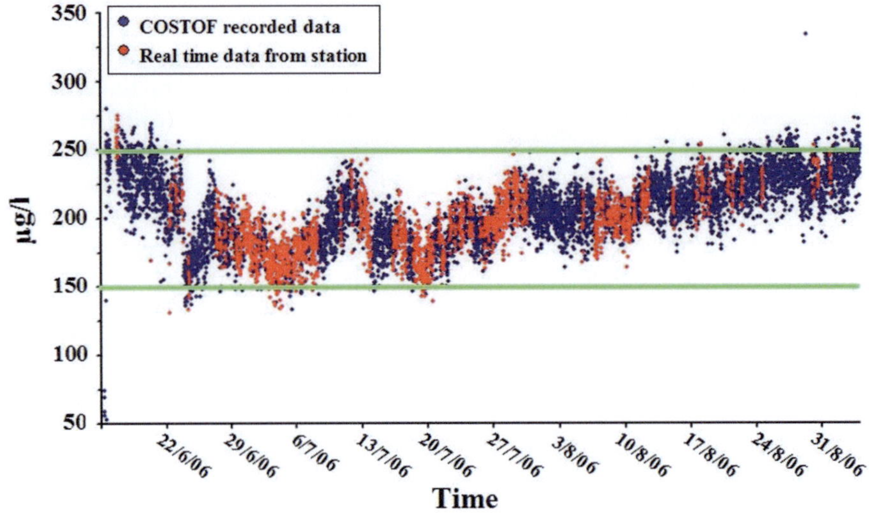

Figure 9.15 TRIOS hydrocarbon fluorometer data from station-2.

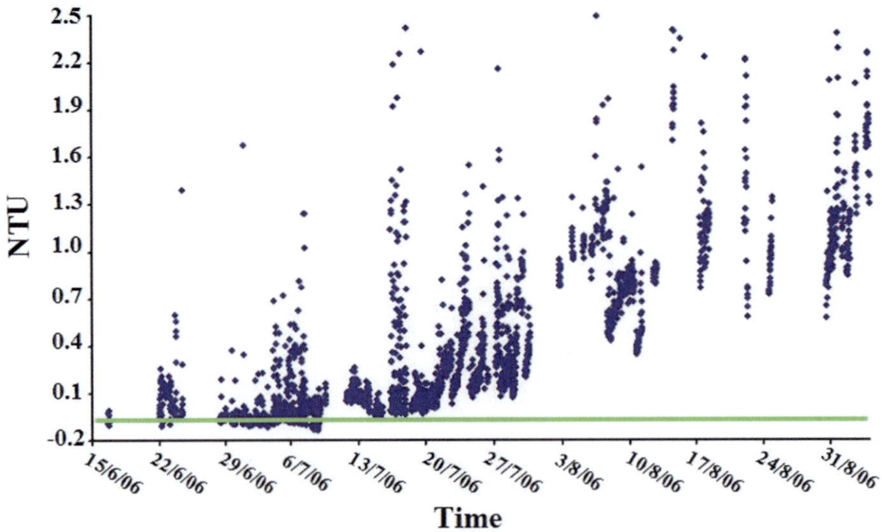

Figure 9.16 Backscatter meter data recorded from station-1.

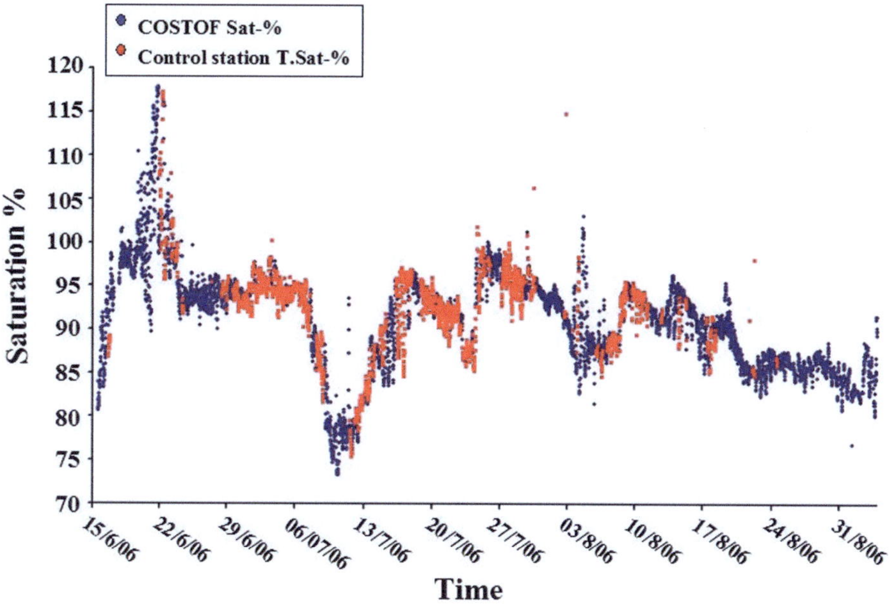

Figure 9.17 Aanderaa probe O_2 record from station-1.

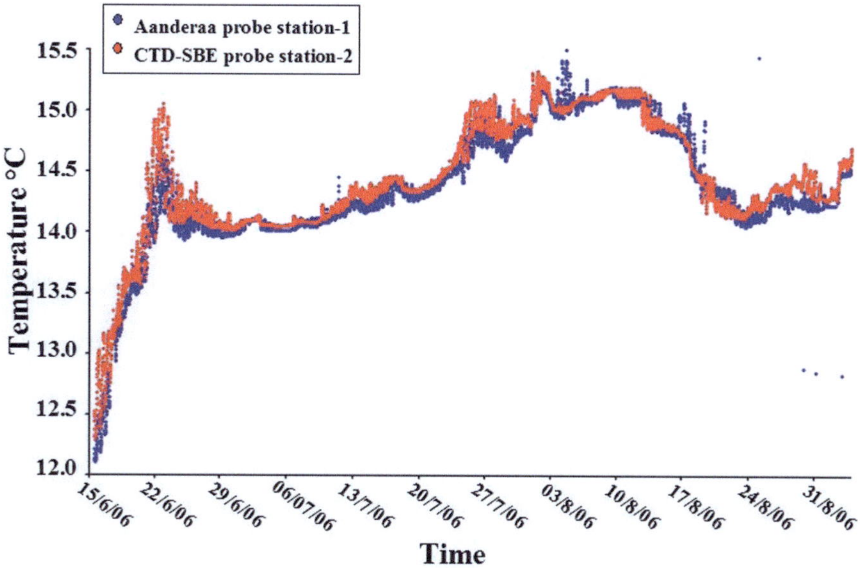

Figure 9.18 Temperature data from the Aanderaa probe on station-1 and from the CTD of station-2.

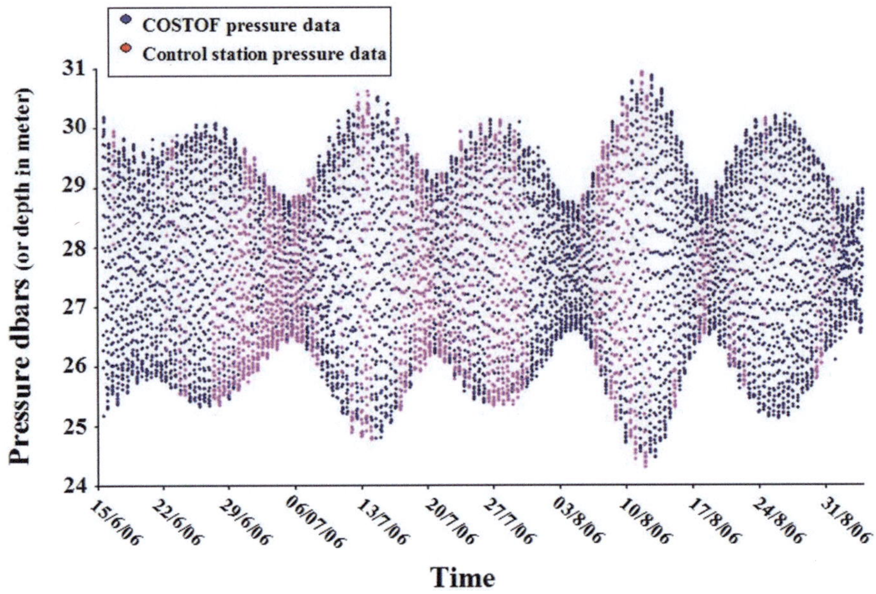

Figure 9.19 CTD immersion data records from station-2.

9.6 Conclusions

The initial review of marine pollution events in the real world and the analyses of conditions and constraints of operation of the monitoring system led to the writing of functional specifications, which were the basis of the ROSE system technical definition. The aim of these "Technical specifications" was to define a system able to meet the full range of operational situations. Effort was put into defining a system combining, as much as possible, elements of general use, with a limited number of parts specific to situations. Major requirements applied to adaptability and modularity of seafloor stations. Previously-built and proved sub-systems were integrated into the system as much as possible. But efforts on design renewal and upgrading have been made, which led to the genuine and innovative concept of anchored floating stations, positioned at a convenient height for measurements above the seafloor.

The "Coastal" demonstration prototype comprised all parts of an operational system. An at-sea demonstration was implemented for a significant duration of 2.5 months and was representative of an operational coastal deployment. Globally, system operation and behavior were satisfying, except on a few points to be upgraded and for which remedial solutions have been devised.

Rates of error free transmissions with floating ROSE stations were lower than with bottom laid stations. This drawback was attributed to interferences between the modem acoustic beam and station structure, when the acoustic path from station to buoy is close to

Figure 9.20 ROSE "Great depth" version.

the horizontal. From tank tests carried out after recovery of the system, solutions have been devised to remedy the problem.

Although Messengers are intended for distant deep-sea stations, coastal demonstration was an opportunity to test in a real environment the complete sequence of Messenger functions, which had been separately proven by tests on shore. Biofouling development disturbed the Messenger release but solutions to secure the process were defined.

The station sensor payload had to fulfill two types of measurements: to detect and as much as possible quantify emissions of pollutants, and to measure pertinent environmental parameters. A set of sensors was gathered for the demonstration and gave satisfying results. Glass windows of optical sensors were efficiently protected against biofouling development by a local chlorination device. While a wide range of environmental parameter sensors is available, proof tests of other pollutant sensors than hydrocarbon fluorometer, such as pH sensor or SBSE (Stir Bar Sorptive Extraction) sampler, should be performed in real and operational environment.

Beyond polluting wreck monitoring, the ROSE system is efficiently used as a mobile observatory for environmental or geohazard monitoring, and for this occasion, a "Great depth" version was developed (Figure 9.20). A good example is the Var canyon turbidite and sediment fluxes monitoring realized in 2009 and 2010 with the downlooking ADCP profiler (Blandin et al., 2007; Vangriesheim et al., 2007; Ruhl et al., 2011).

List of abbreviations

ADCP	Acoustic Doppler Current Profiler
ASSEM	Array of Sensors for Seabed Monitoring (European project)
CLSI	Contact Less Serial Interface
COSTOF	COmmunication STOrage Front end
CTD	Conductivity - Temperature - Depth
GSM	Global System for Mobile (Communications)
GPRS	Global Packet Radio Service
MMI	Man Machine Interface
ROSE	Réseau acoustique Orienté Surveillance d'Epaves - Acoustic Network for Wreck Monitoring
ROV	Remotely Operated Vehicle
SBSE	Stir Bar Sorptive Extraction
VHF	Very High Frequency

Acknowledgment

The ROSE system development was carried out with support from the French Research Ministry within the framework of RITMER, Research and Technical Innovation Network on marine accidental pollutions.

References

Beranzoli L., Calore D., Favali P., Marvaldi J. and Nicot M. (2004) ORION-GEOSTAR-3: A prototype of seafloor network of observatories for geophysical, oceanographic and environmental monitoring. In: Proceedings of the 14th International Offshore and Polar Engineering Conference, Toulon, France.

Blandin J., Person R., Strout J.M., Briole P., Etiope G., Masson M., Golightly C.R., Lykousis V. and Ferentinos G. (2003) ASSEM: A new concept of observatory applied to long term SEabed Monitoring of geohazards. In: Proceedings of OCEAN'2003, San Diego. USA.

Blandin J., Vangriesheim A., Legrand J., Coail J.Y. and Leildé B. (2007) Real time transmission of current and turbidity data from the near bottom Var canyon system. Extended Abstract, Martech Second International Workshop on Marine Technology, Barcelona, Spain.

Delauney L. and Compère C. (2009) An example: Biofouling protection for marine environmental sensors by local chlorination. Marine and Industrial Biofouling 4, 119–134.

Delauney L., Compère C. and Lehaître M. (2010) Biofouling protection for marine environmental sensors. Ocean Science 6, 503–511.

Labat J. and Macchi O. and Laot C. (1998) Adaptive decision feedback equalization: Can you skip the training period? IEEE Transaction on Communication 46(7), 921–930.

Laot C., Glavieux A. and Labat J. (2001) Turbo equalization: Adaptive equalization and channel decoding jointly optimized. IEEE Transactions on Communication 19(9), 1744–1752.

Lurton X. (2002, 2010) *An Introduction to Underwater Acoustics: Principles and Applications*, Chapter 7. Chichester and Heidelberg, Springer Praxis, Heidelberg.

Ruhl H.A., André M., Beranzoli L., Çağatay M.N., Colaço A., Cannat M., Dañobeitia J.J., Favali P., Géli L., Gillooly M., et al. (2011) Societal need for improved understanding of climate change, anthropogenic impacts, and geo-hazard warning drive development of ocean observatories in European Seas. Review Article. Progress in Oceanography 91, 1–33.

Vangriesheim A., Khripounoff A. and Mas. V. (2007) Current and turbiditic events observed in the VAR canyon (Mediterranean Sea). European Geosciences Union, General Assembly, Vienna, Austria.

10 Construction of the DONET real-time seafloor observatory for earthquakes and tsunami monitoring

K. Kawaguchi[1], S. Kaneko[1], T. Nishida[2] and T. Komine[3]

10.1 Introduction

The Japanese islands are located at a quadruple junction of tectonic plates. Along these plate boundaries, a large number of earthquakes occur every year, supplied with energy by tectonic plate activity. Japan has a sophisticated seismological observatory network on land to estimate the location, magnitude, mechanism and other information about the earthquakes in support of research for earthquake procedures. However, the land observatory network is insufficient to cover the seismic activity below the seafloor. Expansion of the seismological observatory network to the seafloor is essential to properly interpolate the hypocenter distribution below the seafloor, and this expansion enables the highly accurate numerical results for plate boundary earthquake activities. To answer this demand, a unique development program for a submarine cabled real-time seafloor observatory network named DONET (Dense Ocean-floor Network system for Earthquakes and Tsunamis) was undertaken. The DONET was targeted at developing the infrastructure technologies of large-scale real-time seafloor research and surveillance for earthquake, geodetic and tsunami observation and analysis. This program began in 2006 on the To-Nankai region in the Nankai trough as the object area for these observations. The Nankai Trough is located at the boundary of the Philippine Sea plate and Eurasian plate, where mega-thrust earthquakes

1 Japan Agency for Marine-Earth Science and Technology, Department of Oceanfloor Network System Development of Earthquakes and Tsunamis, Kanagawa, Japan

2 Ocean Cable & Communications, Kitakyushu, Japan

3 Nichiyu Giken Kogyo Co., Ltd., Saitama, Japan

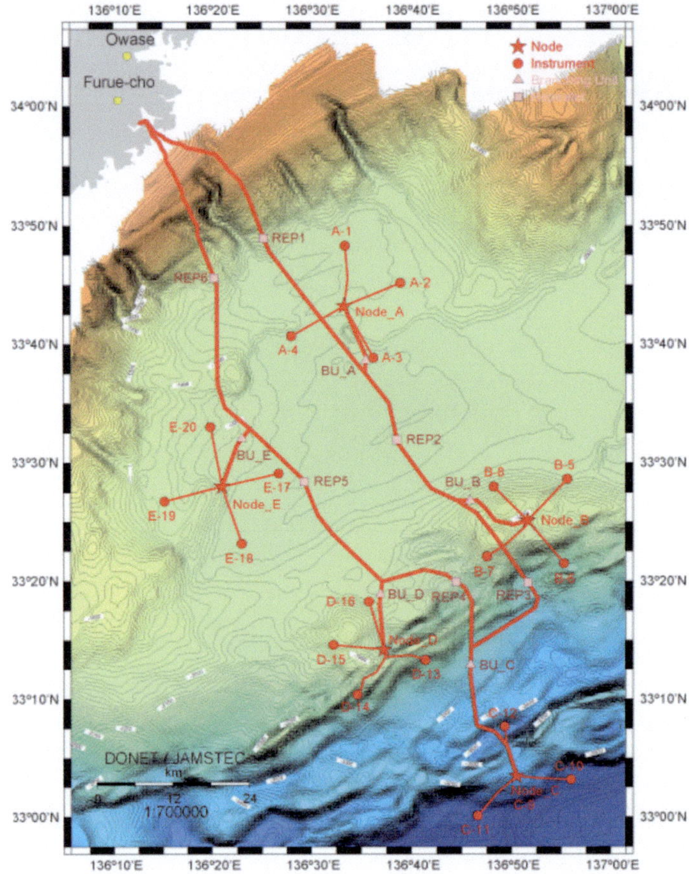

Figure 10.1 DONET backbone cable system and observatory arrangement at Nankai trough. In the DONET project, the 20 sets of earthquake and tsunami observatories are planned to be deployed on the seafloor at the interval of 15–20km to cover the anticipated focal region of the To-Nankai earthquake.

occur frequently with intervals of 100–150 years. The last earthquake to hit this region was in 1946 and the probability of another large earthquake is in the order of 60% that it will occur within the next 30 years. Reinforcement of earthquake surveillance capabilities for this region are urgently needed in order to mitigate the impact of this potential disaster.

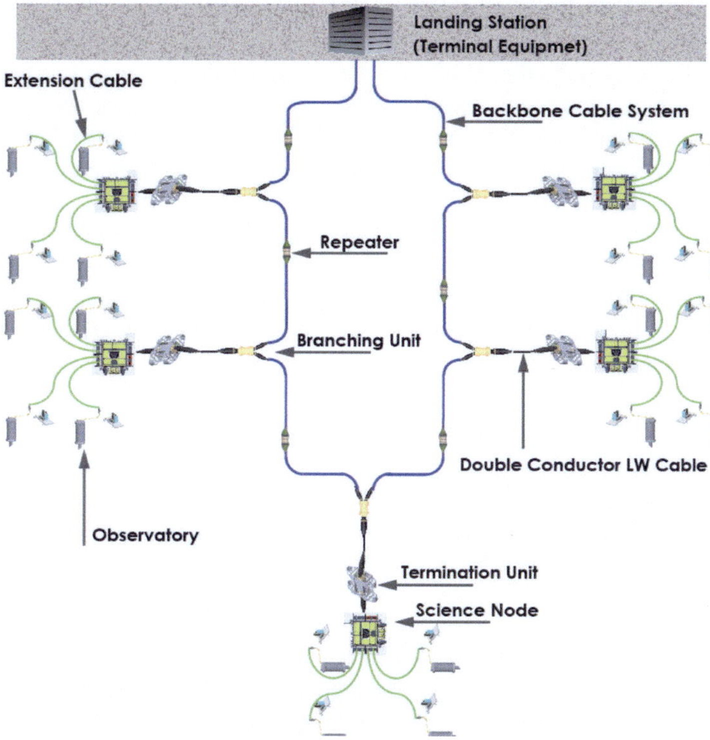

Figure10.2 A system design concept of the DONET seafloor observatory network.

10.2 System overview

The DONET seafloor observatory network is a most suitable subsea infrastructure for the seismogenic zone monitoring. The initial plan schedule to install 20 sets of complex cabled monitoring observatories on the seafloor at an observatory interval of 15–20km as depicted in Figure 10.1. Observation performance equal to the onshore seismological observatory network can be expected to be obtained from DONET. In addition to the initial plan, another 20 sets of preliminary observatory interfaces are prepared in the network by the consideration of the improvement of observation capability in the future.

One of the difficult challenges with underwater technology is keeping up the large-scale observing system over an extended period of time (20–30 years) in seawater. In the development of a submarine cabled observatory network, the number of observatories in the system is very important in order to secure the performance of earthquake observation but, on the other hand, it has a large influence on the total system reliability, because the components of the observatory are the trickiest part to ensure for high reliability. A novel system design concept is necessary for the observatory network development to meet two conflicting demands such as "high reliability system design" and "state-of-art measurement".

It should be able to replace, maintain and enlarge the components of an observatory network during the operating lifetime, and it should also be able to repair an internal or external system failure such as cable abrasion, seabed movement, fishing activity, manufacturing defect, etc. To achieve these requirements, the DONET proposes a system topology that consists of three major components with different system reliability: the high reliability backbone cable system; the maintainable science node; and the expandable observatory. Figure 10.2 shows the planned system design to be implemented in DONET (Kawaguchi et al., 2007; Kawaguchi et al., 2008).

10.3 Backbone cable system

The backbone cable system provides the power feed line and the communications channel to the apparatus on the seafloor. The system consists of several types of submarine cables, repeaters, branching units (BU), and termination unit (TU: science node interface). The constituent of the backbone cable system introduces existing submarine telecom cable technologies as much as possible to secure high reliability. The design of the backbone cable system was done in a manner to preclude any primary system failure in its operational lifetime. Both end of the backbone cable system will be on land and connected to the constant current DC power supplies with different polarity to have the redundancy of power feeding channel. The backbone cable system allows loading up to 3kW (3kVDC/1A) electric power during operation and managing six pairs of optical fiber lines for data communications (one for BU control and five for node communications channel). The six optical amplifiers (repeaters) are located at intervals of every 40–80km of optical fiber length to achieve the long-distance data transmission in the designed backbone cable system length of 320km. These repeaters fit into the coherent optical time-domain reflectometry (C-OTDR) optical fiber trouble location tracking. The five BUs equipped in the backbone cable system have the function to branch the power feed line and the communications channel toward the TU. A pair of exclusive optical fiber communication channels in the backbone cable system controls the high voltage switching relay in the BU for managing the power feeding path to the science node (through the TU) to realize a safe subsea construction and management of unexpected status of network components. A specially manufactured double conductor submarine cable (DC-LWC) was developed for the branching trunk of the BU to achieve the bidirectional electrical and fiber optical connection with the TU. The performance of this newly-designed submarine cable suits the recommendation that the ITU-T telecommunication submarine cable fits to this component for a highly reliable DONET backbone cable system. The TU is an interface with the science node and equipped with each end of DC-LWC. Two pairs of communications channels are allocated between the TU and terminal equipment on landing station (pier-to-pier connection) by in-

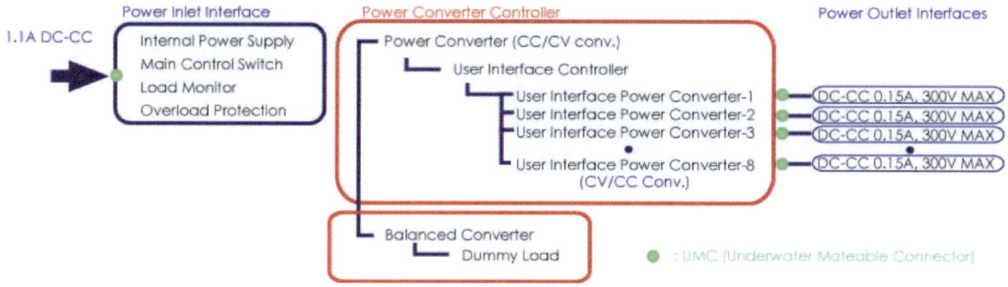

Figure10.3 A block diagram of the DONET power distribution system in the science node. The DONET power distribution system is a balanced constant current DC to constant current DC converter. The system receives 1.1A constant DC power from land and distributes 0.15A constant DC power to eight power outlet interfaces for observatories. Maximum output of each outlet is 45W.

dependent routes (ring topology) to ensure the reliability of data collection. Then each TU is equipped with a high voltage (up to 3KVA connection) optical fiber and electrical hybrid underwater mateable connector (UMC) for science node interface.

10.4 Science node

The science node is a device with the function of a hub that connects the observatory with the backbone cable system. The science node can be attached to or removed from the backbone cable system with an optical fiber and electrical hybrid/high voltage UMC. Each science node is equipped with eight standard (up to 1KVA connection) hybrid UMCs for observatory interface, and manages the power distribution, data transmission and time synchronization controls for versatile, precise and flexible observations on the seafloor. The power distribution system manages the divergence of power from the backbone cable system to observatories. It receives 500 watts of constant current DC power supplied from terminal equipment at a landing station and distributes 40 watts maximum secondary power to each observatory interface as the occasion demands. A constant current DC power system was selected for secondary power output to avoid a general electrical system failure and to achieve high power transmission efficiency through 10km-long submarine cable connections between the science node and the observatory. The power distribution system has a balanced converter that equalizes the science node power consumption constantly while corresponding to the change of secondary load as illustrated in Figure 10.3. This function is effective in protecting the observation network from the unstable condition of a secondary load (such as fluctuation of observatory power consumption, addition/deletion/

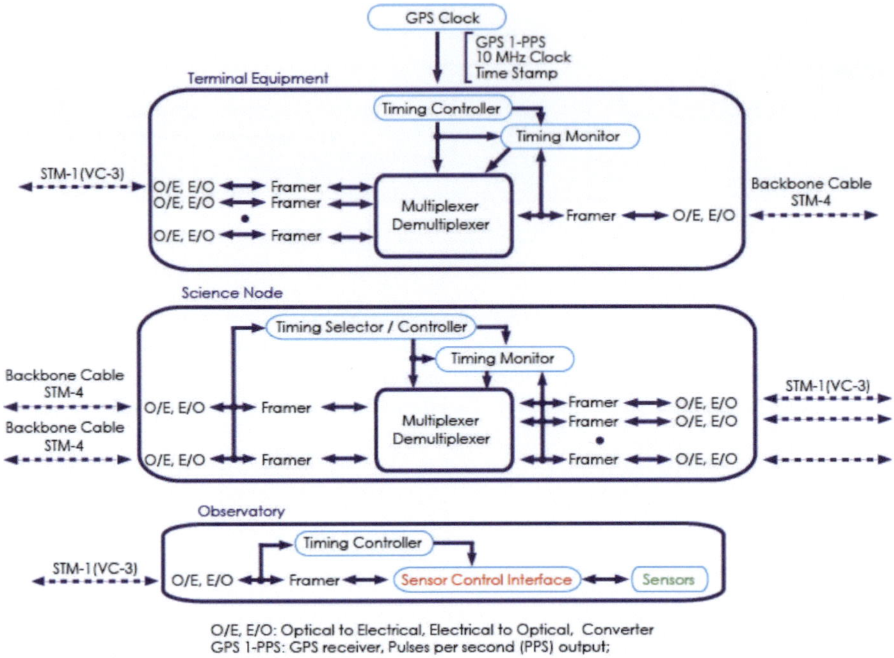

Figure 10.4 A block diagram of the DONET precise time synchronization system in the science node. The DONET precise time synchronization system distributes the time stamp and GPS 1PPS timing information to the science nodes and observatories. Each time information is embedded in the section overhead of the STM (Synchronous Transfer Mode) network management protocol and realizes the less than 1 microsecond of time synchronization accuracy in the network components.

failure of the observatory, problems with the power feeding line, etc.). The data transmission control system manages the distribution of a downlink signal (command) from terminal equipment, collecting and unlinking the data from observatories and sharing precise timing and time stamp information in the network. A SONET/SDH (Synchronous Digital Hierarchy) network management protocol is used for communicating digital information in this system over optical fiber. Data transmission capacity between terminal equipment on land and a science node on the seafloor is approximately 600 Mbit/s. A 50 Mbit/s bidirectional data transmission capability is secured between the science node and each observatory. Precise timing and time stamp information are embedded in the SDH section overhead for time synchronization control. The time synchronization control system can provide the less than 1 microsecond of time synchronization accuracy among the components of the network system (Figure 10.4). All science node components are integrated in the titanium alloy pressure housings to take it to 5000m water depth seafloor and mounted on a titanium chassis to stand up to long-term seawater environment use (Figure 10.5).

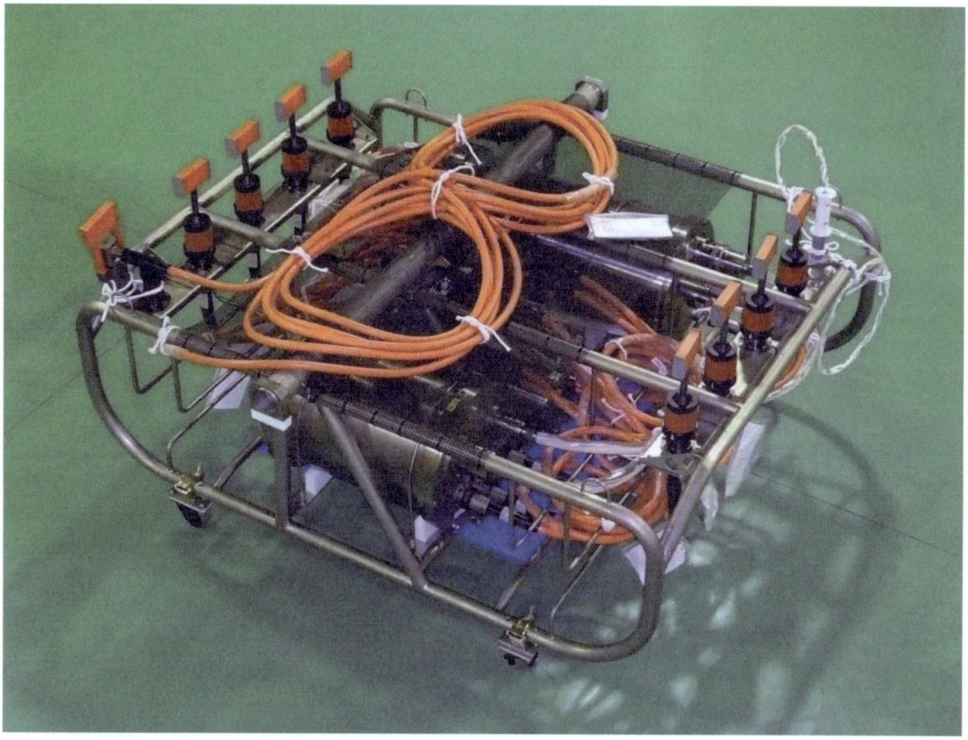

Figure 10.5 DONET science node. The DONET science node consists of the power distribution system, data transmission and time synchronization system, eight standard hybrid UMCs (Underwater Mateable Connector), a high voltage hybrid UMC, titanium alloy pressure housing and chassis. The weight of the system is 4.1kN in air and 1.8kN in pure water.

10.5 Observatory

The goal of the DONET project is to develop a novel real-time and long-term observation infrastructure for monitoring earthquake and tsunami on the seafloor. The variety of sensors are planned to be put together to cover a broad measurement range on the seafloor. The standard DONET observatory is composed of a sensor package and an extension cable system. The sensor package consists of a ground motion sensing system and pressure sensing system to aim for precise earthquake, geodetic and tsunami observation. The ground motion sensing system is made up of a broadband seismometer, a strong motion accelerometer, a gimbals stage and backup data storage. The constituents of the ground motion sensing system are assembled in a titanium alloy pressure vessel to bring the system to the seafloor (Figure 10.6). If there is an unexpected system failure, data in storage can be uploaded to terminal equipment simultaneously with real-time data when the system failure is restored. The measurement of environment control is essential to obtain high-quality

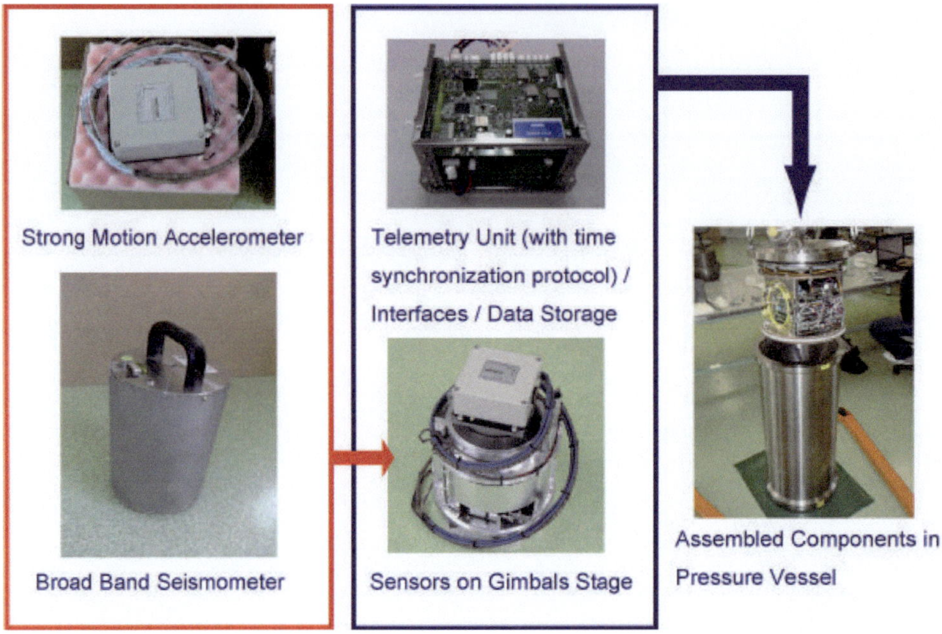

Figure 10.6 DONET observatory component: ground motion sensing system. The DONET ground motion sensing system consists of two different types of seismometers, a telemetry unit and a gimbals stage. All components are assembled in the titanium alloy cylindrical pressure vessel to spend long periods of time in seafloor sediment. The weight of the system is 1.5kN in air and 0.7kN in pure water.

earthquake and geodetic observations on the seafloor. Burying the ground motion sensing system into the sediment layer on the seafloor is expected to reduce the background noise from seafloor current, water temperature variation and other environmental effects (Araki et al., 2008; Kaneko et al., 2009). The pressure sensing system is composed of a pressure gauge, a differential pressure gauge, a hydrophone and a thermometer. All the components of the pressure sensing system are put in an oil-filled container to protect the sensors from damage by the infiltration of seawater (Figure 10.7). The observatories deployed on the seafloor are connected to one of the five science nodes in the backbone cable system with a point-to-point link as star formed topology (Figure 10.2). A very thin electrical/fiber optical hybrid submarine cable called the "extension cable" was originally designed to secure a pair of power distribution lines and bidirectional data communication channel between any observatory and science node. The standard hybrid UMCs, fitted on both ends of the 10km length of extension cable, make possible the maintenance or replacement of sensor packages on the seafloor as the occasion demands.

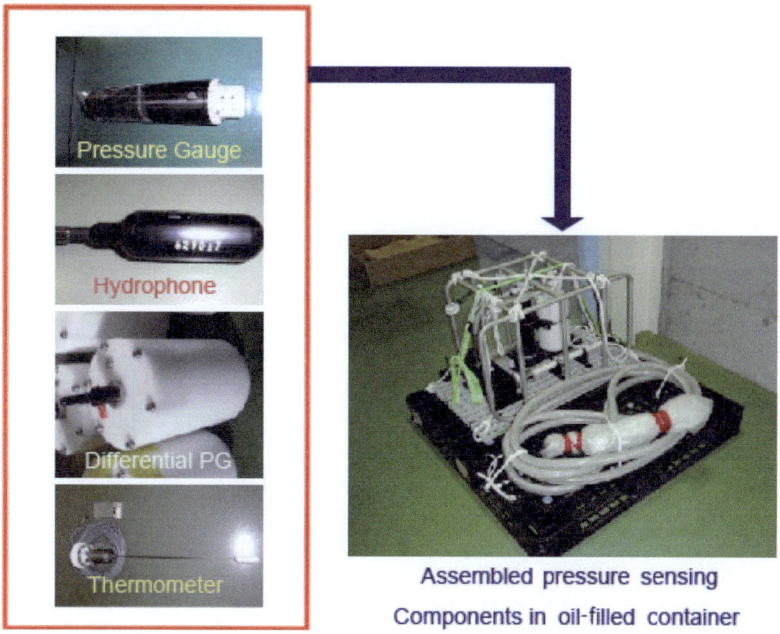

Figure 10.7 DONET observatory component: pressure sensing system. The DONET pressure sensing system consists of three different types of pressure sensors and a thermometer. All system components are put in an oil-filled container to protect the sensors from the seawater environment. The weight of the system is 0.6kN in air and 0.2kN in pure water.

10.6 Scenario

Figure 10.8 shows a scenario of the submarine cabled observatory network construction. The construction starts from the backbone cable installation by a cable laying ship. The several types of submarine cables, repeaters, branching units and termination unit were taken on the cable laying ship and installed on the seafloor according to the planned RPL (Route Position List). The TUs that become an interface with the science node are equipped with an acoustic marker to enable easy ROV approach on the deep seafloor. The science node is loaded on the ROV and carried to the target TU on the seafloor with the help of an acoustic marker. The ROV places the science node beside the TU and makes a connection using the UMC. Following the science node installation, the observatory construction and extension cable laying operation is carried out by the same ROV. The observatory is placed in a pre-installed cylindrical casing that is deployed approximately 10km from the science node. This casing is prepared by the gravity corer seafloor sampling system in advance and the inside sediment is removed with a suction pump system on ROV just before the observatory is constructed. A 10km length of extension cable was laid on the seafloor using a ROV-equipped cable laying system. Both ends of the extension cable have UMC to

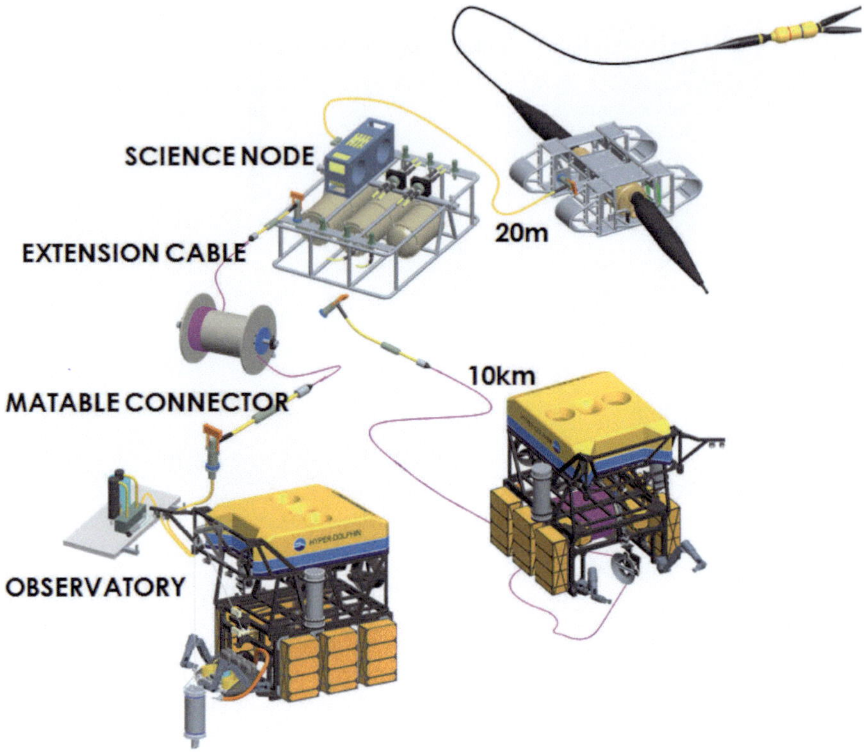

Figure 10.8 Operation scenario of observatory construction by ROV. The DONET backbone cable system (submarine cables, repeaters, BUs and TUs) is installed onto the seafloor by conventional cable-laying methods from a cable ship. The science node and observatory are constructed by the ROV on the seafloor after the installation of the backbone cable.

make a connection between science node and observatory. After connecting the extension cable, the observatory will be started up remotely from the terminal equipment on a landing station.

10.7 ROV for observatory construction

To implement the construction of a DONET observatory, the ROV should be able to endure the large variation in buoyancy that emerges during network component installation, and has to have an ability to control laying the extension cable between two exact locations 10km away with very slight position error. The scientific work-class ROV "Hyper Dolphin" in JAMSTEC had a modification under the DONET project installing the cable

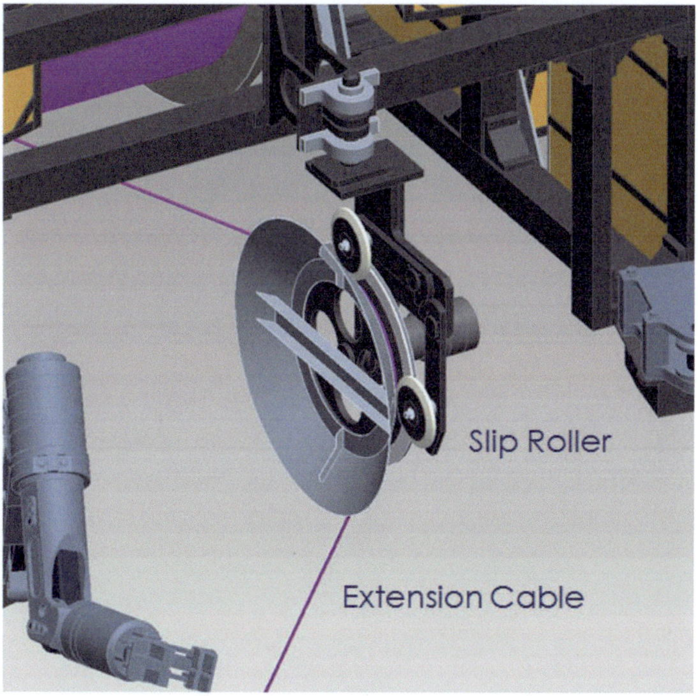

Figure 10.9 Extension cable management system. This controls the cable payout speed voluntarily. The slip roller mechanically manages the cable payout tension to 300N maximum.

laying system so that the requirements of construction could successfully be executed in practice (Kawaguchi et al., 2001, 2006, 2009; Bird, 2002; Shepherd et al., 2007).

The cable laying system consists of a tension controlled extension cable management system, a cable bobbin elevator and a variable buoyancy control system (VBCS). The tension controlled extension cable management system (Figure 10.9) is able to control the cable payout speed voluntarily to manage the reasonable cable slack corresponding to laying over rough and undulating seafloor terrain. The slip roller and bobbin break mechanically manage the cable payout tension maximum of 300N. This tension control is suited for the extension cable breaking strength of 1kN, which is a design value to prevent unexpected restriction of a ROV with the extension cable in the sea. The cable bobbin elevator (Figure 10.10) makes it possible to equip and release the cable bobbin together with a 10km length of extension cable in air and in water. It is designed to generate 10kN of pulling torque, which is sufficiently powerful to lift the cable bobbin of 6.5kN in the air. When in operation, the cable bobbin is fixed to the chasse of the cable laying system with the pair of stab rods.

The VBCS is a system able to adjust the buoyancy of the ROV by flooding or draining seawater to the pressure resistant water tank. The "Hyper Dolphin" has the capacity to

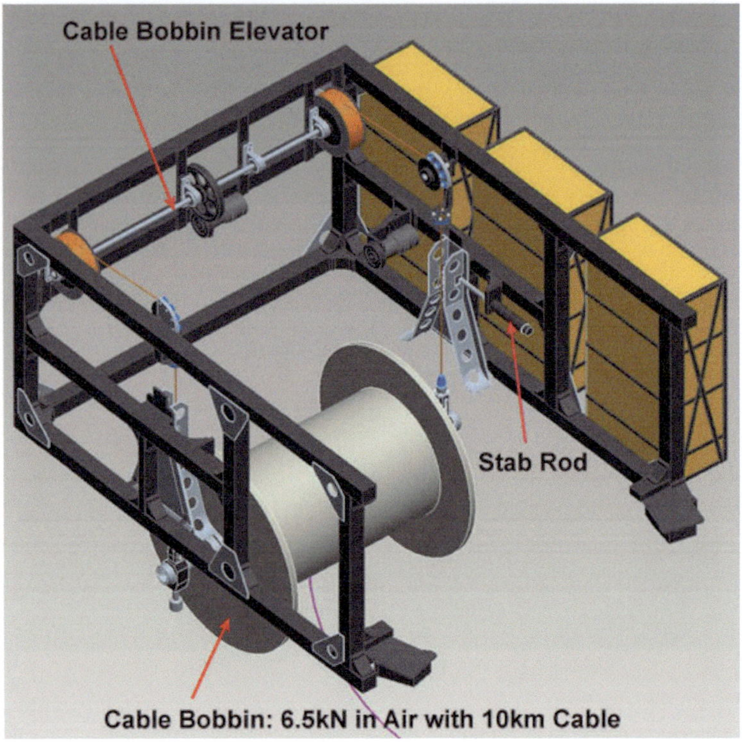

Figure 10.10 Cable bobbin elevator. The cable bobbin elevator make it possible to lift the cable bobbin up and down with a 10km length of extension cable in air and water. The mechanism is used for the bobbin set up on board, or suspending/resuming the 10km distance cable laying operation on the seafloor.

manage approximately 1.5kN of buoyant variation using vertical thrusters. However, this is not sufficient for the cable laying operation where buoyancy variation is 2.2kN during the operation. The VBCS equipped a pair of 50L volume pressure resistant water tanks to compensate the 1kN of buoyancy additionally in water to maintain the mobility of the ROV in cable laying operation as shown in Figure 10.11. A high pressure water pump and solenoid valves control the volume of water in pressure tanks up to 4500m depth of pressure environment referring to the level sensor output.

Beyond the cable laying operation, the ROV is also able to perform the science node installation, observatory construction and cable recovery operations by equipping it with some additional bolt-on accessories contained within the cable laying system as seen in Figure 10.12. The cable laying system and all additional equipment for observatory construction are actuated by hydraulic pressure supplied by the ROV "Hyper Dolphin" by hydraulic power take-off for user payloads.

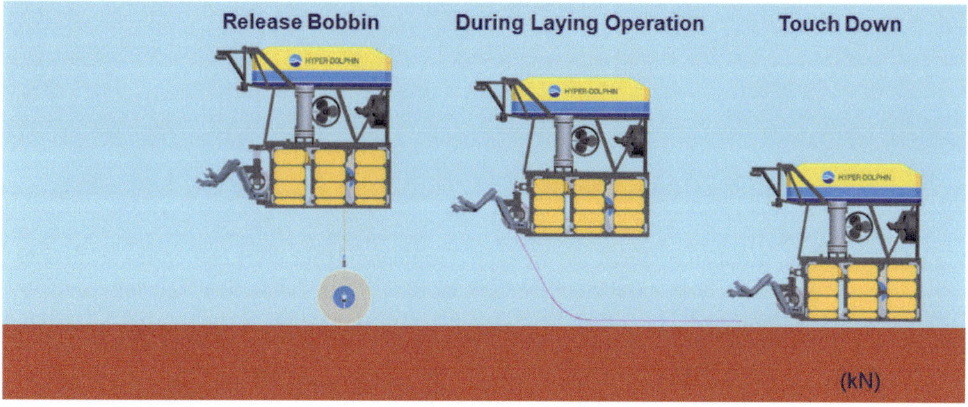

(kN)	Release Bobbin	During Laying Operation	Touch Down
Buoyancy Variation	2.2	0.9	0
with VBCS	1.2	0	0
Adjust Initial Condition	0.7	0	-0.5

Figure 10.11 A consideration of buoyancy variation of ROV in operation. The ROV "Hyper Dolphin" has the ability to manage approximately 1.5kN of buoyancy in water using vertical thrusters. However, this is not sufficient to realize the planned subsea construction scenario of DONET as shown here. The assistance of VBCS (Variable Buoyancy Control System) is indispensable to managing the large buoyancy variation during the operation and to put the theory into practice.

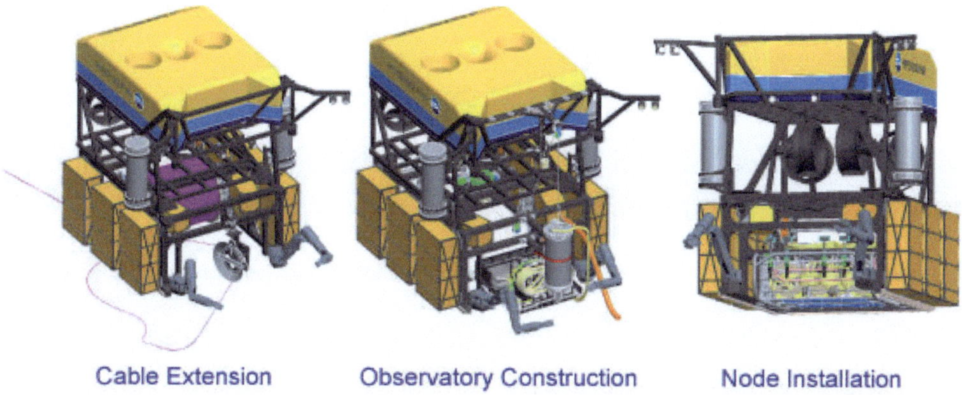

Figure 10.12 Additional uses of the cable laying system. It is designed to use 10km length of extension cable, and it also able to implement other operations such as extension cable recovery, node installation, observatory construction and installation by adding a few mechanical modifications.

Figure 10.13 Node installation. From left to right: Termination Unit (TU) laid on the seafloor by cable laying ship; science node installed on seafloor by ROV; ROV connecting high voltage hybrid UMC from science node to TU.

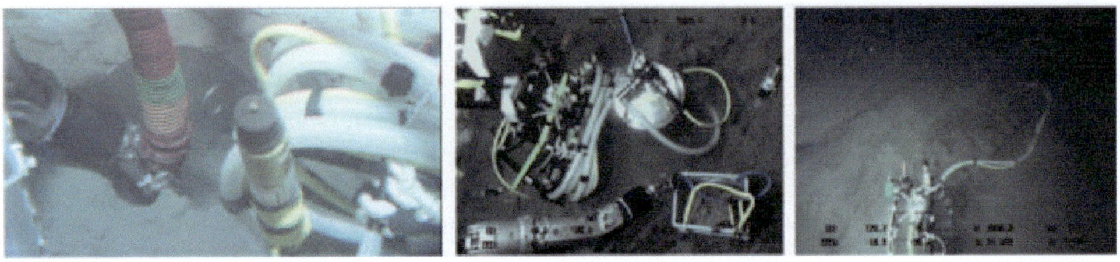

Figure 10.14 Observatory construction. From left to right: burial casing conditioning by DOROTHY suction pump system; fitting the ground motion sensing system into the burial casing; pressure sensing system on seafloor.

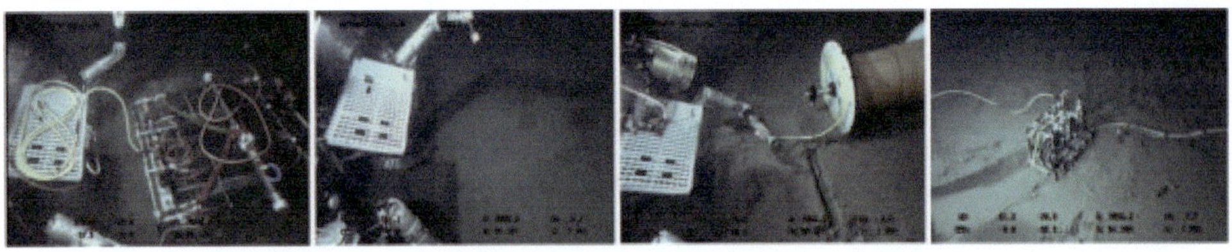

Figure 10.15 Cable laying operation. From left to right: connecting extension cable end UMC to the user port on science node; extension cable laying in operation; releasing the cable bobbin to the seafloor from the cable extension system and taking final extension of extension cable from the side wall of cable bobbin; connecting the final extension UMC to observatory interface on pressure sensing system.

10.8 DONET construction

The DONET backbone cable system installation was conducted from January to early March 2010. For the backbone cable system, 300km lengths were laid on the seafloor by a conventional cable laying method using a cable laying ship. The subsea construction of the world first precise seafloor earthquake and tsunami observatory was started from March 2010, when R/V Natsushima and ROV "Hyper Dolphin" scheduled eight ROV dives for the DONET observatory construction in the expedition NT10-04.

In the first dive of the expedition, the ROV placed a science node in the 2000m depth of DONET observation area "A" and connected it to the TU using UMC (Figure 10.13). Immediately after the connecting operation, the physical condition of the installed science node was confirmed from the node control interface at a landing station.The first DONET observatory (consisting of a ground motion sensing system and a pressure sensing system) was set up at observation site "A-3" during the second dive. The ground motion sensing system was fitted into a burial casing that had been driven into the seafloor sediment in advance of expedition. The pressure sensing system was put on the seafloor sediment about 10m away from of the burial hole (Figure 10.14). During day 2 of setting up the observatory, the "Hyper Dolphin" connected the observatory "A3" to the science node "A" using a 10km length of extension cable system (Figure 10.15). It took approximately 10 hours to lay the cable between the two target points. Figure 10.16 shows a cross section of the as-laid extension cable laying route. In this case, the substantial distance from node to observatory is 9452m from the cross section map and it takes 9800m of extension cable

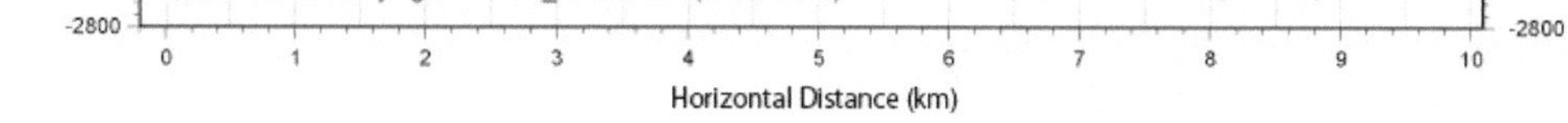

Figure 10.16 A result from the extension cable laying operation. The figure shows the cross section of as-laid extension cable route. Calculated true distance from science node to observatory on the bathymetry map is 9478m and the actual cable payout is 9800m from marking on extension cable. The calculated average cable slack is less than 4% from this operation. This number is reasonable compared with the average design cable slack of 8%.

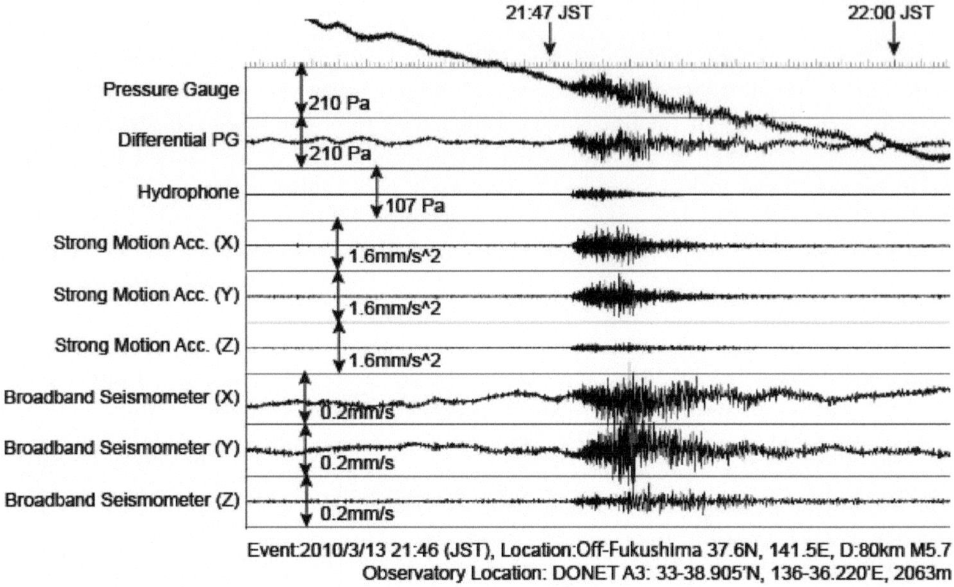

Figure 10.17 Initial data from the observatory. This figure shows a seismic event observed as the first trial observation developed. The hypocentre location of the event is approximately 800km from the observatory. The record shows the observatory has good S/N ratio for seismic observation.

with cable slack. This means the average cable slack of this operation is less than 4% of the physical cable length. This number is very reasonable for any type of cable laying operation, including the novel DONET cable laying method by ROV. Once the cable laying operation was completed, the UMC connections operations were performed and the observatory was successfully booted-up from a landing station and started to acquire a data set from the observatory. Figure 10.17 shows initial data from the observatory. The quality of the data satisfied the requirement of DONET project sufficiently.

10.9 Summary

DONET is a development program for a novel surveillance infrastructure for earthquakes and tsunamis. This program spanned a time frame from April 2006 to March 2010. DONET successfully created a new observation architecture, sensing system and field survey capabilities along with all the related technologies and new techniques for seafloor observatory construction in this duration and it was demonstrated in field operation. The final stage of DONET concluded with the construction of a network started in January 2010, with the first observatory successfully launched in March 2010. This observatory produced a very precise data set that has a level of quality never before collected from the seafloor. In addition, at the time of writing, 19 observatories are waiting in storage for seafloor deployment in the near future. The data from DONET will be made available to the Japan Metrological Agency, local government, and world research communities (on a joint research basis) real-time in the near future.

Acknowledgment

The contract research "Development of Dense Ocean Floor Network System for Earthquakes and Tsunamis" is supported by the Ministry of Education, Culture, Sports, Science and Technology, Japan.

The authors wish to thank the Captain and crew of R/V `Natsushima`, the "Hyper Dolphin" operation team, for ROV modification, also the SOSI (Sound Ocean Systems Inc.) and ISE (International Submarine Engineering Ltd.) for VBCS integration.

References

Araki E., Kawaguchi K., Kaneko S. and Kaneda Y. (2008) Design of deep ocean submarine cable observation network for earthquakes and tsunamis. In: Proceedings of OCEANS'08 MTS/IEEE KOBE-TECHNO-OCEAN '08 (OTO'08), April 8–11, Kobe, Japan.

Bird L.E. (2002) ROV based tool sled for the placement of fiber optic cable between benthic instrument nodes. In: Proceedings of OCEANS'02 IEEE Biloxi, October 29–31, Biloxi, USA.

Kaneko S., Araki E., Kawaguchi K., et al. (2009) Installation method of high-quality seismic observation in the seafloor. In: Proceedings of OCEANS'09 IEEE, May 11–14, Bremen, Germany.

Kawaguchi K., Nishida T. and Obana S. (2001) A thin fiber cable laying system for mobile deep seafloor observation. In: Proceedings of OCEANS 2001, IEEE, November 5–8, Hawaii, USA.

Kawaguchi K., Hirata K., Nishida T., Obana S. and Mikada H. (2002) A new approach for mobile and expandable real time deep seafloor observation: Adaptable Observation System. IEEE Journal of Oceanic Engineering 27, 182–192.

Kawaguchi K., Araki E. and Kaneda Y. (2007) A design concept of seafloor observatory network for earthquakes and tsunamis. In: Proceedings of International Symposium on Underwater Technology 2007/International Workshop on Scientific Use of Submarine Cables and Related Technologies 2007, April 2007, Tokyo, Japan.

Kawaguchi K., Kaneda Y. and Araki E. (2008) The DONET: A real-time seafloor research infrastructure for the precise earthquake and tsunami monitoring. In: Proceedings of

OCEANS'08 MTS/IEEE KOBE-TECHNO-OCEAN '08 (OTO'08) 2008, April 8–11, Kobe, Japan.

Kawaguchi K., Kaneko S., Nishida T. and Komine T. (2009) Cable laying ROV for real-time seafloor observatory construction. In: Proceedings of OCEANS'09, IEEE, May 11–14, Bremen, Germany.

Shepherd K., Tamburri K. and Mills R. (2007) Observatory cable laying system. In: Proceedings of OCEANS'07, IEEE, Sep. 29-Oct. 4, Vancouver, Canada.

11 GEOSTAR-class observatories 1995–2012: A technical overview

F. Gasparoni[1], F. Furlan[1], F. Bruni[1], F. Zanon[1], P. Favali[2], L. Beranzoli[2], G. Marinaro[2], A. De Santis[2] and H.W. Gerber[3]

11.1 Introduction

From the scientific point of view, the deepwater environment represents the ultimate frontier for Earth observation and understanding fundamental processes (solid earth studies, as well as oceanographic, climatic and environmental investigations) (Kopf et al., 2012). Development and operation of seafloor observatories, defined as "unmanned stations, capable of operating for long-term at seafloor, supporting the continuous and stable operation of a number of instrumented packages related to various disciplines" (a more detailed but conceptually similar definition is proposed by NRC (NRC, 2000)), are now recognized as the essential approach to achieve full-time presence at deep seafloor and overcome the main limitations of the traditional ship-based approach, intrinsically episodic and inadequate to provide data at the temporal and spatial scales required (Favali and Beranzoli, 2006; Favali et al., 2010; Lampitt et al., 2010).

The distinction of in-situ investigations from other kinds of research (like ship cruises) is more evident in deep-sea science because "to be in-situ" raises significant technological challenges (Gasparoni et al., 1998; Ollier et al., 2002). The ocean bottom is remote, hostile, corrosive to delicate instruments and highly variable in temperature and pressure. Although some aspects are similar to space exploration (hostile environment, remoteness, etc.), others are rather different: energy from the sun is not available and efficient commu-

1 Tecnomare SpA, Venezia, Italy

2 Istituto Nazionale di Geofisica e Vulcanologia, Roma, Italy

3 Beuth Hochschule für Technik Berlin, Germany

nication links are not possible, making long-term operation in deep water very challenging as planetary activities. The first problem to be solved to establish a deep-sea observation capability is in fact to gain "the possibility to be there" in exactly the right place and for the required period of time (Beranzoli et al., 2002b; Favali et al., 2004).

Technology, and technology innovation in particular, plays a fundamental role in the development of advanced solutions capable of answering the challenging requirements of seafloor observatory science, introducing new capabilities or allowing old functions to be performed with greater efficiency (Ruhl et al., 2011, Aguzzi et al., 2012). Transfer of experiences from the oil&gas industry applications offers a unique opportunity of cross-fertilization, bridging the gap between the practice of offshore technology and the possibility of developing engineered solutions for scientific investigation (Favali and Beranzoli, 2006).

This contribution intends to provide a chronological and logical history of the GEOSTAR-class seafloor observatories (Favali and Beranzoli, 2009b), mainly focusing on the technological aspects.

11.2 The origins: ABEL and DESIBEL

Recognizing these needs and technological challenges, since 1989 the European Union (EU) has promoted a specific action within the R&D Framework Program, namely the Marine Science and Technology Programme (MAST), where important studies and projects were developed with EU support. In addition, dedicated international workshops and conferences were organized, contributing to the establishment of a European network among the different parties involved (academic and scientific institutions, industries, etc.) (Beranzoli et al., 2000b).

Within the initiatives promoted by the EU, two project studies in particular, carried out in the framework of the EU MAST-2 Programme, may be considered the origin of GEOSTAR concept: ABEL and DESIBEL.

The former project was a feasibility and financial study aimed at identifying the scientific requirements, possible technological solutions and opportunities for the development of an Abyssal BEnthic Laboratory (ABEL) (Berta et al., 1995). The basic requirements of the corresponding study can be summarized as follows: to ensure the possibility of carrying out in-situ scientific, multidisciplinary, autonomous seafloor observations and experiments, at water depths up to 6000m.

Awarded to Tecnomare, and carried out between 1992 and 1993, the study proved the feasibility of the concept of a benthic laboratory, capable of operating both in autonomous and controlled mode for periods of several months up to one year. The proposed configuration was a network of cooperating stations, capable of being reconfigured according to specific mission requirements.

The system, shown in Figure 11.1, includes a main station (main Benthic Investigation Laboratory) devoted to the execution of the most complex tasks, with a number of secondary fixed stations (Satellite Stations) acting as nodes of the measuring network and a mobile vehicle (Mobile Station) that extend capabilities of the fixed stations enabling

Figure 11.1 Tecnomare concept of Abyssal Benthic Laboratory (1993).

possibility of surveys in the investigation area and even interaction with the fixed stations. ABEL architecture also includes a dedicated module to deploy and recover the stations, as well as a shore station.

The ABEL concept represents the equivalent at deep seabed of an onshore multidisciplinary meteorological or geophysical laboratory. Significant analogies may be identified also with respect to past and ongoing studies of planetary stations.

The approach used to overcome limitations of systems presently in operation and to address technological development was based on three key elements:

- extension of the operating capabilities of the instrumented bottom stations, to ensure adequate support to the multiplicity of the scientific packages foreseen
- surface-assisted deployment and recovery, for accurate and controlled execution of the marine operations
- maximum interaction with the scientific or technical user during all phases of the mission, to allow remote control and effective operability of the stations.

All the above proposed technical solutions were translated into practice with the development of GEOSTAR.

The second study (DESIBEL) was aimed at investigating methods for deployment and intervention on future benthic stations (Rigaud et al., 1998). Within this latter study four concepts were investigated, namely:

- an active docking system with a mobile hook (LOMOS)
- an active docking system with a special ROV (REMORA)
- a light scientific ROV (ROV 6000)
- a free swimming vehicle (FREE MODULE).

For each concept, engineering studies were conducted as well as cross comparisons, mainly based on simulations of a variety of operational conditions. In particular, LOMOS turned out to be the most suitable solution where heavy payloads needed to be managed, such as the advanced benthic stations previously identified in ABEL study.

In parallel, several international conferences and workshops reconfirmed the need to "join forces" to achieve – even with different scientific objectives – the realization of multidisciplinary sea-bottom observatories and to extend at a global scale the existing land-based networks of permanent observatories of Earth processes (Frugoni et al., 2006). It was argued that such an approach would allow a significant overall cost reduction and contribute greatly to the development a new generation of "carriers" of scientific packages that are indeed required to advance the present understanding of a great variety of Earth processes.

11.3 GEOSTAR

In order to further proceed with the development process started with the ABEL feasibility study, seven scientific and technological European organizations[4] joined their efforts in the GEOSTAR (GEophysical and Oceanographic STation for Abyssal Research) project (Beranzoli et al., 1998; 2002a; Jourdain, 1999), aimed at the development of the prototype of an innovative deep sea observatory capable of carrying out long-term geophysical, geochemical and oceanographic observations at abyssal depths (4000m) (Berta et al., 1995; Gasparoni et al., 2002).

The concept proposed (and implemented in a two-phase project) is shown in Figure 11.2.

The observatory is characterized by original and innovative technological solutions such as:

- the open frame in light, non-magnetic alloy
- active devices for the deployment of specific packages (like the seismometer and the magnetometers)
- a dedicated data acquisition and mission management system, based on custom-built low power hardware

4 Istituto Nazionale di Geofisica, Tecnomare, CNR - Istituto per la Geologia Marina, Technische Universitat Berlin, IFREMER, ORCA Instrumentation, University of Newcastle Upon Tyne

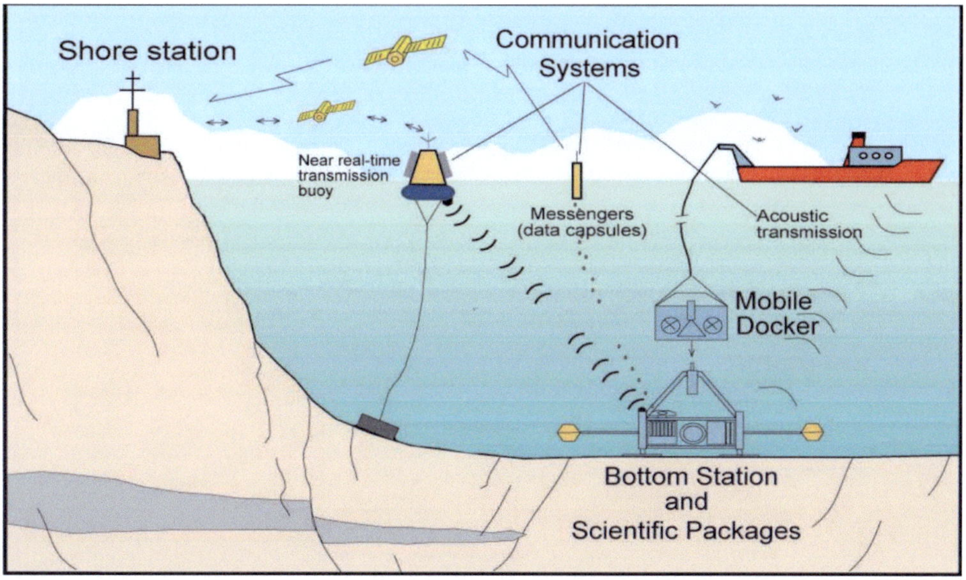

Figure 11.2 GEOSTAR concept (as proposed to the EU MAST III programme, 1995).

- autonomous mission capabilities, including power management and self-diagnostics
- possibility of being reconfigured according to different mission requirements and sites
- a dedicated deployment and recovery vehicle (Mobile Docker or MODUS), derived from LOMOS concept
- multiple possibility of interfacing with external devices (communication systems, deployment system) for continuous control of system status both during the deployment phase and during the mission.

The frame configuration is one key aspect of GEOSTAR's success (Beranzoli et al., 1998). It is a stand-alone autonomous unit, based on a four-legged aluminum open frame supporting all the scientific packages and the mission payload (such as the vessels housing the data acquisition and control system, the battery pack, the communication systems, etc.). For quicker and more reliable deployment and recovery the frame is equipped with a docking cone (Figure 11.3) on which the mechanical connector that mates with the Mobile Docker is mounted (Gerber and Clause, 2001; Clauss and Hoog, 2002).

The single frame, "heavy in water" concept provides important advantages over concepts based on multiple sensor packages managed by ROVs and underwater junction boxes:

Figure 11.3 GEOSTAR Bottom Station frame (right) and mobile docker (left). Scientific payload not mounted. The magnetometer boom on the left of the Bottom Station is shown in the extended position; the seismometer module is partially visible at the left of the two pressure vessels.

- single operation required for deployment (and subsequent recovery)
- no volume reserved for buoyancy and recovery equipment
- increased reliability (no active device at seafloor)
- stable at seafloor
- insensitive to variations of payload
- the same vehicle (the Mobile Docker) can manage several Benthic Stations.

The open frame allows an easy and effective installation and access to the mission payload and gives a virtually unlimited possibility of reconfiguration/extension according to new or different requirements (Gasparoni et al., 1998). This characteristic was verified during the operational life of GEOSTAR (refer to Table 11.1, summarizing GEOSTAR payload configuration in the six missions carried out so far) (Beranzoli et al., 2000a, b; 2003).

The relative availability of volume and resources (electrical power, interfaces, acquisition and processing power) stimulated scientists to not limit their attention to commercial off-the-shelf sensors only, but also to conceive special, innovative instrumented packages specifically for GEOSTAR (Beranzoli et al., 2002a). Technology development work was

Parameter	Mission 1		Mission 2		Missions 3 and 4		Missions 5 and 6	
	Sensor	**Sampling rate**	**Sensor**	**Sampling rate**	**Sensor**	**Sampling rate**	**Sensor**	**Sampling rate**
Broad-band seismometer	Guralp CMG-1T	20 samples/s	Guralp CMG-1T	100 samples/s	PMD/Eentec EP-300-DT	100 samples/s	Guralp CMG-40	100 samples/s
Accelerometer							Guralp CMG-5T	100 samples/s
Scalar magnetometer	GEM Systems GSM-19L	1 sample/ min	GEM Systems GSM-19L	1 sample/ min	GEM Systems GSM-19L	1 sample/ min		
Vectorial magnetometer	INGV prototype	1 sample/ 10 s	INGV / Tecnomare prototype	1 sample/ 10 s	INGV / Tecnomare prototype	1 sample/ 10 s		
Gravity meter			CNR-IFSI prototype	1 sample/s	CNR-IFSI prototype	1 sample/s	CNR-IFSI prototype	1 sample/s
Hydrophone					OAS E-2PD	100 samples/s	OAS E-2PD	100 samples/s
Pressure							Paroscientific 8CB4000-I	1 sample/ 15 s (MISSION) 1 sample/5 s (EVENT)
ADCP	RDI Workhorse 300 kHz	1 profile/h	RDI Workhorse 300 kHz	1 profile/h	RDI Workhorse 300 kHz	1 profile/h	RDI Workhorse 300 kHz	1 profile/h

Parameter	Mission 1		Mission 2		Missions 3 and 4		Missions 5 and 6	
	Sensor	**Sampling rate**	**Sensor**	**Sampling rate**	**Sensor**	**Sampling rate**	**Sensor**	**Sampling rate**
CTD	Sea Bird SBE-16	1 sample/h	Sea Bird SBE-16	1 sample/h	Sea Bird SBE-16	1 sample/h	Sea Bird SBE-16 plus	1 sample/h
Turbidity	Chelsea Instruments Alphatracka II	1 sample/h	Chelsea Instruments Alphatracka II	1 sample/h	Chelsea Instruments Alphatracka II	1 sample/h	Wet Labs ECO-BBRTD	1 sample/h
pH / Eh	Wolfson Sensor Group prototype	(*)	Systea prototype (AMT sensors)	(**)	Tecnomare prototype (AMT sensors)	1 smp/6 h (calibration every day)		
Current meter			FSI 3D-ACM	2 samples/s	FSI 3D-ACM	2 samples/s	Nobska MAVS-3	5 samples/s
Automatic water sampler			McLane RAS-48-500	2 × 500 ml samples/ 10 days	McLane RAS-48-500	2 × 500 ml samples/ week		
Precision tilt meter	Applied Geomechanics 716	1 sample/h	Applied Geomechanics 716	1 sample/h	Applied Geomechanics 716	1 sample/h		
Compass			Falmouth Ostar	1 sample/h	Falmouth Ostar	1 sample/h	Falmouth Ostar	1 sample/h
Inertial Measurement Unit (Accelerometer + gyro)							Landmark LMRK10IMU-150-12-100	200 samples/s

* Automatic analyser also including H_2 and H_2S electrodes; developed but not installed for the mission

** Developed but not installed for the mission

Table 11.1 GEOSTAR payload and sampling rates.

Figure 11.4 Typical GEOSTAR mission setup; note the payload arrangement on the open frame.

therefore not simply addressed to manage standard sensors, but also to develop customized underwater versions of sensors originally designed for different applications (this is the case with the vectorial magnetometer and gravity meter (Iafolla and Nozzoli, 2002)), to support the design of completely new packages (as with the automatic chemical analyzer) or simply to make available a platform where packages designed and developed by other institutions could be operated (Favali et al., 2002) (Figure 11.4, Figure 11.5 and Figure 11.6).

Particular attention was paid in the observatory design in order to take into account the specific requirements of the various scientific packages, in terms of mounting constraints and minimization of possible interferences. This applies in particular for the magnetometers and the seismometer, whose specific requirements do not allow a direct mounting on the Bottom Station frame, as the resulting measurements would be affected by noise and disturbances induced by the station itself and devices mounted on it. As regards the magnetometers, it had been decided to mount the glass spheres housing the sensors at the end of long booms (approx. 2.5m) hinged at two opposite corners of the Bottom Station frame. These booms are automatically extended when at the seabed, after a remote command by the surface operator; this ensures that the sensors are adequately located far from the Bottom Station structure that is the source of electronic noise.

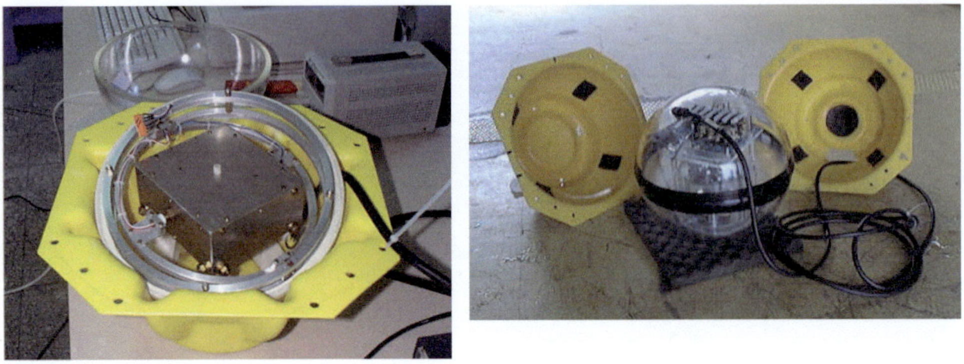

Figure 11.5 Gravity meter (left) during customization work and laboratory tests; (right) Vectorial magnetometer integrated into a glass sphere.

Figure 11.6 (Left) Prototype chemical analyzer engineered by Tecnomare. (Center) Experimental seismometer derived from a space prototype integrated to GEOSTAR frame during wet test in IFREMER basin. (Right) GEOSTAR junction box providing interfaces for payload.

Another active device has been conceived to manage the seismometer, in order to release it from the station and couple with the seafloor (Figure 11.7 and Figure 11.8). Again, this actuation is commanded by the surface operator. At recovery the instrument hangs below the station, suspended on a rope.

Both actuation systems have been designed and manufactured specifically from this application; basically they consist of modified Acoustic Releases directly interfaced to observatory electronics. Finally, all sensors whose measurements could be affected by an

Figure 11.7 Different arrangements for the seismometer. From left to right: Guralp seismometer first arrangement (GEOSTAR mission 1); same instrument with an upgraded arrangement (GEOSTAR mission 2 and SN1 first mission); arrangement for PMD seismometer (GEOSTAR missions 3 and 4, SN2 mission 1, SN4 mission 1); Guralp seismometer final arrangement in titanium sphere (SN1 missions 2 and 3).

Figure 11.8 Seismometer management systems (left) used in GEOSTAR, SN2, SN3 and SN4; (right) used in SN1.

excessive tilt of the Bottom Station once at seafloor are provided with suitable levelling systems: motorized and remote controlled for the seismometer, passive (gimbals) for the gravity meter and vectorial magnetometer.

The "heart" of GEOSTAR is the Data Acquisition and mission Control System (DACS), on which the most advanced functionalities depend. Long autonomy, high reliability, capability to manage a wide range of sensors and devices, capability to manage large quantities of data are the basic requirements that triggered the development of a dedicated hardware for:

- payload management and control
- mission management
- power management and control
- technical status parameters monitoring

This hardware has been designed to meet the requirements of complex and multidisciplinary instrumented systems operating at sea, taking into account the necessity to operate in critical conditions (reduced volume, limited quantity of energy, hostile environment) and according to standardization criteria constituting one of the peculiar characteristics of GEOSTAR-class observatories (Table 11.2).

The architecture consists of different low power microprocessor units working in parallel (Figure 11.9). Number and type of units are selected case by case according to the complexity of the tasks and functionalities required by the experiments. Typical architecture of a GEOSTAR-class observatory includes:

- one Mission Control Unit, in charge of system configuration and supervision, interface with the communication systems, technical status monitoring
- a number of Payload Management Units, in charge of scientific packages data acquisition and control, data processing, data storage.

Weight (kN)	25.4 (in air), 14.2 (in water)
Dimensions (mm)	3500 × 3500 × 3300 (magnetometer booms retracted)
Design depth (m)	4000
Material	Aluminum 5083 (frame); titanium grade 5 (vessels); stainless steel (docking pin)
Data acquisition and mission control	4 boards (32 bit microcontroller MC68332)
Data storage	Hard disks, CompactFlash
Power supply	24 VDC, 3000 Ah Lithium-thionyl chloride
Power consumption (mA)	70 (idle mode), ~300 (mission mode)
Status parameters	Voltage, current, temperature, heading, tilt x/y, water intrusion, echo sounder

Table 11.2 GEOSTAR main characteristics (data refer to the last version).

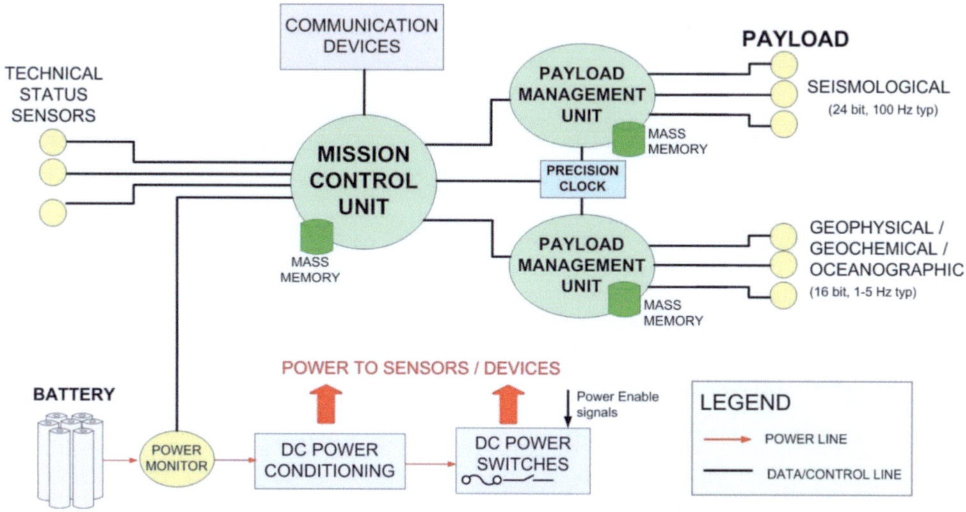

Figure 11.9 Reference architecture of a GEOSTAR-class observatory DACS.

The DACS is completed by auxiliary boards dedicated to power regulation, power switch (allowing switch on or off each instrumented package according to the preprogramed strategy or the occurrence of significant events like failures, energy loss, etc.), status monitoring (technical parameters like battery voltage, current, internal temperature).

A low power precision clock (rubidium) with long-term stability of 10-9 can be added to meet the requirements of seismological monitoring.

GEOSTAR deployment and recovery are ensured by a dedicated deep-sea vehicle (MODUS) – basically a special, simplified version of a Remotely Operated Vehicle – which is able to handle heavy payloads (up to 30 kN) with a weight of less than 10 kN (Figure 11.10). A comparison with the numbers of the few available deep-sea ROVs capable of operating at depths greater of 4000m (30–50 kN weight, 1–2 kN max payload), provides a clear idea of the efficiency and cost-effectiveness of the GEOSTAR concept. Unlike a standard ROV, MODUS has no free swimming capabilities and operates suspended from its electro-mechanical umbilical cable (25mm diameter, 1.8 kN/km weight in water) providing both power and fiber optic telemetry (Gerber et al., 2002). Electrical thrusters ensure mobility on the horizontal (x–y) plane, while the winch of the support vessel regulates its descent/ascent (z-axis). By means of visual (TV cameras) and instrumental (sonar) systems MODUS is capable of locating the predetermined installation area or find GEOSTAR for the subsequent recovery (Clauss et al., 2004) (Figure 11.11). An operation range of 5% of the water depth has been verified.

Figure 11.10 GEOSTAR–MODUS connection principle, based on docking pin (on GEOSTAR) and latching device (on MODUS).

MODUS guidance and control is ensured by a Surface Unit including monitors, PC with human-machine interface, joysticks for steering, and video recorders (Gerber and Claus, 2005)

Interaction with the user during the mission is one of the key functionalities for a seafloor observatory, enabling data transfer to the end users as well as full control of the mission.

Generally speaking, three basic configurations for a seafloor observatory can be identified, as regards the levels of interaction with the remote operator (and associated architecture of the communication infrastructure):

- autonomous operation; no connection apart from possibility of episodic access (e.g., from a ship of opportunity) where the observatory is provided with acoustic telemetry
- near-real-time connection (NRTCS); remote accessibility via underwater acoustic telemetry and a moored relay buoy connected to shore via radio or satellite, with limited capacity (in terms of quantity and bandwidth of transmission due to the acoustic telemetry)
- real-time connection to a shore station (via power and communication cable); enables "permanent" operation with full integration of the observatory in larger monitoring networks.

Levels of interaction are implemented according to the installation site location, available infrastructures and mission requirements. During the development of GEOSTAR-class observatories, all these configurations were implemented and operated in the field; in particular, GEOSTAR adopted the NRTCS scheme, with increasing levels of complexity and associated functionalities in the six missions carried out (see Table 11.3). SN2 operated in

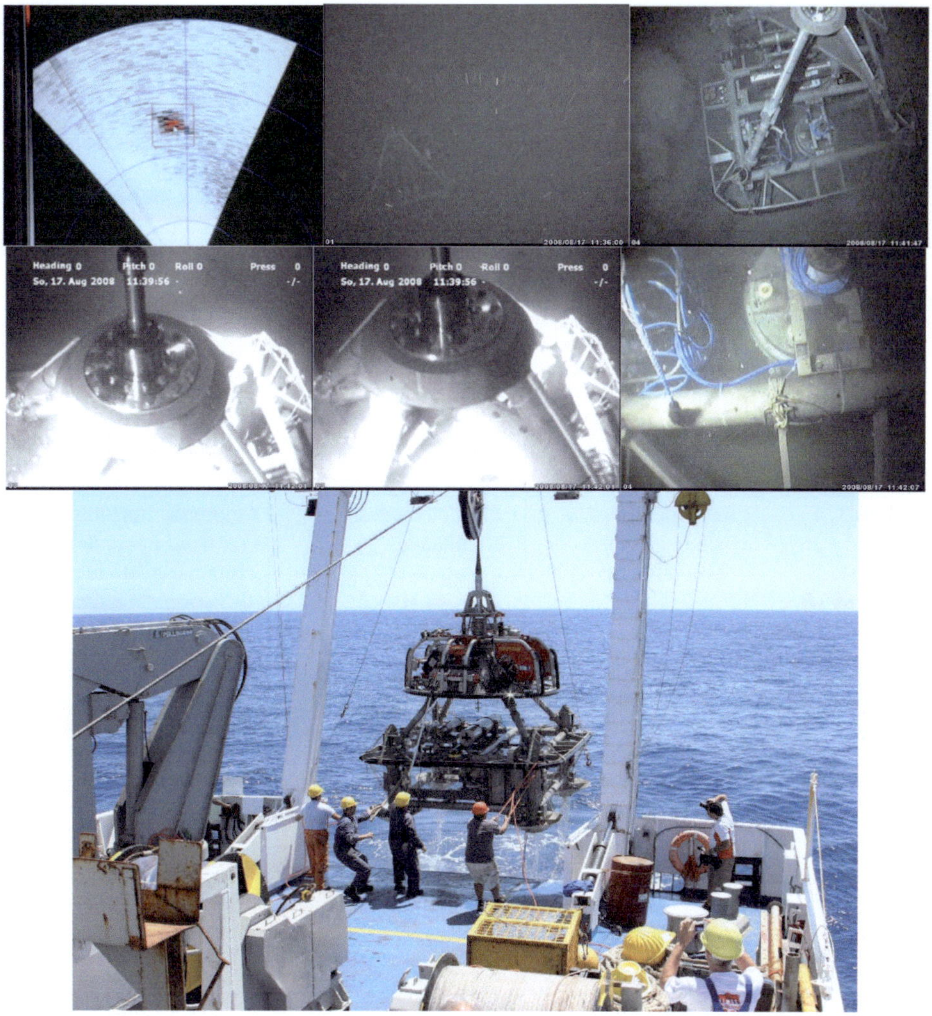

Figure 11.11 Typical recovery sequence of GEOSTAR (from top to bottom, left to right): observatory detected by the sonar, visual contact established, close approach, MODUS manually guided over the docking cone, MODUS lowered by the winch operator, docking complete, observatory onboard

autonomous mode, SN1 and GMM operated in autonomous mode but were also configured as a cabled observatory, SN3 and SN4 worked in near-real-time connection.

An essential component of the NRTCS architecture is a buoy moored in the immediate vicinity of the observatory and configured to operate as a communication relay between the observatory and the onshore remote operator (Marvaldi et al., 1998; Beranzoli et al., 2004). GEOSTAR buoy evolution is illustrated in Figure 11.12 and Table 11.4.

		Activated from surface	**Activated from seafloor**
Mission 1 (GEOSTAR 1) Adriatic Sea	Acoustic link • ship of opportunity	• Mission control commands • Summary data files upload	No functionality implemented
	Data capsules		ARGOS Messenger release (programmed or on event)
Mission 2 (GEOSTAR 2) Tyrrhenian Sea (offshore Ustica Island)	Acoustic link • shore station (via relay buoy) • ship of opportunity	• Mission control commands • Summary data files (mission, technical status) upload	No functionality implemented
	Data capsules		ARGOS Messenger release (programmed or on event)
Missions 3, 4 (ORION/ GEOSTAR 3) Tyrrhenian Sea (Marsili Seamount)	Acoustic link • shore station (via relay buoy) • ship of opportunity	• Mission control commands • Summary data files (mission, technical status) upload • Wave forms (seismic, hydrophone) upload	Automatic transmission • (at fixed intervals) summary data files (mission, technical status)
Mission 5 (NEAREST) Gulf of Cadiz (offshore Portugal)	Acoustic link • shore station (via relay buoy) • ship of opportunity	• Mission control commands • Summary data files (mission, technical status) upload • Event information (pressure, seismic) upload	Automatic transmission • (at fixed intervals) summary data files (mission, technical status) • (on event) event information (pressure, seismic)
Mission 6 (NEAREST) Gulf of Cadiz (offshore Portugal)	Acoustic link • shore station (via relay buoy) • ship of opportunity	same as Mission 5	Automatic transmission • (at fixed intervals) summary data files (mission, technical status) • event catalog (pressure, seismic) • (on event) event information (pressure, seismic)

Table 11.3 GEOSTAR-operator interaction levels.

Figure 11.12 GEOSTAR buoy evolution: (left) mission 2; (center) missions 3 and 4; (right) missions 5 and 6.

Configuration	GEOSTAR 2 (mission 2)	ORION (missions 3,4)	NEAREST (missions 5,6)
Power	Primary lithium-thionyl chloride, 28 V 312 Ah	Lead-acid, 24 V 40 Ah recharged by 2×110 Wp photovoltaic panels	Lead-acid, 24 V 40 Ah recharged by 3×125 Wp photovoltaic panels
Communication (underwater segment)	12 kHz multimodulation acoustic modem	12 kHz multimodulation acoustic modem	12 kHz multimodulation acoustic modem
Communication (surface segment)	Inmarsat Mini-M VHF radio link	VHF radio link Iridium	Globalstar
Payload	Technical status parameters	Technical status parameters	Meteo station Barometric sensor Buoy attitude Technical status parameters

Table 11.4 GEOSTAR buoy main technical specifications.

Figure 11.13 ARGOS Messengers (GEOSTAR mission 2 configuration) (left) during deployment; (right) released.

A second communication system was specifically developed for GEOSTAR, based on a set of releasable capsules called "Messengers", capable of transferring data by ARGOS satellite telemetry once arrived at the sea surface (Figure 11.13). Two types of Messengers were used:

- Expendable Messengers, released periodically (depending on the mission duration) or under particular conditions (i.e., in case of failure detection in the observatory); up to 32 Kbytes of data can be stored and then transferred through satellite telemetry
- Storage Messengers, released on external request (e.g., by operators on a ship), and storing up to 40 Mbyte of data.

11.3.1 GEOSTAR mission 1 (Adriatic Sea)

From the beginning, the GEOSTAR project was conceived as a two-step development process. The first phase (1995–1998) was carried out under the framework of EU Marine Science and Technology (MAST-3) program and was aimed at demonstrating the technology of the concept in shallow water and with a limited but essential set of functionalities implemented. The demonstration included a short mission (<1 month) in shallow water. The results of this represented a go/no-go point for the continuation of the project to the second phase, aimed at completing the technology development and carrying out the first deep-sea long-term mission with a complete payload and full capabilities implemented (Favali et al., 1998).

Due to the limited duration of the shallow water mission, the observatory was configured to operate in autonomous mode with the possibility of being interrogated from the ship of opportunity via vertical acoustic telemetry (12 kHz multimodulation acoustic modem ensuring up to 2400 bit/s). For the same reason, no dedicated surface logistics was developed (MODUS was managed through a rope and power/telemetry umbilical simply fastened to it); temporary pressure vessels for the DACS and battery pack were developed (in aluminum, 200m design depth); a smaller battery pack was used. The configuration included also three ARGOS Messengers (two expendable, one storage).

After an extensive phase of wet test in the IFREMER deep basin and TUB Berlin water circulation basin, GEOSTAR was deployed in August 1998 in the Adriatic Sea (offshore Ravenna) at the depth of 42m, and operated continuously for 450 hours. This mission had the goal of demonstrating all the capabilities of the system in real conditions. For the purpose of the technology demonstration, three main objectives were to be fulfilled:

- verify GEOSTAR capability to manage the scientific payload (and in particular the most demanding ones: seismometer and magnetometers)
- verify capability of the dedicated vehicle MODUS to manage GEOSTAR deployment and recovery procedures
- verify the possibility to interact with GEOSTAR during the mission, through the communication systems adopted.

Results obtained led to the conclusion that these objectives were fully achieved; all system capabilities and marine operations were successfully verified in real conditions, and important feedback obtained about the system enhancement in view of the second phase of the project (Gerber et al., 1999; Beranzoli et al., 2003).

The acoustic telemetry link ensured the ability to interact with the observatory during all phases of the mission (to check status of the system while at seabed, collect data, command Storage Messenger release, stop mission and lock seismometer masses), and at the same time represented an essential back-up of the cable link used during deployment. An Expendable Messenger was automatically released during the mission and its data successfully received via ARGOS.

Significant data regarding seismic events, magnetic field variations, water current and water characteristics of the area were collected and made available to scientists for proper analysis, demonstrating GEOSTAR's capability of operating as a multidisciplinary observatory.

Marine operations were carried out by the Italian R/V Urania (a medium-sized oceanographic ship, 1115 t gross tonnage, 61.3m overall length, 11.1m wide), that proved perfectly suitable to handle MODUS and the associated operational procedures (Figure 11.14).

11.3.2 GEOSTAR mission 2 (Southern Tyrrhenian Sea)

The goal of the second phase of the GEOSTAR project (EU GEOSTAR 2, 1999–2001) was to complete the technological development of the observatory and provide a full-scale demonstration of the concept during a long-term (6–12 months) scientific mission in a deep-sea site.

Work on the observatory was basically limited to an upgrade of its main subsystems in order to manage the extended payload and the new functionalities required. The temporary aluminum DACS and battery vessels used in mission 1 were replaced with new titanium vessels rated 6000m depth; the seismometer management system was also optimized; the number of ARGOS Messengers was extended to five (two Storage, three Expendable). The same approach was adopted for MODUS, whose hydrodynamic design was optimized and navigation payload increased (with the addition of sonar, colour camera and altimeter) (Gerber et al., 2002; Clauss et al., 2004).

Figure 11.14 GEOSTAR recovery onboard R/V Urania at the end of the first mission (September 1998).

The two remaining subsystems necessary to complete the development of the GEOSTAR concept were the object of a dedicated work:

- the near-real-time communication system (Marvaldi et al., 2002), based on the development of a surface buoy (moored in the vicinity of the observatory deployment site) managing the operation of a satellite link (surface part) and an acoustic link (underwater part), and allowing a remote operator to interact with the observatory during the mission
- the observatory handling system, based on a dedicated electro-mechanical umbilical cable and winch, extending GEOSTAR's operativeness up to 4000m depth.

Mission 2 started in September 2000 and concluded in April 2001 with the system recovery. Deployment and recovery operations were managed by the same ship used for mission 1 (Figure 11.15).

The mission site selected was located in the abyssal plain of the Southern Tyrrhenian Sea, offshore Ustica Island, Italy. The site originally selected was about 3000m deep, but

Figure 11.15 GEOSTAR recovered onboard R/V Urania at the end of mission 2 (March 2001).

due to technical problems with the umbilical termination during the preliminary test and two aborted tentative GEOSTAR's deployment, the water depth was reduced to approximately 2000m, while maintaining the full significance of the mission. After detailed investigation work by the manufacturer, the cause of the problem was identified in that, although formally rated at 4000m, the umbilical termination could not withstand pressures greater than 200 bar; the problem was then fixed and the handling system made available for subsequent applications (not limited to GEOSTAR-class observatories management and with a track record of 3700m depth reached).

The mission made it possible to demonstrate and confirm all system capabilities in deep-sea conditions.

Approximate 4150 hours of data (corresponding to 173 days of full operation) were collected with an acquisition efficiency of 99.6%.

During the mission a continuous set of geophysical, oceanographic and environmental parameters was acquired with a single reference time. Moreover, discrete water samples were collected for subsequent laboratory analysis. The complete dataset was downloaded from the observatory hard-disks just after the recovery; however, during the mission scientists and engineers could get real-time data access to the observatory from shore via the NRTCS (one interrogation per day) or via the periodic release of Messengers data capsules (one release per month). This allowed the execution of complete checks on the system functionality, as well as the start of the scientific data analysis while the mission was still ongoing. Quality of data collected was high, as demonstrated by the scientific literature so far produced (De Santis et al., 2006; Etiope et al., 2006; Iafolla et al., 2006).

11.3.3 GEOSTAR missions 3 and 4 (Southern Tyrrhenian Sea)

The third and fourth GEOSTAR missions were carried out within the framework of ORION-GEOSTAR 3 EC project (2002–2005). The technical goal of ORION (Ocean Research by Integrated Observation Networks) was the development of a seafloor network for scientific research and early warning of major hazard events (e.g., earthquakes and volcanic eruptions) (Favali et al., 2006). The basic idea was therefore to extend the effective operational radius of the single observatory in order to better cover the area of interest, transferring at seafloor the standards and rules established for the traditional shore-based communication networks. For this purpose, ORION was conceived as a series of interconnected stations, hereafter referred as to "nodes"; each node being network-accessible from the others to exchange data and commands (Gerber and Clauss, 2005).

The network configuration implemented in the project (see Figure 11.16) is based on two satellite stations configured as hosts (i.e., nodes that do not forward messages to other networks) and a main station configured as a gateway (i.e., a node that forwards messages to other networks) (Beranzoli et al., 2004).

The main node (gateway) was an upgrade of GEOSTAR, while the two satellite nodes (named SN3 and SN4) were specifically developed, maintaining most of the peculiar characteristics of GEOSTAR, to ensure maximum standardization and interoperability. SN4 (originally foreseen in the ORION experiment) on request from the European Commission was developed in this project but used in the framework of a parallel project (ASSEM – Array of Sensors for long-term SEabed Monitoring of geo-hazards) for a long-term mission, to demonstrate compatibility and integrability of GEOSTAR technology in other European seafloor networks (Rolin et al., 2005).

Although the network developed in the project was limited to three nodes, its standard and modular architecture allow expansion and reconfiguration according to the specific requirements of future applications. The nodes communicate through horizontal acoustic telemetry, while existing shore networks (phone network, Internet, etc.) are interfaced with the ORION seafloor gateway through a near-real-time communication buoy, supporting the operation of a vertical acoustic telemetry (to communicate with the gateway) and a surface (satellite or radio) link.

New acoustic protocols were specifically developed according to ORION's telemetry architecture. Moreover, GEOSTAR DACS was upgraded to manage the acoustic links together with the new payload and to implement new functionalities aimed at extending the level of interaction of the underwater network with the remote user, such as: automatically send periodical messages to shore (scientific and diagnostic data summaries, or autono-

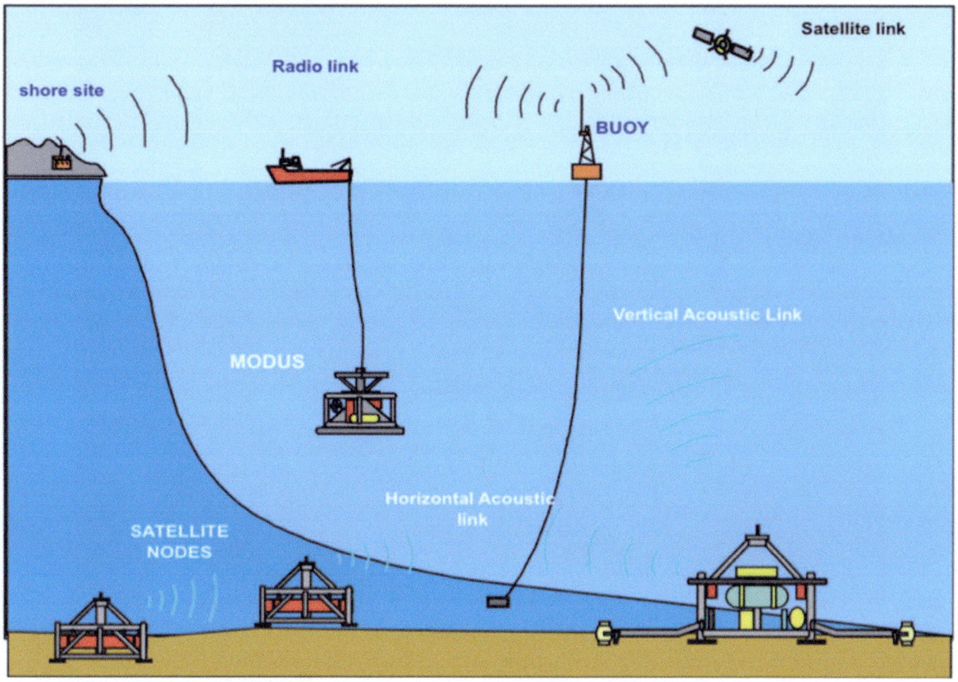

Figure 11.16 ORION concept (as proposed to EU Fifth Framework Programme, 2002).

mous alarm messages); manage messages from the satellite nodes; process hydrophone data to detect events; allow remote user to interrogate any node in the network; and send commands or request data.

Technological solutions developed in the project were demonstrated and validated in a pilot experiment dedicated to long-term continuous geophysical and oceanographic monitoring of Marsili seamount (Southern Tyrrhenian Sea), one of the largest underwater volcanoes in Europe.

GEOSTAR and SN3 were deployed from R/V Urania in December 2003 (Figure 11.17, Figure 11.18, Figure 11.19), at a depth of about 3350m.

The satellite node was placed at approximately 1300m from GEOSTAR.

From the start of mission 3, the system was affected by technical problems in the vertical acoustic telemetry that basically precluded any type of communication between the gateway and the surface buoy. For this reason, it was decided to continue the mission, leaving the observatories to operate in autonomous mode and anticipate recovery to fix the problems and proceed with a new deployment. Then GEOSTAR and SN3 were recovered in the period 23–30 April 2004, disembarked and stored in Catania INFN (Istituto Nazionale di Fisica Nucleare) workshop for the necessary technical interventions.

Figure 11.17 GEOSTAR onboard R/V Urania ready for deployment (mission 3, December 2003).

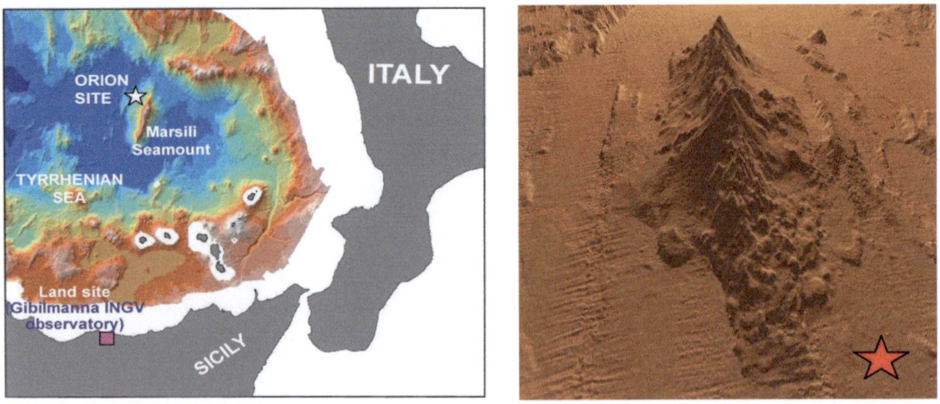

Figure 11.18 ORION mission site. (Left) geographic location of the site. (Right) 3D digital image of the Marsili Seamount seen from NW. Observatory network deployment site indicated by the star. Maps redrawn from Marani et al. (2004).

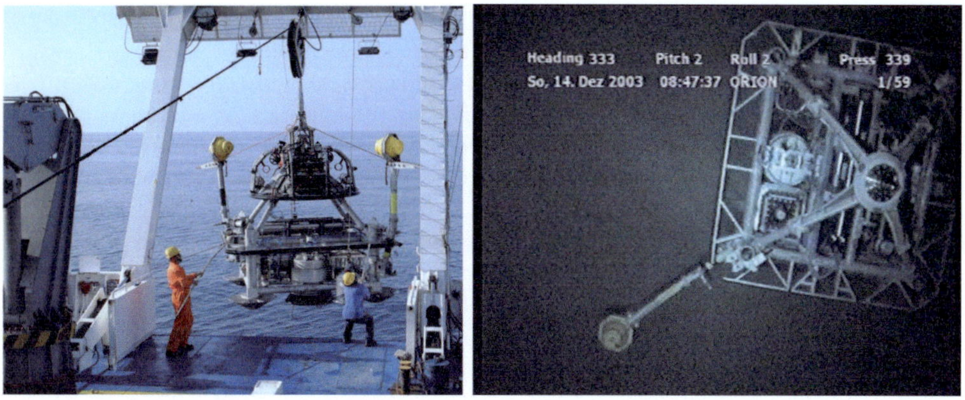

Figure 11.19 (Left) GEOSTAR deployed from R/V Urania, December 2003. (Right) GEOSTAR seen from MODUS immediately after completion of deployment phase and vehicle disconnection.

The observatories were redeployed in mid-June 2004, approximately in the same sites of the first mission, and the fourth mission started. The two observatories were left working until battery discharge and were finally recovered in April 2005.

This time, it was possible to demonstrate all the functionalities implemented: periodic messages from the observatories were received at the shore station and the remote operator was able to make direct interrogations. However, after two months of correct operation, the communication system evidenced new problems (to the buoy acoustic transducer and the radio link) and finally the buoy mooring line was broken.

In spite of the communication problems, GEOSTAR (and SN3) operated reliably throughout the mission duration, and analysis of data recovered indicated that both observatories were able to produce and transmit all the scheduled messages.

The two consecutive missions 3 and 4 demonstrated the validity of the ORION concept and provided many significant technical results. First of all, GEOSTAR operativity (observatory + intervention system + ship + procedures) was fully demonstrated at 3350m (the previous record was 1950m from the Ustica mission 2). With the same battery pack, GEOSTAR worked for 10,257 hours (corresponding to about 427 days of mission), reaching and significantly exceeding the original target of a 1-year operation.

Based also on the experience gained with the SN1 first mission (Monna et al., 2005), the original seismometer deployment procedure implemented in GEOSTAR-class observatories could now be considered well proven and qualified. During the mission period the seismometer recorded many (about 900) local, regional and teleseismic events. The high quality of seismic recordings confirmed the validity of the installation procedure (as developed and already tested in the GEOSTAR projects) and good ground coupling of the sensor.

The automatic chemical analyzer finally made its first mission, providing good quality data. The analyzer recorded more than 260 pH continuous data in parallel to the automatic water sampler that collected 38 samples for on-shore laboratory analyses on dissolved gas in water, cations and anions, minor and trace elements, radionuclides.

From the scientific point of view, for the first time, the Marsili volcano seamount was the object of a long-term monitoring activity. During the missions, an enormous quantity of data was acquired and made available for scientific evaluation (De Santis et al., 2007; Beranzoli et al., 2009). Data analysis evidenced significant geophysical and oceanographic time variations (Fuda et al., 2006). In particular, magnetic data allow the estimation of some conductivity structure at different depths under the Marsili volcano and gravimetric data show relevant signal patterns at low frequency (Vitale et al., 2009); seismic data show local activity with recurring events. Significant correlations between recorded time series could be related to the activity and structure of the volcano. GEOSTAR missions provided the starting point for a more ambitious activity to study the Marsili volcano, with objectives that overcome the pure scientific relevance of the phenomenon, and involve social and economic aspects such as natural hazard management (understand and monitor the risk of a possible eruption and associated risk of a catastrophic tsunami) and the renewable resource (the volcano as an offshore geothermal energy source).

Problems with the NRTCS buoy (mooring line rupture, unreliable operation of the vertical acoustic modem) were related to failures of commercial "off-the-shelf" products. Thus the validity of the concept was not affected and the problems occurred could be easily solved with a more careful specification and selection of the products.

11.3.4 GEOSTAR mission 5 (Gulf of Cadiz)

GEOSTAR missions 5 and 6 were carried out under the framework of the EU project NEAREST (Integrated observations from NEAR shore sourcES of Tsunamis: towards an early warning system), whose objectives included the development and test of an operational prototype of a near field tsunami warning system (Chierici et al., 2008). The prototype is designed to operate in tsunami generation areas for detection-warning purposes as well as for scientific measurements. The reference area of the project was the Gulf of Cadiz (offshore Portugal; Figure 11.20). The key elements of NEAREST concept are shown in Figure 11.21.

The tsunami detector is hosted inside the seafloor observatory, and includes a pressure sensor, a seismometer and two accelerometers. The tsunami detection procedure is based on a trigger on pressure and seismic events:

- seismometer: trigger on local strong earthquakes
- pressure: detection of sea level anomalies (tsunamis wave), triggering on processed sea level data compared to assigned threshold.

Pressure data are processed inside the observatory in real time and by means of an original tsunami detection algorithm conceived and implemented by INGV (Istituto Nazionale di Geofisica e Vulcanologia), IRA-INAF (Istituto Nazionale di AstroFisica–Istituto di RadioAstronomia) and CNR-ISMAR (Consiglio Nazionale delle Ricerche–IStituto di scienze MARine), and capable of detecting centimetric tsunami waves. The tsunami detector sends a near-real-time automatic alert message to surface when a seismic or a pressure signal exceeds a selectable threshold indicating a strong local earthquake or a tsunami wave event.

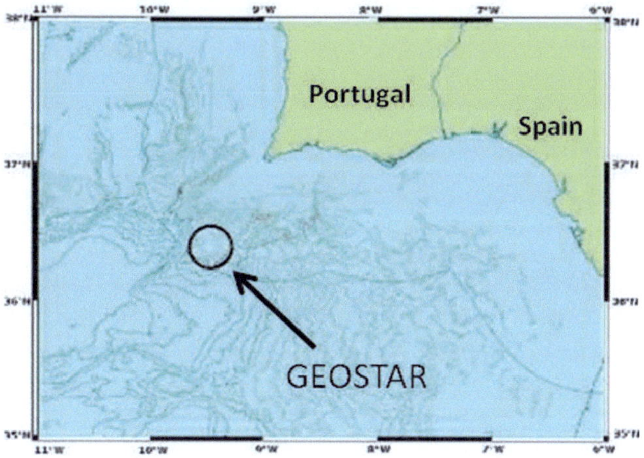

Figure 11.20 NEAREST installation site.

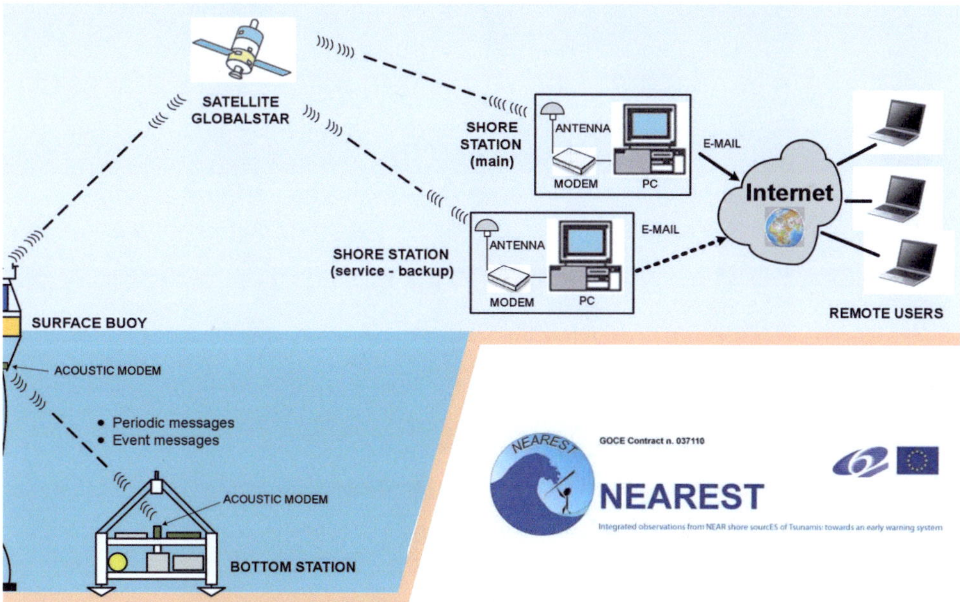

Figure 11.21 NEAREST concept.

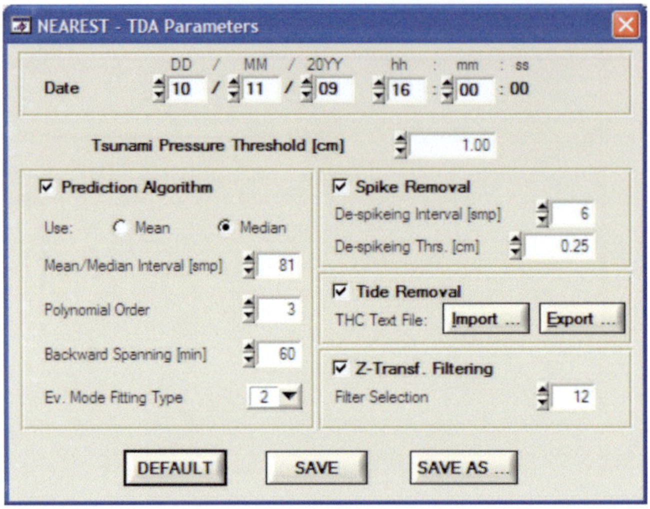

Figure 11.22 Configuration of the tsunami detection algorithm.

Figure 11.23 GEOSTAR ready for installation in Cadiz Gulf (mission 5, August 2007).

The algorithm is based on real time pressure data analysis, consisting of spikes and tide removing, low pass filtering and linear prediction: the output is then compared to a given pressure threshold allowing the detection of anomalous events (Chierici et al., 2008). Different configurations of the algorithm may be adopted, modifying a configuration file at mission start (Figure 11.22 shows the configuration panel of the tsunami detection algorithm). The algorithm can be reconfigured at any time, provided that the acoustic and satellite links are operational.

The Shore Station acts as a "Warning Center", in charge of collecting, integrating and evaluating data recorded at the sea bottom.

Thanks to its unique characteristics, GEOSTAR was selected to host the tsunami detector and be reconfigured and upgraded for the NEAREST experiment. Part of the payload used in previous missions (magnetometers and chemical analyzer) was removed, new instruments like a pressure sensor and Inertial Measurement Unit were integrated, old sensors (unavailable or not suited for the application) were replaced with new ones (seismometer and current meter) and the processing capability was improved with a new powerful CPU board dedicated to the real-time tsunami data processing.

In parallel, the buoy was configured according to project requirements, fixing the problems experienced in the previous mission: new mooring line, new low power electronics based on the standard GEOSTAR hardware, new instrumentation payload (meteorological station, GPS, satellite modem) and power supply (batteries and photovoltaic panels) (Figure 11.23).

GEOSTAR and the buoy were installed above an active, potentially tsunamigenic structure, the Marques de Pombal Structure at a depth of 3200m in August 2007. Mission 5 was therefore in operation.

During the experiment, all the sensors and software worked properly with the exception of a malfunctioning of the acoustic communication system located on the surface buoy that basically precluded any remote access to the observatory. Only direct interrogations from ship of opportunity, bypassing the buoy, were possible. In addition, the buoy suffered another failure to the mooring in November 2007; nevertheless, position data continuously transmitted by the ARGOS beacon allowed the prompt organization and execution of the recovery intervention.

The observatory was recovered in August 2008, one year after deployment (Figure 11.24). Subsequent analysis of scientific and technical data indicated a successful execution of all the mission tasks, including data acquisition and storage, automatic processing of pressure and seismometer data, automatic production and transmission of the data messages. However, due to the problems with the communication buoy the system was not fully able to demonstrate the feasibility of tsunamis warning detection and transmission.

11.3.5 GEOSTAR mission 6 (Gulf of Cadiz)

Following the results of mission 5, the decision was taken to organize an additional mission within the project and in synergy with the LIDO (Listening to the Deep-Ocean environment) Demonstration Mission funded by the EC project ESONET (European Seas Observatory NETwork) Network of Excellence, in order to get a complete demonstration of the communication chain between the seafloor abyssal station configured for the tsunami detection and the shore station.

Figure 11.24 GEOSTAR seen by MODUS during the final approach before docking (August 2008).

Figure 11.25 GEOSTAR before installation in Cadiz Gulf (mission 6, November 2009).

From	Subject	Received
flavio.furlan@tecnomare.it	NEAREST GEOS DATA FILE of 23/11/09 ore 18:00:00	lunedì 23/11/2009 22.41
flavio.furlan@tecnomare.it	NEAREST BUOY EVENTS CATALOG FILE of 22/11/09 ore 18:00:00	domenica 22/11/2009 20.42
flavio.furlan@tecnomare.it	NEAREST BUOY STATUS FILE of 22/11/09 ore 18:00:00	domenica 22/11/2009 20.42
flavio.furlan@tecnomare.it	NEAREST BUOY DATA FILE of 22/11/09 ore 18:00:00	domenica 22/11/2009 20.42
flavio.furlan@tecnomare.it	NEAREST GEOS STATUS FILE of 22/11/09 ore 18:00:00	domenica 22/11/2009 20.41
flavio.furlan@tecnomare.it	NEAREST GEOS DATA FILE of 22/11/09 ore 18:00:00	domenica 22/11/2009 20.41
flavio.furlan@tecnomare.it	NEAREST BUOY EVENTS CATALOG FILE of 22/11/09 ore 12:00:00	domenica 22/11/2009 14.42
flavio.furlan@tecnomare.it	NEAREST BUOY STATUS FILE of 22/11/09 ore 12:00:00	domenica 22/11/2009 14.41
flavio.furlan@tecnomare.it	NEAREST BUOY DATA FILE of 22/11/09 ore 12:00:00	domenica 22/11/2009 14.41
flavio.furlan@tecnomare.it	NEAREST BUOY EVENTS CATALOG FILE of 22/11/09 ore 06:00:00	domenica 22/11/2009 12.42
flavio.furlan@tecnomare.it	NEAREST BUOY STATUS FILE of 22/11/09 ore 06:00:00	domenica 22/11/2009 12.42
flavio.furlan@tecnomare.it	NEAREST BUOY DATA FILE of 22/11/09 ore 06:00:00	domenica 22/11/2009 12.41
flavio.furlan@tecnomare.it	NEAREST GEOS DATA FILE of 22/11/09 ore 06:00:00	domenica 22/11/2009 12.41
flavio.furlan@tecnomare.it	NEAREST BUOY EVENTS CATALOG FILE of 22/11/09 ore 00:00:00	domenica 22/11/2009 4.42
flavio.furlan@tecnomare.it	NEAREST BUOY STATUS FILE of 22/11/09 ore 00:00:00	domenica 22/11/2009 4.42
flavio.furlan@tecnomare.it	NEAREST BUOY DATA FILE of 22/11/09 ore 00:00:00	domenica 22/11/2009 4.42
flavio.furlan@tecnomare.it	NEAREST GEOS STATUS FILE of 22/11/09 ore 00:00:00	domenica 22/11/2009 4.41
flavio.furlan@tecnomare.it	NEAREST GEOS DATA FILE of 22/11/09 ore 00:00:00	domenica 22/11/2009 4.41
flavio.furlan@tecnomare.it	NEAREST BUOY EVENTS CATALOG FILE of 21/11/09 ore 18:00:00	sabato 21/11/2009 22.42

Figure 11.26 Automatic messages received from GEOSTAR and buoy during mission 6.

To fix the previously-occurring problems, a new model of acoustic modem was adopted and the buoy mooring line completely redesigned. Minor upgrades were finally implemented to GEOSTAR and the buoy (Figure 11.25).

The payload was the same as previous missions, with the addition of a stand-alone hydrophone, powered by a dedicated battery pack.

The new deployment cruise took place with the R/V Sarmiento de Gamboa in November 2009.

This time, the GEOSTAR-buoy-shore station communication link worked properly, demonstrating the validity of the concept. Messages automatically produced by GEOSTAR and the buoy were correctly received at the Shore Station and dispatched to the end users, according to the scheme already shown in Figure 11.21; Figure 11.26, Figure 11.27 and Figure 11.28 provide examples of the email delivered with the messages attached, the converted data (binary to spreadsheet) and a typical communication log respectively.

Operation of the communication link was, however, interrupted at the end of December 2009, due to occurrence of severe damages to the buoy instrumentation (probably caused by extreme weather conditions or ship collision). The GEOSTAR mission continued until July 2010 (when the mission was automatically stopped and the observatory put in IDLE mode to preserve the necessary energy to keep the rubidium clock working until recovery).

GEOSTAR was finally recovered in June 2011. Again, download and subsequent analysis of data proved the correct operation of the observatory during the mission; all the scheduled tasks were executed and, in particular, pressure events were detected.

With the conclusion of mission 6, the first European tsunami warning system based on simultaneous acquisition and processing of seismic and pressure data has been qualified and is now available for possible operational application.

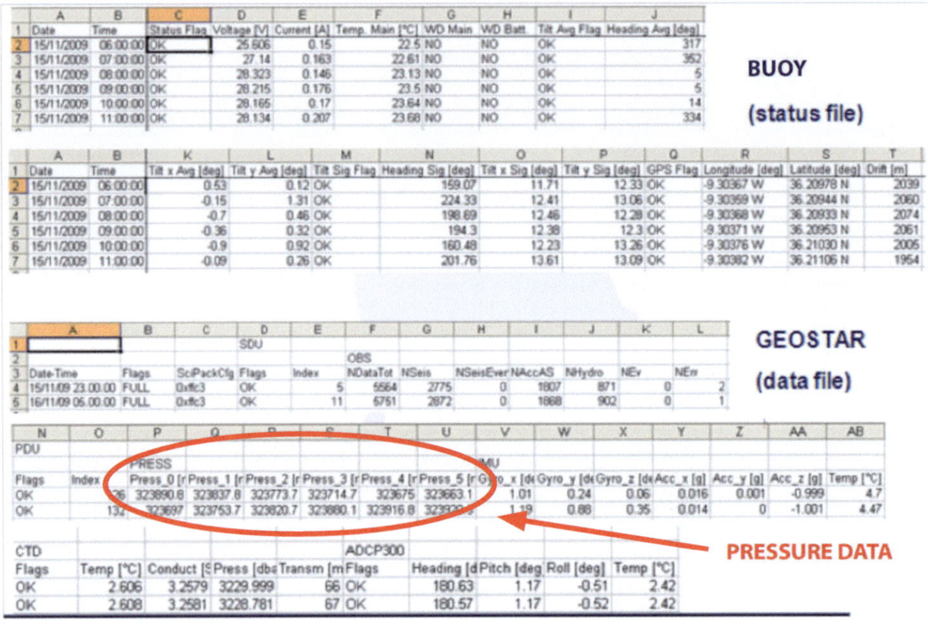

Figure 11.27 Examples of data messages of GEOSTAR and buoy during mission 6.

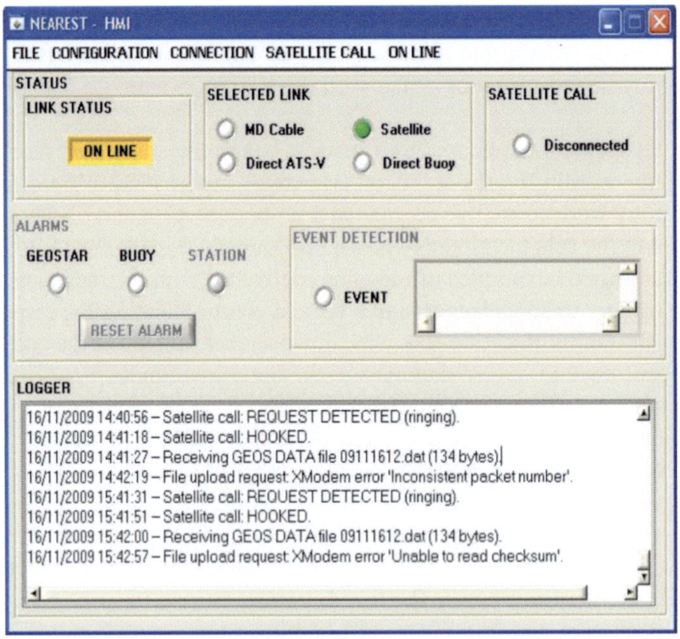

Figure 11.28 Example of satellite calls log during mission 6.

11.4 SN1

Since the early phases of GEOSTAR development, it appeared evident that the concept offered significant opportunities of exploitation. This conviction was the origin to two parallel initiatives: (1) extending the onshore Italian seismic network to the offshore environment (leading to the development of SN1; (2) establishing the first application of a seafloor observatory in a polar environment (leading to the development of MABEL). The technical solution developed for both applications consisted of an optimized version of GEOSTAR, making the resulting observatory easier to handle but at the same time ensuring the highest level of standardization with GEOSTAR. Peculiar characteristics may be summarized as follows:

- smaller frame, fully compatible with Modus (Figure 11.29 allows comparison between GEOSTAR and SN1 dimensions)
- same deployment/recovery procedure and surface logistics (cable, winch)
- same seismometer installation device, positioned in the centre of the frame
- new battery pack, based on 12 V 480 Ah modules specially developed by SAFT and fitting into 200mm internal diameter vessels; this will become a standard for all GEOSTAR-class observatories
- same data acquisition and mission management electronics, reconfigured to fit into 150mm internal diameter vessels; this will become a standard for all GEOSTAR-class observatories.

Basically, the new architecture maintains the same functionalities of GEOSTAR, apart from the capability to host the Argos Messenger container and the extendable booms for the magnetometers.

Figure 11.29 Comparison between GEOSTAR (left) and SN1 (mission 1 center; mission 2 right).

In particular, SN1 maintains the open architecture, fully reconfigurable concept that is one of GEOSTAR's most peculiar characteristics; this made possible SN1 evolution from a battery-powered autonomous version mainly devoted to seismological monitoring, to a cabled version real-time connected to shore and supporting a fully multidisciplinary payload. Evolution of SN1 scientific payload is summarized in Table 11.5.

Sensor	Mission 1	Mission 2	Mission 3	Sampling rate
Triaxial broad-band seismometer	Guralp CMG-1T	Guralp CMG-1T	Guralp CMG-1T	100 Hz
CTD	Seabird SBE37-SM	Seabird SBE37-SM	Seabird SBE37-SM	1 sample/h
Gravity meter	IFSI prototype	IFSI prototype	IAPS prototype	1 Hz
Current meter	FSI 3D-ACM	FSI 3D-ACM	Nobska MAVS-3	2 Hz
Scalar magnetometer		Marine Magnetics Sentinel 3000	Marine Magnetics Sentinel 3000	1 sample/h
Vectorial magnetometer			INGV prototype (3-axes)	0.5 Hz
Low frequency hydrophone	OAS E-2PD	OAS E-2PD	OAS E-2PD	100 Hz
Low frequency hydrophone			SMID DT-405D(V)1	2 kHz
Inertial measurement unit			Landmark LMRK20-AHRS 150-02-100	200 Hz
Absolute pressure gauge			Paroscientific 8CB4000-I	4 to 60 samples/min
Differential pressure gauge			SCRIPP Institution of Oceanography, University of San Diego UCSD	100 Hz
ADCP			RDI, Workhorse Sentinel 600 kHz	1 profile/hour
Bioacoustics			4 hydrophones SMID TR-401(V)1	96 kHz
Compass			Falmouth Ostar Compass	1 Hz

Table 11.5 SN1 payload and sampling rates.

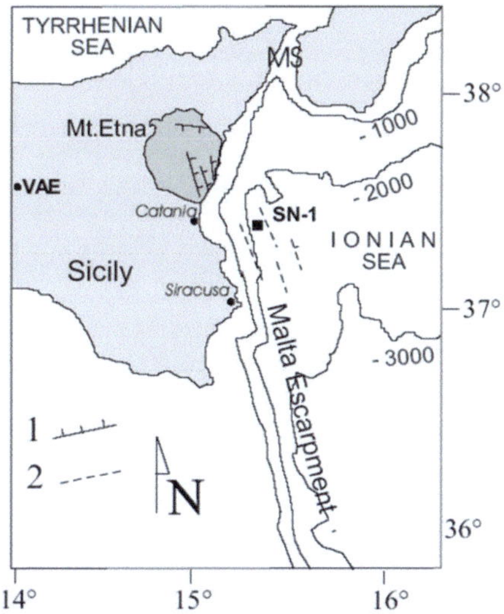

Figure 11.30 Tectonic sketch of Eastern Sicily (Ionian Sea) offshore area; SN1 installation site is indicated by the black square. MS=Messina Strait (redrawn from Monna et al., 2005).

11.4.1 SN1 mission 1 (Ionian Sea)

The SN1 project was developed in the Framework of the 2000–2003 Program of the Italian National Group for Defense against Earthquakes (GNDT) (Beranzoli and Favoli, 2005). The main objective of the project was to deploy a seafloor observatory in the Ionian Sea abyssal plain (Figure 11.30), a few tens of kilometers off the Eastern Sicily coasts and to integrate it into the existing onshore seismic network operated by INGV. The Ionian area facing Eastern Sicily is recognized as the site of important seismogenic underwater structures, the most important of which is the Ibleo-Maltese structure that is considered responsible for the most disastrous earthquakes of the area: Catania (1693, max. MCS intensity XI) and Messina (1908, max. MCS intensity XI).

In operation from October 2002 to May 2003, SN1 successfully completed the first mission at 2105m (about 25km east from Catania).

For this mission SN1 was configured to operate in autonomous mode (i.e., with internal data recording and battery power). The observatory was also provided with a vertical acoustic modem that allowed periodic interrogations from a ship of opportunity during the mission (Figure 11.31).

Deployment operations were carried out by the crane barge Mazzarò (Figure 11.32), demonstrating feasibility of GEOSTAR-class observatory management by a ship of opportunity. During the mission, high-quality seismic, gravimetric and environmental data were collected, confirming the correct operation of the observatory data acquisition and mission management system (Monna et al., 2005; Sgroi et al., 2007).

Figure 11.31 SN1 loading before the 2002 deployment.

Figure 11.32 Selection of images of SN1 deployment (above) and recovery (below).

11.4.2 SN1 mission 2 (Ionian Sea)

Following the success of the first mission, it was decided to upgrade SN1 to become a cabled observatory, in view of a new deployment and connection to the 25km electro-optical cable installed in the same site by INFN (the Italian National Institute for Nuclear Physics). The main purpose of this was to support a scientific pilot experiment of natural neutrino detection in deep sea (NEutrino Mediterranean Observatory, NEMO Project) (Favali et al., 2006b, 2011).

The main peculiarity of the cable design is that 20km off-shore it is spliced into two separate tails, each about 5km long. Each tail is terminated into a frame equipped with two ROV-mate connectors (Ocean Design, 8 way hybrid); thus, two powerful independent infrastructures are available for the connection of seafloor experiments. Thanks to an agreement between INGV and INFN, the Northern Branch was reserved to SN1, while the Southern Branch is dedicated to support the NEMO pilot experiment detectors.

At the shore end, the cable is terminated in the INFN-LNS (Laboratori Nazionali del Sud) laboratory located in the Catania harbor.

The overall system configuration is shown in Figure 11.33.

The observatory upgrade was carried out in 2003–2004. The technical approach followed for this work was to add the new functionalities while maintaining the old ones; this resulted in a hybrid configuration, allowing SN1 to be powered from shore and communicating in real-time with the Shore Station located in the LNS-INFN laboratory inside

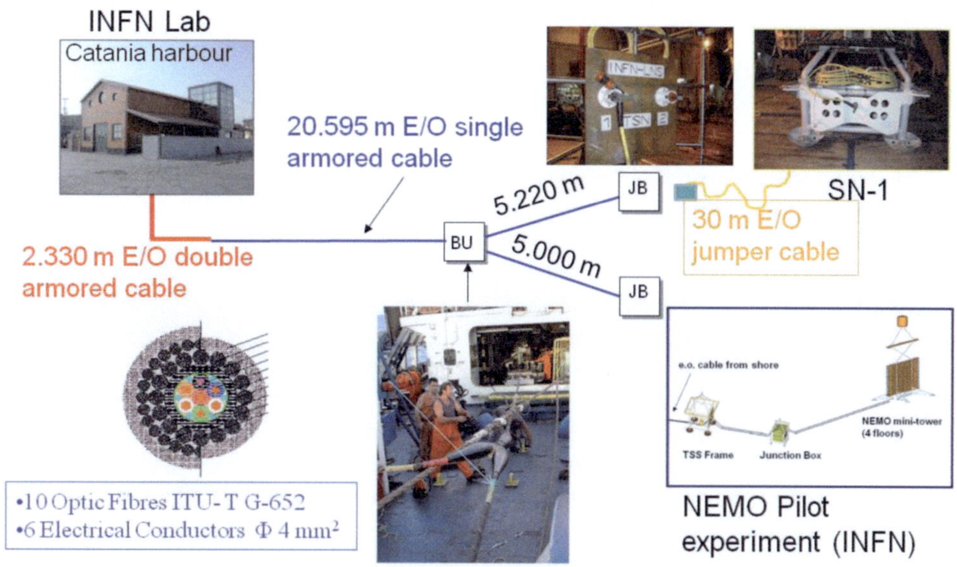

Figure 11.33 SN1 configuration for mission 2 (BU=branching unit, JB=junction box).

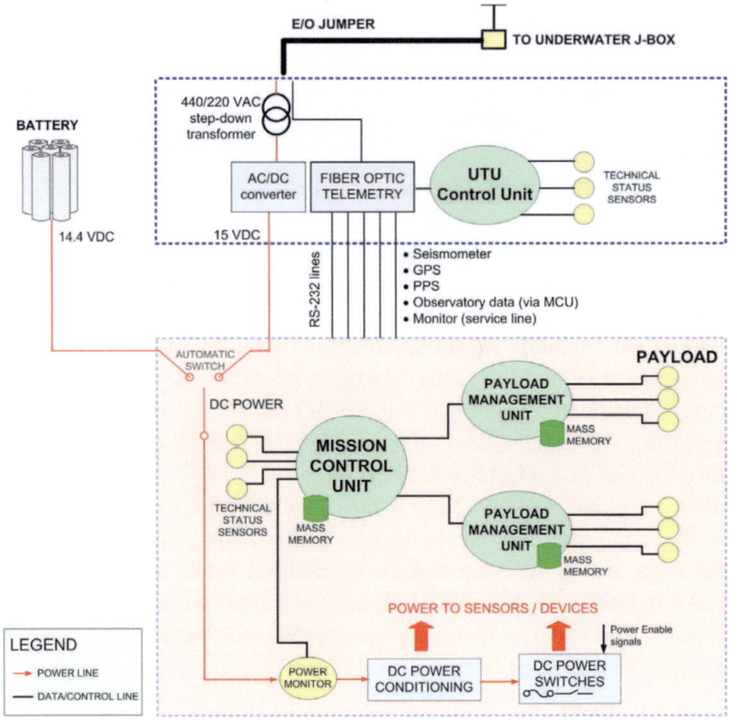

Figure 11.34 SN1 DACS architecture (mission 2).

Catania harbor, but at the same time having the possibility to be operated in autonomous mode (Favali et al., 2006).

SN1 architecture for the second mission is shown in Figure 11.34. The standard battery pack, connected to a switch that automatically determines the highest voltage source, ensures a temporary back-up in case of loss of power from shore. Note also the addition of a new unit (UTU) to the DACS architecture, managing external power and fiber-optic (FO) telemetry. New functions were implemented in the existing hardware and software; in particular, a third operational mode (REAL-TIME mode) was created in addition to the standard ones (MISSION mode and IDLE mode) always implemented in any GEOSTAR-class observatory. Basically, when in REAL-TIME mode, the payload data lines were directly switched to the FO telemetry, bypassing the internal storage. The SN1 Shore Station was then integrated to INGV land-based networks.

In January 2005, the observatory was deployed at the same site as the previous mission and connected to the submarine cable. Marine operations were carried out by the C/V Pertinacia and the observatory connection to the junction box was performed by a work class ROV equipped with manipulator (Figure 11.35).

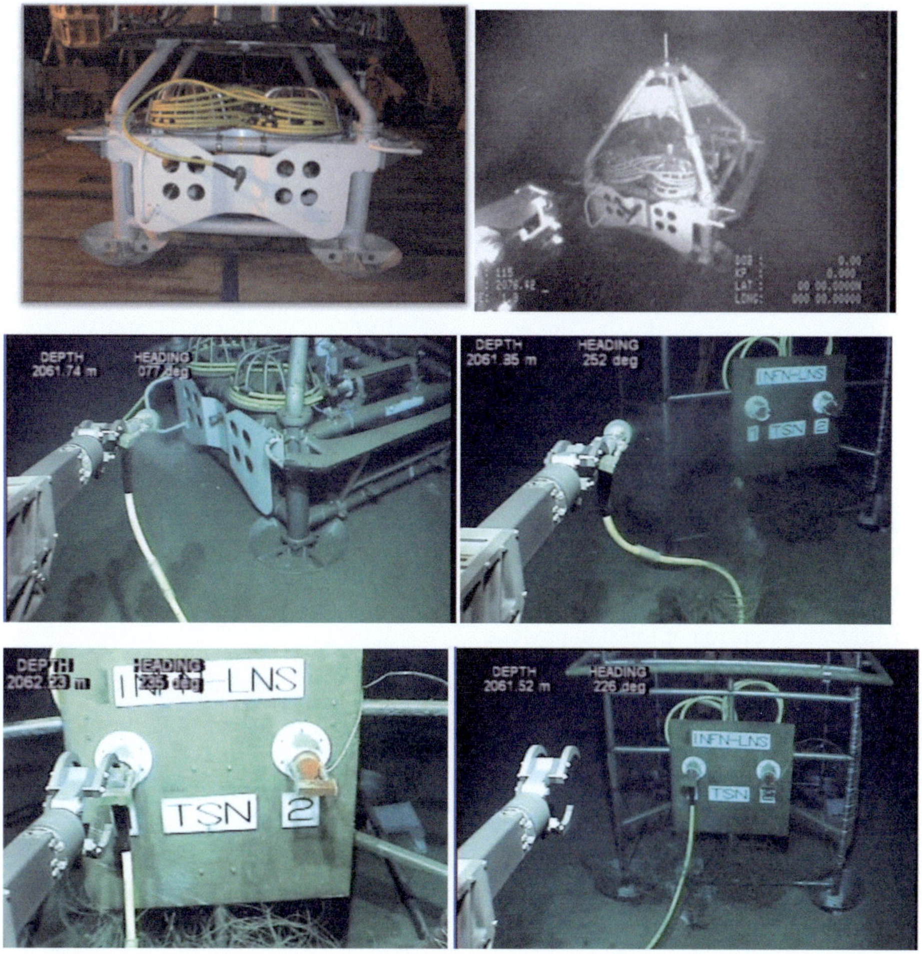

Figure 11.35 SN1 connection to the underwater junction box (SN1 mission 2, January 2005).

SN1 operated satisfactorily until recovery, which occurred in May 2008. Most of the time, it operated connected in real-time to shore, apart from a few periods of stand-by caused by damage to the umbilical cable in the shore vicinity. Integration of SN1 data with the Italian Seismic Network was also successfully verified.

At recovery the observatory was found in good condition, confirming the suitability of the concept to constitute the basis of permanent monitoring networks. With this mission, SN1 became the first real-time seafloor observatory in Europe and one of the few in the world. It was also the first seafloor observatory operative in one of the "key-sites" planned in the EC project ESONET.

11.4.3 SN1 mission 3 (Ionian Sea)

Thanks to the success of previous missions, a unique infrastructure was developed (observatory, cable, junction box, shore station) and the SN1 site had been selected as one of the nodes of the forthcoming European large-scale research infrastructure EMSO (European Multidisciplinary Seafloor and water column Observatory) (Favali and Beranzoli, 2009a), the network of seafloor and water column observatories recommended by ESFRI (European Strategy Forum on Research Infrastructures) addressing the long-term monitoring of environmental processes related to ecosystems, climate change and geo-hazards (Favali et al., 2011). To meet the challenging requirements of this initiative, SN1 was significantly upgraded (support obtained under the framework of EC project ESONET NoE) as briefly described below. With a marine operation similar to that carried out in 2005, in June 2012 SN1 was deployed again, connected to the submarine cable and finally, after successful completion of technical tests, commissioned for action. Since then SN1 has been providing real-time data to the Catania Shore Station and INGV seismological seismic network. SN1 is part of the first operative node in real-time of EMSO (NEMO-SN1, Western Ionian Sea; Favali et al., 2012).

The most significant technical aspects of the new observatory configuration (shown in Figure 11.36) are:

- significant extension of mission payload
- evolution from the previous hybrid configuration to a fully-cabled configuration
- new shore station architecture and functionalities.

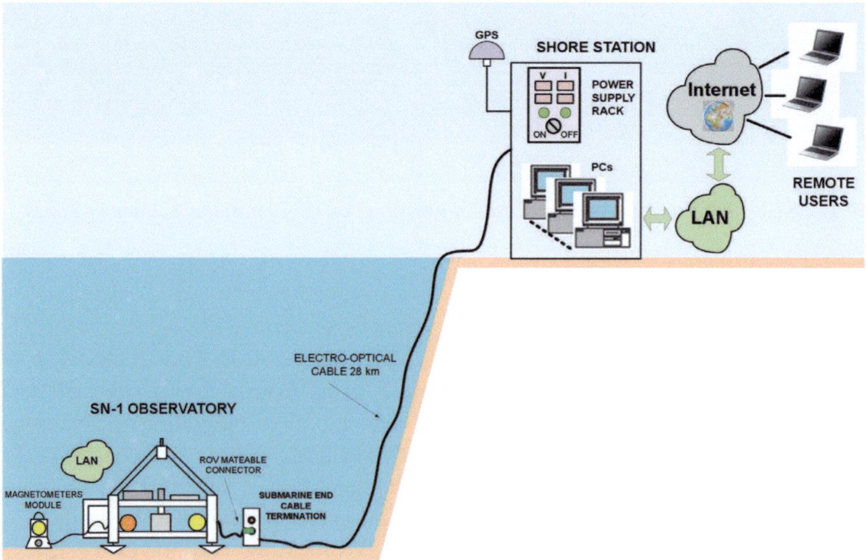

Figure 11.36 SN1 concept (mission 3 ongoing).

As regards the payload, SN1 is now capable of supporting a complete set of sensors for seismological, geomagnetic, gravimetric, accelerometric, oceanographic, hydro-acoustic and bio-acoustic measurements, most of which were not present in the previous configuration (Favali et al., 2012). One of the most significant features is the installation of acoustic sensors used for the passive acoustic detection of cetaceans to localize and fully track them. Monitoring marine mammals can help researchers to better understand their population trends in relation to climate changes and human impact. Thanks to their broad bandwidth, the hydrophones can detect a large variety of marine mammals. The system is also equipped with a tsunami detector (working on the same principle as the prototype developed and operated in GEOSTAR missions 5 and 6 and based on the simultaneous measurement of the seismic and bottom pressure signals and a new high performance tsunami detection algorithm) (Chierici et al., 2012).

The new configuration maintains the mechanical frame, the deployment and recovery procedure, the seismometer installation procedure and the interface with the electro-optical cable. Apart from the new internal arrangement (for the additional payload and devices), the most significant improvement is the adoption of a separate module for magnetometers designed to be handled by the ROV in charge of the observatory connection to the junction box.

Transition to a fully cabled architecture meant the complete redesign of the data acquisition and mission control system. A new DACS architecture was defined and implemented, removing all the Payload Management Units, the internal mass memories, batteries, acoustic telemetry and relevant interfaces.

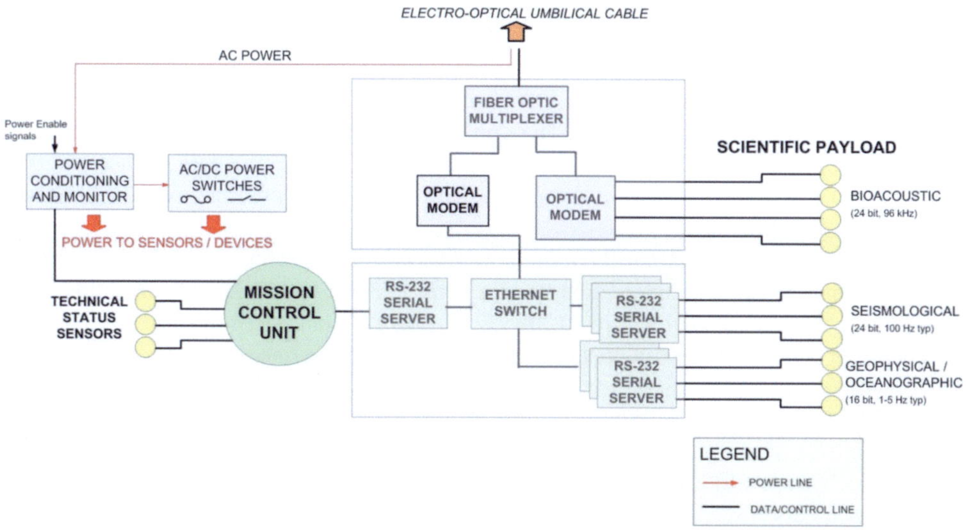

Figure 11.37 SN1 DACS architecture (mission 3 ongoing).

Figure 11.38 SN1 during final integration and dry test (2011).

The new control and telemetry system includes a TCP/IP-based network consisting of two local area networks, one onshore linking data acquisition and control computers and one offshore connecting the sensors. The communication system is designed to make it easy for observatory users to access the instruments and acquire the data onshore. Serial servers (i.e., Ethernet to RS-232 converters) allow transparent communications between topside and subsea instruments.

Figure 11.37 provides details about the new electrical architecture of the observatory.

A peculiarity of the communication system is the use of two redundant optical fibers to improve system redundancy and reliability. Furthermore, the system employs four separated CWDM (Coarse Wavelength Division Multiplexing) frequencies: two for acoustic data downlink, one uplink shore to sea for observatory control and one downlink for geophysical data.

The new SN1 Shore Station is installed, like the old one, in INFN-LNS (Istituto Nazionale di Fisica Nucleare - Laboratori Nazionali del Sud) workshop at Catania harbour. It hosts the land termination of the cable, the onshore data acquisition system, the power supply for underwater instrumentation and the GPS antenna for time synchronization. A radio link up to 92 Mbps to LNS-INFN is available for data connectivity. From there, a high-speed

Ethernet link (1 Gbps) to the Internet is used for data access by scientific users and the general public. The Shore Station is capable of sustaining the intense data rate received from deep sea (about 50 Mbps) and distributing it over the Internet. High-rate data from deep sea are acquired by dedicated PCs, named Data Servers (DS).

A Control PC is dedicated to mission configuration settings (diagnostic alarm threshold, enable/disable sensors, etc.), managing SN1 operative modes IDLE – MISSION and remote control access through a client application.

A dedicated machine is equipped with RS-232 expansion cards for acquisition and data storage of oceanographic and geophysical payload.

Data Servers for acoustic data (ADS) are equipped with professional audio cards capable of sustaining the underwater hydrophones data stream. A first analysis code is implemented for real-time data, recording, visualization and listening of acoustic data. This code also provides real-time statistical measurements of the acoustic background, such as sound pressure density spectrum, that can be used in off-line analysis to locate acquisition time with presence of biological sounds.

Networking connectivity to authorized remote users is made through VPN (virtual private network).

The power system is designed to deliver sufficient power to the observatory providing isolation and fault protection, tripping off the power supply in case of ground fault or overcurrent. It can deliver up to 1 kVA and an adjustable output voltage up to 500 VAC. The isolated power is delivered to the observatory via the $4mm^2$ conductors inside the umbilical. Other features include: automatic and manual voltage ramp control (soft start), visualization and RS-232 transmission of power status parameters (insulation resistance, currents and voltages). All relevant equipment – observatory, PCs, communication systems, etc. – operates from a dedicated UPS (Uninterruptible Power Supply).

11.5 MABEL (SN2)

The goal of the MABEL (Multidisciplinary Antarctic BEnthic Laboratory) project was to develop and operate a multidisciplinary observatory for the continuous and long-term measurement of geophysical, oceanographic and chemical parameters in Antarctic sea waters (Calcara et al., 2001).

In the Polar regions, the peculiar advantages of the seafloor observatory approach are even more evident, considering the hostile environment and logistic difficulties as well as the perspective of studies in these areas. In particular, Antarctica is scientifically considered to be of strategic importance for the comprehension of many complex phenomena that are not only related to regional processes but, more importantly, to the condition, dynamics and sustainability of the whole planet.

From the technical point of view, the logistical issues forced the adoption of solutions ensuring minimization of costs; in particular, a new deployment procedure was defined that did not use the MODUS and associated equipment (electro-optical umbilical and dedicated winch).

Sensor	Manufacturer and model	Sampling rate
Seismometer	PMD/EENTEC EP300-DT	100 samples/s
CTD	SeaBird SBE 16 Seacat	1 sample/h
Transmissometer	Chelsea Instruments Alphatracka II	1 sample/h
Current meter	FSI 3D-ACM	2 samples/s
Chemical analyser (pH, Eh)	INGV/Tecnomare prototype (with AMT sensors)	1 sample/2 days
Automatic water sampler	McLane RAS 48-500	500ml sample/8 days

Table 11.6 MABEL (SN2) payload and sampling rates.

Weight (kN)	16.3 (in air), 9.5 (in water)
Dimensions (mm)	2900 (L) × 2900 (W) × 2900 (H)
Design depth (m)	4000
Data acquisition and mission control	2 boards (32 bit microcontroller MC68332)
Data storage	Ruggedized Hard Disk (120 GB), 2 × CompactFlash
Power supply	12 Vdc, 1920 Ah Lithium-thionyl chloride 24 Vdc, 890 Ah Lithium-thionyl chloride
Power consumption (mA)	<70 (idle mode), <180 (mission mode)
Status parameters	Voltage, current, temperature, heading, tilt x/y, water intrusion, echo sounder

Table 11.7 MABEL (SN2) main characteristics.

Most of the work was therefore focused on the adaptation of the GEOSTAR-class observatory concept to the challenging and peculiar logistics and environmental conditions. In this respect, the qualification tests represented a significant part of the activities carried out (Cenedese et al., 2004).

As regards the mechanical layout, MABEL shares with SN1 the same frame design. The most peculiar aspect characterizing MABEL's mechanical design is the modification of the upper cone, which makes possible the MABEL deployment using a simple rope and standard winch (Figure 11.39). Basically, the solution consisted in integrating a lifting point for a standard acoustic release into the cone, avoiding any interference to the subsequent recovery by MODUS; the problem was solved by designing a special ring, mounted inside the cone below the docking pin. The solution is fully reversible, i.e., the ring can be

Figure 11.39 MABEL (SN2) deployment in the Weddell Sea, Antarctica (December 2005).

removed, allowing MABEL to return to the original configuration; at the same time other GEOSTAR-class observatories can be equipped with the same device.

The main drawback of this solution was the control of MABEL status sensors (echo sounder, etc.) on periodic acoustic interrogation during deployment.

As regards mission management, most of the efforts were dedicated to minimize power consumption and ensure operation in cold conditions. Further upgrades of the well-proven GEOSTAR hardware led to an average power consumption of less than 180 mA @ 12 VDC (Table 11.7).

To further increase mission autonomy, a second 24 VDC battery pack was developed, in addition to the standard 12 VDC, 1920 Ah pack. The additional 24 VDC, 890 Ah pack (developed by SAFT and featuring the same modular architecture and dimensions of the 12 VDC pack) was dedicated to the chemical analyzer and the precision clock.

Qualification of the MABEL system and procedures in polar environment was object of specific activities during the project. A first phase of tests was carried out in 2002 at HSVA Hamburg Large Ice Model Basin (ARCTECLAB) (Figure 11.40) with the financial support of the European Commission Human Potential and Mobility Program. Work included:

- deployment and recovery sequences in cold temperatures (water about 0°C, air −15°C)
- execution of simulated missions in cold water (including a 3-day mission at 5m depth on the bottom of the ice tank)
- acoustic communication tests in cold water
- chemical analyzer operation in cold temperatures.

Tests demonstrated the capability of the observatory to operate in Antarctic conditions.

MABEL DACS and seismometer were then subject to dedicated qualification tests in cold climatic chamber (Electrolux, Italy) down to –20°C, successfully.

In addition, the standard hard disks were replaced with rugged versions, with extended temperature range.

Figure 11.40 MABEL (SN2) qualification tests in HSVA Basin, Hamburg (2002).

11.5.1 MABEL (SN2) mission 1 (Weddell Sea, Antarctica)

MABEL deployment was scheduled during the 2005–2006 cruise of R/V Polarstern in the Weddell Sea (Antarctica). The selected site was located at a water depth of approximately 1874m, 60 miles offshore Neumayer German Station. Operations were carried out on 5 December 2005. A first tentative deployment was aborted at approximately 200m due to a water alarm inside DACS, halting the operation, and recovering MABEL on board. The DACS vessel was checked and a problem of condensed water discovered in proximity to the water detector. DACS was then closed again in a controlled environment (0°C and dry atmosphere). A few hours later, the deployment procedure was started again. During the first part of the descent, the observatory's functionalities were checked by periodic interrogations by acoustic telemetry; close to the seabed, reliability of transmission becomes too poor due to excessive noise from ship propellers.

Touchdown occurred ca. 3 hours from launch. The mission was started automatically on 6 December 2005, 16.00 UTC with the seismometer release.

On 1 January 2006, Polarstern returned to the deployment site, allowing successful execution of some interrogations with the acoustic modem. The observatory was correctly found in mission and all functionalities verified. Summary data records (containing average hourly technical and scientific data) relevant to four different mission days were also recovered.

After one year (Polarstern cruise 2006–2007) a first attempt to recover MABEL failed due to bad weather conditions and the inadequate positioning of the MODUS winch onboard Polarstern.

At that time, MABEL was interrogated via acoustics and found in idle mode, as expected; several summary messages (scientific and technical) were successfully recovered before leaving the observatory in place. Data collected made it possible to get a reliable and significant picture of the mission.

Another recovery cruise was organized two years later (Polarstern cruise 2008–2009); this time (16 December 2008) MABEL was successfully recovered (Figure 11.41, Figure 11.42) and all data made available to scientists for analysis (Gerber and Clauss, 2009).

For the first time ever, a deep sea multidisciplinary observatory was installed and successfully operated in the extreme conditions of polar waters. Polarstern proved to be perfectly suitable for MABEL management. Polar conditions proved to be critical for the operation of some commercial scientific sensors (namely, the automatic water sampler and the current meter); problems experienced may, however, be easily overcome with a more careful selection and qualification phase of the observatory payload.

11.6 SN3

With the successful results obtained from GEOSTAR mission 2, in mid 2001 the technological development of the observatory could be considered complete and ready for the execution of new scientific missions. From the technical point of view, a new challenge was identified: extend the observation capability of a seafloor observatory (basically limited to

Figure 11.41 (Left) MODUS during final approach to MABEL (SN2). (Right) MABEL successfully recovered.

Figure 11.42 MABEL (SN2) on board R/V Polarstern (December 2008).

Figure 11.43 SN3. (Left) on board R/V Urania during the December 2003 deployment cruise (photo allows comparison with GEOSTAR). (Right) SN3 with ASTRA ready for deployment.

the deployment point), making possible monitoring over a whole area and at the same time maintaining a near-real-time access to the data.

To reach this goal, the concept of a "network of seafloor observatories" (derived from the ABEL study) was proposed, based on a main observatory operating as a gateway to the underwater network, and satellite observatories operating as nodes of the network, ensuring the coverage of the area of interest (Beranzoli et al., 2004).

The ORION-GEOSTAR 3 project (2002–2005) offered the opportunity of implementing and demonstrating this concept; in parallel to the GEOSTAR and buoy upgrades, two new observatories were developed (SN3 and SN4).

As regards the mechanical design, SN3 maintains the architecture and dimensions of SN1 and SN2. However, the frame was customized to host ASTRA (Automated Sensor Burial Tool for Set-Up of Subsea Seismic Networks), a special module developed within the project and aimed at carrying out a remotely controlled burial of a seismometer inside sediment, to improve the seismometer-seabed coupling. For this purpose, the frame design was modified including a removable section, allowing the observatory to be configured in two alternative ways (Figure 11.44):

a) mounting the standard seismometer assembly, adopted in all GEOSTAR-class observatories

b) mounting ASTRA.

Figure 11.44 Comparison between SN3 configurations; (left) with ASTRA; (right) with a standard seismometer.

In the "standard" configuration, SN3 was used for the two consecutive ORION missions, while in the second configuration SN3 was used for the execution of shallow water demonstration tests of ASTRA.

SN3 electronic (hardware/software) architecture derives directly from the standard GEOSTAR-Class observatory architecture shown in Figure 11.9. In this case, three CPU boards were adopted: the Mission Control Unit and two Payload Management Units (one dedicated to the seismometer, the other to the hydrophone).

Specific functionalities were implemented to make SN3 operate as a "satellite" of GEOSTAR from the network point of view.

First of all, capability to manage a horizontal acoustic communication link was implemented. This link allows SN3 to send its data to the Shore Station via GEOSTAR (which acts as a "gateway") and the relay buoy; the ORION system design also allows contact between SN3 and the surface (ship of opportunity or the Shore Station via GEOSTAR, buoy and radio relay link), to send commands or download data, thus using the acoustic network at its full potential.

The mission software was upgraded, so to the standard functionalities of all GEOSTAR-class observatories (related to data acquisition and storage, mission management, status monitoring) was added the following capabilities required by the project objectives:

- capability of automatic generation and transmission of periodic messages

- real-time processing of hydrophone data and automatic detection of events by means of a standard Short Term Averaging/Long Term Averaging (STA/LTA) triggering algorithm, implemented in the dedicated CPU unit managing the hydrophone
- implement a new mission data structure named Event Message produced in case of event detection
- possibility to download seismometer or hydrophone wave forms, corresponding to any period of the mission: this is particularly useful, e.g., in the case of occurrence of a seismic event.

The main characteristics of SN3 are shown in Table 11.8, and the payload and sampling rates in Table 11.9.

Weight (kN)	14 (in air), 8.5 (in water)
Dimensions (mm)	2900 × 2900 × 2900
Design depth (m)	4000
Data acquisition and mission control	3 CPU boards (32 bit microcontroller MC68332)
Data storage	Hard disks, CompactFlash
Power supply	12 VDC, 1920 Ah Lithium-thionyl chloride
Power consumption (mA)	120 (idle mode), ~350 (mission mode)
Status parameters	Voltage, current, temperature, heading, tilt x/y, water intrusion, echo sounder

Table 11.8 SN3 main characteristics.

Parameter	Missions 1 and 2	
	Sensor	Sampling rate
Broadband seismometer	PMD/EEntec EP300-DT	100 samples/s
hydrophone	OAS E2PD	100 samples/s

Table 11.9 SN3 payload and sampling rates.

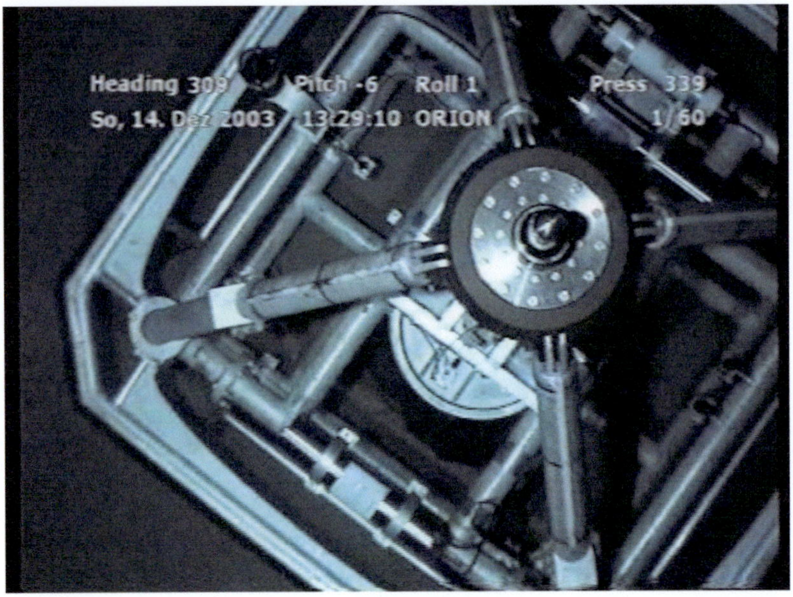

Figure 11.45 SN3 seen by MODUS immediately after deployment at seabed.

11.6.1 SN3 missions 1 and 2 (Southern Tyrrhenian Sea)

SN3 missions were carried out in parallel with GEOSTAR's third and fourth missions (Figure 11.45). For details on the installation site, see Section 11.3.3.

From the technical side, the two missions provided further confirmation of the maturity and soundness of the GEOSTAR-class observatory concept, in particular as regards distinctive aspects such as the effectiveness of the deployment and recovery procedures, the quality of the seismometer management procedure, the reliability of the data acquisition and mission management hardware and software.

In spite of technical problems with the vertical acoustic telemetry system that affected the entire first mission and part of the second, networking and near-real-time communication were demonstrated: SN3 was able to periodically (every 6 hours, i.e., 4 times a day) send summary technical messages and data messages to surface through the communication path. These are short packets, compatible with the low bandwidth of the acoustic link but at the same time sufficiently exhaustive about the health status of the observatory.

SN3 was also able to reply to the acoustic commands/queries issued from the Shore Station, that are relayed from the radio link, the buoy and GEOSTAR in turn; this operation is quite complex, requiring many hops and some time to complete.

Besides the technical results, SN3 gave its significant contribution to the scientific mission, providing a set of seismological data that complemented those collected by GEOSTAR in its parallel mission. However, efficiency of data collection was smaller, due to technical problems with the seismometer which caused a higher number of data packets to be lost by the acquisition system and a sudden failure of the sensor during the first mission.

11.7 SN4

SN4 was originally conceived to be the third node of the ORION experiment. In this scenario, SN4 would have operated as a satellite node of an underwater seismological network (including also GEOSTAR and SN3) and consequently it would have the same architecture and functionalities already described for SN3. Instead, the role and configuration of SN4 were changed at the specific request of the EU commission to integrate one of the ORION nodes into the shallow water experiment that the parallel EU project ASSEM was going to develop in the Corinth Gulf (Greece) (Rolin et al., 2005).

The new mission requirements imposed a significant revision of the observatory configuration; notably, a new mechanical architecture was studied to comply with the logistic constraints of the experiment, particularly as there was no possibility of using MODUS for the deployment and recovery of the observatory. The problem to solve was to develop a new version of a GEOSTAR-class observatory, maintaining most of its distinctive characteristics but at the same time making it manageable in a different way both from the logistic (installation, recovery) and logical (communication interface) point of view. Different concepts were studied, including pop-up configurations (basically an evolved OBS); the final choice was again to rely on the potentialities offered by the single frame, open architecture. SN4 was therefore designed as a "heavy" (although significantly lighter than the sister versions) observatory, deployable with a simple rope and acoustic release and with provision to host different recovery systems (adaptable according to the logistic facilities available; details will be given in the description of the missions) (Figure 11.46).

De facto, SN4 represents the smallest GEOSTAR-class observatory maintaining full compatibility with the original seismometer management system. Standardization also involves:

- data acquisition and control hardware, including the precision clock
- payload supported
- battery pack
- acoustic telemetry
- the basic mission management functions implemented, including
  - acquisition from all scientific packages and status sensor
  - preparation and continuous update of hourly data messages
  - management of bidirectional communications via hydro-acoustic telemetry link
  - actuation of commands received (e.g., data request, system reconfiguration, restart)
  - complete data back-up on internal memory.

The main technical features and scientific payload of SN4 are summarized in Table 11.10 and Table 11.11.

Weight (kN)	6.6 (in air), 1.5 (in water)
Dimensions (mm)	2000 × 2000 × 2000
Design depth (m)	600
Data acquisition and mission control	3 boards (32 bit microcontroller MC68332)
Data storage	Hard disk, CompactFlash
Power supply	12 VDC, 1920 Ah Lithium-thionyl chloride
Power consumption (mA)	120 (idle mode), ~450 (mission mode)
Status parameters	Voltage, current, temperature, heading, tilt x/y, water intrusion, echo sounder

Table 11.10 SN4 main characteristics (data referred to the last version).

Parameter	**Mission 1**		**Missions 2 and 3**	
	Sensor	**Sampling rate**	**Sensor**	**Sampling rate**
Broad-band seismometer	PMD/EEntec EP300-DT	100 samples/s	Guralp CMG-40T	100 samples/s
Hydrophone	OAS E2PD	100 samples/s	OAS E2PD	100 samples/s
CH_4	Capsum METS	1 sample/s	Franatech METS	1 sample/s
CH_4 with pump			Franatech METS + SeaBird SBE-5T	1 sample/s (pump ON for 5 min every 30 min)
CTD			SeaBird SBE-16plus	1 sample/10 min
Turbidity			WET LABS Echo-BBRTD	1 sample/10 min
Current meter			NOBSKA MAVS-3	5 sample/s
Dissolved oxygen			Aanderaa Optode 3830	1 sample/s

Table 11.11 SN4 payload and sampling rates.

Figure 11.46 SN4: (left) 2004 configuration; (right) 2009 configuration.

11.7.1 SN4 mission 1 (Corinth Gulf)

Parallel to the ORION project that led to the development of SN3 and the Marsili pilot experiment, the EU project ASSEM was aimed at developing a seafloor network for the continuous monitoring of marine geo-hazards (Rolin et al., 2005). Two experiments were planned, the first in an area with a slope instability risk (offshore Norway), the second in an area characterized by an active fault (Corinth Gulf, Greece) representing the most active extensional basin in Europe, with high rates of margin uplift (several mm per year). The array of measurement nodes developed for the Corinth Gulf experiment included pore-pressure sensors, tilt meters and extensometers (Figure 11.47). Integrated to this network, SN4 provided continuous monitoring of seismic activity as well as methane release from the seafloor. For details about the scientific payload adopted, together with data acquisition rates, see Table 11.11.

Since SN4 is deployed jointly with other EC ASSEM nodes, it also carries a hydroacoustic modem capable of communicating towards a surface transducer (from a ship, or attached to a buoy). This acoustic link allows for a bidirectional communication to be established with a remote operator using a transducer from a ship: this was the case during deployment operation; SN4 could also communicate acoustically with a buoy placed near it at the sea surface. In fact, one of the main targets of the ASSEM project was to test and demonstrate the feasibility of a network of seabed nodes, communicating to the surface through acoustic modems, and delivering data summaries during their operation to a ded-

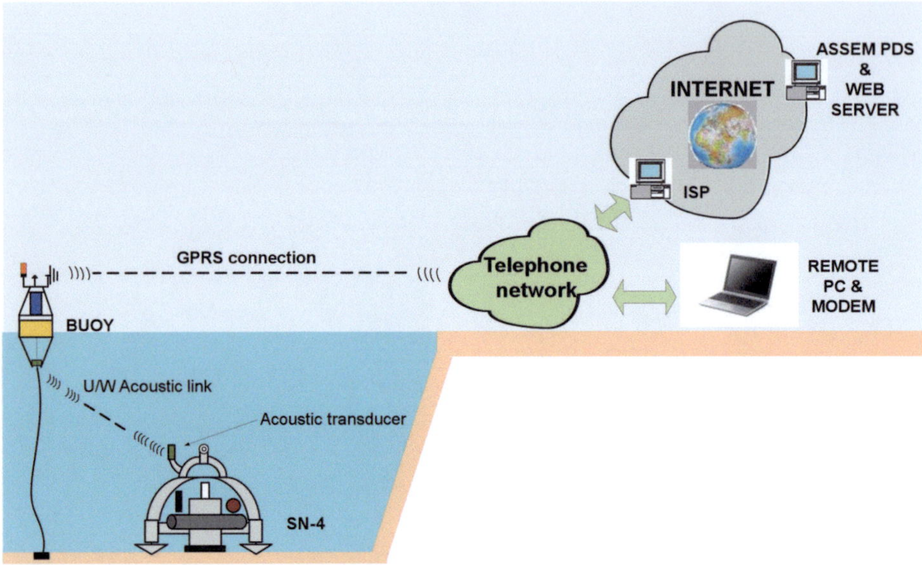

Figure 11.47 SN4 concept (mission 1).

icated "server" in order to publish relevant data on a dedicated website. This purpose was accomplished by using different communication media: in the case of SN4, the acoustic link reaches a surface buoy, whereas a GPRS modem instigates a phone call to an Internet Service Provider in order to transfer all the data collected by the buoy during time. Finally, summary data are collected on a Permanent Data Server so that they can be published on a dedicated website.

Moreover, SN4 could be reached with a PC connected to a phone line, and interrogated to check for system health and to retrieve relevant data, e.g., seismic waveforms.

For this mission, an assisted recovery was planned, based on an underwater intervention by an available ROV or manned submersible; for this purpose, SN4 was equipped with a sling terminating with a ring (both clearly visible in Figure 11.48). The recovery procedure consisted in engaging the ring with the manipulator and attaching it to a rope deployed from the same ship.

SN4 was installed on 20 April 2004 (379m depth) and recovered on 24 November 2004. The observatory operated uninterruptedly all of the time (approx 5230 hours), carrying out all the tasks programmed.

During the period, the relay buoy was operative (until approximately mid July 2004), the networking worked and the effectiveness of bidirectional link from SN4 to ASSEM PDS and to remote PC for data retrieval was proved.

On 28 April 2004 immediately after the occurrence of a significant earthquake in the area, the communication infrastructure allowed for the first time the retrieval of the seismic waveform from the Tecnomare laboratory (see Figure 11.49).

Figure 11.48 (Left) Underwater photograph of SN4 in Corinth Gulf taken from NCMR THETIS ROV. (Right) SN4 being recovered onboard R/V AEGEO (note the seismometer hanging below the frame).

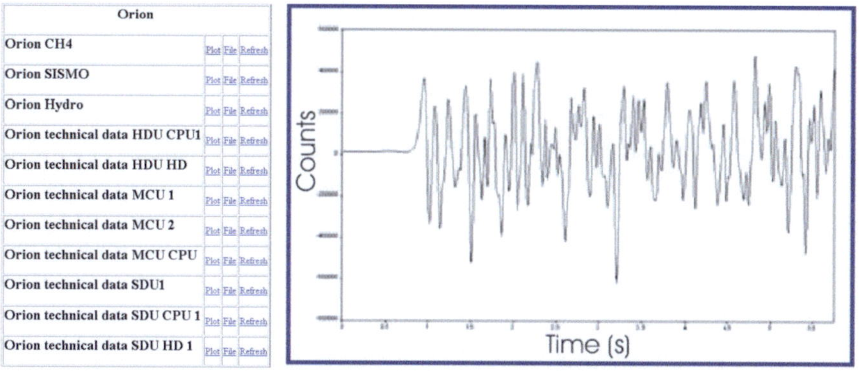

Figure 11.49 (left) SN4 technical summary data page as seen from ASSEM PDS dedicated webpage; (right) acoustically retrieved waveform of 28 April 2004, Corinth Gulf earthquake (M4.6).

Reliable operation of SN4 was confirmed just before recovery, when the observatory was successfully interrogated from the ship and the transmission of autonomous messages at the programmed times verified.

Good quality seismic data (98.5%) and hydrophone data (100%) were recovered, allowing scientific analysis from INGV seismologists. Also, 100% of methane sensor measurements were correctly acquired, but with no practical value due to an almost instantaneous drift of the sensor leading to meaningless data. This experience, associated with the outcomes deriving from the parallel operation of GMM offshore Patras, evidenced a gap in the long-term reliability of the underwater methane sensor technology.

11.7.2 SN4 missions 2 and 3 (Marmara Sea)

In the framework of EU project ESONET NoE, SN4 was selected for the execution of a demonstration mission in Marmara Sea, recognized as a seismic gap that will be probably filled in during the next decades by a large (M ≥7) earthquake along the North Anatolian Fault (NAF) system (Gasperini et al., 2012b). In this scenario, long-term multidisciplinary observatories play an essential role for their unique capability to continuously monitor natural processes that are either very episodic, or statistically require long time series to be detected. Again, a GEOSTAR-class observatory was selected, being the only well-proven technology available in Europe fully meeting the requirements of the application.

Payload was significantly enhanced compared to the previous mission, aiming at better quantifying the temporal relations between fluid expulsion, fluid chemistry and seismic activity along the NAF (Gasperini et al., 2012a). In particular, a new broadband seismometer was selected and integrated with gas and oceanographic sensors allowing identifying local signals related to the fluid expulsion events and eventual local or distant earthquakes that may influence gas migration and seepage processes (Marinaro et al., 2008).

Two methane sensors working in parallel were adopted (following feedback from previous experiences and laboratory qualification and test phases), one directly exposed to the environment and the other connected to a small pump flushing fresh water in front of a sensor membrane, in order to eventually reduce biased signals induced by water turbulence effects. The sensor mounting arrangement is shown in Figure 11.50 (the two methane sensors are visible in the foreground, attached to the internal side of the bumper).

In this application, the hydro-acoustic telemetry link was used only to communicate with a ship of opportunity, to check for system status during descent towards the seabed and for periodical interrogation; hence, the networking functionalities developed for the ASSEM experiment were disabled.

While the deployment procedure remained unchanged (rope terminated with acoustic release), the recovery procedure was redesigned to allow SN4 recovery without any underwater intervention (either by diver or by ROV) and, consequently, reduce economic and logistic efforts. For this purpose, a recall buoy canister was integrated in the observatory, equipped with 400m of recovery rope. Accordingly, the total weight in water of the observatory was reduced to 0.15 kN (about 150kg) by installing four glass spheres on the frame and optimizing the mechanical design. In this way, it was possible to carry out all the marine operations by a light ship of opportunity (the 32m R/V Yunus, owned and operated by Istanbul University).

The site selected for the mission (approx 166m depth, coordinates 40.73 N, 29.40 E) is on the offshore extension of the active NAF, which has been the source of many destructive earthquakes and presents with continuous seismic activity and methane degassing. SN4 operated for one year, in two consecutive missions of about 6 months each.

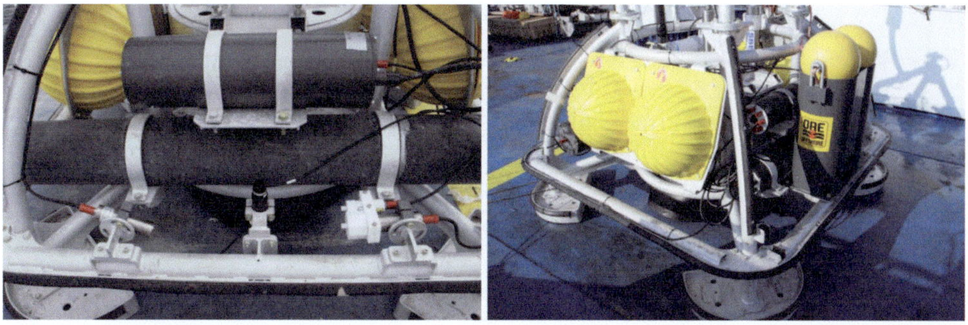

Figure 11.50 (Left) Methane sensors arrangement (the one fitted with a flow-through pump is visible on the right). (Right) Detail of the frame with two buoyancy spheres and recovery system based on a pop-up buoy and rope canister.

Figure 11.51 (Left) SN4 during deployment in Marmara Sea (2009). (Right) the mission site.

SN4 mission 2 was carried out between October 2009 and March 2010 (Figure 11.51). SN4 and its payload operated with complete reliability over the whole period, corresponding to 3863 hours (almost 161 days) acquisition time.

At the end of the mission, SN4 was recovered to download data and replace the battery pack and the methane sensors. Servicing of the observatory was carried out at ITU pier at Tuzla (Turkey).

Figure 11.52 (Left) SN4 during recovery from mission 2. (Right) R/V Yunus that managed SN4 deplyoment and recovery.

Then, SN4 was redeployed in the same site and mission 3 was carried out between March and September 2010. Unfortunately, during this period, the observatory was trawled by fishermen, capsized and moved from the deployment site. This fact affected the significance of data produced by some sensors (in particular the seismometer). Nevertheless, recovery operations, although more complex, were successful and the observatory suffered only minor damage (Figure 11.52).

SN4 missions 2 and 3 represent the longest monitoring of temperature + gas + seismicity at seabed, ever done (Marinaro et al., 2011). A significant number of CH_4 peaks were detected (with frequency about 1 peak every 2 days), showing significant correlations with other parameters (temperature, pressure, dissolved oxygen, turbidity) and patterns similar to those observed during past missions 1 and 2 of the GMM observatory (operating in a gas-bearing pockmark in the Patras Gulf, Greece; see next section). Broadband seismometers recorded low-frequency signals in correspondence with these events, possibly related to vibrations induced by gas seepage. This time, methane sensor technology proved to be mature for long-term applications in seafloor observatories.

Once again, synoptic observation from multidisciplinary sensors proved to be fundamental for a better comprehension of complex and poorly understood phenomena. Redundancy of critical sensors is also opportune especially in case of long-term, autonomous missions where the remote operator may have little (or no) control of the system status.

Summing up, SN4 proved to be a highly cost-effective and efficient observatory, operable with very limited and simple logistics and capable of providing high-quality scientific data; its robustness was also proved during the unexpected phases of the third mission.

For future applications, SN4 can be reconfigured to operate as a cabled observatory, ensuring permanent real-time monitoring of the Marmara Sea and thus the study of relationships between fluids and seismicity.

11.8 GMM

GMM (Gas Monitoring Module) is a light observatory specifically designed for long-term gas monitoring at the seafloor. Gas seeps, either offshore or onshore, reflect deep hydrocarbon generation processes and may provide useful information on the nature of the exploitable natural gas.

On the other hand, seeps may also represent hazards for humans and buildings, because of the explosive properties of methane; gas in marine sediments and onshore soil can, then, damage building and infrastructures by gas-pressure build-up or by degradation of geotechnical properties of ground foundations. Not least, seeps are a source of greenhouse gas for the atmosphere; offshore seeps may release large amounts of methane that can enter the atmosphere, especially if the depth of the seep is shallower than 200–300m (Etiope et al., 2005).

In the offshore environment, detection of gas seepage is much more complicated than onshore and, so far, it has been accomplished through techniques based on rough "sniffers" to detect hydrocarbon anomalies in near-bottom waters, or by direct, expensive and time-consuming sediment sampling and analysis.

The present approach to study the occurrence of methane in seawater was then based on the combination of the peculiar characteristics of GEOSTAR-class observatories (the single-frame architecture, the multiparametric approach, the custom-developed data acquisition and mission control hardware and software) with the use of a new generation of solid-state methane sensors available on the market.

Figure 11.53 GMM: (left) 2004 configuration; (right) 2010 configuration.

The GMM frame is based on a light circular aluminum tripod. The feet are oversized to prevent settlement into the sediment. Each foot can hold a steel ballast to increase stability, if necessary. Design allows modification of frame height by the installation of flanged extension tubes bolted directly to the feet (Figure 11.53).

GMM electronics perform the following tasks: acquisition from all scientific packages and status sensors; preparation and continuous update of hourly data messages, transmitted on request; scientific payload management (switch on/off of individual sensors according to command from the remote operator); processing of methane data to detect occurrence of events (sudden variations of methane concentration); management of commands received (e.g., data request, system reconfiguration, restart); back-up of data in internal mass memory (Marinaro et al., 2004).

GMM main characteristics are summarized in Table 11.12.

GMM has been developed in the framework of the European Commission ASSEM project and since then has been used in the Patras Gulf (two consecutive missions, 2004–2005) and (after upgrade work for payload extension) in Katakolo harbor (2010–2011).

11.8.1 GMM missions 1 and 2 (Gulf of Patras)

The scientific goal of the first two GMM missions was the long-term monitoring of an active pockmark located in the Gulf of Patras (Corinth Shelf, Greece), 40m water depth and 1.5km distance to shore (Marinaro et al., 2006) (Figure 11.54).

GMM scientific payload and sampling rates for the Patras Gulf missions are summarized in Table 11.13. At that time, no experience of long-term operation of methane sensors was available anywhere in the world; this fact led to the decision to adopt three methane sensors mounted in series ("revolver" type configuration), including a master and two back-up sensors (normally powered off and activated in case of failure or bad functioning of the master). Alternatively, the observatory could be configured to operate a back-up ("auxiliary") sensor in parallel to the master one, allowing for data comparison. The methane sensors mounting arrangement is clearly visible in Figure 11.58, with the three sensors

Weight (kN)	1.5 (in air); 0.7 (in water)
Dimensions (mm)	1500 (diameter) × 1550 (height)
Design depth (m)	1000
Data acquisition and mission control	1 board (32 bit microcontroller MC68332)
Mass memory	Compact Flash
Power supply	12 V, 960 Ah Lithium-thionyl chloride
Power consumption (mA)	80 (idle mode); ~150 (mission mode)
Status parameters	Voltage, current, temperature, pressure, acceleration, water intrusion

Table 11.12 GMM main characteristics.

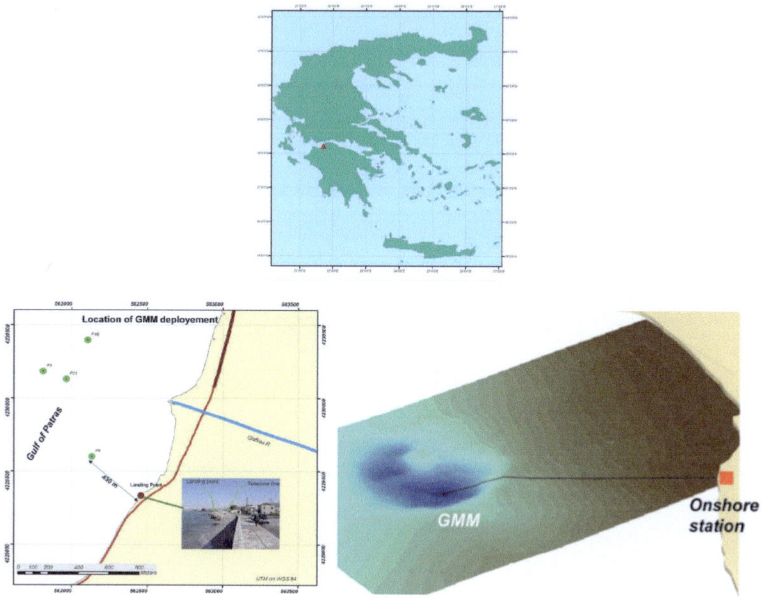

Figure 11.54 GMM installation site (missions 1 and 2).

Parameter	Missions 1 and 2		Mission 3	
	Sensor	**Sampling rate**	**Sensor**	**Sampling rate**
CH_4	3 × Capsum METS	1 sample/s	Franatech METS	1 sample/5s
CH_4 with pump			Franatech METS + SBE-5T	1 sample/5 s (pump ON for 5 min every 30 min)
H_2S	AMT GmbH electrode microsensor	1 Hz for 30s every 10 min	AMT GmbH electrode microsensor	1 sample/5s
CTD	SeaBird SBE-37-SI Microcat	1 sample/10 min	SeaBird SBE-37-SI Microcat	1 sample/min
Turbidity			WET LABS Echo-BBRTD	1 sample/5s
Current meter			FSI 3D-ACM	2 sample/s
Dissolved oxygen			Aanderaa Optode 3830	1 sample/min

Table 11.13 GMM payload and sampling rates.

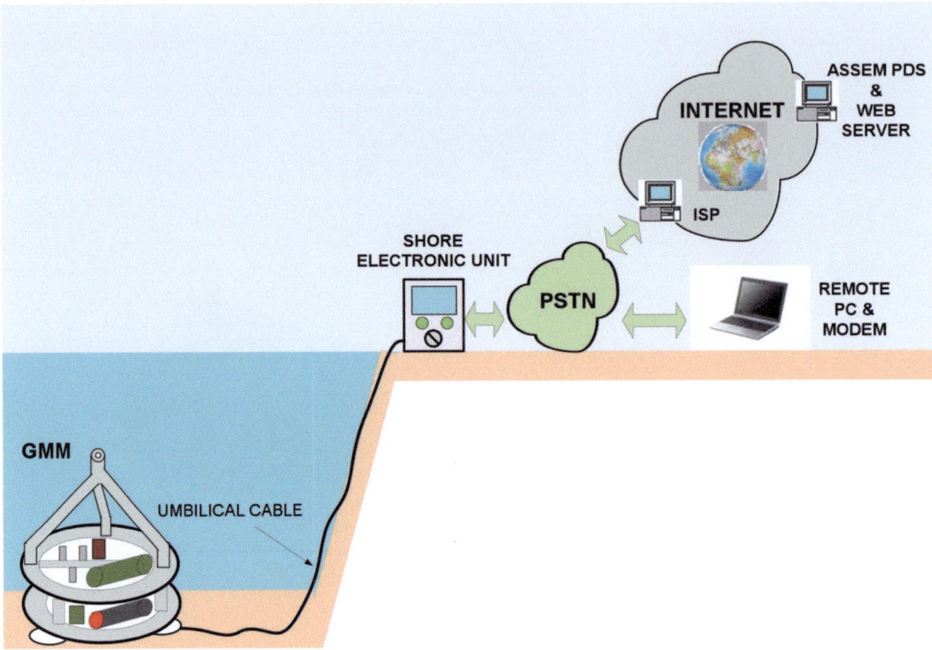

Figure 11.55 GMM cabled concept (missions 1 and 2).

placed horizontally on the top of the frame. Methane sensors were associated to an H2S sensor and a CTD (conductivity, temperature and depth).

Due to the favorable site characteristics, the observatory was simply lowered to the seafloor with a rope and positioned in the desired place by divers.

In addition to the scientific goal of the mission, the ASSEM project had the technical goal of demonstrating the feasibility of a network infrastructure connecting several seabed observatories (called "nodes"), sharing common communication protocols and interfaces and furthermore allowing for a centralized data collecting, publication and storing facility. A remote server PC of the ASSEM PDS (Permanent Data Server) located in the IPGP (Institut de Physique du Globe de Paris) was devoted to collecting and archiving datasets coming from offshore observatories via different communication paths somehow (acoustic modems, phone line, GSM, GPRS, ...) connected to the internet. Moreover, IPGP facilities were in charge of publishing datasets to a dedicated website. The solution developed for GMM connection to the ASSEM network differs from the one adopted for the rest of the observatories. Thanks to the proximity to shore, a cable connection was selected for GMM, while the other ASSEM nodes (deployed in deeper sites) were connected via underwater acoustic/surface GSM telemetry, managed by a moored relay buoy (Figure 11.54).

The underwater cable served to connect GMM to a Shore Unit equipped with a phone line modem providing a remote telemetry link to the Greek public telephone network. The configuration is shown in Figure 11.55.

Figure 11.56 (Left) GMM ready for deployment in Patras Gulf; (center) Shore Unit; (right) umbilical cable.

For safety reasons, GMM was not powered through the umbilical cable. Instead, it was provided with a 12 V, 960 Ah battery pack (half of the standard 1920 Ah battery pack developed for the other GEOSTAR-class observatories) ensuring six months of autonomous operation.

Periodically during the mission, GMM was able to initiate a telephone internet connection through a local Internet Service Provider (a dedicated account was created for this purpose): the GMM Shore Unit was equipped with a modem with embedded TCP/IP features, making easier the work of GMM microcontroller (who itself has not the processing power to properly handle TCP/IP stack complexities, while acquiring data at the same time).

This way, during the whole mission period GMM was able to send data packets to the remotely accessible ASSEM PDS on a daily basis. Datasets consisted of summary technical (status parameters) and scientific measures which got published on the project's dedicated web server.

Furthermore, having a dedicated phone line and number, the GMM seabed observatory could be easily contacted in real-time (e.g., for diagnostics, data download and mission/sensors reconfiguration), just by "calling" it with another phone line modem connected to a remote PC.

This link allowed early detection of a failure relating to the methane sensors; this led to the decision to stop the mission and recover the observatory for the necessary servicing operations. GMM was therefore recovered at the end of September 2004 and redeployed one day later, after CH_4 and H_2S sensor replacement. In the second mission, GMM operated until mid-January 2005.

Summarizing, GMM worked throughout six months in the two consecutive missions (April–July 2004 and September 2004–January 2005). The combined monitoring period amounted to 201 days (4824 hours of data acquisition in total). The data acquisition and control system worked without failure throughout the monitoring period, allowing data transmission and control in near-real-time via cable and modem links. This represented

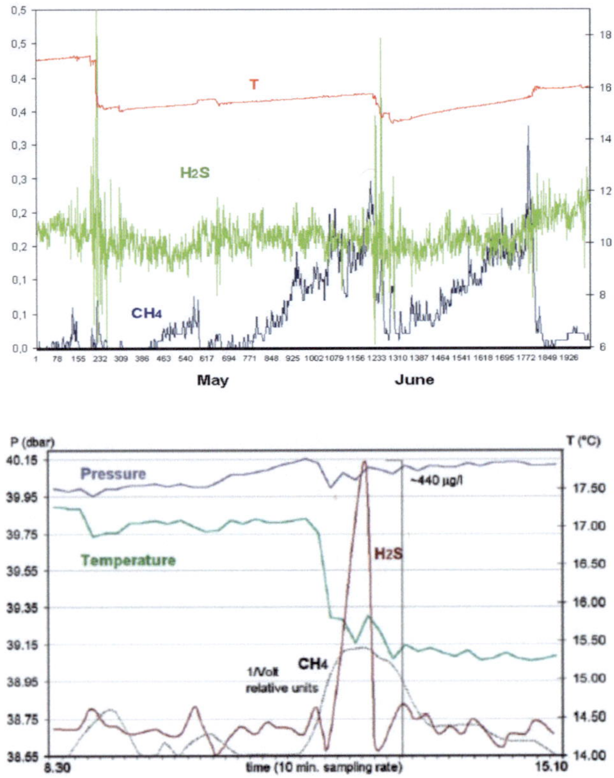

Figure 11.57 Examples of gas seepage events detected by GMM (Marinaro et al., 2006).

the first long-term monitoring ever done on gas leakage from pockmarks by means of CH_4+H_2S+T+P sensors. The results show frequent temperature T and pressure P drops associated with gas peaks; in particular, over 60 events occurred in 6.5 months, likely due to intermittent, pulsation-like seepage (Marinaro et al., 2006) (Figure 11.57). Decreases in temperature in the order of 0.1–1°C (up to 1.7°C) below an ambient T of ca. 17°C (annual average) were associated with short-lived pulses (10–60 min) of increased CH_4+H_2S concentrations. This seepage "pulsation" can either be an active process driven by pressure build-up in the pockmark sediments, or a passive fluid release due to hydrostatic pressure drops induced by bottom currents cascading into the pockmark depression. Redundancy and comparison of data from different sensors were fundamental to interpret subtle proxy signals of temperature and pressure which would not be understood using only one sensor.

11.8.2 GMM mission 3 (Ionian Sea)

The "pioneering" experience of the Patras Gulf missions evidenced some critical aspects of the methane sensors (concerning in particular long-term stability, repeatability and cross-sensitivity), limiting their immediate transfer and operational use in long-term applications. Feedback from the mission allowed sensor manufacturers to upgrade the product. In addition, INGV and Tecnomare organized a dedicated qualification phase in laboratory, testing the two commercially-available underwater methane sensors (Franatech METS and Contros Hydro-C) in controlled conditions, in order to verify their reaction time and their response to temperature variations and water turbulence, either in the presence or in the absence of methane in solution. On the basis of the results obtained, adoption of a pump and a flow-though chamber, able to provide a constant water flow in contact with the sensitive membrane, was then recommended to avoid bio-fouling on the membrane (hence to increment long-term autonomy) and to reduce eventual signal variations induced by changes in water currents (Marinaro et al., 2011).

In the framework of EU HYPOX (in situ monitoring of oxygen depletion in hypoxic ecosystems of coastal and open seas, and land-locked water bodies) project, GMM was recently used in the Katakolo harbor (Ionian Sea, Greece). This area is heavily affected by intense gas seeps, posing a severe hazard for local tourist activities and at the same time providing a unique natural laboratory to study seepages and their impact in the oxygen budget in the seawater.

For this mission, GMM payload was extended (see Table 11.2) and the observatory configured to operate completely in autonomous mode (battery powered, internal data storage, no communication with external devices/users). Detection of gases (O_2, CH_4 and H_2S) is associated with physical-chemical factors, i.e., temperature, pressure and conductivity. Gas

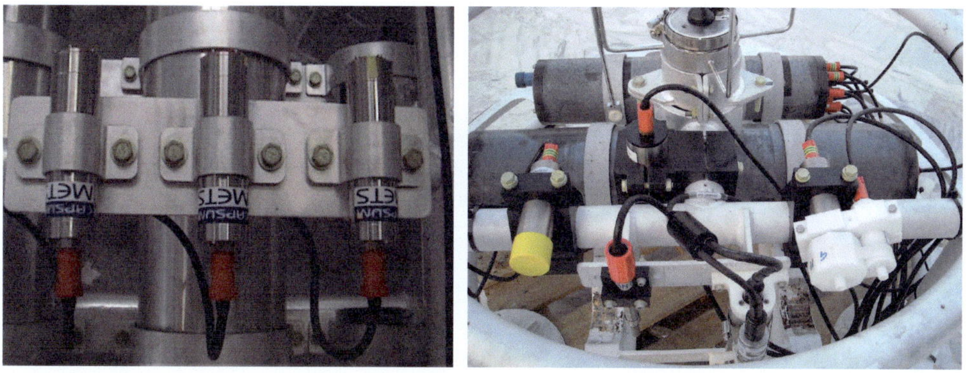

Figure 11.58 Comparison between methane sensor arrangement in GMM. (Left) 2004 configuration (three sensors in series). (Right) 2010 configuration (two sensors, of which the one fitted with flow-through system is visible on the right).

Figure 11.59 (Left) GMM deployment in Katakolo harbor. (Right) GMM recovery after conclusion of mission 3.

detection is based on the use of oxygen, methane and hydrogen sulphide sensors commercially available. This time two methane sensors were adopted (see Figure 11.58), operating in parallel; one was fitted with pump and flow-through chamber, while the other was directly exposed to the environment, thus allowing performance comparison during time.

GMM was deployed on 22 September 2010 and recovered on 17 January 2011 (Figure 11.59).

The observatory successfully operated over the whole mission period (101 days), with 100% data acquisition efficiency. Preliminary analysis shows periods of O_2 decrease (hours) associated with enhanced CH_4 events. Short-term events of T and P drops are associated to CH_4 peaks (as observed in other seepage sites). Data from the two methane sensors show good correlation and absence of drift.

11.9 Conclusions

A fleet of seafloor observatories has been qualified during 18 missions at water depths of up to 3350m, duration up to 3.5 years and operation in autonomous, acoustic linked or cabled (real-time) configuration.

Basic data on these missions (sites, depths, duration and reference projects) are provided in Table 11.14. GEOSTAR and the five derived observatories are characterized by an innovative and alternative concept with respect to other ongoing applications, that proved to be technically sound, cost-effective and suitable to serve the challenging goals of seafloor observatory science.

Full operativeness was achieved (equipment, procedures, personnel, logistics) and high quality scientific data collected.

System	Mission	Site and depth	Period	Project
GEOSTAR	1	Adriatic Sea (42m)	August 1998, 450 h	GEOSTAR (1995–1998)
	2	Tyrrhenian Sea (1950m)	Sep 2000–Apr 2001	GEOSTAR 2 (1999–2001)
	3	Tyrrhenian Sea (3350m)	Dec 2003–Apr 2004	ORION-GEOSTAR 3 (2002–2005)
	4	Tyrrhenian Sea (3350m)	Jun 2004–Apr 2005	ORION-GEOSTAR 3 (2002–2005)
	5	Gulf of Cadiz (3200m)	Aug 2007–Aug 2008	NEAREST (2006–2010)
	6	Gulf of Cadiz (3200m)	Nov 2009–Jun 2011	NEAREST (2006–2010) ESONET (2007–2011) LIDO-DM
SN1	1	Ionian Sea (2105m)	Oct 2002–May 2003	GNDT (2000–2003)
	2	Ionian Sea (2105m)	Jan 2005–May 2008	NEMO
	3	Ionian Sea (2105m)	Jun 2012-on	EMSO (2008 on)
SN2	1	Weddell Sea (1874m)	Dec 2005–Dec 2008	PNRA–MABEL (2005–2009)
SN3	1	Tyrrhenian Sea (3350m)	Dec 2003–Apr 2004	ORION-GEOSTAR 3 (2002–2005)
	2	Tyrrhenian Sea (3350m)	Jun 2004–Apr 2005	ORION-GEOSTAR 3 (2002–2005)
SN4	1	Corinth Gulf (379m)	Apr 2004–Nov 2004	ASSEM (2002–2004)
	2	Marmara Sea (167m)	Oct 2009–Mar 2010	ESONET (2007–2011)
	3	Marmara Sea (167m)	Mar 2010–Sep 2010	ESONET (2007–2011)
GMM	1	Corinth Gulf (40m)	Apr 2004–Jul 2004	ASSEM (2002–2004)
	2	Corinth Gulf (40m)	Sep 2004–Jan 2005	ASSEM (2002–2004)
	3	Ionian Sea (<10m)	Sep 2010–Jan 2011	HYPOX (2009–2011)

Table 11.14 Mission data.

Acknowledgments

The authors wish to thank Renato Campaci, Stefano Cenedese, Roman Chomicz, Vincenzo Ciccarelli, Daniele Calore, Felice Da Prat, Heiko de Vries, Wilfried Langner, Peter Longerich, Andrea Marigo and Simone Virgilio for their essential contribution to the successful technical development of the fleet of GEOSTAR-class observatories.

Photos used in this chapter have been taken from the huge archive produced in many years of work. Special thanks are due to Luigi Innocenzi (author of most of them).

References

Aguzzi J., Company J.B., Costa C., Matabos M., Azzurro E., Mànuel A., Menesatti P., Sardà F., Canals M., Delory E., et al. (2012) Challenges to the assessment of benthic populations and biodiversity as a result of rhythmic behaviour: Video solutions from cabled observatories., in: Gibson, R.N., R.J.A. Atkinson, J.D.M. Gordon, R.N. Hughes, D.J. Hughes, I.P. Smith (Eds), Oceanography and Marine Biology: An Annual Review 50, 235–286.

Berta M., Gasparoni F. and Capobianco M. (1995) Abyssal Benthic Laboratory (ABEL): A novel approach for long-term investigation at abyssal depths. J. Marine Systems 6, 211–225.

Beranzoli L., De Santis A., Etiope G., Favali P., Frugoni F., Smriglio G., Gasparoni F. and Marigo A. (1998) GEOSTAR: A Geophysical and Oceanographic Station for Abyssal Research. Phys. Earth Planet Int. 108(2), 175–183.

Beranzoli L., Braun T., Calcara M., De Santis A., Di Mauro D., Etiope G., Favali P., Frugoni F., Montuori C., Palangio P. (2000a) GEOSTAR, an observatory for deep-sea geophysical and oceanographic researches: Characteristics, first scientific mission and future activity. Mem. Soc. Geol. It. 55, 491–497.

Beranzoli L., Braun T., Calcara M., Calore D., Campaci R., Coudeville J.-M., De Santis A., Etiope G., Favali P., Frugoni F. et al. (2000b) European seafloor observatory offers new possibilities for deep-sea study. EOS, Trans. AGU, 81(5), 45–49.

Beranzoli L., Braun T., Calcara M., Calore D., Campaci R., Coudeville J.-M., De Santis A., Di Mauro D., Etiope G., Favali P., et al. (2002a) GEOSTAR: Geophysical and oceanographic station for abyssal research. Elsevier Oceanography Series 66, 307–315.

Beranzoli L., Favali P. and Smriglio G. (Eds) (2002b) *Science-Technology Synergy for Research in the Marine Environment: Challenges for the XXI Century*. Developments in Marine Technology, 12, Amsterdam: Elsevier.

Beranzoli L., Braun T., Calcara M., Casale P., De Santis A., D'Anna G., Di Mauro D., Etiope G., Favali P., Fuda J.L., et al. (2003) Mission results from the first GEOSTAR observatory (Adriatic Sea, 1998). Earth Planets Space 55, 361–373.

Beranzoli L., Calore D., Favali P., Marvaldi J. and Nicot M. (2004) *ORION-GEOSTAR-3: A Prototype of Seafloor Network of Observatories for Geophysical, Oceanographic and Environmental Monitoring*. ISOPE-Int. Soc. Offshore and Polar Engineers, Proc. 14th Int. Offshore and Polar Engineering Conf., Toulon, May 23–28.

Beranzoli L. and Favali P. (2005) SN1 – The first Italian seafloor observatory for seismic monitoring. In: M. Maugeri (Ed.), *Seismic Prevention of Damage: A Case Study in a Mediterranean City*. Advances in Earthquake Engineering 14, 367–376. Southampton (UK) and Boston (USA):

Beranzoli L., De Santis A., Calcara M., Ciafardini A., De Caro M., Favali P., Frugoni F., Iafolla V., Lo Bue N., Marinaro G., et al. (2009) Multiparametric seafloor exploration: The Marsili Basin and Volcanic Seamount case (Tyrrhenian Sea, Italy). Third IASME/WSEAS International Conference on Geology and Seismology (GES'09), WSEAS Press, 153–157.

Calcara M., Beranzoli L., Braun T., Calore D., De Santis A., Etiope G., Favali P., Frugoni F., Gasparoni F., Montuori C. and Smriglio G. (2001) MABEL: A multidisciplinary benthic laboratory for deep sea, long-term monitoring in the Antarctica. Terra Antarctica 8(2), 115–118.

Cenedese S., Calcara M., D'Anna G., Evers K.-U., Favali P. and Gasparoni F. (2004) MABEL – The first seafloor observatory for multidisciplinary long-term monitoring in Polar environment. ISOPE-Int. Soc. Offshore and Polar Engineers, Proc. 14th Int. Offshore and Polar Engineering Conf., Toulon, May 23–28.

Chierici F., Zitellini N., Favali P., Beranzoli L., Pignagnoli L., Embriaco D., Carrara G., Marinaro G., Lo Bue N., Monna S., et al. (2008) Tsunami warning prototype in the frame of the EC NEAREST project, Rendiconti online. Soc. Geol. It. 2, 1–3.

Chierici, F., P. Favali, L. Beranzoli, A. De Santis, D. Embriaco, G. Giovanetti, G. Marinaro, S. Monna, L. Pignagnoli, G. Riccobene et al. (2012) NEMO-SN1 (Western Ionian Sea, off Eastern Sicily): A cabled abyssal observatory with tsunami early warning capability. ISOPE-Int. Soc. Offshore and Polar Engineers, Proc. 22nd Int. Offshore and Polar Engineering Conf., Rhodes, July 17–22.

Clauss G. and Hoog S. (2002) Deep sea challenges of marine technology and oceanographic engineering. In: L.P. Beranzoli, P. Favali and G. Smriglio (Eds) *Science-technology Synergy for Research in Marine Environment: Challenges for the XXI Century*. Developments in Marine Technology 12. Amsterdam: Elsevier, pp. 133–142.

Clauss G., Hoog S., Stempinski F. and Gerber H.W. (2004) Advanced deepwater intervention with MODUS – Latest results from model tests and full scale operations. In Proceedings of 14th International Offshore and Polar Engineering Conference (ISOPE 2004) Vol II, Toulon (France), pp. 377–386.

De Santis A., Di Mauro D., Cafarella L., D'Anna R., Gaya-Piqué L.R., Palangio P., Romeo G. and Tozzi R. (2006) Deep seafloor magnetic observations under GEOSTAR project. Ann. Geophys. 49 (2/3), 681–694.

De Santis A., Di Mauro D., Cafarella L., Palangio P., Beranzoli L., Favali P. and Vitale S. (2007) Extending magnetic observations to seafloor: The case of GEOSTAR and ORION missions in the Adriatic and Tyrrhenian Seas. Publ. Inst. Geophys. Pol. Acad. Sc. C-99 (398), 114–123.

Etiope G., Marinaro G., Favali P., Furlan F., Cenedese S. and Gasparoni F. (2005) New technologies for methane leakage monitoring from seafloor – description and first operational results. Poster presented at the 7th Offshore Mediterranean Conference, Ravenna (Italy), 16–18 March, 2005.

Etiope G., Favali P., Fuda J.L., Italiano F., Launbenstein M., Millot C. and Plastino W. (2006) The Benthic Boundary Layer: Geochemical and oceanographic data from the GEOSTAR-2 observatory. Ann. Geophys. 49 (2/3), 705–714.

Favali P. and Beranzoli L. (2006) Seafloor observatory science: A review. Ann. Geophys. 49(2/3), 515–567.

Favali P. and Beranzoli L. (2009a) EMSO: European Multidisciplinary Seafloor Observatory, Nuclear Instruments and Methods. Physics Research Section A: Accelerators, Spectrometers, Detectors and Associated Equipment 602, 21–27, doi:10.1016/j.nima.2008.12.214.

Favali, P and Beranzoli L. (2009b) GEOSTAR technology. ESONEWS, The Newsletter of the European Seas Observatory Network 2(3), 1–8.

Favali P., Smriglio G., Beranzoli L., Braun T., Calcara M., Etiope G., Frugoni F., Millot C., Fuda J.-L., Marani M.P., et al. (1998) GEOSTAR – Scientific goals of the project and results of the first test phase. IEEE-OCEANS '98 Conference Proceedings, vol. 2, 1088–1090, doi: 10.1109/OCEANS.1998.724404.

Favali P., Smriglio G., Beranzoli L., Braun T., Calcara M., D'Anna G., De Santis A., Di Mauro D., Etiope G., Frugoni F., et al. (2002) Towards a permanent deep sea observatory: The GEOSTAR European experiment. In: L. Beranzoli, P. Favali and G. Smriglio (Eds) *Science-technology Synergy for Research in Marine Environment: Challenges for the XXI Century*. Development in Marine Technology 12, Amsterdam: Elsevier, pp. 111–120.

Favali P., Beranzoli L., Calcara M., D'Anna G., Etiope G., Frugoni F., Lo Bue N., Marinaro G., Monna S. and Montuori C., et al. (2004) Single-frame multiparameter platforms for seafloor geophysical and environmental observations: Projects and missions from GEOSTAR to ORION. OCEANS '04. MTTS/IEEE TECHNO-OCEAN '04, Vol. 4, 2000–2007, doi: 10.1109/OCEANS.2004.1406450.

Favali P., Beranzoli L., D'Anna G., Gasparoni F., Marvaldi J., Clauss G., Gerber H.W., Nicot M., Marani M.P., Gamberi F., Millot C. and Flueh E.R. (2006) A fleet of multiparameter observatories for geophysical and environmental monitoring at seafloor. Ann. Geophys. 49 (2/3), 659–680.

Favali P., Beranzoli L., D'Anna G., Gasparoni F. and Gerber H.W. (2006) NEMO-SN1: The 1st "real-time" seafloor observatory of ESONET. Nuclear Instruments and Methods in Physics Research Section A: Accelerators, Spectrometers, Detectors and Associated Equipment 567(2), 462–467, doi:10.1016/j.nima.2006.05.255.

Favali P., Person R., Barnes C.R., Kaneda Y., Delaney J.R. and Hsu S.-K. (2010) Seafloor observatory science. In: J. Hall, D.E. Harrison and D. Stammer (Eds) Proceedings of the OceanObs'09: Sustained Ocean. Observations and Information for Society Conference 2, Venice, Italy, 21–25 September 2009, ESA Publication WPP-306 ISSN:1609-042X, doi: 10.5270/OceanObs09.cwp28.

Favali P., Beranzoli L., Italiano F., Migneco E., Musumeci M. and Papaleo R. (2011) NEMO-SN1 observatory developments in view of the European Research Infrastructures EMSO and KM3NET. Nuclear Instruments and Methods in Physics Research Section A: Accelerators, Spectrometers, Detectors and Associated Equipment 626–627, S53–S56, doi:10.1016/j.nima.2010.04.139.

Favali P., Chierici F., Marinaro G., Giovanetti G., Azzarone A., Beranzoli L., De Santis A., Embriaco D., Monna S., Lo Bue N., et al. (2012) NEMO-SN1 abyssal cabled observatory in the Western Ionian Sea. IEEE J. Oceanic Engineering, doi: 10.1109/joe.2012.222.4536.

Frugoni F., Beranzoli L., Favali P. and Basili A. (Eds) (2006) Special issue dedicated to Giuseppe Smriglio. From land networks to seafloor observatories. Ann. Geophys. 49(2/3), 515–879.

Fuda J.L., Millot C., Hoog S. and Gerber H.W. (2006) Analysis of ADCP data above a bottom observatory. Ann. Geophys. 49(2/3), 715–728.

Gasparoni F., Calore D., Campaci R. and Marigo A. (1998) GEOSTAR – Development and test of an innovative benthic station for long term observations at abyssal depths. In: Proceedings of the IEEE Conference OCEANS '98 (on CD-ROM).

Gasparoni F., Calore D. and Campaci R. (2002) From ABEL to GEOSTAR: Development of the first European deep-sea scientific observatory. In: L. Beranzoli, P. Favali and G. Smriglio (Eds) *Science-technology Synergy for Research in Marine Environment: Challenges for the XXI Century*. Developments in Marine Technology 12. Amsterdam: Elsevier, pp. 143–159.

Gasperini L., Polonia A., Del Bianco F., Etiope G., Marinaro G., Favali P., Italiano F. and Çağatay M.N. (2012a) Gas seepage and seismogenic structures along the North-Anatolian Fault in the Eastern Marmara Sea. Geochem. Geophys. Geosyst. 13(10), Q10018, ISSN:1525-2027, doi:10.1029/2012GC004190.

Gasperini L., Polonia A., Del Bianco F., Favali P., Marinaro G. and Etiope G. (2012b) Cold seeps, active faults and the earthquake cycle along the North Anatolian Fault system in the Sea of Marmara (NE Turkey). Bollettino di Geofisica Teorica ed Applicata 53(4), 371–384, doi: 10.4330/bgta0082.

Gerber, H.W., G.Clauss, 2001. Development and operational optimization of MODUS - the GEOSTAR subsea docking station, Proceedings of OMAE 2001 Conference, Rio de Janeiro (Brazil), 3-8 June 2001.

Gerber H.W. and Clauss G. (2005) Space shuttle MODUS – Key system for the installation of networks of Benthic Stations. In: Proceedings of OMAE 2005 Conference, Haikidiki (Greece).

Gerber H.W and Clauss G. (2009) MABEL: Recovery operation of the first long-term heavy benthic laboratory in the deep sea of Antarctica. In: Proceedings of the 28th OMAE International Conference on Ocean, Offshore and Arctic Engineering, 31 May–5 June 2009, Honolulu (Hawaii, USA), paper 80251.

Gerber H.W., Favali P., Smriglio G., Gasparoni F., Marvaldi J. and Coudeville J.-M. (1999) GEOSTAR benthic observatory: Technological results. J. Acoust. Soc. Am. 105(2), 1169, doi: http://dx.doi.org/10.1121/1.425545.

Gerber H.W., Clauss G. and Hoog S. (2002) MODUS: Remotely operated carrier for abyssal research – Experiences in the Mediterranean Sea. In: Proceedings of ISOPE 2002 Conference, paper No.2002-JSP-320.

Iafolla V. and Nozzoli S. (2002) Gravimeter for deep sea measurements. In: L. Beranzoli, P. Favali and G. Smriglio (Eds) *Science-technology Synergy for Research in Marine Environment: Challenges for the XXI Century*. Developments in Marine Technology 12. Amsterdam: Elsevier, pp. 183–197.

Iafolla V., Nozzoli S., Fiorenza E. and Milyukov V. (2006) Deep-sea gravity measurements: GEOSTAR-2 mission results. Ann. Geophys. 49(2/3), 695–704.

Jourdain J.Y. (1999) First trial of GEOSTAR, the geophysical and oceanographic European station for abyssal research. In: G. Ollier (Ed.) EC Project Information Booklet EUR18885, p.31.

Kopf A., Camerlenghi A., Canals M., Ferdelman T., Mevel C., Pälike H., Roest W., Ask M., Barker-Jørgensen B., Boetius A., et al. (2012) *Deep Sea and Sub-Seafloor Frontier*. White Paper European Commission, pp. 58.

Lampitt R.S., Favali P., Barnes C.R., Church M.J., Cronin M.F., Hill K.L., Kaneda Y., Karl D.M., Knap A.H., McPhaden M.J., et al. (2010) In situ sustained Eulerian observatories. In: J. Hall, D.E. Harrison and D. Stammer (Eds) *Proceedings of the OceanObs'09: Sustained Ocean*. Observations and Information for Society Conference 2, Venice, Italy, 21–25 September 2009, ESA Publication WPP-306, doi:10.5270/OceanObs09.pp.27.

Marani M.P., Gamberi F. and Bonatti E. (Eds) (2004) *From Seafloor to Deep Mantle: Architecture of the Tyrrhenian Back-arc Basin*. Mem. Descr. C. Geol. d'It., LXIV.

Marinaro G., Etiope G., Gasparoni F., Calore D., Cenedese S., Furlan F., Masson M., Favali P. and. Blandin J. (2004) GMM-a gas monitoring module for long-term detection of methane leakage from the seafloor. In: G. Etiope and P. Favali (Eds) *GEM-Geologic Emissions of Methane from Lands and Seafloor: Mud Volcanoes and Observing Systems*. Environmental Geology 46/8, 1053–1058, doi:10.1007/s00254-004-1092-2.

Marinaro G., Etiope G., Lo Bue N., Favali P., Papatheodorou G., Christodoulou D., Furlan F., Gasparoni F., Ferentinos G., Masson M. and Rolin J.-F. (2006) Monitoring of a methane-seeping pockmark by cabled benthic observatory (Patras Gulf, Greece). Geo-Mar. Letters, doi: 10.1007/s00367-006-0040-4.

Marinaro G., Etiope G., Favali P., Beranzoli L., Gasperini L., Gasparoni F., Furlan F., Geli L., Henry P. and Çağatay M.N. (2008) SN4 seafloor observatory in Marmara Sea. Rendiconti on-line Soc. Geol. It. 2, 1–3.

Marinaro G., Etiope G., Gasparoni F., Furlan F., Bruni F. (2011) Gas seepage detection and monitoring at seafloor. Paper presented at the 10th Offshore Mediterranean Conference, Ravenna (Italy), 23–25 March 2011.

Marvaldi J., Blandin J., Podeur C., Coudeville J.M., Antoine J., Barbot D., Fellmann D. and Rhodes D. (1998) GEOSTAR – Development and test of a communication system for deep-sea benthic stations. In: *Proceedings of the IEEE Conference OCEANS '98* (on CD-ROM).

Marvaldi J., Aoustin Y., Ayela G., Barbot D., Blandin J., Coudeville J.-M., Fellmann D., Loaëc G., Podeur C. and Priou A. (2002) Design and realisation of communication systems for the GEOSTAR project. In: L. Beranzoli, P. Favali and G. Smriglio (Eds) *Science-technology Synergy for Research in Marine Environment: Challenges for the XXI Century*. Developments in Marine Technology 12. Amsterdam: Elsevier, pp. 161–181.

Monna S., Frugoni, F., Montuori, C., Beranzoli, L. and Favali, P. (2005) High quality seismological recordings from the SN1 deep seafloor observatory in the Mt. Etna region. Geophys. Res. Lett. 32, doi:10.1029/2004GL021975.

NRC (National Research Council) (2000) *Illuminating the Hidden Planet. The Future of Seafloor Observatory Science*. Washington D.C.: National Academy Press.

Ollier G., Favali P., Smriglio G. and Gasparoni F. (2002) Perspectives and challenges in marine research. In: L. Beranzoli, P. Favali and G. Smriglio (Eds). *Science-technology Synergy for Research in Marine Environment: Challenges for the XXI Century*. Developments in Marine Technology 12. Amsterdam: Elsevier, pp. 3–9.

Rigaud V., Semac D., Nokin M., DESIBEL Team, Tietze G., Amann H., Goetz V. and Pascoal A. (1998) New methods for deep-sea interventions on future benthic laboratories. DESIBEL project: Final results, comparison of concepts and at sea validation. In: *Proceedings of OCEANS '98*, IEEE Conference Nice (France) (on CD-ROM).

Rolin J.-F., Blandin J., Lykousis V., Strout J.M., Etiope G., Favali P., Briole P., Ballu V., Papatheodorou G., Ferentinos G., Cathy D. and Masson M. (2005) Common issues between cabled and non-cabled observatories in ASSEM project. OCEANS 2005 – Europe 2, 872–877, doi: 10.1109/OCEANSE.2005.1513171.

Ruhl H.A., André M., Beranzoli L., Çağatay M.N., Colaço A., Cannat M., Dañobeitia J.J., Favali P., Géli L. and M. Gillooly, et al. (2011) Societal need for improved understanding of climate change, anthropogenic impacts, and geo-hazard warning drive development of ocean observatories in European Seas. Progress in Oceanography 91, 1–33, doi:10.1016/j.pocean.2011.05.001.

Sgroi T., Beranzoli L., Di Grazia G., Ursino A. and Favali P. (2007) New observations of local seismicity by the SN1 seafloor observatory in the Ionian Sea, off-shore Eastern Sicily (Italy). Geophys. J. Int. 169, 490–501, doi:10.1111/j.1365-246X.2007.03348.x.

Vitale S., De Santis A., Di Mauro D., Cafarella L., Palangio P., Beranzoli L. and Favali P. (2009) GEOSTAR deep seafloor missions: Magnetic data analysis and 1D geoelectric structure underneath the Southern Tyrrhenian Sea. Ann. Geophys. 52(1), 57–63.

Part III

World-wide recent and ongoing projects and programmes

12 The two seafloor geomagnetic observatories operating in the western Pacific

H. Toh[1] and Y. Hamano[2]

12.1 Introduction

The need for seafloor observatories has been intensifying for decades. It has long been pointed out that a "land-based" view alone cannot provide the true image of the Earth because the continents occupy only 30% of the Earth's surface. Disciplines in solid Earth geophysics, such as seismology, strongly require an "ocean-based" view of the Earth as well as the "land-locked" view in order to know the true internal structures of our planet. Some of the disciplines such as geodesy or geomagnetism may circumvent the difficulty since remote sensing of potential fields by low-Earth-orbit (LEO) satellites (e.g., Gaya-Piqué et al., 2005) is possible in those disciplines. The "ocean-locked" view, however, is indispensable even for disciplines in which the Earth-observing satellites play dominant roles in providing the bulk of necessary data, especially when temporal variations of geophysical phenomena such as the geomagnetic secular variations are of particular interest.

Multidisciplinary efforts aimed at constructing seafloor observatories (e.g., GEOSTAR; Beranzoli et al., 1998, 2003) resulted in successful pilot data acquisition even in the hostile environment of the seafloor which had prevented us from obtaining meaningful geophysical observations for a long time. In North America, a broadband ocean bottom seismic observatory was built in Monterey Bay, which was originally uncabled (Romanowicz et al., 2003, 2006). However, it has recently been connected to a submarine cable in order to realize real-time data transmission from the seafloor (Romanowicz et al., 2009; http://www.

1 Data Analysis Centre for Geomagnetism and Space Magnetism, Graduate School of Science, Kyoto University, Sakyo-ku, Kyoto, Japan

2 Department of Deep Earth Structure and Dynamics Research, Japan Agency for Marine-Earth Science and Technology, Kanagawa, Japan

mbari.org/mars/). On the other hand, it should also be noted here that those attempts have mostly been made and repeated in coastal regions such as the Mediterranean Sea rather than in open oceans. This is partly due to the fact that the coastal seafloor observatories can be serviced at better maintenance rates and hence they tend to result in better data recovery rates. However, this does not necessarily mean that the geophysical data are acquired in areas of high data demand. Seafloor geophysical observatories in deep and open oceans are still necessary in this sense.

Now that there have been a miniconstellation of magnetic satellites (Ørsted, CHAMP (Challenging Mini-Satellite Payload), and SAC-C (Satelite de Aplicaciones Cientificas-C)) in space and the next long-life LEO magnetic satellite has been launched (SWARM; Olsen et al., 2006; Friis-Christensen et al., 2006), the importance of comparative studies of geomagnetic data by ground networks and those by magnetic satellites is increasing. Because the four successive satellite missions cover over a full solar cycle, the long-term monitoring of the gradual change not only in the Earth's main field but also in the external field of the magnetospheric and/or ionospheric origin is now possible. This chapter reports on how the world's first seafloor geomagnetic observatory NWP (North-Western Pacific) is producing data that are compatible with those obtained by the geomagnetic satellites (Ørsted: Neubert et al., 2001; Olsen, 2007; CHAMP: Reigber et al., 2002) and the existing geomagnetic observatory network (e.g., as for the OHP geomagnetic network, see Fukao et al., 2001; Shimizu and Utada, 1999).

Toh and Hamano (1997) began to develop a prototype geomagnetic observatory for long-term deployments at remote open oceans in coordination with the OHP project. The instrument, the first version of the Seafloor EM Station (SFEMS), was further improved (Toh et al., 1998) by adding a magnetotelluric (MT) variograph consisting of a combination of a three-component fluxgate-type magnetometer and a two-component voltmeter with a pair of mutually orthogonal electric dipoles of approximately 5m long to sense the seafloor geoelectric field. In-house silver-silver chloride electrodes (the TOK electrodes in Perrier et al., 1997) were adopted for stable measurements of the geopotentials during the long-term deployments. The SFEMSs are designed to continuously measure not only the geomagnetic but also the geoelectric field more than one year by one-minute sampling. In the course of its development, it was found that attitude monitoring is crucial to realize the absolute observation of the geomagnetic vector field by the unmanned systems (Toh et al., 2004). We therefore decided to equip accurate tilt meters and a fiber-optical gyro for orientation detection in addition to original EM sensors such as an absolute scalar Overhauser-type (Overhauser, 1953) proton precession magnetometer (OHM) and the vector fluxgate magnetometer. The final version of SFEMSs is now operating at NWP since August 2001 and at WPB since June 2006 (Toh et al., 2006).

So far, slightly less than 2000-day-long time-series of seafloor EM fields have been obtained at NWP. Precise attitude data make the seafloor data nearly conform to those on land. Moreover, the absolute scalar geomagnetic measurement is now possible even on the seafloor by an Overhauser scalar absolute magnetometer taking advantage of nuclear magnetic resonance. The state-of-the-art technology is now utilized not only in space but also on the seafloor. One of the main points of this chapter is to show how well the slow and smooth temporal variation of the magnetic field, i.e., the geomagnetic secular variation, coincides at both locations. Close agreement with satellite data collected more than

500km above the seafloor will be demonstrated. In the following, this chapter outlines the long-term geomagnetic observation on the seafloor being conducted by the two SFEMSs in the western Pacific. Specifically, (1) instrumentation of the geomagnetic observatories on the deep seafloor, (2) deployment/recovery procedures of SFEMSs at the time of recent sea experiments, (3) description of obtained EM time-series, and (4) results of analysis of the geomagnetic secular variations found in the observed geomagnetic data will be illustrated. Finally, the detected geomagnetic secular variation and future perspectives such as application of the seafloor data to regional geomagnetic field modeling and research on ocean dynamics are discussed and concluded.

12.2 Instrumentation at sea

Outer views of the SFEMS are shown in Figures 12.1(a) and (b). The SFEMS is a pop-up, self-contained system that can monitor the geomagnetic and geoelectric field continuously, together with auxiliary measurements of environmental parameters. Figure 12.1(c) shows the sensors and electronics that measure the EM fields and the environmental parameters. The real attitude on the seafloor is usually very stable as shown in Figure 12.1(d). The continuous monitoring of the EM field and the environmental parameters is achieved by a preset sampling rate apart from detection of the instrument's azimuth on the seafloor. The environmental parameters include temperature and the instrument's attitude, i.e., horizontal tilts and the orientation of the SFEMS on the seafloor. The tilt variations and the azimuthal measurements with respect to the geographical north turned out to be essential, especially for detection of the geomagnetic secular variation. The azimuthal measurements alone are conducted intermittently on the seafloor, viz., once in a couple of months with duration of days, due partly to the limitation of power supply and to magnetic noises that were produced by a fiber-optical gyro (indicated by 'Gyro' in Figure 12.1(a)) at the time of the measurements. The measuring frame of the instruments' attitude was carefully aligned to that of EM measurements prior to the SFEMSs' deployments at sea.

The geomagnetic field can be continuously monitored by a combination of an Overhauser-type absolute scalar proton precession magnetometer ('OHM' in Figure 12.1(a)) and a fluxgate-type vector magnetometer (Ocean Bottom ElectroMagnetometer; OBEM). The geoelectric field sensed by the mutually orthogonal electrode dipoles is also recorded by OBEM. All the EM data and the environmental parameters are archived not only by each client sensor such as OHM, OBEM and Gyro but also by an interface circuit (I/F) being equipped with a master clock of sub-ppm precision and with a large capacity for digital storage. Every measurement timing is controlled by the master clock of I/F.

Recovery of the SFEMS is normally enabled by acoustic release. The SFEMS is equipped with an acoustic transponder for this purpose with a release code unique to each SFEMS. The SFEMSs are designed to survive on the seafloor as deep as 6000m for more than three years in terms of the acoustic release. However, the lifetime of the SFEMSs is constrained by the capacity of power supply from lithium primary cells (Batt) attached to each SFEMS rather than that of the acoustic transponder. Power saving schemes built in the

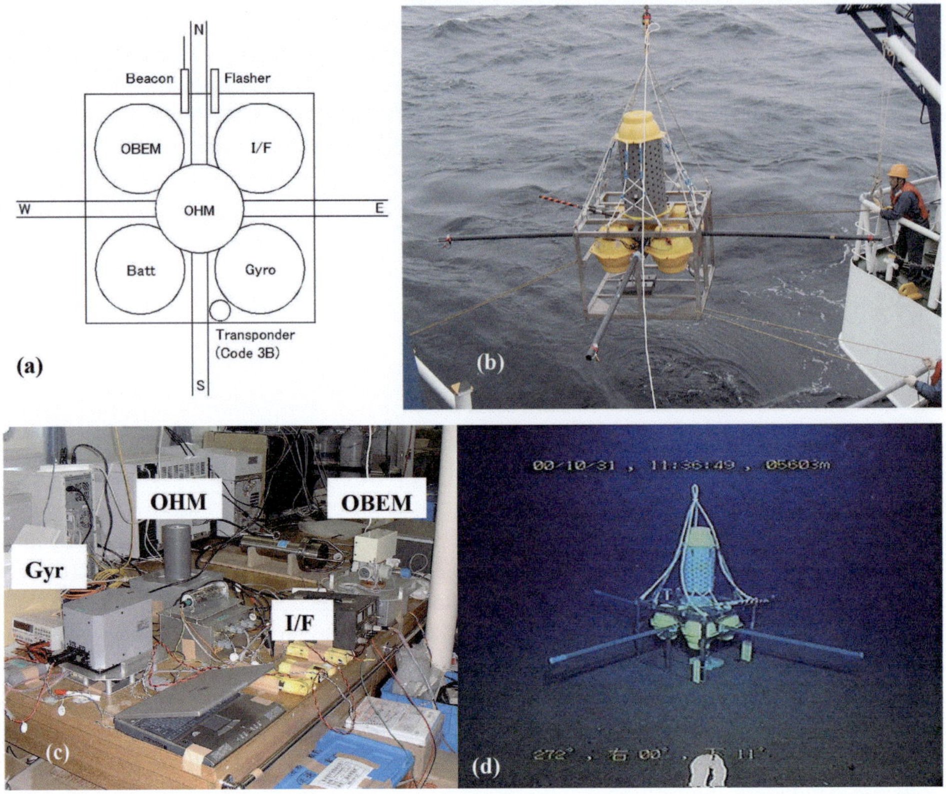

Figure 12.1 (a) A schematic top view of the SFEMS. It mainly consists of five buoyant glass housings that contain EM sensors and electronics. The perpendicular dipoles are for geoelectric measurements. In addition, the SFEMS is equipped with an acoustic unit (transponder) for release and a radio transmitter and flashing light for prompt discovery at the sea surface. (b) The SFEMS just before launch from R/V Kairei. (c) The EM sensors and electronics inside the SFEMS. (d) The SFEMS in operation on the seafloor. The height of the instrument is approximately 2m. (Photos are by courtesy of JAMSTEC.)

electronics of the SFEMSs have made it possible to elongate the SFEMSs' actual lifetime on the seafloor by more than two years even with a battery package with capacity of only several hundred *Ahs*.

Although the SFEMS weighs approximately 400kg in air at the time of deployment, its weight in water is only 50–60kg. After releasing their non-magnetic lead weights, they weigh less than 300kg in air at recovery. They are also equipped with a VLF radio transmitter and a flashing light for easier spotting of the instruments on the surface. As for physical dimensions and sensor resolutions of the SFEMS, refer to Table 12.1. Detailed instrument packaging is also given in Toh et al. (2006).

The most outstanding feature of the SFEMS is the first application of OHM to the seafloor geomagnetic observation. The OHMs are known to be the only sensors for practical use in long-term absolute magnetic measurements at this moment because of their low power consumption rate. Specifically, the OHM attached to each SFEMS can measure one-minute values of the absolute geomagnetic total force with a resolution of 0.01 nT for as long as two years by a 18V–120Ah lithium primary cell package, which had been impossible by ordinary proton precession magnetometers. Faster sampling is also possible at the cost of lifetime. However, it cannot be faster than 5 sec since the present Overhauser sensors adopted for the SFEMSs are not the so-called "proton oscillator" that can create continuous proton precessions. It takes at least a few seconds (the relaxation time of the proton precession in the geomagnetic field) per reading. Refer to Primdahl (2002) for details of the OHM sensor such as its principle. The OHM sensor and other relevant electronics are housed in separate glass spheres (see Figures 12.2(a) and (b)) to shut out possible noise from the electronics. The sensor sphere is further placed on top of a PVC cylinder as shown in Figures 12.1(b) and (d) for final noise protection and to avoid surface effects of the seafloor to the OHM sensor.

12.3 Seafloor experiments

Since earlier seafloor experiments have been already described in Toh et al. (1998, 2004, 2006), we will focus on three recent seafloor experiments in this section. All of the three cruises involved were conducted by a combination of R/V Kairei and ROV Kaiko 7000II operated by JAMSTEC. They are KR05-08, KR06-06 and KR07-08 research cruises, among which only KR06-06 was organized to deploy an SFEMS for the first time to WPB. Establishment of another long-term EM observation site on the seafloor in the middle of the Philippine Sea plate enabled monitoring of dynamics in the Earth's outer core at a

L × W × H [m]	1.28 × 1.28 × 1.85
Weights [kg]	385 (in air), 50 (in water), 285 (in air without lead ballast)
Magnetic resolutions	0.01 nT (for both OHM and OBEM)
Magnetic accuracy/stability	0.2 nT (OHM), 4 nT/year (OBEM)
Electric resolutions	0.3 μV (~ 60 nV/m by 5m electrode span)
Electric field precision	0.1 μV/m
Attitude precisions	5 arc seconds (tilt), 20 arc seconds (azimuth)
Digital storage	2 GB flash memory
Submersible depth [m]	6000

Table 12.1 Physical dimensions and sensor resolutions of the SFEMSs.

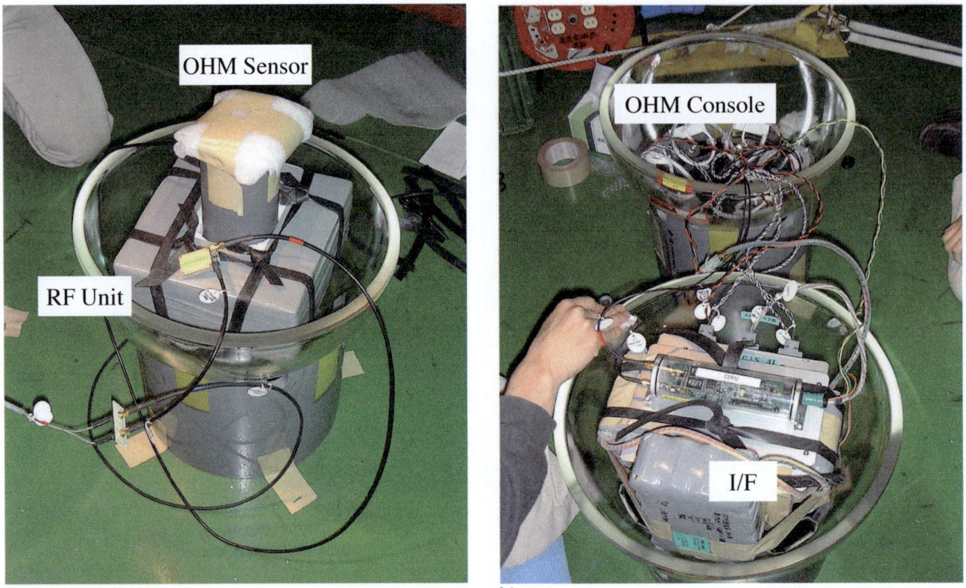

a) b)

Figure 12.2 (a) The OHM sensor cylinder containing chemical liquid abundant in free radicals. The RF unit generates the RF field necessary to cancel the electric spins in the sensor cylinder (Primdahl, 2002). (b) The other electronics of the OHM are assembled into the OHM console and contained in another 17-inch pressure-tight glass housing together with I/F so as to avoid affecting the sensor unit.

regional scale. The combined KR05-08 and KR07-08 cruises were dedicated to carrying out maintenance necessary for continuation of the long-term seafloor observation at NWP.

Prior to those three seafloor experiments, EM data were available only at NWP although the total length of the obtained EM time-series was more than 1300 days. A pair of KR05-08 and KR07-08 cruises at NWP has added another 21-month-long EM time-series, yielding the total length slightly less than 2000 days at NWP. In 2005, however, the EM data have a gap of approximately five months from February through July due to an unsuccessful proposal for a research cruise in 2004 (we were not able to do any sea experiments between KR03-08 and KR05-08). This experience has made us elongate the lifetime of the SFEMSs by as long as two years. As for a list of available seafloor geomagnetic data, refer to http://wdc.kugi.kyoto-u.ac.jp/caplot/index.html.

Figure 12.3 shows a site map depicting the relative position of the two long-term seafloor EM sites, and a schematic cartoon illustrating the procedure of the SFEMS replacement. By making use of the ROV, it is now possible to install a new SFEMS to a seafloor position very close to the old SFEMS, say within several tens of meters, and conduct simultaneous side-by-side EM measurements for a few days per sea experiment before the recovery of the old SFEMS. Table 12.2 lists the positions of the two SFEMSs estimated by acoustic ranging conducted at the time of each deployment cruise.

a)

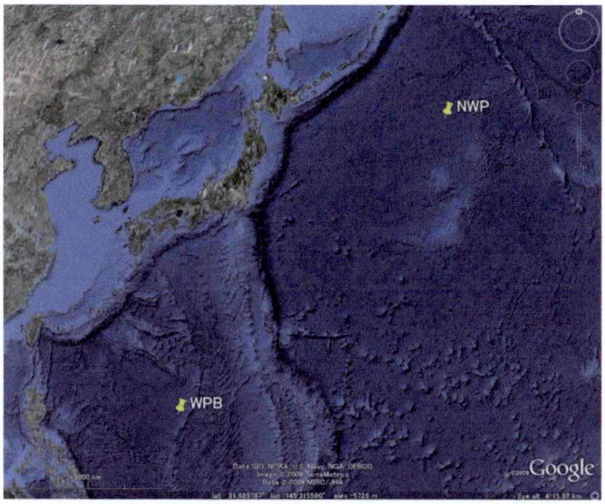

b)

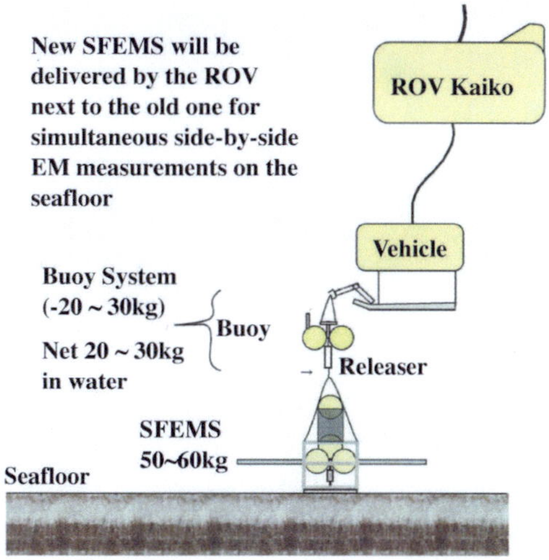

Figure 12.3 (a) The site map of the SFEMSs operating in the western Pacific. (b) The schematic figure of the replacement operation of the SFEMSs on the seafloor using the ROV.

Site name	Positions [WGS84]	Seafloor age	Duration
NWP	41 06' 08" N, 159 57' 47" E −5580 m	129 Ma (Nakanishi & Winterer, 1998)	August 2001 ~
WPB	19 19' 27" N, 135 06' 41" E −5702m	49 Ma (Salisbury et al., 2006)	June 2006 ~

Table 12.2 Positions of NWP and WPB.

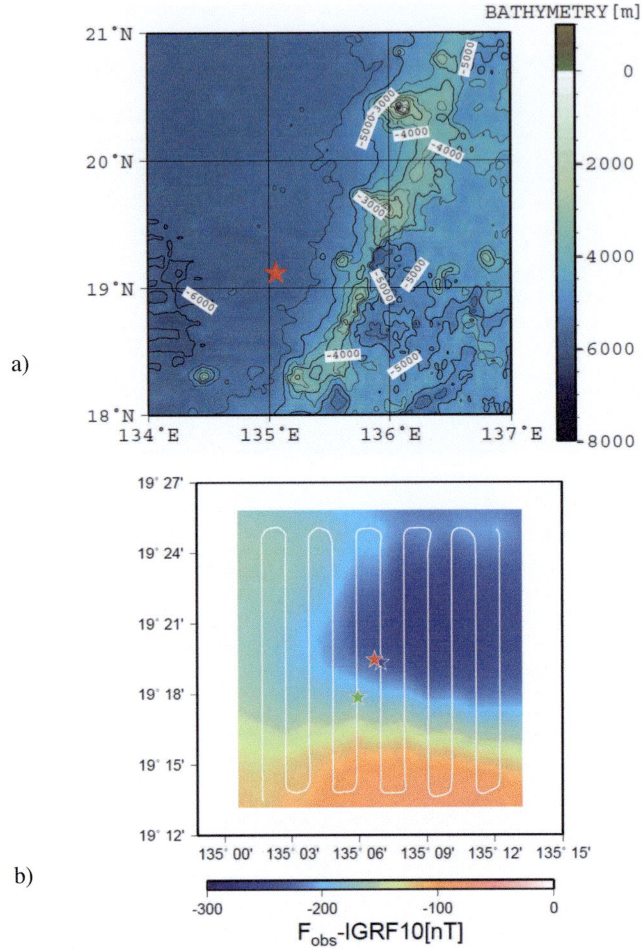

Figure 12.4 (a) Bathymetry around WPB (red star). (b) Result of the magnetic survey in the vicinity of WPB. The blue and green stars are, respectively, the drop site of the SFEMS at the sea surface and the location of WP-1 where a borehole seismometer (Suyehiro et al., 2002) was installed. The white line shows the ship's track at the time of surface tow of an absolute proton precession magnetometer in order to yield the magnetic anomaly map.

Mapping of magnetic anomalies in the vicinity of the long-term observation site is one of the indispensable procedures for establishing seafloor observatories in addition to a detailed swath mapping around the site and acoustic ranging to determine the instruments' positions on the seafloor. Figure 12.4 shows the result of the magnetic mapping in the vicinity of WPB as well as that of the swath mapping. It is evident that WPB sits on a negative anomaly whose spatial gradient seems small.

12.3.1 Observed time-series on the seafloor

As noted above, the recent three cruises have added 21 months of data at NWP and commenced the long-term EM observation at WPB. Because the long-term seafloor observation at NWP started at the beginning of August 2001, the new input of the data lengthened the total duration of the obtained EM time-series for approximately six years with a data gap of five months from the end of February through the middle of July 2005. We give a brief report on our latest dataset at NWP here in this section in order to outline what the raw seafloor data look like.

Figure 12.5 shows the attitude data of the SFEMS at NWP. It is evident from the azimuthal measurements conducted on the seafloor on three different days with a 6-month interval (Figure 12.5(a)) that the X-axis of the SFEMS's measuring frame pointed between 1.5° and 2° clockwise from the true north during the 2005–2007 deployment. This was further confirmed by an in-situ measurement of the instrument's azimuth using a precise gyroscope of ROV Kaiko7000II at the time of the replace operation of the SFEMSs on the seafloor. Because the three azimuth measurements on the seafloor were very consistent with each other and the precision of the fiber-optical gyro used is known to be 0.2°, the standard error of the azimuthal average is less than 11 arc seconds by taking all the azimuth readings of 4320 (3 days by one reading per minute) on the seafloor. The tilt variations in

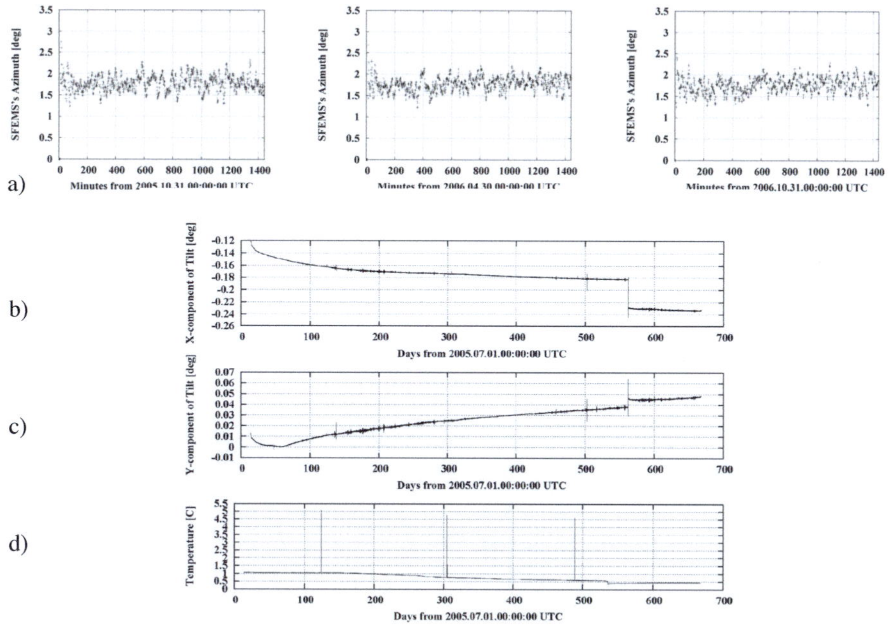

Figure 12.5 The attitude (a–c) and temperature (d) data observed at NWP from July 2005 through April 2007. (a) Azimuthal measurements conducted on the seafloor on three different days with a 6-month interval. (b) Tilt variations for the X-axis of the measuring frame. (c) Same as (b) but for the Y-axis. Tilts are positive downward. (d) Temperature variations within the OBEM sphere.

Figures 12.5(b) and (c) contain spikes and jumps due to large earthquakes that occurred mostly along trenches located several hundred kilometers away from NWP, whereas the three spikes seen in the temperature variations in Figure 12.5(d) correspond to the three azimuthal measurements conducted on the seafloor. The temperature variations show those within the OBEM pressure-tight glass housing. The temperature inside the OBEM glass sphere normally increases at the time of the azimuthal measurements since the fluxgate magnetometer is switched to a continuous mode during the occasional azimuthal measurements to record as many geomagnetic data as possible at a cost of temporarily enhanced power consumption rate.

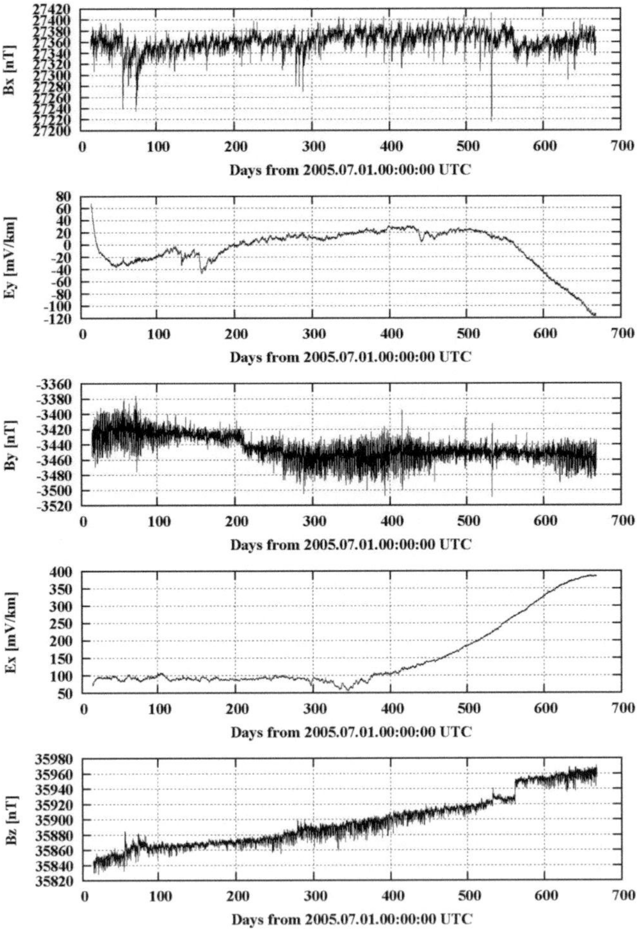

Figure 12.6 Five EM components observed at NWP from July 2005 through April 2007. They are, respectively, the northward geomagnetic (Bx), the eastward geoelectric (Ey), the eastward geomagnetic (By), the northward geoelectric (Ex) and downward geomagnetic (Bz) components from top to bottom.

The observed 5-component EM data are shown in Figure 12.6. They are truly "raw" data in the sense that no corrections for tilts, azimuth, timings and electrode drifts have been made. However, the almost northward orientation of the SFEMS (see Figure 12.5(a)) as well as its small absolute values of tilt averages (Figures 12.5(b) and (c)) have made the raw time-series similar to the corrected time-series. The decreasing By and increasing Bz components give a very good agreement with what has been observed as the geomagnetic secular variation at NWP (Toh et al., 2006).

12.4 The geomagnetic secular variation contained in the time series

Figure 12.7 compares the geomagnetic total force synthesized from the vector geomagnetic data by the 3-component fluxgate magnetometer of the SFEMS with the absolute geomagnetic total force observed by the scalar Overhauser proton precession magnetometer. It is evident that the observed difference is stable for over half a year on the seafloor, and only shows a small systematic trend of a few nanoteslas. This trend could arise from the instrumental drift of the fluxgate magnetometer used, although it is still within the required

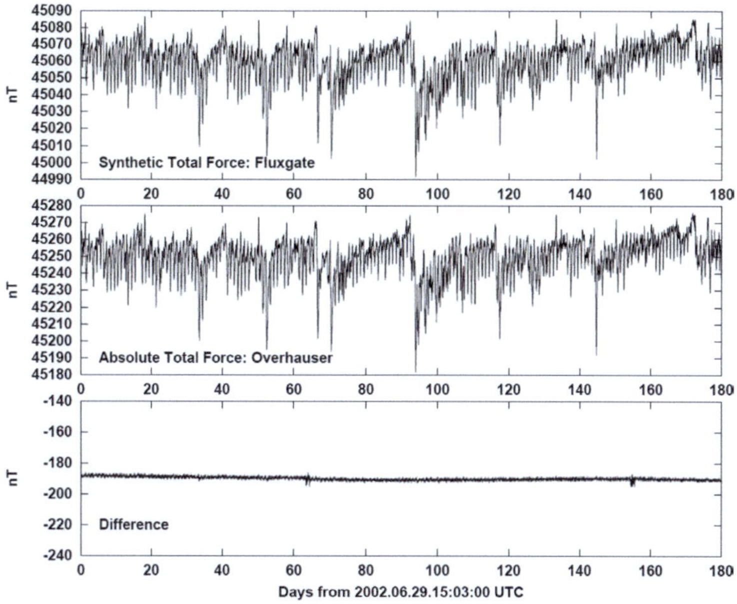

Figure 12.7 Comparison of synthetic total force (top) with the absolute geomagnetic total force (middle). The difference between the two (bottom) shows quite stable variations with a nearly constant offset of approximately 190 nT. The two disturbances seen at the end of August and November 2002 are caused by magnetic noises at the time of in-situ Gyro operations. Reproduced from Toh et al. (2006).

stability of 5 nT/year by INTERMAGNET. The nearly constant difference between the synthetic and absolute total force motivated us to compare the vector geomagnetic field observed on the seafloor more with those estimated by the global models in the hope of detecting the vector secular variation for the first time on the seafloor.

Raw three-component magnetogram longer than one year from July 2002 is shown in Figure 12.8 (upper left). Initial tilt variations (Figure 12.8, upper right) are normally larger than variations in the latter half of the observation. This is because each newly-deployed instrument needs several months to settle down on the seafloor. However, it is possible to monitor the tilt variations precisely enough as shown in Figure 12.8. Once the precise tilt information is given, it is straightforward to rotate the inclined measuring frame back into the horizontal plane. The result of the tilt correction is also shown in Figure 12.8 (bottom three).

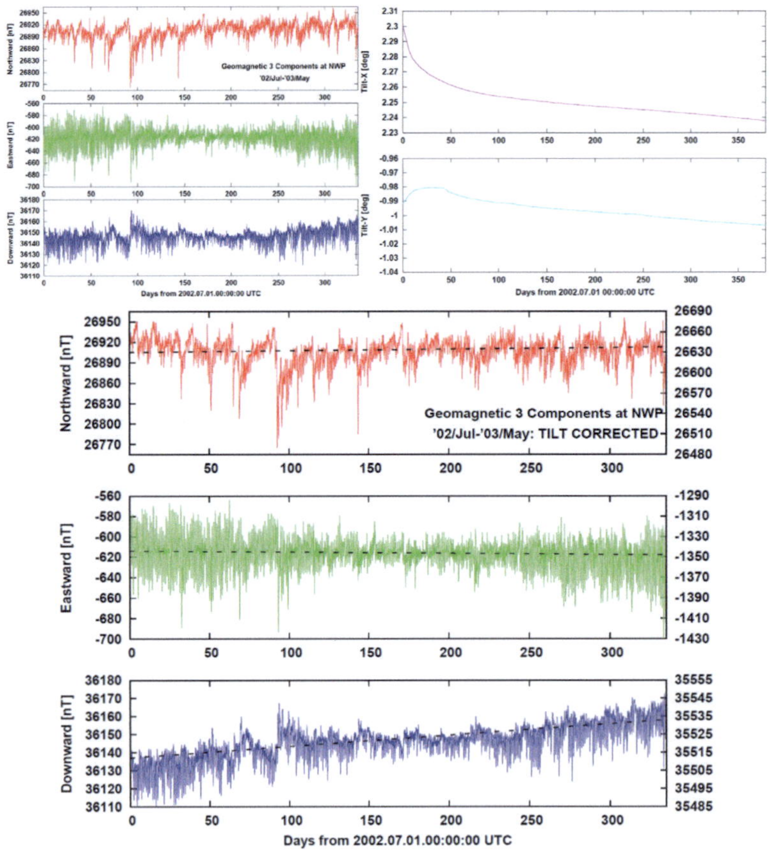

Figure 12.8 The tilt-uncorrected vector geomagnetic field observed on the seafloor (upper left), simultaneous tilt variations (upper right) and the tilt-corrected geomagnetic three components (lower three). The dashed lines in the corrected data are the prediction of the geomagnetic secular variations at NWP using IGRF-11 (International Association of Geomagnetism and Aeronomy, Working Group V-MOD, 2010).

12.5 Discussion

Figure 12.8 clearly shows that the tilt-corrected geomagnetic three components possess almost linear trends that coincide well with the predicted vector geomagnetic secular variations. This implies that detection of the secular variation for each geomagnetic component is now realized even on the seafloor by the SFEMSs. A preliminary study on the detected geomagnetic secular variation has shown that most of the increase in the vertical component with time can be attributed to that of the westward drift of the equatorial dipole (Toh et al., 2010). This is compatible with the concept of the so-called "Pacific Dipole Window" (PDW; e.g., McWilliams et al., 1982) that implies the dominance of geomagnetic dipole components not only in the spatial distribution but also in temporal changes.

The construction of regional geomagnetic reference field models is important in the sense that:

1. It can make a significant contribution to the arguments on existing geomagnetic anomalies (e.g., PDW) of regional scale.
2. Density of geomagnetic observatories in a specific portion of the Earth's surface is a function of time and can be occasionally enhanced for some temporal duration.

Toh et al. (2007) derived a regional geomagnetic reference field over the western Pacific (RGRF-P) using data from NWP and geomagnetic stations on oceanic islands established/maintained by OHP. Although the regional geomagnetic model, derived for the first time in the area of sparse data both in space and time, has revealed the presence of PDW, a detailed description of the field modeling is out of the scope of this chapter and will be explained elsewhere in this volume (Toh and De Santis, 2014).

Another important topic that has not been included in this chapter is possible tsunami warnings by our long-term seafloor EM stations in the western Pacific. We have observed a few examples of EM field generation by large-scale barotropic motions of sea water columns associated with tsunamis so far (e.g., Toh and Hamano, 2007). The very long geomagnetic time-series acquired at NWP was found to contain very clear tsunami-induced magnetic signals at the time of the 2011 Tohoku earthquake off the Pacific coast, even though NWP was located 1500km from its epicenter. It therefore is quite feasible to use the SFEMSs as tsunami sensors. The SFEMS as a tsunami detector has several advantages over the existing tsunami sensors such as pressure gauges on the seafloor. While the conventional tsunami sensors are mostly scalar, the SFEMSs are essentially vector sensors because they are measuring the intrinsically vector EM field. This could enable the SFEMSs to find directions of tsunami propagation as well as detection of tsunami arrival. Application of the SFEMSs to tsunami monitoring is our next major improvement of the present version of the SFEMSs. Integration of the SFEMSs into the existing tsunami monitoring systems such as NOAA's DART system (Wei et al., 2008) is preferable in this respect. A more detailed report on this issue is a matter of another research paper and out of scope of the present paper.

12.6 Conclusions

The two seafloor geomagnetic observatories are now operating in the western Pacific. The long-term geomagnetic observation by the SFEMS in the Northwest Pacific Basin implies that the geomagnetic secular variation of the vertical component is arising mostly from the westward drift of the equatorial dipole. This means that the vector geomagnetic secular variation can now be detected even on the deep seafloor by the present version of the SFEMSs, although it is still subject to further instrumental improvement in the future. A combination of the SFEMSs at NWP and at WPB is expected to reveal core dynamics in regional scale by resolving difference in the vector geomagnetic secular variations at each site.

Two important research topics need more thorough descriptions. They are the regional geomagnetic field modeling making use of the presence of the seafloor observatories and the ground observation network, and detection of tsunami-induced EM signals captured by the SFEMSs (Toh et al., 2011). The former can address the present arguments on the existing geomagnetic anomalies such as PDW.

The integration of the SFEMSs to the global tsunami monitoring systems is one of the most promising research topics pertinent to the straightforward application of the SFEMSs to disaster mitigation. Early tsunami warning systems is increasingly needed since the occurrence of the great Sumatra earthquake in 2004 that generated extremely hazardous tsunamis. The SFEMSs (Toh et al., 2011) are quite new as tsunami sensors in the sense that they can monitor tsunami propagation directions by single site observation. This is obviously impossible by conventional scalar tsunami sensors.

Acknowledgments

We are indebted to the skilful help of the captains and crew of R/V Kairei and ROV Kaiko 7000II of Japan Agency for Marine-Earth Science and Technology at the time of the sea experiments. We also thank Mr T. Motobayashi and Dr T. Goto for preparation of the SFEMSs since their help is indispensable for successful deployment and recovery of the instruments. Support given by Earthquake Research Institute, University of Tokyo throughout this study is greatly acknowledged. This work was partly supported by Grant-in-Aid for Scientific Research, Ministry of Education, Culture, Sports, Science and Technology, Japan (No. 18204038).

References

Beranzoli L., De Santis A., Etiope G., Favali P., Frugoni F., Smriglio G., Gasparoni F. and Marigo A. (1998) GEOSTAR: A GEophysical and Oceanographic STation for Abyssal Research. Phys. Earth Planet. Inter. 108, 175–183.

Beranzoli L., Braun T., Calcara M., Casale P., De Santis A., D'Anna G., Di Mauro D., Etiope G., Favali P., Fuda J-L., Frugoni F.,Gamberi F., Marani M., Millot C., Montuori C. and

Smriglio G. (2003) Mission results from the first GEOSTAR observatory (Adriatic Sea, 1998). Earth Planets Space 55, 361–373.

Friis-Christensen E., Lühr H. and Hulot G. (2006) Swarm: A constellation to study the Earth's magnetic field. Earth Planets Space 58, 351–358.

Fukao Y. et al. (2001) The Ocean Hemisphere Network Project (OHP). In: B. Romanowicz et al. (Eds) *OHP/ION Joint Symposium Workshop Report*, 13–29.

Gaya-Piqué L.R., De Santis A. and Torta J.M. (2005) Use of CHAMP magnetic data to improve the Antarctic geomagnetic reference model. In: C. Reigberet al. (Eds) *Earth Observation with CHAMP – Results from Three Years in Orbit*. Berlin: Springer.

International Association of Geomagnetism and Aeronomy, Working Group V-MOD (2010) International Geomagnetic Reference Field: The eleventh generation. Geophys. J. Int. 183, 1216–1230, doi: 10.1111/j.1365-246X.2010.04804.x.

McWilliams M.O., Holcomb R.T. and Champion D.E. (1982) Geomagnetic secular variation from ^{14}C dated lava flows on Hawaii and the question of the Pacific non-dipole low. Philos. Trans. R. Soc. London, Ser. A. 306, 211–222.

Nakanishi M. and Winterer E.L. (1998) Tectonic history of the Pacific-Farallon-Phoenix triple junction from Late Jurassic to Early Cretaceous: An abandoned Mesozoic spreading system in the Central Pacific Basin. J. Geohpys. Res. 103, 12453–12468.

Neubert T., Mandea M., Hulot G., von Frese R., Primdahl F., Jørgensen J.L., Friis-Christensen E., Stauning P., Olsen N. and Risbo T. (2001) Ørsted satellite captures high-precision geomagnetic field data. Eos, Trans. Amer. Geophys. Union 82, 81–88.

Olsen N. (2007) Ørsted. In: D. Gubbins and E. Herrero-Bervera (Eds) *Encyclopedia of Geomagnetism and Paleomagnetism*, pp. 743–746.

Olsen N., Haagmans R., Sabaka T.J., Kuvshinov A., Maus S., Purucker M.E., Rother M., Lesur V. and Mandea M. (2006) The *Swarm* End-to-End mission simulator study: A demonstration of separating the various contributions to Earth's magnetic field using synthetic data. Earth Planets Space 58, 359–370.

Overhauser A.W. (1953) Polarization of nuclei in metals. Phys. Rev., 92, 411–415.

Perrier F.E., Petiau G., Clerc G., Bogorodsky V., Erkul E., Jouniaux L., Lesmis D., Macnae J., Meunier J. M., Morgan D., Nascimento D., Oettinger G., Schwarz G.., Toh H., Valiant M.J., Vozoff K. and Yazici-Cakin O. (1997) A one-year systematic study of electrodes for long period measurements of the electric field in geophysical environments. J. Geomag. Geoelectr. 49, 1677–1696.

Primdahl F. (2002) Resonance magnetometers. In: P. Ripka (Ed.) *Magnetic Sensors and Magnetometers*. Boston and London: Artech House, pp. 267–304.

Reigber, C., Luhr H. and Schwintzer P. (2002) CHAMP mission status. Adv. Space Res. 30, 129–134.

Romanowicz B., Stakes D., Uhrhammer R., McGill P., Neuhauser D., Ramirez T. and Dolenc D. (2003) The MOBB experiment: A prototype permanent off-shore ocean bottom broadband station. Eos, Transactions, American Geophysical Union 84(34), 325, 331–332.

Romanowicz B., Stakes D., Dolenc D., Neuhauser D., McGill P., Uhrhammer R. and Ramirez T. (2006) The Monterey Bay broadband ocean bottom seismic observatory. Ann. Geophys. 49, 607–623.

Romanowicz B., McGill P., Neuhauser D. and Dolenc D. (2009) Acquiring real time data from the broadband ocean bottom seismic observatory in Monterey Bay (MOBB). Seism. Res. Lett. 80, 196–204.

Salisbury M.H., Shinohara M., Suetsugu D., Arisaka M., Diekmann B., Januszczak N. and Savov I.P. (2006) LEG 195 synthesis: Site 1201 – A geological and geophysical section in the West Philippine Basin from the 660-km discontinuity to the mudline. In: M. Shinohara et al. (Eds) *Proc. ODP, Sci. Results* 195, 1–27.

Shimizu H. and Utada H. (1999) Ocean hemisphere geomagnetic network: Its instrumental design and perspective for long-term geomagnetic observations in the Pacific. Earth Planets Space 51, 917–932.

Suyehiro K., Araki E., Shinohara M. and Kanazawa T. (2002) Deep sea borehole observatories ready and capturing seismic waves in the western Pacific. Eos, Trans. Am. Geophys. Union 83, 624–625.

Toh H. and De Santis A. (2014) Modelling of regional geomagnetic field based on ground observation network including seafloor geomagnetic observatories, this volume.

Toh H. and Hamano Y. (1997) The first realtime measurement of seafloor geomagnetic total force – Ocean Hemisphere Project Network. J. Japan Soc. Mar. Surv. Tech. 9, 1–13.

Toh H. and Hamano Y. (2007) Present status of seafloor electromagnetic stations in the northwest Pacific. Oral presentation (JSS016) at the XXIV General Assembly of the International Union of Geodesy and Geophysics, Perugia, Italy, July 2–13.

Toh H., Goto T. and Hamano Y. (1998) A new seafloor electromagnetic station with an Overhauser magnetometer, a magnetotelluric variograph and an acoustic telemetry modem. Earth Planets Space 50, 895–903.

Toh H., Hamano Y., Ichiki M. and Utada, H. (2004) Geomagnetic observatory operates at the seafloor in the Northwest Pacific Ocean. Eos, Trans. Amer. Geophys. Union 85, 467–473.

Toh H., Hamano Y. and Ichiki M. (2006) Long-term seafloor geomagnetic station in the northwest Pacific: A possible candidate for a seafloor geomagnetic observatory. Earth Planets Space 58, 697–705.

Toh H., Kanezaki H. and Ichiki M. (2007) A regional model of the geomagnetic field over the Pacific Ocean for epoch 2002. Geophys. Res. Lett. 34, L09308, doi:10.1029/2007GL029341.

Toh H., Hamano Y., Goto T. and Utada H. (2010) Long-term seafloor electromagnetic observation in the Northwest Pacific may detect the vector geomagnetic secular variation. Data Sci. J. 9, IGY100-IGY109, doi:10.2481/dsj.SS_IGY-004.

Toh H., Satake K., Hamano Y., Fujii Y. and Goto T. (2011) Tsunami signals from the 2006 and 2007 Kuril earthquakes detected at a seafloor geomagnetic observatory. J. Geophys. Res. 116, B02104, 10pp., doi:10.1029/2010JB007873.

Wei Y., Bernard E.N., Tang L., Weiss R., Titov V.V., Moore C., Spillane M., Hopkins M. and Kanoglu U. (2008) Real-time experimental forecast of the Peruvian tsunami of August 2007 for U.S. coastlines. Geophys. Res. Lett. 35, L04609, doi:10.1029/2007GL032250.

13 The DELOS project: Development of a long-term observatory in an oil field environment in the Tropical Atlantic Ocean

P.M. Bagley[1], K.L. Smith[2], B.J. Bett[3], I.G. Priede[4], G. Rowe[5], J. Clarke[6], A. Wall[6], H.A. Ruhl[7], D.M. Bailey[7] and B. Bazica[8]

13.1 Introduction

The deep-sea environment into which oil company operations are gradually extending is generally poorly understood, with surveys regularly discovering new habitats and communities of animals previously unknown to science. There is inevitably a lack of historical data which can be used as a basis for baseline knowledge and prediction. It is, however, apparent that all deep-sea environments support a wide range of animals that contribute significantly to global biodiversity and ecosystem function. Hitherto, only a limited number of deep-ocean sites have been monitored for periods exceeding 5 years (Smith et al., 2001; Billett et al., 2001). At these sites, important annual cycles have been observed with considerable variability from year to year and changes in dominant fauna over decadal time scales (Smith et al., 2001, 2008; Billett et al., 2001; Ruhl et al., 2008). In the Pacific,

1 Aker Solutions, Aberdeen, UK
2 MBARI, Moss Landing, California, USA
3 National Oceanography Centre, University of Southampton, Southampton, UK
4 University of Aberdeen, Oceanlab, Aberdeen, UK
5 Texas A&M University, Galveston, USA
6 BP, UK
7 Glasgow University, UK
8 National Institute of Fisheries Research, Angola

evidence suggests a linkage between the surface climate and the deep sea, and major climatic events such as the El Niño Southern Oscillation have influenced the deep-seafloor epibenthic megafauna community (Ruhl and Smith, 2004). Similar relationships between climate and deep-sea communities have been observed in the Atlantic. In an oil production area, such spontaneous changes need to be distinguished from any anthropogenic influences imposed on the deep-sea environment.

Here we discuss the development of DELOS (Deep ocean Environmental Long-term Observatory System) to provide long-term environmental monitoring data to both science and the oil industry (BP) in the vicinity of an oil production field in 1400m water depth. This represented a unique partnership between science and industry which should prove beneficial to both parties. The oil industry requires a high level of quality in design and manufacture to ensure that DELOS is safe for use in an area of active extraction, and fit for purpose. We discuss the process from design through to installation.

13.1.1 Science and oil industry collaboration

A series of scientific meetings were set up to discuss the feasibility of establishing a long-term monitoring station within BP operating areas. The first meeting (September 2004) concluded that the establishment of a long-term monitoring station in the Atlantic Ocean would, after a minimum 3–4 year time span, begin to allow discrimination of patterns of local change that could be compared with global changes. An objective should be to achieve a 25-year time span of operation to allow monitoring for the lifetime of the production in the area. This would provide long-term environmental monitoring but also make a major

Name	Affiliation	Expertise
Dr Ken Smith (Chairman)	MBARI, USA	Deep-ocean biology and processes
Professor Gill Rowe	Texas A&M University, USA	Marine biology and ecosystem function
Professor Monty Priede	Oceanlab, University of Aberdeen, UK	Deep-ocean biology
Jim Clarke	BP, UK	Project manger, local environmental impact
Anne Walls	BP, UK	Environmental manager
Bomba Bazica	National Institute of Fisheries Research, Angola	Angolan fishery science
Dr Phil Bagley	Oceanlab, University of Aberdeen, UK	Ocean instrumentation and engineering
Dr Henry Ruhl	NOC, UK	Deep-ocean biology and processes
Dr David Bailey	University of Glasgow	Marine biology

Table 13.1 The DELOS steering committee.

contribution to science. At the conclusion of the first meeting, Oceanlab (University of Aberdeen) were tasked with developing a number of design options to satisfy the scientific requirements and the practicalities of deploying a platform in the vicinity of BP offshore operations. A further meeting in April 2005 discussed and enhanced both the scientific objectives and the engineering design.

The science meetings comprised a committee of deep-ocean biologists, engineers, and oil industry environmental experts (Table 13.1) with the remit to control all aspects of the DELOS (Deep-ocean Environmental long-term Observatory System) project from scientific quality through to data access and future plans.

13.1.2 Rational

By establishing long-term monitoring of the deep-sea physical environment and biological activity in that environment, it should be possible to develop an understanding of the ecological processes taking place and to differentiate between natural variation and changes arising from the presence of oil industry facilities. Two platforms are deployed, one within 50m of a seafloor well (near field), and a second 16km from any seafloor infrastructure (far field). These platforms are situated in the tropical Atlantic Ocean in an oil and gas exploration region termed Block 18 at a depth of 1400m. The platforms will be deployed for 25 years and serviced every 6 months by ROV. The DELOS scientific objectives are to:

- Determine long-term natural environmental conditions at a deepwater site in Block 18
  - Compare variations observed at near- and far-field monitoring sites
  - Increase understanding of mechanisms linking climate change to deep-water ecology
  - Measure and monitor deep-sea biological communities
  - Understand the pace of recovery from any unforeseen impacts
  - Differentiate between natural and man-made changes providing a linkage between marine biodiversity and climate change
- Determine long-term effects of the monitoring platform itself on natural processes
  - Understanding on the reefing effect of large fixed structures in the deep-water environment;
  - Contributing to understanding of potential effects of sub-sea equipment in general.

13.2 System description

Each platform comprises two parts: the seafloor docking station that is deployed on the seafloor at the start of the monitoring program and remains for the duration of the 25 years; a number of observatory modules that are designed to perform specific environmental monitoring functions. Each observatory module carries enough battery and storage capacity for autonomous operation for at least six months. Towards the end of the 6-month deployment

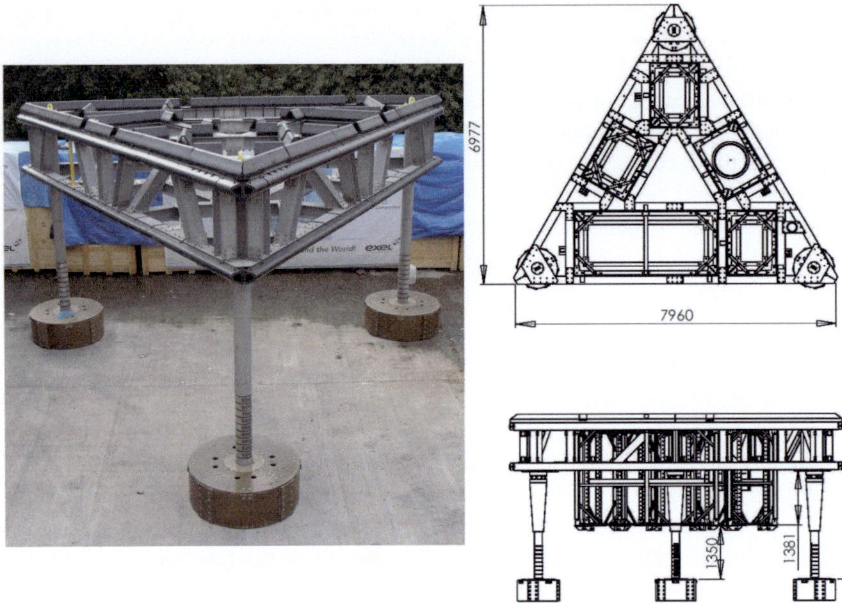

Figure 13.1 The DELOS seafloor module.

period each platform requires ROV (Remotely Operated Vehicle) intervention to recover observatory modules to the surface for brief periods of time for service, calibration and data download.

13.2.1 Seafloor docking stations

The seafloor docking stations consist of a robust glass fiber triangular framework designed to withstand long-term seafloor deployment and periodic service by industrial ROV (Figure 13.1). Glass fiber is used to eliminate any corrosion effects. The standard techniques of deploying sacrificial anodes to sub-sea metal structures to reduce corrosion could not be used because of the potential effect of the sacrificial anode on seafloor biological processes. The large sacrificial anode volume required to protect a steel structure for 25 years' deployment may result in altered seafloor chemistry and bias any results. Therefore, the frame was constructed from sections of glass/vinyl ester resin composite material with joints glued (3M Scotchweld 9323), and bolted together with Super Duplex (25% chrome content stainless steel) doubler plates (Hartley, 2008). The frame was designed by 2HOffshore Ltd, UK from the Oceanlab concept, and the glass fiber sections were manufactured and the framework fabricated by Excel Composites Ltd, UK.

13.2.1.1 Glass fiber material testing

Pultruded glass/vinyl ester composite material exhibited the potential strength, robustness and stability for long-term deployment in the deep ocean. However, no data were available for its performance under the deployment conditions. Therefore, a research program was initiated to investigate glass fiber performance after prolonged immersion in sea water. The research (conducted at the National Physics Laboratory, UK) involved immersing composite material tokens in pressurized, heated seawater baths for a conditioning time period. Raising the temperature of the composite material in this manner simulates ageing, thus an 8-week elevated temperature immersion is equivalent to 10 years deployed in the deep ocean (Hemmerich, 1998). At the conclusion of the conditioning period, composite material tokens were subjected to three mechanical tests (sheer, tensile, flexural) to characterize the aged performance of the material (2Hoffshore Ltd, 2006; Mera, 2007). Some tokens were conditioned for longer periods of up to 24 weeks and confirmed the predicted data trend. Results show little evidence of serious degradation in composite material properties. The accelerated ageing process may not take into account all material degradation processes; however, extrapolating these results show the GRP composite material to have a predicted period to "end of life" well in excess of the design deep-sea deployment duration.

13.2.1.2 Structure analysis and modelling

Mathematical modelling of the DELOS structure allows the designers to test the integrity of their design against a number of predicted load cases before manufacture. Issues raised can be addressed at the design stage rather than post manufacture. Particularly important in the oil industry, modelling can also enable the structure to be analyzed for safety during transport, installation and use. The availability of the aged material data also allows end-of-life analysis to be performed.

Modelling of steel structures is routine in the oil industry. However, glass fiber has varying properties in different directions and at the bolted joint sections there could be load-dependent relative slip between the joint members resulting in nonlinear behavior. In steel structures, local stress points tend to be smoothed out due to its ductile nature; however, glass fiber does not exhibit this behavior and local stress points may result in failure (Octa Engineers, 2006a). Therefore, two modelling approaches were adopted: a global frame analysis to identify the behavior and stability of the frame (Octa Engineers, 2006b), and a localized model to detail the performance of important joints, and stress points (Octa Engineers, 2006c) The global model was constructed using Structural Analysis Computing Software [8] and enabled critical load conditions to be identified and provided results that were fed into the more detailed localized model. The localized models were constructed using ANSYS finite element analysis software (Clarke, 1996). Both models enabled a good picture of structural behavior to be developed (e.g., Figure 13.2), allowing a suite of load cases to be assessed:

8 SACS (Structural Analysis Computing Software), Version 5.2, Service Pack 3.

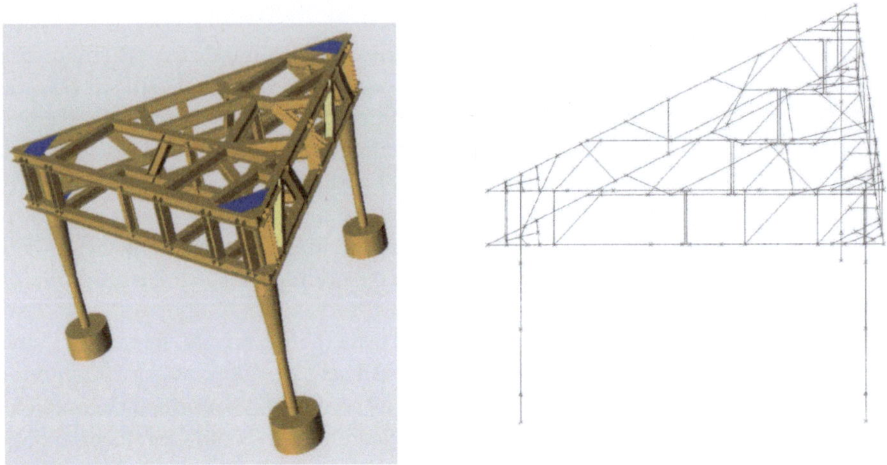

Figure 13.2 Visual representation of DELOS docking station model used for structural analysis.

- The end-of-life performance of the structure, using the end-of-life material properties from the material testing: to determine if the structure would survive recovery from the seafloor after a 25-year installation
- Lift analysis of the docking station with as-new material properties to determine stability of the structure during mobilisation and installation crane lift operations
- Remotely Operated Vehicle (ROV) impact analysis using end-of-life material properties to determine "no-go" areas for ROV operations
- Transportation analysis: due to the size of the docking station the docking frame had to be transported in two halves. This analysis took account of the reduced structural integrity of the frame when in its transportation state
- Installation analysis: during installation when the completed frame is deployed over the side of the ship through the wave-affected area (splash zone), wave slam effects on the structural elements were assessed
- Lift analysis of DELOS instrument modules using end-of-life material properties.

In all modelled load cases the analysis was guided by the EUROCOMP [15] design code handbook for composite material structures, and in each case the design either passed or surpassed the design code requirements (Octa Engineers, 2006a, 2006b, 2006c).

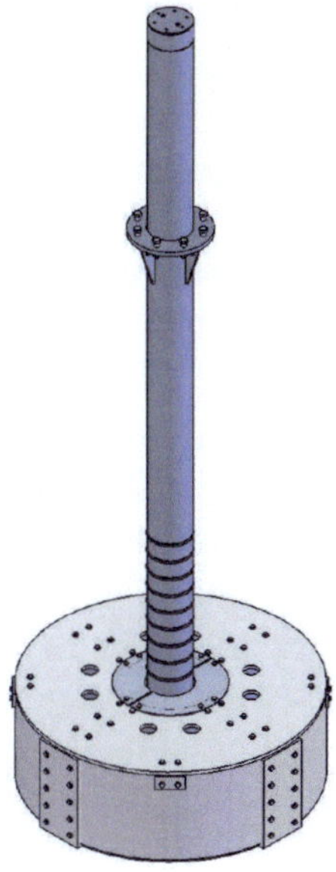

Figure 13.3 DELOS leg and mud mat design.

13.2.1.3 Foundation design

The foundation stability of the DELOS structure over the long term is an important element in the design process. The DELOS structure must be able to withstand ROV operations, including potential impacts, sustain instrument module removal and insertion, and remain in the same location without sinking or slipping for 25 years. During the design phase, the use of piling techniques was considered but rejected because of the potential disturbance to the seafloor. Therefore, a mud mat design was chosen for its foundation stability and reduced impact on the seafloor (Figure 13.3).

The mud mat resembles an inverted can, with small holes in the top surface to allow water to escape during insertion into the seafloor, resulting in little sea-floor disturbance. However, once inserted, the mud mat operates in a similar manner to a suction pile. The holes in the top surface are not large enough to allow easy extraction and therefore vacuum forces exist, the mud mat firmly fixing in the sediment.

The size of the mud mat and the sheer stress of the sediment determine the load capacity of the mud mat. The sheer stress data were available from the block 18 geotechnical survey [16] and foundation calculations were performed by BP (Guo, unpublished data). A number of load cases were considered and mud mats designed to maintain foundation stability against these forces. During calculations, a safety factor of 3 was built into the calculations to ensure no foundation instability for the life of the project. A mud mat of 1.5m diameter and skirt penetration of 0.5m will give each mud mat a:

- Vertical bearing capacity of approximately 4 tonnes (each DELOS frame has three mud mats giving a total vertical bearing capacity of 12 tonnes). Vertical bearing capacity is the static vertical load the mud mats are able to withstand before sinking will occur
- Horizontal (sliding) capacity of 0.49 tonnes (1.47 tonnes for the frame having three legs). Sliding can occur from sea current forces acting on the structure, slowly pushing the structure in the direction of the prevailing current. The maximum side load experienced by the DELOS frame due to sea current is approximately 0.4 tonnes for a 100-year current (maximum current magnitude expected in a 100 years)
- Static overturning stability of 2.2 tonnes, dynamic (peak) overturning stability of 9.1 tonnes. The potential for the DELOS frame to be overturned by ROV impact was assessed. Water drains relatively slowly from the small holes in the top surface of the mud mat, resulting in a significantly larger dynamic than static overturning stability.

The bearing pressure of the mud mats themselves on the seafloor may affect the sediment animal community. Initially, the pressure from the mud mat will be borne by changes in pore water pressure which will dissipate over time. Analysis using a simple linear elastic model shows the unfactored pressure due to one mud mat of 7 kPa dissipates to a tenth of the value (0.7 kPa) at a radial distance of 0.5m beyond the mud mat diameter and 2.8m below it (Guo, unpublished data).

13.2.1.4 Hydrodynamic modelling

The aim of the shape and form of the DELOS structure was to reduce its impact on the seafloor. Ideally, the structure should be as small and as light as possible, but when the scientific and engineering constraints are considered a realistic design evolves. The structure dimensions were determined by the scientific requirement for a wide view image of the seafloor and instrumentation that looked outward from the structure to relatively undisturbed seafloor areas. Given these dimensions, engineering design dictates a robust structure that can withstand ROV operations and maintain structural stability for 25 years. Once the design had been modelled and proved safe for long-term deployment its performance in a hydrodynamic environment was assessed. Of particular interest was the avoidance of sediment resuspension around the DELOS docking station legs.

Measurements of the sediment particle size and sediment sheer stress from the Block 18 pre-installation survey[9] were fed into a hydrodynamic analysis around the DELOS mud mat and leg to determine if any sediment surface resuspension would occur at observed sea current velocities. The analysis assumed an installed mud mat and leg with currents ranging from $5cms^{-1}$ to $20cms^{-1}$ (Figure 13.4).

Figure 13.5 shows the output of the hydrodynamic modelling for four heights above the sediment surface. The output is a color intensity chart with green/yellow representing the nominal current velocity of approximately $5cms^{-1}$, blue representing current velocities less than normal, and red greater than normal. The output shows elevated current velocities on the front edges of the leg and in the turbulent flow behind it. Analysis of the sedi-

9 A. Hill. DELOS Foundation Design – Technical memorandum #1 (Rev2). BP Exploration & Production Technology 2006 (Restricted commercial).

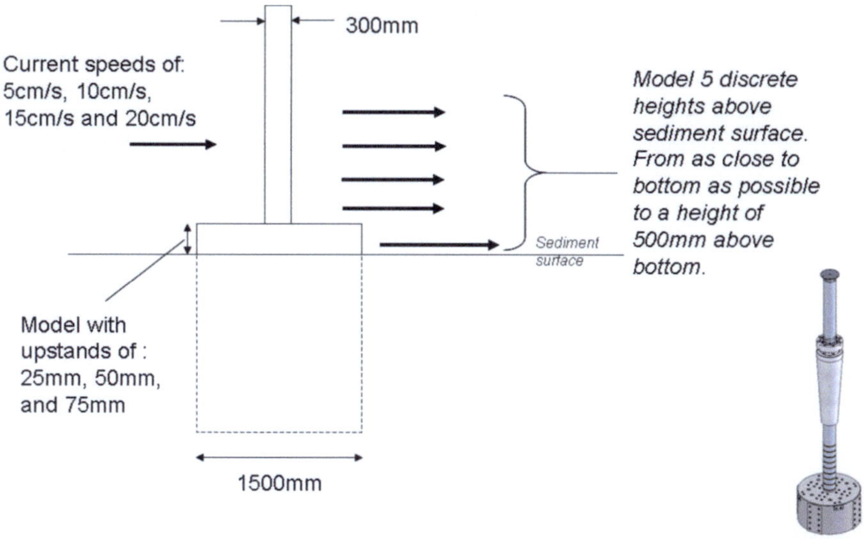

Figure 13.4 Hydrodynamic modelling specification. Assume current is approaching the leg from the left at four current speeds ranging from 5 cm/s to 20 cm/s. Analyse currents at 5 distinct heights above the sediment surface. Assuming mud mat and leg dimensions as shown.

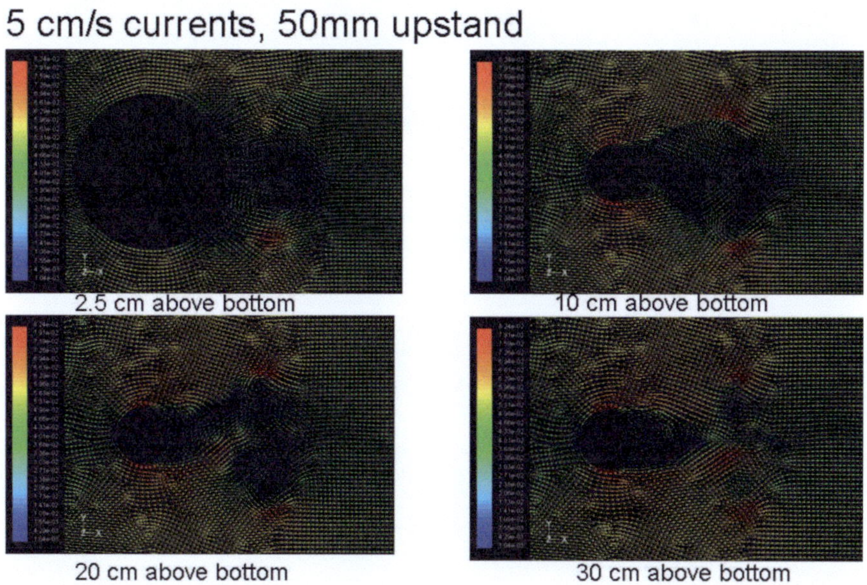

Figure 13.5 Hydrodynamic modelling current vector results given a 5cm/s linear current approaching from the left hand side of the output. The colours represent current velocities. Maximum current velocity (red) = 8.2 cm/s, minimum current velocity (blue) 0.1 cm/s

ment particle size and sheer stress shows that fine superficial sediment (0.002mm diameter particles, representing 30% of the superficial sediment) may start to resuspend at current speeds greater than 20cms^{-1}. The hydrodynamic analysis shows that turbulent current velocities exceeding 20cms^{-1} occur when a sea current of 15cms^{-1} or greater exists (Kaufmann and Smith, 1997). Annual current data from Block 18 show currents occasionally exceed 15cms^{-1} in the months of September, October, and November (see footnote 9). During these periods, it is possible some sediment resuspension may occur around the DELOS legs.

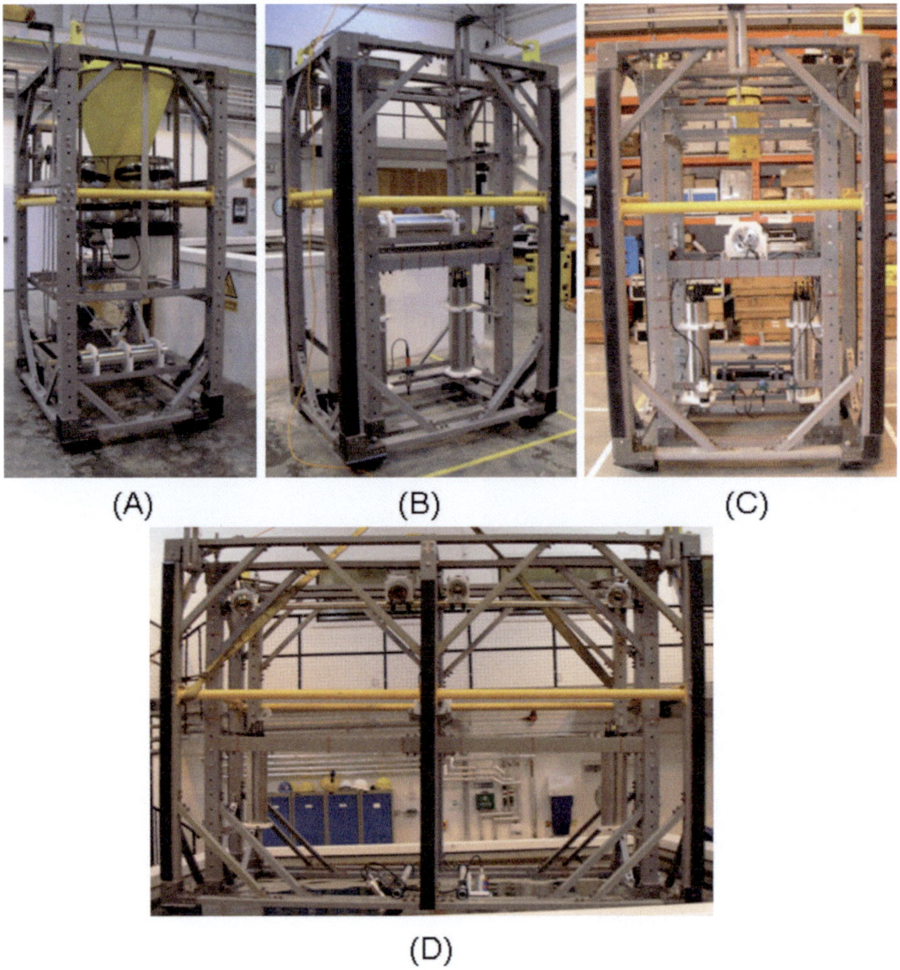

Figure 13.6 One set of DELOS Observatory modules. (A) Sediment trap module; (B) acoustic module; (C) oceanographic module; (D) Camera module.

13.2.1.5 Observatory modules

Each observatory module contains a suite of related instrumentation housed within either a standard size or over-size observatory module frame (Figure 13.6). For ease of data offload and instrument programing, a controller is also situated within each frame. The controller deals with different interface issues for each of the sensors and instruments, enabling a standard programing interface to be presented to the user. It also acts as a backup, storing a copy of each sensor's data in a central location.

Observatory modules are designed to be placed into the sea-floor docking module by ROV. Each module has sufficient battery and memory capacity to operate for a maximum of 6 months. All instrumentation, including batteries and memory, will be confined within the dimensions of the standard observatory module frame designs. During transportation or storage, the frame protects the instruments within and also allows easy access for crane or forklift operations. Table 13.2 shows the observatory module makeup for each of the near and far-field platforms. The observatory module makeup is identical for each platform apart from the sediment trap module in the near field. The requirement to float the sediment trap 100m above the seafloor (see sediment trap module below) would prevent normal routine industrial ROV operations around the near field well, and thus a sediment trap is not used at the near field site.

13.2.1.6 Camera module

The camera modules contain two camera systems, a close view and wide view camera system (Figure 13.6d).

Close view camera

The close view camera takes time-lapse close-up photographs of the seafloor and associated fauna (Figure 13.7). In a relatively unstudied area, such as the West African continental slope, it is vital to obtain good-quality high-resolution images of the indigenous fauna. The close view stills camera will give the flexibility required to correctly identify both invertebrates and fishes. It is anticipated that these high-quality images will also have considerable educational and public outreach potential.

Near-field platform observatory modules (50m from sub-surface well)	**Far-field platform observatory modules (16km from any seafloor infrastructure)**
Camera module	Camera module
Oceanographic module	Oceanographic module
Acoustic module	Acoustic module
Guest module 1	Sediment trap module
Guest module 2	Guest module 1

Table 13.2 The DELOS observatory module makeup.

Figure 13.7 Examples of typical close view camera images. (Left) Cusk eel *Holcmycteronus asuamses*. (Right) Baird's smooth head *Alepocephalus bairdii*.

Wide view camera

The wide view camera takes time-lapse photographs of a large area of the seafloor (Figure 13.8). These observations enable a visualisation of seasonal seafloor sedimentation processes, passing animals and disturbance events over a large $20m^2$ area. This scale of observation is essential to categorize any patterns of long-term change in the benthic environment over the 25-year life span of the project. This imaging set up has been used to excellent effect in the deep Pacific Ocean (e.g., Smith et al., 2006). The synergy between the close view camera and the wide view camera will allow us to assess large-scale and long-term patterns of diversity and community change with accurate identification of putative species.

Both camera systems utilize a 5M pixel digital stills camera (Kongsberg Ltd.) capable of taking in excess of 3000 images over a 6-month deployment. The cameras were chosen for the image quality and proven use in deep-sea conditions; however, they were designed to operate with a ROV and thus use too much power for long-term autonomous deployment. Therefore, an Oceanlab controller was used to schedule the time-lapse photography, only applying power to the cameras and strobes for the duration of the photograph and image storage. The use of an external controller also allows time-lapse photographs to be biased towards expected periods of higher animal activity if required.

The wide view camera requires a greater area of illumination of the seafloor (approximately $20m^2$) and thus the standard strobe system was modified to incorporate a further two slave strobes systems. These slave flashes trigger in sympathy with the main camera flash, utilizing a bespoke blue/green spectrum optical slave circuit.

Figure 13.8 An example of a view camera time lapse images of a 20m^2 area of the sea. Circled area is a holothurian moving across the field of view during three consecutive time lapse photographs.

13.2.1.7 Oceanographic module

A suite of oceanographic instruments is essential for any long-term monitoring station. They provide background measurements to characterize the environment for all other observation modules in the docking station. Each oceanographic module houses (Figure 13.6c):

- 300kHz Acoustic Current Doppler Profiler (Teleydyne RDI) to provide current profiles of water column above DELOS
- Transmissometer (Wet labs C Star): measures the total particle load in the water column (this includes organic matter/sand/sediment/etc.)
- Fluorometer (Wet labs) measures phytopigments which can be related at times to the organic matter content (Chlorophyll a) of the particle load identified by the transmissometer (i.e., the "fresh" material or "food" arriving on the seafloor)
- Local seabed current meter (Aanderaa) measures sea currents close to the seafloor
- Conductivity, temperature, and pressure sensors (Aanderaa)
- Oxygen sensor (Aanderra) measures dissolved oxygen levels available to the local seafloor community.

Each oceanographic sensor is controlled by a central controller that takes readings at a preprogramed schedule, and logs the data in a central location for easy user access. Where internal sensor memory is available, identical data are stored on board the instrument for backup purposes.

13.2.1.8 Acoustic module

The acoustic module contains both active and passive acoustic sonars (Figure 13.6b).

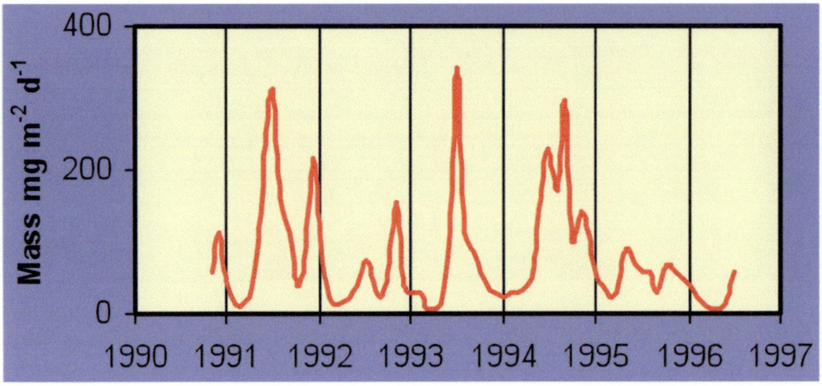

Figure 13.9 An example of long-term sediment trap data from the deep Pacific Ocean.

Passive sonar

A passive bioacoustic sensor will monitor the natural sounds generated by animals within its detection range, as well as the background noise level. This system will allow passing vocalizing cetaceans to be identified (from characteristic sound spectra) and counted. This is achieved through the use of an omni-directional deep ocean hydrophone (Sensor Technology) coupled to a hard disk based on a broadcast quality recorder (Sound Devices). Recording quality (bit depth), bandwidth, and gain can be programed as required by the application. Block 18 is a relatively unstudied area and, as such, knowledge of acoustic spectra and intensity was lacking. As an initial setting a 24 bit depth sampling at 32 kHz was used, resulting in a continuous maximum recording duration of 200 hours that was time lapsed over the 6-month deployment.

Active sonar

High-frequency active sonar systems enable fish movements to be observed at a lower resolution but at much greater range than photographic systems. This module contains an active sonar system to record movements of fish with suitable target strength, and sedimentation processes to ranges of up to 150m from the DELOS platform. We utilize a 675kHz mechanically scanned visualization sonar (Kongsberg mesotech) controlled by an Oceanlab single board PC-based controller. The controller enables the normally tethered active sonar to operate autonomously for 6 months without a human interface. At the time of manufacture, a single board PC operating Windows XP required power levels not compatible with a battery-based system designed to operate for 6 months. Even in a low power sleep state (Microsoft sleep condition S3/S5) the single board PC required excessive power, and therefore an independent microcontroller was used to power the single board PC from a user predefined schedule. Once active, the single board PC then deals with all the interface issues to enable the correct operation of the active sonar.

The active and passive sonars operating in conjunction will potentially be able to monitor fish reaction to acoustic disturbance events.

13.2.1.9 Sediment trap module

Phytodetritus from plankton in the surface layers falls to the seafloor in seasonal pulses (Rice et al., 1986; Billett et al., 1983; Massion and Raybould, 2006) (Figure 13.9). This input of material is the major source of energy for the deep-sea community. A sediment trap collecting and periodically storing this fallout enables the composition and quantity of this energy input to be measured. However, deploying a sediment trap close to the seafloor will increase sampling bias from benthic boundary layer horizontal sediment transport currents. Therefore, to represent an unbiased record of phytodetrial fallout the sediment trap is deployed a minimum of 100m above the DELOS seafloor platform.

As all DELOS observatory modules require ROV intervention every six months, the sediment sampler (McLane) must initially be retained within the dimensions of the observatory module for ease of ROV handling. However, once the module has been installed in the DELOS seafloor docking station and ROV intervention has been completed, the sediment trap module must be able to deploy the sediment sampler 100m above the seafloor, and to recover it back into its module just prior to the next DELOS service.

This is achieved by attaching floats to the sediment sampler so that it is positively buoyant and then connecting it to a subsea winch (modified Inter Ocean VPS winch) via 100m of Dynema rope. The subsea winch can be operated by a number of different command inputs for the purposes of backup. Normally, the winch would haul and veer under program control at set intervals. Should the schedule need to alter due to weather or service interval changes, the winch can be actuated via an acoustic command from the surface vessel. Should either of these systems fail, the winch can be manually operated by a ROV manipulator arm. These data, along with seafloor photographs, transmissometer and fluorometer (in the Oceanographic module data), will build up a comprehensive picture of food input into the deep ocean site.

13.2.1.10 Guest module

Guest modules will initially be left empty and will be available for use in the future. Currently, we have applications from both the science community and BP to use some of these modules for new research.

13.2.1.11 ROV module

The science package offered by the DELOS system is enhanced by adding vertical water column data to the DELOS instrument suite. This is achieved by adding a small data logging instrument package on the local ROV tether management system to enable profiling data to be added to the overall scientific database. During normal ROV operations, the tether management system would be deployed vertically through the water column, enabling oxygen, conductivity temperature and depth measurements to be taken. Positional data from the ROV field service vessel will be merged to provide input to a positional database.

The far field DELOS platform fully populated with observatory modules is shown in Figure 13.10.

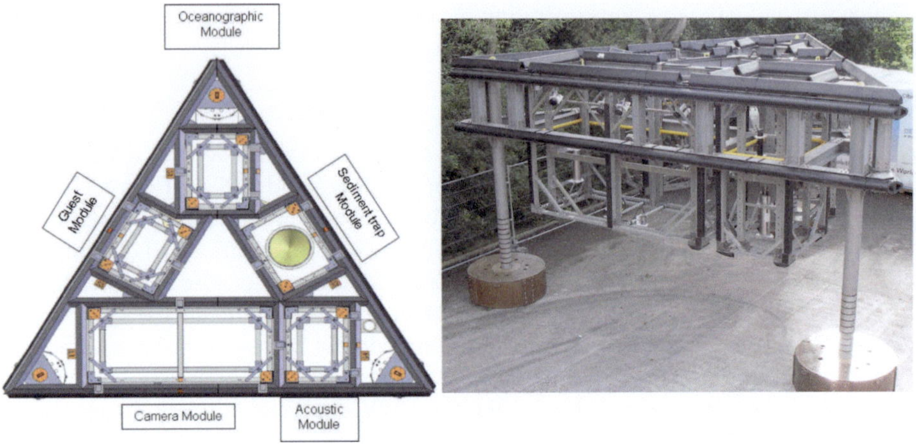

Figure 13.10 The far field DELOS platform fully populated with observatory modules. The location of observatory modules is also shown.

13.2.1.12 Battery packs

Each module is powered by a battery pack capable of operating the module for the deployment duration between services. Initially, bespoke rechargeable lithium ion battery packs were used. However, as the period between service intervals increased from 6 to 12 months, the quiescent current required to power the lithium ion battery protection circuitry (over charge, over discharge, over current, and over temperature) became a significant factor in the life of the batteries. Therefore, the lithium ion batteries were replaced with bespoke alkaline primary battery packs. The alkaline batteries have a reduced low temperature performance when compared to lithium ion cells, but have a superior self-discharge and a high capacity.

13.3 Installation

The DELOS was installed in early February 2009 using the Field Service Vessel (FSV) Oceanteam Bourbon 101. The vessel has full dynamic positioning, and two heave-compensated 100-tonne cranes. Subsea operations were controlled by Hercules class ROVs operated by Subsea 7. Initially, a ROV acoustic survey was completed to map the topography of the seafloor and to determine the installation location at both the near and far field sites. Both platforms had to be deployed at similar depths to comply with the scientific specification.

Each DELOS seafloor platform was lowered by crane towards the seafloor. Once near the seafloor, the ROV stabilized the DELOS platform and the ship manoeuvred into position through a defined safe working area. Once in position, the ROV rotated the platform to the desired orientation, and the crane lowered the platform to the seabed. A second ROV observed the mud mat penetration and confirmed that there was little seabed disturbance. The mud mats penetrated to within 50mm of the mud mat top surface. Once installation was completed at the near field site, the platform was left to settle whilst the second platform was installed at the far field site 16km away.

Once both platforms were installed, the ship returned to the near field site to install the observatory modules. In turn, the camera, oceanographic, and acoustic modules were lowered toward the seafloor. Once near the seafloor, the ROV stabilized the module whilst the ship manoeuvred through the safe working area to the DELOS platform location. A second ROV spotted the appropriate platform slot for the module, and while the heave compensated crane lowered the module, the first ROV guided the module into the slot. When all the modules were installed, a final ROV observation was made of all installed modules, mud mat penetration and the overall level (observation of bulls eye to determine pitch and roll) of the platform. Finally, the ship returned to the far field site where the remaining modules were installed (camera, oceanographic, acoustic, and sediment trap). The sediment trap module was only installed at the far field site. Installation was similar to other modules; however, if required, the sediment sampler could be deployed manually by the ROV.

13.4 Periodic service

The battery-powered observatory modules are designed to operate for six months. They are recovered to the surface using a similar procedure to that described above but in reverse. Once at the surface they are serviced (batteries charged, data offloaded, cleaned) and calibrated as required and redeployed to the seafloor. Should any observatory module require a repair that cannot be carried out "on site", the module would have to be returned to shore for service. The delay will result in no data collection from that module until it can be returned and redeployed. However, because the program utilizes industrial ROV time of opportunity, delays for instrument replacement may only be for days or weeks.

13.5 Results

The DELOS platforms were installed and have been operational since February 2009. The platforms have been serviced every six months as planned and a summary of these data is available on the DELOS website (http://www.delos-project.org/). These long-term data sets will over time build into an invaluable resource that will allow science to investigate and describe long-term change in the environment within the vicinity of an oil production field. First results from the DELOS system have been reported by Vardaro et al. (2013).

13.6 Discussion

The move in ocean science to record data in the oceans over long time scales has resulted in a number of proposals and active systems to fulfill this challenge (Taylor, 2009; Tunnicliffe et al., 2008; Priede et al, 2003; Katz, 2009; Rapidis, 2009). However, significant challenges exist in the long-term reliability, and resistance to bio-fouling of sensors and subsea systems. The DELOS provides a "stepping stone" between these fully cabled long-term real-time systems and shorter-term research carried out currently. DELOS is providing scientific data every six months after each service interval. Returning each observatory module to the surface every six months has the detrimental effect of interrupting data gathering for the service period. However, it also offers the opportunity to service the instruments, perform antifouling procedures, repair or replace damaged sensors and to update the suite of instrumentation as technology changes and as scientific demands alter.

Placing the DELOS within the vicinity of an oil production field has the advantage of long-term scientific access to oil-related infrastructure such as ROVs and field service vessels. This infrastructure will be in place for the life of the oil field (approximately 25 years), ensuring continued access to the DELOS platforms.

The opportunity to work in collaboration with oil companies in this way has the potential in the future to allow further environmental platforms to be located in the oceans around the world for a relatively low cost when compared to installing purpose-built cabled observatory networks.

References

2HOffshore Ltd. (2006) GRP material test specification. Doc No 1844-SPC-0002 Rev02. 6 February 2006.

Billett D.S.M., Lampitt R.S., Rice A.L. and Mantoura R.F.C (1983) Seasonal sedimentation of phytoplankton to the deep-sea benthos. Nature 302(5908), 520–522.

Billett D.S.M., Bett B.J., Rice A.L., Thurston M.H., Galeron J., Sibuet M. and Wolff G.A. (2001) Long-term change in the megabenthos of the Porcupine Abyssal Plain (N E Atlantic). Progress in Oceanography 50(1–4), 325–348.

Guo Y. Three-dimensional numerical simulation of hydrodynamics of flow around a marina structure. Unpublished data.

Hartley J. (2008) Composites the right choice for subsea ocean observatory. Composite Technology, December 2008, 54–56.

Hemmerich K.J. (1998) General aging theory and simplified protocol for accelerated aging of medical devices. Medical Plastics and Biomaterials 5(4), 16–23.

Hill A. (2006) DELOS Foundation Design – Technical memorandum #1 (Rev2). BP Exploration & Production Technology 2006 (Restricted commercial).

Katz U.F. (2009) Status of the KM3NeT project. Nuclear Instruments & Methods in Physics Research Section A – Accelerators, spectrometers, Detectors and Associated Equipment 602(1), 40–46.

Kaufmann R.S. and Smith K.L. (1997) Activity patterns of mobile epibenthic megafauna at an abyssal site in the eastern North Pacific: Results from a 17-month time-lapse photographic study. Deep-Sea Research Part I – Oceanographic Research Papers 44(4), 559–579.

Massion G. and Raybould K. (2006) MARS: The Monterey accelerated research system. Sea Technology 47(9), 9.

Mera R.D. (2007) Mechanical properties of GRP subjected to water immersion at temperature and pressure. Progress report no. 3 – Final report. National Physical Laboratory E06010252/3 (Restricted commercial).

Octa Engineers (Designs) Pvt Ltd. (2006a) Structural analysis and design of DELOS docking station: Executive summary of DELOS docking station. Doc No. 06-003-DC-1800-104 Rev 1, 06/12/2006.

Octa Engineers (Designs) Pvt Ltd. (2006b) Structural analysis and design of DELOS docking station:- Global FE analysis of DELOS frames. Doc No. 06-003-DC-1800-103 Rev 0, 01/12/2006.

Octa Engineers (Designs) Pvt Ltd. (2006c) Structural analysis and design of DELOS docking station:- Structural design report for DELOS docking station. Doc No. 06-003-DC-1800-102 Rev 0, 01/12/2006.

Priede I., Solan M., Mienert J,. Person R., Van Weering T.C.E., Pfannkuche O., O'Neill N., Tselepides A., Thomsen L., Favali P., Gasparoni F., Zitellini N., Millot C., Gerber H.W., De Miranda J.M.A and Klages M. (2003) ESONET-European seafloor observatory network. In: 3rd International Workshop on Scientific Use of Submarine Cables and Related Technology, 263–265.

Rapidis P.A. (2009) The NESTOR underwater neutrino telescope project. Nuclear Instruments & Methods in Physics Research Section A – Accelerators, spectrometers, Detectors and Associated Equipment 602(1), 54–57.

Rice A.L., Billett D.S.M., Fry J., John A.W.G., Lampitt R.S., Mantoura R.F.C. and Morris R.J. (1986) Seasonal deposition of phytodetritus to the deep-seafloor. In: Proceedings of the Royal Society of Edinburgh Section B: Biological Sciences 88, 265–279.

Smith K.L., Kaufmann R.S., Baldwin R.J. and Carlucci A.F. (2001) Pelagic-benthic coupling in the abyssal eastern North Pacific: An 8 years time series study of food supply and demand. Limnology & Oceanography 46(3), 543–556.

Smith K.L., Baldwin R.J., Ruhl H.A., Kahru M., Mitchell B.G. and Kaufmann R.S. (2006) Climate effect on food supply to depths greater than 4,000 meters in the northeast Pacific. Limnology and Oceanography 51(1), 166–176.

Smith K.L., Jr., Ruhl H.A., Kaufmann R. S. and Kahru M. (2008) Tracing abyssal food supply back to upper-ocean processes over a 17-year time series in the northeast Pacific. Limnol. Oceanogr 53(6), 2655–2667.

Ruhl H.A. and Smith, K.L. Jr. (2004) Shifts in deep-sea community structure linked to climate and food supply. Science 305, 513.

Ruhl H.A., Ellena J.A. and Smith K.L. Jr. (2008) Connections between climate, food limitation, and carbon cycling in abyssal sediment communities. PNAS 105(44), 17,006–17,011.

Taylor S.M. (2009) Transformative ocean science through the VENUS and NEPTUNE Canada ocean observing systems. Nuclear Instruments & Methods in Physics Research Section A – Accelerators, spectrometers, Detectors and Associated Equipment 602(1), 63–67.

Tunnicliffe V., Barnes C.R. and Dewey R. (2008) Major advances in cabled ocean observatories (VENUS and NEPTUNE Canada) in coastal and deep sea settings. In: 2008 IEEE/OES US/EU-BALTIC International Symposium, 66–72.

Vardaro M.F., Bagley P.M., Bailey D.M., Bett B.J., Jones D.O.B., Milligan R.J., Priede I.G., Risien C.M., Rowe G.T., Ruhl H.A., Sangolay B.B., Smith, K.L. Jr., Walls A., & Clarke J., (2013) A Southeast Atlantic deep-ocean observatory: first experiences and results. *Limnology & Oceanography: Methods* 11: 304–315.

14 Sub-sea environmental observatory integrated with the KM3NeT neutrino telescope infrastructure in the Mediterranean Sea

A. Holford[1] (on behalf of the KM3NeT Consortium), H. van Haren[2], J. Craig[1] and I.G. Priede[1]

14.1 Introduction

The concept of fiber optic cabled environmental observatories on the seafloor with real time data transmitted to shore is now being developed in a number of projects around the world (NEPTUNE 2012, MARS 2008, DONET 2008, ESONET 2012). However, the first attempt at deploying an optical cabled system in the deep sea was not directed towards study of the marine environment but to astronomy. In order to detect high-energy neutrinos originating from cosmic sources it was realized during the 1970s that the detector would have to be much larger than could be built in any laboratory. The solution was to instrument a large volume of seawater in the deep sea with photo-detectors that could monitor the flashes of light generated by high-energy particles passing through the planet. This could only be achieved through the use of the then novel underwater glass fiber optic cables, and the first such undersea observatory attempting to use the new cables was the DUMAND (Deepwater Underwater Muon and Neutrino Detector) project off Hawaii (DUMAND 2003). Sub-sea observatory elements were tested in 1982 and a fiber optic cabled system was deployed in 1993, only to be abandoned in 1995 owing to technical difficulties. With a location close to the equator (latitude 19.5°N) DUMAND would have been capable of scanning the entire universe every 24 hours as a result of the rotation of the Earth (Roberts,

1 University of Aberdeen, Oceanlab, Aberdeen, UK.

2 Royal Netherlands Institute for Sea Research, Den Burg (Texel), The Netherlands.

1992). This concept has now been replaced by international projects to build two neutrino telescopes, one in Antarctica (using the southern polar ice-cap as an alternative to seawater) and a second observatory in the northern hemisphere. Such a dual system can monitor the entire sky continuously in contrast to a single observatory at the equator which relies on rotation of the Earth to gain complete coverage.

In the absence of a continental ice cap in the Arctic, the northern hemisphere telescope will be built in seawater. The Mediterranean Sea (locations at latitudes 36–43°N) has been chosen as an appropriate site for the European KM3NeT project and overcomes some of the problems of open ocean location that frustrated DUMAND. KM3NeT will be complementary to the US ICECUBE (ICECUBE, 2014) project in Antarctica. With a large sub-sea infrastructure, including power and data transmission, it is proposed that KM3NeT should also encompass earth–sea science studies at the chosen location in the Mediterranean Sea. This will complement and augment the large quantity of marine research that has already taken place in this region investigating phenomena such as bioluminescence (production and emission of light by a living organism as the result of a chemical reaction).

In this chapter we consider the integration of oceanographic, climatological, geological and biological applications into the astro-particle physics infrastructure.

14.1.1 Neutrino astronomy

Our knowledge of the universe has primarily been derived from the observation of electromagnetic waves throughout the spectrum from long wavelength radio waves, the light spectrum (infrared to ultra violet), to very short wavelength X-rays and gamma rays. Traveling in straight lines, electromagnetic waves provide an accurate image of the universe but are susceptible to absorption so there are inevitably parts of universe that cannot be viewed by earth-bound telescopes. Cosmic rays, which are mostly composed of protons or atomic nuclei minus their orbiting electrons, provide an additional means of understanding the cosmos beyond the solar system. However, due to their positive charge, their paths are affected by the galaxy's magnetic fields and therefore these particles cannot be used to establish the source of cosmic rays. Neutrinos are weakly-interacting particles which travel in straight lines, penetrate even the interiors of stars and enter our atmosphere unhindered at very close to the speed of light. This makes neutrinos an ideal candidate for detection by a telescope to accurately map the universe. High-energy neutrinos originate from supernova remnants, pulsars and micro-quasars in our Galaxy and from extragalactic sources such as active galactic nuclei and γ-ray burst emitters. As a neutrino traverses through the mass of the Earth, interactions may occur whereby a muon is released and emerges upwards from the seafloor in the antipodal location traveling at more than the speed of light in closely the same direction as the parent neutrino. The passage of the muon through water produces Cherenkov radiation which at a fixed point consists of a short flash of light analogous to an electromagnetic version of a sonic boom. The detection of Cherenkov light is achieved using an array of light detectors housed in glass spheres known as optical modules (OM), deployed in the deep sea. Knowing the arrival time and the position of the OMs enables the reconstruction of the muon track to a precision of a few tenths of a degree.

The European Union agreed to fund the design study and preparatory phase of a cubic kilometer neutrino telescope, KM3NeT, in the Mediterranean Sea. Its infrastructure will be

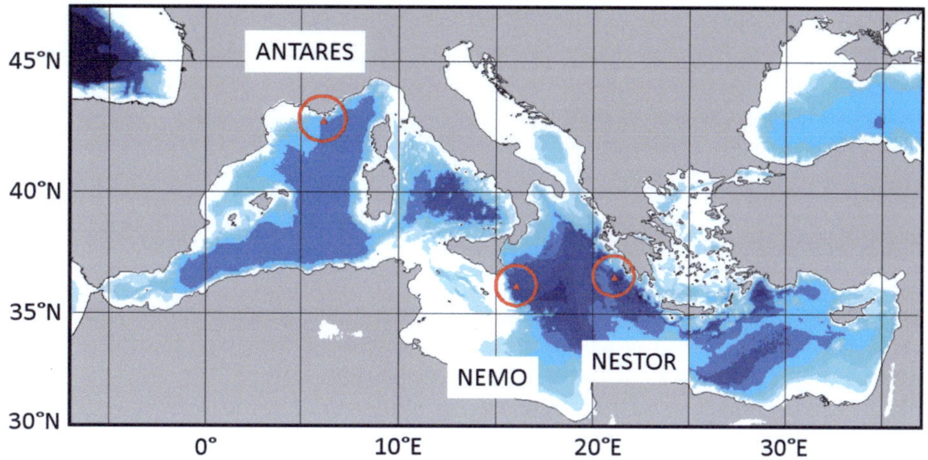

Figure 14.1 The Mediterranean Sea showing the location of the three KM3NeT neutrino telescope pilot projects ANTARES (2475m depth) in the Ligurian Sea, NEMO (3350m) and NESTOR (4500–5200 m) in the Ionian Sea. The circles indicate 100km radius around each location that could be instrumented via the Km3Net junction box. Shading indicates depths at 1000m intervals.

shared with multidisciplinary undersea observatories, making continuous long-term measurements in the area of oceanography, climatology, geophysics, geohazards and marine biological sciences.

14.2 Scientific case for a cabled infrastructure in the Mediterranean Sea

With the transparency of its water the Mediterranean Sea provides the large mass necessary in which to house a large array of light sensors to detect Cherenkov light emissions. The geographic location is perfect since the area of the sky covered includes the center of our Galaxy and, due to the rotation of the Earth, a telescope situated at a latitude (λ) 36° and 43° north can observe upwards-going neutrinos from most of the sky (about 3.5 π sr). Declinations below $-90°+\lambda$ are always visible, while those above $+90°-\lambda$ are invisible. Most of the Galactic plane, including the Galactic center, can be observed for most of the sidereal day.

In addition, the Mediterranean deep sea is of prime interest for marine and earth sciences research. The earth–sea sciences infrastructure, which has been included in the design, will enable the development of a permanently cabled deep-sea observatory with potential for important discoveries. This infrastructure will form part of a wider global environmental monitoring system for investigating climate change issues and enabling the provision of early warning systems regarding earthquakes and tsunamis.

Studies into the possibility of constructing a neutrino telescope in the Mediterranean Sea have been ongoing for more than a decade. Three pilot projects, ANTARES, NEMO and NESTOR, situated at different locations identified in Figure 14.1 have been used to explore different configurations and techniques.

14.3 KM3NeT conceptual design

Based on the experience gained from the three pilot studies, a generic conceptual design (KM3NeT, 2008) has been proposed for the Mediterranean Sea cubic kilometer neutrino telescope, KM3NeT, and its marine science observatories.

This neutrino telescope will be composed of a number of vertical structures known as Detection Units, anchored to the sea bed and kept vertically upright with buoys. Each detection unit consists of a series of mechanical structures, called storeys, supporting the necessary sensors, electronic components, power and data line interfaces. In contrast to previous designs in which each OM housed one large photo-multiplier tube (PMT), KM3NeT is evaluating designs with multiple small PMTs within each glass sphere, including the high-voltage bases and data acquisition interfaces. Therefore the complete detector may have in the order of 100,000 PMTs. Each detection unit "string" or "tower" is likely to have 20 storeys spaced at intervals of 40m with the top buoy at over 800m above the anchor on the seafloor. A practical telescope will have over 150 such detector units with spacing of 150–200m between them depending on the final configuration chosen. An electro-optical

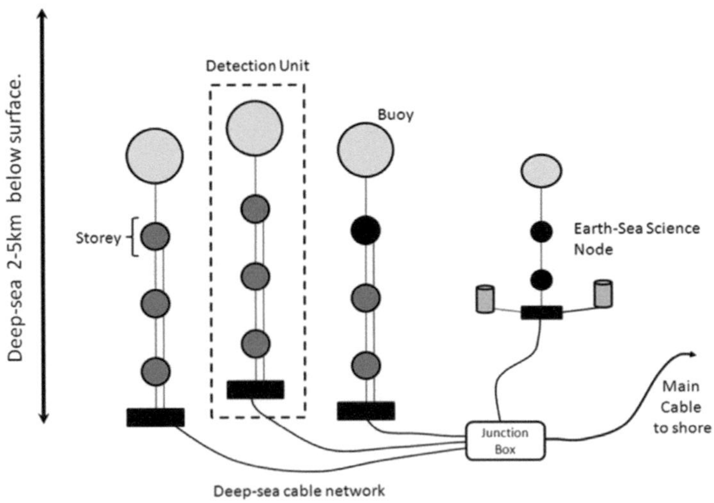

Figure 14.2 Schematic diagram of a KM3NeT junction box to which are connected detection unit strings anchored to the seafloor and an earth–sea science node. In practice, there are many more detection units and storeys as well as multiple earth–sea science nodes.

cable connected via the anchor to a deep-sea cable network runs through the detection unit supplying power and retrieving data from each storey. This network typically contains one or more junction boxes and electro-optical cables, through which all data are transmitted to shore and configuration data sent to the detector. A shore station equipped with substantial computing power collects the data, applies online filter algorithms and saves the data onto mass storage devices.

Earth–sea science nodes are connected to the neutrino telescope cable infrastructure, thus providing electrical power and data links to a series of sensors, moorings and platforms. These may be positioned under the sea-floor, on the sea-floor and reaching up through the water column to the mixed layer below the sea surface as shown in Figure 14.2. Some small earth–sea science sensors may be incorporated into the neutrino telescope detection units.

These nodes are to be positioned at optimal places around the detector providing environmental data relevant to the neutrino telescope and also allowing the marine science communities to perform continuous long-term monitoring experiments and implement earthquake and tsunami early warning systems. Assembly of the telescope sub-sea infrastructure will probably take a number of years. The system could start functioning with a small number of detector units and gradually be expanded over time. Planning of the earth–sea science nodes should allow for possible changes to, or extension of, the telescope infrastructure.

Typically the earth–sea science junction boxes will have connections to one or two mooring strings with profilers or fixed sensors, a suite of seafloor sensors such as seismo-

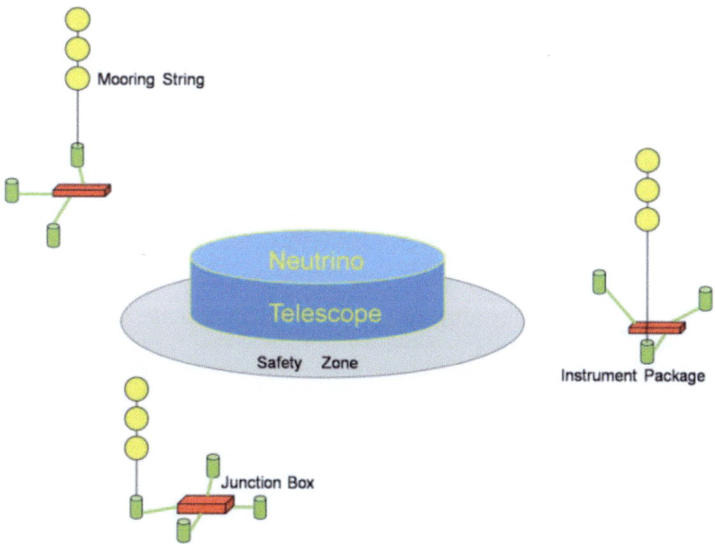

Figure 14.3 KM3NeT conceptual design with the volume of water instrumented for the Neutrino telescope surrounded by a safety zone outside which are located the earth–sea science nodes, each comprising a junction box, a mooring string with sensors in the water column and sensors on the seafloor.

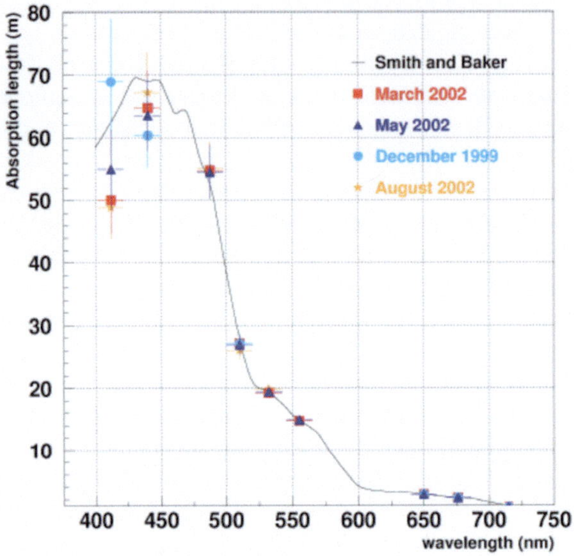

Figure14.4 Average absorption length as a function of wavelength, for four seasons at the NEMO site. Measured using WETLABS AC9 sensor ~25cm path length.

graphs, cameras and acoustic instruments built into modular frames that can be inserted in fixed platforms on the sea-bed (Figure 14.3). Data will be transmitted to shore in real time for processing and dissemination, whilst instruments may be controlled via commands sent from the shore station.

14.3.1 Site criteria

The Mediterranean Sea offers optimal conditions to host an underwater neutrino telescope. Several sites have been identified for such a telescope, with the following criteria driving the final choice:

- Optical properties of sea water for light in the wavelength range 350 nm to 550 nm;
- Distance to the coast for ease of deployment, risk reduction and cable costs;
- Sufficient depth to reduce background noise from atmospheric muons;
- Level of bioluminescence;
- Rates of bio-fouling on optical surfaces;
- Rates of sedimentation;
- Sea current velocities at those depths.

Careful studies of all three candidate sites have been carried out in order to identify the most suitable one. These results are discussed in the following sections.

14.3.2 Water optical properties

The optical properties of water are important when deciding where to construct an underwater neutrino telescope. Much data have already been gathered from the pilot sites and the results are discussed in the following sections.

14.3.2.1 Light transmission

The study of deep-sea water optical properties has been carried out at each site through a long-term programme across all seasons (Riccobene et al., 2007, Capone et al., 2002, Anassontzis et al., 2010). Seawater absorbs and scatters light depending on the water temperature and salinity, as well as the characteristics and concentration of the suspended particulates. These parameters vary between marine sites and possibly as a function of time. Measurements of absorption length show that typically in the Mediterranean Sea (Figure 14.4) there is a maximum at wavelengths around 450 nm and values are close to those recorded for optically pure sea water (Smith & Baker, 1981).

14.3.2.2 Optical background

Decay of radioactive elements in the water and bioluminescence produced by organisms can influence the background counting rate of OMs.

14.3.2.2.1 Potassium 40

The dominant radioactive particle in natural sea water is potassium 40 (^{40}K) and the following 40K decay channels contribute to the production of optical noise:

$$^{40}K \rightarrow {}^{40}Ca + e^- + \bar{\nu}_e \quad (BR = 89.3\%)$$

$$^{40}K + e^- \rightarrow {}^{40}Ar + \nu_e + \gamma \quad (BR = 10.7\%)$$

A large fraction of electrons produced in the first reaction is above the threshold for the production of Cherenkov light. The second reaction produces photons with energies of 1.46 MeV and therefore can lead through Compton scattering to electrons above the Cherenkov light production threshold. The intensity of Cherenkov light from 40K radioactive decays depends mostly on its concentration in sea water. Since the salinity in the Mediterranean Sea has small geographical variation, this noise effect is largely site independent.

14.3.2.2.2 Bioluminescence

Bioluminescence forms an important part of the optical background of the deep sea. It can be categorized into two types: a steady background glow produced by bioluminescent bacteria, and intermittent flashes produced by bioluminescent organisms. The organisms that produce the bioluminescent flashes range in size from sub-millimeter to several meters. Flash intensities range from 10^9 to 10^{13} photons flash^{-1} (Priede et al., 2008) Emission maxima of most species range from 450–490 nm (Herring, 1983), corresponding to maximum light transmission in seawater. Although it is estimated that up to 90% of deep sea animals are bioluminescent (Herring 2002, Haddock et al., 2010), background flash rates are very low (Widder et al., 1989) . However, defensive bioluminescent responses can be stimulated when planktonic organisms are carried by currents past underwater topography (Craig et

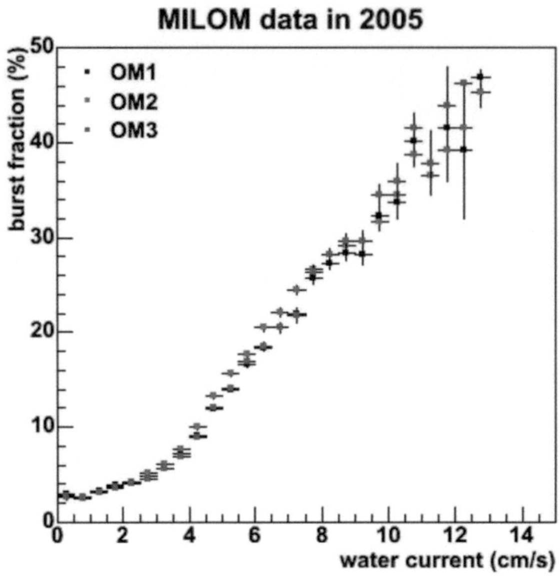

Figure 14.5 Correlation between water current velocity and bioluminescence burst rates around optical modules at the ANTARES site. The burst fraction is the fraction of time with count rates exceeding 120% of the baseline rate.

al., 2011a) and structures. Flow of 5.5 ± 3.4 mm s^{-1} can be sufficient to generate the necessary shear force (Hartline et al., 1999).

Priede et al. (2008) modeled the frequency of bioluminescent events resulting from organisms impinging on a submerged sphere, such as the OM of a neutrino telescope (Eq. 14.1):

$$impacts\ s^{-1} = \pi\left(\frac{\phi_{sphere}}{2} + \frac{\phi_{animal}}{2}\right)^{2} \times v \times \rho \qquad (14.1)$$

where $Ø_{sphere}$ is the sphere diameter, $Ø_{animal}$ the animal diameter, ρ the density (bioluminescent sources m^{-3}), and v the current velocity. If it is assumed that each impact results in a bioluminescent response, the flash rate is directly proportional to the density of bioluminescent animals, the water current velocity and the cross-sectional area of the sphere and the organism.

Flashes, or bursts, of light have been detected by the array lines of OMs installed at prototype neutrino telescopes. At the ANTARES telescope site, data from the MILOM line showed a correlation between sea-current velocity and the bioluminescent burst rate (Naumann-Godo et al., 2007, Bertin 2009) (Figure 14.5).

Individual flash characteristics vary between species, but do share a common pattern of intensity. Typically, after stimulation there is a delay, followed by a rapid rise in intensity

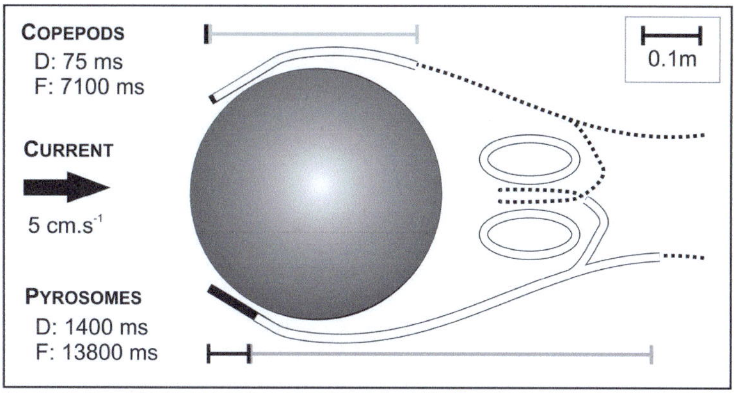

Figure 14.6 Expected pattern of stimulated bioluminescence around a KM3Net detector unit, 43cm-diameter glass sphere. Hypothetical bioluminescence field produced by a copepod and a pyrosome advected by a current of 5 $cm.s^{-1}$. When an organism encounters and is stimulated by the sphere there is a delay (D – solid bar) before start of the flash duration (F – open bar). The mean flow is indicated by dotted lines. The result is an asymmetric stimulated bioluminescent field with most photons emitted downstream.

and a slower decay. Craig et al., (2009). showed that although organisms impinge on an OM on the upstream side, most light would be produced on the downstream side as a result of the flash delay and further stimulation in the eddy vortices (Figure 14.6).

The number of bioluminescent organisms in the water column ($N.m^{-3}$) has been measured using profiler systems based on an ultra-low-light camera focused downwards on a rectangular mesh. The system is mounted on a frame and records during vertical descent through the water column from the sub-surface to the seafloor. The mesh stimulates bioluminescent flashes which are counted and converted to a density with depth. Values in the Mediterranean Sea are an order of magnitude lower than previous data recorded in the Atlantic Ocean and show a general decrease with increasing depth (Priede et al., 2008; Craig et al., 2009, 2011) (Figure 14.7). Seasonal changes have been detected but these are not significant at depths >2500m where the neutrino telescope is likely to be installed.

14.3.2.3 Deep-sea currents

Deep-sea currents have been monitored at all three sites over long periods of time. At the Toulon site measurements were performed during the investigation phase using autonomous mooring lines. Since integration of the instrumentation line in 2005, real-time measurements have been performed continuously. The current speed average was ~5$cm.s^{-1}$ with peak flow occasionally exceeding 35$cm.s^{-1}$ (Aguilar et al., 2006; van Haren 2011) .

Deep-sea current measurements in the Capo Passero region have been recorded using stand-alone current-meter moorings. Current intensity and direction have been monitored

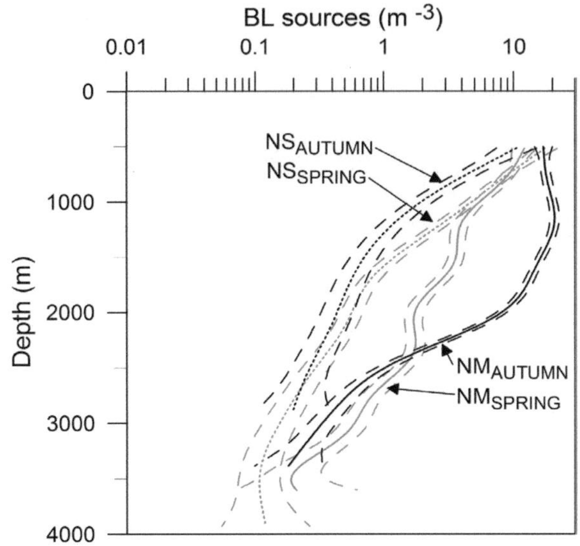

Figure 14.7 Density of bioluminescent animals as a function of depth in the Ionian Sea during different seasons. Curves (GAM, generalized additive modeling, smoothers) show the depth trend at the NEMO (NM) and NESTOR (NS) sites, in autumn 2008 and spring 2009, detected using the ICDEEP profiler. Dashed lines: 95% CL.

almost continuously since 1998 to within 500m of the seabed. Data for the period July 1998–December 1999 have been analyzed and the results presented in (Migneco et al., 2004). This analysis shows that deep-sea currents are unaffected by depth and were stable in both direction and intensity with an average value of about 3cm.s^{-1} and peaks not exceeding 12cm.s^{-1}. These results were confirmed for the period August 1999–August 2002.

At the Pylos site the deep-sea currents have been monitored since 1989 with different moorings and stand-alone current-meters (Aggouras et al., 2006). These studies show that velocities rarely exceed 6cm.s^{-1}. In general, the flow at the 4500m site is northward and is below 4cm.s^{-1} 90% of the time and at the 5200m site it is southward and substantially weaker, with a speed below detection threshold 95% of the time.

14.3.2.4 Sedimentation

The rate of sedimentation in these three regions has been obtained using sediment traps deployed at each site in different years. Records at the Capo Passero site (KM3NeT, 2008), and the Toulon site (Amram et al., 2003) indicate that the highest mass flux values occurred during the late winter/spring and the autumn early winter periods as shown in Figure 14.8. The mass fluxes were recorded at the Pylos site during 2006–2008 and low sedimentation rates confirmed the oligotrophic character of the area (Stavrakakis & Lykousis, 2011).

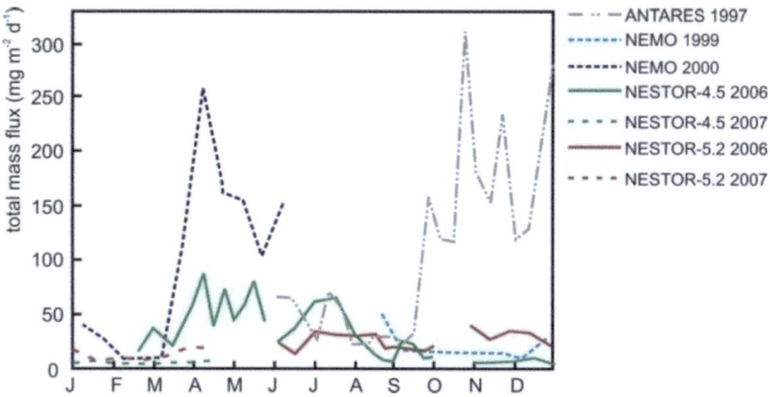

Figure 14.8 Downward total mass fluxes of particulate matter as recorded from sediment traps at the three KM3Net neutrino telescope sites.

14.3.2.5 Biofouling

Fouling is classically considered to progress in five main stages (DeLauney et al., 2010):

1. Formation of a primary film by adsorption of organic and inorganic macromolecules immediately after immersion.
2. Transport of microbial cells to the surface and the immobilization of bacteria on the surface.
3. Attachment of bacteria to the substratum through extracellular polymer production, forming a microbial film.
4. Development of a more complex biological community with the presence of multicellular species, microalgae plus debris, sediments, etc.
5. The final stage is characterized by attachment of larger marine invertebrates such as barnacles, mussels and macro-algae.

Biofilms may form on any solid structures in a pelagic environment and in the case of a neutrino telescope can lead to degradation of the transparency of the surface of the OMs and presence of bioluminescent bacteria in the film will likely increase optical noise. Studies at the ANTARES site (Amram et al., 2003) indicated that growth of fouling is very slow at depth with loss of light transmission for a vertical glass surface estimated to be ~2% after one year. Fouling appears not to progress beyond stage 3 on this time scale and at the depths of interest for neutrino telescopes. A piezophilic luminous bacterium Photobacterium phosphoreum strain designated ANT-2200 has been isolated from 2200m depth at ANTARES and it was demonstrated that the light output from this organism is five times higher at 22 MPa pressure than at atmospheric pressure (Ali et al., 2010). It is assumed that the presence of this bacterium on aggregates around the telescope can contribute to bioluminescent background but this and other marine luminous bacteria may also be capable of growing on the biofilm coating OMs. In the Ionian sea at the NEMO site a maximum of potentially luminescent bacteria was found between 900 and 1200m depth (De Domenico, 2003).

Research at the NESTOR site has concentrated on the metabolic activity of bioluminescent bacteria in the water column, and on the influence of various materials on the settlement of bacteria colonies and biofouling. A colonisation experiment has been developed, providing surfaces for bacterial growth and a protective mechanism preventing wash-off during retrieval. It has been deployed at various depths ranging from 1500m to 4500m carrying sample surfaces of aluminium, titanium, glass, limestone and slate facing both upwards and sideways. After 155 days during 2007 there was no evidence of macro-fouling and only a loosely adhered biofilm could be observed. Scanning Electron Microscopy of surfaces revealed presence of attached bacteria and molecular biological techniques detected some bacteria at all depths (Bellou et al., 2011).

14.3.2.6 Distance offshore

Distance offshore is a major determinant of capital cost of a neutrino telescope; the further offshore the greater the cost of cable. Ship time required for installation and operations also increases owing to longer passage times and greater time loss if bad weather intervenes. Costs increase with increasing depth of installation, winches with longer cables become heavier, support vessels consequently increase in size and time required for any sub-sea operations become longer. The basic engineering of the telescope hardware also becomes more challenging with increasing depth. The Mediterranean Sea offers a choice of sites with different distances offshore and depths. NESTOR offers several locations from a depth of 3000m at 20km offshore to 5200m depth at 60km offshore, ANTARES has a depth of 2475m at 45km offshore and NEMO 3350m at 100km offshore. All the factors relating to costs, telescope performance, seafloor slope, depth and distance offshore can be weighted to select an optimal location in a GIS (Geographic Information System) environment (Niedzielski et al., 2009).

14.3.3 Scientific opportunities in the Mediterranean Sea

In general, observations in the deep sea have until now been made by autonomous measuring systems, deployed for up to a year and requiring recovery in order to retrieve the data. Data storage and battery energy capacity have limited data sampling rates in such systems to once every 10 minutes or every hour and there are inevitable breaks in data collection.

The proposed KM3NeT deep-sea infrastructure, permanently cabled to the shore, will enable continuous data to be collected without interruption over long periods. Data capture rates will be orders of magnitude faster than used hitherto and will allow new phenomena to be investigated with sampling rates of the order of 1 Hz. For instance, it will be possible for the first time to investigate phenomena such as internal waves and short time-base oscillations in the water column using sensors distributed throughout the telescope array. Real-time tracking of bio-acoustic emissions or vertical migrations of organisms will also be possible. Both the spatial and temporal scale of measurements will be transformed and real-time availability will revolutionise data applications. The system will also provide continuous vigilance in the face of transitory hazardous events such as earthquakes, slope failures and tsunamis.

Sensors placed on the seafloor and in the water column that are cabled to shore will ensure efficient tracking of environmental change on longer daily, annual and decadal time scales. The earth–sea science infrastructure within KM3NeT will form the basis for the Mediterranean section of the EU plan for long-term monitoring of the ocean margin environment around Europe. It is part of the Global Monitoring for Environment and Security (GMES) system and will complement oceanographic networks such as GOOS (Global Ocean Observing System), EuroGOOS and DEOS (Dynamics of Earth and Ocean Systems), and will be multidisciplinary, with stations monitoring the rocks, sediments, bottom water, biology and events in the water column.

The KM3NeT infrastructure will incorporate reconfigurable junctions boxes to which new associate sciences instruments can be connected. For example, it will be feasible to connect elements of the new Ocean Tracking Network that will be capable of tracking fishes and other animals equipped with implanted transmitters (OTN, 2014).

Geologically, the Mediterranean Sea is at a pivotal point between the African and Eurasian tectonic plates and some authors have argued that it contains some of the oldest ocean floor on the planet, recognisable as remnants of the Tethys Sea that was an embayment of the single proto-continent, Pangaea. The present day Mediterranean area is still tectonically active with 23 active volcanoes in Italy and six in Greece associated with subduction of the African plate beneath Eurasia. Vesuvius on the mainland and Etna on the island of Sicily are the best known in Italy and the island of Santorini in Greek waters erupted most recently in 1950. Hydrothermal activity is also seen around volcanic islands. Earthquakes occur regularly in this region, particularly in the eastern basin despite relatively slow motion of the tectonic plates. The Hellenic trench and the parallel Mediterranean ridge which extends across the Ionian Sea to Cyprus are important sea-floor features associated with subduction and accretion (Ambriola et al., 2012). Emissions of methane from the sea-floor due to deep sea mud volcanoes also occur in the Eastern Mediterranean. Slope failures have been detected capable of generating turbidite flows and tsunamis.

Biologically, the Mediterranean Sea has been re-colonised after the formation of the modern ocean basins with limited time for evolution of new species but very special conditions and partial isolation of basins that give opportunities for endemic species to appear. Colonisation and re-colonisation have occurred from the Atlantic Ocean. It is estimated that 50.2% of species are of Atlantic origin, 16.8% Atlantico-Pacific and 4.4% of Indo-Pacific origin (Emig & Geistdoerfer, 2004) and of all species in the Mediterranean 28.6% are endemic or unique to this area. Owing to the direction of colonisation and stagnation events in the east there is a general impoverishment of fauna from west to east. It is estimated that the Mediterranean Sea contains about 8500 species of which only half are generally represented in the Eastern Basin (Bianchi & Morri, 2000). The east is therefore impoverished not only in productivity but also in biodiversity. Despite apparent sparseness of life in the Mediterranean it contains between 4 and 18% of world marine species in 0.32% of the world ocean volume. Numbers of species decrease rapidly with depth so that at 1000m there is 8% of the shallow water number and only 3% at 2000m.

The Mediterranean supports a wide range of fishing activity including pelagic (near surface) such as tuna, sardine and anchovy, demersal species (bottom living), crustacean and molluscs. Generally, the Mediterranean is less productive than other seas reflecting its oligotrophic status. However, there is evidence that production of fish has increased over time. The Western Basin has moved from below the world average to above average production in sea shelf fishery production. This increase may be attributable to fertilisation by run-off from human activity on land. The main increase has been due to small inshore pelagic fishes whereas bottom living fishes have declined. So whilst catches have increased, the quantity of big valuable fish has decreased.

There are 19 species of cetaceans in the Mediterranean Sea of which eight are considered of common occurrence, and only one species of pinniped, the Mediterranean Monk Seal (*Monachus monachus*) which is listed as endangered. Colonies are now confined to the Alboran and Aegean Seas. Public interest in cetaceans of the Mediterranean Sea is great and there is support for research activity. Recently, a Cuvier's beaked whale, *Ziphius cavirostris* (Tyack et al., 2006), was recorded to a depth of 1900m during a dive lasting 85 minutes in the Ligurian Sea. This is the world record for deep diving in mammals.

14.3.3.1 Physical oceanography

An array of the order of 10,000 sensors in a cubic kilometer of deep sea such as KM3NeT offers the opportunity to study water motions in the ocean in unprecedented temporal and spatial detail. Specifically, scientific questions can be addressed on 3D-variations of physical processes on meso-scale (10km) horizontally and smaller scales including those of nonlinear internal wave propagation. This will not only renew our knowledge of some of the relevant physical processes of the ocean, but also of their effects on the distribution of suspended geological, chemical and biological materials.

The redistribution of matter, momentum and heat in the ocean interior depends crucially on turbulent mixing to overcome the otherwise stable vertical density stratification. The impact of this mixing affects not only redistribution of fine grained sediment and possibly associated pollutants, but also life, due to the flux of nutrients up and into the photic zone. It also affects the distribution of plankton in the deep sea and, crucially, the large-scale meridional overturning circulation. It is thought that a major part of the mixing is generated via breaking of (nonlinear) waves in the interior and especially above sloping bottoms. These waves are supported by the self-same density stratification.

Other processes that may be important for the redistribution of materials are small- and meso-scale eddies, which show rather large, preferentially downward vertical motions of up to several 10^{-2}m s^{-1}, or 1000m day^{-1}, so that near-surface materials can be transported to the bottom within a day (van Haren et al., 2006). Such eddies can be associated with meso-scale meandering variations in boundary or along-frontal currents. Recent custom-made high-resolution temperature T-sensors that can be moored at a specific site down to 6000m whilst measuring at a rate of 1 Hz show details of such internal waves up to 40m high and breaking as depicted in Figure 14.9 (van Haren et al., 2005, 2009) .

Most present-day moored physical oceanographic (internal wave) observations are limited by the following dimensions: 10m vertically, typically 10km horizontally and 10–30 minutes in time when sampled over ranges of 500m vertically and durations of one year. An exception was a tri-moored internal wave experiment, in the 1970s (Briscoe, 1975) . This limitation is due to battery and memory constraints as the measurements are made by

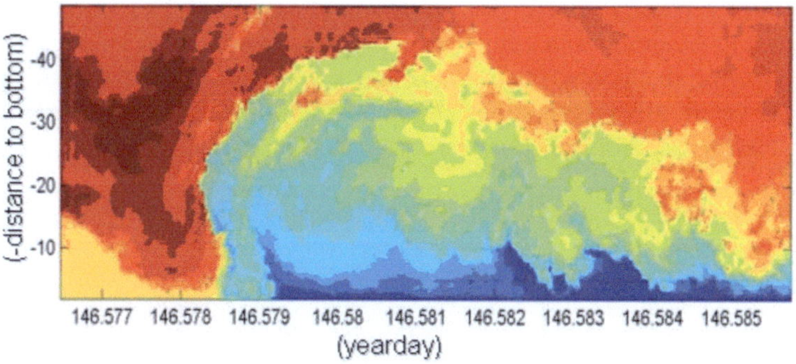

Figure 14.9 Example of data obtained from a vertical array of temperature sensors. Fourteen minutes detail of a backwards breaking, 40m-high wave measured using 100 NIOZ-2 temperature sensors at a depth of 550m near the top of Great Meteor Seamount, North Atlantic Ocean. Temperature ranges from 12.4°C (blue) to 14.2°C (red). Such vigorous process is unlikely to occur in the deep Mediterranean interior but, at a slower rate by about a factor of 3–10, such waves can occur near deep topography.

self-contained instrumentation, while financial constraints restrict the number of simultaneously operated moorings to about 5–10. Within the set-up of KM3NeT, novel oceanographic possibilities arise by sampling at least one decade faster and across one order of magnitude smaller horizontally over the above ranges for the period of at least 10 years. It will enable the operation of an internal wave/small-scale processes antenna, which will resolve amplitude and phase distributions in so far unresolved detail.

In the deep Mediterranean Sea, where the neutrino telescope is planned, the weak stratification supports internal waves having typical periods of 1–24 hours with typical vertical and horizontal wavelengths of 10–1000m. These are very slow waves indeed, but even to properly sample linear, sinusoidal waves, a sampling rate is required that is at least 10 times faster than the shortest period available. If one wants to study internal wave induced mixing via nonlinear steepening and eventual wave breaking into large, not yet even dissipative turbulence scales, one needs to sample 100 times faster, or about once per 10–100s (Figure 14.9 is sampled at 1 Hz for waves of period 20 minutes, occurring once every 12.4 hours; semidiurnal lunar tide).

We envisage the use of the photon detectors PMTs to compare variations in their data with those obtained from conventional oceanographic instruments. It would be beneficial if high-precision T-sensors are installed on each PMT. The high sampling rate of such T-sensors, typically 1 Hz, allows sampling of the larger turbulence overturning scales but does not compare with the somewhat cruder vertical and horizontal spacing between PMTs of 15 and 100m, respectively. In addition, some of the PMT lines will be equipped with acoustic current meters, to resolve directly the flow field. Furthermore, the plan is to equip at least one line in the earth–sea science section outside the neutrino array with some 100–200 densely vertically spaced (1m) high-sampling rate (1 Hz) temperature sensors. These sensors will be moored in conjunction with turbulence velocity sensors and a 300 kHz ADCP covering the lower 100–200m of the water column and focusing on near-bottom (nonlinear internal wave) motions. Another line will hold a continuously profiling yoyo-CTD which will monitor the entire vertical density structure at cm-scale.

14.4 Infrastructure management and operation

There will be need for joint operations management of the earth–marine science infrastructure and the neutrino telescope which would be responsible for coordinating deployment, maintenance, data sharing between the two communities and emergency situations.

14.4.1 Neutrino telescope operations

The telescope will have an operations control center situated either in the shore station or at one of a limited number of remote centers and be operated by a small group of operators on a 24-hour, seven day a week basis. Technicians will be based locally for maintenance operations and emergency situations. The operations center will provide continuous monitoring of the status of all major components such as optical cable, junction boxes, detection units, power supply, data links and local environmental variables. The system will alert the operators when anomalies are detected and predefined procedures will be in place to resolve these problems.

14.4.2 Marine science observatory operations

It is expected that neutrino telescope operations will have priority over the marine science operations, except when agreed in advance, in order to limit any interference.

Three categories of observatory operation are recognized:

- Operational and Civil Protection
  - Earthquakes
  - Tsunamis
  - Oceanography – GOOS (Global Ocean Observing System) contribution
- Ocean & Geosciences Research
- Engineering trials.

In the first category, a high standard of reliability and real-time availability is required with significant down-time being unacceptable. In the second category, a certain amount of service interruption is acceptable but continuity is a priority albeit at low data rates or via data buffering. In the third category, performance and availability are negotiable.

The earth–marine science user community will need some kind of coordinating body with operational staff tailored according to the size and complexity of the earth–marine

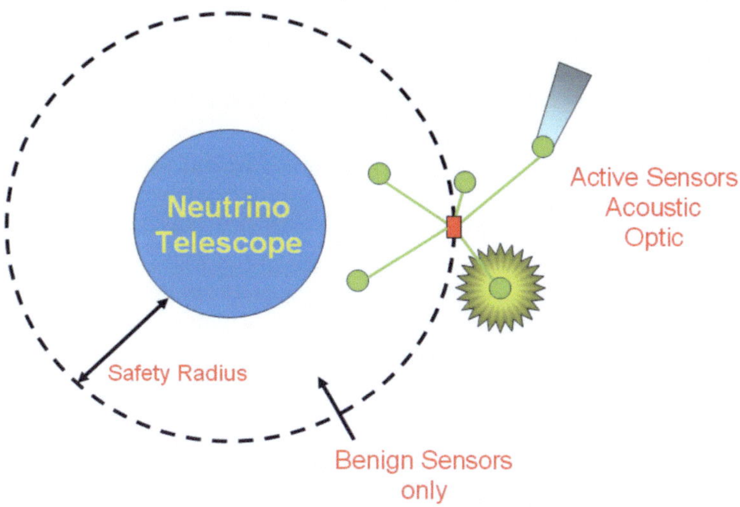

Figure 14.10 Schematic plan of the KM3Net neutrino telescope and an earth–sea sciences junction box on the seafloor. Active sensors with sonars and strobe lights must be outside a safety radius so that telescope functions are not affected. Benign sensors might be placed with that radius.

science nodes. This could be managed according to the rules and methods established in ESONET NoE (ESONET, 2012).

The management team will ensure efficient integration between the KM3NeT associated science communities, environmental agencies and organizations at the national, regional and international level including ESONET, EMSO, GOOS and ORFEUS, thus maximizing dissemination and use of data.

14.4.3 Safety requirements

The experience gained during the integration of the associated science node at the ANTARES site demonstrated the need to define a safety radius between the telescope and the science node to reduce the cost of sub-sea intervention whilst minimising the risk of collisions during such interventions.

A safety zone around the neutrino telescope has therefore been agreed between the astrophysics and the marine science communities. The position of an associated-science node relative to the array, as shown in Figure 14.10, will need to take into account this safety distance and the optimum position for the best scientific results. This will have the added advantage of ensuring independent deployment, operations and maintenance of both installations as the marine science observatories will be serviced regularly every 12 months whilst the neutrino telescope will be serviced on a needs-be basis. This safety zone is defined to avoid interference between the earth–sea sciences activity and the astrophysics infrastructure; it lies within a general exclusion zone protecting the overall subsea infrastructure.

14.4.4 Data management and access

In the marine sciences standardized methods for information management are becoming established to ensure better accessibility and traceability of datasets and ultimately to increase their use for societal benefit. A further important aspect is the connection of ocean observatory effort into larger frameworks including the Global Earth Observation System of Systems (GEOSS) and the Global Monitoring of Environment and Security (GMES) (Ruhl et al., 2011).

The neutrino telescope data acquisition system will filter the data in real-time before archiving and distributing it to users around Europe and beyond. If possible, an existing data center will be used; however, if necessary, a data center similar to those used by astronomical observatories may be implemented for KM3NeT with a dedicated data manager.

Management of the earth–marine science data was made possible by the use of the ESONET NoE data management system (DMAS with reference to the NEPTUNE concept) allowing access to:

- Generic sensor data by a large community for model assimilation and warning for civil protection
- Specific experiment data by their principal investigators
- Images and associated data by the public at large
- Test data by engineering teams.

14.4.5 Public relations and outreach program management

It is important that the observatory results are made available in a simple and visual manner to the citizens of the European Community and beyond. In order to achieve this, all KM3NeT observatory activities will be publicized on a regular basis and links to outreach programs made available at all the institutes involved in the project. All KM3NeT results will be posted on the KM3NeT website. The site will be split in two sections: one for the layman and a more advanced one for scientists. Real-time data will be displayed on the website, whenever possible, for the general public to access. An aim of the layman's site is to get young people interested in all aspects of KM3NeT science and allow the general public to understand how funds are being used to enhance scientific and technical knowledge. The advanced site will aim at generating scientific and technological knowledge among European scientists.

14.4.6 Users and stakeholders

It is envisaged that access to the system and its data will be granted to:

- Scientists from member institutes
- External scientists from within the EU
- Scientific institutions external to the EU
- EU government departments and organizations.
- Non-government organizations.

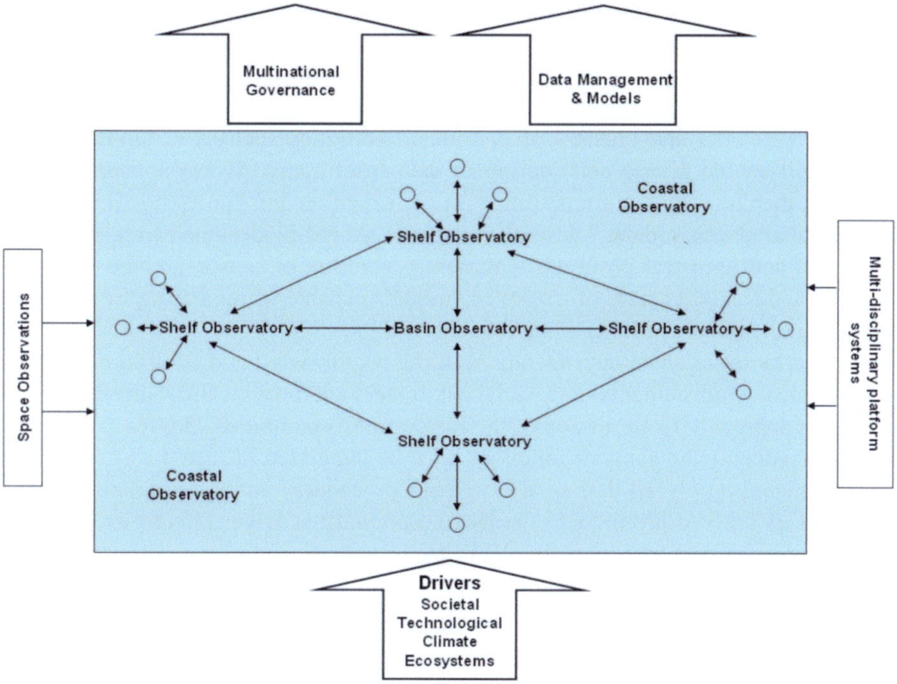

Figure 14.11 The CIESM (Mediterranean Science Commission, or Commission Internationale pour l'Exploration Scientifique de la Méditerranée) scheme for an integrated system of Mediterranean marine observatories into which KM3Net facilities should be incorporated.

14.5 Marine observatory integration in the Mediterranean

At a workshop held in La Spezia in January 2008 (CIESM, 2008) the Mediterranean Science Commission (2012) stated that the Mediterranean is characterized as a miniature ocean and was therefore an ideal model to study oceanic processes and land–ocean–atmospheric interaction. Geological records have shown that this ecosystem amplifies climatologic signals and makes this an ideal test bed for climatic studies.

The Mediterranean environment is facing extreme pressure from an ever-increasing population, climate change and over-exploitation. Recent observations have shown large-scale changes to deep ocean circulation, heat and regional climate, sea level rise, deterioration in water quality, increased number of algal blooms and the collapse of regional fisheries. Such changes have a profound impact on the ecosystems.

Sustainable development will depend more and more on intelligent management of the marine environment in order to protect marine ecosystems and minimize the impact of climate change whilst maintaining the economic benefit to the region. As a result, the formulation of policies must be based on informed decisions which in turn will depend on intelligent support systems, relying on real-time sensing platforms and numerical modeling.

The concept of an Integrated Mediterranean Marine Observatory, as depicted in Figure 14.11, includes instrumentation placed on satellites, sea-buoys, moorings, ships, drifters, profilers, gliders and coastal systems such as arrays of weather stations and marine sensors such as ADCPs.

Complex networks and clusters of distributed activities such as ESONET NoE and KM3NeT will enable access and sharing of data from many diverse sources. Together, this data can then be used and transformed into information and knowledge increasing the economic value of ocean data. These observatories should be designed to provide appropriate spatial and temporal coverage with new generation of sensors, adaptive sampling capabilities and linked to state of the art 3D models capable of predicting future changes and assessing the effects of mitigating actions.

These multipurpose observatories represent the way forward and combining these data with economical, environmental and social parameters will provide the required integrated management approach (Commission of the European Communities, 2007).

Data management and analysis functions must be integrated in such a way as to rapidly and systematically apply all data quality control procedures and disseminate the marine data with all its derived products to the user community at large. This in itself will be a major challenge for data providers involving high volumes of data consisting of disparate data requiring rapid integration and presented in a user-friendly format.

The KM3NeT sub-sea telescope infrastructure will form an important component of the Integrated Mediterranean observatory system. The decision to build in the Mediterranean Sea and the precise site chosen are determined by the requirements of neutrino astronomy; the challenge for the Earth–Sea–Science community is to make the best use of this opportunity to intensively instrument part of the Mediterranean Sea.

Acknowledgments

This work was support by the EU Framework programme KM3NeT projects. Except where otherwise indicated all images are copyright KM3NeT and are used with permission.

References

Aggouras G. et al. (2006) LAERTIS, a multidisciplinary station. Nuclear Instruments and Methods in Physics Research A567: 468–473.

Aguilar J.A. et al. (on behalf of the ANTARES Collaboration) (2006) First results of the Instrumentation Line of the deep-sea ANTARES neutrino telescope. Astroparticle Physics 26, 314–324.

Ali B.A., Garel M., Cuny P., Miquel J-C., Toubal T., Robert A. and Tamburini C. (2010) Luminous bacteria in the deep-sea waters near the ANTARES underwater neutrino telescope (Mediterranean Sea). Chemistry and Ecology 26, 57–72.

Amram P. et al. (on behalf of the ANTARES Collaboration) (2003) Sedimentation and fouling of optical surfaces at the ANTARES site. Astroparticle Physics 19, 253–267.

Anassontzis E.G. et al. (2010) Water transparency measurements in the deep Ionian Sea. Astroparticle Physics 34, 187–197.

Bellou N., Colijn F. and Papathanassiou E. (2011) Experimental settlement study in the Eastern Mediterranean deep sea (Ionian Sea). Nuclear Instruments and Methods in Physics Research A626–627, S102–S105.

Bertin V. (on behalf of the ANTARES Collaboration) (2009) Status and first results of the ANTARES neutrino telescope. Nuclear Instruments and Methods in Physics Research A604, S136–S142.

Bianchi C.N. and Morri C. (2000) Marine biodiversity of the Mediterranean Sea: Situation, problems and prospects for future research. Marine Pollution Bulletin 40, 367–376.

Briscoe M.G. (1975) Preliminary results from the trimoored Internal Wave Experiment (IWEX). Journal of Geophysical Research 80, 3872–3884.

Capone A., Digaetano T., Grimaldi A., Habel R., Lo Presti D., Migneco E., Masullo R., Moro F., Petruccetti M., Petta C., Piattelli P., Randazzo N., Riccobene G., Salusti E., Sapienza P., Sedita M., Trasatti L. and Ursella L. (2002) Measurements of light transmission in deep sea with the AC9 transmissometer. Nuclear Instruments and Methods in Physics Research A487, 423–434.

CIESM (2008) Towards an integrated system of Mediterranean Marine Observatories. No. 34. In: R. Briand (Ed.) CIESM Workshop Monographs. CIESM: Monaco.

Commission of the European Communities (2007) An Integrated Maritime Policy for the European Union. COM(2007) 575 final Brussels, 10.10.2007. 16 pages. http://ec.europa.eu/maritimeaffairs/policy/index_en.htm. Accessed August 31, 2012.

Craig J., Jamieson A.J., Heger A. and Priede I.G. (2009) Distribution of bioluminescent organisms in the Mediterranean Sea and predicted effects on a deep-sea neutrino telescope. Nuclear Instruments and Methods in Physics Research A602: 224–226. doi:10.1016/j.nima.2008.12.043

Craig J., Jamieson A., Bagley P. and Priede I.G. (2011) Seasonal variation of deep-sea bioluminescence in the Ionian Sea. Nuclear Instruments and Methods in Physics Research A626–627, S115–S117.

De Domenico M., Scarfì S., Leonardi M., Raffa F., De Luca M. and De Domenico E. (2003) Microbial communities temporal variations in a pelagic site offshore Cape Passero (Southern Ionian Sea). Biologia Marina Mediterranea 10, 994–997.

Delauney L., Compère C. and Lehaitre M. (2010) Biofouling protection for marine environmental sensors. Ocean Science 6, 503–511.

DONET (2008) Dense Oceanfloor Network System for Earthquakes and Tsunamis. http://www.jamstec.go.jp/donet/e/donet/index.htm. Accessed June 5, 2014.

DUMAND (2003) DUMAND at the University of Hawaii. http://www.phys.hawaii.edu/~dumand/. Accessed June 5, 2012.

Emig C. and Geistdoerfer P. (2004) The Mediterranean deep-sea fauna: Historical evolution, bathymetric variations and geographical changes. Carnets de Géologie CG2004, A01.

ESONET (2012) European Seas Observatory Network. http://www.esonet-emso.org/. Accessed August 31, 2012.

Haddock, S.H.D., Moline, M.A., and Case, J.F., 2010. Bioluminescence in the sea. Annual Review of Marine Science 2(1), 443–493.

Hartline D.K., Buskey E.J. and Lenz P.H. (1999) Rapid jumps and bioluminescence elicited by controlled hydrodynamic stimuli in a mesopelagic copepod, Pleuromamma xiphias. The Biological Bulletin 197, 132–143.

Herring, P.J. (1983) The spectral characteristics of luminous marine organisms. Proceedings of the Royal Society of London. Series B, Biological Sciences. 220: 183–217.

Herring P.J., (2002) The Biology of the Deep Ocean. Oxford: Oxford University Press.

IceCube (2014) South Pole Neutrino Observatory. http://www.icecube.wisc.edu. Accessed June 5, 2014.

KM3Net (2008) KM3NeT Conceptual Design Report for a Deep-Sea Research Infrastructure Incorporating a Very Large Volume Neutrino Telescope in the Mediterranean Sea. Available from: http://www.km3net.org/CDR/CDR-KM3NeT.pdf. Accessed June 5, 2014.

MARS (2008) MARS comes alive, In Annual Report 2008, MBARI Monterey Aquarium Research Institute, pp 4–8. Available from: http://www.mbari.org/news/publications/ar/2008ann_rpt_lowres.pdf. Accessed June 5, 2014.

Migneco E. (2004) NEMO: Status of the project. Nuclear Physics B (Proceedings Supplements) 136, 61–68.

Naumann-Godo, M. et al. (on behalf of the ANTARES Collaboration) (2007) Current status of the ANTARES neutrino telescope. Nuclear Physics B (Proceedings Supplements) 172, 36–40.

NEPTUNE (2012) NorthEast Pacific Time-Series Undersea Networked Experiments. http://neptunecanada.ca/. Accessed August 31, 2012.

Niedzielski T., Priede I.G. and Holford A. (2009) On the optimal siting of cubic kilometre scale neutrino telescope infrastructure on the deep-sea floor. Marine Geophysical Researches 30, 217–227; doi:10.1007/s11001-009-9078-9

OTN (2014) Ocean Tracking Network. http://www.oceantrackingnetwork.org/. Accessed June 6, 2014.

Priede I.G., Jamieson A., Heger A., Craig J. and Zuur A.F. (2008) The potential influence of bioluminescence from marine animals on a deep-sea underwater neutrino telescope array in the Mediterranean Sea. Deep Sea Research Part I 55, 1474–1483; doi:10.1016/j.dsr.2008.07.001

Riccobene, et al. (2007) Deep seawater inherent optical properties in the Southern Ionian Sea. Astroparticle Physics 27, 1–9.

Robert, A. (1992) The birth of high-energy neutrino astronomy: A personal history of the DUMAND project. Reviews of Modern Physics 64, 259–312.

Ruhl H.A., André M., Beranzoli L., Çagatay M.N., Colaço A., Cannat M., Dañobeitia J.J., Favali P., Géli L., Gillooly M., Greinert J., Hall P.O.J., Robert Huber R., Johannes Karstensen J., Lampitt R.S., Larkin K.E., Lykousis V., Mienert K., Miranda J.M., Person R., Priede I.G., Puillat I., Thomsen L. and Waldmann C. (2011) Societal need for improved understanding of climate change, anthropogenic impacts, and geo-hazard warning drive development of ocean observatories in European Seas. Progress in Oceanography 91(1), 1–33; doi:10.1016/j.pocean.2011.05.001.

Smith, R.C. and Baker K.S. (1981) Optical properties of the clearest natural waters (200–800 nm) Applied Optics 20, 177–184.

Stavrakakis S. and Lykousis V. (2011) Interannual mass flux variations of settling particles in the NESTOR basins, SE. Ionian Sea (E. Mediterranean), Greece. Nuclear Instruments and Methods in Physics Research A626–627, S99–S101.

The Mediterranean Science Commission (2012) http://www.ciesm.org/index.htm. Monaco. Accessed August 31, 2012.

Tyack P.L., Johnson M., Aguilar de Soto N., Sturlese A. and Madsen P.T. (2006) Extreme diving of beaked whales. The Journal of Experimental Biology 209, 4238–4253.

van Haren H. (on behalf of the ANTARES collaboration) (2011) Meso- and small-scale vertical motions in the deep Western Mediterranean. Nuclear Instruments and Methods in Physics Research A626–627, S84–S86.

van Haren H., Groenewegen R., Laan M. and Koster B. (2005) High sampling rate thermistor string observations at the slope of Great Meteor Seamount. Ocean Science 1, 17–28; SRef-ID: 1812-0792/os/2005-1-17.

van Haren H., Laan M., Buijsman D.-J, Gostiaux L., Smit M.G. and Keijzer E. (2009) NIOZ3: Independent temperature sensors sampling yearlong data at a rate of 1 Hz. Journal of Oceanic Engineering 34, 315–322; doi:10.1109/JOE.2009.2021237.

van Haren H., Millot C. and Taupier-Letage I. (2006) Fast deep sinking in Mediterranean eddies. Geophysical Research Letters 33, L04606 ; doi:10.1029/2005GL025367.

Widder E.A., Bernstein S.A., Bracher D.F., Case J.F., Reisenbichler K.R., Torres J.J. and Robison B.H. (1989) Bioluminescence in the Monterey Submarine Canyon: Image analysis of video recordings from a midwater submersible. Marine Biology 100, 541–551.

15 ANTARES neutrino telescope and deep-sea observatory

V. Bertin[1], J. Brunner[1], J. Carr[1], P. Coyle[1], C. Curtil[1], J-J. Destelle[1], A. Deschamps[2], S. Escoffier[1], K. Graf [3], C. Gojak[4], J. Hößl[3], R. Lahmann[3], D. Lefèvre[5], C. Lévèque[6], C. Tamburini[5], J-P. Schuller[7] and H. van Haren[8]

ANTARES Collaboration[9]

15.1 Introduction

The ANTARES detector consists of a multidisciplinary undersea observatory associated with a neutrino telescope. The neutrino telescope, with 12 mooring lines holding light detectors, was completed in May 2008 and is destined for research in the field of astroparticle physics, in particular in neutrino astronomy. Instruments for research in marine and Earth

1 CPPM – Centre de Physique des Particules de Marseille, CNRS/IN2P3 et Université de la Méditerranée, Marseille, France

2 GéoAzur, CNRS/INSU, IRD, Université de Nice Sophia Antipolis, Observatoire de la Côte d'Azur, Sophia Antipolis, France

3 ECAP – Erlangen Centre for Astroparticle Physics, Erlangen, Germany

4 DT INSU, Division Technique de l'INSU, La Seyne sur Mer Cedex, France

5 Aix Marseille Université, CNRS, Université de Toulon, IRD, MIO UM 110, 13288, Marseille, France

6 IFREMER, La Seyne-sur-mer Cedex, France

7 Direction des Sciences de la Matière – Institut de Recherche sur les lois Fondamentales de l'Univers, Gif-sur-Yvette, France

8 Royal Netherlands Institute for Sea Research (NIOZ), Den Burg, the Netherlands

9 The results presented in this chapter are the work of the whole ANTARES Collaboration whose members are given in Ageron et al., (2009).

science are distributed on the 12 optical detector lines and are located on a further 13th line specifically dedicated to monitoring the sea environment. An essential feature of the ANTARES infrastructure is the permanent connection to the shore with the capacity for high-bandwidth acquisition of data. In the short term, a significant extension of the multidisciplinary activities is planned and, in the longer term, a major extension of all aspects of the undersea observatory is envisioned.

The development and construction of the undersea observatory has taken several years. The first tests were carried out in 1996 with autonomous mooring lines. The permanent cable link to the shore was operational from November 2002, and during 2003 the first prototype mooring lines were connected for long periods of time to the shore. Oceanographic data are available from the instruments on the site since 2003 with an increasing number of instruments being operational as more of the detector was constructed.

15.2 Science objectives

The combined neutrino and marine observatory has enormous potential for discoveries in a wide range of scientific domains. The unique properties of the neutrino as a messenger from distant and hidden locations in the cosmos offer breakthroughs in many aspects of our understanding of the Universe. Visible light astronomy mainly observes stars like the Sun while the extended multi-wavelength astronomy with radio wave to gamma-rays has discovered a multitude of different sources such as active galactic nuclei, gamma-ray bursts and microquasars. Neutrino telescopes extend further the discovery capability of this new astronomy and will yield essential information on the nature of known sources but also give possibilities of discovering hitherto unknown sources where only neutrinos can emerge from an environment dense in matter. Neutrinos are unique among the messengers of the universe because of their weak interaction with matters. They can only be detected via their rare interactions with nuclei in the vicinity or inside the equipped detector volume, which has therefore to be larger than any other existing particle detector or telescope. The undersea location and the permanent connection to the shore provides the opportunity to install sensors for sea parameters giving continuous long-term measurements. Such data are currently non-existent and the novel potential will create opportunities for discoveries and innovation in many sea science areas.

15.3 Technical description of neutrino telescope and observatory

15.3.1 Stages in construction of the detector

In 1996, the ANTARES Collaboration started R&D activities towards the construction of a neutrino telescope in the deep sea. As neutrinos are detected via short flashes of optical photons (see Section 15.3.2), it is important to know the optical properties of the potential

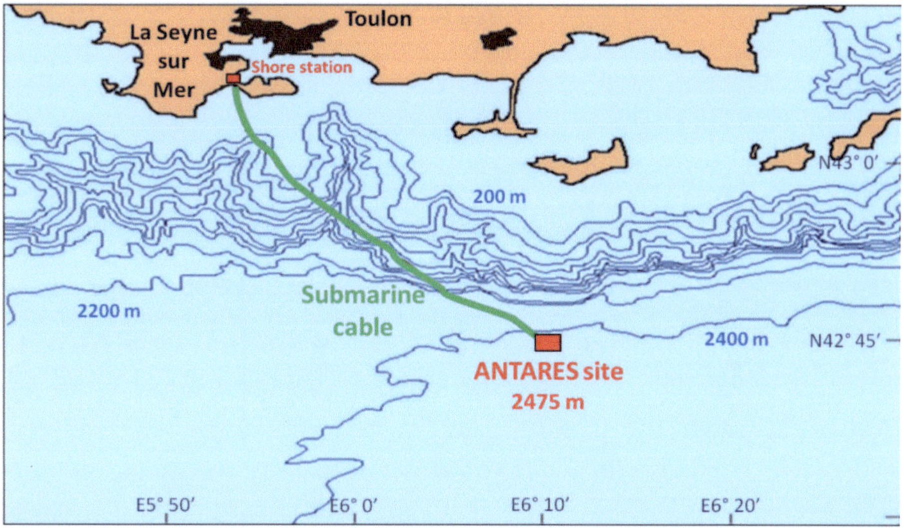

Figure 15.1 Location of the ANTARES site south-east of Toulon in France.

installation site. The first operations were the deployment, operation and recovery of autonomous mooring lines to measure the properties of the water and environment at the ANTARES site which is located off the French coast south-east of Toulon at 42° 48'N, 6° 10' E as shown in Figure 15.1. In these site evaluation surveys (Aguilar et al., 2000; Amram et al., 2000, 2003) comprising more than 60 line deployments, extensive measurements were made of light background from bioluminescence, bio-deposition, sedimentation and light transmission.

The earliest test line in the ANTARES program which was connected to a readout system on shore was deployed in November 1999 and is referred to as the "Demonstrator Line". This line used an old, existing undersea cable donated by France Telecom, connecting the line to a recording station in Marseille. Located on a site near Marseille at a depth of 1200m, the line was operated for a few months and proved various concepts of the design, in particular the acoustic positioning system, and included reconstruction of cosmic ray muons with seven optical sensors.

In 2001, the construction of the actual detector started with the deployment of a new cable between the final site and the shore station in La Seyne-sur-Mer. This cable, the present Main Electro-Optical Cable (MEOC), was deployed on the seabed in November 2001, initially with only a loop-back container at the sea end. In November 2002 the end of the cable was brought to the surface, connected to the Junction Box (JB) and redeployed. Since this date, the battery-operated slow control in the junction box has sent to shore measurements of various parameters in the housing, showing perfect operation of the MEOC and JB system for more than five years.

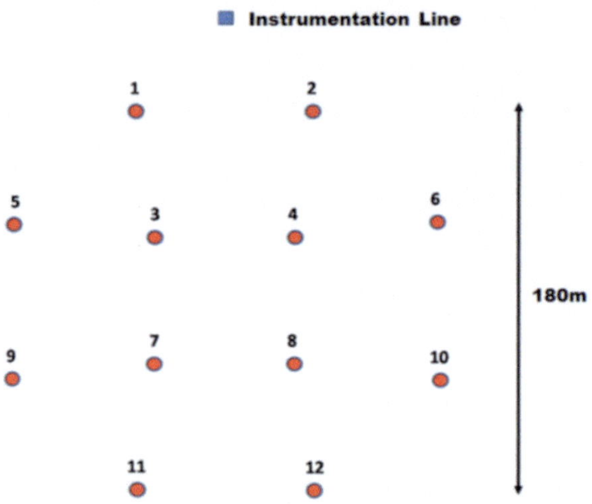

Figure 15.2 Seabed layout of the ANTARES detector. The system has 12 neutrino telescope lines with optical detectors and one instrumentation line. The lines of the neutrino telescope were put into operation between March 2006 and May 2008 and the two versions of the instrumentation line in March 2005 and December 2007 respectively.

During 2003, prototypes approaching the final technology of the detector were operated. Two lines were deployed and connected between November 2002 and March 2003: the Mini Instrumentation Line (MIL), a test instrumentation line, and the Prototype Sector Line (PSL), a short optical detector line with 15 optical sensors. These lines, which were operated in situ until May and July 2003 respectively, again proved the validity of various aspects of the design but indicated as well certain problems with loss of optical transmission in the line cables and leaks in the cables and containers. Nevertheless, the PSL was able to measure the counting rates in the optical detectors, in particular the rates of bioluminescence, over a period of ~4 months. After the experiences of the MIL and PSL lines, some significant changes were made to the detector design and a line incorporating these changes, the Mini Instrumentation Line with Optical Modules (MILOM) was deployed in the sea on 18 March 2005 with the connection on 12 April 2005. The successful operation of this line for several months in 2005 is described in Aguilar et al. (2006). The final detector lines were deployed and connected between February 2006 and May 2008. The lines were deployed using the ship CASTOR and connected on the seabed at different dates using the submarine Nautile or the Remote Operated Vehicle (ROV) Victor of IFREMER. The first detection line was deployed on 14 February 2006 and connected two weeks later. A second detection line was put into operation in October of the same year. A further three lines were connected in January 2007, with another five connected in December 2007. The apparatus reached its complete configuration with the last two lines being deployed in early 2008 and connected on 29 May 2008. The final layout of the line anchors on the seabed is illustrated in Figure 15.2.

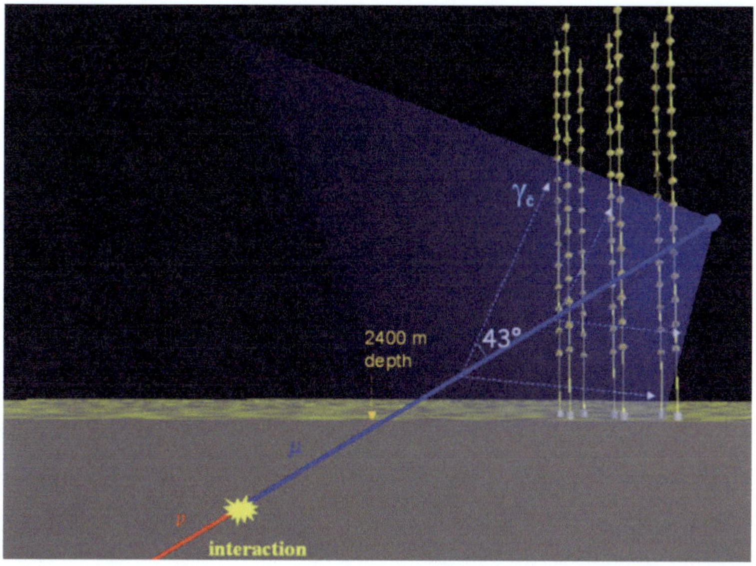

Figure 15.3 Principle of detection of high energy neutrinos in an underwater neutrino telescope. In this illustration, the neutrino makes an interaction in the rock yielding a muon with a direction close to that of the neutrino which emerges from the rock and then gives Cherenkov light in the sea water. This light is subsequently detected by the array of light sensors of the neutrino telescope leading to a measurement of the muon direction and consequently the neutrino direction.

15.3.2 Design of the neutrino telescope

A deep sea neutrino telescope detects neutrinos by observing the Cherenkov light produced by muons crossing the detector. These muons originate from rare interactions of neutrinos in the sea water around the detector or in the rock below it. A matrix of light detectors, in the form of photomultipliers in glass spheres – optical modules (OM) (Aguilar et al., 2006) – is deployed near the seabed. The trajectory of the muon track is reconstructed from the detection timing of the Cherenkov photons as well as from the positions of the photodetectors. The signal from Cherenkov photons occurs within a time window of few microseconds, which helps to distinguish it from various background light sources which exhibit correlations only on much larger time scales. An indirect search for neutrinos can then be performed by selecting the upward-going muons produced by neutrinos which have passed through the entire planet and interacted in the vicinity of the detector. The direction of the incoming neutrino, being almost collinear with the secondary muon, can be determined with an accuracy reaching 0.2° for high energy neutrinos above 10 TeV. Due to its size and the spacing of the photomultipliers, the ANTARES detector has a low energy threshold of ~20 GeV for reconstructed neutrinos. The "effective area" (equivalent area for a telescope with 100% neutrino detection efficiency) increases strongly with the neutrino energy and reaches ~1m2 for PeV energy neutrinos. Figure 15.3 illustrates the principle of neutrino detection with the undersea telescope.

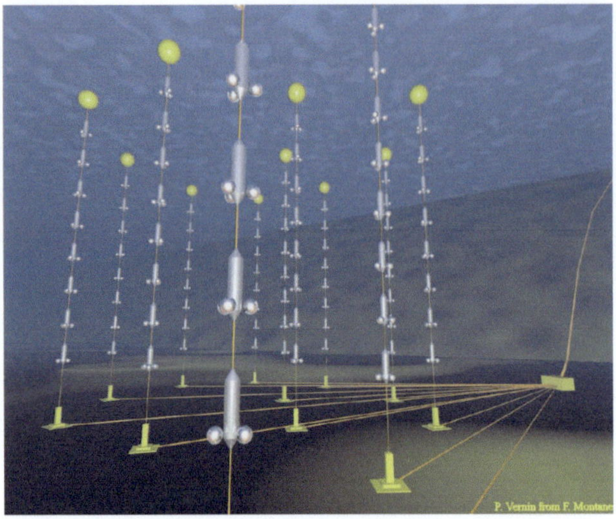

Figure 15.4 Artist's impression of the neutrino telescope, showing the detector lines, the seabed interlink cables, the junction box and the MEOC cable to the shore. For clarity, the number of storeys per line is reduced and items are not drawn to scale.

The Neutrino Telescope consists of 12 mooring lines, each with a total height of ~480m. All lines have a similar design being weighted to the seabed and held nearly vertical by syntactic-foam buoys at the top. An artist's impression of the layout of the neutrino telescope is shown in Figure 15.4, while Figure 15.5 illustrates the components within a typical neutrino telescope line. Each line has a total of 75 photodetectors, grouped in triplets. The seabed at the site is at a depth of 2475m and the OMs are positioned at depths between ~2400m and ~2000m.

The buoy at the top of the line is floating freely and so each line moves under the effect of the sea current, with movements being a few meters at the top for a typical sea current of 5cm s^{-1}. The positions of the OMs are measured with a system of acoustic transponders and receivers at discrete positions on the line and on the seabed, together with tilt meters and compasses on each storey of the line. The positioning system gives a real-time measurement, typically once every two minutes, of the position of every OM with a precision better than 10cm in space.

The default readout mode of ANTARES is the transmission, for each optical module, of the time and amplitude of any photomultiplier (PMT) signal above a threshold of 1/3 of the electronic signal corresponding to one incident photon. Time measurements are referenced to a master reference clock signal distributed to each storey from shore. All signals are sent to shore and treated in a computer cluster of several hundred cores to find photon hit patterns in the PMTs corresponding to muon tracks or other physics events producing light in the water. The grouping of three optical modules in a storey allows local coincidences to

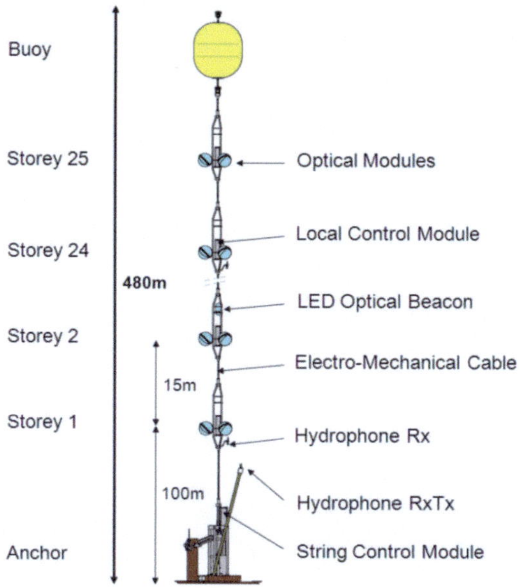

Figure 15.5 A line of the neutrino telescope of the ANTARES detector. For clarity only 4 of the 25 storeys are shown and the lengths of cable between storeys are not to scale.

be made for pattern finding. The front end electronics also allows the sampling of the full waveform of the signal with 128 samples separated by ~2ns, enabling special calibration studies of the electronics.

The readout architecture of the detector has several levels of multiplexing of the photomultiplier signals. The first level is in the Local Control Module (LCM), an electronics container in each storey of the detector where the analogue electrical outputs of the PMTs are digitized in a custom-built ASIC chip, the Analogue Ring Sampler (ARS) before being treated by a data acquisition card, containing an FPGA and microprocessor, which outputs the multiplexed signals of the three local optical modules on an Ethernet optical link. The links from five storeys, forming a Sector, are combined in an Ethernet switch in the Master Local Control Module (MLCM) at every fifth storey and the combined link output is sent on a particular wavelength to a dense wavelength division multiplexing system in an electronics container, the String Control Module (SCM), at the bottom of each line. In the SCM the outputs from the five MLCMs along the line are multiplexed on to one pair of optical fibers. These fibers are then connected to the junction box on the seabed via interlink cables. In the junction box the outputs from up to 16 lines are gathered onto a 48-fiber electro-optical submarine cable, the Main Electro-Optical Cable (MEOC), and sent to the experiment's shore station in the town of La Seyne-sur-Mer. The optical links between each MLCM and the shore station are connected using only passive components. The shore

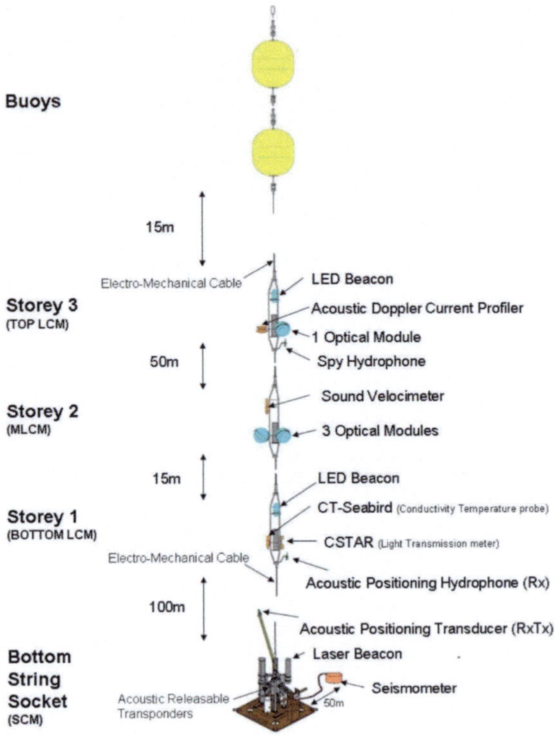

Figure 15.6 The first instrumentation line "MILOM". The elements are not drawn to scale.

station houses a computer cluster for the onshore event filtering (Aguilar et al., 2007) and relevant data are sent via a network data link to be recorded at a computer center in Lyon.

The electrical supply system has a similar architecture to the readout system. The submarine MEOC supplies ~4400 V, 10 A AC to a transformer in the junction box. The 16 independent secondary outputs from the transformer provide ~500 V, 4 A AC to the lines via the interlink cables. At the base of each line, a String Power Module (SPM) shares the same container as the SCM. The SPM distributes 380 V DC to the MLCMs and LCMs in the line, each of which contains a Local Power Box (LPB) to provide the various low voltages required by the electronics cards in each LCM.

15.3.3 Instrumentation line

The 13th line of the ANTARES detector is an instrumentation line which has functions related to calibration of the neutrino telescope but contains several instruments for marine and Earth science. The instrumentation line has evolved in time from the MILOM, which was operational from March 2005 to June 2007, to the Instrumentation Line of 2007 (IL07) which has been operational since December 2007.

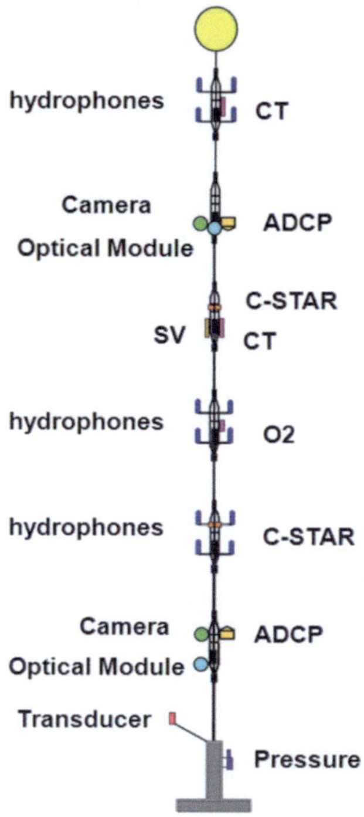

Figure 15.7 The instrumentation line "IL07". Elements are indicated schematically; not drawn to scale.

Operational for 18 months, the MILOM, which is shown schematically in Figure 15.6, was the first ANTARES complete instrumentation line. The line structure consisted of an instrumented releasable anchor (BSS for Bottom String Socket) and three storeys located respectively at 100m, 117m and 169m above the seabed. Two buoys, located at the top, maintained the line almost vertical. The MILOM was equipped with four optical modules: a triplet of OMs on the second storey and a single additional OM fixed on the top storey. Three intense light sources were located on the line being used mainly for OM timing calibration: the Laser Beacon located on the BSS and two LED Optical Beacons attached to the bottom and top storey, respectively. In order to allow the reconstruction of the line shape geometry, the MILOM is equipped with biaxial tiltmeters and compasses located in the electronics container of every storey, and with two acoustic positioning modules: an emission/reception (RxTx) module with its transducer on the BSS and a reception (Rx) module with its hydrophone on the bottom storey.

In addition to these instruments mainly intended for calibration of the neutrino telescope, the MILOM hosted various environmental devices: an acoustic doppler current profiler (ADCP) to monitor the intensity and direction of the underwater flow; a sound velocimeter to record the local value of the sound velocity; a CT probe to measure the conductivity and temperature of the sea water; a transmission meter to monitor the light attenuation of the water; a Spy Hydrophone to record the acoustic activity from the positioning beacons, surface or biological noise; and a broadband seismometer installed in the seabed sediment 50m away from the MILOM anchor to monitor the seismic activity at the site.

The electronics layout, power supplies and readout architecture of the MILOM line was the same as that of the neutrino telescope lines described in the previous section.

In June 2007 the MILOM instrumentation line was recovered and upgraded to the IL07 instrumentation line which was put into operation in December 2007. This new line has six storeys compared to three in the MILOM, so providing space for many other instruments. A significant addition in the new line is the acoustic detection system, described in a following section, consisting of 18 acoustic sensors on the IL07 and the same number on the top of Line 12 of ANTARES. Other important additions are a dissolved oxygen sensor, widely used by physical oceanographers to characterize mixing and ventilation of water masses, and two cameras, continuously connected in order to record images of bioluminescent organisms.

The mechanical structure and general architecture of the IL07 line is similar to that of the MILOM and the neutrino telescope lines. A schematic layout of the instruments on the line is shown in Figure 15.7 and details of the instruments are given in Table 15.1.

15.3.3.1 ADCP

Acoustic Doppler Current Profilers (ADCPs) measure, without moving parts, currents in a large water column in front of the device using ultrasounds waves reflected by particles in suspension which move with the current. These particles reflect the sound with a frequency shifted by the Doppler effect according to their velocity, which is the same as that of the water mass.

The instrument chosen is the Workhorse Monitor ADCP from RD Instruments[10] working at 300 kHz. The use of 300 kHz has the advantage to reduce the "noise" in the ADCP due to the acoustic system used in ANTARES for the detector positioning and vice versa. In the deep sea the low density of particle limits the precision of the measurements. For this reason, the IL07 line has two devices mounted at different levels, one directed downwards and the other directed upwards, in such a way as to monitor the maximum distance of water column.

The Workhorse Monitor ADCP can measure water current velocities in the range 0–5m s^{-1} with a resolution that depends on settings: a measurement range of 150m is achieved, with cells 4m in height and with 5mm s^{-1} resolution on the velocity measured in each cell. The direction of the water velocity is determined with a precision of half a degree. The instrument communicates via a serial port (RS-232) at a baud rate in the range 1200–115,400 (depending on the connection cable length). It is powered with 48V DC voltage; power consumption goes from ~50 mW in standby to 115 W during operation (acoustic signal transmission). The ADCP setup used for the data recorded on the IL07 is 50 cells of 250cm depth with measurements typically taken each minute.

10 Teledyne RD Instruments, http://www.rdinstruments.com/monitor.html

Storey	Height above seabed	Device type	Manufacturer	Model	Measured parameters
6	305m	6 hydrophones	HTI	HTI-90-U	sound level, transients
		CTD	SEABIRD	SBE 37-SMP	conductivity, temperature
5	290m	Optical Module	ANTARES	custom	light level
		ADCP	Teledyne RD	Workhorse	sea current
		Camera	AXIS	AXIS221	images
4	210m	C-Star	WETLABS		water transparency
		SV	GENISEA/ECA	QUUX-3A(A)	sound velocity
		CT	SEABIRD	SBE SI	conductivity, temperature
3	195m	6 hydrophones	Erlangen	custom	sound level, transients
		O_2 probe	AANDERAA	Optode 3830	oxygen level
2	180m	6 hydrophones	HTI	HTI-90-U	sound level, transients
		C-Star	WETLABS		water transparency
1	100m	Optical Module	ANTARES	custom	light level
		ADCP	Teledyne RD	Workhorse	sea current
		Camera	AXIS	AXIS221	images
BSS	0	Pressure sensor	GENISEA/ECA		pressure
		Transponder	IXSEA	RT661B2T	acoustic positioning

Table 15.1 Details of the instruments on the line IL07.

15.3.3.2 CTD

The line has two different models of Conductivity-Temperature-Depth (CTD) probes from Sea-Bird Electronics[11]; the SBE 37-SMP and the SBE SI. Both instruments have high accuracy with conductivity resolution10^{-4} S/m for a measurement range 0–7 S/m and temperature resolution 10^{-4}°C for a measurement range –5 to +35°C. The construction uses

11 Sea Bird Electronics, http://www.seabird.com/products/profilers.htm

titanium and other non-corroding materials to ensure long lifetime with minimum maintenance, making the devices perfect for long duration, fixed site deployment. Conductivity is measured in an "internal-field conductivity cell" that, according to SBE, is unaffected by external fouling and permits the use of anti-fouling devices to inhibit internal fouling. Temperature and conductivity measurements are very stable in time (DC ~ 3 10^{-4} S/m per month, DT ~ 2 10^{-4}°C per month). The devices are powered with 12V DC.

For the CT on IL07 – storey 4, the sampling rate is typically of one measurement each hour, while for the CTD on IL07 – storey 6 and the CTD connected on Line 4 – storey 24, the sampling rate is typically of one measurement every 2 minutes.

15.3.3.3 Sound velocimeter

The IL07 line uses the same model of sound velocity probes as the neutrino telescope lines which are principally intended for the calibration of the acoustic positioning system. The devices are model QUUX-3A (A) manufactured by the GENISEA Company[12]. This device has a titanium case rated for 6000m depths and is suitable for long-term resistance to corrosion. It performs underwater measurements of sound speed velocity in the range 1400–1600m s^{-1} with an accuracy of 0.1m s^{-1}.

All the sound velocimeters have a sampling rate typically of one measurement every 2 minutes.

15.3.3.4 Water transmission

Water optical properties are monitored by measuring the water attenuation coefficient with two C-STAR instruments built by WetLabs[13]. The light attenuation coefficient c is calculated by the transfer equation: T = e-cx where x is the path length (25cm) of the water volume measured.

In the C-STAR, an LED light source provides light that is focused and collimated by an aperture and lens that transmit the light within a given narrow bandwidth. The light passes through a beam splitter where part of the light is monitored by a reference detector and used in a feedback circuit to take into account the variations in the LED source in time or the variation in the instrument's internal temperature. The remaining light enters the sample volume after passing through a first pressure window, transits the sample volume and enters the receiver part after passing through another pressure window. The light passes through additional focusing optics to be finally measured by a silicon photodiode detector which converts the amount of received light to a corresponding 0–5 V analogue output signal proportional to the amount of light received.

The instruments used in IL07 have monolithic housings in hard anodized aluminium, rated for 6000m depth. The devices are mounted horizontally with both pressure windows protected from biofouling by the housings. The overall dimensions of the devices are 47 × 6.4 × 9.3cm.

Data from the C-STAR have been available from April 2009 and have a sampling rate of one measurement every 10 minutes.

---

12 GENISEA/ECA, http://pagesperso-orange.fr/genisea/sound_velocimeter.htm.

13 WET Labs, http://www.wetlabs.com/products/cstar/cstar.htm.

15.3.3.5 Oxygen monitor

The oxygen monitor on IL07 is an Oxygen Optode model 3830 from AANDERAA[14] which is an optical sensor based on dynamic fluorescence quenching. In this device, a specially-designed chemical complex is illuminated with a blue LED and emits in return a red luminescent light with a lifetime that directly depends on the oxygen concentration of the medium.

In the Optode the lifetime measurement is made by a phase shift detection of the returning red luminescence. The foil is excited with blue-green light modulated at 5 kHz. The decay time is a direct function of the phase of the received red light; it is used directly for oxygen detection, without calculating the decay time.

Unlike electrochemical oxygen measurements, Optode sensors do not consume or remove oxygen from the water during the measurement. So the measurement does not affect the oxygen concentration of the medium. The oxygen concentration response of this sensor is exponential, yielding highest sensitivity at low concentrations.

The manufacturer's published performance specifications for the Aanderaa Optodes are: range 0–500μM (concentration) and 0–120% (air saturations); resolution <1μM (concentration) and 0.4% (air saturation); accuracy <8μM or 5% (whichever is greater, concentration) and <5% (air saturation); response time <25 seconds.

The sampling rate of the Optode on IL07 storey 3 is typically one measurement every 4 minutes.

15.3.3.6 Camera

Since 1996, many studies have been done to estimate the optical background level on the ANTARES site and more specifically the bioluminescence contribution (Amram et al., 2000). The early measurements found the background rates to be fully satisfactory for the detector operation; however, during the first year of running the detector, at the beginning of 2006, an unexpected bioluminescence activity was seen by the Optical Modules of the ANTARES detector. This phenomenon is known to be episodic; however, the biological sciences community began to take a great interest in it, since deep waters are expected to be relatively poor in living organisms. As a result, it was decided to connect two cameras on the IL07 line in order to record some living bioluminescent organisms. This system is expected to be able to detect the organisms which lead to the bursts of bioluminescence seen in Figure 15.25 but not the bioluminescence contribution to the baseline.

The installation uses very different concepts to observe deep sea bioluminescence compared to conventional methods using submarines or landers which have a rather short observation window. The following features characterize the camera installation on the IL07:

- integrated image treatment
- self-triggering
- permanent 100 Mbit/sec link to the shore
- more than 2000m below sea surface.

---

14 AANDERAA, http://www.aadi.no/Aanderaa/Products/Sensors/OxygenOptode/default.aspx.

Figure 15.8 Storey 5 of IL07 during deployment, an ADCP (yellow) photomulptiplier (foreground) and a sphere with the camera (background) are visible.

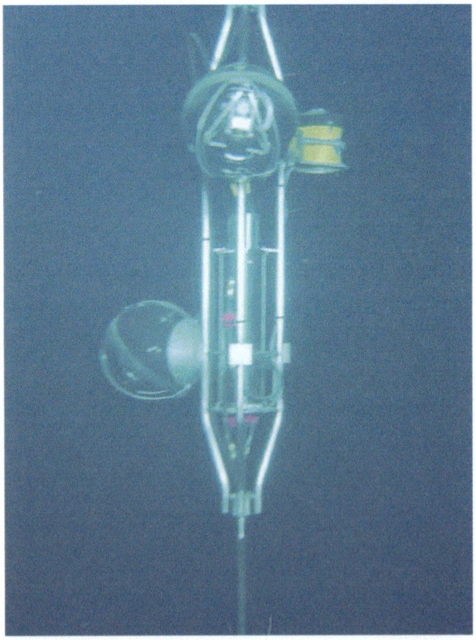

Figure 15.9 Storey 1 after deployment. The camera is installed in the top sphere. (Image © IFREMER)

This combination makes these cameras unique in the world. The image treatment and self-triggering enables the cameras to activate themselves on the faintest bioluminescence flashes by avoiding storing mostly uniformly-black images. The permanent link ensures access to the camera and uploading of images and movie sequences in real time. No other system with similar features has been installed at such a depth so far.

The cameras are standard video surveillance devices (model AXIS 221), designed for low light levels. Their sensitivity is found to be better than 10^{-5} lux. Many parameters of the cameras can be tuned remotely, such as the exposure time and the trigger level. Due to their large viewing angle (90 degrees) they cover several cubic meters of water around them. Currently, both cameras operate with an exposure time of only 0.1 seconds, sufficiently short to register movies.

The cameras are protected from sea water and pressure by a glass sphere as used also for the ANTARES photomultipliers. Together which each camera an optical module with an active photomultiplier has been installed to allow correlating image sequences with photo-multiplier signals. The self-triggering system controls an adjacent light source designed to illuminate the scene after an interesting event triggers the camera. Due to the vicinity of the neutrino telescope which is sensitive to very low light levels only infrared light sources could be used in the present installation.

The acoustic detection devices are explained in Sections 15.3.5 and 15.3.6.

15.3.3.7 Oceanographic instruments on neutrino telescope lines

As well as the instruments on the IL07 line, there are several other sensors installed on the standard neutrino telescope lines: a seismograph (Guralp) on line 12; an oxygen analyser (IODA) on line 12; a current meter Aquadopp from Nortek[15] on line 5 and a SV-CTD sensor (from GENISEA and FSI) on line 4. All these devices are located within 100m of IL07.

15.3.3.8 Mounting of instruments on lines

All instruments mounted on the instrumentation line are fixed to the same mechanical structure designed for optical module mounting. Figure 15.10 shows storey 1 of the MILOM line holding a pressure sensor and a C-Star and Figure 15.11 shows storey 3 of the same line holding the ADCP and an optical module. The instruments on the IL07 line are mounted in the same way and Figure 15.19 gives a sketch of the mounting frame holding hydrophones.

15.3.4 Other instruments in a deep-sea observatory

15.3.4.1 Deep-IODA$_{6000}$

Quantifying organic matter re-mineralization along the whole water column is essential for assessing the role of the micro-organisms in the global carbon cycle in relation to global change. The AOU, derived from in situ temperature, salinity and oxygen measurements, provides an integrated time scale of months to years for oxygen consumption; however, to

15 NORTEK http://www.nortek-as.com/products/CurrentMeter/aquadopp-3d-current-meter.

Figure 15.10 Storey 1 of MILOM line showing a CTD, a C-Star and an optical beacon.

Figure 15.11 Storey 3 of MILOM line showing and ADCP, an optical module, a flotation sphere and an optical beacon.

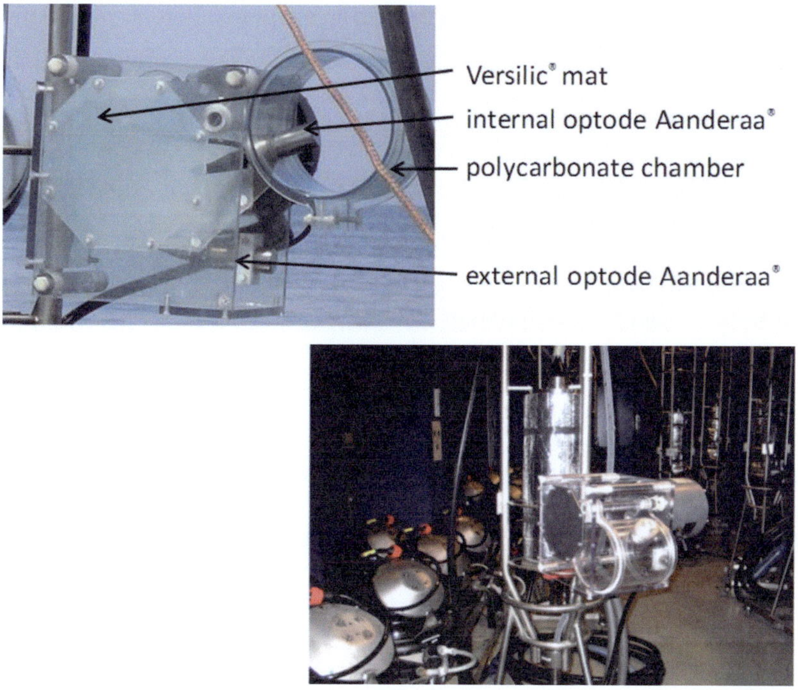

Figure 15.12 (Top) Image of $IODA_{6000}$ before deployment indicating the main components. (Below) Image of $IODA_{6000}$ during assembly on line 12.

constrain the biological activity it is necessary to assess different time scales using complementary approaches for oxygen dynamics.

For this purpose, a new instrument has been developed, the in situ Oxygen Dynamic Auto-sampler ($IODA_{6000}$)[16]. Shown in Figure 15.12, the $IODA_{6000}$ consists of an equipressure system which aims to measure the oxygen concentration and the oxygen dynamics in shallow or deep waters, up to 6000m depth. $IODA_{6000}$ consists of a 5L-chamber in polycarbonate equipped with an internal Aanderaa Optode that samples the seawater by a slow rotation. The seawater sample is enclosed between two Versilic® mats during a period of time (between 12 hours and 5 days according to the biological activity) in order to measure the oxygen dynamics (consumption in the deep waters; consumption and production in surface waters). All the mechanical and electronics parts are mounted in silicon oil so as to be in equipressure with the external environment. An external Aanderaa Optode is also fitted on the $IODA_{6000}$ in order to assess to the time integrated biological activity. A prototype version ($IODA_{6000}$ v4.0) has been deployed with the ANTARES line 12 in the framework

16 K. Arnaud, S. Beurthey, M. Billault, D. Lefèvre, P. Payre, A. Robert, C. Tamburini, APO ANR POTES, 05-BLAN-0161-01.

of the FP7 EuroSITES project which aims to form an integrated European network of nine European deep-ocean (>1000m) observatories.

15.3.4.2 Seismometer

Submarine slope instabilities are well known all around the Mediterranean Sea. The ANTARES site, located south of the Maures Massif, is a relatively unexposed area but active sediment transfer systems are found on either side of the location, westward on the Ligurian Margin (e.g., Var river delta) and East of Toulon, in a zone comprising the Rhone delta. Large volumes of sediments have been destabilized in the past and triggered tsunamis. For instance, a landslide occurred in 1979 during the construction of a jetty offshore of Nice Airport, causing several deaths at the construction site and on the Antibes Cape from the tsunami which followed. Among other causes, strong motion produced by seismic events can trigger such slope destabilisation. Seismic activity (with offshore earthquakes of magnitude >6 along the Ligurian margin and in the Ligurian oceanic basin) is the main component included in the seismic risk assessment along the coast of south-east France. It may, at least in part, be explained by the strong variation of crust thickness from the alpine domain to the Ligurian basin, seen at the Earth's surface as a strong topographic gradient between the Alps (3000m high) and the Ligurian oceanic basin (3000m depth). Despite offshore earthquakes being poorly located, they lie near the base of the continental slope, possibly reactivating ancient faults. As well as earthquakes, unstable slopes close to shore represent a serious geohazard.

The Ligurian Site (comprising the ANTARES Neutrino telescope site, the continental slope offshore of Nice and the area in between) was designated as a test site for new instruments and technologies for measurement at the seafloor and data transmission. Test deployments already performed found applications at other sites of the ESONET European network[17]. The seafloor monitoring stations (with seismometers and accelerometers) deployed by KOERI[18] in the Sea of Marmara in 2009 were developed from collaboration between Guralp manufacturer and GeoAzur laboratory and tested at the ANTARES site.

The ANTARES infrastructure is indeed a unique opportunity for long-term tests of new sea-bottom geophysical sensors. The instrument deployed on the ANTARES site, installed and connected to the MILOM during the operation of this first instrumentation line, and currently to the detector Line 12, is a broadband seismometer developed by Guralp (CMG 3T) offering a real-time transmission of the three components of the ground velocity in the bandwidth from 0.0028 to 50 Hz. The measurements are complemented with those of a differential pressure sensor in order to reduce the noise induced by the motion of the sea water mass. Figure 15.13 shows the installation of the Guralp seismometer on the ANTARES site by IFREMER's ROV Victor prior to its burying into the seabed sediment. Figure 15.14 illustrates an example of a seismic event (earthquake in Mozambique) recorded by the sensor on the ANTARES site in February 2006.

17 ESONET, http://www.esonet-emso.org/

18 KOERI, Kandilli Observatory and Earthquake Research Unit, Bogazici University, www.koeri.boun.edu.tr/

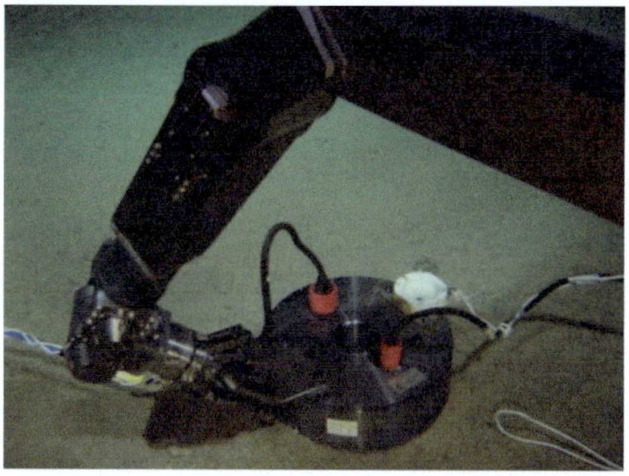

Figure 15.13 Installation of the Guralp broadband seismometer on the ANTARES site by IFREMER's ROV Victor (Image ©IFREMER).

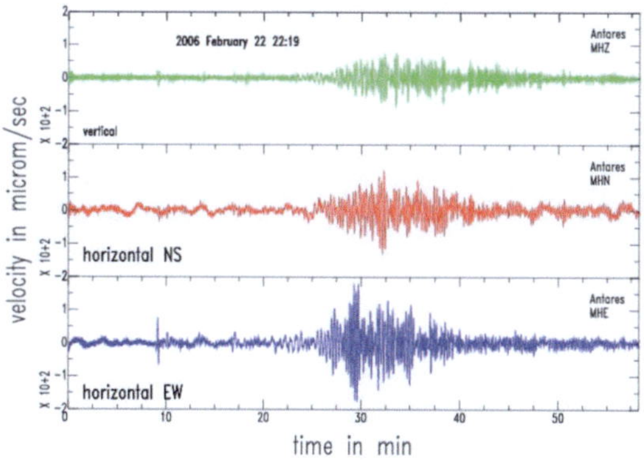

Figure 15.14 Example of a seismic event (earthquake in Mozambique, Mw=7.5) recorded by the Guralp broadband seismograph on the ANTARES in February 2006. The noise is much lower on the vertical component, but the P, S and surface waves can be identified.

15.3.5 Acoustic positioning system

The reconstruction of muon tracks in the neutrino detector is based on precise measurements of the arrival times of Cherenkov photons at the optical modules and requires knowledge of their positions with a precision corresponding to the dimensions of the PMT. The positions of the OMs are obtained with a High Frequency Long Base Line acoustic positioning system which is supplemented with tiltmeter and compass sensors on each storey of the lines. The shape of each line is reconstructed by performing a global fit based on the physical properties of the line and all the available information from the sensors. The

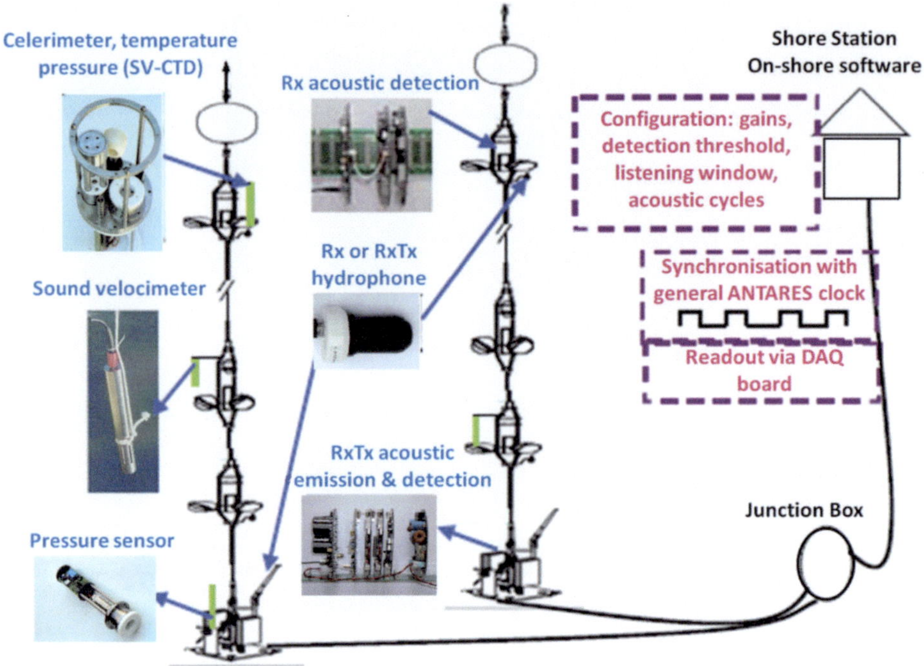

Figure 15.15 Schematic of the elements in the acoustic positioning system.

relative positions of the OMs are then deduced from this reconstructed line shape and from the geometry of the storey holding the three optical modules.

The development and construction of the acoustic positioning system was performed by the company GENISEA/ECA. The system uses measurements of the travel time of 40 to 60 kHz acoustic pulses between receiving hydrophones placed at different heights on the detector lines and emitters fixed at the bottom of each line. Due to the slow movements of the lines in the water flow, the receiving hydrophones can be considered as fixed points on a time scale of one minute during which time different measurements are made throughout the detector. The fixed acoustic beacons are able to receive an acoustic signal in order to determine their own spatial positioning with respect to the other line anchors. In the full system, many individual emitter-hydrophone distances are measured and the position of each hydrophone is obtained by triangulation using all available distances.

Figure 15.15 shows the different parts of the relative positioning system. Each ANTARES detector line is equipped with an acoustic emitter-receiver (RxTx module) fixed on its anchor and receiving hydrophones (Rx modules) fixed on few storeys. There are five Rx modules per line; one is placed on the bottom storey and one on the top storey. The others are distributed in order to obtain a larger density of hydrophones in the top third of the line, where the maximum displacement from a vertical line shape is expected. The RxTx mod-

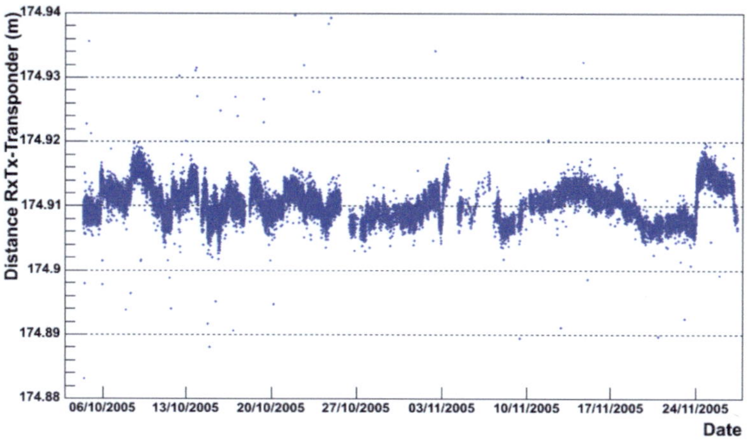

Figure 15.16 Distance measured by the acoustic positioning system between the RxTx module and the autonomous transponder, both being fixed a few meters above the seabed. The measurement is displayed as a function of time, during two months of operation.

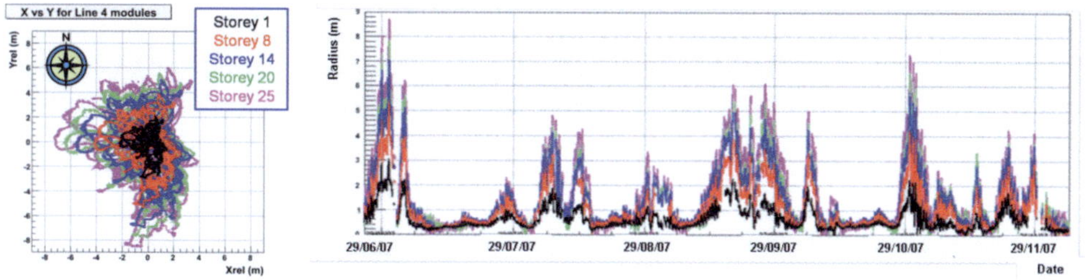

Figure 15.17 (Left) Displacement in the horizontal plane for the hydrophones on five storeys in line 1. (Right) Time dependence of the radial displacement of the hydrophones.

ule is composed of a transducer (emitting and receiving hydrophone) placed at the top of a pole on the line anchor and six electronic boards integrated in the SCM container. It emits the acoustic signals in emission mode and acts as an Rx module in reception mode. An Rx module is composed of a receiving hydrophone placed on the storey and three electronic boards in the LCM container.

The positioning of the whole detector is performed by successively sending acoustic sinusoidal wave packets of a few ms duration (typically 2 ms) from the RxTx module of each line. Several discrete acoustic frequencies between 40 and 60 kHz are used in turn to better differentiate the different acoustic emissions and to avoid interference resulting from fast successive emission of acoustic waves with the same frequency. Detection of the arrival time of the acoustic signal on the piezoelectric element of each receiving hydrophone (Rx module) is done by comparing the threshold amplitude to the numerically filtered signal . The gain of the preamplification as well as the detection threshold is set for each receiver depending on the emission cycle due to the emission frequency and to sound attenuation with propagation on variable distances. The system gives the sound time travel between the emitter and the receiver for each couple emitter-receiver. Knowing the sound velocity profile, the distance between emitter and receiver is deduced from the travel time. Positions of all hydrophones and transducers are then computed from distances using triangulation principles based on a least-mean square minimization.

An example of the precision of the system is given in Figure 15.16 which displays the acoustic distance measured between the RxTx transducer attached on the MILOM anchor and the autonomous transponder. The acoustic system measurements show a resolution of a few mm and a stability of about 1cm over a distance of 174.91m, during two months of operation.

Figure 15.17 gives examples of the line movements reconstructed after the full triangulation procedure. The first plot shows the x–y displacement in the horizontal plane of the five hydrophones at different heights along a line, and the second the absolute value of this displacement as a function of time from July to December 2007.

15.3.6 Acoustic detection system

ANTARES comprises an acoustic detection system which is called AMADEUS (ANTARES Modules for the Acoustic Detection Under the Sea). The system extends the ANTARES detector by a dedicated array of custom-designed acoustic sensors for broad-band recording of signals with frequencies ranging up to 125 kHz. The project was conceived in order to perform a feasibility study for a potential future large-scale acoustic detector for extremely high energy neutrinos from the cosmos. For the detection of such neutrinos with macroscopic kinetic energies exceeding 1017eV (0.02J) a promising approach is the use of acoustic pressure waves produced by the particle cascades that evolve when a neutrino occasionally interacts with a water molecule. One advantage of acoustic waves is the attenuation length of the order of 1km for the peak spectral density of the generated sound waves around 20 kHz (Fisher and Simmons, 1977). AMADEUS is excellently suited for extensive acoustic background studies, including signal correlations on several length scales as well as sourcc localisation. It can be used as a multipurpose device for studies of neutrino detection techniques, position reconstruction, and marine research.

15.3.6.1 System description

Acoustic sensing was integrated into ANTARES in the form of Acoustic Storeys which are modified versions of standard ANTARES storeys, replacing the PMTs by acoustic sensors and using custom-designed electronics for the digitization and pre-processing of the analogue signals. AMADEUS consists of six Acoustic Storeys, three situated on IL07 and on Line 12, respectively. Each Acoustic Storey comprises six acoustic sensors that are arranged at distances of roughly 1m from each other and hence the system comprises a total of 36 sensors. A sketch of the complete ANTARES detector, with the acoustic parts highlighted, is shown in Figure 15.18 and Figure 15.19 shows the design of a standard acoustic storey.

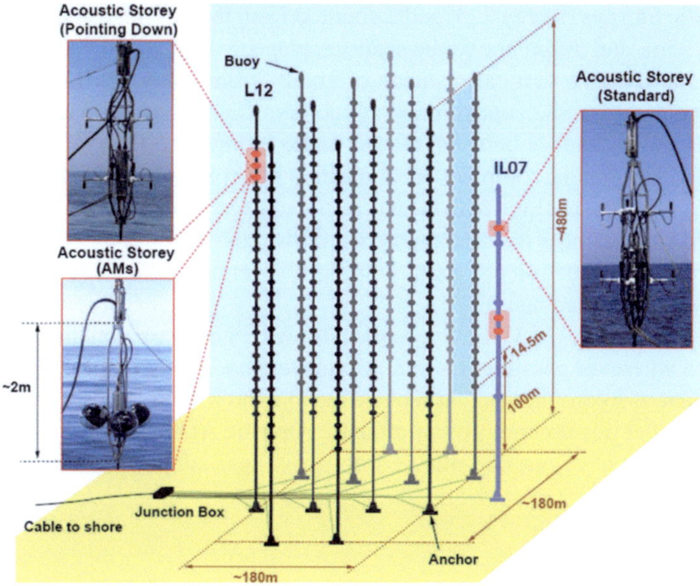

Figure 15.18 A sketch of the ANTARES detector. The six Acoustic Storeys of the AMADEUS system are highlighted and their three different setups are shown.

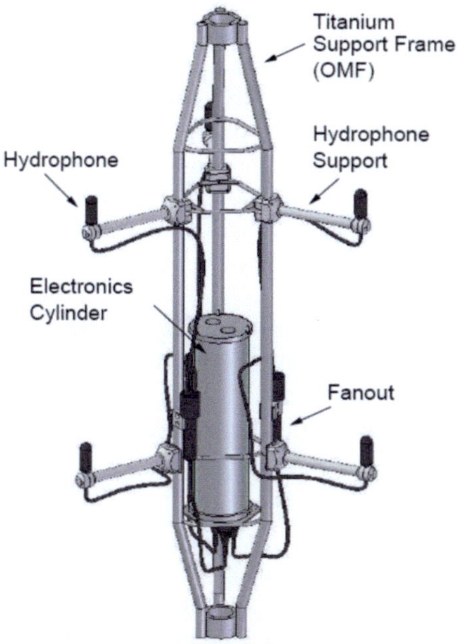

Figure 15.19 Drawing of an Acoustic Storey with hydrophones.

The Acoustic Storeys (storey 2, 3, and 6 counted from the bottom) on the IL07 are located at 180m, 195m, and 305m above the seafloor, respectively (as indicated in Table 15.1). Line 12 is anchored at a vertical distance of about 240m from the IL07, with Acoustic Storeys (storey 21, 22 and 23) positioned at heights of 390m, 405m and 420m. With this setup, the maximum distance between two Acoustic Storeys is 340m.

The AMADEUS design combines local clusters of acoustic sensors (the Acoustic Storeys) with large cluster spacing. With each acoustic cluster, suppression of random noise by requiring local coincidences and reconstruction of the arrival direction of an acoustic wave is possible. From the direction reconstruction from several individual storeys, the position of a source can then be reconstructed.

For the integration of AMADEUS into the ANTARES detector, existing hard- and software was used wherever possible. Design efforts were necessary in three basic areas: first, the development of hydrophones that replace the Optical Modules of standard ANTARES storeys; second, the development of an offshore acoustic ADC and pre-processing board; third, the development of on- and offline software. These subjects will now be discussed in more detail.

15.3.6.2 Acoustic sensors

Two types of acoustic sensors are used in AMADEUS: hydrophones and Acoustic Modules (AMs). In both cases, the sensors are based on piezo-electrical ceramics that convert pressure waves into voltage signals, which are then amplified for readout (Anton et al., 2006). For the hydrophones, the ceramics and amplifiers are coated in polymer plastics. In the case of the AMs, two piezo elements are glued to the inside of a sphere identical to those used for the Optical Modules of ANTARES. The latter non-conventional design was inspired by the idea to investigate an option for acoustic sensing that can be combined with a PMT in the same housing. All acoustic sensors are tuned to be sensitive over the whole frequency range of interest from 1 to 50 kHz with a typical sensitivity around −145 dB re. 1V/μPa or 0.05V/Pa (including preamplifier) and to have a low noise level. The three Acoustic Storeys on the IL07 house hydrophones only, whereas Storey 21 of line 12 holds AMs. In Storey 22 of line 12, the hydrophones were mounted with their cable junction, where the sensitivity is largely reduced, pointing upwards. This allows for investigations of the directionality of background from ambient noise, which is expected to come mainly from the sea surface. Three of the five storeys holding hydrophones are equipped with commercial hydrophones and the other two with hydrophones developed and produced at the Erlangen Centre for Astroparticle Physics (ECAP).

15.3.6.3 Offshore electronics and acoustic data acquisition

For the digitisation of the acoustic signals and for feeding them into the ANTARES data stream, the Acoustic Digitisation board (AcouADC board) was designed. One board processes the differential signals from two acoustic sensors, which results in a total of three such boards per storey. All other components of an electronics container are identical to the standard ANTARES components. In the analogue part of the AcouADC board, each signal is amplified in two stages by one of 12 adjustable factors between 1 and 562. An analogue filter suppresses frequencies below ~4 kHz and above ~130 kHz. The high-pass part cuts into the trailing edge of the low frequency noise of the deep-sea acoustic background (Urick, 1983) and thus protects the system from saturation. The low-pass part efficiently suppresses frequencies above the Nyquist frequency of 250 kHz for the sampling rate of 500 kSamples per second (kSps). In standard mode, the sampling rate is digitally reduced to 250 kSps, corresponding to a down sampling by a factor of 2.

The digital part of the AcouADC board digitises and processes the acoustic data. It is highly flexible due to the use of a micro controller (μC) and a field programmable gate array (FPGA) as data processor. The μC can be controlled with the onshore control software and is used to adjust settings of the analogue part and the data processing. Furthermore, the μC can be used to update the firmware of the FPGA in situ. The digitization is done by one 16-bit ADC for each of the two input channels. The digitised data is read out by the data acquisition card of the electronics container which then handles the transmission of the data to the onshore data processing servers. For the time synchronisation between Acoustic Storeys, the standard ANTARES clock board with 50 ns resolution is employed. The dynamic range achieved for the standard gain factor of 10 is from about 5 mPa to 5 Pa in peak-to-peak amplitude of an acoustics signal. The former value is given by the pressure equivalent of the sensor inherent noise over the frequency range from 1 to 100 kHz, the second marks the system saturation level for this amplification and the sensor sensitivity.

15.3.6.4 Onshore data processing

AMADEUS follows the same "all data to shore" strategy as ANTARES; the offshore data arrive via the TCP/IP protocol at a Gigabit switch in the ANTARES shore station, where the acoustic data are separated from the standard ANTARES data and routed to a dedicated acoustic computer cluster. The cluster currently consists of four servers of which two are used for data filtering. The filtering has the task of reducing the raw data rate of about 1.5TByte/day to about 10GB/day for storage. Currently, three filter schemes are implemented (Neff, 2007): a minimum bias trigger which records 10 sec of continuous data every 60 min; a threshold trigger for strong signals; and a cross correlation trigger, which searches for the expected signal of a neutrino which is bipolar in time. The filtering requires coincidences of a predefined number of acoustic sensors on each storey and can be extended to require coincidences between storeys on the same line. All parameters can be freely adjusted. It is also possible to move parts of the filter into the FPGA of the AcouADC board, thereby implementing an offshore trigger which reduces the size of the data stream sent to shore. Just like ANTARES, AMADEUS can be controlled via the Internet from essentially any place in the world. Data are centrally stored and are available remotely as well.

15.3.6.5 Ambient noise measurements

The ambient noise level in the frequency range from about 200 Hz to 50 kHz in the deep sea is assumed to be mainly determined by the agitation of the sea surface, i.e., by waves, spray and precipitation (Lurton, 2004).

To verify these assumptions, the correlation between daily averages of weather observables (especially the surface wind speed) and the RMS of the ambient noise recorded by AMADEUS was investigated (Lahmann, in press). Weather data recorded at the Cap Cepet

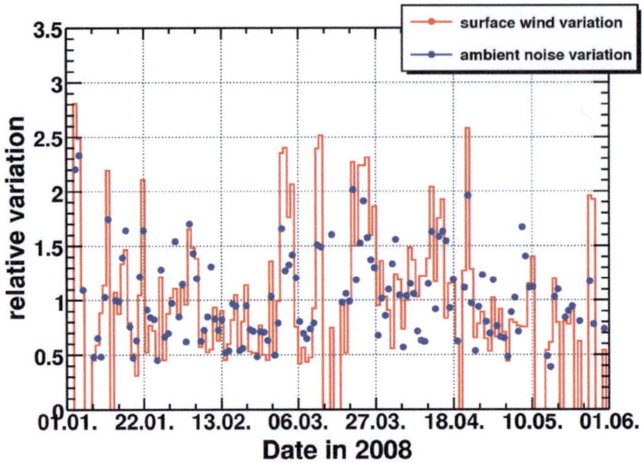

Figure 15.20 RMS noise recorded by a representative hydrophone (dots) and wind speed (solid line) as a function of time for the first half of the year 2008, showing the correlation between the two quantities. Both quantities are displayed relative to their mean over the time period, which is normalized to a value of 1.

weather station (43°5' N, 5°56' E), about 30km north of the ANTARES site, were used. The result of this study is shown in Figure 15.20. Each point in the figure represents the daily average of the RMS noise of 10s samples of data recorded every hour, calculated by integrating the power spectral density (PSD) in the frequency range from 1 to 50 kHz. The correlation coefficient between the wind speed and the recorded noise is found to be about 75% for the data taken during the year of 2008, indicating strong correlation between the two observables. The results are consistent with those reported for other deep-sea sites (Kurahashi and Gratta, 2007; NEMO Coll, 2008).

15.3.6.6 Position reconstruction of sources

Position reconstruction of acoustic point sources is done by first reconstructing their direction from individual storeys and then combining the reconstructed directions from three or more storeys. A prerequisite for this is the knowledge of the position of the Acoustic Storeys. With the six acoustic sensors of a storey, a complete reconstruction (position and three angles) of each Acoustic Storey can be done using the signals emitted by the ANTARES acoustic positioning system. To find the direction of point sources, a beam forming algorithm is used. It is designed to reconstruct plane waves, which for the geometry of a storey is a reasonable assumption for sources with distances above 100m (Richardt et al., 2009; in press). The left plot of Figure 15.21 shows an exemplary signal as recorded by the topmost Acoustic Storey of the IL07. Sensors 6 and 7, 8 and 9, 10 and 11, respectively, are attached to the same vertical tube of the storey; hydrophones with even numbers are located at the bottom (Figure 15.19). Signals arrive at the hydrophones positioned at the

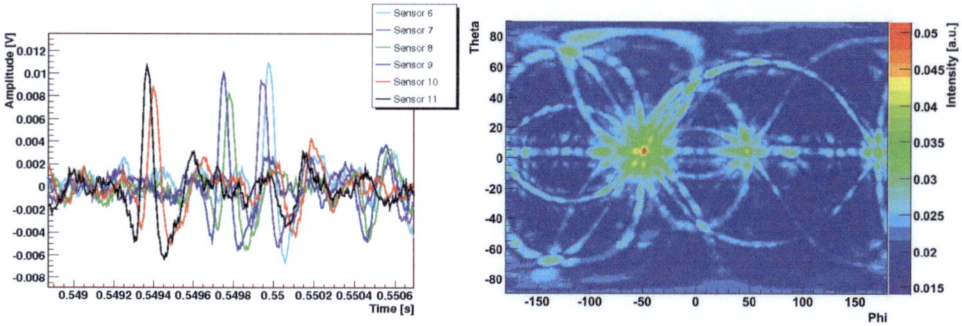

Figure 15.21 Reconstruction of the direction of an exemplary source. (Left) Signals recorded with the hydrophones of the topmost storey of the IL07. (Right) Result of the beam forming algorithm.

same rod almost at the same time, indicating that the direction of the source is close to horizontal. The right plot of Figure 15.21 shows the result of the beam forming algorithm. The intensity at a given solid angle corresponds to the probability that the source is located in that direction. Bands of increased intensity come from potential source directions where the signals from only two or three hydrophones coincide. The most probable direction, at q≈5° and f≈50° is consistent with the intuitive interpretation of the left plot.

15.4 Sample data from ANTARES detector

15.4.1 Data available from the neutrino telescope lines

The principle objective of the neutrino telescope system is to detect neutrinos from the distant cosmos. Figure 15.22 gives an example of one recorded neutrino event. The goal of the experiment is to identify distant astronomic objects such as active galactic nuclei or microquasars which are expected to emit neutrinos. These objects would then appear with a concentration of events from a particular location in the sky. The downwards orientation of the PMT in the detector is to concentrate on upward-going events where the neutrino has traversed the Earth and Figure 15.23 shows the angular distribution of reconstructed events recorded in data taken in 2007 and 2008. The upward-going events in this distribution are dominated by neutrino events with the downward-going tracks due to cosmic ray muon tracks which penetrate the sea water to the depth of the detector. The arrival direction of the upward-going events is shown in Figure 15.24 in a reference system of galactic coordinates. As yet in the ANTARES data the events detected are consistent with being produced by cosmic ray interactions in the atmosphere of the Earth.

In the neutrino detector the amount of light produced by the signal particles is tiny compared to the background light from others sources in the sea. At the depths of ANTARES

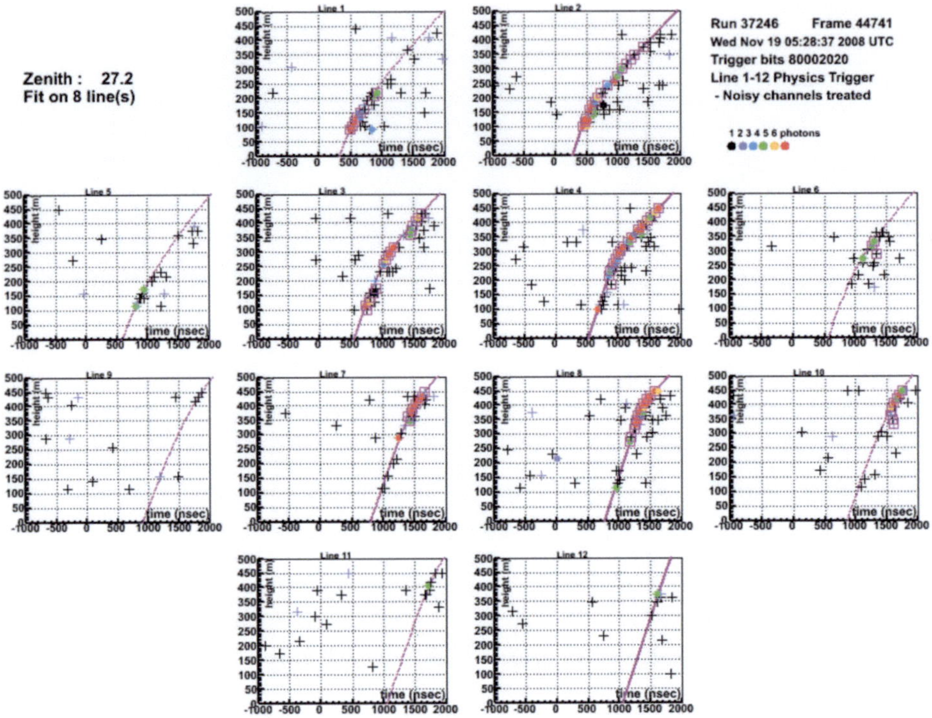

Figure 15.22 Example display of a neutrino event detected in November 2008. A series of 2D plots, one for each detector line, showing on the y axis the vertical position of the hit OM and on the x axis the time of arrival of the light at the OM. In these plots the dots correspond to detected photons and the lines the reconstructed track.

ambient light from the Sun is totally negligible and the dominant sources of light are from radioactive decay of isotopes in the sea water, principally ^{40}K, and bioluminescence from living organisms. Figure 15.25 gives a typical record of the counting rate in the three optical modules in one storey over a period of a few minutes. The constant level is mainly due to ^{40}K decay and the peaks due to impulses of light from different living species. The data show large variations with time. Figure 15.26 gives an example of time variations during the month of July 2007. This figure shows two measures of the light count rate, the baseline which is the constant level in Figure 15.25 and the burst fraction which is the fraction of time where there are peaks greater than 20% higher than the baseline. Although the baseline is relatively constant the burst fraction varies very much with time. The data in Figure 15.26 are for one particular month but the situation can be very different for other periods. It is well known that the bioluminescence is a function of agitation of the living organisms and Figure 15.27 demonstrates that the fraction of bursts is a strong function of the sea current speed.

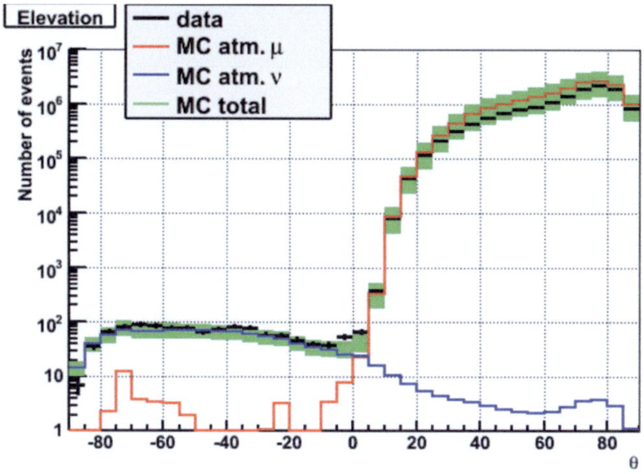

Figure 15.23 Distribution of reconstructed tracks as a function of q, where q is the angle of the track to the horizontal. Values of negative q correspond to upward-going tracks and are dominated by neutrino events. The black points are the data and the coloured histograms the results of Monte Carlo simulations as indicated in the figure legend.

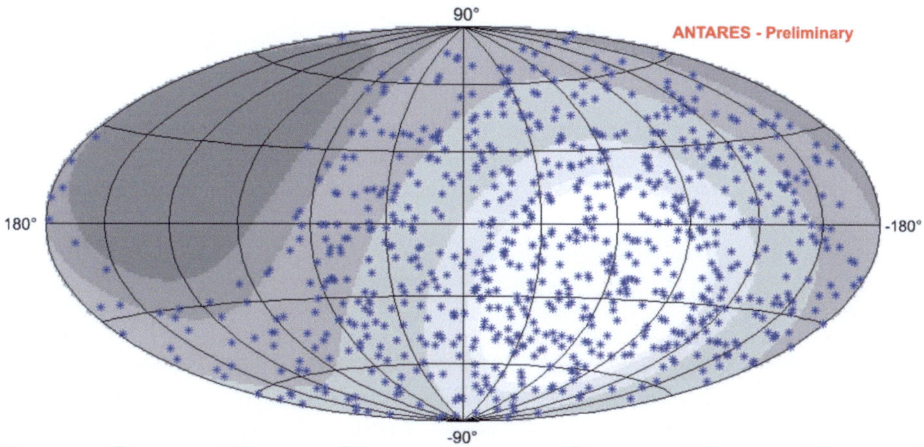

Figure 15.24 Distribution of observed neutrino events in data taken during 2007 and 2008. The coordinate system is that of the Galaxy, with the centre of the Milky Way at the origin and the disk of the galaxy along the x axis. (In this plot the measurements have a small smearing to avoid any scientific conclusions being drawn.)

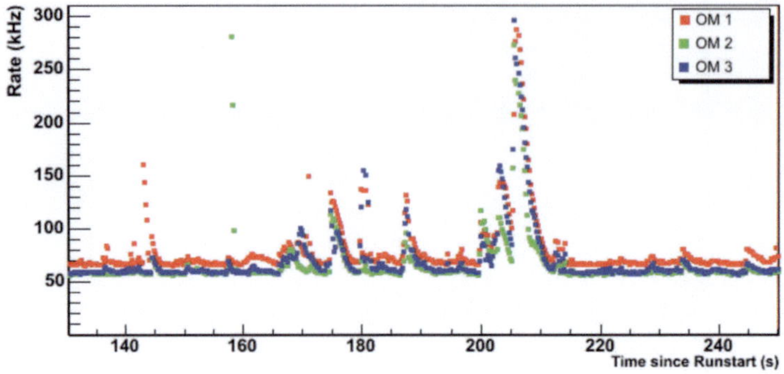

Figure 15.25 Typical example of counting rate in three optical modules of a detector line storey. The three colours indicate the individual rates of each of the OM with the small difference in the base line counting rate being due to slight differences in detector efficiencies.

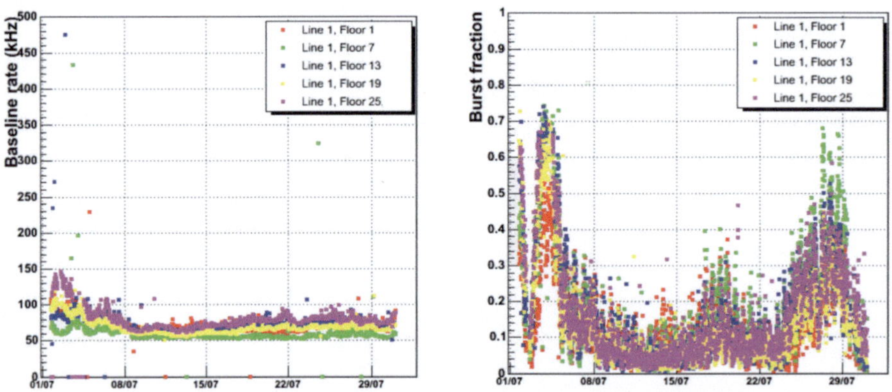

Figure 15.26 Further example of counting rate in optical modules. On the left, the average baseline rate and on the right the burst fraction during the month of July 2007. The six colors indicate the individual rates of OM on different storeys of one line, with storey 1,100m from the seabed and storey 25, 450m from the seabed.

These data on bioluminescence activity have now been recorded on the ANTARES site with measurements at depths between 2400m and 2000m for a period of four years. Figure 15.28 shows the mean counting rate, averaged between the baseline and the bursts, between May 2005 and June 2009. In 2006 and 2010, the underwater neutrino telescope ANTARES detected an unusual phenomenon: the bioluminescence of deep-sea organisms suddenly increased, revealing an unexpected connection between biological activity – bioluminescence – and the motion of water masses in the deep ocean (van Haren et al., 2011). Convective motion in the Gulf of Lion provides deep waters with oxygen and nutrients that fuel in carbon and energy the biological activity (Tamburini et al., 2013).

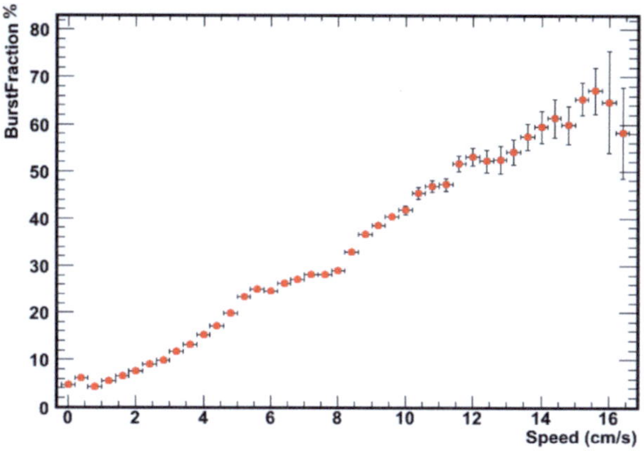

Figure 15.27 The fraction of bursts of bioluminescence as a function of sea current speed. The results are from a year of data taken from summer 2005 to summer 2006.

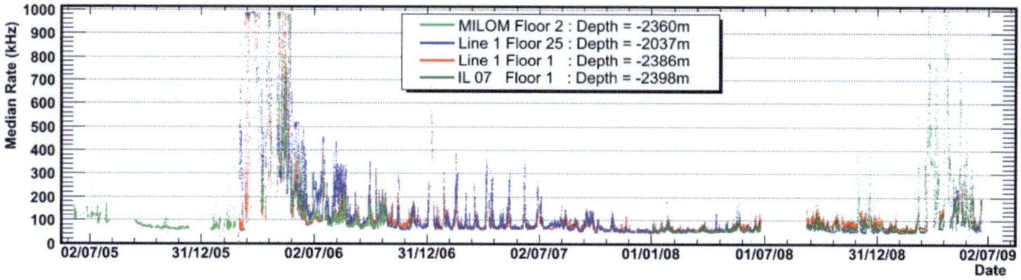

Figure 15.28 Mean counting rate in a single OM recorded on the ANTARES site from May 2005 to June 2009. The data are from different OMs in different lines according to the colour code in the legend.

In addition to the optical modules and position calibration devices, the neutrino telescope lines contain a number of instruments which provide measurements of the environmental parameters and some of these data will be included in the plots in the following section.

15.4.2 Data available from the environment instrumentation in the system

The cabled underwater observatory provides a wonderful opportunity for many sea science areas. As a result of the connection of ANTARES infrastructure to shore, all environmental data are transmitted in real time. Besides monitoring the deep-sea environment, real-time acquisition allows control of data quality and of relevant acquisition parameters settings.

Most of the instruments available on ANTARES are used by both the physicist community and by the sea science community, while others were developed specifically for oceanographic or biological research. Another key issue of the multidisciplinary observatory is the long-term and real-time data measurements: a crucial feature for marine sciences. Data acquisition parameters of oceanography devices are set from shore and so can be modified. Typical sampling rates vary from 1–30 minutes depending on the devices. Some examples of data recordings over short- and long-term periods are given as illustrations of the data available with ANTARES.

15.4.2.1 Sea water current

Water currents are of major importance in understanding ocean circulations with implications for global changes. Since March 2005, the sea current on the site has been measured almost continually with different instruments: the ADCP on the MILOM line, the two ADCPs on the IL07 line and in addition an Aquadopp device on line 5 of the neutrino telescope. ADCP current meters measure current speed and direction by transmitting high frequency sound waves and by measuring the Doppler frequency shift of the return backscattered signal. On the ANTARES infrastructure, ADCP instruments at frequencies of 300 kHz are used, giving properties of the sea water current along a distance of 150m in the water column. A first example of the sea current data is given in Figure 15.29 where the current velocity and direction are plotted for a period of one month. The inertial oscillation of the current is clearly seen with a periods of ~17 hours. The distribution of the magnitude of the sea current velocity in the same period is shown in Figure 15.30.

Figure 15.31 shows a selection of some of the current measurements during a 4-year period. A polar plot of the current directions is given on Figure 15.32 relative to the geography of the site where it can be seen that the dominant flow is east–west with occasional reverses.

15.4.2.2 Oceanographic processes in the deep western Mediterranean Sea

Besides underwater flow, an ADCP also registers variations in acoustic reflections. Typically, a 300 kHz ADCP is sensitive to such reflections off particles larger than about 0.003m. In combination with optical information from photomultipliers, the acoustic data may provide unique information on variation in deep-sea life. In the spring of 2006, the ADCP on the MILOM line measured downward vertical currents of amplitudes up to 0.03m s^{-1}. These currents were accompanied by enhanced levels of acoustic reflection by a factor of about 10 and by horizontal currents exceeding 0.35m s^{-1} (Figure 15.33). These observations coincided with saturation of the telescope's photomultipliers, which is due to high levels of bioluminescence. They follow severe winters when near-surface waters are cooled and dried, so that they sink to the bottom thereby creating vigorous vertical turbulent mixing. Subsequently, the large difference in density between the interior and coastal waters generates a topographic boundary current with associated, about 20-day periodic horizontal and vertical current variations that may transport material downwards. This demonstrates the capabilities of the ANTARES telescope to register and monitor oceanographic phenomena that have not been studied before in detail. The telescope could thus be used as a warning system for events such as intense deep dense-water formation, which occur irregularly

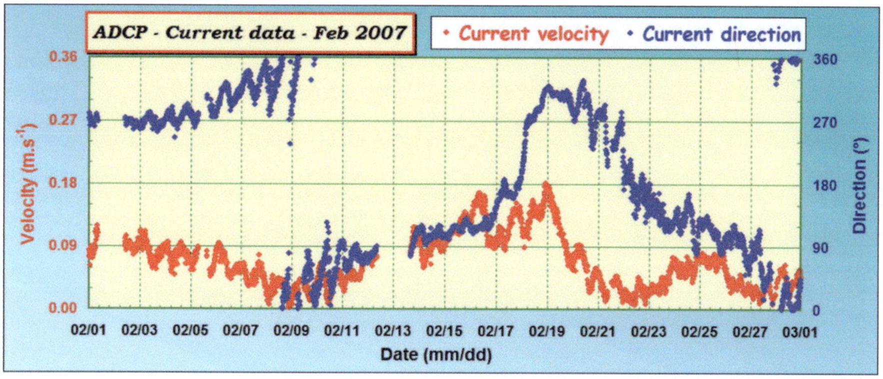

Figure 15.29 Measurements of sea current speed direction recorded with the ADCP on the MILOM line and during a period of one month.

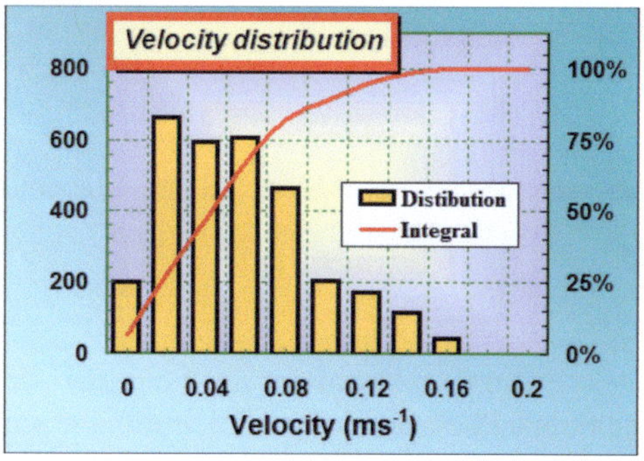

Figure 15.30 The distribution of the current velocity in the same data period as Figure 15.29. The red line represents the integrated content of the histogram and helps to conclude that, for 5% of the time, the current was below 8cm s^{-1}.

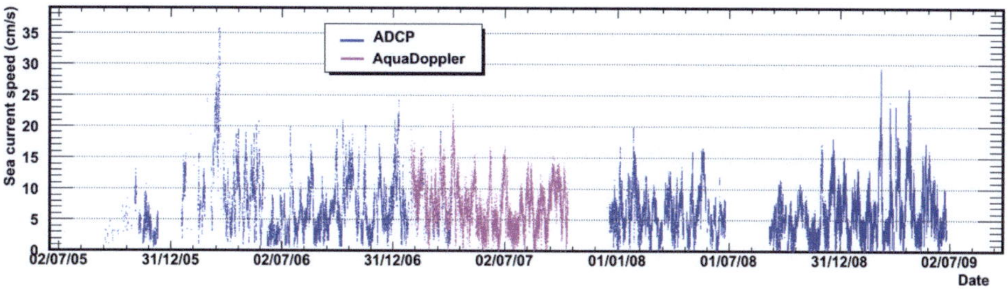

Figure 15.31 Measurements of sea current speed since March 2005. The data are mainly from the ADCPs on the instrumentation lines but between June and December 2007 are from the Aquadopp device on line 5 of the neutrino telescope.

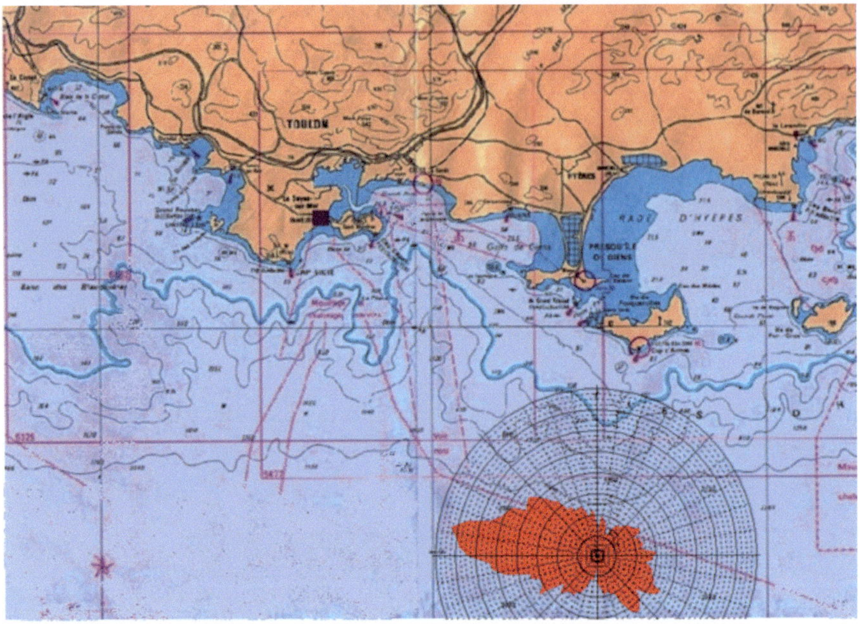

Figure 15.32 Sea current measurement on the ANTARES site. The polar plot shows the relative proportion of the time the current is in a given direction.

once every few years. These events are important for the replenishment of fresh materials to the deep Mediterranean Sea, and oceans in general, but these processes have been studied only to a limited extent because of the rapidity of the process together with their rare occurrence.

15.4.2.3 Internal waves in the deep western Mediterranean Sea

The slowly varying vertical motions (Figure 15.33) are accompanied by short-term (<1 day periods) "noise-like" rapid motions, mainly observed in the vertical currents having typical amplitudes (0.01 m s^{-1}). This is not instrumental noise, but measurements of gravity waves in the deep interior that are supported by the vertical density stratification. Generally, such "internal waves" can exist between the vertical component of the local Coriolis parameter due to the rotation of the Earth: the inertial frequency, of about 17h period at ANTARES, and the buoyancy frequency, which depends on the density stratification (van Haren et al., 2014). In general, density stratification is so weak in deep seas that typical buoyancy periods are 4–5h. However, as an aftermath from newly-formed deep dense waters in winter, near bottom density stratification may not only decrease due to turbulent mixing, but especially also increase by a factor of 10 locally, so that the buoyancy period increases to up to 1 h, as observed (Figure 15.34). This is because newly-formed waters are most dense and spread below older waters that become uplifted and partially mix with the underlying waters. Such high-frequency internal waves have not been observed before in the deep

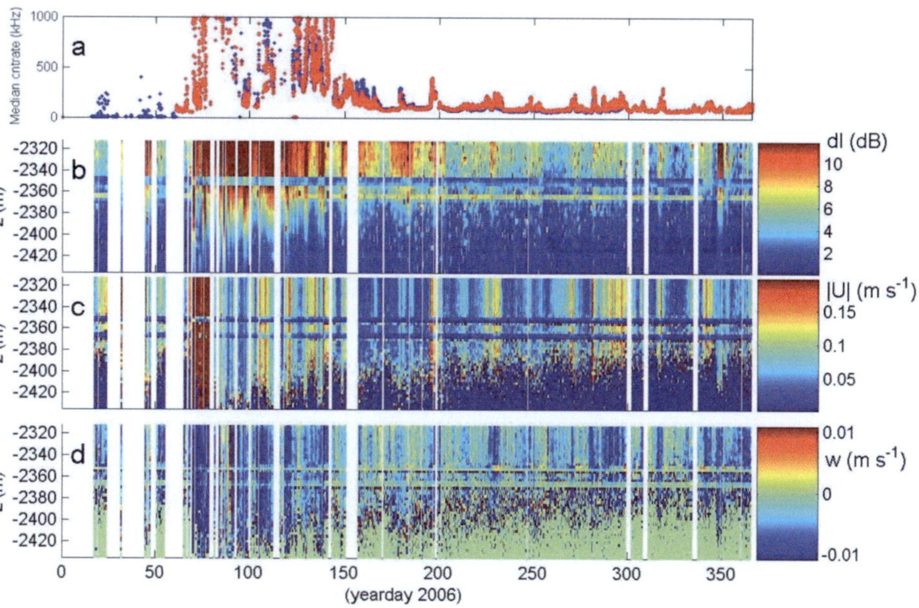

Figure 15.33 (a) Optical baseline data observed on two different lines: 50m below the ADCP on the MILOM line (blue) and from a storey on IL07 (red). (b–d) Raw acoustic ADCP data, time-depth series of 2006. Vertical white lines indicate no data due to data transmission interrupt. (b) Acoustic reflection, limited between [0, 12] dB. Higher levels indicate more suspended particles, e.g., during the period between days 70 and 100, but also periodically ~20 day periods later in the year. (c) Current amplitude. The colour coding is restricted between [0, 0.2] m s^{-1} for display purposes. Large currents occur in spring, and periodically later on. (d) Vertical current, between [–0.01, 0.01] m s^{-1}. Downward currents dominate, especially in early spring, and in ~20-days periods later on. In panels b–d, the good data range reach about 2390m, except between days 70 and 100 when acoustic reflections are large enough to provide non-zero signal. The two horizontal lines at 2350 and 2365m are bad data due to direct sound reflections off storeys 2 and 1, respectively.

Mediterranean Sea. Although they occur anywhere in the sea and oceans, their relevance for deep-ocean mixing is still under investigation. Observations by the ANTARES telescope provide a unique opportunity to add new details on their 3D propagation.

15.4.2.4 Temperature

Data have been available almost continuously from the various CTD devices in the system since March 2005. Temperature and conductivity, the latter used to derive salinity, are two of the most important physical properties of seawater, and both are used as tracers to infer the flow. Figure 15.35 gives an example of the in-situ temperature recordings. Amplitude variations of the in-situ temperature on the ANTARES site are of the order of a few hundredths of a degree Celsius.

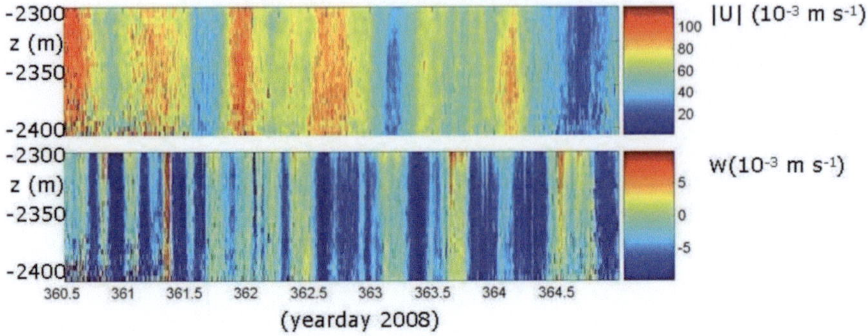

Figure 15.34 Five days detail of raw-ADCP data from IL07 line. (Upper panel) Horizontal current amplitude, dominated by inertial motions of about 17 hour periods. (Lower panel) Vertical current dominated by downward motions and both near-inertial and high-frequency internal waves (up to 1 hour periods: thin vertical lines).

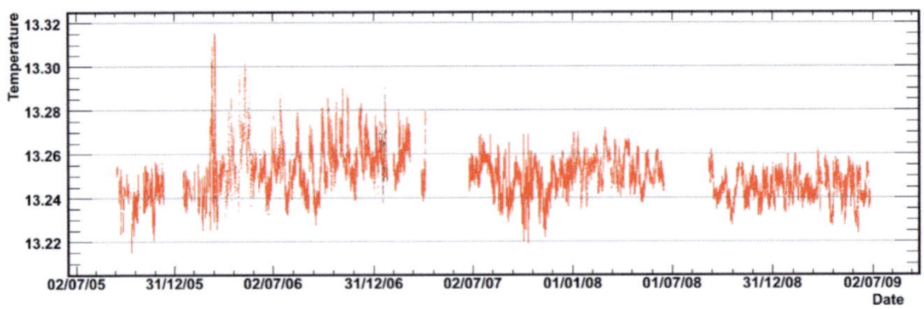

Figure 15.35 Measurement of sea water temperature from March 2005 to July 2009.

15.4.2.5 Oxygen dynamics in the deep waters

The Aanderaa Oxygen Optode on the IL07 line has been deployed with the aim of long-term in-situ monitoring of the oxygen concentration. Combined time series of temperature, salinity and oxygen allow the calculation of apparent oxygen utilization (AOU) in the oceans, which provides a proxy for the biological oxygen consumption since the water mass is no longer in contact with the atmosphere at the ocean surface. AOU is a time integrated estimation of the biological activity with respect to water masses circulation and mixing. Figure 15.36 shows an example of the data of oxygen concentration obtained with the oxygen sensor.

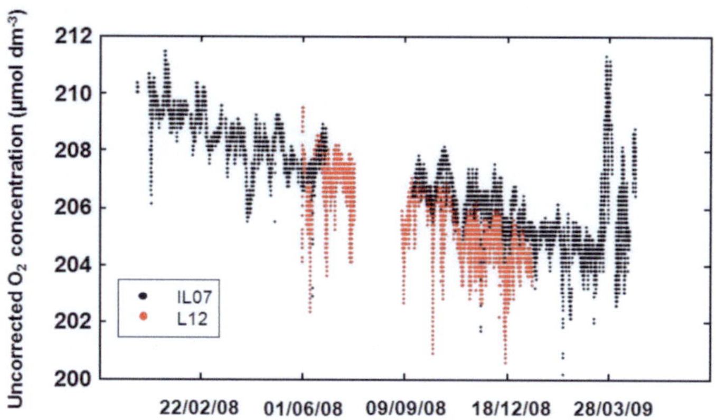

Figure 15.36 Data of in-situ oxygen concentration using Aanderaa Optode Oxygen sensors fitted on the IL07 (black) and L12 (red) lines. These data are uncorrected. These data as a function of time during 2008 and early 2009 exhibit a steady decrease, which may be used to assess in-situ oxygen consumption. The L12 time series exhibits the same trend on a shorter period of observation.

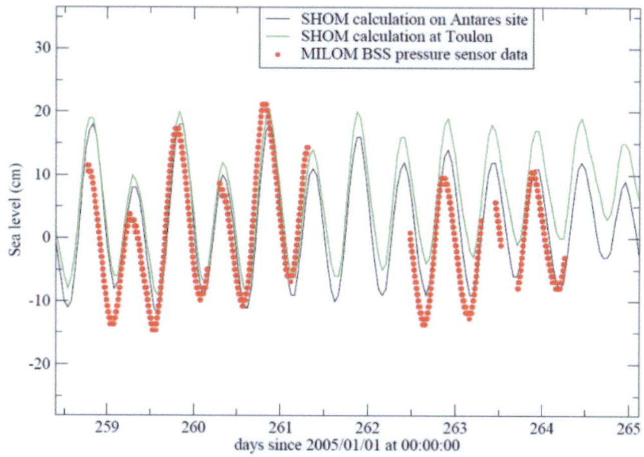

Figure 15.37 Example of pressure measurement data over six days in 2005, compared to tide calculations performed at the Toulon harbor and at the ANTARES site (SHOM data catalogue).

15.4.2.6 Pressure sensor

Pressure is a fundamental hydrological variable, just like temperature and salinity. These three parameters of state characterize sea water physical properties. Understanding the dynamics of sea water includes many topics, one of which is the study of ocean tides. An

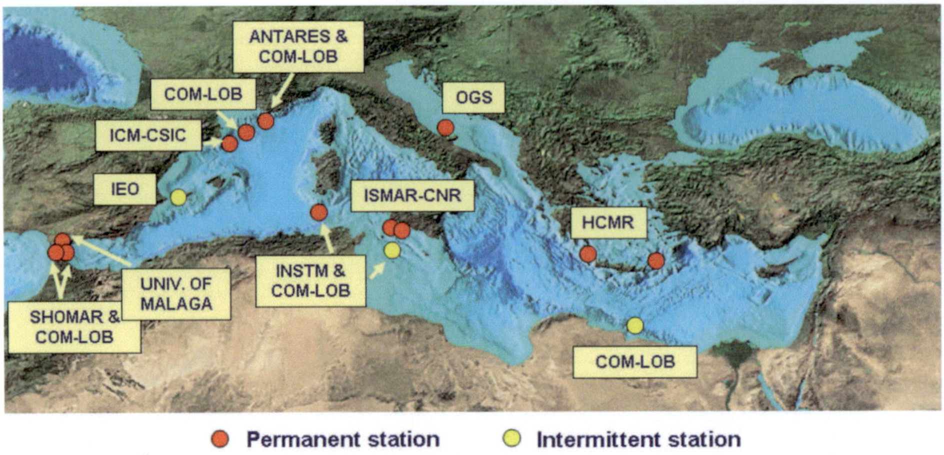

Figure 15.38 Present status of the HYDROCHANGES network, supported by CIESM.

example of pressure data sample recorded on the ANTARES site over six days is given in Figure 15.37. A pressure of 1 mbar is equivalent to a variation of 1cm in the water column. Values are compared to tide calculations computed at the Toulon harbor and at the ANTARES site by the SHOM (Service Hydrographique et Océanographique de la Marine), and a good agreement is found between data and previsions.

Monitoring network

The ANTARES site belongs to the general network of the HYDROCHANGES program19 whose purpose is to elaborate a realistic monitoring strategy to describe long-term hydrological changes in the Mediterranean Sea. Supported by the CIESM (Commission Internationale pour l'Exploration Scientifique de la Méditerranée), the HYDROCHANGES project collects continuous and long-term measurements of temperature and salinity (from CT probes) of deep water in key places in the Mediterranean, as shown in Figure 15.38.

15.4.2.7 BioCamera

Both of the cameras installed to study bioluminescence have been active most of the time since their deployment. More than 400 days of observation have been accumulated and 250 bioluminescence events have been registered so far; on average, one event every three days per camera. Figure 15.39 shows one of the first of such events, taken with an exposure time of 2 sec in December 2007. The field of view is 90 degrees; the distance of the object is at least 20cm which allows estimating its size to be between 5 and 20cm. Figure 15.40 shows an example of a bioluminescence burst taken by one of the photomultipliers adjacent to a

19 HYDROCHANGES, www.com.univ-mrs.fr/spip.php?article44

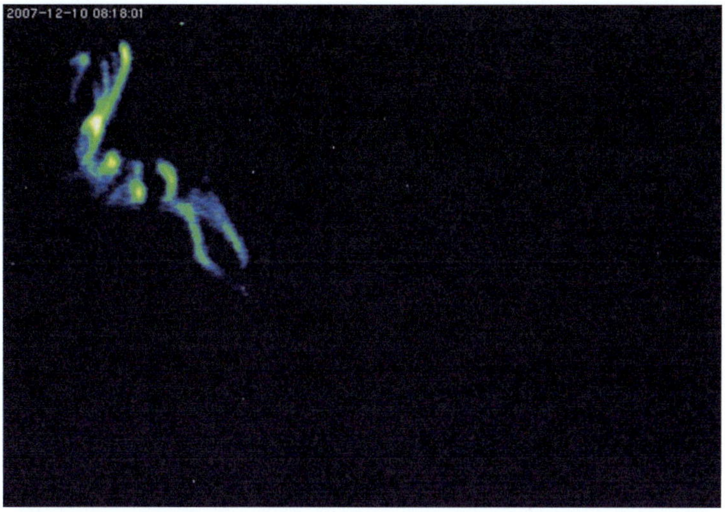

Figure 15.39 Bioluminescence event photographed with the BioCamera (false colours).

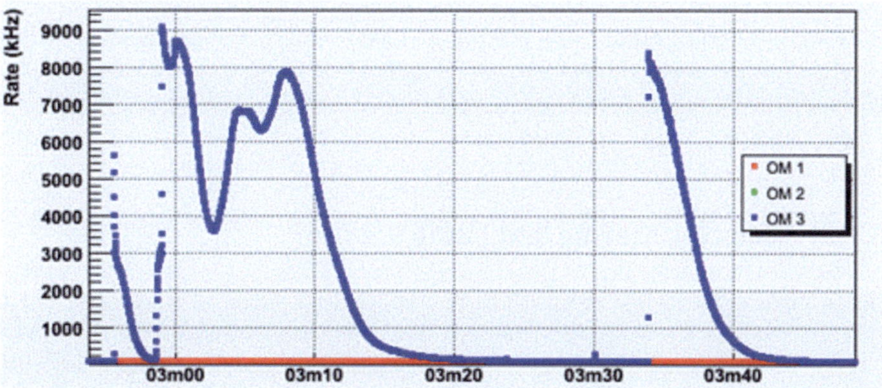

Figure 15.40 Photomultiplier signal of bioluminescence event.

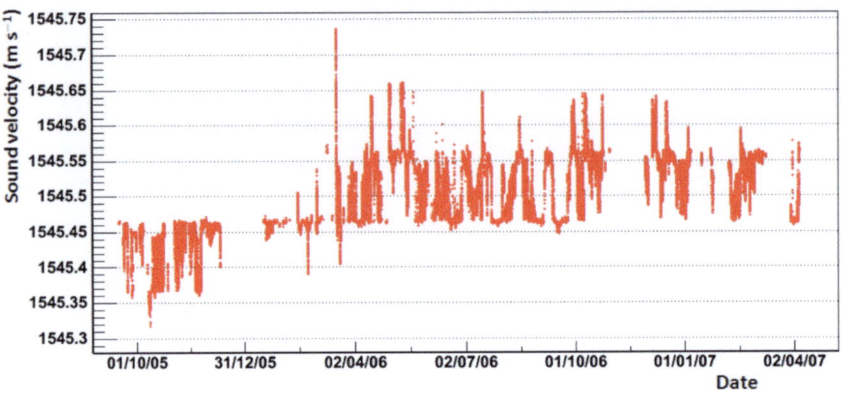

Figure 15.41 Sound velocimeter measurements over 18 months with the MILOM instrumentation line.

camera. The time distribution of the registered events shows statistically significant fluctuations, which means there are more and less active periods in terms of bioluminescence. A visual inspection of the images and movie sequences allows them to be classified. A handful of different patterns can be easily distinguished which most likely originate from different species. A more detailed analysis will be necessary to actually identify these species.

15.4.2.8 Sound velocity

The acoustic positioning requires a value for the sound velocity and the same data can provide useful measurements for other purposes. Figure 15.41 shows a time series of sound velocimeter measurements acquired during the MILOM operations between 2005 and 2007.

15.5 Extensions for Marine and Earth science

Extension plans exist for the ANTARES network on two different timescales. The first project, referred to as the Secondary Junction Box, was recently installed in an initial version. A second project is in the proposal stage for a major expansion to both the neutrino telescope and the marine and Earth science instrumentation

15.5.1 Secondary junction box (SJB) system[20]

The objectives of the secondary junction box project are to provide a system which will be "ready to use" for users of multidisciplinary instruments in terms of electrical power delivery and data transfer to the shore including data recording facilities on shore. The

20 Secondary Junction Box specifications, A080805C, 12/01/11

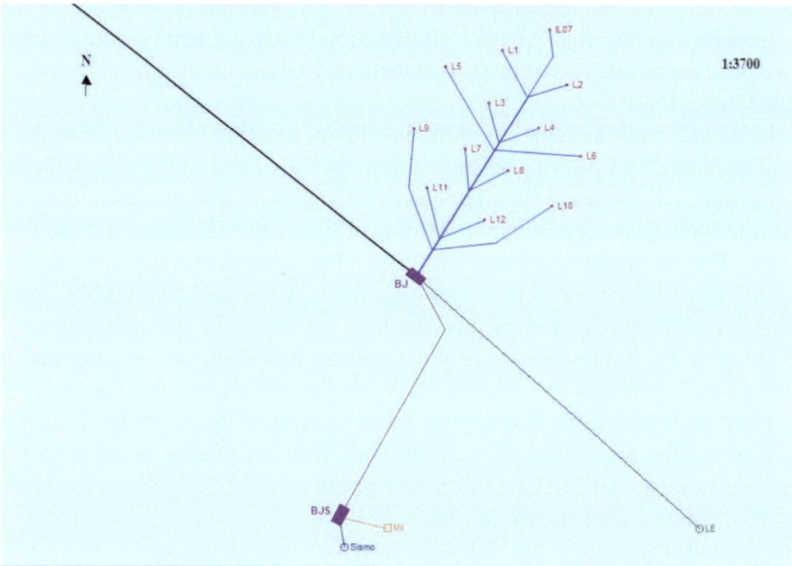

Figure 15.42 Layout of ANTARES site with the secondary junction box network.

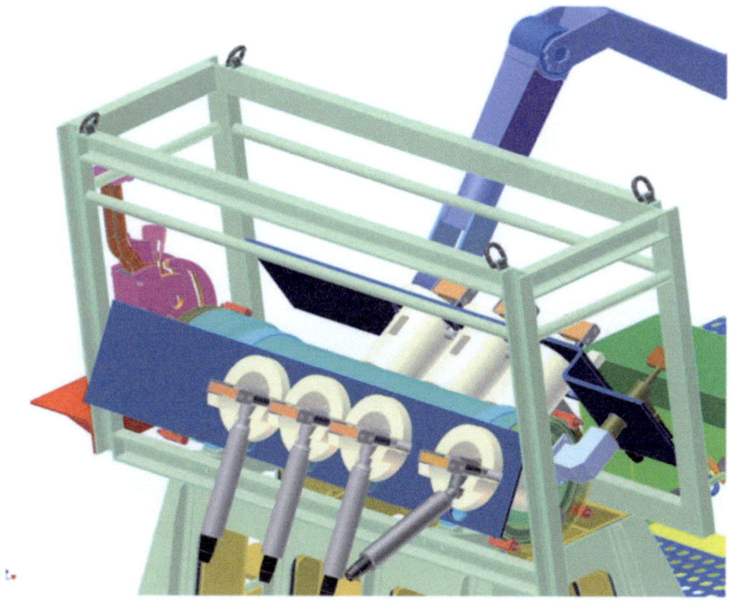

Figure 15.43 Schematic of the secondary junction box system, with a ROV arm behind the device in the process of making a cable connection.(Image © IFREMER)

secondary junction box was constructed by IFREMER and will provide access to general scientific users beyond the ANTARES Collaboration following appropriate proposals. The process to assess proposals and allocate connections to the secondary junction box is managed by the CNRS/IN2P3.

Figure 15.42 indicates the location of the secondary junction box relative to the neutrino telescope. The nature and position of the equipment connected to the shore via the SJB is open to proposals. All proposals must conform to constraint that the operation of the neutrino telescope is not degraded. A cable 400m long links the SJB with the main ANTARES junction box. This secondary junction box offers six general-purpose sockets to connect different equipment which will share operation time, power and bandwidth. Figure 15.43 is a schematic view of the secondary junction box.

On 30 October 2010, the secondary junction box was deployed and connected to the ANTARES primary junction box by the ROV Victor operating from the research vessel Pourquoi Pas? of IFREMER. During the same operation, three initial instrumentation packages were connected to the SJB: an Interface Instrumentation Module (MII); a new generation seismometer GURALP; and a first phase of Deep Sea Net project. Images of these deployed elements are shown in Figure 15.44.

The standard user interface, the "Module Interface Instrumentation" (MII), was developed by the Technical Division of INSU/CNRS and LMGEM under the ESONET merged site call ALBATROSS. This module will allow users several communication protocols and the electrical power required to connect scientific instruments. Two types of port will be available, either a serial link RS232 or Ethernet link 10/100 Mbps on copper. For these two types of interconnection, the available voltage is 48 V. The power available for each connector depends on the power allocated to the MII by the SJB. The integration of a system of data transmission via an acoustic modem is planned for a future version of the MII. This type of transmission will allow the installation of autonomous mooring lines, with low data rates, for short time periods and provides the benefit of real-time connections without the cost of submarine operations. The electronic system is embedded inside a container attached to a frame structure. This structure hosts the wet mateable connectors for ROV operation as well as a release system.

Device type	Manufacturer	Model	Measured parameters
CTD	SEABIRD	SBE 37-SMP	Conductivity, temperature
ADCP	NORTEK	Aquadopp DW	Sea current
Camera	Lyon	Lyon	Images
C-Star	WETLABS	EcoBB	Turbidity
CT	SEABIRD	SBE SI	Conductivity, temperature
O_2 probe	AANDERAA	Optode 3830	Oxygen level
Pressure sensor	PARO Scientifique	Paroscientifique	Absolute pressure

Table 15.2 Instruments installed on the MII.

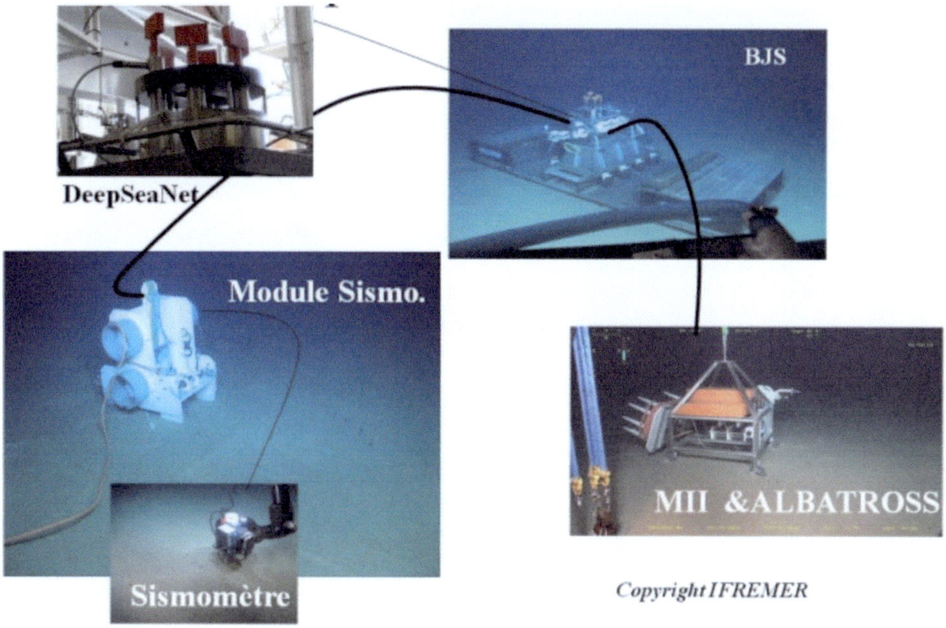

Figure 15.44 Images of the deployed elements of the SJB system.

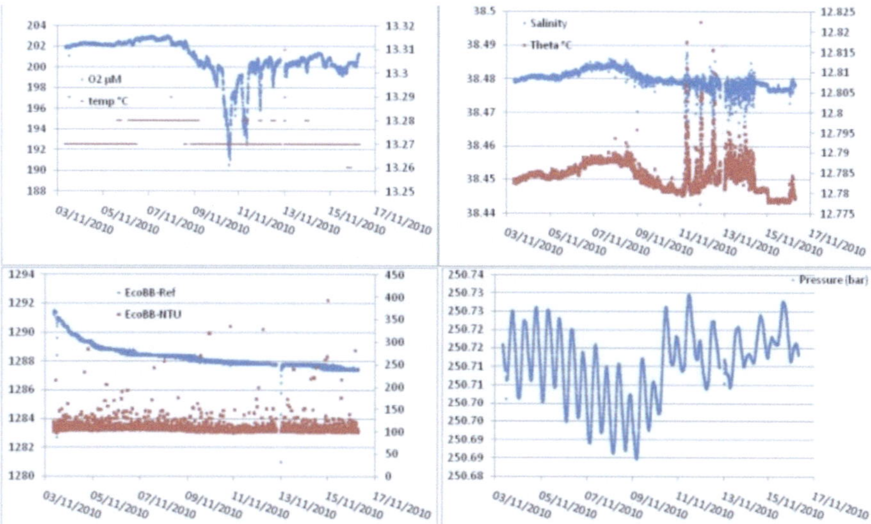

Figure 15.45 First data record with the MII on 3 November 2010.

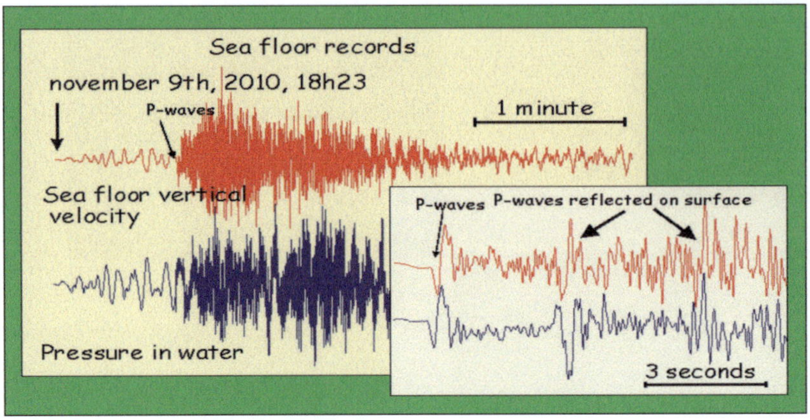

Figure 15.46 Data from the new GURALP seismometer showing a seismic event observed on 9 November 2010.

The currently installed version of the MII hosts seven generic sensors for environmental science as listed in Table 15.2. Some first data are shown in Figure 15.45.

The new GURALP seismometer incorporates a three component accelerometer as well as an absolute and a differential pressure sensor. It is under GEOAZUR management with funding from ANR (MOGLI) and FUI/RATCOM. This seismometer is intended to extend the RESIF broad band seismological antenna towards the seafloor. The pressure sensor completes real-time water level measurements installed on the coast, necessary for the demonstrator for a regional tsunami alert network proposed in the FUI/RATCOM project led by Thales. As an example of the first data from this instrument, a record of an uncommon regional event in the middle of Ligurian Sea is shown in Figure 15.46.

The long-distance network DeepSeaNet[21] will link individual battery-powered instruments (every 50km) using cheap fiber optic cables. One application of this network could be to operate seismic detector offshore from Nice to the ANTARES junction box over distances up to 200km. Deep Sea Net is under IFREMER management with funding from ANR (MOGLI) and FUI/RATCOM. In the current implementation, one optical fiber was plugged on the SJB prior deployment and is connected to the first IP node.

15.5.2 KM3NeT

On a longer timescale, a large deep-sea research infrastructure (KM3NeT[22], is proposed, with a cubic-kilometer-scale neutrino telescope and an extensive network available for marine and earth science experiments; the ANTARES site is one of the candidate sites. Figure 15.47 gives a schematic of how this might be arranged.

21 DeepSeaNet is a work package of the Ligurian Sea project MOGLI (MOuvements sismiques et Gravitaires en mer LIgure) managed by GEOAZUR, UT07-SSC07, Tokyo, April 2007.

22 KM3NeT Conceptual Design Report, which can be found at http://www.km3net.org/CDR/CDR-KM3NeT.pdf, also ISBN 978-90-6488-031-5.

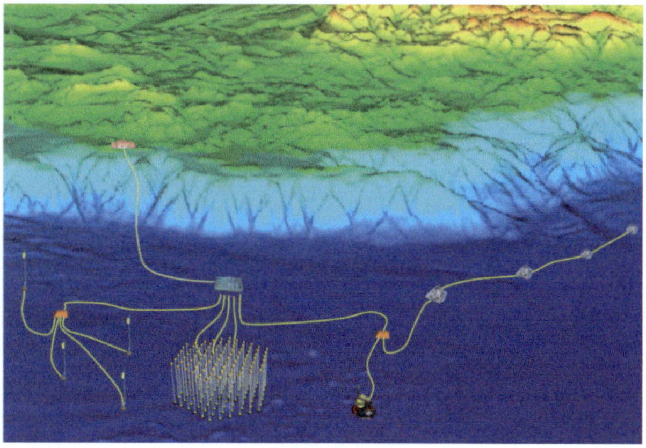

Figure 15.47 Possible long-term extension of the Undersea Observatory on the ANTARES site in the KM3NeT framework, including a cubic-kilometer-scale neutrino telescope and an extensive network for Earth and Marine sciences.

15.6 Summary

After several years of R&D, design and construction the ANTARES Neutrino Telescope was completed in May 2008. Elements of the detector have worked reliably for over five years and much data, both for neutrino astronomy and environmental research, have been collected. This achievement demonstrates the feasibility of long-term permanent scientific observatories in the deep sea.

References

Ageron, M. et al. (ANTARES Collaboration) (2009) Performance of the First ANTARES Detector Line. Astropart. Phys. 31, 277.

Aguilar, J.A. et al. (ANTARES Collaboration) (2005) Transmission of light in the deep sea at the site of the ANTARES neutrino telescope. Astropart.Phys. 23, 131.

Aguilar, J.A. et al. (ANTARES Collaboration) (2005) Study of large hemispherical photomultiplier tubes for the ANTARES neutrino telescope. Nucl. Inst. Meth. A 555, 132.

Aguilar, J.A.et al. (ANTARES Collaboration) (2006) First results of the instrumentation line for the deep-sea ANTARES neutrino telescope. Astropart. Phys. 26, 314.

Aguilar J.A. et al. (ANTARES Collaboration) (2007) The data acquisition system for the ANTARES neutrino telescope. Nucl. Inst. Meth. A570, 107.

Amram P. et al. (ANTARES Collaboration) (2000) Background light in potential sites for the ANTARES undersea neutrino telescope. Astropart. Phys. 13, 127.

Amram P. et al. (ANTARES Collaboration) (2003) Sedimentation and fouling of optical modules at the ANTARES site. Astropart. Phys. 19, 253.

Anton G. et al. (2006) Study of piezo based sensors for acoustic particle detection Astropart. Phys. 26, 3016.

Arnaud K., Beurthey S., Billault M., Lefèvre D., Payre P., Robert A. and Tamburini C. APO and ANR POTES, 05-BLAN-0161-01.

Fisher F.H. and Simmons V.P. (1977) Sound absorption in sea water. J. Acou. Soc. Am. 62, 558.

Kurahashi N. and Gratta G. (2007) Accepted by Phys. Rev. D; preprint arXiv: 0712.1833 [physics.ao-ph.

Lahmann R. for the ANTARES Collaboration. Proc. ARENA08, preprint arXiv:0901.0321. Nucl. Instr. Meth. A. (in press).

Lurton X. (2002) An Introduction to Underwater Acoustics, Principles and Applications Chichester: Praxis Publishing Ltd), Chapter 4.

Neff M. (2007) Studie zur akustischen Teilchendetektion im Rahmen des ANTARES-Experiments: Entwicklung und Integration von Datennahmesoftware, Diplomathesis, Univ. Erlangen-Nürnberg, (FAU-PI1-DIPL-07-003).

NEMO Coll. (2008) Submitted to Deep Sea Research I; preprint arXiv: 0804.2913v1 [astro-ph.

Richardt C. et. al., Proc. ARENA08, to be published in Nucl. Instr. Meth. A. (in press).

Richardt C. et al. (2009) Reconstruction methods for acoustic particle detection in the deep sea using clusters of hydrophones. Astropart. Phys. 31, 19.

Tamburini C, Canals M, Durieu de Madron X, Houpert L, Lefèvre D, Martini S, D'Ortenzio F, Robert A, Testor P and the ANTARES collaboration (2013) Deep-sea bioluminescence blooms after dense water formation at the ocean surface. PLoS ONE 8(7), e67523

Urick R.J. (1983) Principles of Underwater Sound. Los Altos: Peninsula Publishing, Los Altos, USA.

Urick R.J. (1986) Ambient Noise in the Sea. Los Altos: Peninsula Publishing, Los Altos, USA.

van Haren H., Taupier-Letage I. and the ANTARES-collaboration (2011) Acoustic and optical variations during rapid downward motion episodes in the deep north-western Mediterranean Sea. Deep-Sea Res. I, 58, 875-884.

van Haren H. and the ANTARES Collaboration (2014) High-frequency internal wave motions at the ANTARES site in the deep Western Mediterranean. Ocean Dyn., 64, 507-517.

16 NEPTUNE Canada: Installation and initial operation of the world's first regional cabled ocean observatory

C.R. Barnes[1,2], M.M.R, Best[1,3], F.R. Johnson[1,4], and B. Pirenne[1,4]

16.1 Introduction

NEPTUNE Canada (NC) has built the world's first regional cabled ocean observatory in the north-east Pacific Ocean off the coast of British Columbia. The observatory became operational in late 2009 with instruments added to the last node site in 2010–2012 and others replaced or added on an ongoing basis. The observatory, and similar ones being planned elsewhere in the world, is expected to transform the ocean sciences. By introducing abundant power and high bandwidth communications into a wide range of ocean environments, it is now possible to discriminate between and interpret both short- and long-term events, to conduct experiments and receive the data and imagery in real time, and thus work in complex multidisciplinary teams to build a vast database that can be interrogated over the 25 years of the design life of the observatory (Favali et al., 2010; Barnes et al., 2009, 2010a, 2010b, 2011, 2013).

1 previously NEPTUNE Canada, University of Victoria, Victoria, BC, Canada

2 School of Earth and Ocean Sciences, University of Victoria, Victoria, BC, Canada

3 Consultant, and Adjunct in Department of Earth Sciences, Laurentian University, Sudbury, ON, Canada

4 Ocean Networks Canada, Victoria, BC, Canada

It was noted by Walter Munk that the last century of oceanography is marked most by the degree of under-sampling (Munk, 2002). In contrast, cabled ocean observatories will have:

- 24/7/365 presence, variety of sensors, selected locations
- sampling frequencies of subseconds for most parameters
- real-time multidisciplinary, interactive experiments.

All this is possible, and resulting in a vast interactive data archive, with:

- abundant power and high bandwidth communication
- remote control of observatory network and instruments
- real-time high data/imagery return (Gb/sec).

16.2 History of NEPTUNE Canada

In the mid-1990s, scientists through a variety of workshops first conceived of using decommissioned submarine telecommunications for scientific purposes, by cutting the cables at suitable locations and connecting instruments. Japanese scientists adopted such cables in building nine small observatories that positioned seismometers some distance offshore. However, the commercial telecommunication cables were not necessarily positioned at locations of preferred scientific investigations. This concept was advanced as the NEPTUNE Project by US scientists, in particular John Delaney (University of Washington) and Alan Chave (Woods Hole Oceanographic Institution), who argued for the use of new electro-optic cables that would be instrumented using a significant footprint to allow measurements of large-scale Earth/ocean processes.

This was elaborated further through discussions and scientific workshops and resulted in a preferred location for a regional scale cabled ocean observatory over much of the Juan de Fuca tectonic plate off the coast of Oregon, Washington and British Columbia. This is the smallest of the world's 12 main tectonic plates, is adjacent to two well-developed nations with oceanographic facilities and institutions on those coasts, and can be instrumented with acceptable costs and benefits.

In 2000, specialists in Canada invited colleagues in the US to a meeting at Emerald Lake, British Columbia, to consider how a bi-national program could be established. A formal Memorandum of Understanding (MoU) was signed that year with the Institute of Pacific Ocean Science and Technology (IPOST) and the University of Victoria (UVic) representing Canada and the University of Washington (UW), Woods Hole Oceanographic Institution (WHOI), Monterey Bay Aquarium Research Institute (MBARI), and NASA's Jet Propulsion Lab (JPL) at Caltech University representing the US. An Executive Team representing those six institutions then worked over the next three years to establish a bi-national cabled ocean observatory. The US partners raised over \$25M specifically to begin baseline research into the required power and communications systems; over \$1M was raised in Canada to assist in bringing the scientific community together to formulate a scientific program and associated grant proposals (Note: Canadian and US dollar currency rates vary but can be regarded as approximately equal for the purposes of this chapter; most are given in Canadian dollars). For the latter, some basic science requirements were established, including:

- gigabits of bandwidth
- kilowatts of power
- precision timing
- hundreds, perhaps thousands, of attached devices
- 3000m water depth rating
- 25-year design life
- resilient network to achieve moderate operating costs
- over 60TB/yr data flow initially when all instruments deployed
- expandable and extendable: Currently there are 5 nodes, but the system has the capacity to be doubled.

One important outcome of this technological innovation in ocean sciences is that there is a new capacity for rapid multidisciplinary collaboration along with a substantial increase in the speed of analysis and documentation. The main elements of this intellectual revolution can be summarized as:

- using the power and opportunity of the Internet and high bandwidth communications
- promoting large community (team) experiments on complex interdisciplinary problems
- increasing international participation and partnerships
- establishing large interactive databases and time-series
- resulting in rapid publication, communication, imagery
- greater inclusion in educational program and in public outreach
- liberating knowledge to the general public and for public policy debates and new roles for academic research

16.2.1 Concept design and funding phase

In 2003, two notable events occurred. NC received its first major funding ($62.4M) for the observatory infrastructure jointly from the Canada Foundation for Innovation (CFI) and the British Columbia Knowledge Development Fund (BCKDF). The role of IPOST was completed and it was disbanded. UVic led the Canadian consortium of 12 Canadian universities, with partners in government departments (Table 16.1); indeed, UVic was required by the granting agencies to both own and operate the NC observatory.

Participating Canadian universities	
Memorial	Guelph
Dalhousie	Waterloo
UQAR (Rimouski)	Manitoba
Laval	Simon Fraser
UQAM (Montreal)	British Columbia
Toronto	Victoria

Participating Federal government departments (with regional groups)	
Fisheries and Oceans Canada	Institute of Ocean Sciences, Pacific Biological Station
Natural Resources Canada	Geological Survey of Canada Pacific, GSC Atlantic
Environment Canada	Canadian Climate Centre for Modeling and Analysis
Parks Canada	Pacific Rim National Park
National Defense	Esquimalt, Defense Research and Development Canada
National Research Council	Herzberg Institute for Astrophysics

Table 16.1 NEPTUNE Canada university consortium and government partners.

In the US, in late 2003, the National Science Foundation (NSF) formally established the Ocean Observatories Initiative (OOI) with an office and staff in Washington, DC. This effectively replaced the Executive Team arrangement as a planning organization and a MoU was signed between NSF and UVic to jointly develop an interoperable regional cabled ocean observatory. For the proposals developed by OOI, the term NEPTUNE was replaced in the US by Regional Scale Nodes (RSN) as a component of the US coastal, regional and global observatories. This MoU was revised in 2010 and is formally between Ocean Networks Canada and OOI.

Unfortunately, in the US, funding for major projects from the Major Research, Engineering and Facilities Construction (MREFC) account was restricted or frozen for several years between 2003 and 2008. It was hoped that the OOI and NC funding would be concurrent, resulting in some joint Request for Proposals (RFPs) and subsequent cost sharing and savings. With the approval for major US OOI funding, a formal contact was arranged in July 2009 between NSF and Ocean Leadership Inc., which manages the OOI program. Consequently, there will be a significant time difference between the final installation of NC and RSN, in 2009 and probably 2015, respectively.

Funding source	$M
Canada Foundation Innovation	39.9
BC Knowledge Development Fund	38.5
Other government agencies	8.3
University of Victoria	2.0
In-kind support: Industry	17.0
In-kind support: Partner labs	4.5
CFI/NSERC/BCKDF/CANARIE Operations Funding	37.0
Total	147.2

Table 16.2 NEPTUNE Canada funding awarded 2003–2012.

With the delay of the US funding and the inability to secure shared cost reductions, NC applied to the Canadian granting agencies for additional funding and received an additional $20M ($8M CFI, $8M BCKDF, and $4M in-kind support) in late 2006, for a total of $78.4M plus nearly $20M in in-kind support. In 2007, NC received interim operating funds ($13.2M) from Natural Sciences and Engineering Council of Canada (NSERC), CFI and BC Ministry of Advanced Education for a total of cash and in-kind support of $123M (Table 16.2). Also in 2007, UVic established a not-for-profit agency, Ocean Networks Canada (www.oceannetworks.ca), to manage, promote and develop both the NC and the coastal VENUS observatory networks. ONC was successful in securing a two-year award from CFI to cover their operating costs ($24M for 2010–2012); it was also successful in gaining $13.2M over five years for a federal Centre of Excellence in Commercialization and Research (Innovation Centre; www.oceannetworks.ca/technology-services) to support commercialization and education/outreach opportunities spawned by the observatories. In late 2010, the federal government announced a new funding mechanism for Major Science Initiatives. ONC received one of the five successful awards for a 5-year operating grant continuation ($32.9M for 2012–2017, with equivalent provincial and other support now arranged).

Following the establishment of the science requirements noted above, much work went into the detailed concept design in both joint and separate Canadian and US efforts. Several science workshops in each country were held. One in particular was organized by the US National Science Foundation in San Francisco in 2004 to determine the location anywhere in the world for the first regional cabled ocean observatory to be funded by NSF. The workshop recommended that it be the Juan de Fuca tectonic plate region, which was accepted by NSF.

The workshops held by NC resulted in the identification of many important sites for scientific investigation, about 10 of which extended onto the Pacific and Explorer tectonic plates adjacent to the Juan de Fuca plate. An RFP for the subsea wet plant was issued in 2005 after much detailed work and after assessment by two review panels of the responses (averaging 1000p. each), a contract was awarded to Alcatel-Lucent Submarine Networks (A-LSN; then ASN) to design, manufacture and install the backbone and spur cables, repeaters, branching units, and nodes (power and communications centers that reduce the 10kV DC to 400V DC). This process resulted in an important reality check on what was commercially and technically feasible within the research funds that were awarded. The scope of the observatory was reassessed and reduced to five node sites with an additional trawl resistant frame (TRF) at the Middle Valley site, which would be ready for a node component once additional funds were secured at a future date (Figure 16.1). The objective was achieved of locating nodes at key sites near the coast (Folger Passage), on the continental slope (Ocean Drilling Program (ODP) site 889 and Barkley Canyon), on the abyssal plain (ODP 1027), and at the ocean spreading ridge between the Juan de Fuca and Pacific plates (Endeavour and Middle Valley). The 800km backbone cable and 120km of spur cables thus extended across the entire northern part of the Juan de Fuca plate. The initial hope to include sites further to the north at Nootka and onto the Explorer and Pacific plates was not possible but will remain a priority to be reassessed as an opportunity to seek additional infrastructure funds becomes available. A successful application to the latest CFI competition (2012–2013) is allowing a diversified expansion of the instrumentation and sites at Endeavour.

16.2.2 Infrastructure design, testing and installation phase

It is important to note that the location of the nodes (Figure 16.1) was based on scientific priorities, with A-LSN then proposing the best cable route to connect them. Using a cable loop rather than one or more straight lines from shore has the advantage of redundancy,

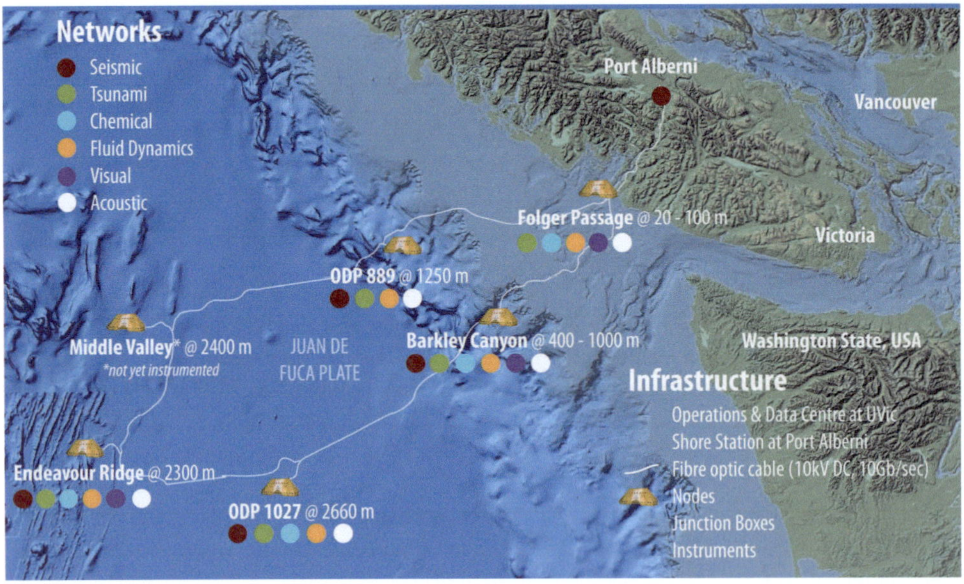

Figure 16.1 NEPTUNE Canada cabled ocean observatory showing 800km backbone cable route, node locations, and Port Alberni shore station. Note that sites ODP 889 and 1027 have recently been renamed as Clayoquot Slope and Cascadia Basin, respectively.

so that if there is a break in the cable, information can still flow back to the shore station without loss of data. The design also allows for future expansion of both additional nodes or cable and additional instruments. The present system of five nodes can be expanded to 10 nodes, either on the existing backbone cable loop or by an expansion geographically by splicing and adding new cable; there is abundant power and high bandwidth communications so that additional instruments can be accommodated in future years.

The architecture proposed by A-LSN and adopted by NC is a ring and branch topology, which achieves the desired functionality for both power distribution and communications and allows a measure of fault tolerance. The backbone ring components are constructed exclusively from components designed and qualified for use in commercial subsea telecommunications systems, leveraging many years of design experience and high reliability of these components. A key infrastructure component is the shore station and UVic was fortunate to purchase an existing one (1100m^2; then renovated and re-equipped) at Port Alberni. The backbone cable forms a continuous loop from and to the shore station (Figures 16.1, 16.2, 16.3) and was installed with repeaters, branching units and spur cables over an 11-week period in late 2007 by A-LSN, with over eight weeks of post-lay burial work using another cable ship. The cable was buried where possible to 1.1m depth across most of the two continental shelf transects down to about 1500m to limit the possible interactions with trawl gear and ship anchors. This NC topology differs from the three separate straight-line cables from two shore stations adopted by the US OOI/RSN where other factors such as power input and location were influential.

Figure 16.2 Shore station at Port Alberni.

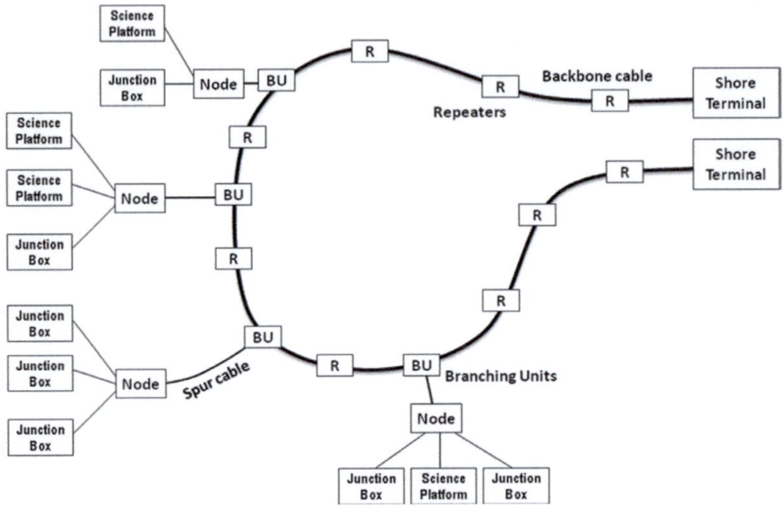

Figure 16.3 Generalized observatory configuration.

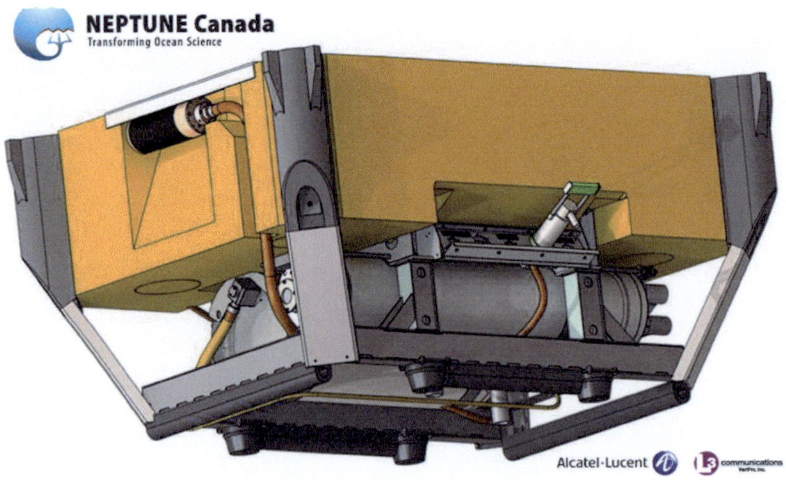

Figure 16.4 Generalized view of a node that controls the communication systems and steps the voltage from 10kV DC down to 400V DC to supply the junction boxes.

Figure 16.5 The first node, protected by its trawl-resistant frame, deployed on the NC observatory at Barkley Canyon in July 2009.

Specialists from A-LSN, with subcontractor L3 MariPro, Santa Barbara, designed and manufactured the trawl-resistant frame for the nodes and the node module that fits into it (Figures 16.4, 16.5). The nodes lie at the end of spur cables, each up to 20km in length, off the backbone cable. Network modifications and upgrades can be made by replacing the nodes, while the backbone remains unaffected. Key details of the power architecture are:

- power feed equipment at Shore Station: Output 10kV DC, 8 Amp
- modified telecom branching units
- medium voltage converter at each node: Output 400V DC, 25 Amps.
- low voltage DC-DC converters at nodes and junction boxes: Output 15, 24, 48V DC

Each node provides six interface ports for connection of science instrument arrays or extensions. Each port provides dual optical Ethernet links and up to 9kW of electrical power at 400V DC. Wetmateable connectors make the connections between node and extension cables. Fourteen junction boxes (Figure 16.6) designed and built for NC by OceanWorks International, Vancouver, BC, support up to 10 instruments each and can be daisy-chained where necessary. They accommodate both serial and 10/100 Ethernet instruments, and provide a variety of voltages (400V, 48V, 15V and 12V).

Most of the major technology challenges during the implementation of this subsea network have now been overcome. A-LSN resolved the three most significant development tasks: the 10kV DC wetmate connectors, the medium voltage converter and the low voltage communications converter. With Ocean Design, Inc. (ODI; now Teledyne ODI, Inc.), A-LSN successfully developed a high voltage wetmate connector that links the spur cable into the node. The backbone equipment has all been qualified and installed; the shore station

Figure 16.6 Junction box in harness, which steps voltage from 400V DC to 48/24/15V DC to the different instruments.

modifications are complete; the junction boxes are all manufactured. However, additional time to complete the development work and formal qualifications extended the installation date of nodes into mid-2009 and consequently pushed the deployment of instruments into the 2009 and 2010 weather windows (May–September). With the repair of a fault in the Branching Unit at Middle Valley and some post-lay burial work, final commissioning and system acceptance occurred by November 2010.

16.2.3 Instrument design, testing and installation phase

The science workshops resulted in the definition of the following main scientific themes:

- plate tectonic processes and earthquake dynamics
- dynamic processes of fluid fluxes and gas hydrates in the sea bed
- regional ocean/climate dynamics and effects on marine biota
- deep-sea ecosystem dynamics
- engineering and computational research.

These themes are being addressed by many specified and approved experiments involving scores of researchers. Overall, the NC observatory is investigating a great diversity of

Figure 16.7 Remotely operated crawler, Jacobs University Bremen, Germany, investigating gas hydrates at Barkley Canyon.

Earth/ocean processes, phenomena and events, with instruments deployed in the subsurface (boreholes), on the seabed, and buoyed through all or part of the water column. About 130 instruments will host several hundred sensors. Larger and more complex instrument packages include those attached to a tethered ROV crawler (Figure 16.7) and a 400m vertical profiler (Figure 16.8).

An award in 2012 of 1.2m Euros to Laurenz Thomsen (Jacobs University Bremen) will facilitate further development of the crawler, with the ability for vertical lift-off from, and movement above, the seabed. The observatory is designed to facilitate real-time observations and experiments that will establish a vast data archive, allowing scientists interactive participation with both the instruments and the archive. The observatory is also designed and located to address a wide spectrum of research questions. Experiments will focus on: earthquake dynamics and tsunami hazards; fluid fluxes in the ocean crust and sediments, including gas hydrate deposits; ocean and climate dynamics, including acidification, nutrient fluxes and impacts on biota; deep-sea ecosystems dynamics; and applied engineering and computer science research. Scores of scientists have been involved in the design of the experiments and the instrumentation; others are mining the abundant database; participation by others is invited (info@oceannetworks.ca). Many abstracts and conference proceedings have been published and initial scientific results are appearing in the primary literature (e.g., Riedel et al., 2009; Swadinsky and Edwards, 2009; Edwards et

Figure 16.8 The 400m vertical profiler by NGK Ocean, Saitama and Tokyo, Japan operating near Barkley Canyon.

al., 2010a, b; Riedel et al., 2010; Rogers et al., 2010; Willoughby et al., 2010; Robert and Juniper, 2012; Rona and Light, 2011; Thomsen et al., 2012 ; Purser et al., 2013). Among the many examples of international collaboration, it must be emphasized that colleagues and institutions in the US have helped considerably in areas such as initial power and communications design work, ship time, engineering consultation, seafloor mapping and bathymetric data, the IODP borehole program, and some instrument development particularly for the Endeavour hot vents. The observatory can be readily expanded in the future through the subsea wet plant, number of nodes, or number and types of instruments as funding allows or as new instruments are added by other researchers and government or commercial partners.

The bench and initial wet testing of the 130 individual instruments is complete, together with integration with the junction boxes on research platforms and frames and DMAS connectivity, to support the various experiments near the main node sites. Highland Technologies, Sidney BC, was contracted for much of this work. More advanced wet testing uses the Saanich node of the Victoria Experimental Network under the Sea (VENUS, www.oceannetworks.ca), also led by UVic. One large instrument pod was installed and tested using the CCGS John P. Tully with ROPOS from late September 2008 to mid-Feb-

Figure 16.9 Instrument pod for test deployment in Saanich Inlet being loaded onto CCGS Tully.

ruary 2009 (Figure 16.9). The pod hosted examples of most of the components of the full deployment in mid-2009 – an extension, an instrument frame, a junction box, and a variety of instruments and their associated whips/cables, including in particular Tempo-Mini designed and built at Ifremer, France with its video camera, iron and oxygen sensors, and anti-biofouling system (www.oceannetworks.ca/sights-sounds/live-video/tempo-mini-vent-camera). Most of the instruments worked well, but others needed adjustments or return to manufacturers for modification. Under an MOU with Ifremer, that institution tested the complex package of instruments on the pod and sent over six engineers to assist with deployment. Substantial modifications were completed prior to deployment at Endeavour in late 2012.

NC contracted Rocketday Arts, Victoria, to redesign the website, which has a Web 2.0 environment, and to establish new systems that accommodated the start of the data flow in late 2009. The website continues to have content added and further structural revisions to make it fully interactive with the user community and the lay public, in part with the development of the Oceans 2.0 system and now incorporated within www.oceannetworks.ca.

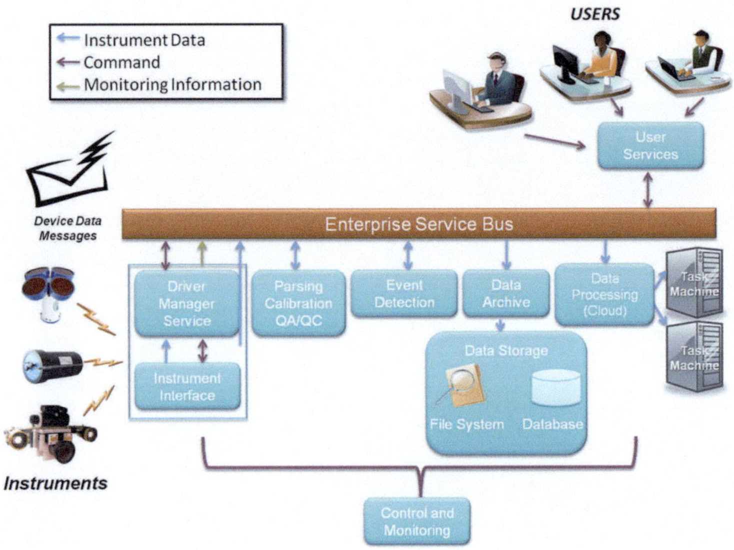

Figure 16.10 Architecture of the DMAS system: the Enterprise Service Bus implements a publish and subscribe model with asynchronous communication between the various elements to be applied to the data.

16.3 Data management and archiving

The need for a software system to support NC's underwater infrastructure is critical. The software serves as an interface between users and the cable observatory and responds to a triple mandate:

- to acquire data from various instruments and sensors underwater
- to provide life-time storage and redistribution capabilities for all data
- to allow duly authorized users to remotely and interactively control experiments.

The main features of the Data Management and Archive System (DMAS) are:

- DMAS mandate
  - Make data available to researchers in (near) real time
  - Store data for long-term time-series studies.
- DMAS features:
  - Allow human interaction with instruments
  - Facilitate automated, routine "survey campaigns"

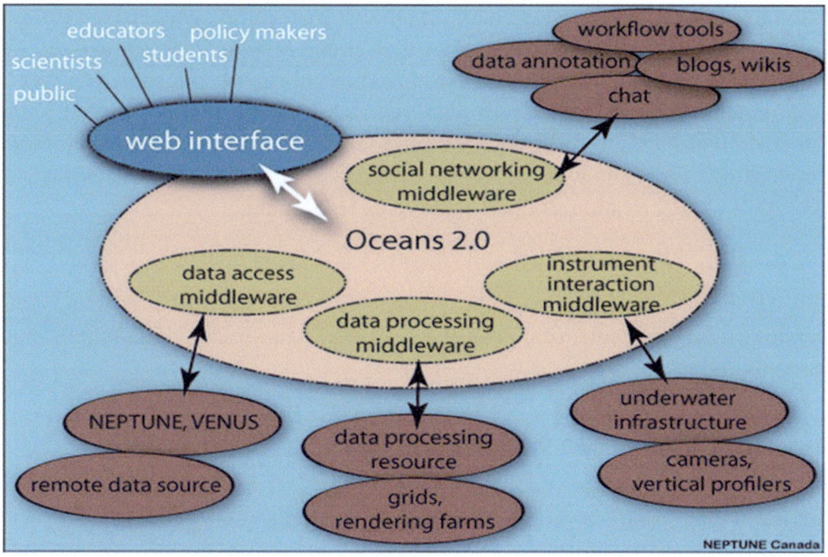

Figure 16.11 Various components are shown of the participative Web 2.0 approach by DMAS using CANARIE funding.

- Facilitate automated event detection and reaction definition
- Store scalar data in data warehouse for easy x-correlation
- Store complex data in fast file system
- User access mostly through the web/specialized client software.

The DMAS (Pirenne, this volume) is being developed in-house, using the Agile development methodology to deliver frequent, incremental versions of the system. This approach was first applied to VENUS, with data flowing since 2006, and subsequently to NC to establish the system specifications, understand the issues, develop the software, and train operations staff on observatory control and monitoring. As a safeguard against a damaging earthquake affecting Vancouver Island and the data archive, NC has completed an arrangement to have a completely mirrored data archive located at the University of Saskatchewan, Saskatoon. No data are discarded; some large data streams are transmitted to other affiliated institutions, such as the seismic data, which are sent in real-time to IRIS based at the University of Washington, Seattle.

CANARIE funding permitted DMAS to adopt a Service-Oriented Architecture (SOA), using Web services to expose the functionality of DMAS' various components. An internal messaging bus allows various functional components to interact through the publish and subscribe paradigm, using Java as a programming language (Figure 16.10).

Additional CANARIE support is enabling DMAS to develop a modern environment for both professional and lay users: data access, data processing and experimentation control within a Web 2.0 environment (Figure 16.11). This will allow the users, on top of data and instrumentation access, to perform data visualization and analysis online with either default or custom processing code, while simultaneously interacting with each other. These social networking aspects lie within NC's new Oceans 2.0 environment, which was funded at $1.4M (2008–2010) to address three areas of development:

- Interoperability between VENUS and NC and a few other selected data centres
- A Web 2.0 environment that will allow users to search and discover data, work with colleagues online on a scientific project involving ocean-related data and interact with instrumentation
- Set up and integrate an underwater HDTV camera with high-quality compression and with controls available in the Web 2.0 environment.

Following an international Interoperability Workshop at UVic in September 2008, pilot projects for interoperability were initiated with selected observatories.

The NC observatory includes a network backhaul link from the Port Alberni shore station to UVic's data centre, mostly funded through a CANARIE award, with Shaw Business Solutions providing a dedicated 10Gbps/10yr service. All the equipment additions (power feed equipment and laser-based communication and security systems) and physical modifications at the Port Alberni Shore station, acquired for NC in 2004, have been completed and the 10Gbps connection to the new UVic Data Centre is fully operational.

16.4 Challenges for NEPTUNE Canada

Establishing the NC cabled ocean observatory network has proven to be extremely challenging on many fronts. Space constraints do not permit a detailed review of all challenges, but Table 16.3 shows many of the principal elements that are involved in building such an observatory with the elements arranged in approximate sequence of encounter.

Separate from these principal elements are several challenges that have been considered by the NC Executive Committee (Tables 16.4, 16.5), with advice provided by various advisory committees (Table 16.6). These include a range of issues that require effective internal and external relationships and communication in order to achieve the maximum efficiency for the observatory and the maximum return on investment (ROI) of public funds (Table 16.4).

Another set of challenges are those that are technical in nature and can be expected since this type of region cabled ocean observatory is the first of its kind in the world (Table 16.5). These embrace a wide range of topics ranging from the initiation of the main RFP for the wet plant to the in-house development of the DMAS. Several of these have been addressed in partnership with key contractors such as Alcatel-Lucent.

Additionally, there have been many special issues experienced in building the NC observatory network (Table 16.6). Some involve interaction with other users of the marine

• Vision, articulation, concept • Science priorities, experiments • Ownership and liability • Funding proposals/O&M costs • Science requirements: engineering, DMAS • Engineering: network design, power and communications, wet plant, shore station, backhaul • Permits and rights of way • Route surveys, node sites, GIS • Operation/data centre • Risk assessment	• DMAS/Cybersecurity: in-house development, evolving technologies, distributed databases, storage • Education and outreach • Special stakeholders: First Nations, fishers, navies • Communications: science/publication, communities, media, partners • Partnerships: institutions, funders, foundations, international • Socio-economic benefits • Time, contingency, renewal, expansion costs (all in short supply)

Table 16.3 Principal elements in building a regional cabled observatory.

- Translate vision(s) into realistic plans
- Establish accurate installation/operating costs
- Establish collegial team of management specialists
- Maintain support of scientific community (key users)
- Attract interest of other users (industry, policymakers, educators, public, media)
- Expand user community with new researchers/students
- Fully utilize potential of data archive and time series
- Accommodate to real-time data flow, respond to natural events as they occur; develop truly multidisciplinary teams/science; improve data management
- Develop vision/plans for progressive expansion and sophistication of network and experiments

Table 16.4 Special challenges: Internal and external relationships.

- Design of RFP and wet plant contract (c. $50M)
- Network design, manufacture, installation (Alcatel-Lucent)
- Sea-bed mapping of cable route and node locations
- Cable deployment in hostile environments; burial issues
- Infrastructure technology development: nodes/medium voltage converter; junction boxes; connectors
- Modifications and development of science instruments and platforms: serial vs. Ethernet-ready; vertical profiler; crawler; AUV docking
- Building the Data Management and Archive System (DMAS): interactive communications and observatory control
- Establishing the shore station, data backhaul, operation/data centre

Table 16.5 Challenges: Selected technical issues.

• National security: consultations with Canadian/US navies; *Cybersecurity Committee*
• Commercial fisheries: competing use of ocean environment; *Fisheries Advisory Committee*
• Coordination with international scientific community; *Science Planning Committee and Science Users Committee*
• Archiving data, time-series, and real-time data use by diverse users; *DMAS Advisory Committee*
• Public education and outreach: critical to long-term advocacy and impact, different funding sources
• MSI operating funds: $12M/yr; successful CFI and provincial awards; successful CECR award
• Management oversight and continuous upgrades: *Ocean Networks Canada with external Board of Directors*

Table 16.6 Challenges: Special issues and oversight procedures.

environment such as the Canadian and US navies regarding homeland security, and with the fishing industry where fishing gear can damage the NC wet plant and instruments unless agreements are developed that ensure that such negative interactions are minimized and mitigated as much as possible. NC has established several advisory committees (Table 16.6) with external representation to help secure policies and solutions. An ongoing issue will be to secure additional and continuing operating funds to sustain the observatory over its expected 25-year life span. Representations have been made at the highest levels to the granting councils and to key federal government agencies to support the establishment of procedures and funding of operating grants for a Major Science Initiatives (MSI) program. After being considered by the Science and Technology Innovation Council and the federal Department of Industry, this new program was established in fall 2010 by the Canada Foundation for Innovation and ONC received one of the five awards in 2012 ($32.9M, 2012–2017, representing 40% of total operating costs).

It would appear inevitable in a multiyear programme of installation and operation for such a complex observatory that delays would occur in the planned schedule, not to mention cost overruns. Over the last few years, for example, several key timeline issues were recognized or encountered that posed challenges in attaining the programme on schedule and on budget (Table 16.7).

The challenges faced by NC that are largely captured in Tables 16.4–16.7 show the diversity and often concurrent nature of these issues, the requirement for forward planning and adequate contingency funds, and the need for a national program to support the operations of major national infrastructure facilities. At the stage of final installation, it may be useful to identify some of the lessons learned by NC, although some of these will be specific to Canada, national funding arrangements, or the style and scope of the observatory facility (Table 16.8). The result of the delay in funding the OOI is that Canada and the US have only loose joint management structures of their combined observatory assets located on the Juan de Fuca tectonic plate. The subsea wet plant arrays are not connected as

- Meet Alcatel-Lucent's schedule for constructing nodes and junction boxes – Q3/06
- Receipt of additional $20M infrastructure funds – Q2/06
- Arrange data backhaul contracts and shore station upgrade – Q3-4/06
- Arrange maintenance contract with cable ship operator – Q3-4/06
- Establish a data and operations centre – Q2/07
- Delay in installing nodes and commissioning of system (from 2007/08 to 2009/10)
- Node delay forced delay in installing instruments (from 2008/09 to 2010/11) and initial data flow (December 2009)
- Transitional operating funds ($13.2M) received from NSERC/CFI/BC Govt. – Q3–4/08
- Initial operational funding secured ($19.3M) for 2010–12 – Q2/10
- Operating funds largely secured for 2012–17 – Q2/12 but an ongoing need

Table 16.7 Challenges: Key timeline issues.

Installation/operation phase

- CFI restrictions: no operational funding; no education/outreach allowed; precise budgeting-RFPs required
- 6-yr offset of initial Canadian Funding (late 2003) and US/NSF/OOI funding (late 2009)
- Challenges to develop a bi-national integrated management of the regional cabled ocean observatory (CFI-BCKDF/UVic vs. NSF/OOI; MOU): c. $250M in total
- Multinational integrated management of operational phase; $30M/yr for 30yrs
- CFI: installation funds only; only in late 2010 has Canada established a programme to provide operating funds for Major Science Initiatives to be managed by CFI
- Downturn in telecom industry at time of wet plant RFP: competitive period for installation
- Canadian funding ahead of US support: potential advantages for Canadian industry
- NEPTUNE Canada to be an important driver for science, science policy, education/outreach, CA*net5 (CANARIE), ocean technology, environmental protection/security, offshore resource development

Table 16.8 Challenges: Selected lessons learned.

was expected some years ago. There will be an opportunity to combine the data from both observatories, assuming that the cyberinfrastructure is fully compatible and interoperable. In 2009, NC organized an Interoperability Workshop with five different observatories in North American and Europe (reported on the NC website) and has demonstrated effective interoperability. OOI/CI used the data from the IOOS program to test interoperability between different data sets. The start of operation dates between NC and OOI are different, being 2009 and anticipated 2015, respectively.

- Transforming the ocean sciences
- Improved understanding of climate change
- Comprehending natural hazard processes: earthquakes, tsunamis, slope failures
- Understanding ecosystem dynamics
- Monitoring ocean pollution/acidification
- Sustaining commercial fisheries
- Understanding the role of gas hydrates
- Industry: ocean technologies, services, data products
- Information to public, decision-makers and politicians
- Educating the next generation
- International partnerships > global networks

Table 16.9 Future opportunities for the NEPTUNE Canada observatory network.

- New scientific knowledge and interpretations
- Development of new technologies
- Renewable and non-renewable resource development
- Climate change driven by the oceans
- Hazard mitigation: earthquakes, tsunamis, storms, slope failures
- Sovereignty and security
- Port security and shipping
- Ocean management and policy formulation
- Education, outreach and public information

Table 16.10 Selected applications of cabled ocean observatories.

16.5 Future opportunities for NEPTUNE Canada

With the final installation of the NC observatory infrastructure, a wide range of opportunities is generated (Table 16.9). Some stem from the new knowledge and scientific interpretations of the data and imagery, focused on the main science themes such as ocean/climate change, ocean acidification, recognizing and mitigating natural hazards, non-renewable and renewable natural resources. There are ample opportunities for commercialization of particular technologies and data services and products, which are more fully detailed in the next section dealing with socio-economic benefits.

16.6 Socio-economic benefits of NEPTUNE Canada

Although the NC project was funded on its scientific merits and innovation, there was a requirement to demonstrate the socio-economic benefits to both Canada and British Columbia. Not only will cabled ocean observatories transform the ocean sciences, there will be a range of applications in various scientific areas of the ocean and related sciences, but also applications in other sectors such as sovereignty, security, transportation, data services, and public policy (Table 16.10).

Using two specific examples for elaboration, the first being maritime emergencies, where cabled ocean observatories applied to both natural and anthropogenic emergencies may be short or long term in duration:

- *Short-term (natural)*: earthquakes, tsunamis, gas hydrate releases, slope failures, storms, and harmful algal blooms
- *Short-term (anthropogenic)*: oil spills, ship wrecks/collisions, and security crises
- *Long-term (natural)*: climate/ocean change, sea-level change, mass extinctions, and coastal flooding
- *Long-term (anthropogenic)*: fisheries collapse, diminished defense capabilities and maritime information systems

The second example is specific to the British Columbia coastal and offshore region. In 2004, the federal government passed the Oceans Act with a related legislation dealing with Canada's Ocean Strategy to assist in implementing the Act. For implementation, the federal government needed to gain the support of several of the provinces. The first to sign the Memorandum of Understanding was British Columbia. Implementation of several elements of the Ocean Strategy is being assisted by the NC and VENUS cabled ocean observatories, including:

- improving the scientific knowledge base for estuarine, coastal and marine ecosystems
- promoting research in monitoring and protection of Marine Protected Areas (Endeavour, Race Rocks, Gulf Islands Conservancy)
- generating data supporting pollution policy development (i.e., acoustic noise, chemical sensors)
- supporting new and emerging opportunities for oceans industries
- supporting ocean business development.

Overall, there are myriad applications that would directly or indirectly contribute benefits to Canada and British Columbia (Table 16.11).

16.7 Summary and invitation

This chapter has reviewed the development of the NEPTUNE Canada, the world's first regional cabled ocean observatory, from concept to operations. With the installation of the backbone cable system, shore station facilities, and the development of a more sophisticated DMAS, NEPTUNE Canada has transitioned from the installation to the operational phase. NC has received about \$100M support for the infrastructure and, more recently, interim operating funding ($13.2M through 2007 to 2010) from the Natural Sciences and Engineering Research Council of Canada (NSERC), CFI and the BC Ministry of Advanced Education. This was followed by operating funds of $19.3M for 2010–2012 from CFI and equivalent support (2012–2017) through CFI/MSI and BC provincial awards. With this support, NC staff has been expanded to nearly 50 members.

Public policy

Environmental
- Seismic detection
- Seafloor stability
- Tsunami detection
- Climate monitoring and prediction
- Conservation and protection; Marine Protected Areas
- Oceans Act

Resources
- Fisheries
- Oil and gas
- Gas hydrates

Public health and safety
- Earthquake/tsunami hazard
- Climate change
- Algal blooms/toxicity
- Oil spills and pollution
- Marine mammals/tourism

Security
- Transportation
- Acoustic detection
- Technology development and testing
- Port security

Economic development

Growth of marine technology industry
Innovative technologies
- Instrumentation; observing systems
- Cyber-infrastructure and data management
- Database mining
- Sub-sea operations
- Security systems

Tourism
- Ocean-related exhibits in aquaria and science centres
- National and marine parks
- Cruise boat industry

Education and public outreach

Internet access to real-time data and video
Bringing the oceans to the classrooms in real-time; student experiments
Bringing oceans to the public/policy makers/politicians through real-time programming (media, science centres, aquaria)
Exhibits in the Canadian Museum of Nature (Ottawa) and the Quebec Museum of Civilization (Quebec City)
Novel expression media (IMAX, Banff Centre, Emily Carr Institute for Art and Design)
Improving public understanding of the complexity of the Earth System (marine biosphere, climate, earthquakes, tsunamis)

Table 16.11 Benefits for Canada.

The NC team has achieved the building and initial operation of the world's first regional cabled ocean observatory. The observatory network is destined to transform the ocean sciences and provide a wealth of new opportunities (Tables 16.9–16.11). Along with several other countries planning cabled observatories over the next decade, the result will be a progressive wiring of the oceans. The observatory is designed to be expandable in its footprint, in the number of nodes and instruments, and provides a magnificent facility for testing prototypes of new technologies monitored and demonstrated in real time. NC and ONC (info@oceannetworks.ca) extend an invitation for new scientific, government and industrial participation, experiments, instrumentation, and data services.

Acknowledgments
We acknowledge the financial support provided for the installation phase primarily from the Canada Foundation for Innovation and the British Columbia Knowledge Development Fund. CFI, the Natural Sciences and Engineering Council of Canada, and the BC Ministry of Advanced Education, together with support for specific aspects from CANARIE and the University of Victoria, have provided funding towards the early operating phase. In-kind support, notably from Alcatel-Lucent Submarine Networks, and other grant awards have come from many partners and contractors noted in the text of this paper. The University of Victoria at all levels has been and remains committed in a myriad of ways to stewarding the observatory network, as part of Ocean Networks Canada, as a national facility and for which the NEPTUNE Canada team and its partners are deeply appreciative. Karen Douglas kindly assisted in the formatting of the final manuscript.

References

Barnes C.R., Best M.M.R. and Zielinski A. (2009) The NEPTUNE Canada regional cabled ocean observatory. Sea Technology 49(7), 10–14.

Barnes C.R., Best M.M.R., Johnson F.R., and Pirenne B. (2010a) Final installation and initial operation of the world's first regional cabled ocean observatory (NEPTUNE Canada). Canadian Meteorological and Oceanographic Society Bulletin 38(3), 89–96.

Barnes C.R., Best M.M.R. and Pirenne B. (2010b) NEPTUNE Canada cabled ocean observatory: Final installation and initial results. Innovation 14(2), 30–33.

Barnes C.R., Best M.M.R., Johnson F.R., Pautet L. and Pirenne B. (2011) Understanding Earth/ocean processes using real-time data from NEPTUNE Canada's widely distributed sensor networks, North-East Pacific. Geoscience Canada 38(1), 21–30.

Barnes C.R., Best M.M.R., Johnson F.R., Pautet L. and Pirenne B. (2013) Challenges, benefits and opportunities in installing and operating cabled ocean observatories: Perspectives from NEPTUNE Canada. IEEE Journal of Oceanic Engineering 28(1), 144–157.

Edwards R.N., Schwalenberg K., Willoughby E.C., Mir R. and Scholl C. (2010) Marine controlled source electromagnetics and the assessment of seafloor gas hydrate. In M. Riedel, E.C. Willoughby and S. Chopra (Eds) *Geophysical Characterization of Gas Hydrates*. Society of Exploration Geophysicists, Chapter 10.

Favali P., Person R., Barnes C., Kaneda Y., Delaney J. and Hsu S-K. (2010) Seafloor observatory science. In J. Hall, D.E. Harrison and D. Stammer (Eds.) *Oceans Observations '09, Proceedings,* Sustained Ocean Observations and Information for Society 2. ESA Publication WPP-306, doi: 10.5270/OceanObs09.cwp.28.

Munk W. (2002) The evolution of physical oceanography in the last hundred years. Oceanography 15, 135–141.

Pirenne B. (2015) Data Management and Archiving System for seafloor observatories: Acquisition, archival, analysis, interoperability. In P. Favali, L. Beranzoli, and A. Desantis (Eds) *Seafloor Observatories*. Praxis Publishing Ltd, Chichester, UK (this volume).

Purser A., Thomsen L., Barnes C., Best M., Chapman R., Hofbauer M., Menzel M., and Wagner H. (2013) Temporal and spatial benthic data collection via Internet operated Deep Sea Crawler. Methods in Oceanography 5, 1–18, doi: 10.1016/j.mio.2013.07.001.

Riedel M., Willoughby E. C., Edwards R.N., Hyndman R.D., Spence G.D., Chapman N.R., Chen M.-A., Novosel I. and Schwalenberg K. (2009) Gas hydrate offshore Vancouver Island, northern Cascadia Margin. In T. Collett, A. Johnson, C. Knapp, and R. Boswell (Eds) *Natural Gas Hydrates: Energy Resource Potential and Associated Geologic Hazards*. American Association of Petroleum Geologists Memoir 89, pp. 433–450.

Robert K. and Juniper S.K. (2012) Surface-sediment bioturbation quantified with cameras on the NEPTUNE Canada cabled observatory. Marine Ecology Progress Series 453, 137–149.

Rogers G.C., Meldrum R., Mulder T.R, Baldwin R. and Rosenberge A. (2010) First observations from the NEPTUNE Canada Seismograph Network. Seismological Research Letters 81(2), 369.

Rona P. and Light R. (2011) Sonar images hydrothermal vents in seafloor observatory. Eos, Transactions American Geophysical Union 92(20) 169–170.

Swidinsky A. and Edwards R.N. (2009) The transient electromagnetic response of a resistive sheet: Straightforward but not trivial. Geophysical Journal International 179, 1488–1498.

Thomsen L., Barnes C., Best M., Chapman R., Pirenne B., Thomsen R. and Vogt J. (2012) Ocean circulation promotes methane release from gas hydrate outcrops at the NEPTUNE Canada Barkley Canyon node. Geophysics Research Letters 39, L16605, doi: 10.1029/2012GL052462.

17 The ALOHA cabled observatory

B. M. Howe, F. K. Duennebier and R. Lukas[1]

17.1 Introduction

Sustained observation of the ocean is difficult. Ocean science requires new and varied ways to observe the ocean, each with its own strengths and weaknesses, in order to advance our understanding and lay the foundations for predictive models and their applications. Recent technological advancements, such as autonomous mobile platforms (floats, gliders and AUVs), satellite transmission and acoustic data links, are providing new data acquisition methods, but sensors, instruments and platforms requiring high power, high data rates or continuous access to the water column and seafloor are as yet out of reach of these technologies. Cabled observatories – such as the ALOHA Cabled Observatory (ACO) – provide remote interactive instrument control, continuous real-time data streams and large amounts of electric power. Cabled and autonomous technologies are complementary in that the cable systems can provide the power and communications interface to subsurface autonomous fixed and mobile platforms using docking stations and acoustic communications but without the routine use of ships, thus significantly increasing the spatial footprint and efficiency of the overall observing system.

Submarine cable systems have been used for science since the 1960s, with much work in Japan since the 1970s. In the 1990s, scientists in the United States began using cables to support their instrumentation and several systems were installed, e.g., LEO-15, ATOC, HUGO, and H2O (Forrester et al., 1997; ATOC et al., 1995; Duennebier et al., 2002; Petit et al., 2002; an historical account is given by Howe et al., 2012). Two technologies opened the door for these observatories: remotely operated vehicles (ROVs) with power, high-resolution video, manual dexterity, heavy lift capability, long continuous work time at the

1 School of Ocean and Earth Science and Technology, University of Hawaii, Honolulu, Hawaii, USA

ocean floor, and wet-mate electrical and optical connectors allowing instrumentation to be attached to the observatory by the ROVs. These developments and projects, and the associated planning efforts, have led to many of the ocean- and Earth-observing systems that are described in this volume.

As the design of new cable systems progress (Favali and Beranzoli, 2006) and experience with test bed systems accumulates, the realities of the design, cost, and complexities – as well as advantages and disadvantages of such cable systems are becoming clearer. The full-scale new cable systems coming on line, NEPTUNE Canada, RSN (Regional-scale Nodes of the NSF Ocean Observatories Initiative), and the Japanese DONET, will provide important contributions to cabled observatory efforts. At the same time, the re-use of existing cables provides a cost-effective complementary path to the deep ocean with minimal compromise in capability.

The ACO is a prototypical example of a deep ocean system that uses a retired fiber optic communications cable. The ACO architecture uses highly reliable existing transoceanic cable systems to provide power and communications bandwidth. The cables are terminated on land at existing cable stations that contain the infrastructure required to power and control the observatories and transmit commands and data. Since the cable is already in place and is designed to operate for well beyond its commercial lifetime, costs of converting retired cable systems to scientific use are substantially lower than for design and installation of new systems.

In the following section the background of the ACO is presented. This is followed by a description of the ACO infrastructure. Section 4 discusses current and future science and engineering topics that the ACO has or can address. Concluding remarks follow, including an epilogue briefly describing the recent installation of the complete node and the current operating status.

17.2 Background

The ACO is located at Station ALOHA (22°45'N, 158°W; see Figure 17.1), about 100km north of Oahu, Hawaii. Since October 1988, the Hawaii Ocean Time-series (HOT) program has investigated temporal dynamics in biology, physics, and chemistry at Station ALOHA, a deep-ocean field site in the nutrient poor North Pacific Subtropical Gyre (NPSG). HOT conducts near monthly ship-based sampling and makes continuous observations from moored instruments to document and study NPSG climate and ecosystem variability over semi-diurnal to decadal time scales (Karl and Lukas, 1996; Karl et al., 2001). The recent work of Dore et al. (2009) on ocean acidification trends and dynamics at ALOHA stands out. Understanding the impacts of climate forcing on the ecosystem (Bidigare et al., 2009) is only possible by having long time-series observations coupled with constantly evolving and improving ocean models.

Long-term change in parameters such as salinity and rare and episodic temperature events at Station ALOHA (Lukas et al., 2001; Lukas and Santiago-Mandujano, 2001; Alford et al., 2011) can be better observed with the new capabilities of the ACO because

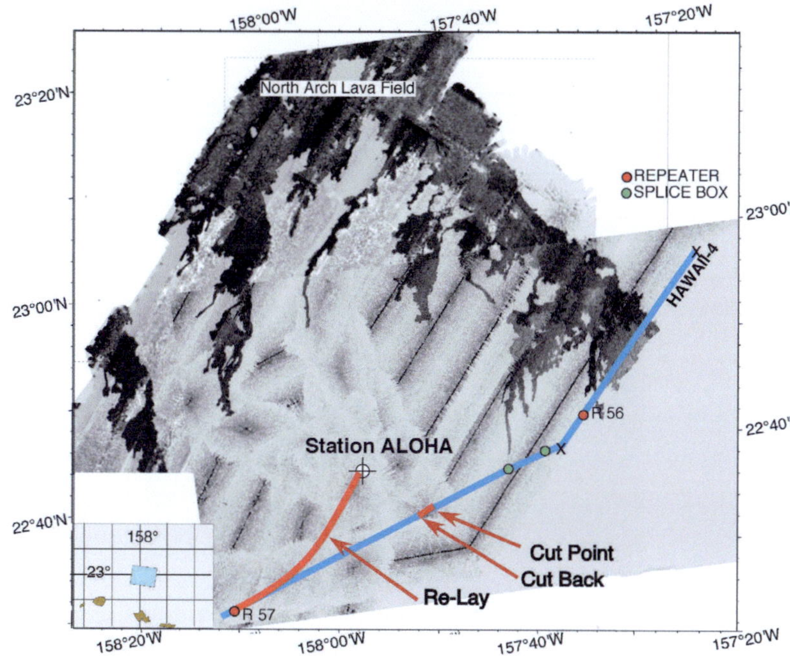

Figure 17.1 Station ALOHA and the ALOHA Cabled Observatory installation. The blue line shows the original cable route, and the red shows the re-laid cable path. The background is EM-120 side-scan swaths taken during site surveys for this project and processed by the Hawaii Mapping Research Group. The dark regions are part of the north-arch lava flows with high backscatter. The flat sediments have low backscatter. The sea cable termination is located at 22° 44.3211' N, 158° 00.3698' W at 4726m water depth.

sustained real-time detection of events will support adaptive sampling within and around the observatory.

In 2002, the ACO had been funded as a Major Research Instrumentation project by the US National Science Foundation (NSF) and was in the process of obtaining permission to use the retired ANZCAN coaxial cable that ran near Station ALOHA. Teleglobe, Inc., the main owner, went into bankruptcy and negotiations failed. HAW-4, a newer first-generation optical cable system, was then retired and made available for the ACO by AT&T. A meeting of cable engineers at Rutgers University in 2003 led to a specific design for cable hardware that would provide data transport through first-generation optical cable systems (Tremblay and Duennebier, 2006). A test of the production hardware in 2006 demonstrated the capabilities by sending 100 Mb/s data from the Makaha Cable Station on Oahu to a cable station in California where it was looped-back to Oahu and recorded error-free, a round-trip path of more than 8000km.

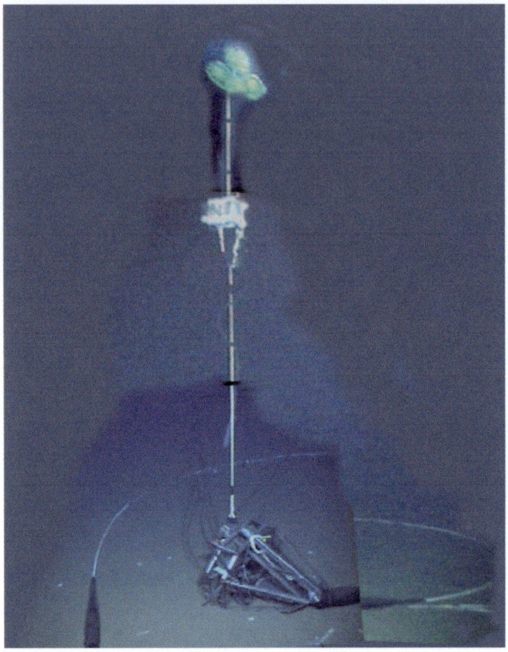

Figure 17.2 The ACO sea cable termination and proof module. The proof module with hydrophone and pressure sensor and the seawater ground are on the short mooring that was deployed in February 2007. ROV *Jason* took this composite image in October 2008

Installation of the ACO was organized in two phases. During the first phase, the HAW-4 cable was cut, both ends terminated, and 20km of cable was relaid to Station ALOHA. This operation was accomplished in February 2007 using the cable ship USNS *Zeus* (Figure 17.1). The cable termination was lowered to the ocean floor with a set of electronics, a hydrophone, and pressure sensor (the "Proof Module") to provide proof of concept and assurance that the system would operate as designed (Figure 17.2). Data began flowing to Oahu shortly after the sensors were in the water and ran continuously (except for brief, planned outages) until the package was recovered in October 2008.

During Phase 2, a cruise to install the ACO general-purpose observatory node planned for October 2007, had to be canceled because of late delivery of faulty (i.e., obviously cracked) titanium pressure cases for the observatory electronics. A rescheduled cruise in October 2008 recovered the proof module and deployed the ACO node (Figure 17.3), but failure of optical dry-mate connectors between the junction box and the observatory module forced its immediate recovery. The plan had been to install the main ACO infrastructure, two hydrophones, an absolute pressure sensor, two acoustic Doppler current profilers (ADCPs), a vertical thermistor string, two temperature/conductivity sensors (CTDs), and an acoustic modem supplied by Woods Hole Oceanographic Institution (WHOI). The therm-

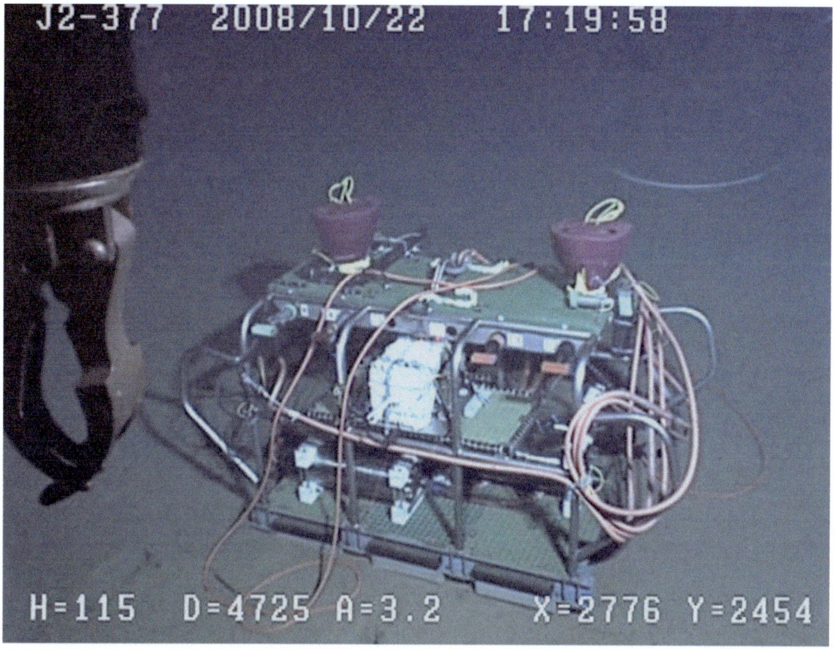

Figure 17.3 The ACO observatory module. ROV *Jason* took this composite photograph in October 2008, just after the module was set on the bottom.

istor string was deployed in an autonomous mode, to record temperatures with sampling for up to three years, but the other systems could not be installed.

Phase 3 consisted of observatory upgrades, testing and deployment. Upgrades were based on lessons learned (e.g., connectors) and new technology (e.g., precise timing). Testing was extensive, facilitated by special test connectors and system modularity. Deployment used the RV Kilo Moana and the ROV *Jason*, with several new instruments including a camera system and a secondary node with a fluorometer and additional CTDs with oxygen sensors.

17.3 Infrastructure

The ACO uses technology from the HUGO and H2O observatories, with improvements based on lessons learned from both (Duennebier et al., 2008), as well as from the October 2008 attempted Phase 2 deployment. In the following subsections we describe the Phasc 3 system elements (Figure 17.4), making brief reference to changes relative to Phase 2. Phase 3 modifications occurred in 2009–2011 and deployment in May–June 2011.

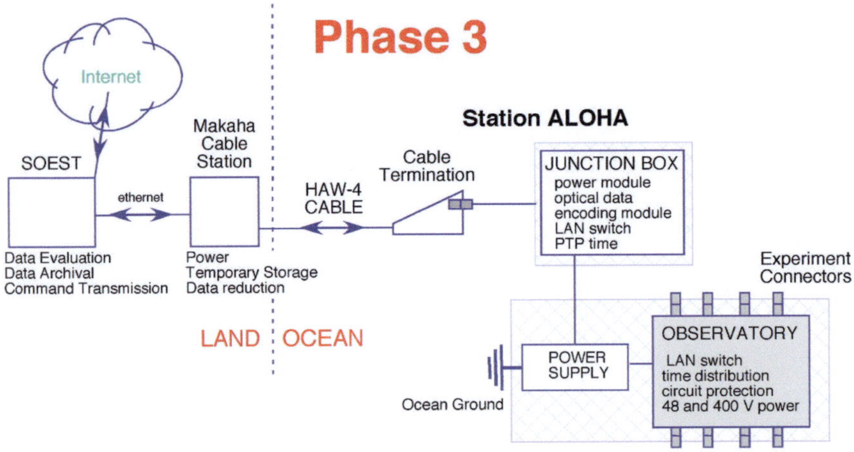

Figure 17.4 The ACO block diagram for Phase 3, deployed in June 2011.

17.3.1 Shore station and cable

At the AT&T Makaha shore station, the shore power system supplies the single conductor sea cable with a constant 1.6 A at a (negative) voltage that varies depending on load, up to 1280 V for the present observatory configuration. Because the cable is single conductor, the sea water is used as the second electrical conductor, with a cathode (positive) just offshore of the shore station and an anode (negative) at the most seaward end of the electrical system. AT&T communications equipment provides the interface between the cable system and the Internet through the University of Hawaii (UH) via a dedicated 3 Mb/s line. The interface includes access to the supervisory functions, such as repeater status. A grandmaster clock synchronized to GPS time provides system time at the sub-microsecond level using the precision time protocol (described below). At the seaward end of the cable is a titanium and fiberglass frame with the cable termination, an Ocean Design, Inc. (ODI) hybrid electro-optical wet mateable connector.

17.3.2 Junction box

In the junction box, an AT&T SL560 regenerator is used to interface to the cable optical fibers and convert between optical and electrical signals. A custom multiplexer ("muldex") board is used to interface these electrical signals with 100 Mb/s optical Ethernet. In essence, the muldex samples the lower data rate Ethernet signal at a higher data rate suitable for the SL560, and vice versa. Details are presented in Tremblay and Duennebier (2006). In Phase 3, the optical Ethernet is connected to a managed industrial Ethernet switch with electrical ports connected to the observatory module and a local time server.

Precise and accurate time is obtained using the IEEE-1588v2-2008 precise time protocol (PTP) that provides sub-microsecond accuracy and precision over Ethernet. Switches following this protocol append packets to the Ethernet stream that contain the measured time-varying latency within the switch. Further, the drifts of all clocks in the system are monitored and delays through the system are measured many times a second. A PTP client connected to the Ethernet switch provides 1-PPS and IRIG-B signals.

The Ethernet, 1-PPS, and IRIG-B are provided on a 12-pin ROV wet mate connector. Normally, this is connected to the observatory module (as shown in Figure 17.4), but it can also be connected, for example, to the hydrophone experiment module (HEM) so the combination becomes a "proof module" as in Phase 1.

Three power modules based on Zener diodes supply local power for the regenerator/muldex, the Ethernet switch/PTP, and 20 W at 48 V. The high voltage is passed through, provided on a 4-pin ROV wet mate connector. Normally, this is connected to the observatory power supply followed by the anode (seawater electrode) to complete the series circuit. If the junction box is operated by itself or with the HEM, then this is connected to the seawater electrode. The 48 V is supplied on the 12-pin connector, which then has the standard "observatory configuration" (see below).

17.3.3 Power system

Submarine commercial telecommunications cables, such as HAW-4, are single conductor, use the low resistance seawater as the return path, and are double-end powered by shore stations on each end. They operate in constant current, constant power mode for several reasons: robust immunity to single electrical ground faults, and power supplies are simple because loads (the repeaters) are constant.

After the cable is cut for scientific use, the power system is supplied from only one end and voltages vary depending on the requirements of the observatory. To deliver power to an experiment at the ocean-end of such a cable as for ACO, the power available from the cable needs to be converted in two ways: (1) the power delivered to the experiment must be isolated from the voltages on the cable; and (2) the power delivered to the experiment must match the amount required. Such conversions are very common in electronic design. The power required and the source voltage define the converter operating point (input voltage and current).

In the more usual (terrestrial) constant voltage configuration, with the converter close to the source, the input voltage is fixed and the input current adjusted as necessary (e.g., Howe et al., 2002). If the convertor requires more power at a fixed voltage, more current is drawn from the source, up to the power handling limits of the converter and delivery capability of the source. Such converters operate as constant power devices. They are designed to handle a relatively wide range of input voltages (2:1 or more) and a much wider range of output currents up to a maximum amount of power. These convertors are quite stable when supplied from a voltage source.

But this constant power mode produces a stability problem when the input power is supplied in the form of a constant current. Assume that the convertor is operating at an input voltage and current and thus power that supplies the needed output power, and are within the limits of the convertor. If the needed power increases slightly, the input power must

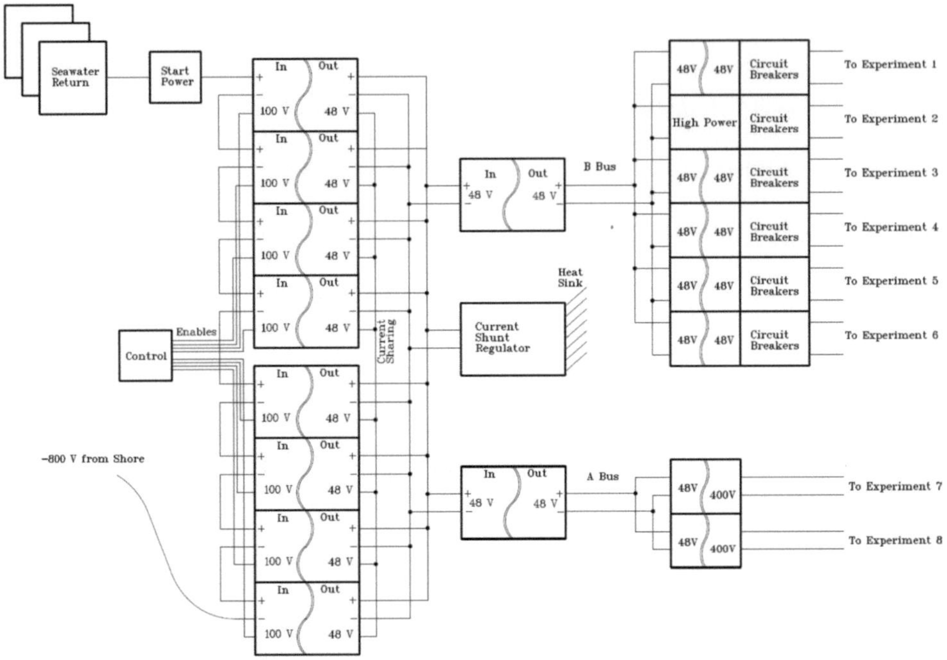

Figure 17.5 The ACO power supply block diagram. The observatory uses a stack of relatively low-voltage converters to extract sufficient power from the 1.6 A constant current from the cable.

also increase. So the input current tries to increase. The current increase can be supplied for a short period by discharging the capacitance but this soon results in a drop in the input voltage to the converter. This in turn causes the converter to try to draw even more current from the cable and the voltage rapidly collapses to a low voltage at the constant current.

This instability can be managed by using active current shunts that consume the excess power if the load power decreases. Sensing the input voltage to the converter controls the amount of power consumed by the shunts. If the voltage is higher than desired, the current to the shunts is increased to return the voltage to the correct level. The power consumed by the shunts is converted into heat and passed into the ocean water around the observatory.

An active shunt can be thought of as a resistor and a field effect transistor (FET) in series. The observatory hotel and user load is in parallel with the active shunt. The FET can be (relatively) instantaneously controlled so the total current going through the observatory load and through the active shunt (FET + resistor load) is the required constant amount. ACO uses several of these shunts in parallel to distribute the heat and provide redundancy. The active shunts must be designed to be able to consume all the power that can be delivered by the converter when the experiment is turned off or not connected (FET "open" fully conducting). The maximum power available to the experiment occurs when the current shunts are consuming zero power (FET "closed" non-conducting). Any attempt to draw more power will result in the voltage collapse described above.

When the input current is fixed, as it is in our case (because of in-line repeaters), the input voltage must increase to increase the available output power. Most commercially-available modules, however, are limited to ~200 W. To increase the available power, we use multiple convertor modules. The inputs are connected in series to increase the voltage drop along the current loop to 800 V, which increases the power that can be delivered from the cable to 1280 W (at 1.6 A). The outputs are connected in parallel and operated at a constant voltage. To be able to connect all the outputs in parallel, of course, the outputs must be isolated from the inputs (Figure 17.5).

With the outputs properly connected in parallel, the convertor sharing circuitry will ensure that each converter delivers the same amount of current as the others. Since the same current passes through the input of each module, each module converts the same amount of power as the others. The ACO power supply uses eight convertors, each with a 100 VDC nominal input and a regulated 48 VDC output. Each convertor is capable of 200 W.

The parallel output reduces the effect of power consumption changes in any one experiment by summing all the experiment currents. It also allows the design of low voltage current shunt regulators to enable a stable operating point. See Harris and Duennebier (2002) for more details on the design of a constant current power supply of this type.

Although the 48 V now is a stable source of power for the experiment modules, other converters are used to provide isolated power to each experiment. Each of these isolated outputs is further protected by circuit breakers programmable to disconnect the experiment power if it exceeds the current specified by the system operator. Because the main power supply can deliver only 1280 watts maximum (8 stacks × 100 V × 1.6 A), any one experiment can obtain the rated power but not every user can get the maximum power (see below).

17.3.4 Observatory module

The observatory module has eight science channels or ports supplying power (two ports at 400 V/300 W, five ports at 48 V/150 W, and one port at 48 V/500 W), 100 Mb/s Ethernet (some ports can accommodate serial RS422 and RS232), and 1-PPS and IRIG B timing to science users. The associated connectors are identical to the one on the junction box, and to those used on MARS, NEPTUNE Canada, DONET, and planned for RSN, so that experiments should be interchangeable. (This informal, de-facto connection interface "standard" was arrived at by consensus during numerous meetings of the various planning and engineering groups over the last decade or more.) From the power supply 48 V user bus, the power on each channel goes through separate, controlled dc–dc converters to provide isolation and on/off switching.

Both sides of the 48 V are monitored by the system to determine possible ground faults. As long as the voltages are near +24 V and –24 V with respect to the Observatory pressure case, none of the supplied circuits are connected to seawater. Multiple microprocessors, each connected to the Ethernet, are used to control the power and communications associated with the individual channels.

17.3.5 Other system aspects

Based on the experiences from the October 2008 deployment attempt, three modifications or upgrades have been performed. The first change is to modularize the mechanical system, separating the junction box pressure case from the observatory frame. With this configuration, the highly reliable junction box can remain in place with the proof module (or any other low power experiment) if the observatory node needs to be recovered. If expansion of the system is required, a new observatory/power supply module can be inserted between the current junction box and the observatory frame without recovery of any in-place modules. Further, testing is simplified.

Secondly, by converting the optical Ethernet (the output of the muldex) immediately to electrical form in the junction box, the need for downstream optical connectors is eliminated. The third change makes use of the new PTP capability (mentioned above) that greatly simplifies system time distribution.

Lastly, several other problems associated with cable and connector assemblies were experienced during the 2008 cruise. One was difficulty in the physical mating of the hybrid optical–electrical connector, connecting the junction box and the sea cable termination. The manufacturer has now provided alignment guides to mitigate this problem. Once the connection is made between the junction box and the sea cable, the connection should (ideally) never need to be broken again, thereby reducing future risk. Further, since 2005 when the original cable/connector assemblies were made, the manufacturer has instituted new practices and testing procedures that will reduce overall risk associated with cable and connector assemblies.

All exposed metal components of the observatory are titanium. In most cases, bulkhead connectors on pressure cases are dry mate. Pressure balanced oil filled (PBOF) hoses are used to connect these dry mate bulkhead connectors to flange mounted ROV wet mate connectors, as well as between wet mate connectors on the observatory frame, the junction box frame, and the sea cable termination. This modularity permits parallel independent testing of the various components.

Data management and archiving, along with interactive control to facilitate adaptive sampling, are essential elements of an ocean observatory and will be developed for ACO during Phase 3. Our experience to date indicates that the efforts associated with data management, archiving, processing, and scientific quality control and product presentation should not be underestimated.

17.4 Research

The motivation for deploying the ACO infrastructure is to conduct research that cannot otherwise be accomplished in the abyssal and overlying ocean. Compelling scientific research questions have been posed that only measurements from the ACO can uniquely and cost-effectively address. There are also engineering research and development issues that can be addressed within the framework of the ACO, some related to observatory infrastructure and some related to sensors. The scientific questions and engineering issues that will be raised by the exploratory measurements made via the ACO are equally important.

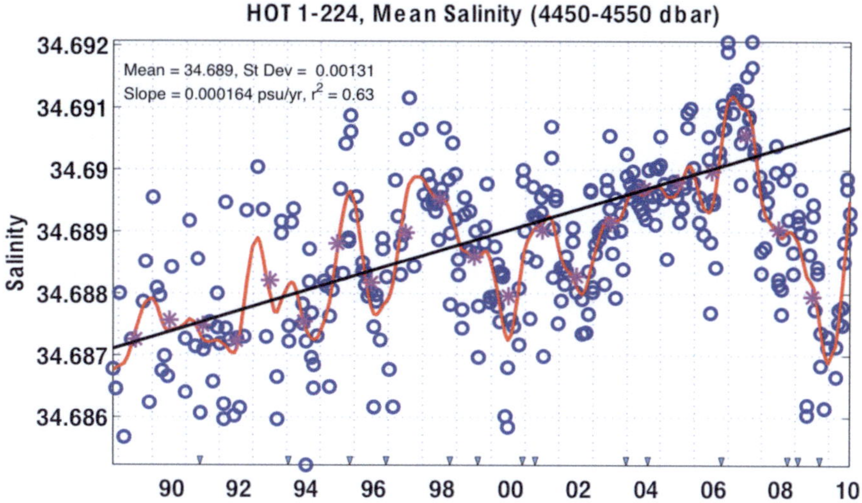

Figure 17.6 Observed deep salinity variation and trend. The inverted triangles near the time axis indicate the start of cold events shown in Figure 17.7.

17.4.1 Research with core measurements

Because the effort and resources that are required to deploy the ACO are significant, it is essential that there be initial research returns from that effort. Given the long lead time for designing, proposing, obtaining funding, and the ship and ROV scheduling needed to add experiments to the ACO, a "core" set of measurements will be made as part of the ACO deployment. These core measurements will serve the multiple purposes as outlined above, and will also provide a long time-series context for subsequent observatory-based experiments. The core measurements are summarized in Table 17.1. An example of what might be expected for the basic core measurement of salinity and temperature is given in Figures 17.6 and 17.7 (updated from Lukas and Santiago-Mandujano, 1996). In Figure 17.6, the increasing trend in abyssal salinity as measured by the HOT program is shown along with the high frequency variations around the trend. The relatively recent freshening of this water mass (Nosse et al., 2012) may be related to Rossby waves associated with the 2007–2009 La Niña. Figure 17.7 shows large amplitude (for the abyss) cold events with sudden onsets. Lukas et al. (2001) attributed these events to episodic flow of abyssal waters from the Maui Deep over the shallower ridge of the Nuuanu Seamounts into the Kauai Deep, where Station ALOHA is located. The long recovery time from cold events is due to relatively slow vertical turbulent diffusion of heat. Note there is no obvious correspondence of salinity variations (Figure 17.6, triangles) with the cold overflow process. The high frequency sampling of the observatory will allow filtering out of internal wave noise in the salinity record that can not be resolved by the sparse shipboard sampling.

Platform/Sensor suite	Sensors	Comments
Thermistor array/ acoustic modem (TAAM) mooring from seafloor to 200m	Seabird SBE37 thermistors 10 equally spaced (2 w/pressure) Wetlabs FLNTU fluorometer (chlorophyll) and backscatter (turbidity) WHOI 10 kHz micro-modem	A physically coupled dual mooring. Battery-operated thermistor array and one fluorometer communicate with the ACO using Seabird inductive modem. The thermistors also record internally. The modem is directly connected to the observatory.
Camera frame CAM	AXIS 214 PZT color video camera 2 LED Multi SeaLites lights Hydrophone	Internet surveillance camera in 17in polished glass sphere.
Observatory node OBS	Conductivity/temperature CT Seabird 37 (un-pumped) 2 Sontek 250 kHz ADCP	On top of node frame; CTD swings outboard for some horizontal separation from the electronics.
Hydrophone experiment module (HEM)	Two hydrophones, bandwidth 0.01–5000 Hz and 24–40,0000 Hz Digiquartz pressure, 16 Hz	On separate sled, can work directly off junction box (as proof module during deployment), or standard science connector on node (switched over after node deployment).
AMM secondary node	2 Seabird 52/43MP $CTDO_2$ Wetlabs FLNTU	Primary sensors Connected to a SIIM, AMM node and then to the OBS.

Table 17.1 ACO core measurements.

The ACO node frame is deployed with two upward looking ADCPs and a MicroCAT CTD connected to the µSEM (micro science experiment module) that is dry-mate connected to the main observatory node pressure case. There are two co-joined 200m-high cabled components (on separate experiment ports) integrated into the thermistor array/ acoustic modem (TAAM) mooring system. One cabled component is a 10-element thermistor array plus one fluorometer, all self-powered with batteries for at least two years but communicating in both directions through the ACO via an inductive modem; the second provides power to and two-way communications with a 10 kHz WHOI micro-modem system. The acoustic micro-modem will communicate with nearby autonomous systems such as the NSF-funded HOT profiler mooring installed in 2012.

The original hydrophone/pressure proof module was reconfigured to use the standard 12-pin connector, so it can connect either to the junction box as a proof module or to an experiment port on the observatory node (the hydrophone experiment module, HEM). It now has two hydrophones separated by 1.2m with combined effective bandwidth of 0.01 Hz to 40,000 Hz; sampling rate can be remotely selected up to 192 kHz. A secondary node with

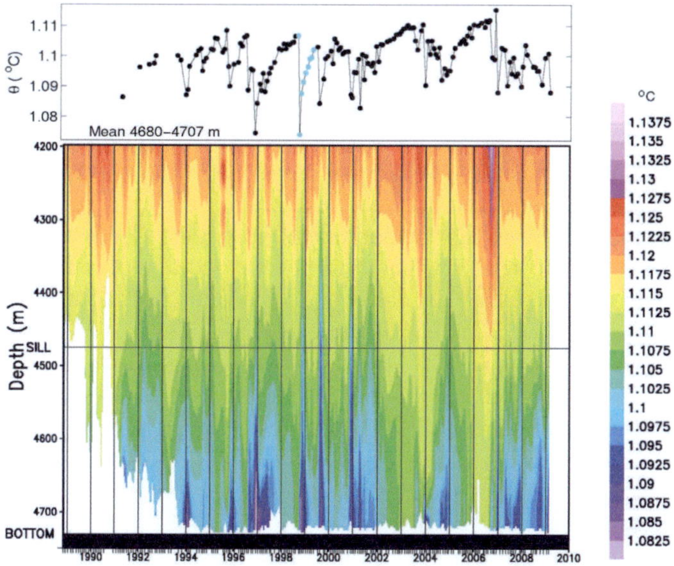

Figure 17.7 The intermittent cold, deep overflow. Near-bottom potential temperature (colors; values given by color bar on right side) at Station ALOHA in the depth-time domain (bottom panel) from Conductivity-Temperature-Depth profiles. Hawaii Ocean Time-series (HOT) project cruises, roughly one month apart, are indicated along the time axis. The sill depth of the Nuuanu Seamounts ridge is indicated by a horizontal line; the ocean bottom is indicated by the solid black area. Potential temperature averaged over the depth range 4680–4707m for individual cruises is shown in the top panel.

two CTDs with oxygen sensors and a fluorometer will sit next to the node. An Ethernet video camera with lights and another hydrophone also sits on a tripod nearby.

An example of an engineering issue to be addressed is the utility of acoustic Doppler measurements to obtain estimates of abyssal current profiles. We simply do not know the concentration climatology of the backscattering particles needed to obtain reasonably accurate deep ocean currents, and thus how far above the seafloor such profiles can be obtained. The two ADCPs will allow us to conduct exploratory measurements for the design of future abyssal current measurements. Optical measurements will help quantify the backscattering particle density, which may be variably related to near-surface productivity events and to strong near-bottom current events that suspend sedimentary materials.

Another engineering issue concerns the calibration stability of conductivity sensors deployed near the bottom. Can the relatively small, but important, salinity signals that are observed in the abyssal ocean (Figure 17.6) be observed with high fidelity via the ACO? Will the accumulation of sediments be an important factor?

An important engineering unknown for future observatory design that needs to be explored is the range and fidelity of acoustic modem communications in the abyssal ocean.

By placing an acoustic modem sufficiently far above the seafloor, received signals can be listened to with the dual hydrophones that are proposed as part of the core sensor suite, while the micro-modem can receive direct signals from other sources, such as ships, gliders and moorings, without the complication of bottom and observatory infrastructure reflections.

The ACO core measurements should be of broad interest, not just related to a current scientific hypothesis; it is intended that these exploratory core measurements inspire new hypotheses that may motivate new experimental proposals. Given the need to site an acoustic modem above the seafloor, the cost-effective addition of inductively communicating temperature sensors to the riser cable will provide the vertical structure of temperature variations around the ACO increasing the spatial footprint of measurements. The cold overflow events observed by the HOT program will be continuously monitored, with the possibility of measuring the full spectrum of thermal responses to overflow events which are likely important (e.g., Figure 17.7; Lukas et al., 2001; see more recently Alford et al., 2011). Given that these sensors are battery powered, their effective and efficient use will require commanding hibernation and awakening adaptively for event sampling, a compelling topic of research in itself.

Bio-optical sensors on the riser cable and on the ACO itself will help to quantify the scattering particle density, which will aid the interpretation of ADCP return intensity variations, and will help understand the CTD calibration changes due either to particle fall events that may be associated with near-surface productivity blooms, or to resuspension of bottom sediments during strong flow events. Video frame grabs will support categorization of particles and enable estimation of currents as the particles are advected within the field-of-view. Particle export events to the abyss will be observed, and their relationship to near-surface productivity blooms will be studied (see Figure 17.8).

The dual hydrophones will provide redundancy and some level of directionality to the acoustic measurements that have already been shown to be of interest to widely separated scientific communities. Along with absolute pressure measurements, the variations of sea level will be measured directly, and the acoustic spectrum will provide useful constraints on the sea surface elevation spectrum. Signals from earthquakes, tsunamis, and human activities will provide information about the solid earth that will be exploited. Marine mammal acoustic signatures are already the focus of several research groups using the data obtained from the proof module deployment (Oswald et al., 2009, and Ou et al., 2012).

Because the acoustic data collected during Phase 1 has proven to be exceptionally rich, we provide some detail here. Data were collected for ~20 months from February 2007 to October 2008. 1.5-minute time samples of 24 kHz were sampled from the data every 5 minutes to provide power spectral densities of the acoustic data from 0.0114 Hz to 10 kHz (6 decades in frequency, 20 octaves). The 12 spectra obtained each hour were averaged to provide hourly spectra. The statistics of the ambient sound and the dependence on wind speed using these spectra are given in Figures 17.9 and 17.10, respectively.

The acoustic data span more than 150 dB in energy level. The log of the number of samples at each frequency with a particular energy level ("sample density") is contoured in Figure 17.9, allowing visualization of measurements from a single hourly observation in the 20-month period to over a thousand observations. The highest energy levels are observed at the microseism peak near 0.2 Hz. Note the relatively low range of spectral values

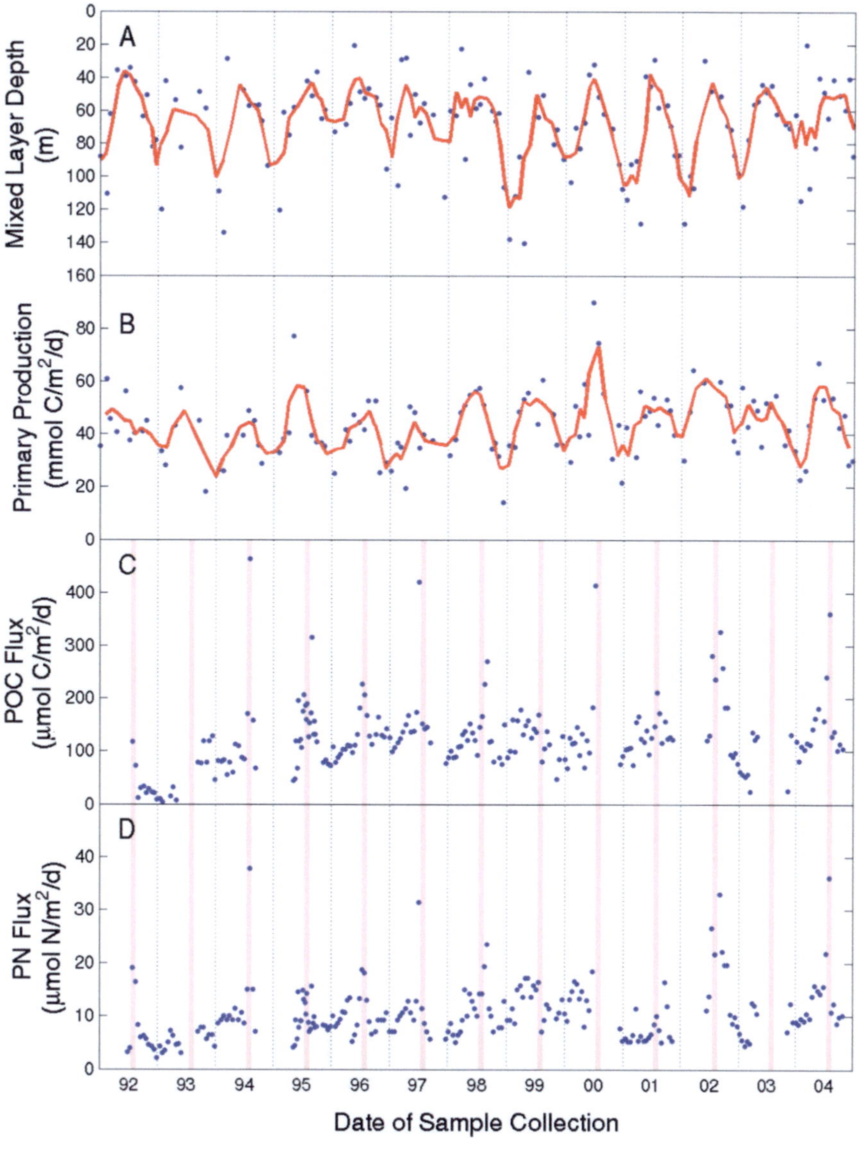

Figure 17.8 Deep export events. A composite of key environmental variables observed at Station ALOHA for the period 1992–2004. [A] Mixed-layer depth, [B] Primary production, [C] POC flux at 4000m, [D] PN flux at 4000m. The data in [A] and [B] were collected on approximately monthly Hawaii Ocean time-series cruises (closed blue circles) and the red trend lines are 3-point mean values for each parameter. The data in [C] and [D] were collected in bottom-moored time-series sediment traps. The lightly shaded period in [C] and [D] corresponds to the period 15 July–15 August when deep sea particulate matter fluxes are at their annual maxima. (Courtesy of D. Karl and R. Letelier.)

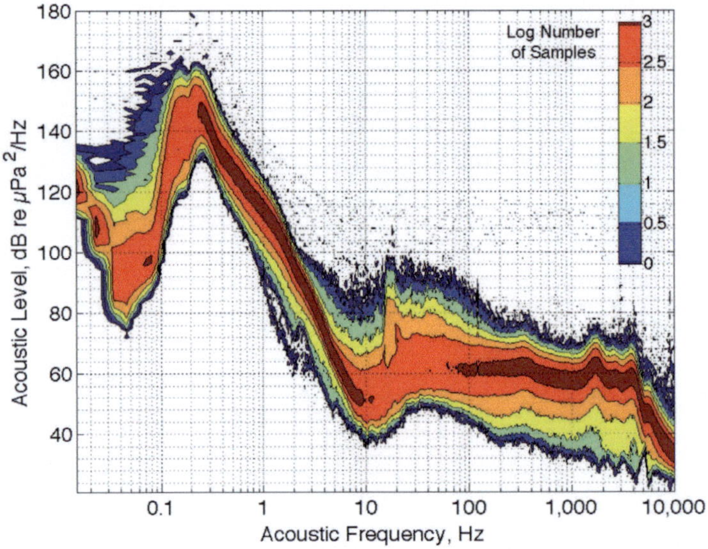

Figure 17.9 ACO acoustic spectrum. Histogram of occurrences of all hourly acoustic data from 0.0145 Hz to 10 kHz. Colors represent the log of the number of hours where each level was observed at each frequency. (From Duennebier et al., 2012.)

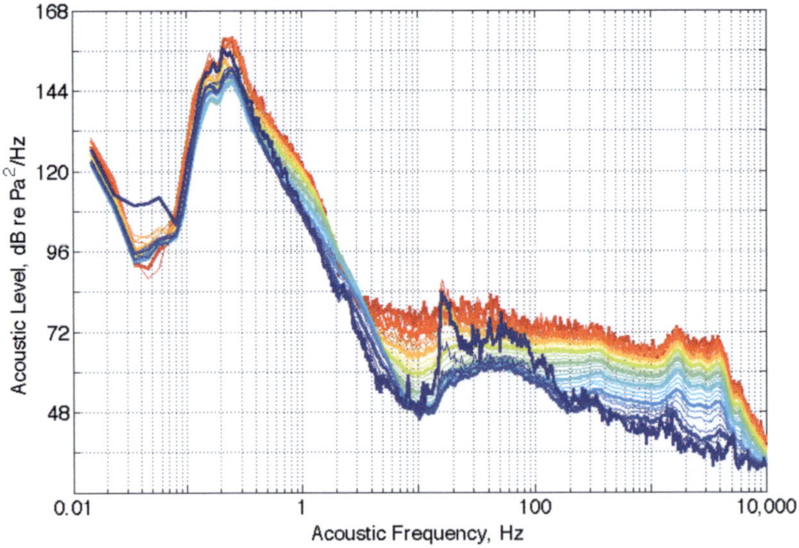

Figure 17.10 ACO acoustic level and wind speed. Sorted spectra are plotted by color in 0.5m/s steps from zero to 15m/s with heavy lines for even wind speeds from zero (dark blue) to 14m/s (red). (From Duennebier et al., 2012.)

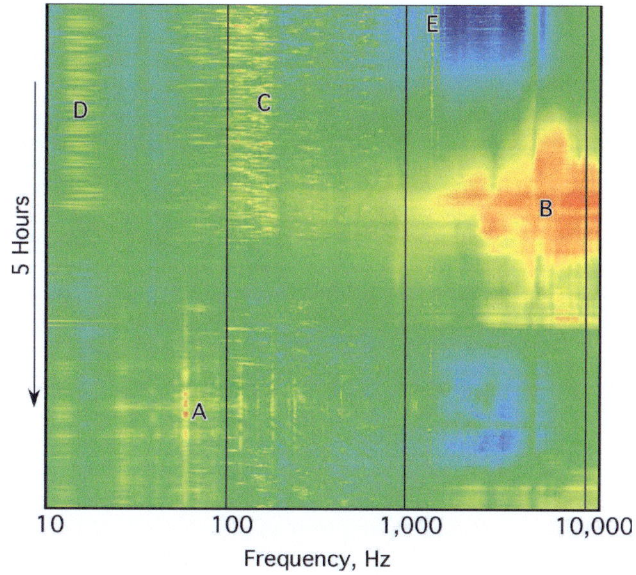

Figure 17.11 Waterfall spectrogram of ACO hydrophone data. Red colors signify more energy relative to blue colors. (A) a ship passage, (B) a rainstorm, (C) humpback whales, (D) fin whales, and (E) minke whale.

between 2 and 6 Hz compared to other frequency bands. A sharp increase in the range of spectral levels above 16 Hz continues to at least 4 kHz. A peak near 16 Hz may be associated with blue whale vocalizations (Diachok and Duennebier, 2009). While the general trend of spectral levels decreases between 14 Hz and 5 kHz, the median level is stable at near 60 dB. Resonances above 2 kHz are the result of the physical size of the sensor being close to the acoustic wavelengths. The ambient background level is above instrument noise level from below 0.02 Hz to about 4 kHz, where the system noise floor is roughly 30 dB re $\mu Pa^2/Hz$.

The hourly-averaged acoustic spectra were sorted by wind speed into 0.5m/s bins, and the median spectral levels at each wind speed were determined and plotted in Figure 17.10. Higher acoustic levels are generally associated with higher winds except for the noise hole around 0.04 Hz. The maximum range of the median acoustic level with wind speed (about 30 dB) is observed near 2-4 kHz, with acoustic levels correlating with wind speeds as low as 1m/s. The variation of acoustic level with wind speed is largest near 10 Hz at wind speeds between 7 and 15m/s, changing at a rate of about 4 dB/m/s. Noise unrelated to the wind speed puts a floor under acoustic levels at lower wind speeds. Detailed analysis of noise levels between 50 and 500 Hz recorded in 1975 by Gaul et al. (2007) shows a strong correlation with high wind speeds that is limited by shipping noise at wind speeds below 2.5m/s. Their analysis yields acoustic levels that are within 2 dB of those presented here near 500 Hz but their noise floor is lower at low wind speeds. This difference may be the result of their selection for times when local shipping is absent, while the data presented here have not been selected.

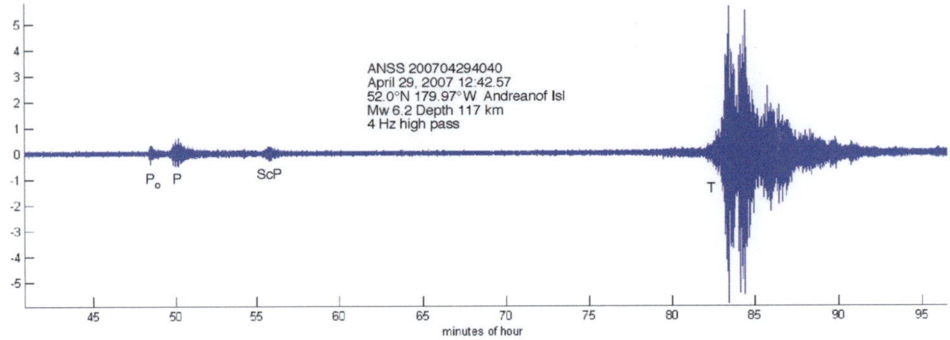

Figure 17.12 An Aleutian earthquake recorded by the ACO proof hydrophone. 4 Hz high pass filtered.

Figure 17.11 shows a representative sample of ambient sound, including various whales, shipping, and a rain event. Research will continue on the surface wave–wave interactions that are observed in the hydrophone data (Farrell and Munk, 2008; Duennebier et al., 2012), as well as seismic signals from various sources (Figure 17.12).

17.4.2 Examples of future research directions

While we do not know the full variety of innovative ideas that will be proposed to take advantage of the ACO infrastructure, we provide some examples of viable future experimental research activities that have been extensively discussed. They illustrate the transformative nature of the observatory approach to deep ocean research.

Mooring sensor systems. A mooring sensor network with profiler has been designed and tested in shallow water (ALOHA-MARS mooring (AMM); Howe et al., 2010). A similar if not identical mooring system will be proposed for installation at ACO to address numerous science goals (Figure 17.13). In the meantime, the "HOT Profiler" mooring project has installed an autonomous battery powered (two 0.8m steel spheres with alkaline batteries) mooring system with a profiler, using an inductive power transfer system similar to the AMM. This mooring was deployed in June 2012 with enough on-board energy to run the profiler and other systems continuously for half a year. It has several means to communicate. First is an acoustic modem that can "talk" with the ACO acoustic modem and hydrophones as well as with the ship during routine ALOHA-HOT visits. The second is via a small, slack tethered surface buoy with GPS, Iridium and FreeWave radio. Ultimately, the cabled mooring system in Figure 17.13 will enable high-resolution vertical profiles of physical, chemical, and biological quantities over the entire water column. This will eventually allow shifting some of the burden of the routine HOT sampling to this automated system (with much improved sampling), freeing the human and ship resources to focus more on new, more difficult and challenging sensors and sampling.

Tsunamis. The absolute pressure sensor in the Phase-1 system had a resolution of better than 5mm of equivalent sea level, enough to detect a small tsunami, although none were

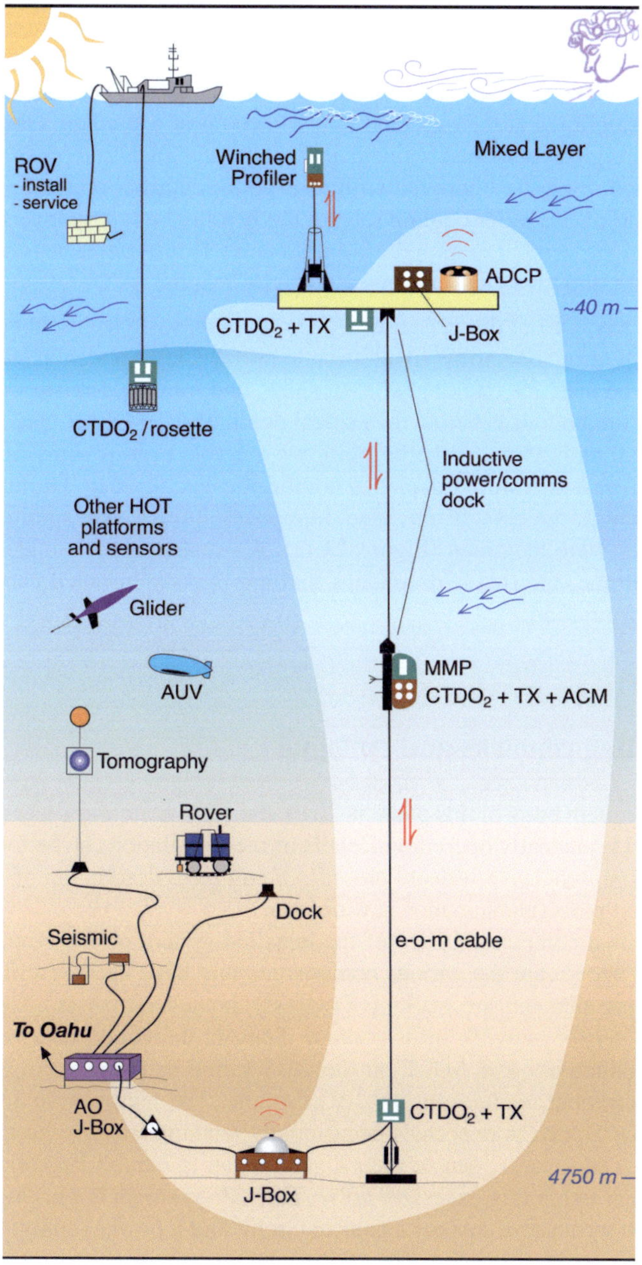

Figure 17.13 The ALOHA mooring sensor network. The lightly shaded area indicates the 2003 proposed moored sensor network. A similar mooring system has been developed as part of the ALOHA-MARS mooring project.

observed during the recording period. A potential problem of using pressure sensors for tsunami detection is that the pressure signal from seismic Rayleigh waves generated by tsunamigenic earthquakes recorded at the ocean floor can be several centimeters of equivalent ocean depth in amplitude at the expected time of arrival of a tsunami. Obscuration of the tsunami pressure signal could lead to misidentification of signals and possible false alarms. Tsunami detection might be improved with near-bottom current measurements, since the horizontal motion of the water column (above the benthic boundary layer) is expected to be far larger than the vertical motion during a tsunami. This hypothesis is one of the motivations for installing both a pressure sensor and ADCP at the ACO.

Hydrophone array. One of the most useful and robust sensor systems that can be installed at the ACO is a hydrophone array. In addition to surface environmental studies discussed above, it would be used for tracking whales, ships, and moving acoustic sources near the ACO. Although the bottom is below the critical depth of the SOFAR channel (varies seasonally between 0 and 200m off the bottom), we are able to hear ships from as far away as 160km and T-phases from earthquakes around the Pacific Rim. During the initial 20 months of recording, the calls of fin, blue, humpback, minke, and sperm whales and as well as porpoises were identified (Figure 17.11). A suitable array is under consideration but, in the meantime, the two hydrophones that are already installed can provide crude directionality information.

17.5 Concluding remarks and epilogue

Since writing the main body of this paper in 2010, the ACO node was successfully installed in June 2011 and is currently operating. Details of the installation can be found in Howe et al., 2011. The comprehensive website provides a complete description of the system and real-time data and plots (aloha.manoa.hawaii.edu).

Figure 17.14 is a composite image of the installed system on the bottom as observed by ROV *Jason*. Shown are the various components that have been described above. The main infrastructure elements are working flawlessly; however, there is mixed success with the sensors and sensor network infrastructure. Sensors delivering data in real time are: temperature, conductivity and ADCP on the OBS frame and two hydrophones (HEM – hydrophone experiment module) on the JBOX frame. The temperature and conductivity sensors on the OBS are (as expected) occasionally contaminated by heat from the electronics beneath. The pressure sensor after a year began to exhibit large step changes and drift. One ADCP is exhibiting excessive noise. The video camera is working, but the lights failed after a few weeks (we suspect a control circuit and a power supply) and the associated hydrophone is not functioning. The 10-element thermistor array extending 200m into the water column is recording data internally but with no connection to the observatory because a dry-mate connector failed during deployment. Ground faults have disabled the WHOI acoustic micro-modem and the AMM secondary node, and thus the science instrument interface module (SIIM) with the two pumped CTDO2s and the fluorometer. The replacement/repair of the non-functional instruments will take place over the next year or

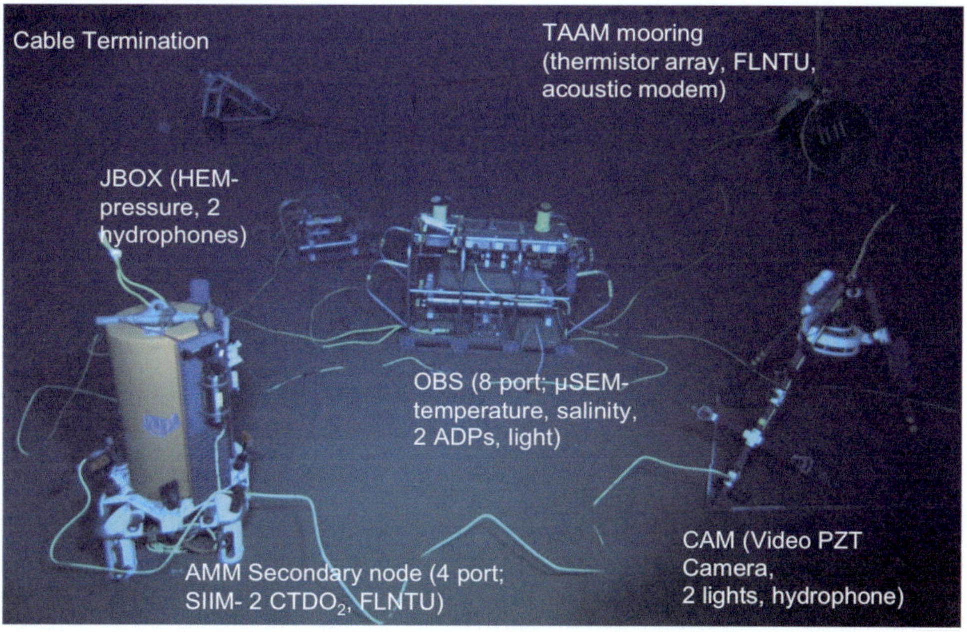

Figure 17.14 The layout of the ACO system and sensors; the bottom depth is 4728m (the foreground is true to scale; the cable termination frame and TAAM mooring anchor in the background are farther away, pasted in here). Image by ROV *Jason*.

so. Learning from these problems will be all-important as we and the community move forward with deep-sea cabled observatory systems.

While too early for science results using the current data, it is clear there is a wealth of information in the data now being collected. For instance, in the one month of video data, roughly 400 events involving 15 or more species of visible macro fauna occurred; the most exciting one was of a deep-sea lizard fish (unsuccessfully) attacking an aristeid shrimp (J. Drazen and A. Fleury, personal communication). ADCP data over the observed 23–70m depth range reveal strong internal tides. The acoustic data continue to reveal a very rich soundscape including earthquakes, shipping, sonars, wind and wave effects, and especially marine mammals.

The value of long, sustained time-series of ocean observations only increases with length. The exceptional record from HOT at Station ALOHA is the perfect example of this. HOT is successful because it maintains the continuity and is constantly improving the high quality base time series, and thus it acts as a magnet for projects with tremendous synergy, e.g., the WHOTS meteorological/upper ocean mooring, the Center for Microbial Oceanography: Research and Education (C-MORE) which uses Station ALOHA as its primary field site, and the ALOHA Cabled Observatory. Over the long-term, the ACO will provide the infrastructure necessary to continue and improve the base time series, while enabling the expansion of the spatial and temporal sampling needed to address ever more difficult questions.

With the successful deployment of the ACO node, the community is expected to respond with new designs for experiments to use the system. A benthic monitoring system to make long-term sediment flux and benthic ecology studies has been proposed by K. Smith (MBARI). The authors and colleagues will be proposing to extend the spatial footprint of the node and mooring systems using bottom HPIES (horizontal electric field (barotropic velocity), pressure, and inverted echosounder) and gliders; the latter will not only sample the ocean directly, but provide "data mule" service for the HPIES and other autonomous instruments in the area. Active and passive acoustics will also be proposed to extend the footprint to basin scale á la the Acoustic Thermometry of Ocean Climate (ATOC) project (Dushaw et al., 2009).

In the longer timeframe we envision the AUV docking stations mentioned in the introduction, with vehicles performing maintenance tasks as well as science missions. For acoustics, the precise and accurate timing will be essential in simultaneously serving navigation, communications, and science needs. Further, the observatory will be used in education and outreach to the public; the real-time data including video and audio are expected to be very useful in this regard.

The ACO will be one of the few sites in the global ocean that can provide sustained low-maintenance observations, and the only one at such water depth, 4728m. As already said, sustained observation of the deep ocean is difficult and the past and continuing problems with connectors are just one reflection of this. Having such a facility, not only for science use but also as an engineering test bed for deep-sea operations, is essential.

The ACO is a proof-of-concept system for cable re-use. The existence of 10s of thousands of kilometers of in-place retired or soon-to-be-retired first-generation fiber optic and other commercial cable systems are a potential resource for marine science. Several of these cables go through regions of interest, such as south of the Aleutian Islands, and across the equator to New Zealand, where they could be effectively utilized in place. In many cases, the precise location is less important than the supply of reliable power and communications. These systems offer potentially lower operational costs and far more data bandwidth and electrical power than autonomous buoy observatories and should be considered by the research and operational oceanography communities.

Acknowledgments

The ALOHA Cabled Observatory project has been and is funded by the National Science Foundation, awards OCE-0652430, OCE-0939570 and OCE-1239637. This project would not have been possible without the cooperation and assistance of Mark Tremblay, AT&T retired, and the team at the University of Hawaii: James Babinec, Grant Blackinton, Brian Chee, Michael Cole, William Doi, Joseph Gum, David Harris, Zensho Heshiki, James Jolly, Fernando Santiago-Mandujano, Kimball Millikan, Jefrey Snyder, and Mario Williamson. The Monterey Bay Aquarium Research Institute (MBARI) provided invaluable assistance with data management software (SSDS and SIAM) and general advice about seafloor observatories; specific individuals included Peter Braccio, Duane Edgington, Steve Etchemendy, Kevin Gomes, Robert Herlion, and Tom O'Reilly. The support of the

University of Hawaii, the Research Corporation of the University of Hawaii, AT&T (including Wayne Yamamoto and team at the Makaha Cable Station), Tyco Telecommunications, Joint Oceanographic Institutions, Inc., the U.S. Navy, the technical team from the MSC SMS Detachment at NWSY, Cheatham Annex, Williamsburg VA, the captains and crews of the R/V *Thompson* and R/V *Kilo Moana* and Matt Heintz and the ROV *Jason* team, and the advice and encouragement of many others is greatly appreciated.

References

ATOC Instrumentation Group: Howe B.M., Anderson S.G., Baggeroer A., Colosi J.A., Hardy K.R., Horwitt D., Karig F., Leach S., Mercer J.A., Metzger K., Jr., Olson L.O., Peckham D.A., Reddaway D.A., Ryan R.R., Stein R.P., von der Heydt K., Watson J.D., Weslander S.L. and Worcester, P.F. (1995) Instrumentation for the Acoustic Thermometry of Ocean Climate (ATOC) prototype Pacific Ocean network. *Proceedings, Oceans '95,* MTS/IEEE, October 9–12, San Diego, CA, 1483–1500.

Alford M.H., Lukas R., Howe B.M., Pickering A. and Santiago-Mandujano F. (2011) Moored observations of episodic abyssal flow and mixing at station ALOHA. Geophys. Res. Lett. 38, L15606, doi: 10.1029/2011GL048075.

Bidigare R.R., Chai F., Landry M.R., Lukas R., Hannides C.C.S., Christensen S.J., Karl D.M., Shi L. and Chao Y. (2009) Subtropical ocean ecosystem structure changes forced by North Pacific climate variations. J. Plankton Res. 31, 1131–1139.

Diachok O. and Duennebier F. (2010) Blue whale vocalizations and the seasonal variability of ambient noise levels at the Aloha Observatory. J. Acoust. Soc. Am. 127, 1784.

Dore J.E., Lukas R., Sadler D.W., Church M.J. and Karl D.M. (2009) Physical and biogeochemical modulation of ocean acidification in the central North Pacific. Proc. Nat. Acad. Sci. USA, 10.1073/pnas.0906044106.

Duennebier F.K., Harris D., Jolly J., Caplan-Auerbach J., Jordan R., Copson D., Stiffel K., Babinec J. and Bosel J. (2002) HUGO: The Hawaii Undersea Geo-Observatory. IEEE J. Ocean Engineering V 27, #2, 218–227.

Duennebier F., Harris D. and Jolly J. (2008) ALOHA cabled observatory will monitor ocean in real time. Sea Technology 49(2), 51–54.

Duennebier F.K., Lukas R., Nosal E.-M., Aucan J. and Weller R. A. (2012) Wind, waves, and acoustic background levels at Station ALOHA. J. Geophys. Res. 117, C03017, 21pp., doi:10.1029/2011JC007267.

Dushaw B.D., Worcester P.F., Munk W.H., Spindel R.C., Mercer J.A., Howe B.M., Metzger K., Jr., Birdsall T.G., Andrew R.K., Dzieciuch M.A., Cornuelle B.D. and Menemenlis D. (2009) A decade of acoustic thermometry in the North Pacific Ocean. J. Geophys. Res. 114, C07021, 24pp., doi:10.1029/2008JC005124.

Farrell W.E. and Munk W. (2008) What do deep sea pressure fluctuations tell us about short surface waves? Geophys. Res. Letts. 35, L19605.

Favali P. and Beranzoli L. (2006) Seafloor observatory science: A review. Annals Geophys. 49, 515–567.

Forrester N.C., Stokey R.P., von Alt C., Allen B.G., Goldsborough R.G., Purcell M.J. and Austin T.C. (1997) The LEO-15 long-term ecosystem observatory: Design and installation. *Proc. IEEE Oceans'97* 2, 1082–1088.

Harris D.W. and Duennebier F.K. (2002) Powering cabled ocean-bottom observatories. IEEE J. Oceanic Engineering 27, 202–211.

Howe B.M., Kirkham H. and Vorpérian V. (2002) Power system considerations for undersea observatories. IEEE J. Oceanic Engr. 27, 267–274.

Howe B.M., Chao Y., Arabshahi P., Roy S., McGinnis T., Gray A. (2010) A smart sensor web for ocean observation: Fixed and mobile platforms, integrated acoustics, satellites and predictive modeling. IEEE J. of Selected Topics in Applied Earth Observations and Remote Sensing, doi: 10.1109/JSTARS.2010.2052022.

Howe B.M., Lukas R., Duennebier F. and Karl D. (2011) ALOHA cabled observatory installation. OCEANS 2011, Kona, Hawaii, 19–22 Sept. 2011.

Howe B.M., Duennebier F.K., Butler R., Lukas R.B. (2012) Scientific uses of submarine cables: Evolutionary development leading to the ALOHA Cabled Observatory. Mains'l Haul, J. Pacific Maritime History 48, 100–119.

Karl D.M. and Lukas R. (1996) The Hawaii Ocean Time-series (HOT) program: Background, rationale and field implementation. Deep-Sea Res. II 43, 129–156.

Karl D.M., Dore J.E., Lukas R., Michaels A.F., Bates N.R. and Knap A. (2001) Building the long-term picture: The U.S. JGOFS time-series programs. Oceanography 14, 6–17.

Lukas R. and Santiago-Mandujano F. (2001) Extreme water mass anomaly observed in the Hawaii Ocean time-series. Geophys. Res. Lett. 28, 2931–2934.

Lukas R., Santiago-Mandujano F., Bingham F. and Mantyla A. (2001) Cold bottom water events observed in the Hawaii Ocean time-series: Implications for vertical mixing. Deep-Sea Res. I 48(4), 995–1021.

Nosse C.T., Santiago-Mandujano F., Lukas R.B., Dore J.E., Weller R.A. and Plueddemann A.J. (2012) Recent strong interannual variation disrupted pycnocline and abyssal salinity trends at Station ALOHA. Eos Trans. AGU, Ocean Sci. Meet. Suppl., PosterID: A0189.

Oswald J.N., Au W.W.L. and Duennebier F. (2011) Minke whale (Balaenoptera acutorostrata) boings detected at the Station ALOHA Cabled Observatory. J. Acoust. Soc. Am. 129, 3353–3360, doi:10.1121/1.3575555.

Ou H., Au W.W.L. and Oswald J. (2012) A non-spectrogram-correlation method of automatically detecting minke whale boings. J. Acoust. Soc. Am. 132, 4, EL317-EL322, doi:10.1121/1.4747816.

Petitt R., Harris D., Wooding B., Bailey J., Jolly J., Hobart E., Chave A., Duennebier F., Butler R., Bowen A.D. and Yoerger D.R. (2002) The Hawaii-2 Observatory. IEEE J. Ocean Engineering 27(2), 245–253.

Tremblay M.D. and Duennebier F.K. (2006) Format and bit rate independent communications over regenerated undersea fiber optic cable systems. *Proceedings of the Scientific Submarine Cable 2006 Conference*, Marine Institute, Dublin, Ireland, 7–10 February 2006, pp.182–185.

18 Next-generation science in the ocean basins: Expanding the oceanographer's toolbox utilizing submarine electro-optical sensor networks

J. R. Delaney and D. S. Kelley[1]

18.1 A grand challenge

For decades, oceanographers have gone to sea to study limited portions of the ocean for short periods of time. In the 1950s, the reach and duration of research in the oceans was extended, utilizing limited bandwidth satellite systems for surficial imaging and communications. For the coming decades, however, the grand challenge in Ocean Sciences is the need to successfully design and aggressively implement novel strategies and innovative infrastructures to dramatically increase our rate of discovery and understanding of the complex interactions operating throughout the volume of the ocean basins (Figure 18.1). The next-generation approaches must be delivered fast enough and well enough to provide confident anticipation of short- and long-term ocean-generated threats to society. These approaches must also provide the basis for predicting, well in advance, positive and negative impacts of potentially nonlinear "tipping points" in our planetary ecosystem resulting from shifts in the dynamic behavior of the oceanic system. In a timely manner, the required distributed infrastructure must be created throughout the oceans and on/in the seafloor to quantitatively characterize linked ocean processes at the required temporal and spatial scales. This infrastructure must include nested scales of experimental arrays from the basin- to the centimeter-scale, and from temporal scales of decades (climate change) to microseconds (earthquakes).

There is a growing urgency to understand ocean complexity and to strategically plan ocean monitoring and its future use by society. By 2030 there will be a billion more people

1 University of Washington, School of Oceanography, Seattle, USA

Figure 18.1 The global ocean is a complex system of many hundreds of interacting processes and it is the 'flywheel' of our planetary life support system. Until the advent of humans as a major environmental agent, this system was dominated by two energy sources: solar energy from without, and geothermal energy from within. Only by establishing a permanent and highly interactive telepresence within this system will we come to understand it well enough to forecast its behavior with fidelity. Electro-optical cables can be the backbone for this next generation capability.

on the planet, with another billion arriving in the following two decades (United Nations, 2004). Over half the population of the planet now lives or works adjacent to coastal environments and the contribution of coastal economies to many nations' overall economic health is substantial. The increased pressures placed on our habitable domains mean that humans will eventually have to learn to "manage" vast portions of the planet. However, we do not currently have the insight, the knowledge, or the wisdom to do so. The next two to four decades must be spent preparing for the time when we will have no choice but to confront this daunting responsibility.

Our focus now must be on exploring the time domain at spatially meaningful scales to understand the behavior of the ocean and to anticipate and sample the unexpected. In this regard, the ocean basins are still the last physical frontier on Earth. Not only do they

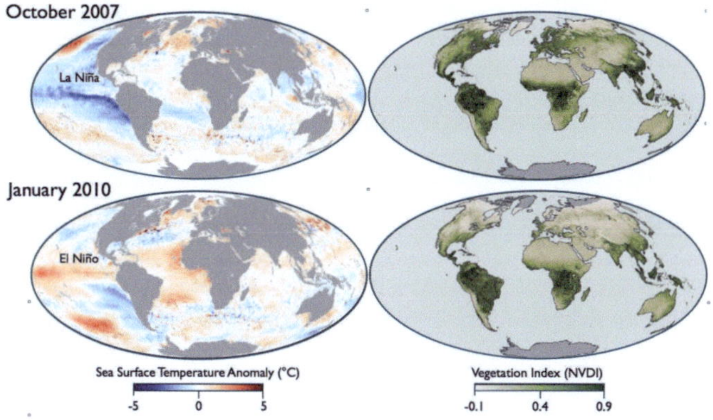

Figure 18.2 Sea surface temperature and vegetation index anomalies in 2007 and 2010. The temperature anomaly is calculated from the departure of average conditions in the top millimeter of the ocean's surface as measured from NASA's Terra satellite. The vegetation index is a scale of "greenness" that is based on the number and type of plants, and how leafy and healthy they are. In 2006–2007, Southern California had the driest winter in 130 years. These images illustrate the profound impact that the oceans have on crop production on the continents. Images are from NASA Earth Observations (NEO) (http://earthobservatory.nasa.gov/GlobalMaps/).

episodically deliver devastating impacts on human society, they harbor potentially vast mineral and bio/microbial wealth that is largely unexplored (National Research Council, 1999). Further, the patterns of ocean circulation and changing sea-surface temperature distributions correlate closely in time with ever-shifting patterns of drought and flooding on the continents (Figure 18.2). It is therefore not surprising that the dynamic behavior of the ocean is linked directly to patterns of food growth and famine on the adjacent continents; we must learn to understand and predict all of these linkages in both time and space to meet the planetary challenges that will face us in the coming decades.

Simple monitoring of the ocean using traditional technologies is no longer sufficient. Moving to the next level of understanding and predictive modeling requires interactive capabilities projecting four-dimensional human telepresence across entire volumes of the ocean from the atmosphere to well within the seafloor. Requirements will include the ability to detect significant behavioral excursions, both expected and unexpected, within the volume of the oceans, and to be able to reconfigure deployed sensing/measurement modalities to quantitatively and adaptively characterize the processes unfolding within these energetic phenomena.

Success in this endeavor will be a computation- and communication-intensive challenge. With success, it will require significant storage/archiving activities, supplemented by transport of vast, ever-growing amounts of data from the natural environment, and from increasingly sophisticated model simulations, to globally distributed users. Scientists, policy makers, and citizens linked by the Internet will become increasingly engaged in our ocean as it becomes more important and accessible to all.

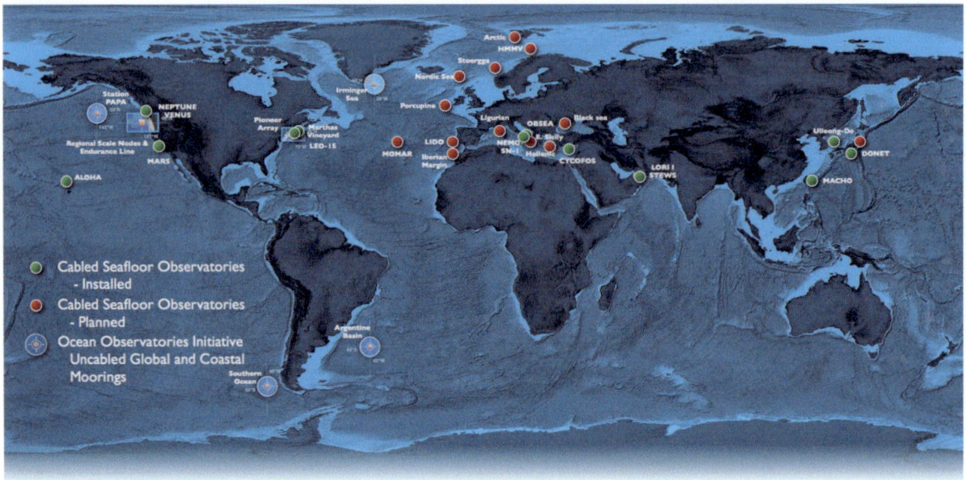

Figure 18.3 Locations of installed (green dots) and planned cabled seafloor observatories (red dots). Many of the sites located around the perimeter of Europe are part of the extensive ESONET (European Sea Floor Observatory Network) that will eventually entail 5000km of fiber optic cable on the seafloor (http://www.oceanlab.abdn.ac.uk/research/esonet.php). Also shown are the locations of high latitude uncabled mooring arrays (Papa, Irminger, Southern Ocean, Argentine basin), the coastal arrays (Pioneer, Washington and Endurance Lines), and the cabled Regional Scale Nodes plate scale observatory that comprise the National Science Foundation's Ocean Observatories Initiative (http://www.oceanobservatories.org/),

18.2 Ocean complexity – natural laboratories

Understanding the complexity, the dynamics, and the resilience of the ocean reservoir (Figure 18.1) is central to achieving the capacity to address increasingly urgent issues of population expansion and environmental degradation. There will be numerous competing priorities for funding over the next several decades such that researchers will not be able to encompass the entire global ocean with equal focus and coverage. Therefore, key representative ocean volumes must be carefully selected for intensive study of interlinked processes comprising the full scope of oceanic complexity (Figure 18.2).

With these goals/priorities in mind, several countries have now installed or are planning to install cabled observatories within the ocean basins (Figure 18.3; Table 18.1). Examples of installed systems include Japan's and Taiwan's early warning seismic and tsunami systems – DONET (Development of Dense Ocean-Floor Network System for Earthquakes and Tsunamis) and MACHO (Marine Cable Hosted Observatory), respectively. Turkey has also installed a local cabled seismic array. These systems are dominantly for operational use. In contrast, major installations in Canada (NEPTUNE Canada) and in the US (Regional Scale Nodes (RSN) as part of the Ocean Observatories Initiative) are focused on science, but they also have operational functions (Figure 18.4) (Delaney et al., 2000; Barnes et al., 2011). These two high-power (10 kV; 8 kW) and high-bandwidth (10 Gb/s) systems were

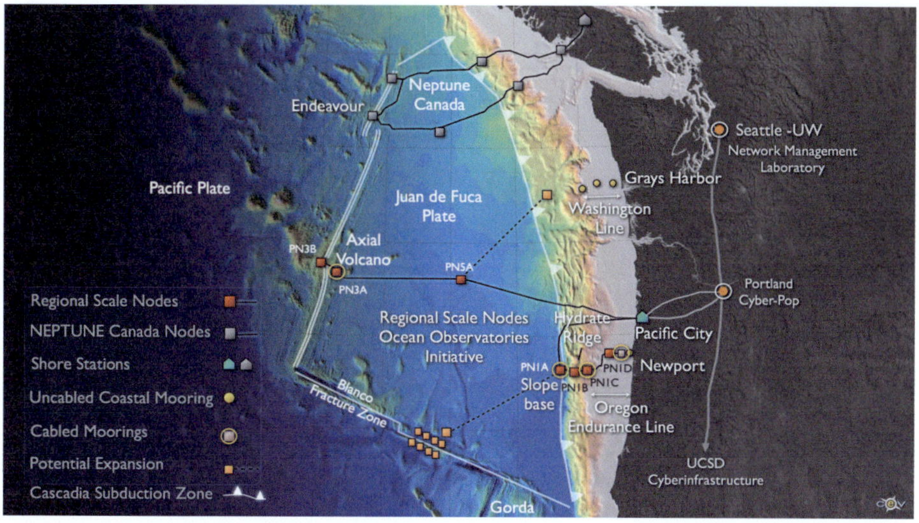

Figure 18.4 Location of the 900km high-power and bandwidth US Regional Scale Nodes cabled observatory in the Northeast Pacific. The backbone cables and seven primary nodes provide power and bandwidth to secondary infrastructure (instruments, junction boxes, medium power nodes, full water column moorings) at key scientific study sites along the Juan de Fuca plate and in the overlying ocean. A shore station located at Pacific City, Oregon, brings the power and communications onshore where data are passed off to the terrestrial fiber. The real-time coupling of the RSN and NEPTUNE Canada through the Internet will provide unprecedented data flow and two-way communications to many hundreds of sensors. In concert, the networked sensor arrays will for the first time allow seafloor and oceanographic measurements to be made of key processes operating at the scale of an entire tectonic plate. Real-time transmission allows rapid response to major perturbation events that include seafloor volcanic eruptions, earthquakes, tsunamis, coastal anoxia events, and harmful algal blooms.

initially proposed under the name NEPTUNE (NorthEast Pacific Time-Series Undersea Networked Experiments; Delaney et al., 2000) as part of a United States–Canada partnership (see Section 18.2.1). When these two systems are fully installed in 2014, they will include 1700km of cable and 14 nodes that span the Juan de Fuca tectonic plate stretching from north of Vancouver Island to near the California–Oregon border and over 400km offshore. These natural laboratories will allow many hundreds of experiments to be conducted simultaneously in a well-documented framework (Delaney et al., 2000; Delaney and Barga, 2009).

Interactive, distributed sensor networks, are exemplified by the two cabled systems now being deployed in the NE Pacific (Figure 18.4), which utilize novel technologies, and create large-aperture "natural laboratories" employing real-time experimental control over the entire "laboratory" volume. Fixed and mobile assets distributed throughout a given volume, all communicating via the Internet at nearly the speed of light, allow constant surveillance of, and response to, a wide spectrum of physical, chemical, and biological processes interacting within key portions of the ocean system. Cabled laboratories extend unprecedented power and bandwidth to a wide range of interactive real-time sensors, instruments, and robots that can carry out routine experiments and next-generation science. These arrays can be programmed to investigate emergent events that include erupting volcanoes, major storms, fish stock migration patterns, or progressive environmental anomalies such as ocean acidification or low oxygen "dead" zones. Other infrastructure now coming online includes instrumented autonomous underwater vehicles with cabled docking stations and gliders that extend the spatial 3D coverage of the networks (Eriksen et al., 2001; Molin and Schofield, 2009). These new capabilities will empower many investigators to push the envelop on creativity and exploration of the time-space domain within a number of natural systems. The following sections provide an overview of the US cabled observatory as an example of this emerging global approach to studying the ocean basins.

Cabled observatory	**Location**	**Status**	**Purpose**
1. DONET: Dense Ocean-Floor Network System[1]	Japan, Nankai Trough – 250km cable	Operating	Monitoring and early warning system for earthquakes and tsunamis
2. MACHO – Marine Cable Hosted Observatory[2]	Taiwan – western Okinawa Trough: 45km, planned 500km cable	Cable installed 2011	Monitoring and early warning system for earthquakes and tsunamis
3. Regional Cabled Observatory[3] (formally known as NEPTUNE North East Pacific Time-series Undersea Networked Experiments)[3]	Southern 2/3 Juan de Fuca Plate – 900km cable	Cable installed 2011	Seismic and tsunami monitoring, volcanic activity, melt migration, hydrothermal vents, gas hydrate formation and methane flux, climate change, anoxia events, ocean acidification; subseafloor biosphere
4. NEPTUNE Canada [4]	Northern 1/3 Juan de Fuca Plate – 800km cable	Operating	Seismic and tsunami monitoring, volcanic activity, melt migration, hydrothermal vents, gas hydrate formation and methane flux, climate change, anoxia events, ocean acidification; subseafloor biosphere
5. MARS – Monterey Accelerated Research System[5]	California – Monterey Bay – 52 km cable	Operating	Ocean acification, marine ecosysems and biodiversity, tidal currents, earthquakes
6. VENUS – Victoria Experimental Network Under the Sea[6]	Vancouver Island– Sannich Inlet and Strait of Georgia – 5km cable	Operating	Ocean mixing, currents, fish and marine mammal movements, sediment dynamics

7. NEMO-SN-1[7]	Southern Italy – 25km cable	Operating	Gravity, earthquakes, currents, ocean noise, marine mammals
8. CYCOFOS – Cyprus Coastal Ocean Forcasting and Observing System[8]	Eastern Mediterranean – Levantine Basin	Operating	Early warning tsunamis, seismic events
9. Chinese Ocean Observatory[9]	South China Sea – 1.1km another planned for 20km	Cable installed 2011	Marine ecosysems and biodiversity, tidal currents, earthquakes
10. Seafloor Observatory (OBSEA)[10]	Barcelona, Italy	Installed, 4km cable	Marine ecosystems, currents, seismicity
11. Martha's Vineyard	Woods Hole MA	Operating	Marine coastal climatology, coastal processes
12. LORI Observatory	Sea of Oman – >20km, 4 nodes	Operating	Seismic and tsunami warning
13. ALOHA	Station ALOHA, 100km north of Oahu	Operating	Climate-ecosystem variability, microseism energy, mammal detection
14. South Korea	South of Ulleong-DO Island, East Sea, 20km cable	Operating	Tsunami warning system
15. LEO-15	Tuckerton New Jersey, 8.1–9.6 km	Operating	Climate-ecosystem variability, coastal marine processes

1 Kaneko et al., 2009 (http://www.jamstec.go.jp/jamstec-e/maritec/donet/);
2 Hsu et al.,, 2007;
3 Delaney et al., 2009 (http://www.interactiveoceans.washington.edu/);
4 Barnes et al., 2007 (http://www.neptunecanada.ca/);
5 http://www.mbari.org/mars/;
6 Taylor, 2009 (http://venus.uvic.ca/);
7 Favali et al., 2006 (http://roma2.rm.ingv.it/en/facilities/seafloor_multidisciplinary_observatories/2/sn-1);
8 http://www.oceanography.ucy.ac.cy/cycofos/ocean-subsea.html;
9 http://www.chinadaily.com.cn/cndy/2011-04/06/content_12276002.htm;
10 Aguzzi et al., 2011 (http://www.obsea.es/);
11 Edson et al., 2000 (http://mvcodata.whoi.edu/cgi-bin/mvco/mvco.cgi);
12 Vall et al., 2011;
13 Howe et al., 2011 (http://aco-ssds.soest.hawaii.edu/index.html);
14 Schleisiek, 2009;
15 Creed et al., 2005

Table 18.1 Representative cabled seafloor observatories installed and operating.

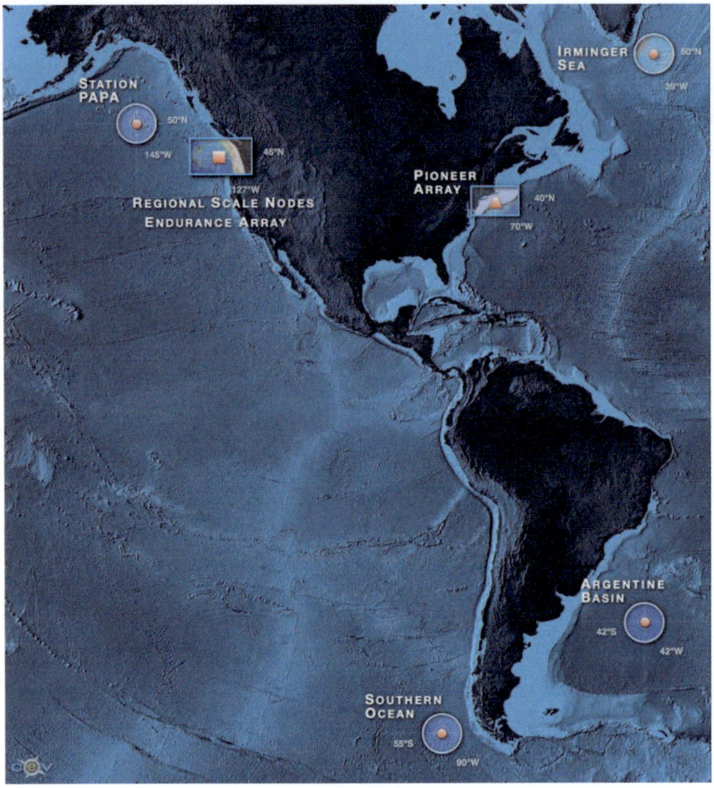

Figure 18.5 Location of infrastructure for the National Science Foundation's Ocean Observatories Initiative (OOI). The OOI includes arrays of uncabled moorings and gliders in the high latitudes – the Southern Ocean off Chile, Argentine Basin, Station Papa and Irminger Sea. In concert, these sites comprise the Global Scale Nodes components of the OOI. Some of the key measurements at these sites are to quantify air–sea exchange of energy and mass in areas dominated by high winds, the uptake of carbon in these areas, and to understand impacts of climate variability on ocean circulation and weather. Coastal Scale Nodes (CSN) sites include an array of uncabled moorings and autonomous underwater vehicles at the Pioneer Array, and uncabled moorings off Washington, cabled and uncabled moorings off Oregon and a fleet of gliders (6) as part the Endurance array. Focused interests for the coastal sites include ocean acidification and anoxia events, and the impact of climate change on coastal ecosystems. The Regional Scale Nodes share coastal sites with the CSN and are unique to the OOI in the capability to provide tens of kilowatts of power, tens of Gb of bandwidth and >30% future expansion with real-time, two-way communication. Its focus includes dynamic processes associated with underwater volcanism and crustal formation, life in extreme environments, methane hydrate formation, and the role that transient events (e.g., earthquakes, volcanic eruptions, and slope failure) play in transfer of mass and energy from the seafloor to the hydrosphere. High power, cabled moorings focus on climate change in the Northeast Pacific, and biogeochemical investigations overlapping with both GSN and CSN.

18.2.1 Development of the US undersea cabled observatory

The concept of a high-power and high-bandwidth underwater cabled observatory was first explored as early as 1987 (Delaney et al., 1987). Over a decade later, this concept was further developed by a consortium led by the University of Washington that finalized the concept of a plate-scale submarine electro-optically cabled ocean observatory to study highly energetic earth and ocean processes from shallow coastal waters off Washington, Oregon, and British Columbia out to the Juan de Fuca spreading center. The hope was that the Canadian system would cover the northern one-third of the Juan de Fuca Plate and the US system would cover the central and southern two-thirds of the plate.

In 2000, the Ocean Science Division of the National Science Foundation (NSF) put forth a proposal for Major Research and Equipment Facility Construction (MREFC) funds from congress to incorporate the NEPTUNE cabled observing systems concept into an even broader ocean science program that included: (1) high-latitude measurements of heat fluxes and greenhouse gas exchange at the air–sea interface with mooring arrays at Station Papa in the Gulf of Alaska, the Irminger Sea, Argentine Basin, and southwest of Chile; and (2) focused inquiries into critical coastal environments where populations are stressing near-shore oceanic systems with mooring arrays off Newport Oregon (Endurance Array – Oregon Line), Grays Harbor Washington (Washington Line), and in the Northwest Atlantic off the coast of New England (Pioneer array) (Figures 18.4 and 18.5). The high-latitude and coastal network of buoys, gliders and autonomous underwater vehicles became known as the Coastal and Global Scale Nodes (CGSN). Linking the marine assets is a Cyber-infrastructure (CI) component to provide infrastructure for data flow, management and distribution, as well as observatory resource management. The entire program was named the Ocean Observatories Initiative (OOI – http://www.oceanobservatories.org/) and it was developed in response to interest in the global ocean from across the oceanographic community.

In anticipation of a US NEPTUNE system coming on line, in 2002 MBARI (Monterey Bay Aquarium Research Institute) proposed the development of a 52km, single node cabled observatory in Monterey Bay, California called MARS (Monterey Accelerated Research System – http://www.mbari.org/mars/). A partial objective was that MARS would be a test bed for NEPTUNE (Figure 18.6). However, other objectives included examination of deep-sea life, impacts of increasing ocean acidity on marine life, and the in situ identification and quantification of both shallow and deep water microorganisms (e.g., Sherman and Smith, 2009; Scholin, 2010). MBARI was awarded ~$10M in 2002 for MARS with funding from the David and Lucile Packard Foundation and NSF and the cable was installed in 2007 (http://www.mbari.org/mars/).

In 2003, representatives of the ocean science community met at an NSF-sponsored cabled observatory workshop and designated the NE Pacific as the site for the first regional cabled ocean observatory in recognition that a representative suite of global ocean processes occur in this area (RECONN, 2003). Also in 2003, NEPTUNE Canada, led by the University of Victoria, announced receipt of federal, provincial and private funding to complete planning and implementation of its network on the northern portion of the Juan de Fuca Plate (http://neptunecanada.ca).

In 2006, based on continued science planning linked to engineering design and development, NEPTUNE U.S. passed the OOI Conceptual Design Review, and in 2007 NEPTUNE U.S. was renamed the Regional Scale Nodes within the OOI (Figures 18.4 and 18.5). As part of a competitive process, in 2007 the University of Washington (UW) became the Implementing Organization (IO) for the RSN and was tasked with developing the design, fabrication and installation of the RSN, as well as operating the cabled network for the first two years subsequent to construction (http://www.ooi.washington.edu/). Woods Hole Oceanographic Institution is the lead IO for the high-latitude moorings and Pioneer Array, Oregon State University for the Endurance and Grays Harbor arrays, and the University of California San Diego is the IO for CI.

In 2009, the Consortium for Ocean Leadership (http://www.oceanobservatories.org/) and the NSF signed a cooperative agreement supporting the construction and initial operation of the OOI with MREFC funding at ~\$385M over 5.5 years. Following the RSN award of ~\$126M, L-3 MariPro was awarded a \$76.8M contract by the UW to design, build and install the major cable components of the RSN. Cable installation began in the summer of 2011 (http://www.interactiveoceans.washington.edu/). Following supplementary and operational funding of nearly \$150M to NEPTUNE Canada, the Canadian system became operational in 2009.

18.2.2 Regional scale nodes

The fiber-optic cables of the RSN will carry electrical power and telecommunications bandwidth into the oceans to serve the needs of science, education, and the community at large (Figure 18.4). The primary infrastructure includes modified telecommunication cables (backbone cables), repeaters, and seven primary nodes (PN) to distribute power from a shore station in Pacific City, Oregon to key sites along the Juan de Fuca plate (Figures 18.4, 18.7, 18.8, 18.9). These sites include the base of the Cascadia subduction zone, which is overdue for a major earthquake, particularly in the southern margin where large earthquakes occur on average at intervals of 240 years (the last one was in 1700) (Goldfinger et al., 2008, 2012). The Cascadia margin where ocean waters are highly influenced by the California current (PN1A), methane seep and gas hydrate sites at Southern Hydrate Ridge (PN1B), areas that experience anoxia events and harmful algal blooms along the Oregon and Washington coasts (RSN and the Endurance Array (PN1A, PN1C and PN1D)), and Axial Seamount, the most magmatically active volcano on the Juan de Fuca spreading center, located >400km off shore (PN3A and 3B)(Figures 18.8 and 18.9). Also included on the network are distributed sensor arrays on the seafloor and sensors, platforms, and crawlers on full water column moorings (Figures 18.8 and 18.9). When fully operational, the cables will provide real-time two-way communications between land and sea. As installed, each node will provide 8 kW of power and 10 Gb/s bandwidth, with the capability to upgrade to a total of 200 kW of power and up to 240 Gb/s bandwidth. These capabilities are distributed from the land-based shore station at Pacific City to subsea terminals (Primary Nodes) at ocean depths as great as 3000m and throughout the water column via state-of-the art instrumented moorings (Figure 18.8). The backbone cable is installed, and the Primary Nodes were installed Summer 2012. All secondary infrastructure will be installed by the end of 2014, with commissioning of the system scheduled for early 2015.

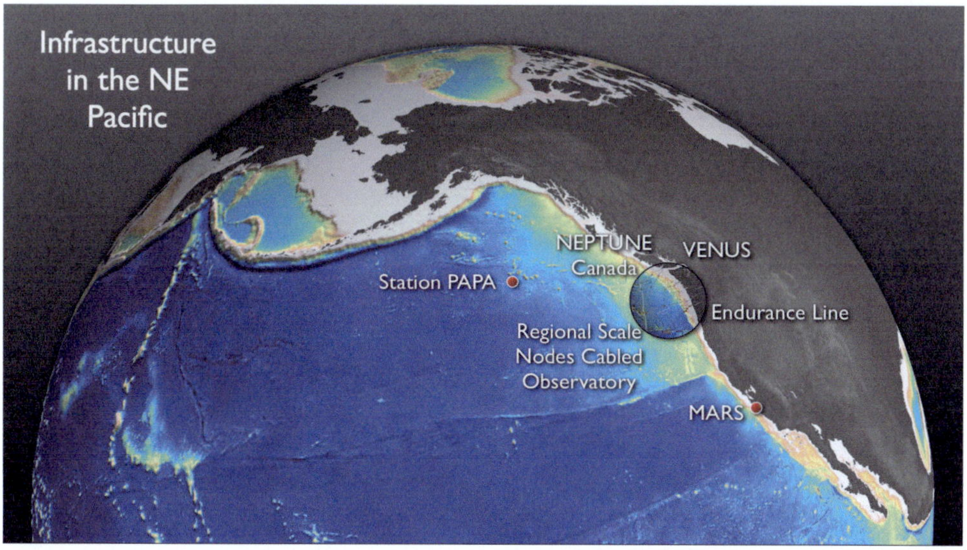

Figure 18.6 Observing infrastructure in the Northeast Pacific. There is a high confluence of cabled and uncabled observatories within the Northeast Pacific that make it possible to study process linkages at ocean basin to centimeter scales. This area is currently the only place where a plate scale observatory with high power and bandwidth is installed. Cabled systems include the RSN with connections to the Endurance array and NEPTUNE Canada (Juan de Fuca Plate), VENUS (Sannich Inlet – Strait of Georgia, Canada), and MARS (Monterey Bay, California). Connections to basin scale processes are provided by Station Papa in the Gulf of Alaska, which includes a surface piercing mooring, two flanking moorings, and gliders, and the cabled ALOHA mooring off Hawaii (not shown).

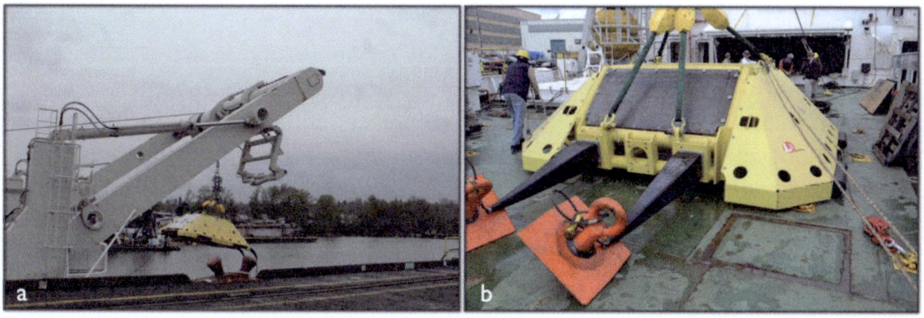

Figure 18.7 (a) An RSN primary node being dock-tested off of the fantail of the cable ship *TE Subcom Dependable* in Portland, Oregon, prior to installation summer 2012. The primary nodes are built by L3MariPro. (b) Primary node with protective housing for deployment in waters shallower than 1500m.

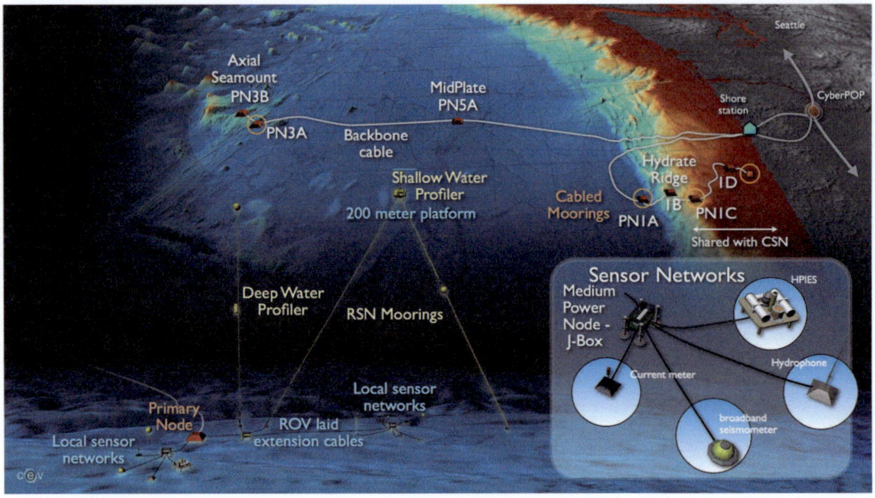

Figure 18.8 Secondary infrastructure on the RSN includes three sites that each host two, high-power (3kW) and bandwidth (1 GigE) water column moorings with a broad suite of sensors. One mooring includes an instrumented deep profiler that traverses from ~3000m to 200m beneath the ocean's surface, while another two-legged mooring hosts an instrumented platform at 200m beneath the ocean's surface and an instrumented winched shallow profiler. Secondary nodes and junction boxes provide power, data transmission, and extension cable ports for sensors on the seafloor and to moorings.

Backbone cables: The network features two major submarine backbone cables. One cable branch extends from the shore station in Pacific City to the base of the Cascadia margin (PN1A), to the Southern Hydrate Ridge study site, and to the coastal scale nodes (CSN) endurance line to the south and another branch extends from the shore station across the Juan de Fuca Plate to the Axial Seamount study site on the Juan de Fuca Ridge (Figures 18.4, 18.8 and 18.9). The backbone cable connects the shore station to the primary nodes, which are distribution centres for extension cables that provide direct access to the specific sites of scientific interest. The RSN backbone cable is supplied by L3 MariPro Inc. It includes ~900km of telecom industry subsea cable with one fiber pair to provide power (10 kV DC, up to 8 A) and communications with initial transmission of 10 Gb/s via fiber-optics between the primary nodes and the shore station. The cable was buried to ~1500m water depth via ploughinging and ROV jetting to avoid aggression by trawling. Cable deployed at shallow depths and in rough topography includes variable degrees of armour. The entire primary infrastructure is monitored and controlled by a network management system (NMS) that allows the communication and power systems to be monitored for activity, status, utilization and it also allows controlled allocations of resources across the system (e.g., power). The NMS can be accessed from the shore station or from remote facilities including the RSN Observatory Network Management Laboratory at the UW.

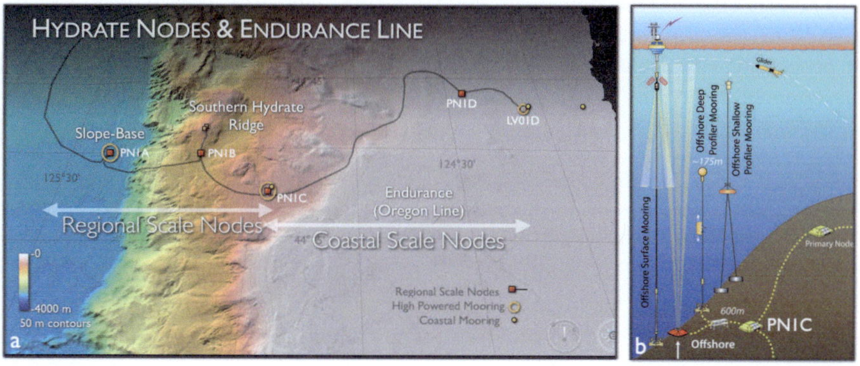

Figure 18.9 (a) An extensive array of cabled and uncabled infrastructure extends >120km west of Newport Oregon, hosting >10 instrumented moorings at water depths of 3000m to 25m, providing unprecedented in situ, real-time measurements of geological, physical, chemical and biological processes within deep to shallow water environments. Southern Hydrate Ridge, one of the best studied methane hydrate systems will host an array of geophysical, chemical and biological sensors to examine processes linkages among earthquakes, fluid flow, release of methane and evolution of seafloor biological communities. (b) Shared infrastructure between RSN and CSN at 600m water depth includes two cabled moorings with wire crawlers, and a surface piercing mooring with satellite communications at Primary Node PN1C. A benthic package will include an upward-looking velocity profilers, CTDs, CO_2 and pH sensors, pressure sensors, and current meters. Gliders, as part of the Coastal Scale Nodes operated by Oregon State University, will traverse offshore from Newport, Oregon to the base of the Cascadia margin and north to Grays Harbor, Washington.

Primary nodes The primary nodes are subsea terminal points for the backbone cable and are also supplied by L3 MariPro Inc (Figures 18.4 and 18.7). The primary nodes have connection points for instruments, medium-power junction boxes, extension cables and moorings. The continuous flow of power and availability of bandwidth allow continuous, real-time interactive science experiments at the seafloor and on moorings that extend from the seafloor to the ocean surface. The main function of the primary nodes is to distribute power and bandwidth to the secondary infrastructure (low power nodes and medium and low power junction boxes, extension cables, instruments and moorings) and also to contain housekeeping functions of the system control, out-of-band communications, and engineering monitors. Using the same kind of equipment used by the subsea telecommunications industry provides the high reliability needed for a system with a planned lifetime of 25 to 30 years.

Primary nodes house two main assemblies: the Base Interface Assembly (BIA) and the Science Interface Assembly (SIA). The BIA is hardwired into the primary node assembly and to the backbone cable, requiring installation and recovery with an industry cable ship. The BIA houses medium voltage converters (MVCs) that convert the 10 kVDC primary level to the 375 VDC level and send power on to the low power nodes and low and medium power junction boxes. The BIAs also contain two backbone expansion ports (10 kV). The SIAs are wetmated within the primary node assembly allowing deployment and recovery by an ROV for ease in maintenance and upgrades. Each SIA houses five wetmatable science ports with 1 GbE and 375 V and two high-bandwidth ports (10 GbE, 375V) with wetmate connectors. Primary nodes deployed in shallow waters (<1500m) are housed in large, trawl-resistant frames similar to those used by NEPTUNE Canada and MARS (Figure 18.7).

Backhaul system: The backhaul system includes two dedicated high-bandwidth terrestrial telecommunication cables leased from Tillamook Lightwave. It provides Internet connectivity from the Pacific City Shore Station to the Internet through a CyberPop (Point of Presence) in Portland, Oregon and data transfer to the OOI cyberinfrastructure in San Diego and the UW. The CyberPop provides some data storage, but predominantly serves as a data-video imagery routing station to distribute two-way information over the Internet in real time. The shore station houses power feed equipment to convert utility power to high-voltage regulated and controlled power for each of the submarine cables that run from the shore station into the ocean. Also included in the shore station is line termination equipment that has the optical drivers and receivers for the submarine cables that terminate in the shore station. A cable landing, beginning in conduits that run 1524m offshore, provides the transition for the submarine cables from the ocean bed through buried conduits to the shore station.

Medium power and low power junction boxes and low power nodes: The medium (MPJB) and low power J-boxes (LPJB) distribute power, communication and timing to instruments. They include eight sensor ports that can handle 12–48 volts DC, with communication capabilities of 10/100 BASE-T, RS232, RS485 and pulse-per-second timing (PPS and 1588). Ports are configured predeployment and include wet- and drymate connectors (Figure 18.8). They are maintainable with an ROV. Low power J-boxes provide up to 50 W per port, with a 200 W per LPJB limit. Medium power J-boxes provide up to 200 W per port for all eight ports. Low power nodes (LVN) provide 48 V and 1-GigE communications capabilities that furnish the correct data and power interface to MPJB and LPJB. Expansion ports provide "daisy-chain" capabilities via extension cables to other LVNs, LPJBs or MPJBs, making the system highly expandable.

Extension cables: Extension cables provide power and communication links between the primary infrastructure, secondary infrastructure and instruments. Extension cables link primary nodes to low power nodes and J-boxes, and to instruments across the RSN. The cabling powers different types of loads with variable communications requirements and will be deployed in a variety of benign and harsh environments (see Sections 18.2.3.1–18.2.3.3). Therefore, similar to the backbone cable, the extension cables include variable degrees of armoring.

18.2.3 Regional scale nodes science, instruments and moorings

The RSN, coupled with NEPTUNE Canada, forms an unprecedented plate scale experiment focused on integrated investigations spanning the subseafloor biosphere to the hydrosphere, the entire ocean water column, and the sea surface–atmosphere interface

(Figure 18.4). A representative suite of natural phenomena that occur throughout the world's oceans and seafloor are found in the Northeast Pacific Ocean. Spatially associated with the Juan de Fuca tectonic plate, the RSN will enable in-depth study and decadal time-series observations of regional oceanography – including biogeochemical cycles, fisheries and climate forcing, tsunamis, ocean dynamics, life in extreme environments, and plate tectonic processes. Dense spatial and temporal sampling at key sites, coupled with interactive capabilities, will allow the Ocean Science community for the first time to make fundamental measurements and exciting discoveries about processes that occur over centimeter to 100s of kilometer distances, and at seconds to decade time scales. This 25- to 30-year experiment will fundamentally change the way we view and study our planet and how we educate and interact with researchers, students and the public.

The continuous, high-quality observations made over the two- to three-decade lifespan of the whole OOI system will provide essential data to improve predictive models of ocean processes. Integration of data from the OOI and the Canadian-shared, 60-year observing Station Papa, and the OOI regional and coastal scale nodes by the OOI cyberinfrastructure, coupled with integration of real-time data from NEPTUNE Canada, VENUS, and MARS are providing unprecedented new insights into how the ocean functions (Figure 18.6; Table 18.1). These data are forming the basis for learning to manage, or at least adapt to, the most powerful climate modulating system on the planet – the global ocean.

18.2.3.1 Ocean processes in the NE Pacific, the regional scale nodes and endurance array

The RSN and associated arrays (e.g., NEPTUNE Canada, PAPA and VENUS) (Figure 18.6) provide an unparalleled opportunity for long-term, interactive water column-observing capabilities that will allow the capture of both intermittent and episodic events that are dynamically and ecologically significant, as well as interannual and interdecadal variability. Sustained arrays of instruments throughout the water column and on the seafloor will lead to breakthroughs in our understanding of the physical, chemical, and biological processes influenced by bifurcation of the West Wind Drift, Alaska Current, and California Current. As part of the OOI system of systems, the plate-scale RSN is the only planned observatory that links global scale studies such as those being conducted at Station PAPA, and coastal investigations such as those along the Washington and Oregon Margin through the OOI cabled and uncabled Endurance array.

Ocean conditions surrounding the RSN are bracketed in the pelagic and open-ocean by the OOI global site at Ocean Station Papa, and on the coastal end by the Endurance Array off Newport, Oregon and Grays Harbor, Washington. The RSN resides in a complex system of currents where wind- and tide-forced motions lead to turbulent mixing that aids transport of chemical and biological species. The Pacific Northwest is one of the most biologically productive regions of the world, and hypoxia and harmful algal blooms are observed with increasing frequency (Grantham et al., 2004). In this region, the impacts of ocean acidification are highly evident with CO2 saturation values and shallowing of acidic waters that were not anticipated for another 50 years based on modeling results (Feely et al., 2008). These complex physical, biological and chemical processes are all intertwined, and respond to forcing on a wide range of spatial and temporal scales. With a vertical dimension provided by a network of instrumented cabled moorings at four initial sites, the RSN and Endurance array will enable interdisciplinary observations of water-column processes offshore of the continental margin in a region strongly forced by air–sea interaction, by shelf–slope interactions with the deep sea, and by coupled atmospheric/oceanic phenomena acting to produce variations in North Pacific circulation over gyre scales.

To better address these problems, the RSN is developing high power (3 kW) and bandwidth (1 Gb/s) full-water column moorings that will include a broad suite of sensors, an instrumented deep profiler, a 200 m platform and an instrumented winched shallow profiler (Figures 18.8 and 18.9; Table 18.2). A two-legged shallow profiler mooring will include an instrumented platform at 200m and an instrumented winched profiler that travels from the platform to the near surface. A single cable mooring will host the instrumented deep profiler that travels from ~ 3000m water depth to 200m. Initially, the paired moorings will be placed near the base-slope intersection on the Cascadia margin due east of Newport Oregon (PN1A), at a 600m site located ~50km shoreward of the deep water mooring (PN1C), and at the base of Axial Seamount (PN3A). The cabled moorings at 600m, and at 80m water depth will be flanked by an uncabled ocean surface piercing mooring with data transmission through satellites to shore (Figure 18.9). In addition, there will be two uncabled moorings utilizing satellite transmission at 25m offshore Newport. The PN1C, 80m and 25m sites comprise the OOI CSN Endurance array as part of the Oregon Endurance Line; the PN1C site is shared between RSN and CSN. Measurements made by the instrumented cabled moorings will be augmented by uncabled moorings west of Grays Harbor, Washington (Washington Line) (Figure 18.4), that extend to water depths of 500m. Increased spatial coverage will be provided by six gliders that traverse west from Newport Oregon >100km offshore and north to Grays Harbor (Figure 18.4). The unprecedented power and bandwidth (Gb/s) capabilities of the cabled moorings allow for a broad suite of sensors (Table 18.2) that include real-time video imaging and acquisition of high bandwidth hydrophone data for biological applications, upward-looking acoustic Doppler current profilers, and in situ pH and CO2 sensors. Each mooring has at least 30% future expansion capabilities.

The deep-water moorings (PN1A and PN3A) have very different oceanographic foci. The mooring at the Slope Base (PN1A) is situated adjacent to the coastal continental slope at the end of the Endurance Oregon Line, and in concert with the northern Washington Line, provides a unique opportunity for investigating a variety of interdisciplinary coastal studies (Figures 18.4, 18.8 and 18.9). The coastal region of the Pacific Northwest is a classic wind-driven upwelling system. However, the presence of the Columbia River plume and the range of trajectories with which it can impinge on the ocean, and the strong variability of the width of the continental shelf, all play strong roles in setting the system's response and behavior (Hickey et al, 2005; MacFadyen et al., 2005). In addition, the aforementioned large-scale systems affect the coastal region by modulating the pycnocline, nutricline and oxycline depths and offshore pressure gradients, which in turn affect the onshore transport of physical, biological and chemical quantities. The presence of internal waves driven by waves and tides, their interaction with the larger-scale currents, and their eventual breakdown into turbulence, are also vital to setting properties in this coastal region. All of these are expected to change strongly over time, but will be well resolved by the measurements at the base of the slope (PN1A), the Endurance Line, and OOI supporting shipboard sampling for calibration and verification (e.g., samples for dissolved oxygen, chlorophyll, nutrients).

In contrast to the margin setting of PN1A, the mooring at the base of Axial Seamount (PN3A) is far from the continental shelf (>400km) and hence represents an open-ocean or pelagic site in the continuum of observing scales represented in the OOI's cabled system (Figure 18.4). Here, large-scale currents including the North Pacific Current, the subpolar gyre and the northern end of the California Current interact. These currents transport heat, salt, oxygen, and biota, all of which are crucial to the region's ecosystem. Their variability arises from forcing as varied as tides and wind (0.5–5 day timescales) to interannual (El Niño) to decadal (Pacific Decadal Oscillation) timescales. Important questions addressed at Axial are what the impacts of long- and short-term forcing changes are on the structure and transport of the large-scale current system, and what their effects are on the ecosystems. Together with the mooring at Ocean Station Papa, these processes will be studied with observing platforms in the water column at these two sites. Other important questions at Axial include understanding by what processes internal tides form and how strongly the tidal currents break down into turbulence at this site. Internal tides are ubiquitous vertical motions formed by tidal currents flowing past areas of steep topography and therefore, both Axial Seamount and the base of the Cascadia margin provide optimal sites to investigate internal tide formation and impact on heat and chemical transport.

18.2.3.2 Cascadia subduction earthquakes, methane hydrates and seeps, novel organisms

Tectonic plate boundaries are the fundamental building blocks of our planet. They modulate formation of the oceanic crust and its eventual destruction, and fluid movement into and out of the oceanic crust. They are also sites of the largest earthquakes on Earth. The small size of the Juan de Fuca Plate, with its close proximity to shore, makes it an optimal site to install cabled arrays directed at studying scientifically and societally important processes associated with plate boundaries, such as earthquakes, submarine landslides, and tsunamis (Figure 18.10). These arrays, coupled with seismic data from NSF's Earthscope and Cascadia experiments and the Canada–Japan CASSIS project (CAScadia SeISmic experiment) and SeaJade projects will yield critical new insights into deformational processes operative along the Pacific Northwest margin and the impacts of these events on emission of methane-rich fluids from accretionary margins.

Cascadia subduction earthquakes: This past decade has witnessed some of the most devastating earthquakes on record. The 2004 Sumatra tsunami generated by the Sumatra-Andaman earthquake (9.1–9.3 magnitude, ~ 300,000 people killed), the 2008 Sichuan China earthquake (7.8 magnitude, 4.8 million homeless), the 2010 Chile earthquake and tsunami (8.8 magnitude, ~ 500 people killed), and most recently the 2011 earthquake and tsunami at Honshu Japan (9 magnitude, recovery cost >$300 billion, >12,000 killed) serve as vivid reminders that our planet is dynamic and that earthquakes generate profound natural disasters resulting in significant human, economic, and property loss.

The Pacific Northwest, bounded by the Cascadia subduction zone, has experienced earthquakes similar to those in Sumatra, China, Chile, and Japan. This zone experiences large earthquakes at intervals varying from about 200 to 1200 years with an average timing of 200 years (Goldfinger et al., 2003, 2008): it is believed to be overdue for a megathrust event. The last Cascadia megathrust rupture event occurred January 26, 1700 as document-

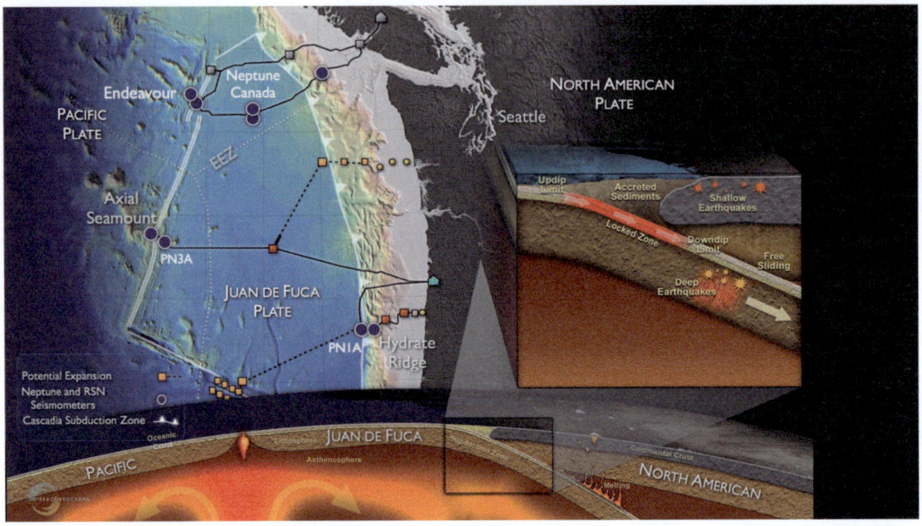

Figure 18.10 The Cascadia margin is overdue for a megathrust event, which historically has included 9 Ma earthquakes. An array of broadband and short-period seismometers and seafloor pressure sensors on the RSN and NEPTUNE Canada provide the beginning of a plate-scale geophysical array important to understanding these highly destructive events. Extension of these arrays, in a fashion similar to the Japanese cabled system DONET, could provide an important early warning system to the Pacific Northwest.

ed in myriad tribal histories and legends (e.g., Thunderbird and Whale) throughout Oregon, Washington, and Canada, as well as by a tsunami that came ashore in Japan (Satake et al., 1996; Ludwin et al., 2005). Extensive paleoseismology studies based on turburdite records over the past 3500 to 10,000 years provide average reoccurrence intervals of great earthquakes of 470–524 years, respectively (Goldfinger et al., 2003, Goldfinger, 2011). It has long been believed that the Cascadia subduction zone was locked; however, recent studies show that there are small- to moderate-sized low-angle thrust events marked by localized earthquakes: some of these events occur near Hydrate Ridge. These earthquakes appear to be associated with roughness on the subducting plate (Trehu et al., 2008). Some subduction zone earthquakes have been documented to initiate at this type of rough zone and then propagate into regions characterized by long, continuous forearc basins (similar to the basin north of Hydrate Ridge) with the magnitude of the earthquake related to the area of the forearc basin (Wells et al., 2003; Song and Simons 2003; Llenos and McGuire, 2007). In addition to the recognition of localized thrust events such as those near Hydrate Ridge, tremor-like seismic events that correlate with aseismic slow slip events are now known to be common along the Cascadia margin (Rogers and Dragert, 2003). These events are believed to represent slip on the plate boundary down dip from the locked zone and they are thought to reflect motion of fluids.

As an initial start to instrumenting the Cascadia subduction zone (and Juan de Fuca Plate), the RSN will install a small array of broadband seismometers with accelerometers, hydrophones, pressure sensors, current meters and short-period seismometers at the base of the accretionary margin west of Hydrate Ridge (PN1A), at the summit of Southern Hydrate Ridge, and at the base of Axial Seamount (Tables 18.2–18.4). While small in scale initially, it is hoped to grow this array in the future to include installation of cabled sensors west of Grays Harbor, and eventually the Blanco Transform Fault (Figure 18.10).

Gas hydrates and methane seeps: It is estimated that gigatons of methane near the surface of the Earth are locked into gas hydrates in the shallow sediments on continental margins (Kvenholden and Lorenson, 2001; Bangs et al., 2011). The hydrates may act as a capacitor in the carbon cycle by slowly storing methane that may be suddenly released into the ocean and atmosphere during slope failure and seafloor erosion (Paull et al., 2003; Maslin et al., 2004; Bangs et al., 2010). For example, erosion of the seafloor offshore of Japan is believed to have resulted in the release of 1.51×10^{11} m^3 of methane (Bangs et al., 2011). The erosional event, which unroofed hydrate-laden sediments hosting overpressured methane, is believed to have resulted in a cascade of events that resulted in the destabilization of kilometer-scale sections of seafloor. hydrate dissaociation, and massive release of methane into the hydrosphere (Bangs et al., 2011). Southern Hydrate Ridge (Figures 18.4 and 18.11), in the central Cascadia accretionary complex, is one of the best-studied gas hydrate deposits. Vigorous seafloor venting and formation of gas rich hydrate deposits near the seafloor have been documented here through ODP drilling during Legs 146 and 204 (Trehu et al., 2004, 2006) and through a series of seafloor studies using submersibles and ROVs (Levin, 2005; Suess et al., 1999; 2001). These studies have provided a good understanding of how gas hydrate is distributed in marine sediments and the processes that lead to heterogeneity in this distribution. The subsurface has been imaged with 3D seismic data (Trehu et al., 2004), which define a focused plumbing system beneath the summit of Hydrate and bottom simulating reflectors at only 120–150m beneath the seafloor (Bangs et al., 2005, 2011). Follow-on high resolution mapping and digital still mosaicking by the RSN team utilizing the ROV's Jason 2 and ROPOS and autonomous vehicle Sentry in 2010 and 2011, show that the most intense methane venting and mat development is currently confined to a 300 × 300m circular shaped area defined by extremely hummocky terrain (Figure 18.11). There is a general tendency for venting to be localized at the summit of the hummocks. Venting of methane and hydrogen sulfide-rich fluids support extensive bacterial mats and clam beds. The area is highly dynamic with numerous extinct and active venting sites and extensive carbonate deposits marking past upflow zones.

Southern Hydrate Ridge is an important observatory site to define the temporal evolution of methane hydrate systems in response to seismic events. It is also an important area to quantify material fluxes from the seafloor and the impacts of methane release on overlying ocean chemistry. It provides an excellent opportunity to understand biogeochemical coupling associated with gas hydrate formation and destruction. To meet these needs, the RSN installed a localized cabled array of instruments in 2013 that includes geophysical, chemical, and physical, and biological sensors (Figure 18.11; Table 18.3).

The real-time interactive capabilities of the cabled observatory are critical to studying gas-hydrate systems because many of the key processes may occur over short time scales.

Winched shallow profiler – 200m below surface

Instrument	Data products
3-wavelength fluorometer	Fluorometric chlorophyll – a concentration (695nm excited by 470nm), Fluorometric CDOM concentration (460nm excited by 370nm). Optical backscatter (red wavelengths)
Seawater pCO_2 sensor	Partial pressure of CO_2 in seawater
Photosynthetically Available Radiation (PAR) sensor	PAR
Downwelling spectral irradiance sensor	Downwelling spectral irradiance. Diffuse attenuation
3-D single point current meter	Turbulent point water velocity. Orientation (pitch, roll, heading)
CTD	Conductivity, temperature, pressure (depth), practical salinity, density
Dissolved oxygen sensor	Oxygen concentration
Optical nitrate sensor	Nitrate concentration
Seawater pH sensor	pH
Spectrophotometer	Optical attenuation, Optical absorption

200m platform

Instrument	Data products
150 kHz Acoustic Doppler Current Profiler (≤300m range)	Velocity profiles. Acoustic backscatter. Orientation (pitch, roll, heading)
Digital still camera, ethernet-based, with lights	Still images of water column biology
5-beam, 600 kHz Acoustic Doppler Current Profiler (≤50m range)	Turbulent velocity profiles. Acoustic backscatter. Orientation (pitch, roll, heading)
Broadband (passive) acoustic receiver (hydrophone) for marine mammal, fish, wind, and rain	Frequency, acoustic pressure waves
CTD	Conductivity, temperature, pressure (depth), practical salinity, density
Dissolved oxygen sensor, optode-based	Oxygen concentration
2-wavelength fluorometer	Fluorometric chlorophyll-a concentration (695nm excited by 470nm), optical backscatter (red wavelengths)

Deep profiler – near bottom to 200m

CTD	Conductivity, temperature, pressure (depth), practical salinity, density
Dissolved oxygen sensor	Oxygen concentration
3-D single point current meter	Turbulent point water velocity. Orientation (pitch, roll, heading)
Spectrophotometer	Optical attenuation. Optical absorption
3-wavelength fluorometer	Fluorometric chlorophyll-a concentration (695nm excited by 470nm). Fluormetric CDOM concentration (460nm excited by 370nm). Optical backscatter (red wavelengths)

Bottom benthic package – 2650m base of cascadia margin (PN1) 2906m base of axial seamount (PN3A)

Instrument	**Data products**
Velocity profiler (300m range)	Velocity profiles, acoustic backscatter, orientation (pitch, roll, heading)
Broadband (passive) acoustic receiver (hydrophone) for marine mammal, fish, wind, and rain	Conductance, susceptance, frequency, acoustic pressure waves, target strength of scatterers
Horizontal electric field and pressure inverted echo sounder. Used for: water column integration of velocity, acoustic travel time, pressure, water column structure	RATT (Roundtrip Acoustic Travel Time), bottom pressure, horizontal electric fields, depth averaged barotropic velocity (GEM), vertically averaged horizontal water velocity (VAHWV)
CTD	Conductivity, temperature, pressure (depth), salinity, density
Dissolved oxygen sensor	Oxygen concentration
Spectrophotometer for optical attenuation and absorption measurements	Optical attenuation, optical absorption, optical backscatter

*Provides sea surface to seafloor profiles of temperature, salinity, specific volume anomaly, and absolute horizontal currents (Luther and Sanford, 2008)

Table 18.2 Instruments on the RSN cabled moorings and associated benthic package.

Figure 18.11 Southern Hydrate Ridge is one of the best-studied methane hydrate sites in the world's oceans. The main RSN node site at the summit of Southern Hydrate Ridge is planned to include a site of vigorous methane bubble release and extensive microbial mats. The entire area is mapped at < 5-m resolution and has been photo-saturated, resulting in a photomosaic documenting the most active seep sites, extensive areas of microbial mat development, and clam beds. Note that the six small "dots" in the center of the image are collapsed holes remaining from the Ocean Drilling Program. The topographic high on the left side of the image is an area of extensive carbonate deposition, with only very minor venting. The insets show the distribution and types of instruments that were deployed in 2013 as part of the RSN. ADCP = Acoustic Doppler Current profiler.

Events such as bubble plume formation, the creation of collapse zones, and increased seepage in response to earthquakes will require adaptive response and sampling capabilities that include full resolution sampling by upward-looking sonars, fluid and gas sampling, long-term duration collection of plume and seep imagery from cameras, and in situ manipulation of chemical sensors which, coupled with flow meters, provide estimates of transient and more stable chemical flux. Future infrastructure ideally would include response capabilities by instrumented autonomous vehicles conducting nested surveys within a fence of cabled instrumented moorings (Gas-Hydrate Observatories Workshop, 2007). Such capabilities would provide full water column chemical measurements and quantification of methane flux through the water column.

18.2.3.3 Linkages among submarine volcanoes, hydrothermal venting, and life in extreme environments: Axial Seamount

For four billion years the global ocean has responded to and integrated the impacts of submarine volcanism. Here, the Earth, the oceans, and life co-evolved (Sullivan and Baross, 2007). Some of the most exciting discoveries in past decades have come from studying the relationships between submarine volcanoes, the gases and fluids they release, and the life that thrives in these novel environments. Over 70% of global volcanism occurs in the oceans with most activity located along mid-ocean ridge spreading centers and it is at this 70,000km long, volcanically active zone where tectonic plates spread apart and where new oceanic crust is formed. Pools of molten material and hot rock within the cores of these volcanoes supply heat to hydrothermal hot springs that emit fluids at >350°C and form metal-rich chimneys that tower up to 45m above the seafloor (Kelley et al., 2002). Within these dynamic environments, magmatically-driven ecosystems host some of the most novel life forms on the planet, which utilize volcanic gases as energy sources, thriving at high pressures and temperatures in the absence of sunlight.

Although earthquakes as destructive processes (landslides, volcanic eruptions, tsunamis) are general knowledge, seismic processes are also important to generation of life itself. For example, seismic events associated with underwater volcanoes result in the release of gases that serve as important energy sources for chemosynthetic microbes (McCollom and Shock, 1997; Lilley et al., 2003). Some of these microbes are genetically the

Winched shallow profiler - 200m below surface

Instrument	**Data products**
Pressure sensor on seafloor, tidal	Seafloor pressure, tides
Low frequency (passive) acoustic receiver (hydrophone) for wind, rain, and seismic t-phases	Frequency, acoustic pressure waves
Broadband seismometer for global/ regional seismic activity	Broadband ground velocity, broadband ground acceleration
3 short-period seismometers for regional/ local seismic activity	Short-period ground velocity
3-D single point current meter	Turbulent point water velocity. Orientation (pitch, roll, heading)
Digital still camera, ethernet-based, with lights	Still images of macrofauna, seeps
Mass spectrometer for dissolved gases in diffuse flow and seep fluids	Dissolved gas concentrations in diffuse flow and seep fluids
Benthic fluid flow sensor	Time series of benthic fluid flow rates
75 kHz Acoustic Doppler Current Profiler (≥500m range)	Velocity profiles, acoustic backscatter, orientation (pitch, roll, heading)
Water sampler, osmosis-based	Physical water samples of diffuse flow and seep fluids for major/trace element chemistry

Table 18.3 Cabled instruments at Southern Hydrate Ridge (~800m water depth).

most ancient of organisms known (Elkins et al., 2008). The highest temperature organisms on Earth have been recovered from submarine vents: one of them was cultured from the Juan de Fuca Ridge and is capable of surviving at temperatures >120°C (Kashefi and Lovely, 2003). However, fewer than 1% of these organisms have been cultured and little is known about their metabolic activity, highlighting the fact that the ocean is still largely unexplored and that profound discoveries will be made at the interface of geological, chemical, and biological studies.

Further advances in understanding these dynamic systems are increasingly dependent on our ability to collect long-term, high-frequency observations using diverse networks of sensors and samplers that are in the right place at the right time. The major volcanic and tectonic events that create the oceanic crust and modulate heat, chemical, and biological fluxes through the seafloor are inherently episodic on decadal timescales and are also short-lived, punctuated events. For example, water column analyses above recently-erupted submarine volcanoes show that significant heat, chemicals, and biological material are ejected into the water column during the formation of megaplumes concomitant and following eruptive events (Lupton, 1995; Delaney et al., 1998). The megaplumes are characterized by anomalous heat-helium content, they are enriched in magmatic gases, the buoyant plumes rise a thousand meters or more above the surrounding seafloor, and they can be 20km across (Lupton et al., 1999). They contain thermophilic microorganisms, not present in background seawater, which require anaerobic conditions (Summit and Baross, 1998). In concert, these observations show that the microbes come from an as yet poorly characterized hot subseafloor environment. They are believed to be injected into the overlying ocean during plume formation as crustal fluids are flushed from depth (Delaney et al., 1998; Summit and Baross, 1998). Significant perturbations can also occur during diking events even if magma does not reach the seafloor, however. For example, measurements of carbon dioxide in high temperature black smoker vents on the Endeavour Segment of the Juan de Fuca Ridge, which were made 3–12 months following melt injection beneath the seafloor, show that the amount of carbon dioxide released during the first few months following the eruption may have rivaled that released by the mature hydrothermal system over an entire year (Lilley et al., 2003). The only way to capture these events, and for the first time make quantitative measurement of important heat, chemical and biological fluxes, is to maintain a long-term monitoring capability at a number of sites with high probability for tectonic and/or magmatic activity.

Axial Seamount on the Juan de Fuca spreading center is an optimal site for a long-term observatory. It is located just one day by ship from the Washington and Oregon coasts and it is one of the main experimental sites on the RSN. It is the most magmatically robust volcano on the spreading center, having erupted in 1998 and most recently in April 2011 (Embley et al., 1999; Chadwick et al., 2006, 2012) (Figure 18.12). It hosts numerous active hydrothermal fields and abundant sites of diffuse flow (Butterfield et al., 1990, 2004). It is one of the best-studied volcanoes along the global mid-ocean ridge spreading center and the only site where in situ pressure sensors have documented up to 13cm/year summit inflation prior to an eruption (Nooner and Chadwick, 2009).

Axial Volcano magmas contain extremely high carbon dioxide concentrations (Helo et al., 2011) supporting a robust subseafloor biosphere (Huber et al., 2003). Axial is the only site where there is greater than decade-long time-series data documenting the evolution

Figure 18.12 An extensive instrument array focused on the ASHES and International District vent fields was deployed within the caldera of Axial Seamount in 2013. Primary Node PN3B will provide the power and data transmission to an array of secondary nodes positioned on the eastern side of the caldera and on the caldera floor. A small array of geophysical sensors will be distributed within and to the side of the caldera to investigate melt migration, seismic events and inflation and deflation of the summit of the volcano. The black outline shows the 3km wide and >8km long eruption that occurred in April 2011. The eruption went undetected until July 2011 because of a lack of a monitoring system at this site. RAS/PPS = remote access in situ fluid sampler and microbial DNA sampler; OSMO = osmotic fluid sampler for diffuse flow sites; BPR/Tilt = bottom pressure sensor and tilt meter; OBS = ocean bottom seismometer; HD = High definition video.

of microbial communities, and vent fluid chemistry and temperature following submarine volcanic eruptions (Huber et al., 2003; Butterfield et al., 2004; Opatkiewicz et al., 2009). Here, individual diffuse flow sites appear to be "microbial islands" with each vent hosting discrete microbial communities that change over time, but which maintain a community structure that is unique to that site, even in vents meters to kilometers apart (Opatkiewicz et al., 2009). The summit of Axial Seamount was significantly transformed during the 2011 eruption with flows reaching nearly 3km across the floor of the caldera, and >8km along the eastern fissure zone. Following the April 2011 event, the axial caldera was filled with murky, volatile and microbial-rich fluids that were injected into the overlying water column. Five months after the eruption, direct observations by the Canadian robotic vehicle ROPOS during the VISIONS11 OOI-UW cruise to Axial showed that numerous "snowblowers" formed along the east and central fissure system. They were characterized by significant egress of methane-rich fluids, and large concentrations of flocculant and microbial material flushed from the subsurface and from the surface of extensive collapse zones (Proskurowski et al., 2012) (http://www.ooi.washington.edu/visions11/file/

Dive+1472+Highlights). The snowblowers may serve as important seeding sites for follow-on microbial and macrofaunal systems.

The high bandwidth and high power fiber optic cables installed at the summit of Axial Seamount, which rises 1500m above the surrounding seafloor will be used to communicate with and power a diverse array of sensors at the summit of the volcano (Figure 18.12, Table 18.4). Real-time high definition video will provide unprecedented views of macrofaunal and microbial communities at the vents and chemical sensors and thermistor arrays will provide information on the environmental conditions in which the biological communities thrive. The instrumentation will also yield information on the impact of flow perturbations associated with eruptive and seismic events on biological communities. Other sensors to be installed include in situ mass spectrometers for fluid – volatile chemistry, broadband and short-period seismometers to monitor earthquake and magma migration activity, temperature and chemical probes in diffuse and black smoker sites, fluid and DNA samplers, and pressure-tilt meters for measurement of preeruptive inflation events and post eruptive deflation (Table 18.4). The in situ DNA samplers initially will focus on in situ filtering and preservation of time series samples in diffuse flow sites. The instrument array at Axial is the largest single in situ experiment in the global ocean focused on long-term measurements of underwater volcanoes with transmission of real-time data and imagery back to shore. All data will be available to the public 24/7/365 and the system as installed will be highly expandable.

18.3 Cabled observatories and amplification with emerging technologies

Ocean Science is rapidly becoming a beneficiary of the convergence of a host of powerful, exponentially evolving, emergent technologies driven by widely diverse technical and economic forces essentially external to the world of ocean research. These cutting edge capabilities include, but are not limited to, very high bandwidth communications, nanotechnology, biotechnology, information technology, in situ genomic analysis, mass spectrometry, computational modeling, imaging technologies, and robotics (Figures 18.13 and 18.14). The Nobel Prize in Physics in 2010 was awarded for work leading to digital imaging that is revolutionizing our capacity to capture both motion and still imagery with a fidelity and visual acuity that rivals or exceeds human vision. More powerful than any single technology will be the progressive integration of these enabling capabilities into highly sophisticated submarine systems designed to conduct challenging remote operations in novel ways.

One area in the oceans where profound discoveries will be made in the coming decades is that of quantifying and characterizing the enormous and diverse microbial and virus communities that thrive within the oceans waters, and in the water-saturated subseafloor environments. New analytical techniques show that we have barely begun to quantify the diversity of organisms in marine environments. It is predicted that the oceans contain 3.6 × 1029 microbial cells (Whitman et al., 1998), yet very few organisms have been fully

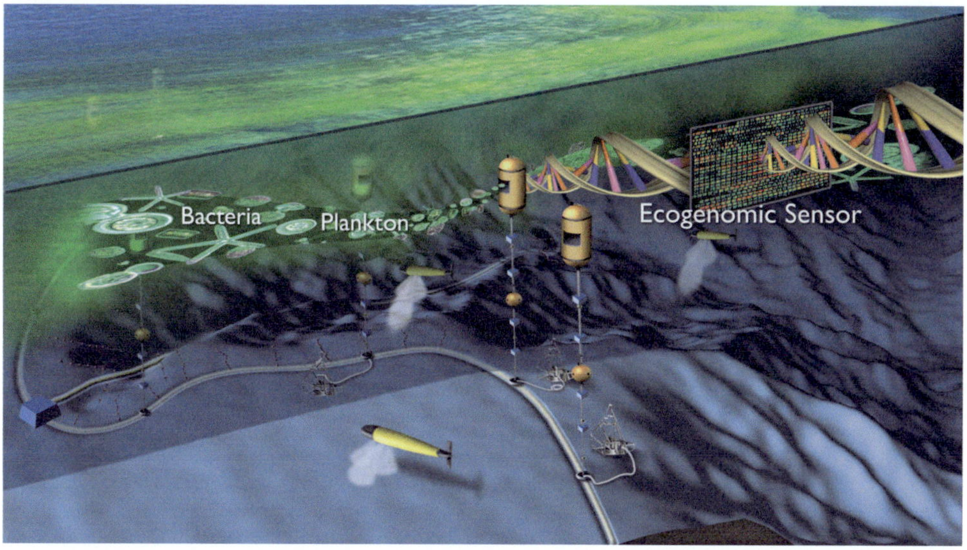

Figure 18.13 A vision of the type of next generation application that might flow from the incorporation into oceanographic research strategies of a rapidly evolving in situ genomic analysis capability within selected volumes of the ocean. Systems such as the Environmental Sample Processor for in situ genomic work (Scholin, 2010) are now being sold commercially (e.g., McLane Research Associates). This capability, connected to cabled observatories, will launch a revolutionary shift in the manner in which we study ocean systems using on-demand genomic analyses of living communities that dominate the base of the oceanic food chain, and can vary widely depending on the environmental conditions.

characterized. The nature of the subseafloor biosphere is largely unknown, as is the diversity and abundance of viruses. New analytical techniques that adopt massive parallel tag sequencing, bring into question the definition of species at the microbial level and indicate that the diversity of microbes may be 1–2 orders of magnitude more diverse than previously thought (Sogin et al., 2006).

The ability to characterize the marine biosphere has substantially benefited by the explosive growth of ecogenomics (a term coined following the completion of the Human Genome Project), and exponential growth of associated technologies. The amount of time for, and the cost of, sequencing a genome have dramatically decreased in the past 10 years, as has the size of the equipment necessary to complete the analysis. Pioneering efforts using derivative capabilities in the ocean have been tested and are expected to be in routine use within five years. For example, the ecogenomic sensor called the Environmental Sampler Processor is now operational for both shallow and deep-sea environments and provides in situ real-time analyses of ribosomal RNA sequences for identification of both bacteria and archaea, as well as numerous invertebrates and harmful algal species (Scholin et al.,

Ashes Vent Field

Instrument	Data products
3 shortperiod seismometers measuring ground motion velocity in 3 orthogonal directions	Velocity, magnitude of earthquakes and location for regional and local seismicity
Thermistor array measuring seafloor diffuse flow temperature and temperature gradients	Temperature of diffuse flow
Osmotic water sampler for major/trace element chemistry of diffuse flow and seep fluids from seafloor	Physical water sample for major and trace chemistry
High definition camera with strobes, pan, and tilt for high definition video imagery	HD video image of vents, diffuse flow, microbial communities, and macrofauna

Eastern Summit

1 shortperiod seismometers measuring ground motion velocity in 3 orthogonal directions	Velocity, magnitude of earthquakes and location for regional and local seismicity
Broadband (passive) acoustic receiver (hydrophone) on the seafloor	Acoustic pressure waves. Target strength of scatterers, for measurement of wind, rain, and seismic t-phases
Broadband seismometer measuring ground acceleration in 3 orthogonal directions.	Velocity and acceleration used to measure global/ regional seismic events, melt migration, harmonic tremor

Central Caldera

1 shortperiod seismometers measuring ground motion velocity in 3 orthogonal directions	Velocity, magnitude of earthquakes and location for regional and local seismicity
Pressure – tilt meter	Measurements of seafloor inflation and deflation due to magmatic processes via measurements of bottom pressure and tilt, X tilt, Y tilt, pressure, orientation (pitch, roll, heading), Seafloor uplift and deflation
Broadband (passive) acoustic receiver (hydrophone) on the seafloor	Acoustic pressure waves, Target strength of scatterers, for measurement of wind, rain, and seismic t-phases

International District

Black smoker fluid chemistry sensor	Temperature, hydrogen concentration, sulfate concentration, hydrogen sulfide concentration, pH
Black smoker fluid temperature-resistivity sensor	Temperature, oxidation-reduction potential, chloride concentration
Mass Spectrometer	Spectrometer scan (partial pressure vs. mass number) for dissolved volatiles in dissuse fluids
In situ remote access water sampler	Physical water sample for major/trace element chemistry, temperature, H2S, pH
Particulate DNA sampler	Physical sample of filtered (2 µm filter) material for microbial DNA
Ethernet Camera with strobes	Digital still imagery of water column biology, diffuse vent site macrofauna and macrofuana
1 shortperiod seismometers measuring ground motion velocity in 3 orthogonal directions	Velocity, magnitude of earthquakes and location for regional and local seismicity
Pressure – tilt meter	Measurements of seafloor inflation and deflation due to magmatic processes via measurements of bottom pressure and tilt, X tilt, Y tilt, pressure, orientation (pitch, roll, heading), Seafloor uplift and deflation

Table 18.4 Cabled instruments at the summit of Axial Seamount (~1540m)

2009; Scholin, 2010). This system has been tested at the summit of Axial Seamount, and it is a candidate system for cabled observatories (http://www.mbari.org/esp/) (Figure 18.13). Other in situ systems include the Autonomous Microbial Genosensor for real-time analyses and identification of active cells using nuclic acid sequence bases amplification (Fries et al., 2007), in situ flow cytometers for analyzing particles, phytoplankton and other micro-organisms (e.g., Olson and Sosik, 2007), and video profilers to image larger organisms such as zooplankton (Stemmann et al., 2008).

Full application of these and other data-intensive sensing techniques within the global ocean would be impossible without the other element recognized by the Nobel Committee for Physics: optical communication. Derivative research has opened the planet to high bandwidth communications. The potential to incorporate semiconductors into more conventional fiber transmission systems holds the potential to dramatically enhance the effectiveness and flexibility of utilizing fiber in the oceans to conduct science and enhance education. The initial vision embodied in Moore's Law regarding the exponential doubling capacity of integrated circuits for computers can be seen as an analogue to other emergent

Figure 18.14 An artist's depiction of a fully automated seafloor analytical laboratory within an active submarine hydrothermal system, operated serially or in parallel by different remote science groups on land using the power and bandwidth provided by industry standard electro-optical cables. Real-time analysis, in situ specimen dissection, and complex experiments can be conducted on materials flowing through the environment of this volcanically-driven ecosystem containing exotic microbial communities. Full operation of such an environmental laboratory will require integration of many emergent technologies. A system of this type, and a host of others, will become commonplace as the emergent technologies are increasingly integrated into next-generation forms of human telepresence in the oceans.

technologies. These capabilities, coupled with in situ genomic, biogeochemical and imaging flow cytometers, will lead to unparalleled insights into linkages among the environment, diversity, and metabolic activity of life within the oceans.

In the hands of innovative scientists working with inspired engineers, many of these technologies will be used to leverage the effectiveness of the many others. The fact that numerous technologies are evolving in an exponential manner may be the basis for suggesting that the ensemble will evolve even more rapidly. The bottom line is that provision of bandwidth and power via submarine cables may optimize next-generation science in the ocean basins by fully capitalizing on a full suite of emergent technologies. The recent use of cables for ocean science is an early phase of a trend toward world-wide networks of cables, both dedicated *and shared*, throughout the global ocean, to vastly enhance human telepresence in a fashion heretofore unimaginable.

18.4 Broader potential – cabled human presence in the sea

Cabled infrastructure is but one piece of the marine research infrastructure needed in the coming decades. Cabled undersea networks may be thought of as the first "interstate" highway that brings growth and discovery by linking and enhancing existing tools such as the US University-National Oceanographic Laboratory System (UNOLS) research vessels, satellites, autonomous underwater vehicles (AUVs), gliders, and unmanned aerial vehicles (UAVs). For example, more effective use could be made of UNOLS's vessels as single research platforms and as service platforms for tending experimental sites within the volumetric footprint of a cabled suite of sensor arrays. Another example might include enhanced use of existing data platforms (ships, satellites, AUVs, gliders, UAVs) for identifying areas worthy of rapid response or sustained observation. Similarly, cabled infrastructure could serve as a "data port" for many mobile assets to plug in and transmit massive amounts of data from the open ocean to be quickly analyzed in the "cloud" computing environment, then returned to the mobile asset and used for near-real time mission updates or validation. Although the initial installation cost of cabled infrastructure is not cheap, when considered over a 25-year lifetime, maintenance costs are low, as are data transmission costs. When viewed as cost per observation, because of the bandwidth necessary to thoroughly interact with complex and dynamic ocean systems, cabled infrastructure is a surprising bargain.

18.5 Global problems require international solutions

Over the coming decades, many nations for many reasons will implement cabled sensing systems in offshore extensions of their territorial seas, continental shelf-slope environments, and Exclusive Economic Zones. As these powerful new forms of integrated marine infrastructure become more sophisticated and the data become routinely accessible, the Internet may emerge as the most powerful oceanographic research tool on the planet.

In response to unexpected but energetic transient events in ocean environments, rapid reconfiguration of key sensor arrays linked to the Internet via submarine electro-optical cables will allow us to image, measure and respond to energetic and previously inaccessible episodic phenomena. These events may include catastrophic natural events such as erupting volcanoes, large submarine mass wasting events, big earthquakes, and giant storms as well as progressive environmental anomalies that include ocean acidification, hypoxia events resulting in "dead zones", and harmful algal blooms

Like the atmosphere, and outer space, the oceans do not belong to any single nation, yet we are all dependent on them. To address some of the changes induced in the global ocean system will require high levels of international collaboration. Because submarine cabled communication systems now bind the international community together, it is fitting that these remarkable products of private industry emerge as an integral part of the solution to one of the major questions confronting our global society: How does the planetary life support system actually work?

Acknowledgments

The authors thank Chris Barnes and an anonymous reviewer for their helpful comments and Mark Stoermer and Hunter Hadaway, Center for Environmental Visualization at the University of Washington, Seattle, WA, for their remarkable graphics, many of which are included in this chapter. The authors also thank the nearly 200 oceanographers that submitted Request for Assistance proposals focused on NEPTUNE science at the request of NSF in 2005. In concert, these proposals formed the inspiration for, and location of, the RSN node sites and exemplified the excitement of the community in science that could be conducted by submarine cabled observatories.

References

Aguzzi J., Manuel A., Condal F., Guillen J., Nogueras M., del Rio J., Costa C., Menesatti P., Puig P., Sardia F., Toma D. and Palanques A. (2011) The new seafloor observatory (OBSEA) for remote and long-term coastal ecosystem monitoring. Sensors 11, 5850–5872.

Bangs N.L., Hornbach M.J. and Berndt C. (2011) The mechanics of intermittent methane venting at Sough Hydrate Ridge inferred from 4D seismic surveying. Earth Planet. Sci. Lett. doi:10.1016/j.epsl.2011.06.022.

Bangs N.L., Hornbach M.J., Moore G.F. and Park J.-O. (2010) Massive methane release triggered by seafloor erosion offshore southwestern Japan. Geology 38, 1019–1022.

Bangs N.L., Musgrave R.J., and Trehu A.M. (2005) Upward shifts in southern Hydrate Ridge gas hydrate stability zone following postglacial warming, offshore Oregon. J. Geophys. Res. 110, doi:10.1029/2004JB003293.

Barletto P., Kelly M., Kelley D.S., Delaney J.R., Ittig B. and Durand C. (2010) Cabled observatory: The operations and maintenance opportunity. SubOptic 2010, Yokohama, Japan.

Barnes C.R., Best M.M.R., Johnson F.R., Pautet L., and Pirenne B. (2011) Understanding Earth/ocean processes in using real-time data from NEPTUNE Canada's widely distributed sensor networks, North-East Pacific. Geoscience Canada 38, 21–30.

Butterfield, D.A., Massoth G.J., McDuff R.E., Lupton J.E. and Lilley M.D. (1990) Geochemistry of hydrothermal fluids from Axial Seamount hydrothermal emissions study vent field, Juan de Fuca Ridge: Subseafloor boiling and subsequent fluid-rock interaction. J. Geophys. Res. 95, 12,895–12,921.

Butterfield D.A., Roe K.K., Lilley M.D., Huber J.A., and Baross J.A. (2004) Mixing, reaction and microbial activity in the sub-seafloor revealed by temporal and spatial variation in diffuse flow vents at Axial Volcano. In: W.S.D. Wilcock, E.F. DeLong, D.S. Kelley, J.A. Baross and S.C. Cary (Eds) *The Subseafloor Biosphere at Mid-Ocean*

Ridges. Geophysical Monograph 144, American Geophysical Union, Washington DC., pp. 269–289.

Chadwick W.W., Jr., Nooner S.L., Zumberge M.A., Embley R.W., and Fox C.G. (2006) Vertical deformation monitoring at Axial Seamount since its eruption using deep-sea pressure sensors. J. Volcan. Geotherm. Res. 150, 313–327.

Chadwick W.W., Jr., Nooner S.L., Butterfield D.A., and Lilley M.D. (2012) Seafloor deformation and forecasts of the April 2011 eruption at Axial Seamount. Nature Geoscience, doi:10.1038/ngeo1464.

Creed E.L., Glenn S., Schofield O.M., Barrier H., Petrecca R.F., Dobarro J.A., McLean S.D., Barnard A.H., Brown K.M., Adams R.S., and Feener S. (2005) *Oceans 2005, Proceedings of MTS/IEEE*, 10.1109/OCEANS.2005.1639828, 657-661.

Delaney, J.R. and R.S. Varga (2009) A 2020 vision for ocean science. In: T. Hey, S. Tansley, and K. Tolle (Eds) *The Fourth Paradigm, Data-Intensive Scientific Discovery*. Microsoft Research, pp. 27–38.

Delaney J.R., Spiess F.N., Solomon S.C., Hessler R., Karsten J.L., Baross J.A., Holcomb R.T., Norton F., Mcduff R.E., Sayles F.L., Whitehead J., Abbott D., and Olso L. (1987) Scientific rationale for establishing long-term ocean bottom observatory/laboratory systems. In: P.G. Teleki, M.R. Dobson, J.R. Moor, and U. von Stackelber (Eds) *Marine Minerals Resource Management Strategies*. pp. 389–411.

Delaney J.R., Kelley D.S., Lilley M.D., Butterfield D.A., Baross J.A., Wilcock W.S.D, Embley R.W., and Summit M. (1998) The quantum event of oceanic crustal accretion: Impacts of diking at mid-ocean ridges. Science 281, 222–230.

Delaney J.R., Ross G.R., Howe B., Chave A.D., and Kirkham H. (2000) NEPTUNE: Real-time ocean and earth sciences at the scale of a tectonic plate. Oceanography 13, 71–79.

Dewey R., Round A., Macoun P., Vervynk J., and Tunnicliffe V. (2007) The VENUS cabled observatory: Engineering meets science on the seafloor. Oceans 07 Conference.

Du Vall K., Ingle S., Snider J., and DiMarco S.F. (2011) Cabled observatories in the Sea of Oman and Arabian Sea. Conf. Proceed. MTS-IEE Oceans'11, 1–6.

Edson J.B., McGillis W.R., and Austin T.C. (2000) A new coastal observatory is born. Oceanus 42, 31–33.

Elkins J.G., Podar M., Graham D.E., Makarova K.S., Wolf Y., Randau L., Hedlund B.P., Brochier-Atmanet C., Kunin V., Anderson I., Lapidus A., Goltsman E., Barry K., Koonin E.V., Hugenholtz P., Kyrpides N., Wanner G., Richardson P., Keller M. and Stetter K.O. (2008) A korarachaeal genome reveals insights into the evolution of the Archaea. Proc. Nat. Acad. Sci. 105, 8102–8107.

Embley R.W., Murphy K.M., and Fox C.G. (1990) High-resolution studies of the summit of Axial Volcano. J. Geophys. Res. 95, 12,785–12,812.

Embley R.W., Chadwick W.W., Jr., Clague D., and Stakes D. (1999) 1998 eruption of

Axial Volcano: Multibeam anomalies and seafloor observations. Geophs. Res. Lett. 26, 3425–3428.

Eriksen C.C., Osse T.J., Light R.D, Wen T., Lehman T.W., Sabin P.L., Ballard J.W., and Chiodi A.M. (2001) Seaglider: A long-range autonomous underwater vehicle for oceanographic research. IEEE J. Ocean. Eng. 26, 424–436.

Favali P. and Beranzoli L. (2006) Seafloor observatory science: A review. Annals of Geophs. 49, 515–567.

Favali P., Beranzoli L., D'Anna G., Gasparoni F., and Gerber H.W. (2006) NEMO-SN-1: the first "real-time" seafloor observatory of ESONET. Nuclear Instruments and Methods in Physics Res. 567, 462–467.

Feely R.A., Labine C.L., Hernandez-Ayon J. M., Ianson D., and Hales B. (2008) Evidence for upwelling of corrosive "Acidified" water onto the continental shelf. Science 320, 1490–1492.

Fries D.P., Paul J.H., Smith M.C., Farmer A., Casper E., and Wilson J. (2007) The Autonomous Microbial Genosensor, an in situ sensor for marine microbe detection. Microsc. Microanal. 13 (Suppl 2), doi:10.1017/S1431927607078816, 2007.

Gas-Hydrate Observatories Workshop (GHOBS) Report, July 18–20 2007, Portland, Oregon, Consortium for Ocean Leadership and Department of Energy.

Goldfinger C., (2011) Submarine paleoseismology based on turbidite records. Annu. Rev. Mar. Sci. 3, 35–66.

Goldfinger C., Nelson C.H., and Johnson J.E. (2003) Holocene earthquake records from the Cascadia Subduction Zone and Northern San Andreas Fault based on precise dating of offshore turbidites. Ann. Rev. Geophys. 31, 555–577.

Goldfinger C., Grijalva K., Bürgmann R., Morey A.E., Johnson J.E., Nelson C.H., Gutierrez-Pastor J., Ericsson A., Karabanov E., Chaytor J.D., Patton J., and Gracia E. (2008) Late Holocene rupture of the Northern San Andreas Fault and possible stress linkage to the Cascadia Subduction Zone. Bull. Seism. Soc. Amer. 98, 861–889.

Goldfinger C., Hans Nelson C., Morey A.E., Johnson J.E., Patton J.R., Karabanov E., Gutierrez-Pastor J., Eriksson A.T., Gracia E., Dunhill G., Enkin R.J., Dallimore A., and Vallier T. (2012) Turbidite event history – Methods and implications for Holocene paleoseismicity of the Cascadia Subduction Zone. U.S.G.S. Professional Paper 161–F, pp. 170.

Grantham B.A., Chan F., Nielsen K.L., Fox D.S., Barth J.A., Huyer A., Lubchenco J. and Menge B.A. (2004) Upwelling-driven nearshore hypoxia signals ecosystem and oceanographic changes in the northeast Pacific. Nature 429, 749–754.

Gray J., cited in: T Hey, S Tansley and K Toll (Eds) *The Fourth Paradigm: Data-Intensive Scientific Discovery*. Microsoft Research. URL: http://research.microsoft.com/en-us/collaboration/fourthparadigm/.

Helo C., Longpre, M-A., Shimizu N., Clague D.A., and Stix J. (2011) Explosive eruptions at mid-ocean ridges drive by CO2-rich magmas. Nature 4, 260–263.

Hickey B., Geier S., Kachel N., and MacFadyen A. (2005) A bi-directional river plume: The Columbia in summer. Cont. Shelf Res. 25, 1631–1656.

Howe B.M., Lukas R., Duennebier F., and Karl D. (2011) ALOHA cabled observatory installation. Conf. Proceed. MTS-IEE Oceans'11.

Hsu S.-K., Lee C.-S., Shin T.-C, Liu C.-S, Huang B.-S., You B., Lin C.-H., Tang D., Chang H.-Y., and Kuo C.-H. J. (2007) Marine cable hosted observatory (MACHO) project in Taiwan, ISEE, 305–307. http://www.asiaone.com/News/Latest%2BNews/Asia/Story/A1Story20110325-270016.html.

Huber J.A., Butterfield D.A., and Baross J.A. (2003) Bacterial diversity in a subseafloor habitat following a deep-sea volcanic eruption. FEMS. Micro. Eco. 43, 393–409.

Kashefi K. and Lovley D.R. (2003) Extending the upper temperature limit for life. Science 301, 934.

Kelley, D.S., Baross J.A., and Delaney J.R. (2002) Volcanoes, fluids, and life in submarine environments. Ann. Rev. Earth Planet. Sci. 30, 385–491.

Kvenvolden K.A. and Lorenson T.D. (2001) The global occurrence of natural gas hydrate. In: C.K. Paull and W.P. Dillon (Eds) *Natural Gas Hydrates: Occurrence, Distribution, and Detection*. Geophysical Monograph 124, American Geophysical Union, Washington D.C., pp. 3–19.

Llenos A.L. and McGuire J.J. (2007) Influence of fore-arc structure on the extent of great subduction zone earthquakes. J. Geophys. Res. 112, B09301, doi:10.1029/2007JB004944.

Levin L.A. (2005) Ecology of cold seep sediments: Interactions of fauna, flow, chemistry and microbes. Ocean. Mar. Bio; An Ann. Rev. 43, 1–46.

Lilley M.D., Butterfield D.A., Lupton J.E., and Olson E.J. (2003) Magmatic events can produce rapid changes in hydrothermal vent chemistry. Nature 422, 878–881.

Ludwin R.S., Dennis R., Carver D., McMillan A.D., Losey R., Clague J., Jonientz-Trisler C., Bowechop J., Wray J., and James K. (2005) Dating the 1700 Cascadia earthquake: Great coastal earthquakes in native stories. Seismological Res. Lett. 76, 140–148.

Lupton J.E. (1995) Hydrothermal plumes: Near and far field. In: S.E. Humphris, R.A Zierenberg, L.S. Mullineaux, and R.E. Thomson (Eds) *Seafloor Hydrothermal Systems: Physical, Chemical, Biological, and Geological Interactions*. Geophysical Monograph 91, American Geophysical Union, Washington D.C., pp. 317–346.

Lupton J.E., Baker E.T., and Massoth G. (1999) Helium, heat, and the generation of hydrothermal event plumes at mid-ocean ridges. Earth Planet. Sci. Lett. 171, 343–350.

Luther D.S. and Sanford T.B (2008) HPIES – measuring water column currents and properties from the seafloor. Ocean Sciences Meeting, Session 153, http://www.sgmeet.com/aslo/orlando2008/viewabstract2.asp?AbstractID=1577.

MacFadyen, Hickey B.M., and Foreman M.G.G. (2005) Transport of surface waters from the Juan de Fuca eddy region to the Washington coast. Cont. Shelf Res. 25, 2008–2021.

Maslin M.M. Owen, Day S., and Long D. (2004) Linking continental slope failures and climate change: Testing the clathrate gun hypothesis. Geology 32, 53–56, doi:10.1130/G20114.1.

McCollom T.M. and Shock E.L. (1997) Geochemical constraints on chemolithoautotrophic metabolism by microorganisms in seafloor hydrothermal systems. Geochim. Cosmochi. Acta 61, 4375–4391.

Molin M.A. and Schofield S. (2009) Remote real-time video-enabled docking for underwater autonomous platforms. J. Atm. Ocean. Tech. 26, 2665–2672.

National Research Council of the National Academies (USA) (1999) *From Monsoons to Microbes, Understanding the Ocean's Role in Human Health*. Washington, D.C.: The National Academy Press.

National Research Council of the National Academies (USA) (2003) *Enabling Ocean Research in the 21st Century: Implementation of a Network of Ocean Observatories*. Washington, D.C.: The National Academy Press.

Nature, News Feature (2010) Human genome: Genomes by the thousands. Nature 467, 1026–1027.

Nooner S.L. and Chadwick Jr., W.W. (2009) Volcanic inflation measured in the caldera of Axial Seamount: Implications for magma supply and future eruptions. Geochem. Geophys. Geosys. 10(2), doi: 10.1029/2008GC002317=5.

Olson R.J. and Sosik H.M. (2007) A submersible imaging-in-flow instrument to analyze nano- and microplankton: Imaging FlowCytobot. Limnol. Oceanogr: Methods 5, 195–203.

Opatkiewicz A.D., Butterfield D.A., and Baross J.A. (2009) Individual hydrothermal vents at Axial Seamount harbor distinct subseafloor microbial communities. FEMS Microbial. Ecol. 70, 413–424.

Paull C.K., Brewer P.G., Ussler W., Peltser E.T., Rehder G., and Clague D. (2003) An experiment demonstrating that marine slumping is a mechanism to transfer methane from seafloor gas-hydrate deposits into the upper ocean and atmosphere. Geo. Mar. Lett. 22, 198–203.

Proskurowski G., Lilley M.D., Baker E.T., Walker S.L., Huber J.A., and Lupton J.E. (2012) Post-eruptive hydrothermal contribution to the water column above Axial Seamount. In: *Proceedings 2012 Ocean Sciences Meeting*, 112, B0775.

RECONN Regional Cabled Observatory Network (of Networks) Report, Cabled Regional Observatory Workshop, 2003 San Francisco, CA, October 7–10, 2003, Report to the National Science Foundation.

Rogers G. and Dragert H. (2003) Episodic tremor and slip on the Cascadia subduction zone: The chatter of silent slip. *ScienceExpress*, 8 May 2003, 10.126.

Satake K., Shimazaki K., Tsuji Y., and Ueda K. (1996) Time and size of a giant earthquake in Cascadia inferred from Japanese tsunami records of January, 1700. Nature 379, 246–249.

Schleisiek K. (2009) Tsunami observatory for South Korea. In: , 10.1109/OCEANSE.2009.5278160.

Scholin C.A. (2010) What are "ecogenomic sensors?" A review and thoughts for the future. Ocean Sci. 6, 51–60.

Scholin C., Doucette G., Jensen S., Roman B., Pargett D., Marin III R., Preston C., Jones W., Feldman J., Everlove C., Harris A., Alvarado N., Massion E., Birch J., Greenfield D., Vrijenhoek R., Mikulaski C., and Jones K. (2009) Remote detection of marine microbes, small invertegrates, harmful algae, and biotoxins using the Environmental Sample Processor (ESP). Oceanography 22, 158–167.

Sherman A.D. and Smith Jr., K.L. (2009) Deep-sea benthic boundary layer communities and food supply: A long-term monitoring strategy. Deep-Sea Res. II. 56, 1754–1762.

Sogin M.L., Morrison H.G., Huber J.A., Wech D.M., Huse S.M., Neal P.R., Arrieta J.M., and Herndt G.J. (2006) Microbial diversity in the deep sea and the unexplored "rare bioshphere". Proc. Natl. Acad. Sci. 103, 12,115–12,120.

Song T-R.A. and Simons M. (2003) Large trench-parallel gravity variations predict seismogenic behavior in subduction zones. Science 301, 630–633.

Stemmann L., Hosia A., Youngbluth M.J., Soiland H., Picheral M., and Gorsky G. (2008) Vertical distribution (0–1000m) of macrozooplankton, estimated using the Underwater Video Profiler, in different hydrographic regimes along the northern portion of the Mid-Atlantic Ridge. Deep Sea Res. II 55(1–2), 94–105.

Suess E., Torres M.E., Bohrmann G., Collier R.W., Greinert J., Linke P., Rehder G., Trehu A., Wallmann K., Winckler G., and Zuleger E. (1999) Gas hydrate destabilization: Enhanced dewatering, benthic material turnover and large methane plumes at the Cascadia convergent margin. Earth Planet. Sci. Lett., 170, 1–15.

Suess E., Torres M.E., Bohrrmann G., Collier R.W., Ricket D., Goldfinger C., Linke P., Heuser A., Sahling H., Heeschen ??, Jung C., Nakamura K., Grienert J., Pfannkuche O., Trehu A., Klinkhammer G., Witicar M.J., Eisenhauer A., Teichert B., and Elvert M. (2001). Sea floor methane at Hydrate Ridge, Cascadia margin. In: C.K. Paull and W.P Dillion (Eds) . Geophys. Monograph. Ser.124, Amer. Geophys. Union, Washington. D.C., pp. 87–98.

Sullivan W.T., III and Baross J.A. (2007) *The Emerging Science of Astrobiology*. Cambridge: Cambridge University Press.

Summit M. and Baross J.A. (1998) Thermophilic subseafloor microorganisms from the 1996 North Gorda Ridge eruption. Deep-Sea Research II 45, 2751–2766.

Taylor S.M. (2009) Transformative ocean science through the VENUS and NEPTUNE Canada ocean observing systems. Nuclear Instruments and Methods in Physics Res. 602, 63–67.

Torres M.E., McManus J., Hammond D.E., de Angelis M.A., Heeschen K.U., Colbert S.L., Tryon M.D., Brown K.M. and Suess E. (2002) Fluid and chemical fluxes in and out of sediments hosting methane hydrate deposits on Hydrate Ridge, OR, I: Hydrological provinces. Earth. Planet. Sci. Lett. 201, 525–540.

Trehu A.M., Bohrman G., Rack F.R., Collett T.S., Goldberg D.S., Long P.E., Milkov A.V., Riedel M., Schultheiss P., Torres M.E., Bangs N.L., Barr S.R., Borowski W.S., Claypool G.E., Delwiche M.E., Dickens G.R., Garcia E., Guerin G., Holland M., Johnson J.E., Lee Y.O.J., Liu C.-S., Su X., Teichert B., Tomaru H., Vanneste M., Watanabe M. and Weinberger J.L. (2004) Three dimensional distribution of gas hydrate beneath southern Hydrate Ridge: Constraints from ODP Leg 204. Earth Planet. Sci. Lett. 20, 541–550.

Trehu A.M., Ruppel C., Holland M., Dickens G.R., Torres M., Collett T., Goldberg D., Riedel M., and Schultheiss P. (2006), Hydrates in marine sediments – lessons from scientific ocean drilling. Oceanography 19, 124–142.

Trehu A.M., Brauniller J., and Nabelek J.L. (2008) Probable low-angle thrust earthquakes on the Juan de Fuca-North America plate boundary. Geology, 36, 127–130.

United Nations Department of Economic and Social Affairs (2004) *World Population in 2030*. New York.

Wells R.E., Blakely R. J., Sugiyama Y., Scholl D.W., and Dinterman P.A. (2003) Basin-centered asperities in great subduction zone earthquakes: A link between slip, subsidence, and subduction erosion? J. Geophys. Res. 108, doi:10.1029/2002JB2072.

Whitman W.B., Coleman D.C., and Wiebe W.J. (1998) Prokaryotes: The unseen majority. Proc. Natl. Acad. Sci. 95, 6578–6583.

19 Technical preparation and prototype development for long-term cabled seafloor observatories in Chinese marginal seas

F. Lu, H. Zhou, X. Peng, J. Yue and P. Wang[1]

19.1 Introduction

Understanding the ocean is critical to humans; long-term observation is the key. The ocean, whose average depth is about 3800m, covers approximately 70% of the terrestrial surface and has a pervasive influence on the global environment and climate; it is therefore critical to understand the mechanism of the complex interactive processes operating in the ocean (Favali and Beranzoli, 2006; Person et al., 2006; Chave et al., 2006; Wang, 2007; Delaney and Barga, 2009; Ruhl et al., 2011). However, the ocean is difficult to understand due to its huge size and harsh environment. Conventional ship-based expeditionary approaches are costly, only provide scientists with piecemeal data, and are constrained by weather conditions. Autonomous instrumented bottom landers with battery packs have been playing a key role in deep-sea investigation for a long time, used during ship-based expeditions from days to months (Person et al., 2006; Favali et al., 2006). Moored-buoy systems, sometimes with vertical profilers, can power bottom landers by fuel generators or solar panels and transmit data by satellites or radio, but they need regular maintenance and their energy and communications capacities are limited. Robotic vehicles, such as human occupied vehicles (HOVs), remotely operated vehicles (ROVs), autonomous underwater vehicles (AUVs), and sea gliders, are powerful tools for accessing remote and hostile environments, such as deep space and the ocean's interior (Bellingham and Rajan, 2007). These manned and unmanned mobile instrumented platforms have already contributed to revealing the diversity and complexity of the ocean, but none of them can stay long-term in the deep sea. As

1 State Key Laboratory of Marine Geology, Tongji University, Shanghai, China

there are few long time-series and high-resolution data, the complex ocean dynamics and earth–ocean interactions on many time and space scales are still poorly understood (Favali and Beranzoli, 2006; Wang, 2007; Delaney and Barga, 2009).

To meet this need, an innovative research facility called "cabled seafloor observatory" has been advocated for observing ocean processes ranging from seconds to decades (Delaney et al., 2000; Chave et al., 2004; Favali and Beranzoli, 2006; Person et al., 2006; Wang, 2007). Cabled seafloor observatories are expected to be "permanent" undersea infrastructure, providing sustained and abundant electrical power and broadband two-way communication bandwidth, similar to the functionalities of the terrestrial power grid and communications network. A large volume of high-resolution data from the full water column, from above the sea surface to below the seafloor, are expected to be collected by many interactive and adaptive in-situ experiments. Moreover, with the sustained energy and communications capacity, cabled observatories can be reinforced by mobile platforms which are more flexible and adaptive (Leonard et al., 2007; Curtin and Bellingham, 2009; Howe et al., 2010). Cabled observatories can be equipped with docking stations for AUVs, sea gliders, seafloor rovers, and on-site ROVs, combining the advantages of fixed and mobile sensor networks. The mobile platforms can expand the spatial coverage of the fixed observatories and the fixed observatories provide the mobile platforms with sustained energy. Cabled observatories are in operation or under construction in Canada, the United States, Europe, and Japan, including large-scale ones, such as NEPTUNE Canada (Barnes et al., 2008), OOI Regional Scaled Nodes (Isern, 2005), ESONET-EMSO (Person et al., 2008; Favali and Beranzoli, 2009), and DONET (Kawaguchi et al., 2008), as well as small-scale ones, such as MARS (Massion and Raybould, 2006), VENUS (Tunnicliffe et al., 2003), ALOHA (Howe et al., 2011), HUGO (Duennebier et al., 2002a), H2O (Duennebier et al., 2002b) , LEO-15 (Schofield et al., 2002), and MVCO (Austin et al., 2002). Decades of high resolution data, covering large undersea areas, are expected to reveal the mechanisms of complex physical, biological, chemical, and geological systems operating within the ocean (Favali and Beranzoli, 2006; Chave et al., 2006; Taylor, 2009).

Long-term cabled seafloor observatories in the Chinese marginal seas are also in the preparation phase, motivated by the science requirements from the science community. The research will be focusing on air–sea interchange, global carbon dioxide circulation, tectonic movement, ocean sedimentology, ocean circulation, ocean acidification, ocean biodiversity, and natural disasters like earthquakes, tsunamis, landslides, typhoons, storm surges, and red tides. A low voltage seafloor observatory using wireless communications, the Xiaoqushan scientific experimental station, has been established in the East China Sea by Tongji University and has been operational since April 2009 (Xu et al., 2011). Chinese marginal seas are characterized by broad shelves and broad slopes under thick sediments. Among all of the Chinese marginal seas, the South China Sea offers a special attraction for Earth scientists because of its location and the well-preserved hemipelagic sediments (Wang and Li, 2009). The South China Sea is the largest marginal sea separating the largest continent (Asia) from the largest ocean (Pacific) and thus functioning as a focal point in land–sea interactions of the Earth system (Wang and Li, 2009). Among the marginal seas in the western Pacific, only the South China Sea has a large deep-sea basin to yield well-preserved immense volumes of calcareous sediment sequences suitable for paleoenvironmental reconstructions (Wang and Li, 2009). The science themes for long-term observation in

Chinese marginal seas are supposed to specifically focus on the following scientific topics:

1. The transport and flux of substances between Yangtze Estuary and East China Sea
2. Observation, prediction, and evaluation of hypoxia zones
3. Monitoring and prediction of natural disasters
4. Fisheries
5. The hydrodynamics of the outer continental shelf and the biosphere
6. The Kowloon reef hydrate zone
7. The hydrodynamics of Bashi Channel
8. Ocean acidification
9. The north slope dynamics of the South China Sea
10. Upwelling.

Based on a systematic research of long-term cabled seafloor observatories, we have developed an experimental observatory prototype for the Chinese marginal seas and three in-situ instrument packages for multidisciplinary observation and prototype testing purposes. The prototype is rated down to 4000m and consists of an undersea station, a junction box, a submarine cable, and a shore station. The submarine cable links the undersea station and the shore station, transmitting 10kVDC/10kW high voltage direct current (HVDC) electrical power and 1Gbps two-way broadband communication bandwidth. There are three instrument packages: a chemical parameter analyzing system (CPAS), a hydrodynamic environment monitoring system (HEMS), and a deep-sea video camera system (DVCS). The junction box and the three instrument packages form a science node as one payload of the undersea station.

In these projects, Tongji University is responsible for the project management, the systematic research of scientific cabled observatories, the system design and test of the experimental observatory prototype, and the instrumentation of the CPAS, including the development of a novel underwater ion chromatography (UIC) unit; Zhejiang University is responsible for developing the junction box and the DVCS; and China Ocean University is responsible for developing the HEMS. In addition, Zhejiang University and Tongji University are both working on HV undersea stations, of different design schemes but similar functionality. Tongji University's first HV undersea station is being developed for an experimental multi-node cabled seafloor observatory which will be deployed in East China Sea around 2013, with a backbone cable of at least 50km.

19.2 System design of cabled seafloor observatories

Cabled seafloor observatories are designed for long-term and real-time ocean observation of complex ocean processes. The designed life of the backbone network of cabled observatories should be no less than 25 years which is the typical requirement for transoceanic telecommunications cable systems. The observatories should be capable of providing various undersea scientific instruments with sustained electrical power and communication bandwidth. Then plenty of in-situ experiments can be well supported, including power hungry and high data rate equipment, such as heavily instrumented platforms, high-definition cameras, benthic seafloor rovers, and robot docking stations.

The infrastructure of a cabled observatory can be generally divided into two critical subsystems: a power system and a communications system. The two subsystems closely interact with each other in the same infrastructure and thus they are strongly correlated (Schneider, 2005). The observatory infrastructure design has particular constraints resulting from the subsea location. It is very expensive to construct and maintain the infrastructure on the seafloor, especially in the deep ocean. The infrastructure must therefore be highly reliable and flexible under low accessibility and low controllability compared with the terrestrial infrastructures. To facilitate regular servicing and replacement, undersea infrastructure components must be both compact and reliable.

19.2.1 Power system

The undersea power system for cabled observatories must be trade-off designed by evaluating relevant technology comprehensively. The principle of the trade-off design is to reduce the overall life cost under the requirements of high reliability and flexibility. Though the detailed designs can differ from one another, the design philosophy and the key components are similar. The submarine cables are used to connect multiple undersea stations located on the seafloor to the terrestrial power system and communication backhaul at the shore stations. Science nodes, consisting of junction boxes and/or scientific instrument interface modules, and instrument/sensor packages, are connected to the junction box interfaces of undersea stations.

Cabled observatories typically use standard single-conductor submarine cables used in transoceanic telecommunications cable systems (with repeaters), in order to reduce the cable cost. Moreover, this selection will also make easier the scientific use of future retired submarine commercial cables. As the optical signals will be gradually attenuated in the fibers, currently in most cases, repeaters are needed in long-distance cable systems of longer than 400km (Harris and Duennebier, 2002). The Erbium Doped Fiber Amplifiers (EDFAs) are widely used to amplify the fiber-optic signals in the submarine cables. The power feed equipment (PFE) located in the shore stations supply constant current HVDC electrical power to the repeaters through the single copper conductor in the cable. This conductor will be used to power both the repeaters and the undersea sensor network.

Cabled observatories are recommended to deliver power using the negative HVDC scheme (Howe et al., 2002). The alternating current (AC) scheme has advantages in voltage level conversion, fault current breaking, and cable insulation life, but it may lead to unacceptably large charging currents and reactive power losses, because typical telecommunication cables have high shunt capacitances and series inductances (Howe et al., 2002). As the compensation cost of cable capacitance and inductance will be very high, generally cabled observatories prefer the DC power delivery scheme. The power line loss can be further reduced by increasing the power delivery voltage, but the maximal voltage is limited by the long-term performance of cable insulations. Considering the typical life of submarine telecommunication cables under the nominal operation voltage of 10kVDC is 25 years, the power delivery voltage is recommended to be 2kVDC to 10kVDC, depending on the observatory's location and the detailed system design. The HVDC electrical power conversion can be realized using the modem power electronics, synthesis from mature and reliable low-voltage high-frequency PWM switching converters (Vorperian, 2007).

Cabled observatories are recommended to use constant voltage sources to power parallel undersea payloads, as this scheme is usually more efficient and has higher power capacity in a networked topology than other schemes (Howe et al., 2002). Moreover, this scheme can more easily handle branching than the constant current source scheme which is used in the submarine telecommunication cable systems. As the standard submarine telecommunication cable has only one conductor, the seawater should be used as the current returning path. Thus, each undersea station and each shore station should be connected to the seawater or to the ground. The sea electrodes of the undersea stations must be positive relative to the transmission line in order to avoid accelerated electrical corrosion, which will lead to costly frequent maintenance. It will be much easier to replace the electrodes of the shore stations. The maximal backbone current is mainly determined by the voltage drop, considering the voltage instability of the power delivery system (Harris and Duennebier, 2002). In the terrestrial grid, the voltage drop from the power source to the load should be less than 5% typically, while the voltage drop can be near 50% in a cabled seafloor observatory, causing much higher energy loss than in a terrestrial grid.

Power monitoring systems (PMS) need to be developed to increase the reliability of the operation of cabled seafloor observatories, usually adopting the techniques used in the energy management systems (EMS) in terrestrial grids. A PMS consists of a human-machine interface (HMI), supervisory control and data acquisition (SCADA), command sending and alert handling, as well as network analysis modules such as power system state estimation, topology identification, and fault location, which should be developed according to the actual configuration of cabled seafloor observatories (Liu et al., 2003; Schneider et al., 2005; Chan et al., 2007).

19.2.2 Communication system

The design of a communications system of a cabled observatory is based mainly on the submarine telecommunications cable systems. The backbone communications system for a typical cabled seafloor observatory is a broadband fiber-optic Ethernet which can be rated up to hundreds of gigabits per second by the Dense Wavelength Division Multiplexing (DWDM) technology. In DWDM systems, a single optical fiber pair can carry tens of optical signals, allowing different wavelengths for different undersea stations. The network infrastructure normally employs Ethernet and optical protocols for in-band data transmission and there can be an out-of-band communications system for the observatory emergency control. The essential functionalities for the observatory operation are provided by the Link Aggregation Control Protocol (LACP, IEEE 802.3) and the Virtual Local Area Networking (VLAN, IEEE 802.1) (Lentz, 2007). Legacy instruments and sensors using serial protocols can be integrated into the cabled observatories by a science instrument interface module (SIIM) for protocol and/or media conversion.

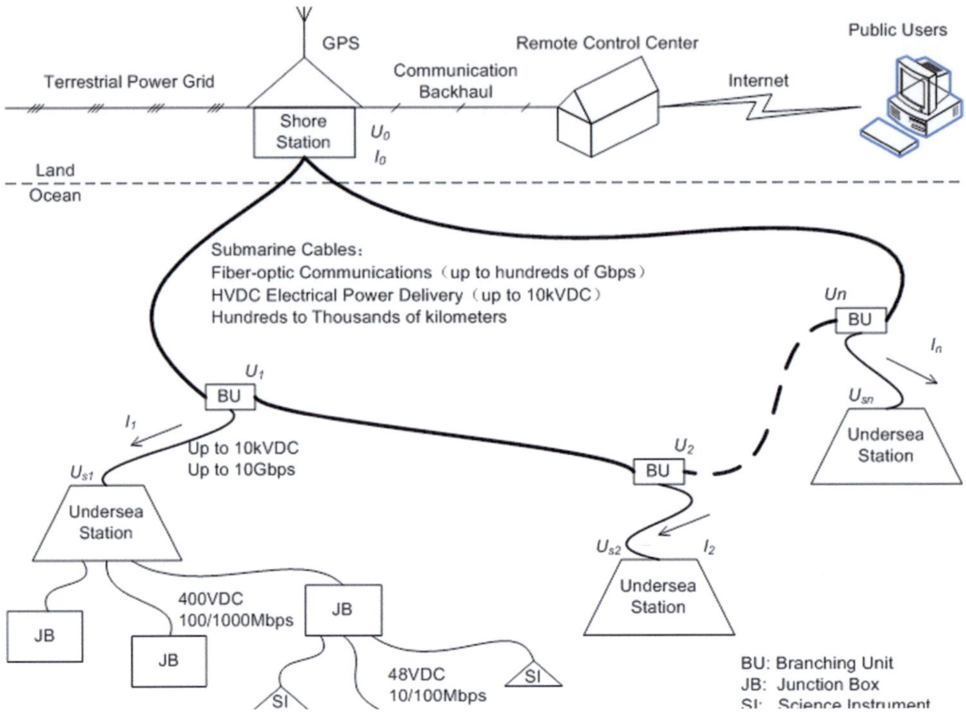

Figure 19.1 System diagram of a typical long-term cabled seafloor observatory network (Lu et al., 2012).

The communications system should provide accurate timing to scientific instruments. In cabled observatories, the Universal Time Coordinated (UTC) is used to timestamp the data stream. Timing can be delivered by Ethernet communications channels, employing Network Time Protocol (NTP, IETF RFC 1305), Simple Network Time Protocol (SNTP, IETF RFC 2030), or IEEE 1588 Precision Time Protocol (PTP) (Lentz and Lecroart, 2009). The accuracy of NTP/SNTP is typically within a few milliseconds and PTP provides much better accuracy within a few microseconds, which is required by applications such as seismology. Accurate timing significantly improves scientific data fusion and mining.

19.2.3 Topology consideration

A multi-node cabled seafloor observatory can be designed in a star, ring, or mesh topology, optimized according to the actual undersea areal coverage. It can be large on a tectonic plate scale with backbone cables as long as thousands of kilometers. The optimal network configuration is based on reliability evaluation, full-lifetime cost assessment, and technical feasibility of construction, maintenance, and operation of the seafloor observatory. The in-service submarine telecommunication cables should also be taken into consideration when planning the scientific cable routes. Generally, ring and mesh topologies are more reliable than other simpler ones, because there are redundant paths delivering electrical power to each undersea station. Moreover, the backbone cable should be segmented for fault isolation purposes, in order to raise the overall reliability of the observatories. Figure 19.1 shows the system diagram of a typical long-term cable seafloor observatory network.

19.3 Prototype development for cabled seafloor observatories

We developed an experimental prototype of a cabled seafloor observatory. The prototype consists of an infrastructure which is essential to the primary functionality of cabled observatories, i.e., providing sustained and ample electrical power and communication bandwidth to various scientific instruments. The key components are an undersea station, a junction box, a submarine cable, and a shore station. Typical system specifications for several regional-scale cabled observatories are adopted (Delaney et al., 2000; Barnes et al., 2008). In this prototype, the submarine cable links the undersea station and the shore station, with the power delivery voltage/power of 10kVDC/10kW and the fiber-optic communication bandwidth of 1Gbps.

The undersea station converts the 10kVDC electrical power to 400VDC electrical power and routes the 1Gbps communication bandwidth to six junction box interfaces, the default bandwidth of each interface being 100Mbps. The junction box converts the 400VDC electrical power to 48VDC electrical power and routes the 100Mbps communication bandwidth to ten scientific instrument interfaces, whose default bandwidth is 10Mbps. Three instrument packages connect to the junction box by SIIMs. The SIIMs are used for LVDC electrical power conversion and/or communication media/protocol conversion, when lower voltage levels or legacy serial protocols are required by conventional scientific instruments and sensors. The prototype is developed with an emphasis on reliability, flexibility, and fault resilience.

19.3.1 Undersea station

The undersea station is designed to process the 10kVDC/10kW electrical power and the 1Gbps fiber-optic communication bandwidth from the backbone cable (Lu et al., 2010). The designed life is 15 years and the depth rating is up to 4000m. The electrical power capacity and the communication bandwidth are upgradable. As technical uncertainties exist, it is the most difficult component in the cabled observatory design.

The undersea station consists of four main parts: a functional node, eight ODI wet-mateable connectors, a cable termination assembly (CTA) and a trawl resistant frame (TRF). The node has two pressure vessels containing electronics for power conversion, data transmission, timing and control. The CTA separates the single conductor and the optic fibers of the incoming submarine cable. The fabricated TRF protects the node, the wet-mateable connectors, and the CTA, mainly against fishery. The CTA and the TRF are deployed together with the cable prior to the node, which is deployed, maintained, and recovered later by a scientific ROV. Figure 19.2 shows the mechatronics integrated design diagram of the

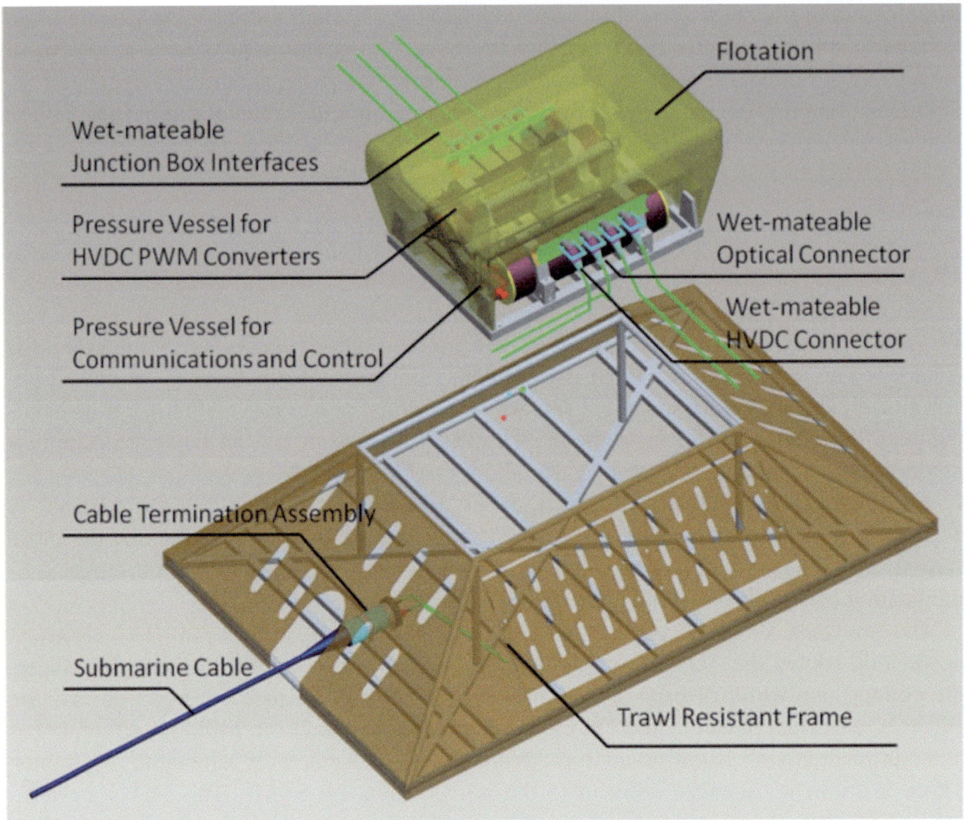

Figure 19.2 Mechatronics integrated design diagram of the undersea station.

undersea station. The CTA may not be necessary when undersea stations are located in a deep-sea environment.

The functional node is designed to be ROV servable, in order to reduce the maintenance costs. As it is too heavy for ROV manipulation, flotation is used to reduce its in-water weight. There are eight ODI wet-mateable connectors installed on the undersea station: one for 10kVDC electrical power, another for optic fibers, and the other six for junction boxes. In the node, one pressure vessel is installed with a 10kV–400V DC–DC PWM converter, which is a critical component in the undersea power system. This converter must be compact, highly reliable, and of high conversion efficiency. There are no COTS for HVDC electrical power conversion with such a compact size. In the undersea station, the 10kV–400V DC–DC converter is synthesized from 25 power converter building blocks (PCBBs), and uses the input-series-output-parallel (ISOP) topology (Vorperian, 2007). The PCBB is a 400V–400V DC–DC converter using the full-bridge topology with a PWM frequency of 60kHz. The HVDC electrical power convertor is inherently redundant and fault-tolerant to a few PCBB failures. Its input voltage is designed to be from 6kVDC to 12kVDC. The other pressure vessel contains communications equipment and the monitoring and control system. These internal loads are powered by 400V–48V/24V/12V/5V DC–DC PWM converters. Each undersea station provides power and Ethernet communications at wet-mateable connected interfaces for junction boxes. These interfaces facilitate flexible connectivity to junction boxes by ROV installation. Each interface is equipped to provide 400VDC electrical power and electric or optic data communications to junction boxes. Constrained by signal attenuation over transmission and connection, the distance from the in-situ experiments to the undersea station can be tens of meters by electric data transmission or several kilometers by fiber-optic data transmission.

The monitoring and control system is designed to make the undersea station more reliable. The voltages and currents of the sources and loads, the switch status of internal and external loads, the external interface ground faults, and the internal health of the pressure vessels are measured to detect, diagnose and isolate faults, and manage internal and external loads according to their priorities (Lu et al., 2010, 2011). The communications system's status, such as the data rates of the junction box interfaces, is also monitored. Whenever an over-threshold fault is detected, the control module will enter into the fault diagnosis and isolation mode. All the thresholds can be modified and reconfigured at the surface control consoles located in the shore stations or remote control centers. The control module also executes commands of the control consoles.

19.3.2 Junction box

The junction box is designed to process the 400VDC electrical power and the 100Mbps communication bandwidth from one of the undersea station's interfaces (Lu et al., 2010). The typical maximal power capacity is 1kW and can be further upgraded to 2kW. The designed life is 10 years and the depth rating is up to 4000m. It has ten scientific instrument interfaces. Each interface provides regulated 400VDC/48VDC electrical power and 10Mbps/100Mbps communication bandwidth. The junction boxes can be connected in a daisy chain configuration. Figure 19.3 shows the junction box with the DVCS tested in MBARI's test tank. The DVCS uses a SIIM to integrate a deep-sea camera and two undersea bulbs for visual observation of big organisms living in the deep sea, with a video server installed in the SIIM. In our current design, the interfaces of junction boxes are standardized and the custom designed SIIMs provide compatible power and communication interfaces to multi-instrument/sensor packages. In the future, the junction boxes are expected to directly serve robotic underwater laboratories and vehicles with compatible interface specifications. In NEPTUNE Canada, junction boxes function like a combination of our junction boxes and SIIMs (Woodroffe et al., 2008).

The junction box consists of two pressure vessels and 10 scientific instrument interfaces (Lu et al., 2010). At present, only two scientific instrument interfaces are installed for the CPAS and the HEMS, using 12-pin ODI wet-mateable connectors. In addition, the DVCS is installed on the same frame as the junction box with SUBCONN dry-mated connectors. Any scientific instrument satisfying the interface specifications, such as mechanical, power, communications, and timing, can be connected to and served by the junction box. In the junction box, one pressure vessel is for the 400V–48V DC–DC PWM converters,

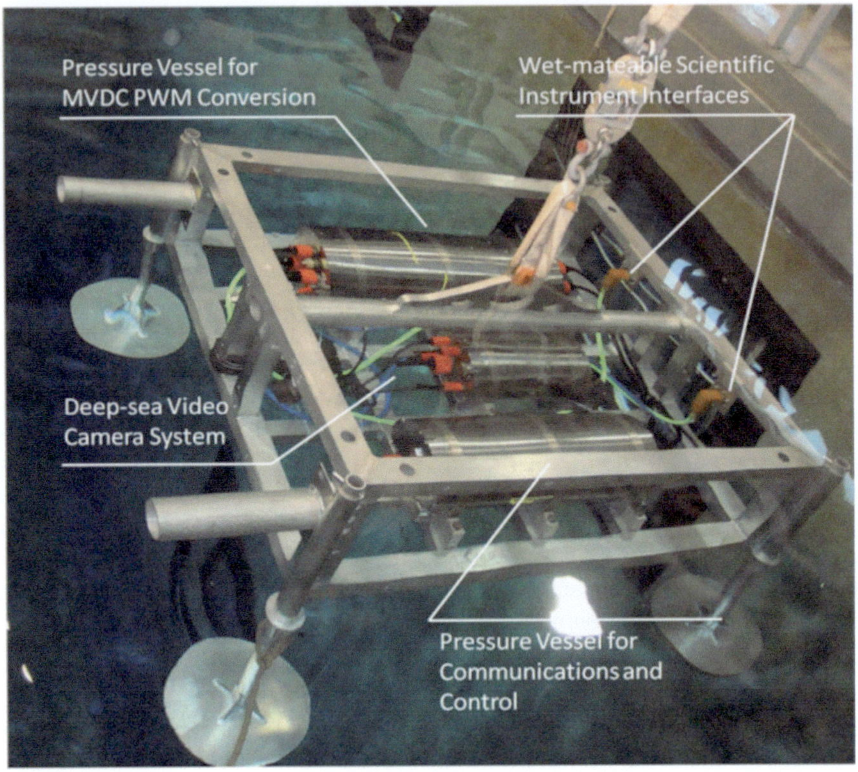

Figure 19.3 The junction box and the DVCS tested in MBARI's test tank.

which are highly reliable COTS components using the input-parallel-output-parallel (IPOP) topology. The other pressure vessel is for the communications, monitoring, and control system. The layer 2 (L2) network switches route the communication bandwidth to supply each instrument package with 10/100Base-T Ethernet. They aggregate variable IP data streams from instrument packages and transmit them to high-speed L2/L3 Ethernet switches in the undersea station. They also distribute Coordinated Universal Time (UTC), at an accuracy of a few milliseconds, from the GPS time server located in the shore station to instrument packages, using the network time protocol (NTP). The monitoring and control scheme is similar to the one used in the undersea station.

19.3.3 Submarine cable

The prototype uses a standard single-conductor submarine telecommunication cable with double armor. The copper conductor has a resistance of 0.96/km and is used to carry the

10kVDC electrical power. The cable has four fiber pairs: two providing in-band communications with a redundant ring, another used for out-of-band control, and the fourth reserved. The in-band communication bandwidth is 1Gbps and the out-of-band system has the data rate of 9600bps. The conductor and fibers are separated in the CTA and connected to the two pressure vessels of the undersea station via an HV power wet-mateable connector and an optic-fiber wet-mateable connector, both of which are manufactured by ODI. Submarine cables have various protection schemes depending on the deployment depth and seabed conditions. In harsh environments, they are recommended to be buried 1–4 meters below the seafloor to avoid damages by outside forces.

In a large-scale observatory network, repeaters have to be used to amplify the fiber-optic Ethernet signals. With the DWDM technology, the data transmission rate in a single commercial transoceanic telecommunication cable can be up to several Tbps, allowing a single fiber pair to serve all undersea stations of a cabled observatory network. Usually, redundant fiber pairs should also be configured in order to increase the overall reliability of the backbone communications system. Each undersea station connects to the backbone cable at a branching unit. In order to isolate faulted undersea stations or cables in a multi-node observatory, the branching units may need to be modified for the capability of backbone cable segmentation (El-Sharkawi et al., 2005). Power switching devices, usually controlled by fiber-optic signals, should be installed in the branching units. When the power monitoring system detects severe faults in an undersea station or a segmented cable, the power switching devices in the respective branching units will act to isolate the fault. Then the remaining system can continue operation. Severe faults might result from HV–MV DC–DC converter failures, pressure vessel leakages, or cable damages. This design strategy can greatly increase the overall network reliability and availability over the life of cabled observatories.

19.3.4 Shore station

The shore station links the undersea power and communications network to the terrestrial power grid and communications backhaul. The shore station has redundant utility AC electrical power line input and high-speed backhaul communications connecting to the remote control center. A cabled seafloor observatory may have one or more shore stations, depending on the network scale and the cable route/landing scheme. The essential power equipment in a shore station is a PFE which converts the utility three-phase 380VAC electrical power to 10kVDC electrical power. Then the HVDC electrical power is delivered to the undersea stations via submarine cables. An uninterruptable power system (UPS) with independent generators is recommended for increasing the reliability of the power feeding.

The essential communications equipment is the L3/L2 network switch routing the two-way broadband data transmission between the land and the ocean. A GPS network time server is used to distribute the coordinated universal time (UTC) to the entire undersea network via the communications system by NTP and PTP (in the future). In addition, standard web services can facilitate interoperability between users and instruments, as well as enable dynamic observatory configuration and virtual instrument networks creation within multiple ocean observatories (St Arnaud et al., 2004). Figure 19.4 shows our general design of the cyberinfrastructure of cabled seafloor observatories.

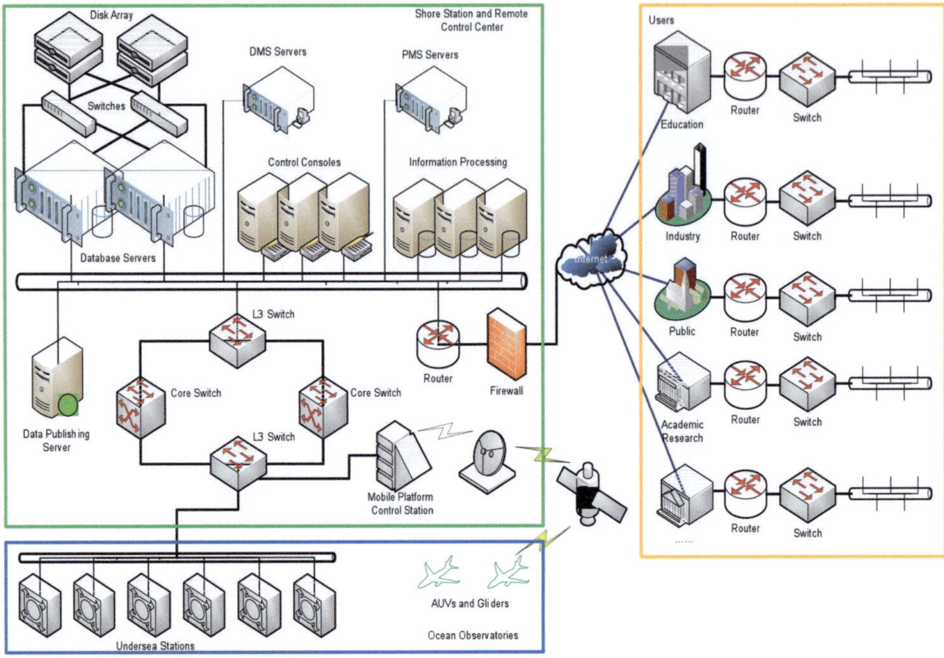

Figure 19.4 General design of the cyberinfrastructure of cabled seafloor observatories.

19.3.5 Reliability engineering

The major design challenge in the prototype design lies in reliability engineering, as the seafloor location poses particular constraints on maintenance (Lu et al., 2011). Deep-sea systems must be highly robust to reduce repair costs. In order to reduce the development costs, high-quality commercial off-the-shelf (COTS) components should be considered first. Custom-made components, such as the HVDC electrical power convertors and the subsea control systems, must be carefully designed and strictly tested to meet the high reliability requirement (Lu et al., 2011). Redundancy and fault tolerance technologies should be used from the most basic circuit to the overall system design. In order to prevent fault propagation, deep-sea components should be designed to possess the capabilities of fault detection, diagnosis, and isolation. In addition, the network infrastructure must also be controllable, expandable, and upgradable to meet the future science needs at interesting locations.

All pressure vessels are manufactured from the same grade of titanium to prevent galvanic corrosion during long-term deployments in the high salinity seawater environment. In order to reduce the aging of electronics, heat dissipation paths from the components to the seawater must be carefully designed in order to keep the internal environment of the pressure vessels cool (Lu et al., 2011).

19.4 In-situ scientific instrumentation

In-situ scientific instruments are essential for measuring chemical, physical, biological, and geological parameters of the ocean (Ding and Seyfried, 2007; Paul et al., 2007; Curtin and Belcher, 2008). Traditional oceanographic sensors and instruments are typically stand-alone and off-line. They are usually designed to consume very low power, in order to be self-contained. In addition, they are usually off-line or have very low data transmission rates. Though the limited power and communications capacity leads to limited functionality, they are still very useful to scientists and have already shown the diversity and complexity of the ocean. Cabled seafloor observatories are powerful research facilities allowing novel on-line intelligent instruments which can now be energy hungry and need high data transmission rates (Scholin et al., 2009; Romanowicz et al., 2009; Sherman and Smith, 2009). Therefore, interactive in-situ experiments can be designed for specific science purposes. By using in-situ scientific instrumentation, the mechanism of complex ocean dynamics, such as interactive processes of varying scales in space and time, is expected to be better understood.

Three scientific instrument packages are developed according to the specifications of the instrument interfaces of the junction box and the requirements of the data management system (DMS) (Lu et al., 2011). They are a chemical parameter analyzing system (CPAS), a hydrodynamic environment monitoring system (HEMS), and a deep-sea video camera system (DVCS), all of which are designed for multidisciplinary oceanographic research and for testing the observatory prototype's functionality. In general, each instrument package consists of a scientific instrument interface module (SIIM) and several undersea sensors or instruments. SIIMs are directly connected to the instrument interfaces of the junction box, which is powered by 48VDC electrical power and 10/100Mbps communication bandwidth.

19.4.1 Scientific Instrument Interface Module (SIIM)

SIIMs are used to connect various types of sensors and instruments to the standardized payload interfaces of the underwater junction box. They have two primary functions: LVDC electrical power conversion and management; and communication protocol or media conversion for data transmission. Thus, SIIMs provide compatible power and communication interfaces between the junction box and the instruments and sensors.

The SIIMs must be designed according to the interface specifications of the junction boxes, and the instruments and sensors. Highly efficient switching DC-DC converters are used in the power modules to save power consumption and reduce heat generation, but additional power line filters should be added after the converters to reduce switching noise, which may affect the measurement accuracies of the sensors (Lu et al., 2011). The soft start circuits are carefully designed to decrease the inrush currents of the instrument packages and the overvoltage protection circuits are applied to protect the system against transient inline voltage spikes, which can damage sensitive components. The SIIM control units monitor all voltage and current of the external sensors and the internal loads, as well as the internal environment parameters, e.g., temperature, humidity, and pressure, of the pressure vessels. The control units execute fault detection, diagnosis, and isolation for fault-tolerance, and also feedback the system health information to the shore stations and the remote operation centers (Lu et al., 2011).

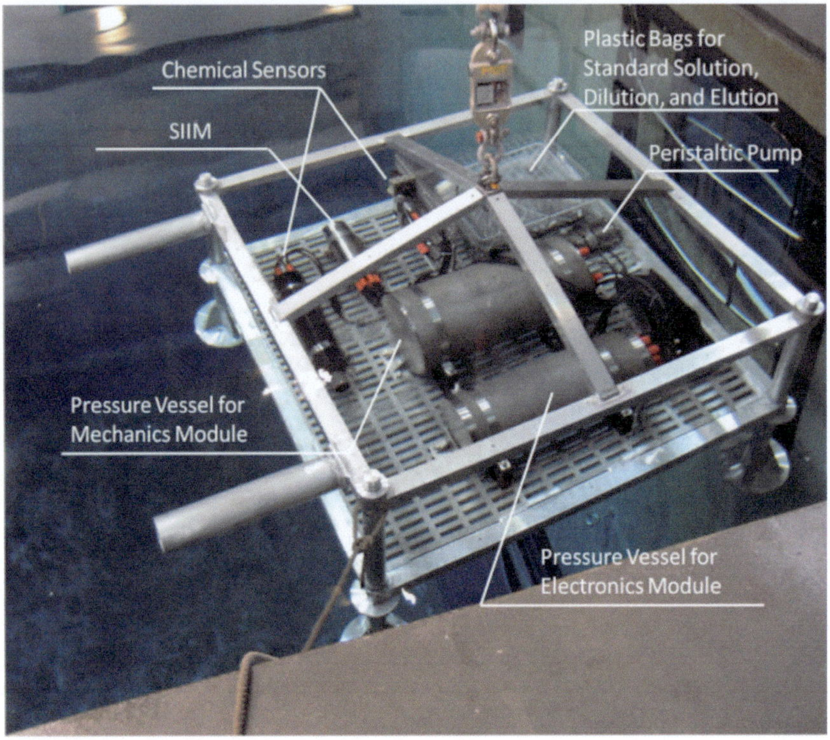

Figure 19.5 The chemical parameter analyzing system tested in MBARI's test tank.

Just as with the power delivery scheme between the junction box and the undersea station, each SIIM also uses two wires (DC+ and DC-) connecting to the instrument interface of the junction box, in order to avoid potential ground faults. The system status and the near-real-time data streams of the instrument packages can be monitored by the computer consoles located at the shore stations and remote control centers. Through the Internet, public users can download interesting science data from the DMS and monitor the status of the in-situ instrument packages from the PMS.

19.4.2 Chemical Parameter Analyzing System (CPAS)

The CPAS consists of an underwater ion chromatography (UIC) unit and several deep-sea chemical sensors. Ion chromatography has the advantage of high precision and can detect multiple anions simultaneously. The UIC is developed for in-situ analysis of the concentrations of anions, such as F-, Cl-, Br-, NO_3^-, SO_4^{2-}, and PO_4^{3-}, in the deep sea, especially in the fields of hydrothermal vents and cold seeps, where the concentrations of Cl^- and SO_4^{2-} are much lower than in common seawater. At present, five chemical sensors are integrated to measure the time-series of concentrations of chlorophyll, dissolved oxygen, nitrate, and methane, as well as the pH values, of the deep-sea water column. Table 19.1 gives the general specifications of the five chemical sensors. Figure 19.5 shows the CPAS tested in MBARI's test tank.

The UIC mainly consists of two titanium pressure vessels and a deep-sea peristaltic pump. One pressure vessel contains the electronic module controlling the anion analysis procedure and processing conductors. The electronic module selects in-coming fluids (seawater, standard solutions, or dilutions) and controls the fluid flow path by setting the

	Chlorophyll A	Dissolved oxygen	Nitrate	Methane	pH
Sensor model	Seapoint Chlorophyll Fluorometer	Aanderaa Oxygen Optode 3975D	Satlantic ISUS V3	Franatech METS	AMT pH-combined Sensor
Detecting range	0–150 μg/L	0–500 μM	0.5–2000 μM	1–500 nM	2–11
Detecting resolution	0.02 μg/L	< 1 μM	0.5 μM	Approx. 0.1 nM	0.01
Detecting accuracy	0.03 μg/L	< 8 μM	± 2 μM	1 nM	0.05
Output signal	Analog	Analog	Analog	Analog	Analog
Output signal range	0–5 Vdc	0–5 Vdc	0–4.096 Vdc	0–5 Vdc	0–5 Vdc
Sampling frequency (period)	0.1 Hz (10 s)	0.1 Hz (10 s)	0.1 Hz (10 s)	0.1 Hz (10 s)	0.1 Hz (10 s)
Water depth	0–6000m	0–6000m	0–1000m	0–4000m	0–6000m
Power supply voltage	8–20 Vdc	7–14 Vdc	6–18 Vdc	9–36 Vdc	9–18 Vdc
Steady-state power consumption	0.18 W	1.0 W	6.5 W	2.8 W	3.0W
Max. power consumption	0.32 W	1.5 W	10 W	4.8 W	4.5W
Working temperature	0–65°C	-5–40°C	0–40°C	0–30°C	−2–38°C
Housing material	Polyurethane	Titanium	Zn electroplating aluminum	Titanium	Titanium

Table 19.1 General specifications of the five chemical sensors.

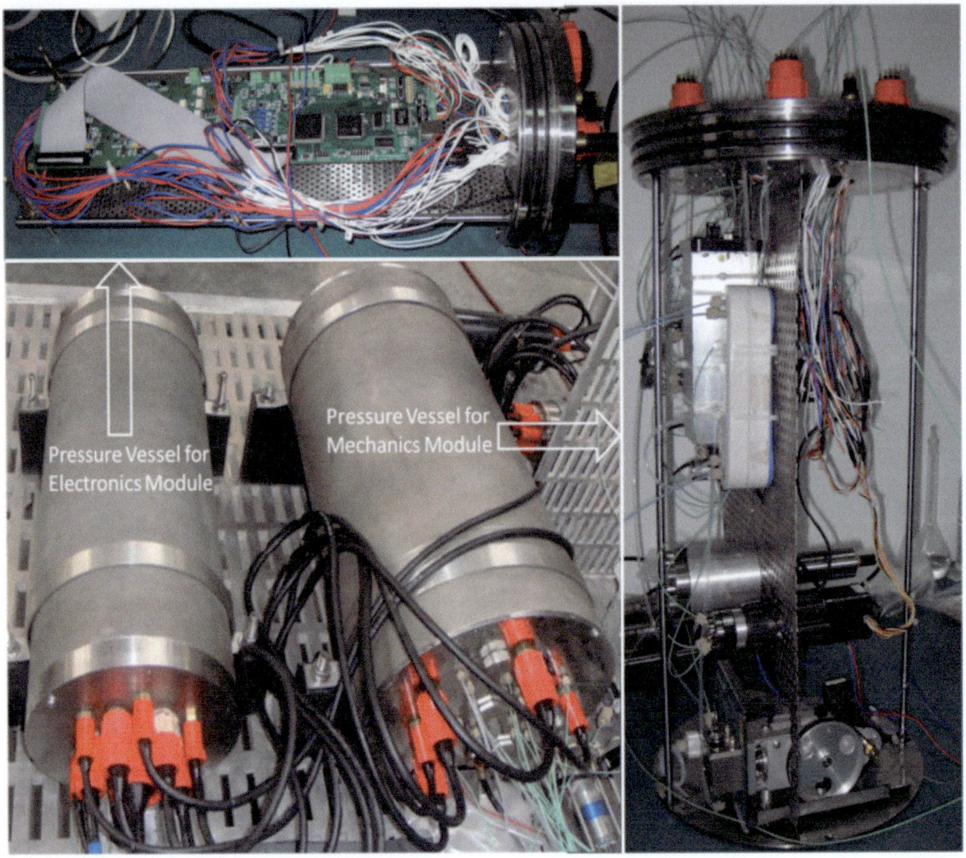

Figure 19.6 The third generation of the underwater ion chromatography (UIC).

positions of relevant valves; therefore the functions of in-situ calibration and dilution can also be realized. In order to derive the anion concentrations, the electronic module processes the raw data of conductors by calculating the peak areas in one analysis and comparing them to the calibration equations. All data streams of command sequences, instrument status, and seawater properties should be processed and packed according to the requirements of the DMS. The other pressure vessel is silicone-oil-filled and contains the mechanic module diluting and analyzing the incoming seawater, as well as calibrating the instrument. The mechanic module mainly contains a chromatographic column, an anion suppressor, a conductivity cell, and a piston pump, as well as tubes and valves. These functional components are connected to several plastic bags filled with eluent, diluent, and standard solutions. A pressure compensator is installed in order to balance the inside and outside pressure of the oil-filled vessel. Thus, these components bear the same internal pressure in the deep sea as on the bench top. The peristaltic pump is used to transfer and elute the incoming seawater through the system. The model of the deep-sea peristaltic pump is KC-11350 from Den-

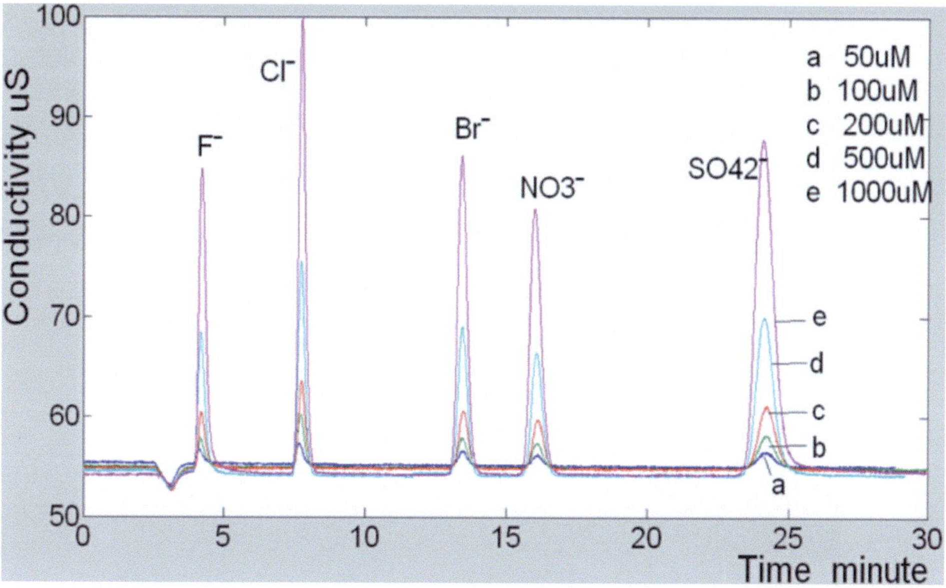

Figure 19.7 Calibration of the UIC using standard solutions (Wu et al., 2010).

mark, rated down to 6000m. The pump has six channels and is controlled by a mini-motor whose nominal speed is 10rpm. The diameter of the fluid tube is 1.5mm to 3mm with a thickness of 0.5mm. Figure 19.6 shows the third generation of the UIC.

The details of in-situ analysis, calibration, and dilution of the UIC can be found in Wu et al. (2010). The standard solutions contain all the anions expected to be measured. In-situ calibration is necessary because the UIC is seriously affected by environmental temperature, fluid pressure, eluent flow rates, and so on. As the Cl^- and SO_4^{2-} in the normal seawater are too concentrated to be measured, the UIC is designed to dilute the incoming seawater by the ultrapure water in the ratio of 1:21 in order to keep the high concentration anions within the detection limit after dilution. There is a micro-motor installed to vibrate the dilutor, making the incoming seawater and ultrapure water fully mixed. Figure 19.7 shows the calibration curves of the UIC using standard solutions.

19.4.3 Hydrodynamic Environment Monitoring System (HEMS)

The HEMS integrates an Acoustic Doppler Current Profiler (ADCP), an Acoustic Doppler Velocimeter (ADV), and a Conductivity Temperature Depth (CTD) instrument using a SIIM. The ADCP, the ADV, and the CTD are products of RDI, Nortek, and Seabird Electronics respectively. The HEMS is used to measure the time-series of salinity, temperature, depth, and the current velocity o the deep-sea environment. Figure 19.8 shows the HEMS tested in MBARI's test tank.

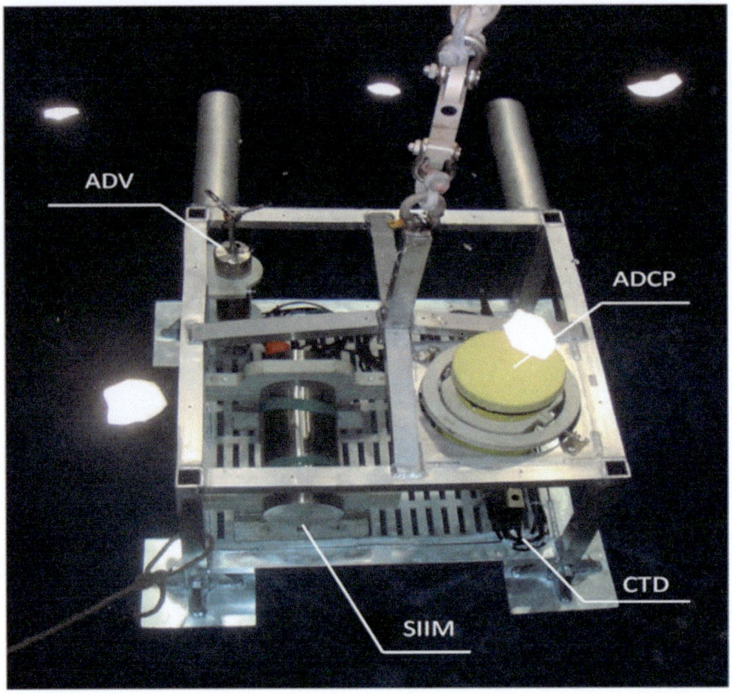

Figure 19.8 The hydrodynamic environment monitoring system tested in MBARI's test tank.

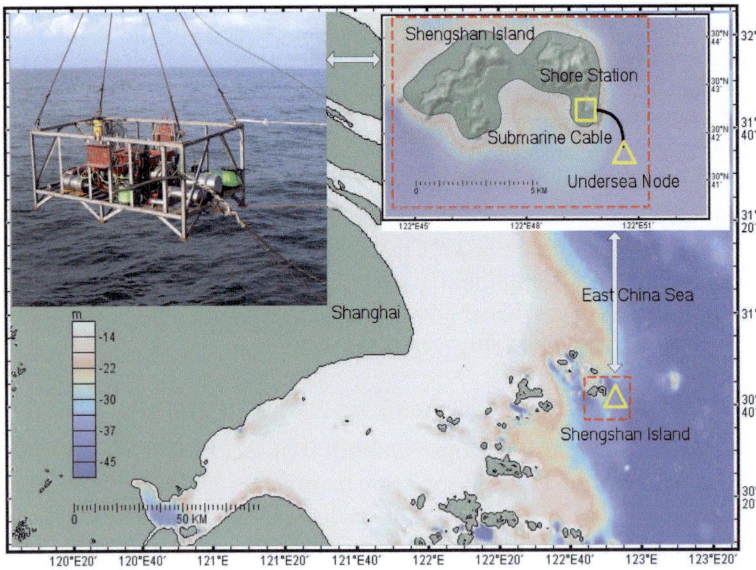

Figure 19.9 The shallow seawater trial of the prototype in the East China Sea (Lu et al., 2012).

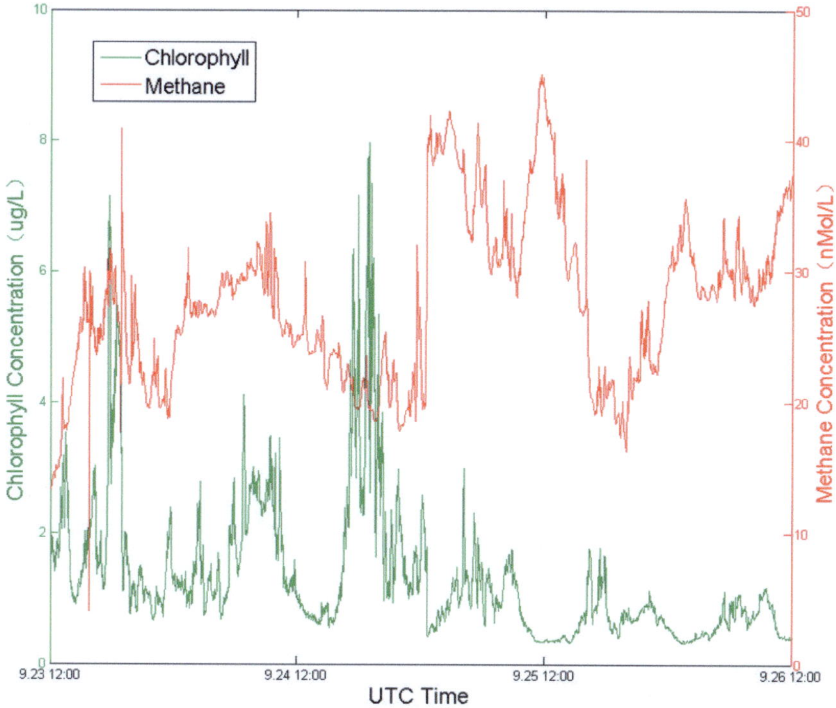

Figure 19.10 Part of scientific data collected in the East China Sea trials.

19.5 East China Sea trials

The prototype was tested in a shallow seawater trial in the East China Sea, near the Shengshan Island, for about three weeks in September 2010. In this experiment, all undersea subsystems were installed on one big frame which was deployed approximately 1km off the island and at approximately 30m water depth. Figure 19.9 shows the shallow seawater trial of the prototype in the East China Sea. Figure 19.10 shows part of the scientifice data collected during the East China Sea trial.

19.6 MARS sea trials

The science node underwent a deep-sea trial on the MARS cabled observatory from April 21 to October 27, 2011, after testing in MBARI's test tank for three days to make sure that the ground fault current and the startup surge current of the science node were within the

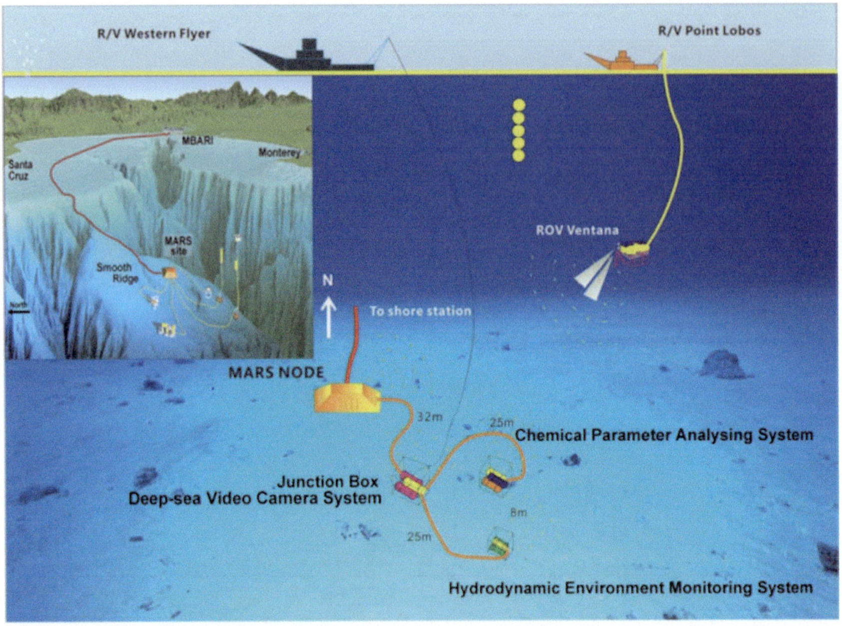

Figure 19.11 Deep-sea trial configuration of the science node and the MARS cabled observatory (Peng et al., 2011).

tolerance limits of the MARS observatory. The science node was deployed by MBARI's research vessels Western Flyer and Point Lobos and ROV Ventana, and deployed on the seafloor of Monterey Bay at a water depth of approximately 890m. MARS provided the science node with 375VDC electrical power and 100Mbps communication bandwidth. Figure 19.11 shows the deep-sea trial configuration of the science node, including the junction box and three instrument packages, and the MARS observatory. Figure 19.12 shows the CPAS, the HEMS, and the junction box on the seafloor in Monterey bay, connected to the sixth port of the MARS cabled observatory by ROV Ventana.

During the experiment, near-real-time scientific and engineering data streams were displayed on the remote computer consoles located at MBARI and stored in the DMS. Figure 19.13 presents the data infrastructure configuration of the science node in the MARS deep-sea trial. Figure 19.14 shows part of science data collected in the MARS deep-sea trial.

The sea trial was successful and the prototype proved to be reliable. The techniques will be applied in long-term scientific observation of the Chinese marginal seas. We also found that special attention should be paid to the issues of biofouling and electromagnetic interference (EMI). The biofouling issue can lead to accelerated performance degradation of many sensors. The EMI issue arises mainly because the switching noise generated by PWM converters can interfere with the analog measurement of seawater properties.

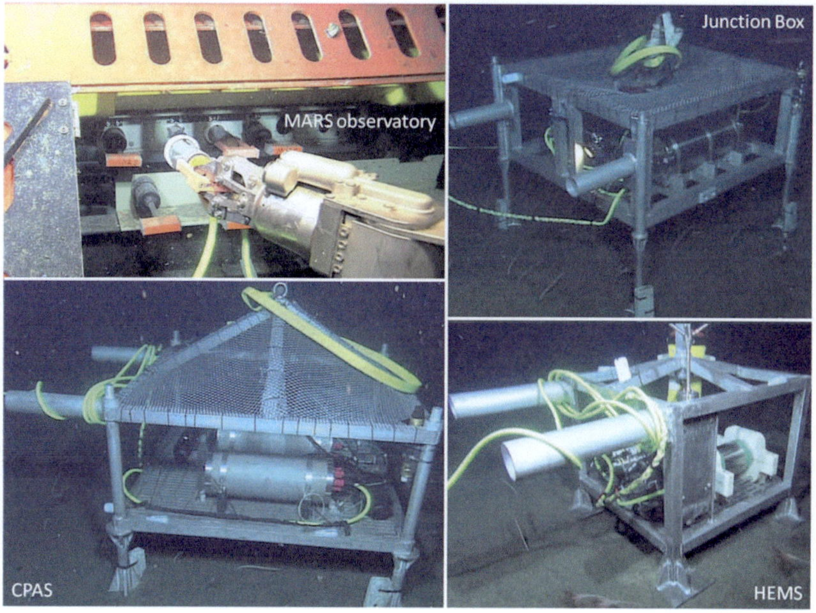

Figure 19.12 The CPAS, the HEMS, and the junction box located on the seafloor of Monterey Bay, connected to the 6th port of the MARS cabled observatory.

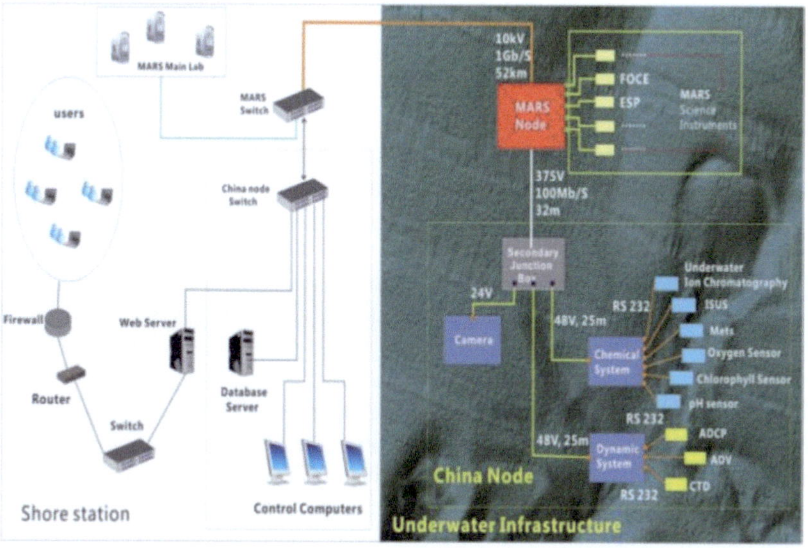

Figure 19.13 The data infrastructure configuration of the science node in the MARS trial (Peng et al., 2011).

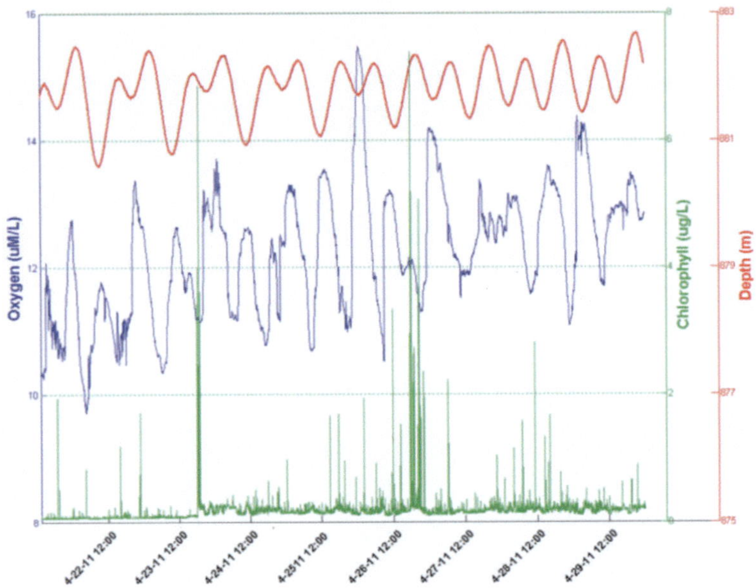

Figure 19.14 Part of the scientific data collected during the MARS deep-sea trial (Lu et al., 2011)

19.7 Concluding remarks

The long-term cabled seafloor observatories merge technologies of the transoceanic submarine telecommunications cables and the PWM switching power convertors. These novel facilities can provide a large number of undersea scientific experiments with sustained and ample electrical power and two-way broadband communication bandwidth. Then long-term observation and real-time experiments in the full water column can be realized and long time-series of high-resolution science data can be collected. The mechanism of the ocean–Earth system is expected to be better understood through the processing and analysis of the four-dimensional information covering large areas over decades.

An experimental prototype of cabled observatories for Chinese marginal seas is developed with emphasis on reliability, flexibility, and fault resilience. The prototype is rated up to 10kVDC/10kW electrical power and 1Gbps fiber-optic communication bandwidth, with depth rating up to 4000m. It consists of an undersea station, a junction box, a submarine cable, and a shore station. Three instrument packages are developed for in-situ multidisciplinary observation: a chemical parameter analyzing system, a hydrodynamic environment monitoring system, and a deep-sea video camera system. The multidisciplinary science node is a payload of the undersea station and consists of the junction box and the three instrument packages.

After laboratory testing, including bench testing, tank testing, and high-pressure testing, the prototype experienced the shallow seawater trial in the East China Sea for about three weeks. The science node experienced the deep-sea trial on MARS for about six months. A stringent test plan was implemented to make sure every component was highly reliable. High-resolution data collected in the sea trial showed the advantage of cabled seafloor observatories over conventional approaches.

Acknowledgments

This work has been funded by the National High Technology Research and Development Program of China and Shanghai Science and Technology Committee. The MARS experiment would not have been possible without the assistance of the team at MBARI: Chris Scholin, Steve Etchemendy, Yanwu Zhang, Mike Pinto, Ken Heller, Knute Brekke, Craig Dawe, Gene Massion, Mike Burczynski, D. J. Osborne, Peter Braccio, James Bellingham, Susan Von Thun, Natalie Fairchild, Katie Loweth, Kim Fulton-Bennett, and the rest of the MBARI team, especially the crews of the R/V Western Flyer and R/V Point Lobos and the ROV Ventana. The authors appreciate the research and engineering cooperation of Tongji University (Bangchun Wu, Zhengwei Wu, Jiangtao Li, Zijun Wu, Bin He, et al.), Zhejiang University (Yanhu Chen, Canjun Yang, Dejun Li, Bo Jin, Ying Chen, et al.), China Ocean University (Peiliang Li, et al.), and Shanghai Jiaotong University (Lian Lian, et al.) throughout the projects. The authors also highly appreciate Dr Yanwu Zhang (MBARI), Dr Mairi Best (University of Victoria), and the anonymous reviewer for the constructive comments that helped improve the quality of this manuscript.

References

Austin T.C., Edson J.B., McGillis W.R., Purcell M., Petitt R A., McElroy M.K., Grant C.W., Ware J. and Hurst S.K. (2002) A network-based telemetry architecture developed for the Martha's Vineyard Coastal Observatory. IEEE Journal of Oceanic Engineering 27(2) 228–234.

Barnes C.R., Best M.M.R., Johnson F., Phibbs P. and Pirenne B. (2008) Building the world's first regional cabled observatory: NEPTUNE Canada. Journal of Ocean Technology 3, 13–18.

Bellingham J.G. and Rajan K. (2007) Robotics in remote and hostile environments. Science 318, 1098–1102.

Chan T., Liu C.C., Howe B.M. and Kirkham H. (2007) Fault location for the NEPTUNE power system. IEEE Transactions on Power Systems 22(2), 522–531.

Chave A.D., Waterworth G., Maffei A.R. and Massion G. (2004) Cabled Ocean Observatory Systems. Marine Technology Society Journal 38(2), 30–43.

Chave A.D., Massion G. and Mikada H. (2006) Science requirements and the design of cabled ocean observatories. Annals of Geophysics 49(2–3), 569–579.

Curtin T.B. and Belcher E.O. (2008) Innovation in oceanographic instrumentation. Oceanography 21(3), 44–53.

Curtin, T.B. and Bellingham J. G. (2009) Progress toward autonomous ocean sampling networks. Deep-Sea Research Part II - Topical Studies in Oceanography 56(3–5), 62–67.

Delaney J.R., Heath G.R. and Howe B.M. (2000) NEPTUNE: Real-time ocean and earth sciences at the scale of a tectonic plate. Oceanography 13, 71–79.

Delaney J.R. and Barga S.R. (2009) A 2020 vision for ocean science. In: T. Hey, S. Tansley and K. Tolle (Eds) The Fourth Paradigm: Data-Intensive Scientific Discovery. Microsoft Research, 27–38.

Ding K. and Seyfried W.E. (2007) In situ measurement of pH and dissolved H2 in mid-ocean ridge hydrothermal fluids at elevated temperatures and pressures. Chemical Review 107, 601–622.

Duennebier F.K., Harris D.W., Jolly J., Caplan-Auerbach J., Jordan R., Copson D., Stiffel K., Babinec J. and Bosel J. (2002a) HUGO: The Hawaii Undersea Geo-Observatory. IEEE Journal of Oceanic Engineering 27(2), 218–227.

Duennebier F.K., Harris D.W., Jolly J., Babinec J., Copson D. and Stiffel K. (2002b) The Hawaii-2 Observatory Seismic System. IEEE Journal of Oceanic Engineering 27(2), 212–217.

El-Sharkawi M.A. Upadhye, Lu S., Kirkham H., Howe B.M., McGinnis T. and Lancaster P. (2005) North East Pacific Time-Integrated Undersea Networked Experiments (NEPTUNE): Cable switching and protection. IEEE Journal of Oceanic Engineering 30(1), 232–240.

Favali P. and Beranzoli L. (2006) Seafloor observatory science: A review. Annals of Geophysics 49(2–3), 515–567.

Favali P. and Beranzoli L. (2009) EMSO: European multidisciplinary seafloor observatory. Nuclear Instruments and Methods in Physics Research Section a-Accelerators Spectrometers Detectors and Associated Equipment 602(1), 21–27.

Favali P., Beranzoli L., D'Anna G., Gasparoni F., Marvaldi J., Clauss G., Gerber H.W., Nicot M., Marani M.P., Gamberi F., Millot C. and Finch E.R. (2006) A fleet of multiparameter observatories for geophysical and environmental monitoring at seafloor. Annals of Geophysics 49(2–3), 659–680.

Harris D.W. and Duennebier F.K. (2002) Powering cabled ocean-bottom observatories. IEEE Journal of Oceanic Engineering 27(2), 202–211.

Howe B.M., Kirkham H. and Vorperian V. (2002) Power system considerations for undersea observatories. IEEE Journal of Oceanic Engineering 27(2), 267–274.

Howe B.M., Chao Y., Arabshahi P., Roy S., McGinnis T. and Gray A. (2010) A smart sensor web for ocean observation: Fixed and mobile platforms, integrated acoustics, satellites and predictive modeling. IEEE Journal of Selected Topics in Applied Earth Observations and Remote Sensing 3(4), 507–521.

Howe B.M., Lukas R., Duennebier F. and Karl D. (2011) ALOHA cabled observatory installation. In: *Proceedings of MTS/IEEE Oceans 2011*, pp. 1–11.

Isern A.R. (2005) National Science Foundation's ocean observatory initiative. Sea Technology 46(6), 55–6005.

Kawaguchi K., Kaneda Y. and Araki E. (2008) The DONET: A real-time seafloor research infrastructure for the precise earthquake and tsunami monitoring. In: *Proceedings of MTS/IEEE Oceans 2008*, pp. 121–124.

Lentz S.T. (2007) The NEPTUNE Canada Communications Network. In: *Proceedings of MTS/IEEE Oceans*, Vol. 1–5, pp. 270–274.

Lentz S. and Lecroart A. (2009) Precision timing in the NEPTUNE Canada Network. In: *Proceedings of MTS/IEEE Oceans 2009 – Europe*, Vol. 1–2, pp. 128–132.

Leonard N.E., Paley D.A., Lekien F., Sepulchre R., Fratantoni D.M. and Davis R.E. (2007) Collective motion, sensor networks, and ocean sampling. In: *Proceedings of the IEEE* 95(1), 48–74.

Liu C.C., Schneider K., Kirkham H. and Howe B. (2003) State estimation for the NEPTUNE power system. In: *Proceedings of IEEE Transmission and Distribution Conference and Exposition* 2, pp.748–754.

Lu F., Peng X., Zhou H., Yue J. and He B. (2012) Design of a prototype system for cabled seafloor observatory networks. Chinese Journal of Scientific Instrument 33(5), 54–60.

Lu F., Zhou H., Peng X., Yue J., He B. and Wu B. (2011) Development of an undersea science node for cabled ocean observatories. In: *Proceedings of MTS/IEEE Oceans 2011*, pp.1–4.

Lu H., Li D., Yang C., Jin B. and Chen Y. (2010) Design and implementation of underwater junction box prototype system for deep seafloor observatory network. Journal of Zhejiang University (Engineering Science) 44, 8–13.

Massion G. and Raybould K. (2006) MARS: The Monterey Accelerated Research System. Sea Technology 47(9), 39–42.

Paul J., Scholin C., Engh G. V. D. and Perry M. J. (2007) In situ instrumentation. Oceanography 20, 70–78.

Peng X., Zhou H., Wu B., Lu F., Wu Z., Yang C., Li P., Li D., Jin B., Feng Z. and Li D. (2011) Test China node on Monterey Accelerated Research System (MARS). Advances in Earth Science 26(9), 991–996.

Person R., Austin Y. and Blandin J. (2006) From bottom landers to observatory networks. Annals of Geophysics 49, 581–593.

Person R., Beranzoli L., Berndt C. et al. (2008) ESONET: A European sea observatory initiative. In: *Proceedings of MTS/IEEE Oceans 2008,* 1–3, pp.1215–1220.

Romanowicz B., McGill P., Neuhauser D. and Dolenc D. (2009) Acquiring real time data from the Broadband Ocean Bottom Seismic Observatory at Monterey Bay (MOBB). Seismological Research Letters 80(2), 197–202.

Ruhl H.A., Andre M., Beranzoli L., Cagatay M.N., Colaco A., Cannat M., Danobeitia J.J., Favali P., Geli L., Gillooly M., Greinert J., Hall P.O.J., Huber R., Karstensen J., Lampitt R.S., Larkin K.E., Lykousis V., Mienert J., Miranda J.M., Person R., Priede I.G., Puillat I., Thomsen L. and Waldmann C. (2011) Societal need for improved understanding of climate change, anthropogenic impacts, and geo-hazard warning drive development of ocean observatories in European seas. Progress in Oceanography 91, 1–33.

Schneider K., Liu C. C. and Howe B. (2005) Topology error identification for the NEPTUNE power system. IEEE Transactions on Power Systems 20(3), 1224–1232.

Schofield O., Bergmann T., Bissett P., Grassle J.F., Haidvolgel D.B., Kohut J., Moline M. and Glenn S. M. (2002) The long-term ecosystem observatory: An integrated coastal observatory. IEEE Journal of Oceanic Engineering 27(2), 146–154.

Scholin C., Doucette G., Jensen S., Roman B., Pargett D., Marin R., Preston C., Jones W., Feldman J., Everlove C., Harris A., Alvarado N., Massion E., Birch J., Greenfield D., Vrijenhoek R., Mikulski C. and Jones K. (2009) Remote detection of marine microbes, small invertebrates, harmful algae, and biotoxins using the Environmental Sample Processor (ESP). Oceanography 22(2), 158–167.

Schneider, K. P., "Analysis of Critical Infrastructure Interactions," University of Washington, 2005

Schofield O., Bergmann T., Bissett P., Grassle J.F., Haidvogel D.B., Kohut J., Moline M., Glenn S. M. (2002) The long-term ecosystem observatory: An integrated costal observatory. IEEE Journal of Oceanaic Engineering 27(2.

Sherman A.D. and Smith K.L. (2009) Deep-sea benthic boundary layer communities and food supply: A long-term monitoring strategy. Deep-Sea Research Part II – Topical Studies in Oceanography 56(19–20), 1754–1762.

St Arnaud B., Chave A.D., Maffei A.R., Lazowska E., Smarr L. and Gopalan G. (2004) An integrated approach to ocean observatory data acquisition/management and infrastructure control using web services. Marine Technology Society Journal 38(2), 155–163.

Tunnicliffe V., Dewey R. and Smith D. (2003) Research plans for a mid-depth cabled seafloor observatory in Western Canada. Oceanography 16, 53–59.

Vorperian V. (2007) Synthesis of medium voltage DC-DC converters from low-voltage, high-frequency PWM switching converters. IEEE Transactions of Power Electronics 22, 1619–1635.

Wang P. (2007) Seafloor observatories: The third platform for earth system observation. Chinese Journal of Nature 29, 125–130.

Wang P. and Li Q. (2009) *The South China Sea – Paleoceanography and Sedimentology*, Chapter 1: Introduction. Springer.

Woodroffe A.M., Pridie S.W. and Druce G. (2008) The NEPTUNE Canada Junction Box – Interfacing science instruments to sub-sea cabled observatories. In *Proceedings of MTS/IEEE Oceans 2008* 1–3, pp.730–734.

Wu B., Peng X., Zhou H., Xiao Y., Lu F. and Wu Z. (2010) Development of a deep-sea chemical analyzer for in-situ anion measurement. In: *Proceedings of MTS/IEEE Oceans 2010*, pp.1–4.

Xu H., Zhang Y., Xu C., Li J., Liu D., Qing R., Luo S. and Fan D. Coastal seafloor observatory at Xiaoqushan in the East China Sea. Chinese Science Bulletin 56(22), 1839–1845.

20 From ESONET multidisciplinary scientific community to EMSO novel European research infrastructure for ocean observation

R. Person[1], P. Favali[2], H.A. Ruhl[3], L. Beranzoli[2], J.-F. Rolin[1], C. Waldmann[4], R. Huber[4], Y. Auffret[1], M. Namık Çağatay[5], M. Cannat[6], J.J. Dañobeitia[7], E. Delory[8], M. Diepenbroek[4], H. de Stigter[9], J.M.A. de Miranda[10], B. Ferré[11], M. Gillooly[12], F. Grant[12], J. Greinert[9], P.O.J. Hall[13], V. Lykousis[14], J. Mienert[11], I. Puillat[1], I. G. Priede[15] and L. Thomsen[4]

1 IFREMER – Institut Français de Recherche pour l'Exploitation de la MER, France
2 INGV – Istituto Nazionale di Geofisica e Vulcanologia, Italy
3 NOCS – National Oceanography Centre Southampton, UK
4 KDM – Konsortium Deutsche Meeresforschunge, Germany
5 ITU – Istanbul Teknik Universitesi, Turkey
6 IPGP – Institut de Physique du Globe de Paris, France
7 CSIC – Consejo Superior de Investigaciones Cientificas, Spain
8 PLOCAN – Plataforma Oceánica de Canarias, Spain
9 NIOZ – Stichting Koninklijk Nederlands Instituut voor Zeeonderzoek, The Netherlands
10 IPMA – Instituto Português do Mar e da Atmosfera, Portugal
11 UiT Universitetet i Tromsø, Norway
12 IMI Irish Marine Institute, Ireland
13 UGOT Goteborgs Universitet, Sweden
14 HCMR Hellenic Centre for Marine Research, Greece
15 UNABN University of Aberdeen, UK

20.1 Introduction

Environmental and climate changes are crucial challenges for sustainable living because of their significant impact on the Earth system and the important consequences for natural resources. Oceans have a primary role in these changes as they regulate heat flux, greenhouse gases and climate whilst harboring many different life forms and resources. Understanding processes in the marine environment is of paramount importance for any prediction of short-, intermediate- and long-term global change.

The oceans are complex dynamic systems on many time and space scales, from episodic catastrophic events to slow trends difficult to discern from the overlying variability of short-term processes. The high resolution, long-time-series collection of multiple variables across a breadth of environments represents the only approach capable of rendering the complexity of these systems. Maintaining observations of key ocean variables – at different observatory locations, over long timescales, with adequate temporal resolution – has been challenging (Lampitt et al., 2010). However, multiparameter datasets enable the enhancement of more multidisciplinary methods, giving strong benefits to many disciplines (e.g., Monna et al., 2014). Seafloor observatories are new tools now available for long-term multidisciplinary data collection (NRC, 2000).

Major international science priorities in the multidisciplinary fields of geoscience, physical oceanography, biogeochemistry, marine ecology, and geo-hazards are being advanced through observatory development and networking (Table 20.1). Moreover, improved availability and consistency in measurements taken across European seas have been stimulating a transformative shift in the ability to conduct cross-cutting science over a wide range of marine habitats (Ruhl et al., 2011).

Globally, a number of large-scale programmes (Favali et al., 2010) have been planned and undertaken to establish permanent seafloor and water-column infrastructures in Europe, Australasia, North America, and East Asia (Figure 20.1). In 1978 Japan started

Geoscience	Interactions with hydrosphere and biosphere, fluid flow and gas seepage through sediments and vents, gas hydrate dynamics, non-living resources, sediment transfer to the deep-sea, and links to climate change
Physical oceanography	Water mass characterization, currents and mixing, thermodynamics, ice cover, climatology, and links to climate change
Biogeochemistry	Global carbon cycle and elemental cycling within the ocean through both physical and biological processes, and climate feedbacks
Marine ecology	Distribution and abundance of sea life, ocean productivity, biodiversity, ecosystem function, living resources, and climate feedbacks
Geo-hazards	Earthquake and tsunami hazards, volcanic hazards, slope instability and failure

Table 20.1 Scientific multidisciplinary fields related to fixed observatories.

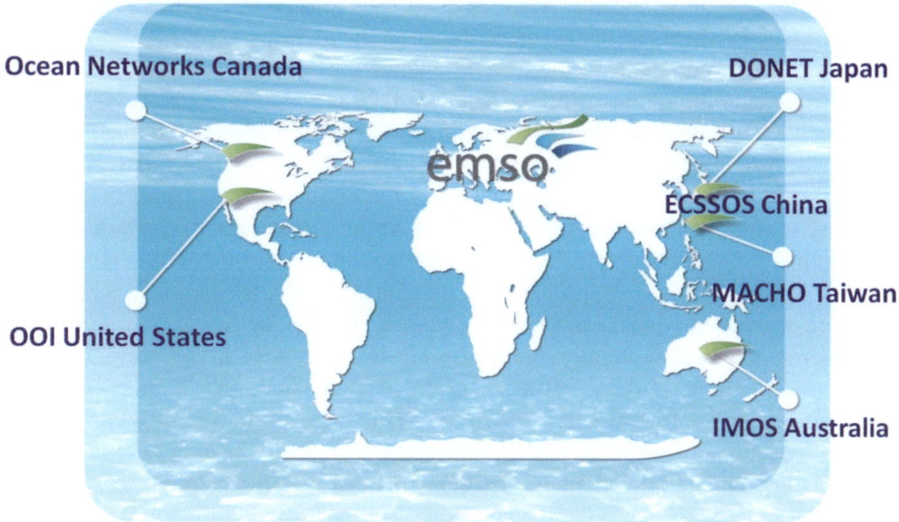

Figure 20.1 Locations of the present seafloor and water-column observatory networks operating or under development in the world.

to develop cabled seafloor observatories focused on real-time monitoring for seismic and tsunami warning; current networks, called DONET 1&2 (Dense Oceanfloor Network system for Earthquakes and Tsunamis), are managed by JAMSTEC (Japan Agency for Marine-Earth Science and Technology). In Taiwan, MACHO (MArine Cable Hosted Observatory) recently started and has deployed a submarine cabled observatory offshore of the eastern part of the island with the main purpose of establishing seismic stations, providing early warning of earthquakes and tsunamis, and monitoring submarine volcanic activity. China is developing ECSSOS (East China Sea Seafloor Observation System; Yu et al., 2012) through a consortium of research institutions. In Canada, ONC (Ocean Networks Canada) includes NEPTUNE Canada (North East Pacific Time-series Underwater Networked Experiments), the world's first regional-scale, multidisciplinary underwater cabled ocean observatory network, and VENUS (Victoria Experimental Network Under the Sea), is a multiplatform, coastal laboratory. The United States has launched the OOI (Ocean Observatories Initiative) aiming at establishing a regional-cabled network (RSN-Regional Scale Nodes), two coastal observatories, and four global high latitude buoys. In Australia, IMOS (Integrated Marine Observing System) is a distributed infrastructure encompassing a broad range of marine observing facilities, with components in Antarctica and New Zealand, tied together with a common data management and access system.

In Europe, the support by the European Commission (EC) of ocean observation dates back to the early 1990s when projects aimed at the development of observatories and observatory network prototypes were funded through the 5th, 6th and 7th Framework Programmes.

The experience gained and the skill achieved by teams of researchers and engineers through many experiments at sea has contributed to the growth of a European marine observatory scientific community across over 14 countries. However, political challenges and persistent fragmentation of disciplinary sectors limited progress towards an integrated understanding of ocean systems. An important effort was thus encouraged by the EC through the funding of ESONET-NoE (European Seas Observatory NETwork – Network of Excellence; 2007–2011). It fostered the integration of the pan-European scientific community made up by researchers and engineers involved in the development and deployment of multidisciplinary ocean observatories, and small medium enterprises (SMEs) specialized in sensors and tools for marine applications. Integration Activities and Joint Research Activities were developed within ESONET to outline a strategy for European research in the marine environment and strengthen the European Research Area in the sector of seafloor long-term observations (Person et al., 2008). In parallel, the elaboration of a European Roadmap of research infrastructures drawn by ESFRI (European Strategy Forum on Research Infrastructures), recognized the value and the maturity of the experience in sea observations and since 2006 included EMSO (European Multidisciplinary Seafloor and water-column Observatory) among the necessary European-scale research infrastructures. These efforts have helped to underpin progress in the environmental sector to more broadly address long-term monitoring of environmental processes (Favali and Beranzoli, 2009) and provide a key supporting element for marine environment policy.

This chapter highlights the ESONET-NoE science and technology major achievements and summarises the results and perspectives of the EMSO research infrastructure in the International framework.

20.2 ESONET and EMSO: Synergic framework

A major objective of the ESONET-NoE was to facilitate the transition from conventional, independently operated, single discipline, short-term, instrumented ocean observations into the EMSO distributed infrastructure of multidisciplinary sustained nodes with a shared science plan and common data services. The heterogeneity of organizational and technical approaches, equipment and methodology used in the observing stations requires the construction of a unique context, shared among the different actors, within which to adopt and develop standards and common tools to enhance interoperability, and assure adaptation to the needs of scientists and generic users. The overall strategy put in place by ESONET was to adopt and foster globally-accepted standards and concepts, to use non-specific approaches and multipurpose set-ups, and to find synergies both in new developments as well as in the use of existing facilities (Puillat et al., 2009).

The most significant test-beds for the integration of methodologies, devices, facilities and infrastructures owned by the partners are represented by the ESONET Demonstration Missions (DM); special Joint Research activities in key sites where EMSO nodes are either operating permanently (cabled systems) or at an early implementation stage with temporary facilities (stand-alone systems) in synergy with regional/national and EC programs and projects. The Demonstration Missions, as described below, have shown the capacity of the community in easily setting up specific research and technology activities and continuing them even beyond the ESONET project context. This outlines an enduring scientific reference frame for scientific collaboration that continues and sustains the EMSO research infrastructure. This framework is supported by the implementation of ESONET-Vi (ESONET-Vision), a major strategic outcome of ESONET, an open association of individuals and research institutions across Europe that will consolidate and promote the synergy between European research communities and the EMSO infrastructure as a service-oriented multi-purpose observatory distributed infrastructure.

The development plan of EMSO, as drafted in the Preparatory Phase project funded by EC (EMSO-PP; 2008–2012), includes multiple aspects of capacity building and retention for science and technology excellence, both in terms of infrastructure and technology management, as well as aspects of financial viability and a legal frame of reference.

The financial and legal issues are also challenging since EMSO will be funded by the participating nations through the respective national funding agencies, and only partially by the EC. This means that EMSO will need a transnational effort to provide Europe with a large-scale monitoring system with a long-term operational life and appropriate organizational structure for helping scientists to answer major scientific questions and policy-makers to design and implement effective environmental and geo-hazard monitoring and mitigation policies.

20.3 Major ESONET-NoE activities and achievements

The ESONET-NoE involved more than 50 partners including research institutions, large companies and SMEs from 14 different European Countries with over 300 scientists, engineers and technicians related to the scientific use of the observatories (Ruhl et al., 2011), data management and technical and implementation issues (Puillat et al., 2012). Relevant partners were also engaged in a set of experiments, namely Demonstration Missions and Test Experiments, which pushed forward the multidisciplinary monitoring capacity.

20.3.1 Demonstration missions and test experiments

ESONET Demonstration Missions and Test Experiments, in synergy with other European and national projects, supported scientific and technological activities and developed EMSO nodes at sites of exceptional global scientific interest.

These activities have integrated and enhanced the European community of marine sciences with respect to basic knowledge tools such as measurement techniques, data, methodologies for data analysis. They represent a significant increase of permanent or nearly-permanent monitoring installation at EMSO nodes. The Demonstration Missions and the Test Experiments were selected and funded by ESONET financial resources according to an internal project call for proposals and peer review. A brief description of the goals and operations developed within the Demonstration Missions is given below showing the large effort and the high-level capacities deployed in support of EMSO. The sites of the Demonstration Missions are shown in Figure 20.2.

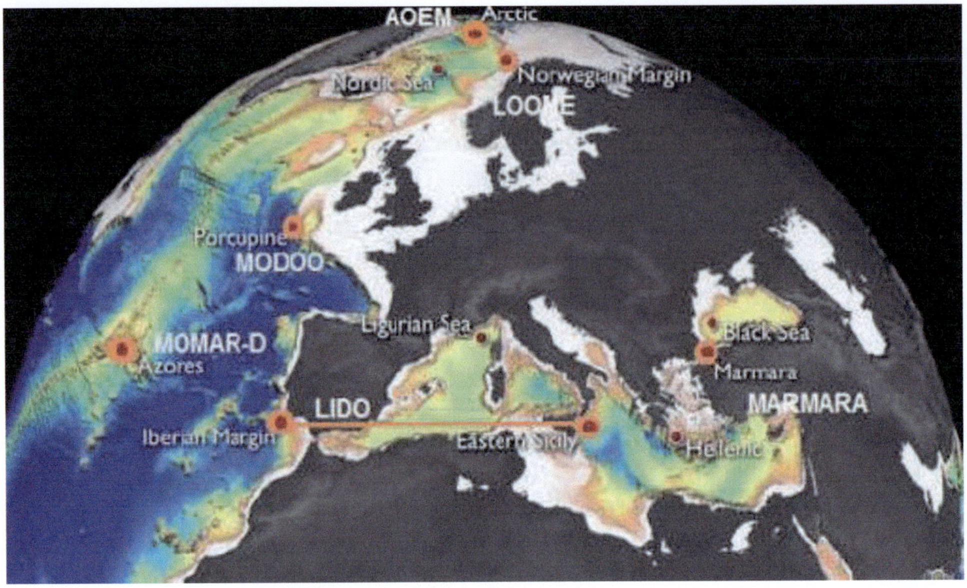

Figure 20.2 Map of the ESONET-NoE demonstration missions.

20.3.1.1 AOEM: Arctic Ocean ESONET Mission: A step towards understanding a key area for climate studies

The Arctic is one of the areas most sensitive to natural and anthropogenic changes. Activities have been focusing on the evaluation of heat and mass transport through the Fram Strait (between Greenland and the Svalbard islands), on the links between climate and surface ocean processes with deep-sea ecosystems, and on methane hydrate dynamics (Ruhl et al., 2011). The activities were conducted within a framework of a sustained cabled observatory network for dissociating hydrate studies in high latitudes. In the eastern Fram Strait, the Hausgarten observatory has been operated by AWI (Alfred Wegener Institut) with repeated sampling and deployment of moorings and freefalling systems since 1999. The observatory comprises a network of 17 permanent stations covering a depth range of 1000–5500m water depth. Recent activities have been conducted within EC projects such as HERMIONE (Hotspot Ecosystem Research and Man's Impact On European Seas) and HYPOX (in situ monitoring of oxygen depletion in hypoxic ecosystems of coastal and open seas, and land-locked water bodies). AOEM, also through its sub-projects MASOX (Monitoring Arctic Seafloor-Ocean Exchange) and ARCOONE (Arctic Operational Oceanography Network in ESONET), addressed important facets related to both technology and real-time environmental data streams to enhance monitoring capacities for climate change. While some AOEM activities focused on optimizing technical solutions for interfaces and data delivery from moorings to the existing cabled node, and establishing common standards for data transfer, other activities were aimed at investigating a causative effect of warming shallow Arctic seas from heat conduction into the seafloor and sub-seafloor sediments. This has potential to perturb the thermodynamic stability zone of methane hydrate, increasing the release of free and dissolved methane gas into the ocean and atmosphere, with consequences for benthic ecology, biogeochemical cycles, and greenhouse gases.

20.3.1.2 LOOME: Long-term observations on mud-volcano eruptions in the Norwegian margin

The long-term observation of a major site of methane emission from the deep European margin has been carried out for the first time at the Håkon Mosby Mud Volcano (HMMV). The HMMV, discovered in 1989, is a cold seep ecosystem located at 1250m water depth on the south-western Barents Sea slope off Norway, an area with a history of seabed slides and tsunamis, and under exploitation for hydrocarbon resources and fisheries. The Barents Sea slope is a target area for sustainable management and monitoring of global change effects. The HMMV is about 1.4km in diameter and reaches up to 15m in height from the seafloor. The volcano sits on top of a giant gas chimney forming a window to the deep sub-surface down to 3000m below seafloor, from where warm, methane-rich fluids are pressed upwards towards the cold seafloor. LOOME acquired a good understanding of spatial scales of geological, hydrographical, chemical and biological parameters of this exemplary mud volcano ecosystem, so that results from the first observation of temporal variation and eruption events immediately led to important findings. Previous work of the partners at HMMV yielded evidence of several eruption events indicated by strong gas bubbling and abrupt temperature changes of almost 10°C within a few days. High resolution bathymetric maps and video observations of the seafloor before and after this event clearly showed changes in the morphology of HMMV. Only continuous observations, recording a wide variety of parameters, can give insight into the mechanics of such eruptions and their early signals, estimating the amount of gas release and the consequences for geochemistry and local communities as well as for seafloor stability. LOOME took advantage of the large amount of data and interdisciplinary knowledge collected at HMMV during previous cruises and developed for the first time a detailed investigation of the temporal variability at an active gas-emitting mud volcano covering the sequence of events before, during, and after an eruption with the analysis of the effect on gas hydrate stability, seafloor morphology and the distribution and colonization patterns of benthic communities. A main goal of the project was the integration of existing technology to establish in a first phase an autonomous non-cabled observatory for monitoring seafloor seismicity, temperature, pore pressure, chemical profiling, sonar detection of gas flares, methane measurements of bottom water, colonization patterns, community structure, and biodiversity.

20.3.1.3 MODOO: MOdular Deep Ocean Observatory: The sustained monitoring of the Porcupine Abyssal Plain for studying biogeochemical processes

The Porcupine Abyssal Plain (PAP), off southwest Ireland, is the location of the longest running, multidisciplinary marine time-series in Europe. Since 1989, regular measurements of a variety of physical and biogeochemical parameters have been undertaken at PAP. From 2002, a full depth multidisciplinary mooring has been in place with sensors taking a diverse set of biogeochemical and physical measurements of the upper 1000m of the water column. Recent activities were undertaken under the MODOO demonstration mission and under EUROSITES (European Open Ocean Observatory Network) EC project, European part of OCEANSITES. MODOO makes use of the existing water-column mooring infrastructure at PAP, which forms part of the EUROSITES network of European Eulerian ocean observatories and developed this within the objectives of ESONET beyond the current state-of-the-art. MODOO also demonstrated the functioning of a prototype observatory based on an advanced version of a BOBO (Benthic Boundary Observatory) lander. This latter was enhanced to host additional sensors and, more importantly, to communicate by inductive link with the mooring via a Data Collection and Dissemination node with real-time data access capabilities. The MODOO concept has thus linked and operated existing stand-alone lander and mooring observatories such that they merge into a single observatory. The MODOO observatory prototype is mobile (or relocatable) as it can be moved to regions where required, and it is modular as its architecture allows other stand-alone systems to be connected.

The science improves our understanding of ocean circulation and links between the surface and deep ocean, and biogeochemical cycling in the water column and through the seafloor/water interface. Short-term variability of the oceans is examined, including physical mixing, ecosystem dynamics and nutrient cycling. This resolution also allows longer-term trends in the Earth's climate to be investigated.

20.3.1.4 MoMAR-D: MOnitoring the Mid-Atlantic Ridge Lucky Strike vent field (off Azores)

Lucky Strike is a large hydrothermal field located in the center of one of the most volcanically active segments of the Mid-Atlantic Ridge. Monitoring this field offers a high probability of capturing evidence of volcanic events, observing interactions between faulting, magmatism and hydrothermal circulation and evaluating their links with the dynamics of hydrothermal fauna. A multidisciplinary approach, from geophysics to microbiology that focuses on fluid circulation, is fundamental.

MoMAR-D Demonstration Mission is a component of the work planned at the Azores EMSO node, which focuses on active processes at ridge hydrothermal systems. Hydrothermal circulation at mid-ocean ridges impacts the transfer of energy and matter from the interior of the Earth to the crust, hydrosphere and biosphere. Seawater circulates through the permeable oceanic crust, exchanges chemicals with the surrounding rocks, and is heated to temperatures reaching 400°C. This hot fluid flows up and is expelled at hydrothermal vents, forming emissions of different physico-chemical properties (from black smokers to diffuse venting). The unique faunal communities that develop near these vents are sustained by chemo-synthetic micro organisms that use reduced chemicals present in the hot fluids as energy sources.

MoMAR-D contributed to the implementation of the Azores EMSO node through a one-year deployment of an acoustically-linked multidisciplinary observing system at the Lucky Strike vent field, with satellite connection to shore.

The SEAMON (SEA MONitoring) technology developed by IFREMER constitutes the basic monitoring system. Two seafloor nodes were deployed, a geophysical node (SEAMON-West) moored in the Lucky Strike lava lake including a pressure probe, an OBS (Ocean Bottom Seismometer), an OBM (Ocean Bottom Magnetometer), T-probes, and a geochemical/ecological node (SEAMON-East) at the Eiffel Tower vent site hosting a videocamera, chemical sensors and CTD/ADCP package. The two nodes were acoustically linked to a surface BOREL (Bouée relais) buoy acting as a data transmission relay and ensuring satellite communication to a land base station. This observatory infrastructure requires a synchronized multidisciplinary dataset, test solutions for sensor interoperability, shore-sensor interactive communication, data management and dissemination. MoMAR-D also developed a public outreach strategy and supported the regional marine protected area policies.

MOMAR-D generates a synergy with on trans-Atlantic collaborations as some instruments and devices (e.g., ridge geochemical and ecological suite, TEMPO-mini) were deployed with the same yearly maintenance pattern of MoMAR on Endeavour, a NEPTUNE Canada site in the Pacific thus allowing inter-comparison studies (Auffret et al., 2009; Vuillemin et al., 2009).

MoMAR-D has been providing the technological know-how and the scientific impetus to move forward and promote more durable monitoring solutions. This includes the extension of real-time monitoring of volcanic, hydrothermal and biological processes to areas nearer the Azores Islands, to assist the Azores region with issues related to volcanic hazards and durable management of environmental resources.

20.3.1.5 LIDO: LIstening to the Deep-Ocean environment with a regional network of multidisciplinary seafloor observatories

The LIDO objective was to enhance the present capabilities of the observatories working in the EMSO sites of the Western Ionian Sea (off Eastern Sicily) and the Iberian Margin (Gulf of Cadiz) by installing additional sensor equipment related to bioacoustics and geo-hazards monitoring.

The LIDO scientific objectives for bioacoustics were to evaluate the human (for example increased shipping) and natural contributions to marine ambient noise and, for the first time, to describe the long-term trends in ambient noise levels, especially from marine mammal populations (migration patterns, presence, and habitat use of key species, such as sperm-, fin- and beaked whales). For geo-hazards, LIDO has continued the experience initiated with NEAREST (Integrated observations from NEAR shore sourcES of Tsunamis: towards an early warning system) EC project in relation to the real-time and near-real-time detection of seismic and seafloor water-pressure signals to improve the early warning capabilities with respect to traditional approaches and for the characterization of potential tsunamigenic sources. Its methodological approach is based on the cross-checking of geophysical, oceanographic and environmental time series acquired on the seafloor and in the water column (Chierici et al., 2012). In the Iberian Margin (Gulf of Cadiz), a GEOSTAR-class seafloor observatory (Favali et al., 2006), with a prototype of a tsunameter, acoustically linked to a surface buoy and through a satellite link from the buoy to shore, was deployed from 2007 to 2011. This installation was the pilot experiment towards a permanent installation. The Iberian Margin is strongly affected by geo-hazards, mostly related to earthquake occurrence and tsunamis; the most important example was the strong tsunamigenic earthquake that destroyed Lisbon in 1755. It is also a crucial area being part of the boundary between Eurasian and Africa plates of great complexity as highlighted by a recent tomographic study (Monna et al., 2012).

One of the most seismically active areas of the Mediterranean is off eastern Sicily in the vicinity of Mt Etna and the MEDI biogeochemical region. It is one of the advanced EMSO sites with the NEMO-SN1 multidisciplinary cabled observatory in operation at a depth of 2100m since 2005. The observatory performs acquisition of geophysical, oceanographic and bioacoustic time-series (Favali et al., 2013).

LIDO established a first nucleus of a regional network of multidisciplinary seafloor observatories contributing to the coordination of high quality research in the ESONET-NoE by allowing the long-term monitoring of geo-hazards and marine ambient noise in the Mediterranean Sea and Atlantic waters. LIDO followed the recommendation of the ICG/NEAMTWS (Intergovernmental Coordination Group for the North-Eastern Atlantic, the Mediterranean and connected seas Tsunami Early Warning and Mitigation System) of IOC/UNESCO (Intergovernmental Oceanographic Commission of United Nations Educational Scientific and Cultural Organization) for the urgent deployment of a tsunami warning system in the related areas with special regard to the definition of transnational seismic and sea level monitoring networks.

20.3.1.6 MARMARA-DM: Looking for a relationship between gas seepage and seismicity in the Marmara Sea (Istanbul Supersite)

The Marmara Sea experiences regular tectonic activity along the North Anatolian Fault, a major continental transform plate boundary (e.g., Görür and Çağatay, 2010). The fault cuts across a natural gas field beneath the Marmara Sea and hydrocarbons seep from the seafloor. The possible relationships between gas seepage and seismicity are of high interest (e.g., Tryon et al., 2010; Gasperini et al., 2012; Tary et al., 2012). After the 1999 Izmit and Düzce earthquakes, the next large earthquake (Mw>7) is expected in the Marmara Sea close to the megacity Istanbul. The Marmara Sea offers the ideal location for seafloor seismogenic observations directed towards risk assessments, because of several factors such as the significant deformation rates (20mm/y), high societal impact (>15 million people), and the numerous fluid vents and related features. The Marmara Sea is thus a unique area to test hypotheses on the relationships between strike-slip deformation, seismic activity, fluid flow and gas expulsion within an active fault zone.

During the MARMARA-DM, the seafloor observatory SN4, together with many innovative instruments (e.g., active acoustics, piezometers, fluid samplers, radon sensor, in addition to OBSs) were deployed as stand-alone systems (Çağatay et al., this volume). The SN4 observatory was deployed in the Izmit Gulf near the termination of the 1999 earthquake rupture to explore relationships between seismicity and gas seepage (Embriaco et al., 2013). Work performed on MARMARA-DM data met difficulties in the precise determination of the depth of offshore earthquakes and in the identification of tremors from noise analysis in the marine environment. Emphasis is now given to focused, small-scale networks and to the identification of repeating earthquakes. These were previously recognized in the Central Basin area, which appears to be the possible hypothetical nucleation site for a rupture on the Istanbul-Silivri segment.

The goal of the MARMARA-DM was to contribute to the establishment of optimized permanent seafloor observatory stations for earthquake monitoring in the Marmara Sea, fully developing the EMSO node to be integrated with the terrestrial monitoring networks.

Three sites are identified as priorities for the future multi-parameter seafloor observatories: the Istanbul-Silivri segment, in the seismic gap immediately south of Istanbul; the Western High/Gas Hydrates area; and the entrance of Izmit Gulf. Several proposals for funding have been prepared on this basis (Çağatay et al., 2011; this volume).

20.3.1.7 Ligurian Sea: Biochemical process of the water column and slope instability monitoring

The Ligurian EMSO node has three specific assets:

1. Being nearly a closed basin, the Mediterranean experiences enhanced coupling of physical, chemical and biological processes in the water column and at the seabed, and is therefore a good natural laboratory for long-term monitoring of these processes.
2. The Ligurian region has a remarkably steep continental slope and active seismicity, which together make it an exceptional case study site for sediment transport, slope failure processes and the related hazards.
3. The Ligurian node is located next to the existing ANTARES neutrino telescope (Astronomy with a Neutrino Telescope and Abyss environmental RESearch) and will share this high-capacity cabled facility with neutrino astronomers.

This test site was selected in order to provide more experience of cabled observatories in the ESONET demonstrating process. A secondary junction box was plugged to the main neutrino telescope infrastructure at a distance avoiding perturbation of the astrophysical photomultipliers. It is able to host instruments of many disciplines, including biogeochemical sensors as well as a broadband seismometer and a bottom pressure recorder contributing to the civil security network of Southern France.

20.3.2 Underwater operation and best practices

Installation and servicing of observatory infrastructure requires experience and physical means of underwater intervention. A review of existing practices in marine operations in Europe shows a wide know-how, from cable laying to ship and ROV operations. It is appropriate here to separate discussion of main cables for cabled observatories vs. other observatory subsea infrastructure.

Concerning the main cable for cabled observatories, it is more than likely that planning, installation, and future maintenance will be contracted out to commercial submarine telecommunications systems and specialized survey companies that make use of existing and proven commercial hardware, jointing methods and techniques. So the project cycle for a scientific system will largely follow the standards of these companies and the steps of commercial undersea cable systems. However, scientific submarine cable systems impose different requirements for route survey data. The most scientifically interesting sites on the seabed tend to be those that pose the most difficulty to the cable route engineer for ensuring the long-term integrity of the deployed cable. In these cases, project schedules can be improved and costs lowered by making use of existing and extensive seabed data from past experiments conducted at well-established sites of scientific interest.

However, direct transfer of methods from offshore industry (e.g., oil and gas) to the installation of underwater scientific modules could be detrimental to the scientific community in terms of equipment availability and installation costs. As opposed to the offshore industry philosophy which offers no compromise to equipment performance, a more flexible approach should be taken by the scientific community in evaluating performance of alternative methods such as smart rigging and lower cost of support ships, versus sea state capability and global cost. Existing facilities: submersibles, ROVs (Remotely Operated Vehicles), AUVs (Autonomous Underwater Vehicles), available within the scientific community will govern the early days of such procedures.

Due to the relatively smaller size of the modules to be deployed in scientific missions compared to offshore industry, and the deliberate choice to perform the operation only during good sea state, deployment/recovery operations are generally not conducted with highly specialized industry ships; rather, oceanographic research vessels are usually used.

Each deployment will be specific, but there are some rules or experience results that can be underlined:

- **Weather conditions and dynamic loads**
  Considering that some scientific operations will not be achieved with highly specialized vessels, with no active heave compensation systems, the cable deployment operations will be done up to Sea State 3. In some cases, active heave-compensation systems can be replaced by passive systems or by smart rigging arrangements, allowing the uncoupling of dynamic loads from surface support before landing on the seabed.

- **Positioning system**
  For all these operations, a precise acoustic positioning system (LBL-Long Base Line or SBL-Short Base Line) is essential. Combined with a transponder on the extremity of the deployment/recovery cable, the placement precision can be on a metre scale.

- **Dredger line and fishing tail on the bottom**
  A cost-effective solution for module recovery by cable is to connect a specific "fishing tail" to the module (this can be prepared on the equipment before landing, or placed by a ROV before recovery operation), that can be dredged directly from the ship by a recovery line equipped with grapnel. This technique needs a good positioning system, with a transponder on the line and on the grapnel. A "dredging area" has to be defined, where there will be no scientific installations, or these will be in a safety position before the recovery operation. The fishing tail should be no longer than 30m, to ease the surface recovery operation. This operation must be achieved in very good weather conditions, with very low ship heave and roll – conditions which are reasonably common during most of the seasons at a number of the EMSO nodes, except those in the Arctic region. It can take a few hours (experience between two and eight hours).

- **Precise orientation with ROV assistance**
  Placing the sensor package at a correct compass heading or measuring the sensor package orientation reliably is necessary when vectorial sensors are used. In this case the use of passive methods, like an orientation "fin" or anchorage line, is not considered reliable, and the intervention of a ROV is needed. Safety procedures must be planned,

with completely separate safe energy sources for lifting devices and ROV deployment. During the entire operation two underwater cables, one for the deployment device, the other for the ROV, are in the water close to each other. This procedure is currently adopted by offshore companies Particular care should be taken to avoid interweaving with the ROV umbilical or tether. The two cables will have to be as similar as possible (diameter and torque properties). It is recommended that the two main lines (lift line and ROV umbilical) be deployed from separate areas of the vessel. The distance between the two cables should be as great as possible, depending on the vessel performance, and with a minimum of 15m. The lift line should be heave compensated, especially from small heave-prone intervention vessels. Compensation includes an active heave-compensated crane or configuration of the lift line in a lazy S, located mid depth, using buoyancy cells to isolate heave motions from load movement below. Dynamic Positioning capability is required on the vessels.

- **Use of a dynamic positioning docking system: MODUS (Mobile Docker for Underwater Sciences)**
  The MODUS system is a specialized mobile shuttle for the deployment, servicing and recovery of benthic stations to/from the seafloor (max 30kN). As one major result of the EC GEOSTAR (Geophysical and Oceanographic Station for Abyssal Research) projects (Favali et al., 2006), MODUS is designed to operate down to 4000m water depth. The complete system comprises a direct winch, an electro-optical cable, a specialized vehicle (MODUS) and an integrated control unit on the ship. MODUS is equipped with sonar, lighting, cameras, altimeter, different sensors for load and acceleration and thrusters for horizontal movements. It has a latch device for remote coupling or decoupling to/from the bottom station equipped with corresponding docking pins. It therefore requires a compatible interface on top of the ocean bottom subsystem.

To deploy and maintain underwater systems as planned in EMSO demands the use of the appropriate marine logistics. Therefore, strict coordination with the ship operators is essential, taking advantage of initiatives like the EC research infrastructure EUROFLEETS (Towards an alliance of European research fleets). This project (now starting a second phase) aims to bring together the European research fleets to enhance their coordination and promote the cost effective use of their facilities.

20.4 Sustained European-scale ocean observations through EMSO infrastructure

EMSO is a pan-European ocean observatory infrastructure, with nodes that cover all the European seas from the Arctic and Atlantic Ocean to the Black Sea passing through the Mediterranean Sea (Figure 20.3). EMSO has been supported within the EC 7th Framework Programme through a preparatory phase project (EMSO-PP), involving 12 countries: Italy (coordinator), France (vice coordinator), Germany, Ireland, Spain, Sweden, Greece, United Kingdom, Norway, Portugal, Turkey and the Netherlands. The EMSO-PP project endorsed the creation of the EMSO-ERIC (European Research Infrastructure Consortium) legal entity to develop and operate ocean observatory infrastructure at the European scale (Favali and Beranzoli, 2009). The ERIC is a legal status adopted by the European Council in May 2009 to facilitate the setting up of projects on the ESFRI (European Strategy Forum on Research Infrastructures) Roadmap.

EMSO is a distributed infrastructure, and potentially expandable, so the number of the nodes and the infrastructure expansion are considered as dynamic parameters. In this respect, the deployment of additional nodes and expansion beyond the initial configuration have to be considered through a process of natural evolution. As an example, the Canary Archipelago has been added to the original list. In fact, PLOCAN (Plataforma Oceánica de Canarias) is a recently incorporated multidisciplinary site which has initiated the progressive implementation of seafloor, water column and air–sea interface monitoring in the Central-Eastern Atlantic Ocean. The PLOCAN station ESTOC (European Station for Time-series in the OCean), with over 15 years of continuous biogeochemical monitoring is an open-ocean station, contributed to EUROSITES' EC project and is currently contributing to FixO3 (Fixed-point Open Ocean Observatories, 2013–2017) EC project and ICOS (Integrated Carbon Observation System, another ESFRI Infrastructure). A seafloor module is planned to monitor volcanic tremor and the seafloor unknown biogeochemistry. The PLOCAN Mobile Observing System is both coastal and open-ocean; three deep-sea gliders are currently in operation. Scientific objectives and operational plans are further detailed in Delory et al. (2011).

The EMSO primary objective is to collect crucial data feeding GEO (Group on Earth Observations), IPCC (Intergovernmental Panel on Climate Change), UNEP (United Nations Environment Programme), and OSPAR (Convention for the Protection of the Marine Environment of the North-East Atlantic) to help form and revise policy and legislation. Furthermore, the EMSO open-ocean data contribute to the Marine Strategy Framework Directive initiated by the EU in 2008, which aims to achieve GES (Good Environmental Status) in the European seas by 2020. The Directive is supporting the development of coherent approaches to assess GES in a comprehensive and holistic manner, thereby supporting an ecosystem-based approach to management.

EMSO aims to supply the required synchronous measurements from fixed locations in four of the five areas defined by the Marine Services as the Global Ocean, the North East Atlantic, the Arctic, and the Mediterranean Sea. EMSO is a logical partner to ICG/NEAMTWS to monitor earthquake-related activities in marine seismogenic areas.

EMSO is the European counterpart of large-scale ocean monitoring programs developed and under development around the world (see the Introduction) and constitutes the sub-sea segment of COPERNICUS, previously known as GMES (Global Monitoring for Environment and Security), the European Programme for the establishment of a capacity for the observation of the Planet, based on in situ measurements, satellite Earth observation, data and services harmonisation.

As with similar environmental research infrastructures global in scope, EMSO is involved in efforts towards interoperation, harmonization (including data management) and is an acknowledged driver for international cooperation. For example, EMSO is working towards marine input into GEOSS (Global Earth Observation System of Systems) as a system involving data policy and compatibility of systems.

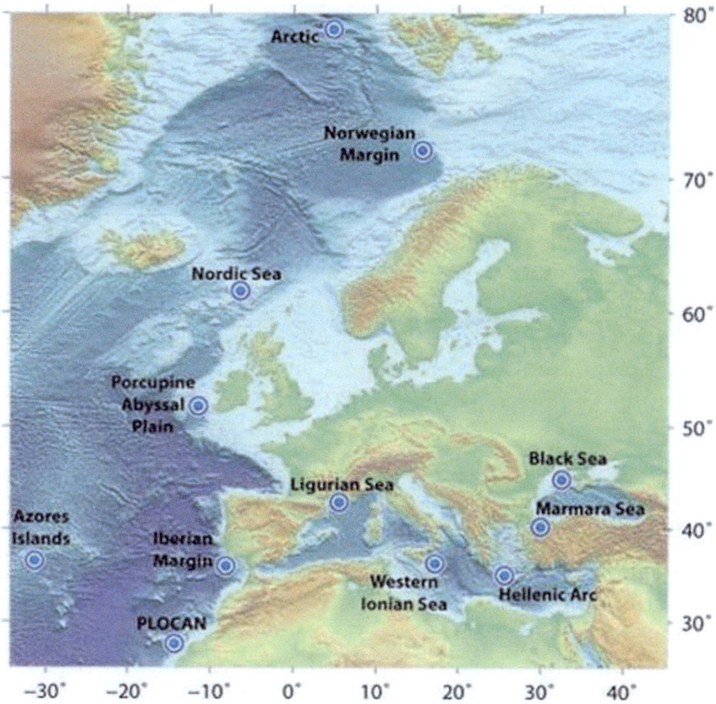

Figure 20.3 The twelve EMSO sites.

20.4.1 Present status of the EMSO nodes

20.4.1.1 Arctic

EMSO observation infrastructure has been established in Hausgarten and Offshore Prins Karls Foreland.

Hausgarten comprises a network of 17 permanent stations covering a watear depth range of 1000–5500m. Repeated sampling and the deployment of moorings and different free-falling systems, which act as observation platforms, has taken place since the summer of 1999. At regular intervals, an ROV is used for targeted sampling, the positioning and servicing of autonomous measuring instruments and the performance of in situ experiments. A 3000m depth-rated AUV further extends the sensing and sampling programs at the node.

Research addresses the study and evaluation of heat and mass transport through the Fram Strait, links between climate, surface ocean processes, deep-sea ecosystems, and methane hydrate dynamics (with co-funding from Statoil-Hydro). Research activities have been also supported by the AOEM ESONET Demonstration Mission, HERMIONE and HYPOX EC projects.

Offshore Prins Karls Foreland is a fluid seep area where methane emissions have been monitored and studied since 2008 through at least one yearly cruise. Scientific studies investigate the possible link between climate-change-induced gas hydrate decomposition and methane release. One hydrographic section has been monitored three times over the last three years, including CH4 analyses. Hydroacoustic surveys with single and multi-beam systems have been repeated yearly. Complementary atmospheric measurements are carried out with Cavity Ring-Down Spectroscopy systems and air sampling.

A phased implementation starting in 2014 will establish the "Fram Array" covering the entire Fram Strait by means of a cabled system (~190km long with 20kV and 4Amp power supply), with the end point at the Hausgarten node. A node will also be installed at the MASOX ESONET Demonstration Mission site, joined to the main Hausgarten cable. Yearly cruises will continue to monitorg seep activity and possible changes. This project is currently under evaluation by the German funding agency.

20.4.1.2 Porcupine Abyssal Plain (PAP)

The PAP EMSO node is a sustained, multidisciplinary observatory in the North Atlantic that has provided key time-series datasets from the 1980s. It includes a single column of strong wire stretching from the sea surface to the seafloor and landers positioned on the seabed. Scientific instruments are attached in clusters like mini constellations all the way down the wire and installed on the landers. The equipment remains deployed for months, transmitting information by satellite or storing information inside the equipment until it is recovered.

Observatory biogeochemical and physical measurements (e.g., temperature, salinity and carbon dioxide) track the effects of climate change on the open ocean and deep-sea ecosystems.

Researchers have been supported within international collaborations such as JGOFS (Joint Global Ocean Flux Study) and BENGAL (High-resolution temporal and spatial study of the BENthic biology and Geochemistry of a north-eastern Atlantic abyssal Locality) and EUROSITES. Subject to funding availability, there is the potential to expand meteorological data buoys (stand-alone moorings) in three sites to monitor cold water coral reefs and slope environments.

20.4.1.3 Azores Islands

The Azores EMSO infrastructure is devoted to the Lucky Strike hydrothermal vent field composed of two nodes (SEAMON East and West). It is acoustically linked to a surface relay buoy (BOREL) to ensure satellite communication with the land base station at IFREMER shore facilities in Brest (France). SEAMON East is dedicated to large-scale geophysical studies in the center of a large lava lake located in the Lucky Strike vent field (Figure 20.4). SEAMON West is deployed at the base of the Eiffel Tower active edifice.

The monitoring infrastructure allows study of geophysical movements of: Earth (seismicity and vertical deformation); water, heat and mineral flow through the vent system; behavior of physical and chemical elements in vent fluids; variations in biogeochemistry and ecological hotspots in the vicinity of vents; links between faunal dynamics; and variations of physico-chemical factors.

Figure 20.4 Systems deployed on the seafloor at the Lucky Strike hydrothermal vent field (Mid-Atlantic ridge); in the background of the figure, one SEAMON array is visible

Scientific activities have been developed under the InterRidge program (Mid-Atlantic Ridge), and MarBEF (Marine Biodiversity and Ecosystem Functioning)-DEEPSETS (Deep-sea & Extreme Environments, Patterns of Species and Ecosystem Time Series), EXOCET-D (EXtreme ecosystem studies in the deep OCEan: Technological Developments), HERMIONE, and CORALFISH (Assessment of the interaction between corals, fish and fisheries, in order to develop monitoring and predictive modeling tools for ecosystem-based management in the deep waters of Europe and beyond) projects among others; and is the site of the MoMAR-D ESONET Demonstration Mission. Yearly maintenance is already scheduled for at least the next five years. A new generation of stations is planned for 2015.

20.4.1.4 Canary Archipelago – PLOCAN

The ESTOC/PLOCAN European station for time-series on the ocean is the most recent node joining EMSO, although its first operation dates back to 1994. It has been part of the EUROSITES network and contributes to BATS (Bermuda Atlantic Time-Series). Ocean physics and biogeochemistry and decadal records of ocean acidification are the main research subjects addressed by the node.

One stand-alone installation at the ESTOC/PLOCAN site and one cabled installation starting with the PLOCAN offshore platform test area (at 100m water depth in a first phase then daisy-chaining to greater depths) are planned and funded by the Regional (Canary) and National governments in Spain.

20.4.1.5 Ligurian Sea

The Ligurian Sea EMSO node is spread over two areas: offshore Nice and Var canyon in the eastern Ligurian Sea and offshore Toulon in the western Ligurian Sea. In the east, the science focus is on atmosphere–sea exchange (DYFAMED – DYnamics of Atmospheric Fluxes in the MEDiterranean Sea), and monitoring of geo-hazards in the Var canyon area and on the Nice unstable slope. In the west, marine geophysical investigations take advantage of submarine cabled monitoring infrastructure developed by the ANTARES program.

The Ligurian Sea is characterized by a wide range of phenomena, such as down welling, coastal upwelling, bioluminescence, particle plumes, nutrient benthic exchange, bottom boundary layer processes, seismicity and slope instabilities. Main projects supporting the previous and recent activities have been JGOFS and EUROSITES.

The planned developments include (a) the deployment of a stand-alone observatory in the eastern Ligurian sea (Var canyon area) starting in 2013, and (b) the cabling of the Nice slope in 2014. The French project MEUST (Mediterranean Eurocentre for Underwater Science and Technology) has the main aim of deploying a second generation shared submarine platform offshore Toulon, within the framework of EMSO and KM3NeT (multi-km3 sized deep-sea Neutrino Telescope).

20.4.1.6 Western Ionian Sea

Since 2005, the area has hosted the first European operational cabled multi-parameter seafloor observatory NEMO-SN1 (Figure 20.5). This comprises a laboratory in the harbor of Catania (land termination of the underwater electro-optical cable), and two data acquisition stations about 25km offshore and at 2100m water depth: a geophysical, oceanographic and bioacoustic observatory and another bioacoustic station. NEMO-SN1 is integrated with land-based networks and transmits real-time seismological data to National Seismological Service Centre managed by INGV in Rome.

Research is focused on: (a) the study and identification of main seismic sources; (b) the generation of tsunami waves and related phenomena (e.g., hydro-acoustic signals; Chierici et al., 2010) by testing a tsunami detection system; (c) the investigation of the deep volcanic activity of Mt Etna; and (d) bioacoustics and ambient noise characterization.

The design, deployment, sustained operation and progressive enhancement of NEMO-SN1 has been supported by projects such as GNDT-SN1 (Italian), LAMS and SIRENA FESR (European Fund for Regional Development), PEGASO (structural funds); and projects funded in the European Framework Programmes, such as ESONET-CA (European Seafloor Observatory NETwork – Concerted Action) and ESONET-NoE (LIDO Demonstration Mission), GENESI-DEC (Ground European Network for Earth Science Interoperations – Digital Earth Community), SCIDIP-ES (SCIence Data Infrastructure for preservation-Earth Science), KM3NeT, NEAREST, TRANSFER (Tsunami Risk ANd Strategies For the European Region) and ENVRI (Common Operations of Environmental Research Infrastructures, cluster of all the ESFRI environmental research infrastructures). This support has, step-by-step, brought the observatory to the international collaboration level.

Further development of NEMO-SN1 will include the hosting of a mooring for physical oceanographic measurements through the water column thanks to the Italian project EMSO-MedIT (structural funds) recently approved in coordination with the FixO3 EC project.

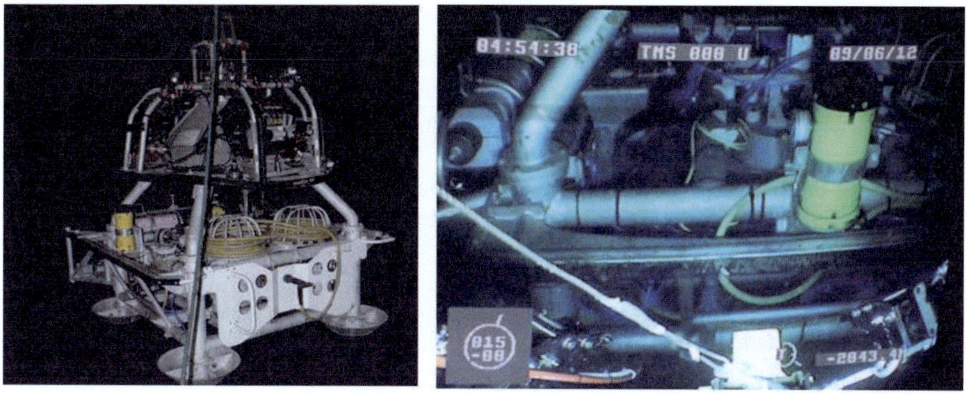

Figure 20.5 NEMO-SN1 multidisciplinary abyssal node during the deployment operations in June 2012. (Left) The observatory ready to be deployed into the water by the deployment/recovery vehicle MODUS. (Right) Node deployed on the seafloor at 2100m water depth (Favali et al., 2013).

The extension of the monitoring area will be achieved by a 100km cable deployed off Capo Passero (south-eastern tip of Sicily) that will become operational in 2014 with new seafloor and water-column observation modules at 3500m water depth.

20.4.1.7 Hellenic Arc

A cabled system for neutrino detection, NESTOR (Neutrino Extended Submarine Telescope with Oceanographic Research), was the first supporting infrastructure for the development of the Hellenic EMSO node. In the area, stand-alone observatories E1-M3A and the Poseidon-Pylos systems, including water-column/seabed seismic and tsunami monitoring systems, were also developed in the framework of EC and national Greek projects; sustained operation is planned in 2013–2014.

Research is driven by the geo-hazards threatening the area (earthquakes, tsunamis and slope instabilities), but climate change issues such as biogeochemical benthic-pelagic interactions and seabed methane fluxes are also being addressed. Constant monitoring of the seabed is also needed to assess the impact and sustainability of oil and gas industry activities.

EUROSITES, HERMES (Hotspot Ecosystem Research on the Margins of European Seas) and HERMIONE, SEAHELLARC (SEismic and tsunami risk Assessment and mitigation scenarios in the western HELLenic ARC), TRANSFER, KM3NeT and FixO3 are EC projects of the 6th and 7th Framework Programmes which have contributed to the acquisition of the necessary skill to develop a permanent cabled infrastructure. This is now funded by the Greek government as a national contribution to EMSO with a budget of €3.7M (structural funds; 2012–2015). The infrastructure, equipped according to the ESONET standards, will provide long-term monitoring at the Poseidon-Pylos site, about 15km offshore and at 1600m water depth.

20.4.1.8 Marmara Sea

The Marmara Sea EMSO node monitors a very crucial area: It is one of the most seismically active zones of the Mediterranean, with high seismic risk affecting the megacity of Istanbul (Görür and Çağatay, 2010). The Marmara Sea is also important from an oceanographic point of view, being the transition between the Black and the Mediterranean Seas. Last but not the least, the area is highly populated and industrialized with consequently heavy environmental problems. Therefore it is essential to establish a seafloor observatory research infrastructure, as node of EMSO, for long-term, real-time observations related to earthquakes, oceanography, climate change and environmental safety. In particular, a continuous multi-parametric monitoring of the North Anatolian Fault which runs beneath the sea aims at obtaining critical knowledge about the processes taking place along the fault during an earthquake cycle. In this regard, the investigation of the relationships between seismicity and fluid activity are of crucial importance, and may possibly lead to the detection of earthquake precursory signals.

As an initial step towards permanent installations, five cabled seismic and oceanographic modules have been installed in the sea as part of the MSBO (Marmara Sea Bottom Observatory) project, managed by KOERI (Kandilli Observatory and Earthquake Research Institute-Boğaziçi University) and funded by Turkish Telecom. A continuation of the activities performed within the ESONET MARMARA-DM will proceed in 2013–2014 with the SN4 seafloor observatory and other stand-alone instruments (OBSs, piezometers, a bubble detector system, etc.) will be redeployed by INGV, IFREMER and ITU in the framework of the EC project MARSITE (New Directions in Seismic Hazard assessment through Focused Earth Observation in the Marmara Supersite).

20.4.1.9 Black Sea

The Black Sea basin is a semi-closed basin confined by active fault systems and high regional tectonics, including high seismicity, and subsequent occurrence of marine natural hazards. In the context of global climate change and of on-going submarine slides in the basin, monitoring and forecasting of marine geo-hazards is of major importance for national authorities in order to adequately plan for protection of population, infrastructure and environment.

Along the Romanian-Bulgarian Black Sea coast, a first nucleus of the EMSO Black Sea permanent seafloor and water-column node has been recently installed thanks to the MARINEGEOHAZARD project funded within the Cross Border Cooperation between Romania and Bulgaria (structural funds; 2010–2013). MARINEGEOHAZARD represents the first major initiative to address in an integrated and coordinated manner a much needed geo-hazard early-warning system for the Black Sea. The main aim of the project is to establish a joint regional early-warning system of natural marine geo-hazards for protection of local communities, environment and assets within the cross-border area. The system consists of five buoys with oceanographic sensors through the water column for water circulation studies, and five seafloor modules for geo-hazard purposes (earthquakes and tsunamis).

Figure 20.6 shows a comprehensive vision of the activities performed at all the EMSO

Figure 20.6 Status of EMSO nodes. Operating nodes are marked by a white star, pictures framed in light blue stay for shallow water sites

nodes, and includes the four shallow water sites used also as test-beds: Koljö Fjord (Sweden), SmartBay (Ireland), MeDON (Marine e-Data Observatory Network) off the island of Molène (France) and OBSEA (Spain).

20.4.2 EMSO data infrastructure

EMSO has an open access policy which requires all data collected by EMSO to be freely accessible. The immediate delivery of data from the oceans directly to the desktops of researchers worldwide turns the observatory network into "gateways to the oceans" accessible to scientists, educators, and the public alike. Consequently, this will increase our understanding of global change, which will lead to socio-economic benefits such as improved climate projections, natural resource management, and human impact mitigation. A European research network of ocean observatories will transform the ability of science to inform European marine government policy and business strategy with a greater certainty and efficiency than has so far been achieved.

The EMSO data infrastructure has been conceived to use the existing distributed network of data infrastructures in Europe and use the INSPIRE and GEOSS data-sharing principles. A number of standards have been set out that will support state-of-the-art transmission and archiving of data with the kinds of metadata recording and interoperability that allow for more straightforward use and communication of data.

In continuation of ESONET efforts, these standards include the Open Geospatial Consortium (OGC) Sensor Web Enablement (SWE) suite of standards, including the OGC standards SensorML, Sensor Registry, Catalogue Service for Web (CS-W), Sensor Observation Service (SOS) and Observations and Measurements (O&M) (Botts et al., 2006).

OGC SensorML is an eXtensible Markup Language (XML) for describing sensor systems and processes. Following on progress from ESONET, EUROSITES, and others, a SensorML profile is being created that can be stored in a so-called Sensor Registry that will act as a catalogue of each EMSO sensor and observatory. This dynamic framework can accommodate the diverse array of data and formats used in EMSO, including the addition of delayed mode data. The main components of the data infrastructure are depicted in Figure 20.7.

Harmonization of data formats is an important task for EMSO, with the cooperation of other marine infrastructures such as EURO-ARGO, in order to enable scientists to effectively make use of the data. It is significant that EMSO faces the greater breadth of data types among this group, and therefore takes the greater initiative in forging this interoperability. An important focus for the data management of these infrastructures was therefore to agree on a common data format (NetDCF), O&M, and to provide the necessary transformation (interface) services. Further, agreement on commonly-used ontologies, especially those for naming the measured variables, had to be reached in cooperation with the marine infrastructures mentioned above. In addition, EMSO has been a leader regarding international standardization and data exchange through intensified cooperation with other environmental infrastructures within an international, multidisciplinary network driven by EC projects such as ENVRI and COOPEUS (cooperation between EU and US research infrastructures, EMSO and OOI are partners in this project). These efforts at standardization and harmonization have made progress across the unprecedented combination of 12 countries, and of course will continue as EMSO nodes have a range of development depending on the status of local activity plans and funding.

Importantly for such an advance in ocean "big data", EMSO is exploiting the power of EGI (European Grid Infrastructure) to create a data infrastructure to serve the wide communities of scientists studying marine mammals, acoustics, oceanography, geology, geophysics, high energy astro-particle physics, and ecology. The distributed computing paradigm of the EU e-infrastructure will be used to provide large CPU and storage capacity. Given this fire hose of "big data", a paradigm shift for many ocean sciences, an important objective is to provide open access and collaborative shared research environments and tools for state-of-the-art analysis approaches.

20.4.2.1 Data management principles

Due to the required distributed sensor infrastructure of EMSO, a distributed data management has been adopted. It is based on already available data infrastructures and archives operated regionally and initially outlined starting with the ESONET-NoE project. This concept and the necessary technological and organizational prerequisites then established continue to be operational for EMSO. The backbone for EMSO data archiving strategy, long-term data preservation, and access is consistent with data centers such as WDC (World Data Centers) and other certified/designated data centers such as the IOC/IODE (IOC/International Oceanographic Data and Information Exchange). The integration of these regional data centers and archives has required considerable effort regarding standardization and harmonization of data, metadata and workflows.

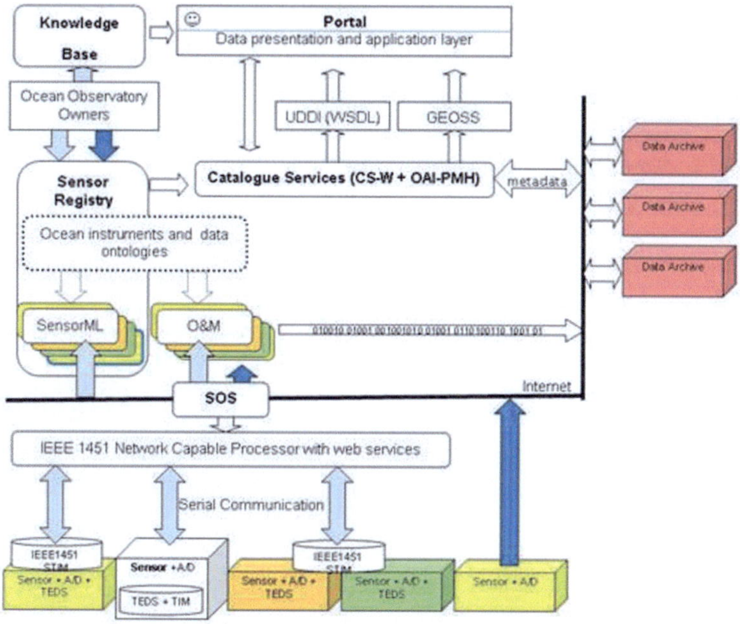

Figure 20.7 Overview of the main standard components used for the data infrastructure of EMSO.

An initial data management plan and data policy for EMSO was established in compliance with the INSPIRE Directive and the European directives regulating access to environmental data, such as 2003/4/EC.

Since the time of the outlining of the ESONET data policy, open access to data has been considered a basic requirement and was strongly favored together with the specification of rules and responsibilities of data providers and data consumers.

Under the ESONET project, the main challenge in data management was to provide a technical architecture based on international standards. Besides common standards for metadata description and exchange such as Open Archives Initiative Protocol for Metadata Harvesting (OAI-PMH) and ISO19139, the core standards of the OGC SWE suite of standards were chosen for implementation. SWE standards represent a generic framework allowing much interpretative freedom. Therefore, to ensure compatibility, specific application profiles have been defined in cooperation with other major European initiatives such as SEADATANET.

Currently, the EMSO information data system is based on the PANGAEA (Data Publisher for Earth & Environmental Science), MOIST (Multidisciplinary Oceanic Information SysTem), and CORIOLIS (in situ data for Operational Oceanography) information data systems, importantly all capable of long-term data archiving services. The above

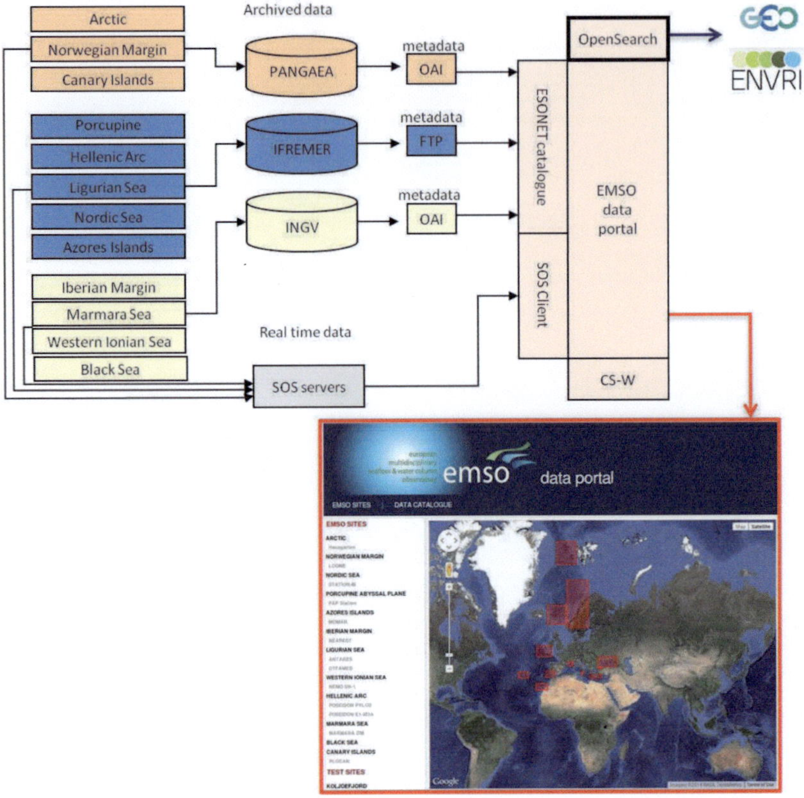

Figure 20.8 The structure of the EMSO data management from the distributed data architecture (left side) to the EMSO data portal (right side).

mentioned standards have been adopted in these systems to implement routines acquiring data from the EMSO nodes. In addition, the data archives provide EMSO data curation procedures including quality control for manual or semi-automatic data uploading into the archive. Each data archive further provides XML metadata files of each data set in either ISO19139 or Global Change Master Directory-Directory Interchange Format (GC-MD-DIF) and offers metadata exchange via the OAI-PMH interface. As quality control and other post-acquisition processing are applied to raw data, these transformed data can be catalogued alongside their raw counterparts.

All the standards described above are also used to integrate the data delivered by the distributed data providers within the ESONET-Vi (community) and EMSO knowledge base which serves as a data portal for the scientific user community as well as the interested public. The data portal offers information about the observatory and sensors through the

ESONET Sensor Registry, and provides access to archived data from the data catalogue, as well as access to real-time EMSO data via SOS interfaces (Figure 20.8). Advanced search query interfaces and simple data visualisation tools allow quick retrieval and analysis of EMSO research data.

The technical platform for the integration of data archives as well as for the public data catalogue is PANFMP (PANGAEA Framework for Metadata Portals). It is a generic and flexible framework for building geoscientific metadata portals, independent of content standards for metadata and protocols (Schindler and Diepenbroek, 2008). PANFMP uses exclusively Open Source Software and, consequently, it was recently released as Open Source and can be downloaded via SourceForge.

The EMSO data catalogue contains data archived by the WDC-MARE (World Data Centre for Marine Environmental Sciences), IFREMER, INGV and HCMR data via the SEADATANET Data Management Infrastructure. Archived data sets cover a broad range of geosciences and include data such as composition of sediments, chemical characterization of sea water and biological information, depending on the scientific focus of the nodes. The majority of data within the data catalogue is available as Open Access data; however for some data sets a moratorium protects the IPR (Intellectual Property Rights) of the responsible scientists as long as detailed investigations are carried out.

20.4.2.2 Interoperability of observatory metadata and sensors

Interoperability is a major challenge for the networking and integration of data and sensors. At the level of technical implementation of ocean observatories, "plug and play" instrument standardization addresses the clear definition of interfaces between the basic building blocks. This is a progressive step in the evolving design of international interoperable systems. In the ESONET project, the IEEE (Institute of Electrical and Electronics Engineers) 1451 standard allows for the envisioned plug and work functionality. Particularly relevant is the possibility for sensors to communicate metadata describing the instrument characteristics in a standard way; the PUCK (Programmable Underwater Connector with Knowledge) MBARI (Monterey Bay Aquarium Research Institute) technology, for instance, can help to automate the configuration process by physically storing information about the instrument within the instrument itself. The information may include metadata descriptions, driver software, GUI (Graphical User Interface) software, or any other information deemed relevant by the observing system. The PUCK standard has been adopted as an OGC standard.

ESONET Sensor Registry: The sensor registry concept, developed out of the work of ESONET-NoE, will form the nucleus of the basic functional blocks of an interoperable EMSO observatory system. In its simplest form it consists of a web interface that allows an observatory operator to insert all necessary metadata to describe system functions. It is necessary that the registry is dynamic in identifying the current status and whereabouts of a given sensor. The prototype sensor registry, from the first definition under the ESONET project, provides a web-based entry module on top of a native XML database (eXist) and additionally offers an OGC catalogue interface to allow information exchange (e.g., with GEOSS). The sensor registry can store observatory and instrument metadata (descriptive data) XML documents according the OGC standard SensorML.

20.5 Conclusions and perspectives

The European scientific and technological community supports the establishment of the EMSO Distributed Research Infrastructure constituted by permanent multidisciplinary seafloor and water-column fixed nodes in the seas around Europe. The implementation of this large-scale research infrastructure has been prepared through the EMSO-PP project and has been developed from the scientific and technical consultations proposed through the ESONET-NoE project.

EU strategy provides the major driver and schedule for EMSO. EMSO will pioneer in delivering multidisciplinary real-time data from the sea, by providing data from the ocean surface through the water column to the benthos and sub-seafloor. It will be partially framed by advancements made on Eulerian (fixed) observatory infrastructures during the ESONET-NoE and EUROSITES projects, and the follow-on project FixO3; but also will compliment broader key infrastructures such as EURO-ARGO, KM3NeT and EPOS (European Plate Observing System).

Indirect effects for the Countries interested in EMSO implementation include scientific and technological innovation, networking opportunities, standardization and interoperability methods and, importantly, the growth of a critical mass of scientists and engineers. Moreover, users of EMSO are not only partners of the nodes or distributed infrastructure, but the entire ESONET-Vi community, which aims at integrating European science by linking geographically scattered complementary research, as well as industrial and governmental elements using data collected by deep-sea nodes. A prognosis from a 2008 ESFRI report predicts that over 70% of the operational costs (personnel, supplies and utilities) end up in the local community as a long-term economic return in all the regions of observatory science. Additionally, the so-called "network effect" generates unexpected new users by making data more readily available and thereby enhancing integration.

Moreover, one of the Core Services of the EMSO-ERIC is defined as the capacity to deliver basic, established, standardized data products and services to International Agencies responsible for Earth monitoring. EMSO-ERIC will therefore not only provide scientific research data but will also supply unprecedented ocean data to COPERNICUS and GEOSS programmes in order to integrate and complement Marine Services of the GMES satellite, sea surface and sub-surface observing systems.

States involved in EMSO will benefit directly from providing services to the distributed research infrastructure during the design and construction/implementation phase. For example, employment opportunities for staff to work on the research infrastructure will be available, and vessels as well as other marine services will be required for maintenance and research to be carried out on a regular basis. EMSO will also be a supporter of SMEs, as innovation will continue to be required to instrument the oceans, a development which is in line with EU priorities.

EMSO research infrastructure will be science driven, based on an up-to-date technology, and service oriented. The participation of the scientific community in the development of the EMSO Science Plan will be boosted by the appointment of members of the EMSO STAC (Scientific and Technical Advisory Committee) by the ESONET-Vi association. This will ensure that technical and management choices are made according to the real need of the scientists.

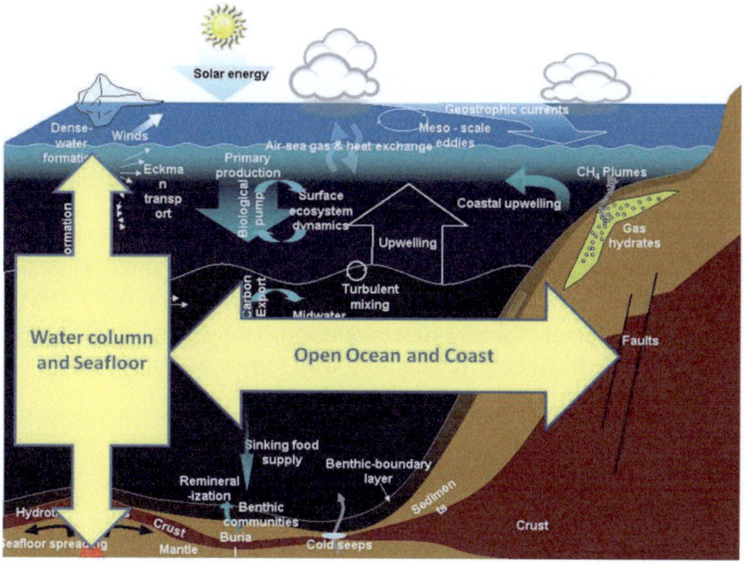

Figure 20.9 Sketch illustrating the idea of infrastructure integration from seafloor to water column, and from open ocean to coast (redrawn from Ruhl et al., 2011).

The EMSO-ERIC operational plan is to constitute a reference European coordination body to perform all the operations necessary to make the observatory nodes scientifically relevant and functional, through a coordinating board and regional technical teams established for each node by the involved ERIC members.

According to an analysis by EMSO-PP, the governance structure of an ERIC could reduce personnel and operating costs by 31% during the construction and by 60% during the operation phase. The centralized management of an ERIC will enhance interoperability and standardization, as well as synchronizing funding of regional observatories. Overall, EMSO-ERIC will increase European competitiveness by representing the counterpart of similar infrastructures at the global scale.

The development and installation of a developing global network of marine research infrastructures will provide an unprecedented and powerful means to understand global physical, biological, chemical, and geological oceanic processes and the complex interactions among geosphere, hydrosphere, biosphere and atmosphere. In the near future, fixed point infrastructures will lead to integration not only from a geographical point of view, connecting all the different infrastructures worldwide, but also looking even more at the ocean as an *unicum,* from seafloor (and sub-seafloor) to the water column and the sea–air interface, and from the open ocean to the coasts, as illustrated in Figure 20.9.

Acknowledgments

We acknowledge the European Commission for its continuous support and funding. We also thank all our colleagues working in the field, often in difficult conditions, and individually acknowledged in most articles of the following bibliography.

References

Auffret Y., Sarrazin J., Sarradin P.M., Coail J.-Y., Delauney L., Legrand J., Dupont J., Dussud L., Guyader G., Ferrant A., Barbot D., Laes A. and Bucas K. (2009) TEMPOMini: A custom-designed instrument for real-time monitoring of hydrothermal vent ecosystems. IEEE OCEANS 2009, Balancing Technology with Future Needs, Bremen, Germany.

Botts M., Percivall G., Reed C. and Davidson J. (2006) OGC Sensor Web Enablement: Overview and high level architecture. Open Geospatial Consortium White Paper, OGC 06-052r2.

Çağatay M.N., Görür N., Geli L., Auffret Y., Rolin J.-F., Henry P., Gasperini L. and Favali P. (2011) The MARDEP Project: The Sea of Marmara Observatory Infrastructure for Multidisciplinary Earthquake and Environmental Research and Monitoring. UT-Underwater Technology Symposium-11, Tokyo.

Çağatay M.N., Geli L., Gasperini L., Henry P., Gürbüz C. and Görür N. (2014) Seafloor observations and observatory activities in the Sea of Marmara (this volume).

Chierici F., Pignagnoli L. and Embriaco D. (2010) Modeling of the hydro-acoustic signal and tsunami wave generated by seafloor motion including a porous seabed. J. Geophys. Res. 115, C03015, doi: 10.1029/2009JC005522.

Chierici F., Favali P., Beranzoli L., De Santis A., Embriaco D., Giovanetti G., Marinaro G., Monna S., Pignagnoli L., Riccobene G., Bruni F. and Gasparoni F. (2012) NEMO-SN1 (Western Ionian Sea, off Eastern Sicily): A cabled abyssal observatory with tsunami early warning capability. ISOPE-Int. Soc. Offshore and Polar Engineers, 22nd Int. Offshore and Polar Engineering Conference, Rhodes, July 17–22, ISBN: 978-1-880653-94-4, ISSN:1098-6189.

Delory E., Hernandez-Brito J. and Llinas O. (2011) The PLOCAN Observatory: A multidisciplinary multiplatform observing system for the Central-Eastern Atlantic Ocean. IEEE OCEANS 2011, doi: 10.1109/Oceans-Spain.2011.6003593.

Embriaco D., Marinaro G., Frugoni F., Monna S., Etiope G., Gasperini L., Polonia A., Del Bianco F., Çağatay M.N., Ulgen U.B. and Favali P. (2013) Monitoring of gas and seismic energy release by multiparametric benthic observatory along the North Anatolian Fault in the Sea of Marmara (NW Turkey). *Geophys. J. Int.* doi: 10.1093/gji/ggt436.

Favali P. and Beranzoli L. (2009) EMSO: European Multidisciplinary Seafloor Observatory. Nuclear Instruments and Methods in Physics Research Section A: Accelerators,

Spectrometers, Detectors and Associated Equipment 602, 21–27, doi: 10.1016/j.nima.2008.12.214.

Favali P., Beranzoli L., D'Anna G., Gasparoni F., Marvaldi J., Clauss G., Gerber H.W., Nicot M., Marani M.P., Gamberi F., Millot C. and Flueh E.R. (2006) A fleet of multiparameter observatories for geophysical and environmental monitoring at seafloor. Ann. Geophys. 49(2/3), 659–680.

Favali P., Person R., Barnes C.R., Kaneda Y., Delaney J.R. and Hsu S.-K. (2010) Seafloor observatory science. In: J. Hall, D.E. Harrison and D. Stammer (Eds), OceanObs'09: Sustained Ocean. Observations and Information for Society Conference 2, Venice, Italy, 21–25 September 2009, ESA Publication WPP-306, ISSN: 1609-042X, doi: 10.5270/OceanObs09.cwp28.

Favali P., Chierici F., Marinaro G., Giovanetti G., Azzarone A., Beranzoli L., De Santis A., Embriaco D., Monna S., Lo Bue N., Sgroi T., Cianchini G., Badiali L., Qamili E., et al. (2013) NEMO-SN1 Abyssal Cabled Observatory in the Western Ionian Sea. IEEE J. Oceanic Engineering 38(2), 358–374, doi: 10.1109/joe.2012.222.4536.

Gasperini L., Polonia A., Del Bianco F., Etiope G., Marinaro G., Favali P., Italiano F. and Çağatay M.N. (2012) Gas seepage and seismogenic structures along the North Anatolian Fault in the Eastern Marmara Sea. Geochem. Geophys. Geosyst. 13(10), Q10018, ISSN: 1525-2027, doi: 10.1029/2012GC004190.

Görür N. and Çağatay M.N. (2010) Geo-hazards rooted from the northern margin of the Sea of Marmara since the late Pleistocene: A review of recent results. Natural Hazards 54(2), 583–603.

Lampitt R.S., Favali P., Barnes C.R., Church M.J., Cronin M.F., Hill K.L., Kaneda Y., Karl D.M., Knap A.H., McPhaden M.J., Nittis K.A., Priede I.G., Rolin J.-F. et al. (2010) In situ sustained Eulerian observatories. In: J. Hall, D.E. Harrison and D. Stammer (Eds), OceanObs'09: Sustained Ocean. Observations and Information for Society Conference 2, Venice, Italy, 21–25 September 2009, ESA Publication WPP-306, ISSN: 1609-042X, doi: 10.5270/OceanObs09, p. 27.

Monna S., Cimini G.B., Montuori C., Matias L.M.M., Geissler W.H. and Favali P. (2012) New insights from seismic tomography on the complex geodynamic evolution of two adjacent domains: Gulf of Cadiz and Alboran Sea. J. Geophys. Res. 118, 1–15, doi: 10.1029/2012JB009607.

Monna S., Falcone G., Beranzoli L., Chierici F., Cianchini G., De Caro M., De Santis A., Embriaco D., Frugoni F., Marinaro G., Montuori C., Pignagnoli L., Qamili E., Sgroi T. and Favali P. (2014) Underwater geophysical monitoring for European multidisciplinary seafloor and water-column observatories. J. Marine System, doi: 10.1016/j.jmarsys.2013.09.010.

NRC (National Research Council) (2000) *Illuminating the Hidden Planet. The Future of Seafloor Observatory Science*. National Academy Press, Washington D.C., p. 135.

Person R., Beranzoli L., Berndt C., Dañobeitia J.J., Diepenbroeck M., Favali P., Gillooly M., Lykousis V., Miranda J.M.A., Mienert J., Priede I.G., Santos R.S. et al. (2008) ESONET: A European Sea Observatory Initiative. OCEANS 2008, MTS/IEEE Kobe Techno-Ocean, 1–6, doi: 10.1109/OCEANSKOBE.2008.4531109.

Puillat I., Person R., Leveque C., Drogou J.-F., Diepenbroek M., Garreau P., Waldmann C. and Auffret Y. (2009) Standardisation prospective in ESONET-NoE and a possible implementation on the ANTARES site. Nuclear Instruments and Methods in Physics Research Section A 602, 240–245, doi: 10.1016/j.nima.2008.12.214.

Puillat I., Lanteri N., Drogou J.-F., Blandin J., Géli L., Sarrazin J., Sarradin P.M., Auffret Y., Rolin J.-F. and Léon P. (2012) Open-sea Observatories: A New Technology to Bring the Pulse of the Sea to Human Awareness. In: M.Marcelli (Ed.), Oceanography. InTech. ISBN: 978-953-51-0301-1. Available from: http://www.intechopen.com/books/oceanography/open-sea-observatories-a-new-technology-to-take-the-pulse-of-the-sea-with-internet-in-the-ocean.

Ruhl H.A., André M., Beranzoli L., Çağatay M.N., Colaço A., Cannat M., Dañobeitia J.J., Favali P., Géli L., Gillooly M., Greinert J., Hall P.O.J., Huber R., Karstensen J. et al. (2011) Societal need for improved understanding of climate change, anthropogenic impacts, and geo-hazard warning drive development of ocean observatories in European seas. Progress in Oceanography 91, 1–33, doi: 10.1016/j.pocean.2011.05.001.

Schindler U. and Diepenbroek M. (2008) Generic XML-based framework for metadata portals. Computers & Geosciences 34(12), 1947–1955, doi: 10.1016/j.cageo.2008.02.023.

Tary J.B., Géli L., Guennou C., Henry P., Sultan N., Çağatay M.N. and Vidal V. (2012) Micro-events produced by gas migration and expulsion at the seabed: A study based on sea bottom recordings from the Sea of Marmara. Geophys. J. Int., doi: 10.1111/j.1365-246X.2012.05533.x.

Tryon M.D., Henry P., Çağatay M.N., Zitter T.A.C., Géli L., Gasperini L,. Burnard P., Bourlange S. and Grall C. (2010) Pore fluid chemistry of the North Anatolian Fault Zone in the Sea of Marmara: A diversity of sources and processes. Geochem. Geophys. Geosyst. 11, doi: 10.1029/ 2010GC003177.

Vuillemin R., Le Rouxv, Dorval P., Bucas K., Sudreau J.P., Hamon M., Le Gall C. and Sarradin (2009) CHEMINI: A new *in situ* CHEmical MINIaturized analyzer. Deep Sea Research Part I: Oceanographic Research Papers 56(8), 1391–1399, doi: 10.1016/j.dsr.2009.02.002.

Yu Y., Xu H., Xu C. and Qin R. (2012) A study of the remote control for the East China Sea seafloor observation system. J. Atmos. Oceanic Technol. 29, 1149–1158, doi: dx.doi.org/10.1175/JTECH-D-11-00115.1.

Websites

ANTARES	http://antares.in2p3.fr
AOEM (ESONET DM)	http://www.esonet-noe.org/Demonstration-missions/AOEM
ARCOONE (ESONET DM)	http://www.esonet-noe.org/Demonstration-missions/
AOEM/ARCOONE	
AWI	http://www.awi.d
BATS	http://bats.bios.edu/
BENGAL	http://en.wikipedia.org/wiki/BENGAL_%28project%29
COOPEUS	http://www.coopeus.eu/
COPERNICUS	
(formerly GMES)	http://copernicus.eu
CORALFISH	http://eu-fp7-coralfish.net/
CORIOLIS	http://www.coriolis.eu.org
DONET	http://www.jamstec.go,jp/donet/
DYFAMED	http://www.obs-vlfr.fr/sodyf/
EC	http://ec.europa.eu
EGI	http://www.egi.eu/
EMSO	http://www.emso-eu.org
ENVRI	http://envri.eu/
EPOS	http://www.epos-eu.org/
ESFRI	http://ec.europa.eu/research/infrastructures/index_en.cf-m?pg=esfri
ESONET-CA	http://www.oceanlab.abdn.ac.uk/esonet/ESONET_fullrep.
pdf	
ESONET-NoE	http://www.esonet-emso.org/esonet-noe/
EURO-ARGO	http://www.euro-argo.eu/
EUROFLEETS	http://www.eurofleets.eu/
EUROSITES	http://www.eurosites.info
EXOCET-D	http://www.ifremer.fr/exocetd/
$FixO^3$	http://www.fixo3.eu/
GENESI-DEC	http://www.genesi-dec.eu/
GEO	http://www.earthobservations.org/
GEOSS	http://www.earthobservations.org/geoss.html
HCMR	http://www.hcmr.gr
HERMES	http://www.eu-hermes.net/
HERMIONE	http://www.eu-hermione.net/
HYPOX	http://www.hypox.net/
ICG/NEAMTWS	http://neamtic.ioc-unesco.org/neamtws
ICOS	http://www.icos-infrastructure.eu/
IEEE	http://www.ieee.org/
IFREMER	http://www.ifremer.fr
IMI	https://www.marine.ie/
IMOS	http://www.imos.org.au

INGV	http://www.ingv.it
INSPIRE	http://inspire.jrc.ec.europa.eu/
IOC	http://ioc-unesco.org/
IODE	http://www.iode.org/
IPCC	http://www.ipcc.ch/
IPMA	https://www.ipma.pt
ITU	http://www.itu.edu.tr
JAMSTEC	http://www.jamstec.go.jp/
JGOFS	http://ijgofs.whoi.edu/
KDM	http://www.deutsche-meeresforschung.de
KM3NeT	http://www.km3net.org
KOERI	http://www.koeri.boun.edu.tr/eng/topeng.htm
LIDO (ESONET DM)	http://www.listentothedeep.org
LOOME (ESONET DM)	http://www.esonet-noe.org/Demonstration-missions/
LOOME	
MACHO	http://macho.ncu.edu.tw
MarBEF-DEEPSETS	http://www.marbef.org/projects/deepsets/index.php
MARINEGEOHAZARD	http://www.geohazard-blacksea.eu/
MARMARA-DM (ESONET DM)	http://www.esonet-noe.org/Demonstration-missions/Marmara
MARSITE	http://marsite.eu/
MASOX (ESONET DM)	http://www.esonet-noe.org/Demonstration-missions/
AOEM/MASOX	
MBARI	http://www.mbari.org/
MEUST	marwww.in2p3.fr/rubrique.php3?id_rubrique=259&id_
parent=7&lang=en	
MODOO (ESONET DM)	http://www.modoo.info/
MOIST	http://moist.rm.ingv.it
MoMAR-D (ESONET DM)	http://www.esonet-noe.org/Demonstration-missions/MoMAR
NEAREST	http://nearest.bo.ismar.cnr.it
NEPTUNE (ONC Network)	http://www.neptunecanada.com/
NESTOR	http://www.nestor.org.gr/
NIOZ	http://www.nioz.nl
NOCS	http://noc.ac.uk
OBSEA	http://www.upc.edu/cdsarti/OBSEA/info
OCEANSITES	http://www.oceansites.org
ONC	http://www.oceannetworks.ca
OOI	http://www.oceanleadership.org
OSPAR	http://www.ospar.org/
PLOCAN (EMSO site)	http://www.plocan.eu/es/
PANGEA	http://www.pangaea.de
PANFMP	http://www.panfmp.org
SCIDIP-ES	http://www.scidip-es.eu/

SEADATANET	http://www.seadatanet.org
SEAHELLARC	http://www.seahellarc.gr/
SmartBay (EMSO shallow water site)	http://www.smartbay.ie/
TRANSFER	http://www.transferproject.eu/
UGOT	http://www.gu.se/
UiT	http://uit.no/
UNABN	http://www.abdn.ac.uk/
UNEP	http://www.unep.org/
UNESCO	http://en.unesco.org/
VENUS (ONC Network)	http://www.venus.uvic.ca
WDC-MARE	http://www.wdc-mare.org/

Part IV

Relevant scientific results with a multidisciplinary emphasis

21 Seafloor observatory for monitoring hydrologic and geological phenomena associated with seismogenic subduction zones

H. Mikada[1] and J. Kasahara[2]

21.1 Introduction

Recently, our seismological knowledge about source mechanisms of plate-boundary earthquakes has revealed the possibility in the detection of spatiotemporal changes in physical properties that may occur at interplate boundaries. Seismological interest is identified to be in seismogenic zones that are deeper than several kilometers from the surface. Although scientific drillings are conducted to identify what is the real cause of physical property changes in the seismogenic zones, there still is a scaling gap between seismological and geological approaches towards the understanding of seismogenic processes. On the other hand, the role of interstitial fluids is studied in much larger spatiotemporal scale. The Nankai Trough, located in the southwestern offshore of Japanese Islands, is the place where the Philippine Sea Plate (denoted as PSP hereafter) begins subducting beneath the Japanese Islands and megathrust earthquakes take place along a low angle interplate boundary (Ando, 1975). The ocean seafloor is a huge conveyer belt that carries sediment layers on its top. At the Nankai Trough, the upper part of the sediments are scraped off to form an accretionary prism zone attached to the overhung island arc materials, while the lower part of the sediments subducts with oceanic basement deep into seismogenic zone (Moore et al., 2001). Fluids included in subducting sediments or in the lower part of sediments on top of oceanic basement are squeezed out and influence the seismogenic processes in conjunction with geochemical dehydration of clay minerals (Hyndman et al., 1995).

1 Department of Civil and Earth Resources Engineering, Kyoto University, Kyoto, Japan

2 Shizuoka University, Shizuoka, Japan

Several scientific drilling experiments were conducted at the Nankai Trough for the purpose of understanding the deformation process and physical properties of deformed sediment layers (Mikada et al., 2002a). Since the drillings were only conducted near the toe or in the shallow part of the Nankai accretionary prism due to the limitations in drilling capability, the interpretation of observational results is for the development of low-angle thrust faults and unconsolidated sedimentary materials at the initial stage of accretionary prism formation. Nonetheless, the role of fluids in piled-up hemipelagic sediments is supposed to be a key to understanding geophysical and geochemical processes underway near the Nankai seismogenic zone (Bangs et al., 2004; Moore et al., 2005; Ienaga et al., 2006). Significantly, the freshening of interstitial water near the décollement was observed not only at the Nankai Trough but at least two other subduction zones, Costa Rica and Barbados, and is considered as scientific evidence for deep-sourced fluid production or dehydration processes in seismogenic zones (Kastner et al., 1993; Brown et al., 2001; Bangs et al., 2004; Moore et al., 2005). In 2001, two long-term hydrogeological observation stations (Advanced CORK; CORK is abbreviation of Circulation Obviation Retrofit Kit; hereafter, it is abbreviated as ACORK) were set up at about 200 kilometers off Cape Muroto (Mikada et al., 2002b). One station is installed about 11 kilometers landward of the Nankai deformation front, while the other is at about 2 kilometers on the seaward side. Presently, these ACORK stations, as well as the other CORK stations, are the only long-term observatories to monitor formation pressures as a function of time in the seismogenic processes at the Nankai Trough. Gulick and Bangs (2004) analyzed seismic reflection profiles and logging data at the frontal thrust zone in the Nankai accretionary prism and discussed that there could be free gas migrating up the fault zones and possibly increased concentrations of hydrate above the base of the hydrate stability zone. Davis et al. (2006) focused on the pressure change observed by the ACORK stations and concluded that non-seismic slip events taking place near the updip limit of seismogenic zone seem to be responsible for the change in formation pressure because of the time synchronization of pressure change with the occurrence of nonseismic events. There seem to be hydrogeological activities that are ongoing, such as fluid circulation, microbial reactions, etc., in the accretionary prism.

When megathrust earthquakes are generated, there should be changes in the seismogenic zone that might appear as thermodynamic changes or geochemical reactions at the inter-plate boundaries (Brown et al., 2001; Moore and Saffer, 2001; Saffer and Bekins, 1998). Since the role of formation and deep-sourced fluids needs to be investigated in detail with respect to the earthquake generation, we need to monitor state variables that describe the relationship with earthquake generation mechanisms by means of long-term observation. Unfortunately, the means to access the deep abyssal environment is limited so the availability of data indicating fluid circulation and related observation remains low. It is observed that the accumulation of crustal stress progresses (Savage, 1995; Ozawa et al., 1999), and long-term observations supported by hypotheses on earthquake generation should be commenced, with enough leading time to avoid any aliasing, before any megathrust earthquakes take place at the plate boundary.

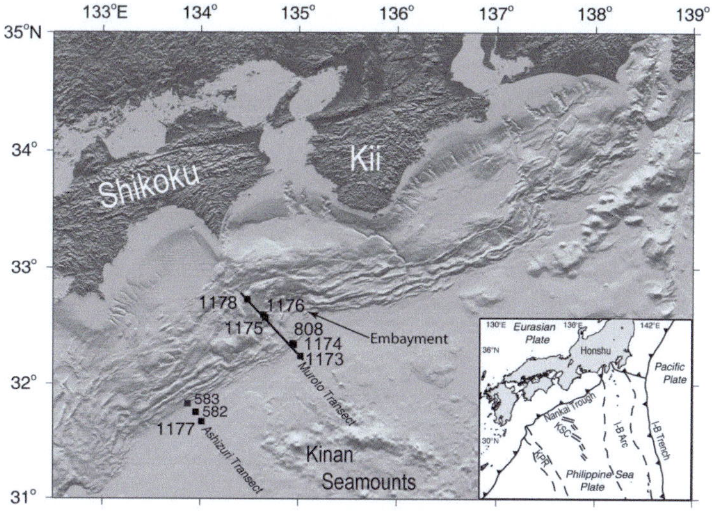

Figure 21.1 Site locations of scientific drilling during Legs 190 and 196 in the Ocean Drilling Program (Mikada et al., 2002; Moore et al., 2005). Numbers indicate the ODP site numbers. The Philippines Sea Plate is subducting in a northwestern direction beneath the Japanese Islands. Two formation pressure monitoring observatories were installed at Sites 808, at the toe of the accretionary prism, and 1173, at a seaward site of the Nankai trough. I-B, KPR, and KSC depict Izu-Bonin, Kyushu-Palau Ridge, and Kinan Seamount Chain, respectively.

21.2 Geophysical and geological settings at the ACORK stations

Figure 21.1 depicts the locations of the drilled sites during legs 190 and 196 of the Ocean Drilling Program (Moore et al., 2005). Figure 21.2 is the interpreted cross-section of the sites shown in Figure 21.1. PSP is subducting at the rate of about 4cm a year in the direction of northwest 310°N beneath the Japanese Islands (Ando, 1975; Sagiya and Thatcher, 1999). The latest GPS observations suggest that the interplate boundary between subducting PSP and overhung Japanese Islands seems completely locked (Mazzotti et al., 2000). The Kinan Seamount Chain on the Shikoku Basin is a fossil axis of seafloor spreading that ceased about 15 Ma (Okino et al., 1994). Two ACORK observatories were installed at Sites 1173 and 808 as indicated by these site numbers in Figure 21.1. The sediment thickness at these sites is about 700m and 1200m, respectively, on top of the subducting basaltic basement (Mikada et al., 2002a). At Site 1173, the ocean basaltic basement is overlain by volcaniclastic facies, massive hemipelagic mudstones of the lower Shikoku Basin facies, then younger hemipelagic mudstones with thin tephra layers of upper Shikoku Basin facies (Moore et al., 2001). Site 1173 is located where no thrust fault is developed due to subduction, while 808 occurs in the middle of the deformation zone (Figure 21.2). The thrust faults develop in the accretionary prism that is landward of the deformation front. A zone of transparent layers, i.e., part of lower Shikoku Basin deposits, is clearly visible

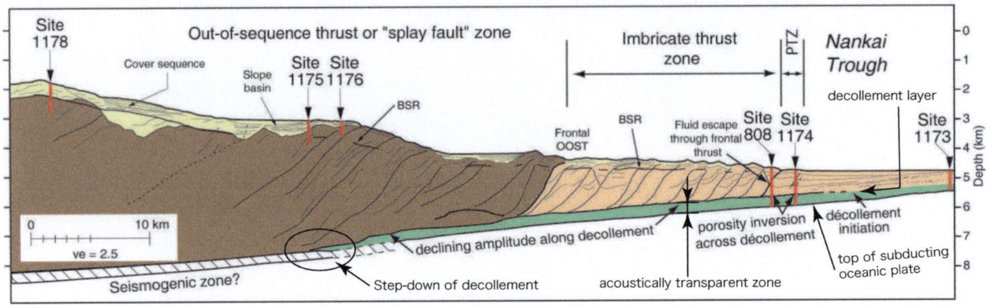

Figure 21.2 Interpreted cross-section along the Muroto transect seismic line shown in Figure 21.1 (modified from Mikada et al., 2002). PTZ and BSR depict protothrust zone and bottom-simulating reflector, respectively. Major observations pertinent to subduction of the oceanic plate and development of décollement are shown in the figure. Seismic reflections show polarity reversal at the décollement layer and at some thrust faults. As the décollement approaches the updip limit of the seismogenic zone, reflection amplitudes from the décollement decline and the step-down of the décollement layer takes place, disappearing or to becoming undistinguishable from the top of oceanic basement. Acoustically transparent zone between the top of oceanic basement and the décollement layer becomes invisible near the updip limit of the seismogenic zone in the figure.

as an acoustically transparent zone between the top of subducting ocean basaltic basement and the décollement layer, i.e., a strong seismic reflector (Mikada et al., 2002a). The log interpretation and the core analysis to the transparent zone, i.e., the lower part of the lower Shikoku Basin facies below the décollement layer, found the porosity reaches a maximum of 40%. These underthrust sediments subduct with the ocean basaltic basement and become seismically invisible in the vicinity of the updip limit of seismogenic zone (Mikada et al., 2002a). High porosity might reflect both anomalous high formation pressure and low permeability of the lower Shikoku Basin facies that results in low acoustic impedance (Tsuji et al., 2005).

Underthrust sediments of the lower Shikoku basin below the décollement seem flattened out in the vicinity of the updip limit of the seismogenic zone. Saffer (2003) compared the sediment materials from some subduction zones, and found that the permeability values for sediments in the Nankai Trough and in Barbados are lower than that in Costa Rica. On the other hand, he found that the accumulation rate of the formation fluid pressures in Nankai and in Barbados is smaller than that of Costa Rica though the sedimentation rates for Nankai and Barbados are larger than that for Costa Rica. This fact suggests that formation fluids are discharged efficiently in Nankai and Barbados under the environment of higher burial rate of low-permeable sediment materials. There is a possibility that the efficient discharge of fluids takes place through the décollement and a series of thrust faults in the accretionary prism in the Nankai Trough.

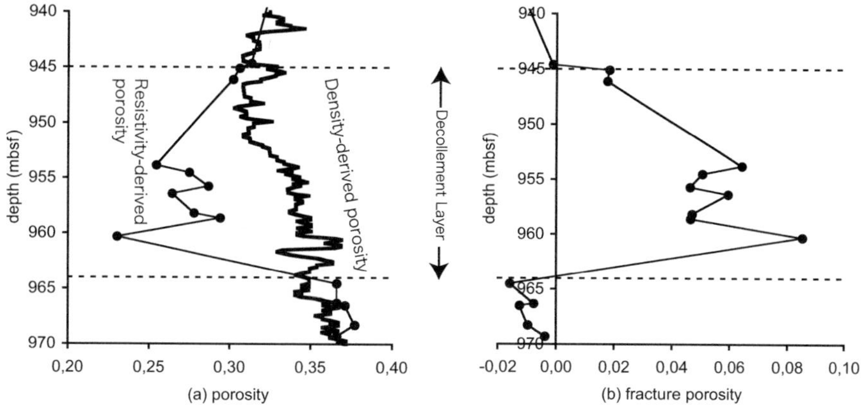

Figure 21.3 Porosities estimated for pores (a) and fractures (b) around the décollement layer at Site 808 (Bourlange et al., 2003). Vertical axis denotes depth in meters below seafloor (mbsf). Solid curve in (a) depicts both density- and resistivity-derived porosities. Resistivity-derived porosity varies between 0.23 and 0.35, while density-derived varies from 0.30 to 0.35. Discrepancies between resistivity- and density-derived porosities are attributed to fracture porosities which augment the permeability. Fractures could also change the permeability depending on fluid pressure. Negative values in (b) indicate estimation error. (Reprinted from Bourlange et al. (2003) with permission from Elsevier.)

Moreover, numerical simulations of fluid movement in the Nankai accretionary prism suggest that it is necessary to understand quantitatively the fluid movement in a spatial scale from the toe of accretionary prism to the seismogenic zone to discuss seismogenic processes including deep fluid production. Bourlange et al. (2003) revisited petrophysical parameters such as porosity, permeability, formation density, cation exchange capacity, and constants in Archie's Law (Archie, 1942) that relate petrophysical properties of porous media with apparent electrical resistivity measurements using data from Logging-While-Drilling that was conducted during Leg-196 at Site 808 and found that fracture porosities become locally several percent in the décollement zone (Figure 21.3). Since fractures in the décollement could contribute to high permeability values especially when fluid pressure rises, they concluded that efficient fluid movement could take place intermittently depending on the fluid pressure. Stauffer and Bekins (2001) also conducted a numerical simulation of fluid movement in the vicinity of the décollement zone in the Barbados subduction zone, and suggested that there are two types of fluid movement: a steady flow and an intermittent flow. They mentioned that the intermittent flow is important to explain quantitatively the freshening of formation water by deep-sourced fluids.

Newberry et al. (2004) tried a genetic analysis using PCR (polymerase chain reaction) amplification for microorganism from core samples acquired at Sites 1173, 1174, 1176, and 1177, indicated in Figure 21.1. They found limited diversity of methane-producing microorganisms with only three gene clusters identified within the and being dominant at

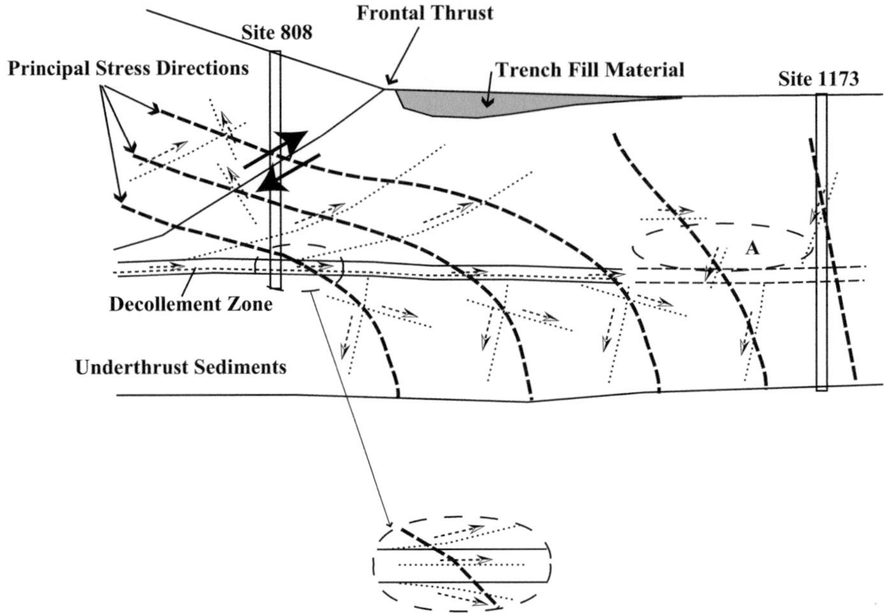

Figure 21.4 Relationship with observed fracture dips at Sites 808 and 1173 (Ienaga et al., 2006). Arrows indicate the inclinations of fracture and slip to be developed by slanted principal stress. Due to heavy mass loading of the prism, vertical normal trend of compaction to the sediments is distorted towards the prism. It is qualitatively indicated that thrust type and horizontal fracturing tend to be formed above and in the décollement zone, respectively. A blow-up image of the décollement zone in the bottom implies that tensile fractures may be developed at the top and base of the décollement zone. An area indicated by "A" denotes a possible zone of tensile field due to spatial changes in principal stress directions. It is necessary to consider possible stress distributions for fluid migrations. (Reprinted from Ienaga et al (2006) with permission from Elsevier.)

all sites. They also found a peculiar decrease in the number of microorganisms at about 400–500m below the seafloor at Sites 1173 and 1174, reflecting high heat flow at these sites. At Site 1176 in the middle of out-of-sequence-thrust zone, members of the genus, which live in general in 48°–73°C environments, are found. Since in-situ temperatures at Site 1176 range between 1°–12°C, the populations may be introduced by past hydrothermal activity or cells that were recently introduced through interstitial fluid flow and not be metabolically active in-situ (Kormas et al., 2003). McNeill et al. (2004) found a principal stress orientation in the same direction of plate convergence using borehole breakout data and Ienaga et al. (2006) found varying dip angles of principal stress axis as a function of locations in the direction of plate convergence. These observations indicate the existence of complex stress orientations in the accretionary prism that could influence the fluid circulations. The depth of penetration by our science drilling is still limited, the relationship of microbes with hydrothermal activity or development of the accretionary prism needs to be discussed in detail in the future when enough data have become available.

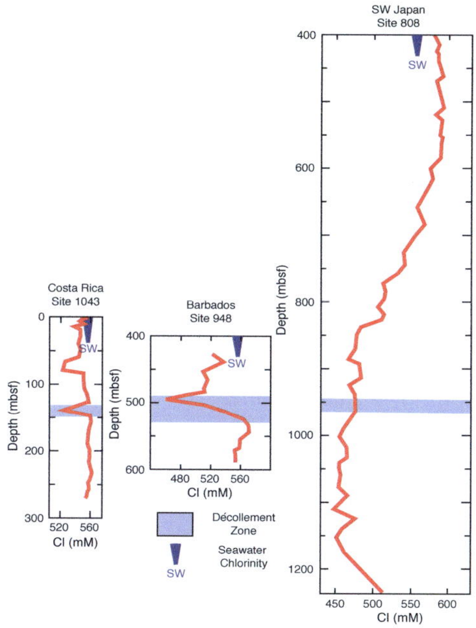

Figure 21.5 Chloride anomalies observed at Costa Rica, Barbados, and Nankai subduction zones (Moore and Silver, 2002). Unlike the chloride anomalies limited in the vicinity of décollement layers for the former two subduction zones, a wide depth zone of anomalous chlorinity was observed at the décollement layer at the Nankai Trough. This observation denotes a relationship between the difference in the chlorinity and that of fluid migrations.

21.3 Freshening observed at the décollement

Moore and Silver (2002) summarized the results from scientific drilling in accretionary prisms conducted at three major seismogenic zones. The décollement zones are identified as a set of layers dominated by horizontal fractures. Although the décollement zones have been recognized as fluid paths, core analyses show the porosity at about 25%, and the permeability ranging 10^{-19}-10^{-18} m^2 for samples from the décollement zones. Also, freshening of formation water was observed in the vicinity of all the décollement zones of the three seismogenic zones (Figure 21.5). In particular, broadening of freshening zone above and below the décollement was observed in the Nankai seismogenic zone. The cause of freshening was suggested as the mixture of deep-sourced fluids that passed through the décollement (Kastner et al., 1993; Saffer and Bekins, 1998).

Saffer and Bekins (1998) did a numerical simulation of the fluid movement from the Nankai seismogenic zone. Their model is based on an assumption that low-chloride fluid is deep-sourced, i.e., produced by the dehydration process (smectite-illite phase transition) in the clay minerals in the sediments near the seismogenic zone with the produced fluids transported through the décollement zone. Even after considering the permeability of 10–16m^2 for the décollement zone, the steady flow, however, is not sufficient to explain the observed freshening in the Nankai Trough and an intermittent supply of low salinity fluid through the décollement that has at least 10^{-13}m^2 of the permeability becomes necessary. They also discussed that at least 20% of weight percent of smectite needs to be included in the sediments subducting into the seismogenic zone and estimated that the intermittent flow took place 80 to 160 years before the measurements of salinity.

Bourlange et al. (2003) pointed out that it is necessary to have the intermittent flow through the décollement at 10^{-12}m^2 permeability in order to explain the pressure dissipation in the low permeable sediments above and below the décollement . They concluded that it is necessary to have a century-scale period to defuse a sudden pressure increase due to the intermittent flow.

Both Saffer and Bekins (1998) and Bourlange et al.(2003) concluded that the intermittent flow is responsible for the most part of freshening of formation fluids in the sediments near the décollement and that the relaxation of pressure causing the last intermittent flow took place in a period of similar duration since the last megathrust earthquake at the Nankai Trough. Jouniaux et al. (1994) confirmed that the permeability of cores, which were sampled during ODP Leg-131, varies with the confined and fluid pressure of the samples. Morgan and Ask (2004) also conducted a laboratory experiment on the cores sampled near the décollement layer during ODP Leg-191 in terms of the stress history of samples and found that the samples over the décollement have experienced in the past the maximum stress of ca. 2.8 times the present stress. Since their observation on the samples above the décollement show static pressure stress due to the sedimentation, stress accumulation near the décollement may be caused by an intermittent augmentation of fluid pressure in the décollement zone. The existence of both steady and intermittent flows in the sediments could be verisimilitude.

Ge and Screaton (2005) tried numerical simulations of intermittent fluid flow during a megathrust earthquake using permeability obtained from core samples of the sediments in the Nankai Trough. Their results indicate the possibility to detect pressure changes at the toe of accretionary prism through the fluid path of the décollement even if the origin of pressure changes are originally caused at a depth of 10km below the surface in the seismogenic zone. They also discussed that the pressure dissipation by intermediate fluid paths intersecting the décollement could lower the signal strength of pressure. In both cases, long-term observations are needed to see how fluid pressure fluctuates during a megathrust earthquake to deepen the present knowledge on fluid circulation in the accretionary prism and its relation to megathrust earthquakes.

21.4 Methane hydrate in accretionary prism

Gulick and Bangs (2004) demonstrated that thrust faults in the accretionary prism near the Nankai Trough are strong seismic reflectors below the bottom simulating reflectors (Figure 21.6). Mikada et al. (2002a) indicated that the log data for a hole penetrating one of the thrust faults show typical gas effects, i.e., abrupt changes such as lower density and

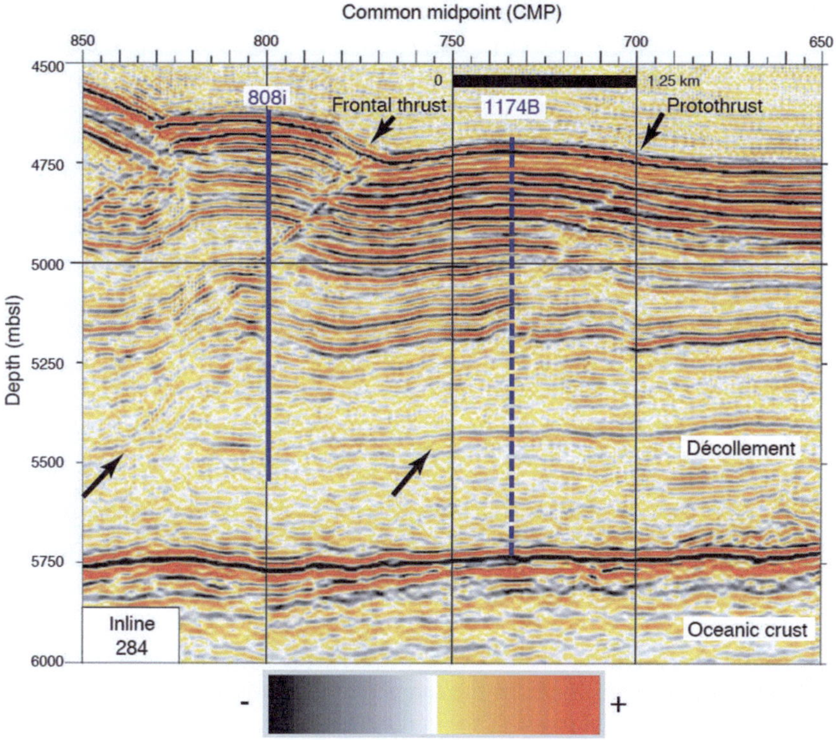

Figure 21.6 Seismic cross section at the toe of the accretionary prism near the Nankai Trough (Gulick and Bangs, 2004). Some parts of the thrust faults appear as negative and strong seismic reflectors below BSR. Also physical property data across the thrust faults at such negative reflectors indicate the existence of low density and high resistivity material inclusion in fluids along the thrust faults.

higher resistivity compared to the surrounding formations. These findings strongly indicate that fluids including gaseous phases migrate along the thrust faults upward to the bottom simulating reflectors or to the seafloor. Suppose the gaseous phases are of methane, they could be deep sourced. Methane hydrates deposits are classified as end-member regimes based on the mechanism that controls gas transport into a methane hydrate stability zone (Trehu et al., 2006).

Brie (1997) demonstrated that the volumetric fraction of gas could be estimated by phase velocity variation of sonic logs. Since abrupt changes in the phase velocity occur when fluids in pore space include gaseous phases, one should relate the magnitude of such changes with the volumetric fraction of gas. His analyses indicate that the change in phase velocity for sonic frequencies is not so drastic to the fraction compared to that for seismic frequencies. He finally used both Vp/Vs ratio and the change in the compressional phase velocity to quantitatively estimate the volume fraction of gaseous phases in the formation fluids. Mikada et al. (2008) analyzed anomalous high sonic velocity observed by sonic logging in free gas zones in the sediments of the eastern and central Nankai Trough and found that the anomalous high velocity could be caused by the inclusion of small methane bubbles in formation water. Banno et al. (2010) confirmed that spherical gas bubbles in water could behave like solids due to high apparent bulk modulus for certain frequencies of sonic signals. According to their analysis, gaseous bubbles in free gas zone in the sediments behave differently from those in hydrocarbon reservoirs. Mikada et al. (2008) concluded that: (1) a certain nucleation process is taking place in-situ in the free gas zone below the bottom simulating reflectors, and (2) the methane comes from both deep-sourced fluids as indicated by Gulick and Bangs (2004) and in-situ nucleation that causes anomalous sonic velocities. Since microbes must be responsible for the in-situ generation of methane bubbles, we now see that it is necessary to understand fluid circulation process and microbial activities in the sediments, as well as the origin of deep-sourced methane, for a full understanding of shallow methane balance. Fluid circulation in the accretionary prism is so deterministic to geological processes of various spatial scales from megathrust earthquakes to the activity of microbes.

21.5 Formation pressure observations

Hyndman et al. (1995) hypothesized that the seismogenic zone is defined by the temperature condition after their estimate of the temperature profile along the Nankai Trough. They advocated that the temperature of the updip limit of the seismogenic zone is at about 100–150°C for the Nankai Trough. Steurer and Underwood (2003) considered that this temperature range could be related to clay mineral diagenesis. The transition of smectite to illite may in particular be strongly related to the shear strength of sediments under such temperatures. They measured the inclusion of smectite in the sediments and concluded that there would not be enough volume of smectite to be transformed to illite in the Nankai seismogenic zone. Moore and Saffer (2001) confirmed that a diagenetic phase transition of clay minerals from smectite to illite takes place at 100–150°C in the mudstone and that this phase transition reduced the shear strength of the material. They also pointed out that a similar diagenetic dehydration from quartz to cristobalite could take place at a slightly lower temperature than that for the smectite-illite transition and that the pressure of formation fluids might influence the shear strength than the clay mineral phase transition. As Brown et al. (2001) and Moore et al. (2005) discussed, the dehydration diagenesis of clay minerals or silicates would freshen the formation fluids but it is not necessary to have such dehydration taking place at depths close to seismogenic zone in the case of the Nankai accretionary prism. To reveal whether or not the freshening is related to deep seismogenic processes, it is essential to understand the whole fluid circulation for both the spatial scale of the accretionary prism and in the time scale of the recurrence period of megathrust earthquakes at a location in the seismogenic zone. If the seismogenic zone is defined by the temperature as Hyndman et al. (1995) proposed, and if the temperature defines the diagenesis of clay minerals (Vrolijk, 1990), the dehydration reaction of any available minerals needs to be revisited to understand whole freshening processes in terms of the seismogenesis.

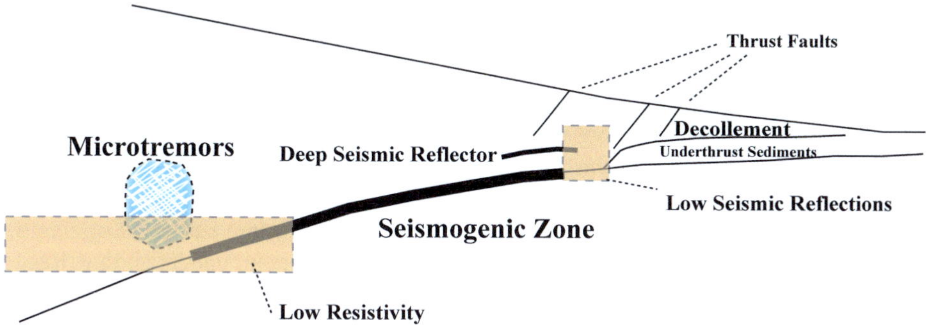

Figure 21.7 Schematic structure around seismogenic zone inferred from seismic reflection, magnetotelluric profiles and microtremor studies (Mikada et al., 2006). Decline of seismic reflection amplitudes is found in the vicinity of décollement step-down and the updip limit of seismogenic zone, while zone of low resistivity and microtremors near the downdip limit. Both phenomena may be explained by the existence of interstitial fluids. However, in-situ data should be provided for further discussions on the fluid circulation in a spatial scale including both accretionary prism and seismogenic zone.

21.6 Discussion

Non-volcanic microtremor activity was found near the downdip limit of the plate boundary in the southwestern part of Japan (Obara, 2002). Although the generation mechanism of such microtremors is not well understood, it has been speculated that interstitial fluids are responsible because of their continuous amplitude variation with time. The necessity to understand the fluid circulation in the accretionary prism has been proposed using freshening data (Kastner et al., 1993), log data (Bourlange et al., 2003), clay mineral analyses (Steurer and Underwood, 2003), formation pressure observation (Davis et al., 2006) and microbe studies (Kormas et al., 2003; Newberry et al., 2004) with respect to seismogenic processes at the Nankai subduction zone. Also, it has become obvious that the models for the Costa Rica and Barbados subduction zones in which fluids of low salinity are supplied through the décollement (Moore and Silver, 2002) may not be suited to explain even the freshening that takes place near the décollement at the Nankai Trough (Moore et al., 2005).

Mikada et al. (2006) summarized studies of the region of low electrical resistivity (Kasaya et al., 2005), deep seismic reflectors (Park et al., 2002), the location of non-volcanic microtremors (Obara, 2002) near the Nankai Trough in a scale of the seismogenic zone 40 to 50km below the surface (Figure 21.7). These observational facts indicating the existence of fluids imply that the deep-sourced fluids may play an important role in the generation of seismogenic zones, at least for the Nankai subduction zone. However, knowledge gaps remain concerning diagenetic processes and fluid paths in the seismogenic zone, except for numerical simulations. Observations on a scale of the seismogenic zone and in a recurrence time-scale of megathrust earthquakes need to be considered in detail to understand and develop predictions for future earthquakes.

The necessity of understanding fluid movement in terms of paths, quantity, and time-scale has been repeatedly mentioned by many authors. The relationship of fluid circulation with the seismogenic processes has only been investigated using numerical simulations to satisfy observations of core data in terms of salinity and microbial activity. Logging data such as resistivity, density, etc., have also contributed to deepen our understanding. Recently, data from long-term observations, such as formation pressure from ACORK and earthquake monitoring, revealed a relationship between regional stress field and non-volcanic tremors. These results emphasize the future needs for the combination of drilling, coring, long-term monitoring, etc., to understand the whole fluid circulation in the sediments that are influenced by regional stresses caused by the subduction of the oceanic plate at the Nankai Trough. As Moore et al. (2005) pointed out, the model indicates that low-salinity fluid produced at seismogenic zone migrates through the décollement layer to freshen the formation water near the toe of the accretionary prism as in Costa Rica and Barbados, but may not be applied directly to the Nankai Trough. This feature is obvious when the freshened zone lies broadly above and below the décollement at the Nankai Trough than in the other subduction zones as shown by Moore and Silver (2002). Phase transition of clay minerals from smectite to illite, proposed by Vrolijk (1990) and Hyndman et al. (1995) with respect to the generation of earthquakes due to the change in shear strength, has not been positively demonstarted due to the unexpectedly smaller smectite content revealed by ocean drilling (Steurer and Underwood, 2003; Moore et al., 2005). Earthquake generation models using phase transition of clay minerals and thermal structure of seismogenic zone need to be revisited to build a model for further reliable and testable hypotheses.

The two-legs' expedition, 190 and 196, of the Ocean Drilling Program was successful in generating information on the contents of clay minerals, the role of microbes in methane production and thermal structure, and with the installation of two long-term hydrogeological observation stations (Mikada et al., 2002a). Fractures located in the sediments were analyzed with respect to regional stress orientations (McNeill et al., 2004; Ienaga et al.; 2006) and variations in local fluid pressure (Bourlange et al., 2003). The existence of ancient high temperature environment in the sediments near the thrust faults indicated by microbe studies (Kormas et al., 2003; Newberry et al., 2004) also suggests that the generation of fractures and faults is strongly related with the dynamics of interstitial fluid movement. Observations in the hydrologically well-isolated sediments at the Nankai Trough revealed that formation pressure variations are likely caused by the accumulation of plate strain by subduction (Kastner et al., 2006). Long-term monitoring of such formation pressure in terms of the recurrence of megathrust earthquakes must be conducted for further understanding of earthquake generation mechanisms.

21.7 Summary

The discussion in this chapter can be summarized as follows:

1. Drilling, coring, and measurements of physical properties through the sediments near the toe of the accretionary prism have been conducted through scientific drilling in the Nankai Trough.
2. Fluids that were considered to be key to understanding the seismogenic processes and the development of the accretionary prism have been confirmed right through the drilling experiments.
3. It is necessary to conduct long-term monitoring of fluid circulations in the accretionary prism to understand processes from the scale of accretionary prism to variations in sediment thickness to understand deep seismogenic processes in at the Nankai Trough.
4. Various problems concerning the fluid circulation are expected to be clarified by future studies. These should provide a quantitative understanding on earthquake occurrence, fluid travel time variation related to earthquake occurrence, pore water pressure variation, and enhanced numerical simulation models for the fluid movement.

Acknowledgments

The authors would like to express their special thanks to Keir Becker, Pierre Henry, J. Casey Moore, Earl E. Davis, Peter Flemings, and Greg F. Moore for profitable discussions. Special thanks should be given to Chris Barnes and the anonymous reviewers for various suggestions in the readability enhancement to the original manuscript, and to Paolo Favali, Laura Beranzoli and Angelo De Santis for their encouragement in the submission of this chapter.

References

Ando M. (1975) Source mechanisms and tectonic significance of historical earthquakes along the Nankai Trough, Japan. Tectonophys. 27, 119–140.

Archie G.E. (1942) The electrical resistivity log as an aid in determining some reservoir characteristics. Petroleum Transactions of AIME 146, 54–62.

Bangs N., Shipley S., Moore G., Gulick S., Kuromoto S. and Nakamura Y. (2004) Evolution of the décollement from the trench into the seismogenic zone inferred from mapping Nankai trough décollement seismic reflections in 3-D, Muroto transect. Geology 32, 273–276.

Banno T., Mikada H., Goto T., Takekawa J. and Onishi K. (2010) Anomalous sonic velocities in the formation containing multi-phase fluids. SEG Expanded Abs. 29, 2659–2663, doi: 10.1190/1.3513393.

Bourlange S., Henry P., Moore J.C., Mikada H. and Klaus A. (2003) Fracture porosity in the décollement zone of Nankai accretionary wedge using Logging While Drilling resistivity data. Earth Planet. Sci. Lett. 209, 103112.

Brie A. (1997) Petrophysical evaluation of sonic logs in shaly sands. In: Proc. 3rd International Well Logging Symposium, Society of Professional Well Log Analysts, Japan Chapter, paper U.

Brown K.M., Saffer D.M. and Bekins B.A. (2001) Smectite diagenesis, pore-water freshening, and fluid flow at the toe of the Nankai wedge. Earth Planet. Sci. Lett. 194, 97–109.

Davis E.E., Becker K., Wang K., Obara K., Ito Y. and Kinoshita M. (2006) A discrete episode of seismic and aseismic deformation of the Nankai trough subduction zone accretionary prism and incoming Philippine Sea plate, Earth Planet. Sci. Lett. 242, 73–84.

Ge S. and Screaton E. (2005) Modeling seismically induced deformation and fluid flow in the Nankai subduction zone. Geophys. Res. Lett. 32, L17301, doi: 10.1029/2005GL023473.

Gulick S.P.S. and Bangs N.L.B. (2004) Negative polarity at the frontal thrust - Is free gas the culprit?: Insights from Nankai accretionary prism off Cape Muroto using seismo-logging integration. In: H. Mikada, G.F. Moore, A. Taira, K. Becker, J.C. Moore and A. Klaus (Eds) *Proc. ODP, Sci. Results, 190/196*, doi: 10.2973/odp.proc.sr.190196.353.2004.

Hyndman R., Wang K. and Yamano M. (1995) Thermal constraints on the seismogenic portion of the southwestern Japan subduction thrust. J. Geophys. Res. 100(15), 15,373-15,392.

Ienaga M., McNeill L.C., Mikada H., Saito S., Goldberg D. and Moore, J.C. (2006) Borehole image analysis of the Nankai Accretionary Wedge, ODP Leg 196: Structural and Stress Studies. Tectonophysics, doi: 10.1016/j.tecto.2006.02.018.

Jouniaux L., Lallemant S. and Pozzi J.-P. (1994) Changes in the permeability, streaming potential and resistivity of a claystone from the Nankai prism under stress. Geophys. Res. Lett. 21, 149–152.

Kasaya T., Goto T., Mikada H., Baba K., Suyehiro K. and Utada H. (2005) Resistivity image of the Philippine Sea Plate around the 1944 Tonankai earthquake zone deduced by Marine and Land MT surveys. Earth Planets Space 57, 209–213.

Kastner M., Elderfield H., Jenkins W.J., Gieskes J. and Gamo T. (1993) Geochemical and isotopic evidence for fluid flow in the western Nankai subduction zone, Japan. In: I.A. Hill, A. Taira, J.V. Firth, et al., (Eds.), *Proc. ODP, Sci. Results 131*, 397–413.

Kastner M., Becker K., Davis E.E., Fisher A.T., Jannasch H.W., Solomon E.A. and Wheat G. (2006) New insights into the hydrogeology of the oceanic crust through long-term monitoring. Oceanography 19(4), 46–57.

Kormas K.A., Smith D.C., Edgcomb V. and Teske, A. (2003) Molecular analysis of deep subsurface microbial communities in Nankai Trough sediments (ODP Leg 190, Site 1176A). FEMS Microbiology Ecology 45, 115–125.

Mazzotti S., Le Pichon X., Henry P. and Miyazaki S. (2000) Full interseismic locking of the Nankai and Japan-west Kurile subduction zones: An analysis of uniform elastic strain accumulation in Japan constrained by permanent GPS. J. Geophys. Res. 105(13),159–13,177.

McNeill L.C., Ienaga M., Tobin H., Saito S., Goldberg D., Moore J.C. and Mikada H. (2004) Deformation and in situ stress in the Nankai Accretionary Prism from resistivity-at-bit images, ODP Leg 196. Geophys. Res. Lett. 31, L02602, doi: 10.1029/2003GL018799.

Mikada H., Becker K., Moore J.C., Klaus A., Austin G.L., Bangs N.L., Bourlange S., Broilliard, J. et al. (2002a) Proc. ODP, Init. Repts., 196 [CD-ROM]. Available from: Ocean Drilling Program, Texas A&M University, College Station, TX 77845-9547, USA.

Mikada H., Becker K., Moore J. C. and Klaus A. (2002b) Leg-196: Deformation and fluid flow processes: Logging while drilling and Advanced CORK in the Nankai Trough accretionary prism. JOIDES Jour. 28 (2), 8–12.

Mikada H., Ienaga M., Goto T. and Kasaya T. (2006) Current research status and meaning of fluid pressure monitoring at the Nankai trough. J. Geography 115(3), 367–382 (in Japanese with English abstract and captions).

Mikada H., Inamori T., Saeki T., Suzuki K. and Onishi K. (2008) Formation evaluation of free gas zone beneath bottom simulating reflector. In: *Proceedings of the 14th Formation Evaluation Symposium of Japan*, Paper J.

Moore G.F., Taira A., Klaus A. and the Leg-190 Scientific Party (2001) New insights into deformation and fluid flow processes in the Nankai Trough accretionary prism: Results of Ocean Drilling Program Leg 190. Geochem. Geophys. Geosyst. 2, doi: 10.129/2001GC000166.

Moore G.F., Mikada H., Moore J.C., Becker K. and Taira A. (2005) Legs 190/196 synthesis: Deformation and fluid flow processes in the Nankai trough accretionary prism. In: H. Mikada, G.F. Moore, A. Taira, K. Becker, J.C. Moore and A. Klaus (Eds) *Proc. ODP, Sci. Results, 190/196* [Online]. Available from: www-odp.tamu.edu/publications/190196SR/synth/synth.html. [Cited 2010-01-02]

Moore J.C. and Saffer D.M. (2001) Updip limit of the seismogenic zone beneath the accretionary prism of southwest Japan: an effect of diagenetic to low-grade metamorphic processes and increasing effective stress. Geology 29, 183–186.

Moore J.C. and Silver E. (2002) Fluid flow in accreting and eroding convergent margins, JOIDES Jour. 28(1), 91–96.

Morgan J.K. and Ask M.V.S. (2004) Consolidation state and strength of underthrust sediments and evolution of the décollement at the Nankai accretionary margin: Results of uniaxial reconsolidation experiments. J. Geophys. Res. 109, B03102, doi: 10.1029/2002JB002335.

Newberry C.J., Webster G., Cragg B.A., Parkes R.J., Weightman A.J. and Fry J.C. (2004) Diversity of prokaryotes and methanogenesis in deep subsurface sediments from the Nankai trough, Ocean Drilling Program Leg 190. Environ. Microbiol. 6, 274–287.

Obara K. (2002) Nonvolcanic deep tremor associated with subduction in Southwest Japan. Science 296, 1679–1681.

Okino K., Shimakawa Y. and Nagaoka S. (1994) Evolution of the Shikoku Basin. J. Geomagn. Geoelectr. 46, 463–479.

Ozawa T., Tabei T. and Miyazaki S. (1999) Interplate coupling along the Nankai trough off southwest Japan derived from GPS measurements. Geophys. Res. Lett. 26, 927–930.

Park J.-O., Tsuru T., Takahashi N., Hori T., Kodaira S., Nakanishi A., Miura S. and Kaneda Y. (2002) A deep strong reflector in the Nankai accretionary wedge from multichannel seismic data: Implications for underplating and interseismic shear stress release. J. Geophys. Res. 107, 2061, doi: 10.1029/2001JB000262.

Saffer D.M. (2003) Pore pressure development and progressive dewatering in underthrust sediments at the Costa Rican subduction margin: Comparison with northern Barbados and Nankai. J. Geophys. Res. 108, 2261, doi: 10.1029/2002JB001787.

Saffer D.M. and Bekins B.A. (1998) Episodic fluid flow in the Nankai accretionary complex: Timescale, geochemistry, flow rates, and fluid budget. J. Geophys. Res. 103(30), 30,351–30,370.

Sagiya T. and Thatcher W. (1999) Coseismic slip resolution along a plate boundary megathrust: The Nankai Trough, southwest Japan. J. Geophys. Res. 104, 1111–1130.

Savage J.C. (1995) Interseismic uplift at the Nankai subduction zone, southwest Japan, 1951–1990. J. Geophys. Res. 100, 6339–6350.

Stauffer P. and Bekins B.A. (2001) Modeling consolidation and dewatering near the toe of the northern Barbados accretionary complex. J. Geophys. Res. 106, 6369–6383.

Steurer J.F. and Underwood M.B. (2003) Clay mineralogy of mudstones from the Nankai Trough reference Sites 1173 and 1177 and frontal accretionary prism Site 1174. In: H. Mikada, G.F. Moore, A. Taira, K. Becker, J.C. Moore and A. Klaus (Eds.), *Proc. ODP, Sci. Results, 190/196* [Online]. Available from: www-odp.tamu.edu/publications/190196SR/211/211.htm. [Cited 2010-01-02]

Trehu A.M., Ruppel C., Holland M., Dickens G.R., Torres M.E., Collett T.S., Goldberg D., Riedel M. and Schultheiss P. (2006) Gas hydrates in marine sediments – Lessons from scientific ocean drilling. Oceanography 19(4), 124–142.

Tsuji T., Matsuoka T., Yamada Y., Nakamura Y., Ashi J., Tokuyama H., Kuramoto S. and Bangs N.L. (2005) Initiation of plate boundary slip in the Nankai Trough off the Muroto peninsula, southwest Japan. Geophys. Res. Lett. 32, L12306, doi: 10.1029/2004GL021861.

Vrolijk P. (1990) On the mechanical role of smectite in subduction zones. Geology 18, 703–707.

22 Modeling of regional geomagnetic field based on ground observation network including seafloor geomagnetic observatories

H. Toh[1] and A. De Santis[2]

22.1 Introduction

We sometimes encounter a situation where we have to model a set of geophysical data acquired within a limited portion of our globe by a rapid and efficient survey such as airborne mapping. The content of the dataset could be a physical quantity that is essentially a potential field like gravity. In order to ensure that the modeled field has appropriate physical properties, it is desirable that the potential, whose spatial gradients correspond to the observed field, fulfills Laplace's equation rather than the modeled field being a mere statistical fit using arbitrary functions. If a vector field, $\vec{A}$, is both irrotational ($rot\,\vec{A} = 0$) and divergence free (div $\vec{A} = 0$), the former means that $\vec{A}$ can be deduced from a scalar potential, ϕ, as $\vec{A}$ =gradϕ, while the latter requires $\Delta\phi = 0$ where $\Delta \equiv$ div·grad. Thus, this kind of vector field is always associated with a scalar potential satisfying Laplace's equation and physically constrained in its own source-free region.

Spherical cap harmonic analysis (SCHA) has been introduced as a tool for regional modeling of any potential fields, since they are solutions of Laplace's equation over a limited portion of a complete sphere (Haines, 2007). Another feature peculiar to SCHs is their seamless transition to spherical harmonics (SHs). Namely, the associated Legendre functions with real degrees used in SCHs converge to those with integer degrees as the aperture of the spherical cap in concern tends to $\pi/2$ (see De Santis, 1992) or π.

1 Data Analysis Centre for Geomagnetism and Space Magnetism, Graduate School of Science, Kyoto University, Kyoto, Japan

2 Istituto Nazionale di Geofisica e Vulcanologia, Rome, Italy

Modeling temporal variations of a potential field as well as its spatial fitting is also important for multidisciplinary purposes. For instance, if a detailed regional model of the geomagnetic secular variation is available for a particular region of the Earth, the model could be utilized as one of the archaeological dating tools (e.g., Pavón-Carrasco et al., 2009).

Although SCHA is not an exact mathematical solution, it has some merits in cases where certain approximations are acceptable. The first appearance of SCHA to geophysical studies can be found in Haines (1985a), in which he introduced two sets of basis functions, defined at the cap edges: one set has the potential itself nil while the other set forces its gradient to disappear. Although those basis functions are mathematically redundant (i.e., each set is complete in itself), they, in turn, ensure rapid convergence of the spherical cap harmonic expansion and reduction of edge effects as well as Gibbs phenomena (see discussion in Section 2.4 of Thébault et al., 2006). Upward continuation of Haines's (1985a) formulation quickly loses accuracy as the altitude enlarges (Haines, 1985b; De Santis and Falcone. 1995). However, it can be applied to a combination of data from not too different altitudes. This, therefore, means that SCHA formulation is practically more efficient in dealing with data collected at one altitude alone.

Recently, Thébault and his co-workers redefined the boundary-value problem (BVP) in a search for the regional geomagnetic potential in order to model data obtained at different altitudes, viz., at geomagnetic satellite orbits and at the Earth's surface (Thébault et al., 2004, 2006). This kind of approach is necessary since it is now an era of geomagnetic measurements by successive long-life low-Earth-orbit (LEO) satellites such as Ørsted (Neubert et al., 2001), CHAMP (Reigber et al., 2002) and SWARM (Olsen et al., 2006). They succeeded in recovering mathematical strictness by solving the BVP within a finite volume, namely, within a frustum of a cone. Their upward and downward continuations, therefore, are now correct within the volume and include not only internal but also external contributions (i.e., external to the cone in concern, not to the Earth's surface), although they had to introduce a combination of special radial functions (see Eqs [18] and [19] of Thébault et al. (2006)) and the Mehler (1868) function for the co-latitudinal direction in addition to SCHs already appeared in Haines's (1985a) formulation. It is worth noting that clarification of the meaning of the internal/external separation for the regional field is one of the important results of Thébault et al. (2006). They have named the new SCHA "Revised Spherical Cap Harmonic Analysis (R-SCHA)". Thébault (2008) further elaborated a two-dimensional (2D) version of R-SCHA (R-SCHA2D) since R-SCHA, by definition, requires data observed at multiple altitudes and becomes unstable when it is applied to single altitude data.

The intent of this chapter is to show to what extent SCHA is capable of describing regional fields by introducing a specific case study in the Pacific region, using SCHA both on land and marine data, where the latter include vector magnetic data from a seafloor observatory (Toh et al., 2004). Below, we describe: (1) in Section 22.2 a brief review of the potential theory; (2) in Section 22.3 formulation of 2D regional modeling problem for the geomagnetic field using SCHA derived by Haines (1985a); (3) in Section 22.4 a regional geomagnetic reference field over the western Pacific (RGRF-P). Finally, in Section 22.5, a comprehensive discussion on the comparison with the emerging R-SCHA is included in order to make clear the difference of the old and new formulations and to summarize the results.

22.2 Potential theory

A good example of potential fields is magnetic induction, $\vec{B}$, exterior to any conductor and magnetic body. Quasi-static Maxwell's equations tell us that $\vec{B}$ is rotation-free since there exist no electric currents flowing outside conductors or magnetic bodies. This means that we can define a geomagnetic scalar potential, V, such that:

$$\vec{B} = -gradV \tag{22.1}$$

It follows that spatial dependence of V obeys the following Laplace's equation because magnetic induction, $\vec{B}$, is known to be always non-divergent in macroscopic physics irrespective to electromagnetic (EM) properties of the medium in which $\vec{B}$ resides:

$$\Delta V = 0 \tag{22.2}$$

The gravity field exterior to any mass distribution is also divergence-free and thus the corresponding gravity potential is subject to Laplace's equation as well. Solutions of Laplace's equation are called "harmonic functions" and determined specifically according to each boundary condition applied to solve Laplace's equation in concern. It goes without saying that linear combinations of the obtained harmonic functions are also solutions of the original Laplace's equation. Thanks to this linear property of the original Laplace's equation, its solutions can be expanded by linear combinations of harmonic functions.

In spherical geometries shown in Figure 22.1, SHs are solutions of Laplace's equation. Gauss (1839) applied SHs for the first time to geomagnetism so as to determine the geomagnetic field and found quantitatively that dipole components are really dominant. His estimate of the internal axial dipole coefficient, $g_1^{\,0}$, was –32350 nT, which is a remarkably good agreement with the present-day dipole coefficient of -29496.5 nT (IGRF-11; IAGA, 2010), if we take into account the insufficient coverage of the geomagnetic observation network as well as the inferior precision of the absolute geomagnetic readings in Gauss' time and the secular variation since then. One of the most important geophysical implications of Gauss' achievement is that his method enabled internal/external separation of the geomagnetic field. This is because the spherical harmonic analysis is always an expansion over a spherical surface and hence the whole three-dimensional (3D) space is inevitably divided into a finite volume within the closed Earth's surface and its exterior. In the case of regional modeling by SCHA, however, the situation becomes somewhat different from the global case, which makes the meaning of the internal/external separation ambiguous. We discuss this issue again later in this chapter.

The determination of any potential field can be regarded as a BVP for Laplace's equation. For example, Thébault et al. (2004) solved a purely Neumann problem at each boundary of a frustum of a cone, namely, giving observed radial components of the geomagnetic field at the top and bottom spherical caps while the θ-components were given on the side boundary. However, they could not include zonal geomagnetic components in the cap's reference frame since they solved the Neumann problem by decomposing it into two inde-

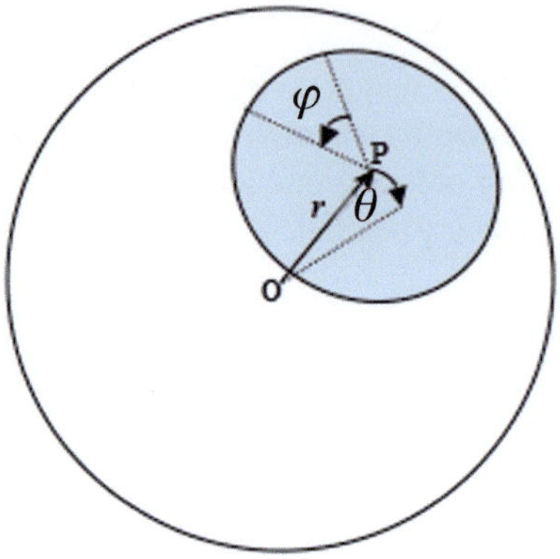

Figure 22.1 The spherical cap on a sphere of aperture θ, radius r and center O together with a spherical coordinate sharing the cap pole, P.

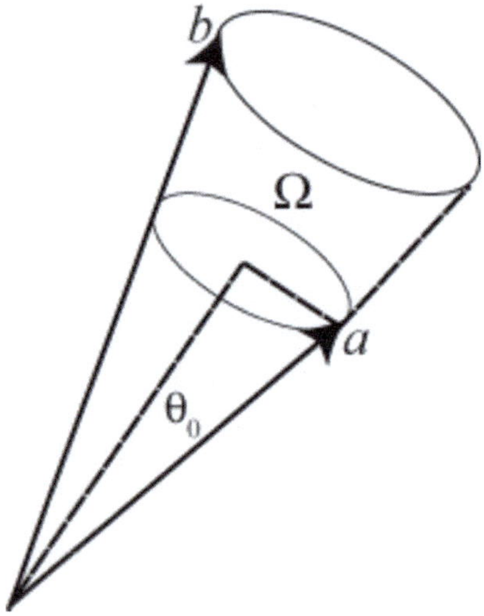

Figure 22.2 A frustum of a cone W defined by the intersection of two spheres of radii a and b, respectively.

pendent sub-BVPs. If they allowed the zonal components, it mixed up the two sub-BVPs and led to violation of the null-flux condition of $\vec{B}$, which is the most important intrinsic property of magnetic induction. To rectify this drawback, Thébault et al. (2006) reformulated their BVP into a mixed BVP. Namely, they solved a Dirichlet problem on the side boundary so as to allow zonal components as well. In either case, the boundary conditions on the radial boundaries were unchanged and the BVPs themselves were solved within a finite volume. This implies that the upward and downward continuations within the frustum of the cone are much more precise than those of Haines's (1985a) formulation, in which radial boundary conditions were given at the Earth's surface and infinity by assuming internal fields alone. The boundary conditions described in Haines (1985a) will be further examined in the next section together with other aspects of his formulation.

Finally, the recent appearance of R-SCHA2D (Thébault, 2008) can be regarded as an R-SCHA counterpart of Haines's (1985a) formulation. Thébault (2008) solved a purely Neumann problem again, whereas he did not suffer from the flux problem this time. This is simply because we are now in a 2D surface, so that it is unnecessary to impose the null-flux condition on the surface integral of $\vec{B}$ over the spherical cap.

22.3 Spherical cap harmonics – 2D case

Laplace's equation (Eq. (22.2)) is usually solved by separation of variables. In the spherical coordinates (r, θ, φ) appropriate to the spherical cap's reference frame (Figure 22.1), Eq. (22.2) becomes:

$$\left\{\frac{1}{r^2}\frac{\partial}{\partial r}\left(r^2\frac{\partial}{\partial r}\right)+\frac{1}{r^2\sin\theta}\frac{\partial}{\partial\theta}\left(\sin\theta\frac{\partial}{\partial\theta}\right)+\frac{1}{r^2\sin\theta}\frac{\partial^2}{\partial\varphi^2}\right\}V=0\,. \tag{22.3}$$

We, therefore, have only to seek solutions of Eq. (22.3) in a form of:

$$V = R\,(r)\;\Theta\;(\theta)\Phi(\varphi). \tag{22.4}$$

Once Eq. (22.3) is decomposed into three ordinary differential equations (ODEs), one for each spherical coordinate, the particular solution for Φ is straightforward when the 2π periodicity in the longitudinal direction is taken into consideration: it is given by $e^{im\varphi}$ where m is integer.

The remaining two ODEs for the radial and co-latitudinal directions are second-order Sturm-Liouville equations. This means that two boundary conditions are required in order to solve the BVP for each direction. In Haines's (1985a) formulation, the two boundary conditions for the co-latitudinal direction are given at the pole and the edge of the spherical cap in concern, respectively. Associated Legendre function, $P_{n_k}^{m}(\cos\theta)$, with real degree, n_k , then appears in order to meet those boundary conditions, while the order, m, remains to be integer as before. Here, k is an integer called "degree index", which is useful for the proper ordering of the real degrees. The reason why we need to introduce real degrees stems from the fact that the aperture of the spherical cap is less than π. It is also noteworthy that $P_{n_k}^{m}$ $(\cos\theta)$, should be of the first kind because the associated Legendre function of the second kind have singularities at the pole.

As for the radial geomagnetic component, the first boundary condition is given at the Earth's surface, while the second boundary condition for r is given at $r\rightarrow\infty$. They result in the following solution for the radial direction:

$$R(r) = a\left(\frac{a}{r}\right)^{n_k+1}, \tag{22.5}$$

where a is the mean radius of the Earth. This implies that his formulation implicitly assumes the geomagnetic potential to be of purely internal origin. What we actually observe at heights is a combination of the internal and external geomagnetic fields. However, Haines's (1985a) formulation still gives a good fit to the observed geomagnetic field of regional scale as long as we stick to the Earth's surface, and provided that the observed geomagnetic data are carefully corrected for the external field.

Haines (1985a) imposed null conditions for both potential, V, itself and its derivative with respect to θ on the cap edge, which is mathematically impossible except for the trivial solution. To circumvent the mathematical difficulty, Haines (1985a) introduced two kinds of basis so that they satisfy separately the null potential condition for $k - m$ = odd and the null gradient condition for $k - m$ = even as follows:

$$V = (r, \theta, \varphi) = 0 \quad \text{for } k - m = \text{odd} \tag{22.6}$$

and

$$\frac{\partial V(r,\theta,\varphi)}{\partial\theta} = 0 \quad \text{for } k - m = \text{even}. \tag{22.7}$$

Each set is a complete solution of the Sturm-Liouville equation, and all basis functions belonging to one set are mutually orthogonal, while this is not usually true when we consider functions from one set to another. Although those basis functions were mathematically redundant, his formulation provided a couple of practical merits for regional modeling of potential fields as described earlier.

The solutions for each tangential direction are combined to give:

$$V = \sum_{m=0}^{\infty}\sum_{k=m}^{\infty} a\left(\frac{a}{r}\right)^{n_k+1} \left\{g_k^m \cos(m\varphi) + h_k^m \sin(m\varphi)\right\} P_{n_k}^m(\cos\theta), \tag{22.8}$$

when Eq. (22.4) and linearity of Laplace's equation are considered. Here, the coefficients, $g^m{}_k$ and $h^m{}_k$, are internal Gauss coefficients for regional expansion. In the next section, a regional geomagnetic reference field based on the regional expansion formula in Eq. (22.8) will be illustrated.

22.4 Regional geomagnetic reference field – A case study

22.4.1 Data

The Pacific region has long been one of the geomagnetically least surveyed areas on the Earth. There are a very limited number of ground stations that continuously measure the vector geomagnetic field in this area. Recently, however, long-term three-component geomagnetic observation has been started by a regional network consisting of mostly oceanic island stations but including one seafloor station (Figure 22.3). (In addition to NWP, another seafloor station, WPB, is now functioning right in the middle of the Philippine Sea.) Refer to a separate report (Toh and Hamano, this volume) for the present status of the geomagnetic observatories at the seafloor in the western Pacific. A regional geomagnetic reference field over the western Pacific (RGRF-P) was constructed for epoch 2002 based on the new geomagnetic data together with that acquired by Intermagnet Magnetic Observatories (IMOs) being operated in the Pacific and its rim.

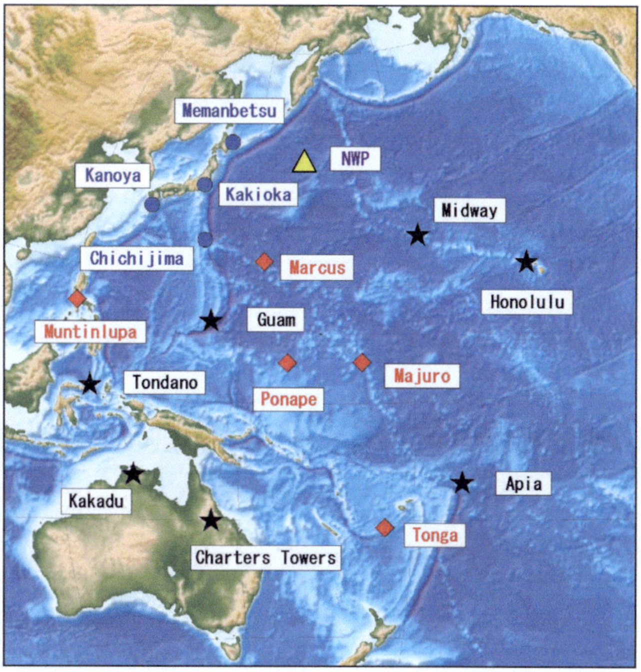

Figure 22.3 Distribution of geomagnetic stations over the western Pacific used in Toh et al. (2007). Diamonds and a triangle are the new island stations and the seafloor station, respectively. Circles are Japanese geomagnetic observatories, while stars denote IMOs

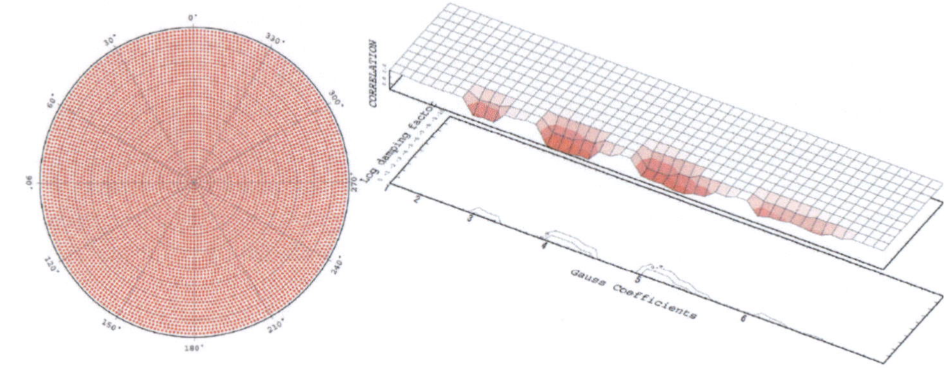

Figure 22.4 (Left) Uniform site distribution for the synthetic test of our SCH expansion code. The observation sites are equidistantly allocated in the spherical cap. (Right) Resultant undulation of correlation between the given and recovered Gauss coefficients in ε-k plane.

Prior to analysis in terms of SCHs, three-component raw annual means of two-year long time-series (from August, 2001 to July, 2003) at each geomagnetic station were corrected for station corrections by a newly-developed method (Toh et al., 2007) which is based on a global estimates of crustal biases (Mandea and Langlais, 2002).

Dipole components (g_1^0, g_1^1 and h_1^1) of IGRF-10 were further subtracted from the corrected annual means because the dipole components have much longer wavelengths than the spatial extent in concern. This type of detrending has been common in analogy with removal of linear trends in time-series analyses.

22.4.2 Synthetic inversion

The determination of the SCH coefficients for each SCH can be formulated into a linear inverse problem. Instead of solving a usual linear inverse problem, we solved the following damped least squares problem with *a priori* information introduced by Gubbins and Bloxham (1985):

$$m = \left\{(WG)^t (WG) + \varepsilon^2 C\right\}^{-1} G^t d \tag{22.9}$$

Here, m and d are the model and data vectors, respectively. m contains the internal model Gauss coefficients to be determined, while d retains the station-corrected and detrended annual means of geomagnetic three components at each station. C, W and G are a model covariance matrix, a data error matrix and a coefficient matrix that projects the model parameters to the data space, respectively. Here we adopted a diagonal matrix for W whose elements are reciprocals of standard errors for each geomagnetic component at each

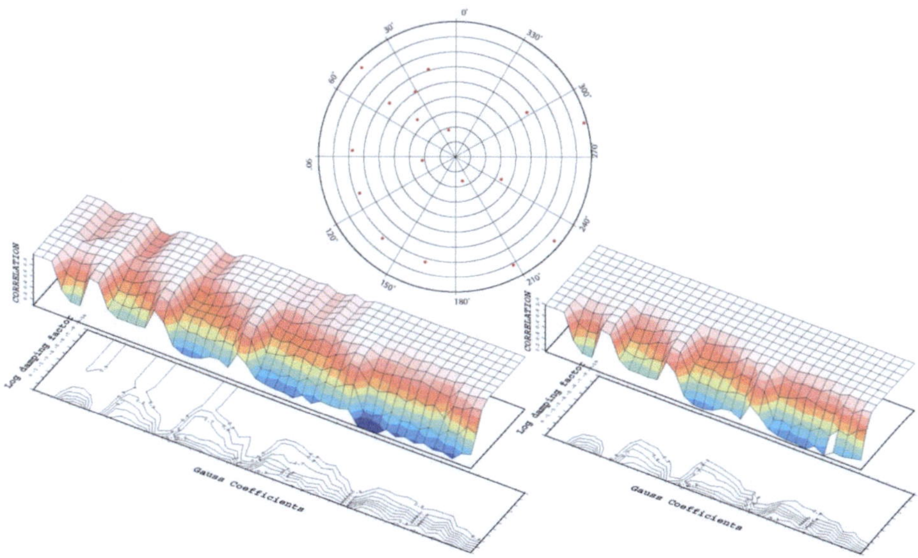

Figure 22.5 Undulation of coherence between the known and reproduced Gauss coefficients in ε-k plane. (Left) Degree-6 expansion shows numerical instability due to spatial aliasing by the biased distribution of the geomagnetic stations shown in the polar plot (middle). The center of the polar plot is placed at 156°E, 15°N in the spherical cap of 45° half-width. (Right) Degree-5 expansion gives adequate fits to all Gauss coefficients, even for the real station distribution and for damping factors smaller than 10^{-5}. Reproduced from Toh et al. (2007).

station. ε is a damping factor of the damped least squares problem and will be determined empirically so as to make the prior information effective. Gubbins and Bloxham (1985) showed that the model variance of Gauss coefficients should at least be proportional to n^{-3} to guarantee convergence of each geomagnetic component on a sphere, where n is the degree of spherical harmonics. Hence, the model covariance matrix, C, was also diagonal, whose elements were, in turn, proportional to $(degree\ n)^{-3}$ for each internal Gauss coefficient belonging to the same degree. This prior information is a rather weak mathematical requirement to stabilize the inversion process in the sense that it only assumes mutually independent model parameters. The weak constraint is sufficient to expand the surface geomagnetic field in our case.

Previous to the final SCH expansion using the real data, several synthetic inversions of known spatial patterns were conducted in order to examine whether the damped least squares solution was stably derived and what was the suitable value for the damping factor. Specifically, a geomagnetic field consisting of a single Gauss coefficient was generated as an input to a newly written SCH expansion code to assess how well the given coefficient was recovered by calculating the spatial coherence between the known pattern and the reproduced pattern. A 3% Gaussian error was added to the generated pattern before each synthetic inversion. First, synthetic inversions with an ideal station distribution of 1° interval were conducted for the Gauss coefficients from g_1^0 through h_6^6. The resultant coherences (Figure 22.4) were found satisfactory for all Gauss coefficients and the damping factor smaller than 10^{-2}. The damping factor was made variable in a range from 10^{-10} to unity. Second, synthetic inversions with the real station distribution were conducted for the same set of Gauss coefficients. Figure 22.5 shows the result of the second synthetic inversion up to degree 5 and 6 with a map of the actual site distribution. It is clear that degree-5 expansion gives adequate fits to all Gauss coefficients for the damping factor smaller than 10^{-5}, whereas there is no damping factor that can satisfy all Gauss coefficients in the case of degree-6 expansion. This means that if the maximum degree is set to 6, Gibbs phenomena governed by spatial aliasing can appear due to the ill distribution of the actual geomagnetic stations, although the degree of freedom of our geomagnetic data nominally allows degree-6 expansion.

22.4.3 The regional geomagnetic reference field over the Western Pacific

The final SCH expansion of the three-component geomagnetic data observed at the 18 stations in the western Pacific was conducted by setting the damping factor and the maximum degree to 10^{-6} and 5, respectively. The aperture of the spherical cap was 45°. The optimized location of the pole was a result of a 1° interval grid search within 155°E–158°E and 12°N–15°N. The geomagnetic field was expanded for the epoch of 2002.08, which converged at RMS 3.86 nT. The result of the final inversion is summarized in Figure 22. 6, where the numerical mesh was centered at (156°E, 15°N).

Except for the unexpectedly large deviation of RGRF-P from IGRF-10 in the radial component at the northern border of the region, the regional model fits to the global model. The large difference in Z-component is probably due to absence of geomagnetic observatories in this part of the cap.

The spatial distribution of the geomagnetic total force in Figure 22.6 clearly shows that there still exists a large non-dipole low which contains both the Midway Island and the Hawaii Island in its center. By comparing the distribution of each geomagnetic component, it is readily known that the non-dipole low mainly stems from the small radial component in the eastern half of the spherical cap. Taking differences between the predicted field values of IGRF-10 and those of RGRF-P leaves negligible residuals as shown in the bottom of Figure 22.6. This means that the presently dominant dipolar fields over the Pacific seen in IGRF-10 are also confirmed by the regionally denser geomagnetic network on the Earth's surface, upon which RGRF-P is based.

It was further revealed that the observed temporal variations of the radial geomagnetic component at several stations (e.g., MUN, TON, NWP etc.) seem to be explained solely by the westward drift of the equatorial dipole. It is noteworthy that the decay of the axial dipole component and increase of the non-dipole components at these stations seem to cancel each other out, leaving the contribution of the equatorial dipole alone in the observed time-series.

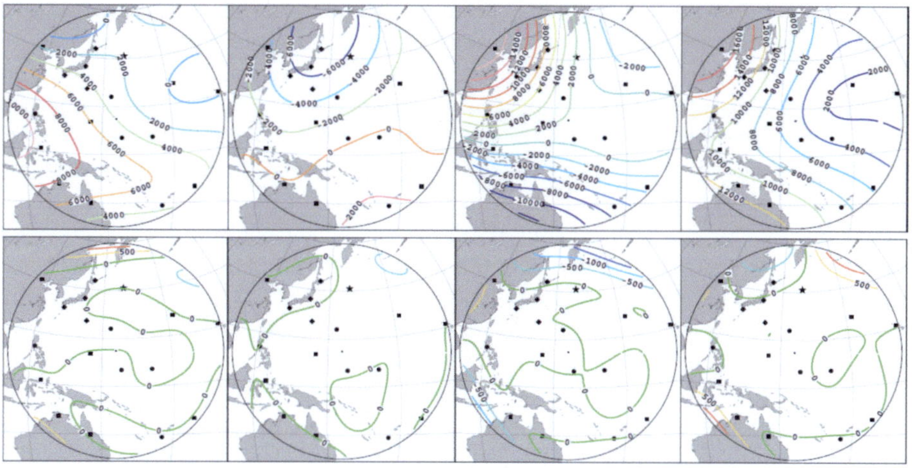

Figure 22.6 (Top) The regional geomagnetic reference field valid at 2002.08 over the western Pacific (RGRF-P) and (bottom) its difference from IGRF-10. The geomagnetic northward, eastward, downward and total force components are respectively shown from left to right. Symbols are the geomagnetic stations. Contour intervals are 2000 (top) and 500 nT (bottom), respectively. Reproduced from Toh et al. (2007).

If both of the non-dipole geomagnetic field and its westward drift over the western Pacific are absent and if the frozen-flux approximation is applicable, it can be translated into very weak core surface flows below the region. The latest core surface flow model based on satellite geomagnetic data (Holme and Olsen, 2006) may support the weak velocity field on the outer core beneath the Pacific Ocean. Furthermore, there are several seismic results (e.g., Zhao, 2004) that suggest the presence of fast *Vp* and *Vs* anomalies at the base of the lower mantle below the north Pacific. This may be consistent with the hypothesis of a cold CMB (Bloxham and Gubbins, 1987) that suppresses the flow field at the top of the outer core beneath the northwest Pacific.

22.5 Discussion

We have already mentioned the drawbacks of Haines's (1985a) SCHA formulation in terms of BVP, although it is still valid and rapid for SCH expansion at one altitude only. The mathematically rigorous version of 2D SCH expansion is now given by Thébault (2008). Since estimation scheme of the geomagnetic secular variation has already been installed in Thébault's (2008) formulation, it is equivalent to Haines's (1985a) formulation in terms of SCH expansion functions. However, upward/downward continuation of these two formulations is of limited use because they are both still 2D, yielding regional geomagnetic potentials basically valid only on the surface of the spherical cap in concern.

Thébault et al. (2006) proposed a complete and 3D version of R-SCHA, which supplements Thébault et al. (2004) by including zonal terms (see arguments in Section 22.2). It is 3D in the sense that R-SCHA3D is valid within a finite volume (a frustum of a cone with two spherical caps for its upper and lower surfaces; Figure 22.2) rather than within a limited portion of a surface. Upward/downward continuation by R-SCHA3D is presently most reliable because the corresponding BVP was solved by rigorously applying boundary conditions on every surface of the frustum of a cone. Separation of the internal and external geomagnetic fields is also nominally possible by R-SCHA3D, though meaning of the derived coefficients is quite different from that of the global case. It is Gauss' (1839) great invention that the geomagnetic field can be separated into those of internal and external origin if the potential field on the Earth's surface is expanded by SHs. The global internal and external Gauss coefficients show up in the regional case to form the regional internal and external Gauss coefficients as mixed sums of the global Gauss coefficients of both origins. This property was first reported by De Santis et al. (1999) together with the meaning of regional averages of the radial geomagnetic component within the spherical cap, and further pursued by Thébault et al. (2006).

The separation is impossible in the case of R-SCHA2D (Thébault, 2008) since spherical caps are not closed surfaces and thus unable to cover any 3D volumes. This is a somewhat disappointing property of R-SCHA2D. However, Thébault's (2008) formulation allows us to give physical meanings to some regional constants. For example, G_0^0 of his formulation can be interpreted as an average of the radial geomagnetic component within the spherical cap as was first pointed out by De Santis et al. (1999). The secular variation of G_0^0 is equivalent to that of $\bar{Z}$ of the region and thus can be used as a measure of, say, the geomagnetic jerk signature (see Figure 8 of Thébault (2008)). The quantity G_0^0 is particularly useful for delineating regional characteristics of a geomagnetic jerk, since some jerks are known to be peculiar to specific regions rather than global phenomena. It is noteworthy that the coefficients, $G_{-1/2}^m$ and $H_{-1/2}^m$, also constitute intriguing regional constants corresponding to averages of the geomagnetic poleward component along the cap edge. They can be used as indicators of the geomagnetic jerk as well as G_0^0.

22.6 Conclusions

It is sometimes required to model physical fields distributed over a limited spatial extent. SCHA is useful especially for cases when the physical fields are potential fields, viz., subject to Laplace's equations. A good example of a field of such kind is the geomagnetic field exterior to the conducting and partly magnetized Earth. In this chapter, construction of a regional geomagnetic reference field over the western Pacific (RGRF-P) was reviewed, following a classical SCHA formulated by Haines (1985a). RGRF-P is based on a newly-acquired geomagnetic dataset observed by a regional network of geomagnetic stations including a seafloor site called NWP. RGRF-P confirmed a small contribution of the non-dipole components to the geomagnetic field in the northwestern Pacific.

A succinct review on recent progress in SCHA was given in order to clarify the merits as well as the demerits of Haines's (1985a) formulation. Recently proposed R-SCHA (Thébault et al., 2004, 2006; Thébault, 2008) consists of 2D and 3D versions (see also a comparison with SCHA in Thébault and Gaya-Piqué, 2008). In both versions, the relation between the regional Gauss coefficients derived by SCHA and the global Gauss coefficients by SHA is clearly given. It is, however, hard to attribute definite physical properties to the separated internal/external coefficients in the case of SCHA. The internal/external separation is, in principle, impossible for R-SCHA2D, although the regional 2D coefficients can be expressed by a combination of the global internal and external coefficients. As for regional 3D cases, the internal/external separation itself is nominally possible. However, unlike global cases, regional internal coefficients do not represent contribution from inside the Earth alone. This is quite a different property of regional expansion compared with global expansion, and should be rigorously distinguished from one another. Finally, it is worth noting that R-SCHA2D adopted boundary conditions so as to form constant terms with clear physical meanings. G_0^0 is equivalent to $\bar{Z}$, while $G^m_{-1/2}$ and $H^m_{-1/2}$ are averages of the geomagnetic poleward component along the cap edge. These constants can be considered as representative of the region in concern at a specific epoch, and thus used for study of the geomagnetic secular variation like jerks.

When, in a hopefully close future, there are more data available from the oceans (from both oceanic islands and seafloor observatories), SCHA will be a powerful mathematical tool to investigate detailed features of the corresponding regional geomagnetic field.

Acknowledgments

We are grateful to Ocean Hemisphere Project Data Center at Earthquake Research Institute, University of Tokyo, and World Data Centres for Geomagnetism (Copenhagen and Kyoto) for having provided us with the geomagnetic data that were indispensable in constructing the regional geomagnetic reference field over the western Pacific. Partial fundings to ADS for this work came from PNRA (ARM and REM projects), Italian Foreign Office (CO-EM bilateral Italy-Japan project) and EC (DS^3F project).

References

Bloxham J. and Gubbins D. (1987) Thermal core mantle interaction. Nature 325, 511–513.

De Santis A. (1992) Conventional Spherical Harmonic Analysis for regional modelling of the geomagnetic field. Geophys. Res. Lett. 19(10), 1065–1067.

De Santis A. and Falcone C. (1995) Spherical cap models of Laplacian potentials and general fields. In: F. Sansó (Ed.) *Geodetic Theory Today*, Springer, Berlin, pp.141–150.

De Santis A., Torta J.M. and Lowes F.J. (1999) Spherical cap harmonics revisited and their relationship to ordinary spherical harmonics. Phys. Chem. Earth 24, 935–941.

Gauss C.F. (1839) Allgemeine Theorie des Erdmagnetismus, In: C.F. Gauss and W. Weber (Herausgeber), *Resultate aus den Beobachtungen des Magnetischen Vereins im Jahre 1838*, Weidmann, Leipzig, p. 157.

Gubbins, D. and Bloxham J. (1985) Geomagnetic field analysis: III. Magnetic fields on the core-mantle boundary. Geophys. J. R. Astron. Soc. 80, 695–713.

Haines G.V. (1985a) Spherical Cap Harmonic Analysis. J. Geophys. Res. 90, 2583– 2591.

Haines G.V. (1985b) Magsat vertical field anomalies above 40°N from spherical cap harmonic analysis. J. Geophys. Res. 90, 2593–2598.

Haines G.V. (2007) Harmonics, Spherical Cap. In: D. Gubbins and E. Herrero-Bervera (Eds), *Encyclopedia of Geomagnetism and Paleomagnetism*, Springer, Dordrecht, pp. 395–397.

Holme R. and Olsen N. (2006) Core surface flow modelling from high resolution secular variation. Geophys. J. Int. 166, 518–528.

International Association of Geomagnetism and Aeronomy, Working Group V-MOD (2010) International Geomagnetic Reference Field: The eleventh generation. Geophys. J. Int. 183, 1216–1230, doi: 10.1111/j.1365-246X.2010.04804.x.

Mandea M. and Langlais B. (2002) Observatory crustal magnetic biases during MAGSAT and Ørsted satellite missions. Geophys. Res. Lett. 29, doi: 10.1029/2001GL013693.

Mehler E.G. (1868) Über die verteilung der statischen electrizität in einem von zwei kugelkalotten begrenzten körper. J. Reine Angew. Math. 68, 134–150.

Neubert T., Mandea M., Hulot G., von Frese R., Primdahl F., Jørgensen J.L., Friis-Christensen E., Stauning P., Olsen N. and Risbo T. (2001) Ørsted satellite captures high-precision geomagnetic field data. Eos, Trans. Amer. Geophys. Union 82, 81–88.

Olsen N., Haagmans R., Sabaka T.J., Kuvshinov A., Maus S., Purucker M.E., Rother M., Lesur V. and Mandea M. (2006) The *Swarm* End-to-End mission simulator study: A demonstration of separating the various contributions to Earth's magnetic field using synthetic data. Earth Planets Space 58, 359–370.

Pavón-Carrasco F.J., Osete M.L., Torta J.M. and Gaya-Piqué L.R. (2009) A regional archeomagnetic model for Europe for the last 3000 years, SCHA.DIF.3K: Applications to archeomagnetic dating. Geochem. Geophys. Geosyst. 10, Q03013, doi: 10.1029/2008GC002244.

Reigber C., Luhr H. and Schwintzer P. (2002) CHAMP mission status. Adv. Space Res. 30, 129–134.

Thébault E. (2008) A proposal for regional modeling at the Earth's surface, R-SCHA2D, Geophys. J. Int. 174, 118–134, doi: 10.1111/j.1365-246X.2008.03823.x.

Thébault E. and Gaya-Piqué L.R. (2008) Applied comparisons between SCHA and R-SCHA regional modeling techniques, G-cubed 9(7), doi: 10.1029/2008GC001953.

Thébault E., Schott J.J., Mandea M. and Hoffbeck J.P. (2004) A new proposal for spherical cap harmonic modeling. Geophys. J. Int. 159, 83–103.

Thébault E., Schott J.J. and Mandea M. (2006) Revised spherical cap harmonic analysis(R-SCHA): Validation and properties. J. Geophys. Res. 111, B01102, doi: 10.1029/2005JB003836.

Toh H., Hamano Y., Ichiki M. and Utada H. (2004) Geomagnetic observatory operates at the seafloor in the Northwest Pacific Ocean. Eos, Trans. Am. Geophys. Union 85, 467–473.

Toh H., Kanezaki H. and Ichiki M. (2007) A regional model of the geomagnetic field over the Pacific Ocean for epoch 2002. Geophys. Res. Lett. 34, L09308, doi: 10.1029/2007GL029341.

Toh H. and Hamano Y. (2014) The two seafloor geomagnetic observatories operating in the western Pacific. This Volume.

Zhao D. (2004) Global tomographic images of mantle plumes and subducting slabs: Insight into deep Earth dynamics. Phys. Earth Planet. Inter. 146, 3–34.

23 Seafloor borehole observatories in the Northwestern Pacific

M. Shinohara[1], E. Araki[2], K. Suyehiro[3] and T. Kanazawa[4]

23.1 Introduction

Seismic imaging of the Earth's interior is useful for understanding mantle structure and dynamics from a regional to a global scale. One of the limitations of the imaging is the uneven distribution of seismic stations, specifically the lack of seafloor stations. The asymmetry and non-uniformity of the existing seismic station distribution makes high-resolution imaging of some parts of the mantle nearly impossible. In addition, seafloor geodesy is an important tool for understanding the dynamics of oceanic plate subduction. Thus, the scientific importance of establishing long-term geophysical stations on the seafloor has been acknowledged (ION/ODP international Workshop, 1995).

Although the oceans cover approximately 70% of the Earth's surface, there are few permanent seismic stations on the seafloor. For high-resolution imaging of the whole mantle, uniform station distribution on the Earth's surface is essential. Seafloor geophysical sensors are needed in many areas but the locations of the seafloor sensors should be carefully selected to optimize imaging. The western Pacific was selected for the installation of seafloor seismic and geodetic sensors by the Japanese Ocean Hemisphere Network Project (OHP: 1996–2002, http://eri-ndc.eri.u-tokyo.ac.jp/en/ohp/index.html). In the western

1 Earthquake Research Institute, University of Tokyo, Tokyo, Japan

2 Japan Agency for Marine-Earth Science and Technology, Yokosuka, Japan

3 Integrated Ocean Drilling Program Management International, Inc., Tokyo, Japan , now at Japan Agency for Marine-Earth Science and Technology

4 Earthquake Research Institute, University of Tokyo, Tokyo, Japan, now at National Research Institute for Earth Science and Disaster Prevention

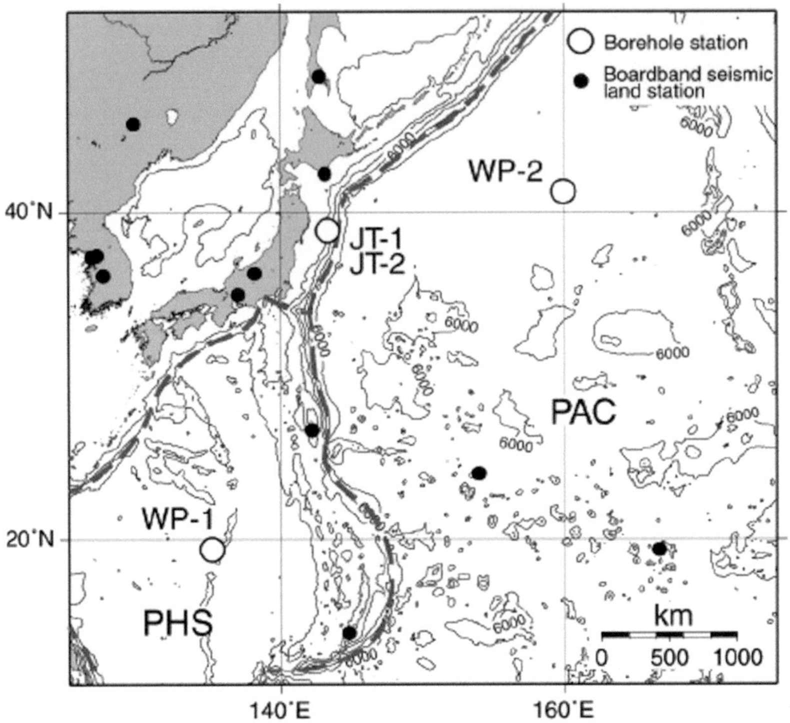

Figure 23.1 Location map of seismic station coverage in the northwest Pacific. Small circles indicate land seismic stations, large circles are seafloor borehole observatories. The JT-1 and JT-2 observatories were constructed in 1999. The WP-1 and WP-2 observatories were installed in August 2000 and April 2001, respectively. Water depths at JT-1 and JT-2 are 2681 and 2194m, respectively. Borehole depths at JT-1 and JT-2 are 1109 and 1084m below seafloor (mbsf), respectively. Water depths at WP-1 and WP-2 are 5721 and 5577m, respectively. Borehole depths at WP-1 and WP-2 are 561 and 460 mbsf, respectively. A few borehole stations can effectively expand the existing land-based network. PAC: Pacific plate; PHS: Philippine Sea plate.

Pacific, the Pacific plate and the Philippine Sea plate subduct below the Asian continent. For resolving problems related to plate subduction, the western Pacific is one of most suitable areas. Based on the results of previous studies using seafloor borehole seismometers, it is becoming clear that seismometers emplaced in boreholes drilled into the oceanic igneous crust experience a low-noise environment. (Kanazawa et al., 1992; Suyehiro et al., 1992; Montagner et al.,1994; Beauduin and Montagner, 1996; Romanowicz et al., 1998; Stephen et al., 1999; Collins et al., 2001). However, to date there has not been observational records covering a span of more than a half year from any seafloor borehole observatory.

For understanding the details of plate subduction, geodetic observations are important. On the Japanese islands, the many geodetic stations that include Global Positioning System (GPS) stations reveal various phenomena related to plate subduction. For example, a slow-slip event on the plate boundary was found after the 2003 Tokachi-oki earthquake by the GPS network (Miyazaki and Larson, 2008). Off Boso peninsula near Tokyo, Japan, slow-slip events on the plate boundary were observed by the GPS network (Ozawa et al., 2003, 2007). In addition, deep, long-period, non-volcanic tremor and slow-slip events in the Nankai trough region were observed by a land-based seismometer and tiltmeter network (Obara, 2002, Ito et al., 2007). These phenomena occurred under the landward slope of trenches and are believed to be related to the dynamics of plate subduction. Seafloor geodetic observations from the landward slope of subduction-zone trenches are required for a better understanding of plate subduction. The instruments employed for geodetic monitoring (e.g., tiltmeters and strainmeters) in marine areas must be installed in boreholes, because these sensors need good coupling to the seafloor.

Four borehole geophysical stations were installed at three sites covering a period from 1999 to 2001 (Figure 23.1). In 1999, two deep-sea borehole stations (JT-1 and JT-2) were installed during Ocean Drilling Program (ODP) Leg 186 (Sacks et al., 2000), and are located immediately above the interplate earthquake generation zone on the landward side of the Japan Trench. The JT-1 and JT-2 stations were installed approximately 50km apart and each station has two types of broadband seismometers, a strainmeter and a tiltmeter. In August 2000, during ODP Leg 191, the borehole seismological observatory WP-2 was successfully installed at ODP Site 1179 in the northwestern Pacific basin (Kanazawa et al., 2001). Borehole seismic observatory WP-1 was installed in the west Philippine basin during ODP Leg 195 in April 2001 (Salisbury et al., 2002). The WP-1 and WP-2 observatories each have two identical broadband seismometers. Together, these observatories cover a large expanse of the northwestern Pacific in areas that comprise major gaps in the coverage provided by pre-existing global seismological observatories. At present, the data from these four borehole observatories have been retrieved by Remotely Operated Vehicle (ROV) visits at an interval of about 0.5–1.5 year (Suyehiro et al., 2002, 2006; Shinohara et al., 2006). In this chapter, we describe the characteristics, layout, and system parameters of the borehole geophysical observatories, and the characteristics of the geophysical data retrieved from them, specifically the ambient seismic noise levels, temporal noise variations, and performance of the observatories. In addition, we discuss the advantages of real-time cabled networks for the future of seafloor geophysical stations.

23.2 Seafloor borehole geophysical observatories

The features seafloor borehole observatories have already been explained in detail (Sacks et al., 2000; Kanazawa et al., 2001; Salisbury et al., 2002). A brief systems-based description of the observatories is given here. The subsea borehole observatories were designed to work as stand-alone systems. All the observatory systems are basically identical except for the sensors and power supply (Figure 23.2). The JT-1 and JT-2 observatories have two kind of broadband seismometers (Guralp CMG-1TD and PMD 2123), tiltmeter (Applied Geomechanics model 510) and strainmeter (Sacks-Evertson type for JT-1 and Sakata-Sacks type for JT-2) (Sacks et al., 1971; Sakata, 1986). For the WP-1 and WP-2 observatories,

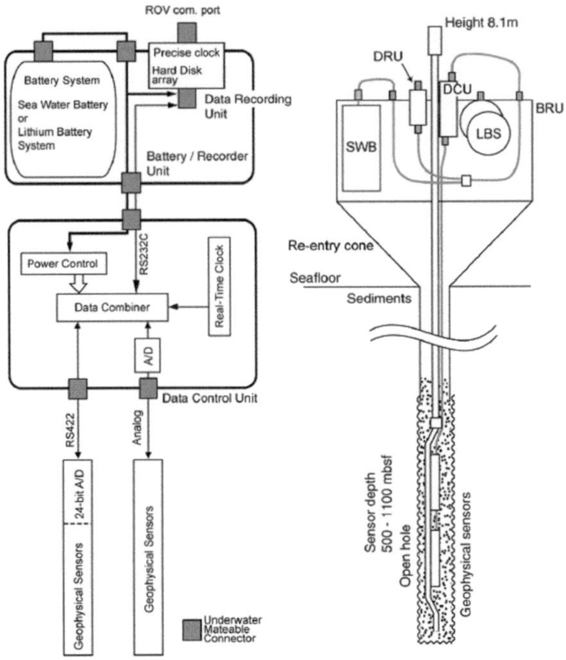

Figure 23.2 Schematic block diagram of the seafloor borehole observatory. Various geophysical sensors are used. The signals from the sensors are digitized at the bottom of the borehole or on the seafloor and are recorded to hard disks on the seafloor. A Sea Water Battery was used for the WP-2 observatory. DRU: data recording unit; DCU: data control unit; BRU: battery/recording unit; SWB: sea water battery; LBS: lithium battery system.

two identical broadband seismometers (Guralp CMG-1TD) were placed near the bottom of the borehole in igneous basement. The geophysical sensors were permanently emplaced in the borehole; they were cemented into the hole in order to improve coupling with the surrounding crust. The signals from the broadband seismometer (CMG-1TD) are digitized downhole in the sensor packages and then sent to the Data Control Unit (DCU) on seafloor. The signals from all sensors except the CMG-1TD are sent to the seafloor in analog form and digitized in the DCU. The DCU combines the digital data from various sensors into a single serial data stream. The data are then telemetered, using RS-232C protocol, to the Data Recording Unit (DRU). The DRU is deployed and recovered by ROV, and provides a communication link to the DCU, and from there to the downhole sensors. The status of the observatory can be checked via an "ROV interface" that enables RS-232C communications between each component of the observatory and the mother ship controlling the ROV. The DRU has four 30 G-byte SCSI hard disks, which is sufficient capacity to store a year's worth of 24-bit data from six channels sampled at 100 Hz.

Another major difference between the observatories is the electrical power supply system. For the WP-1 observatory, power is supplied from a Lithium Battery System (LBS) consisting of two units, each containing 16 large-capacity lithium cells. Eight of these cells are connected serially and two series are connected in parallel, and each unit has a capacity of 2600 Ah. Each battery unit consists of two titanium spheres with an external diameter of 65cm. The weight of each sphere with batteries in air and in water is 210kg and 50kg, respectively. The battery system has sufficient capacity to power the observatory for ~2.5 years. Electrical power for the WP-2 observatory is supplied from a Sea Water Battery (SWB) system (Shinohara et al., 2009), which can supply up to ~24 W with more than 400 kWh capacity. The power is derived from electrolytic dissolution of a magnesium anode in seawater. The SWB system at WP-2 was confirmed to be capable of generating the 6 W of power required by the observatory and to do so for at least 5 years (Shinohara et al., 2009). Power generation of the SWB depends on seawater conditions (salinity, temperature, etc). Although JT-1 and JT-2 also have SWB systems, performance of the SWB system at these sites was inadequate. Therefore, additional lithium batteries were used at JT-1 and JT-2. All the seafloor instruments and the magnesium anode of the SWB are designed to be replaceable by an ROV.

The geophysical sensors for the seafloor borehole observatories were installed from 1999 to 2001 on ODP cruises 186, 191, and 195, respectively. The boreholes at JT-1 and JT-2 have depths of about 1100m with water depths of 2680.7m and 2193.7m, respectively. The sensors at JT-1 and JT-2 were installed at the bottom of the boreholes where the P-wave velocity is about 2km/s and the bulk density is 1.7g/cm^3 (Sacks et al., 2000). The water depths at WP-1 and WP-2 are 5721m and 5577m, respectively. The boreholes at WP-1 and WP-2 were drilled through sedimentary rock into basaltic oceanic basement with a P-wave velocity of about 5km/s and a bulk density of 2.7g/cm^3. The depth of both boreholes is approximately 500m. The seismometers were installed in the basaltic sections of the boreholes.

After completion of drilling, a sensor-string assembly was lowered into the borehole. For the sensor-string assembly, the geophysical sensors were attached to the end of 4.5in-(11.4cm) diameter casing pipe, which ultimately is suspended from the base of the reentry cone. The DCU was attached at the other end of the sensor string assembly, and the sensors and the DCU were connected with separate cables for each sensor. To reduce stress on the cables, the cable has almost neutral buoyancy in seawater. The individual cables were bound to the casing pipe. The 4.5in casing pipe has centralizers that protect the cables from contact with borehole wall. The casing pipe is also used as a conduit for pumping cement to grout the sensors. After the deployment of the sensor string assembly at the targeted depths, cement was pumped from the drilling ship. For example, the seismometers were grouted at depths of 561m below seafloor (mbsf) and 460mbsf in basement rock 19m and 15m above the bottom of the boreholes at WP-1 and WP-2, respectively. After cementing of the sensors was complete, the Battery/Record Unit (BRU) was assembled and lowered through the moon pool of the drilling ship. After emplacement on the re-entry cone, the BRU was released from the drill pipe via acoustic signal. Finally, the drill pipe was decoupled from the sensor string assembly and recovered. One limitation of this installation procedure is that the DCU could not be connected to the BRU on the seafloor, because a TV camera was needed for re-entry of the sensor string assembly.

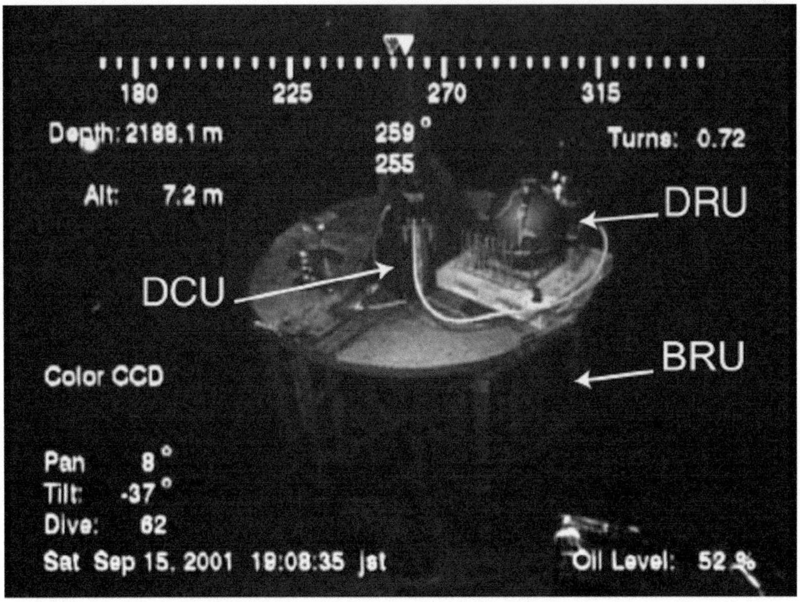

Figure 23.3 A ROV view of the JT-2 station during the maintenance of the system (September 15, 2001). The power is supplied from lithium batteries in the titanium spheres indicated as the DRU. Top of the battery frame is also used as an ROV landing pad.

After successful installation of the JT-1 and JT-2 observatories, an ROV activated both observatories in September 1999, approximately one month post installation. The observatories were not activated at the time of installation because the installation procedure precluded applying power at that time. The first dives at JT-1 and JT-2 were carried out by the ROV Dolphin 3K (depth capability of 3km) belonging to Japan Agency for Marine-Earth Science and Technology (JAMSTEC). During the first visit, operations necessary for starting borehole observation were carried out: connecting batteries to the DCU, initializing borehole sensors, and deployment of the DRU. After activation, system maintenance and data recovery were performed using the JAMSTEC ROVs; the Dolphin 3K, the Hyper Dolphin (depth capability of 3km), and the Kaiko (depth capability of 11km) (Figure 23.3). Visit intervals to JT-1 and JT-2 are approximately 1 year. In 2009, the Hyper Dolphin visited the JT-1 and JT-2, and discontinued observations. Records from the seismometers and tiltmeter were obtained throughout the observation period. Although the observation periods vary by sensor, both seismological and geodetic data were obtained at JT-1 and JT-2. Observations were sporadic due to power limitations. Data recording durations from the tiltmeters are longest, because geodetic observations have precedence. Tilt data of 14 months (July 2002–Jan. 2003, June 2003–Oct. 2003, July 2004–Oct. 2004, May 2006–

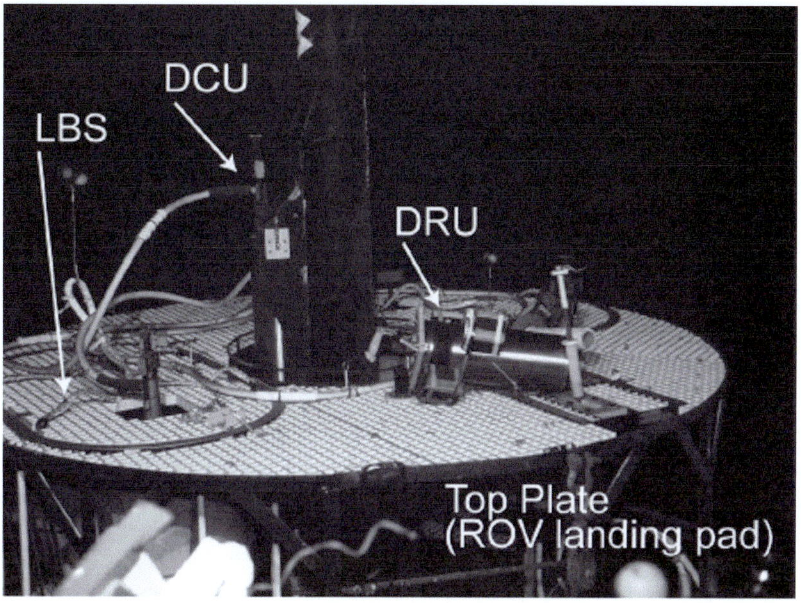

Figure 23.4 A view of the WP-1 station from the ROV Kaiko 7KII (June 16, 2006). The DRU is not in its proper position due to recovery operations. The power is supplied from lithium batteries. The WP-1 observatory was decommissioned in 2006.

Aug. 2006) have been retrieved from the JT-1. The JT-2 has the tilt data of 15 months (Sept. 2001–Dec. 2001, June 2002–Sept. 2002, June 2003–April 2004).

In March 2002, the ROV Kaiko visited the WP-1 site to activate the observatory, and long-term observations were started using the first LBS unit. Both seismometers at WP-1 were confirmed to be working properly via an RS-232C link to the JAMSTEC mother ship R/V Kairei. One seismometer was then deactivated to decrease power consumption. In October 2002, the Kaiko revisited the WP-1 site and approximately 6 months of continuous data were obtained. The Kaiko visited the WP-1 site again in May 2003 and about 7 months of continuous data were retrieved. During the third visit, the Kaiko changed an electric connection and the second LBS unit was put on line to supply the power to the system. Unfortunately, the Kaiko was lost after the third visit and the dive to the WP-1 was postponed until 2006. In June 2006, the new JAMSTEC-owned ROV Kaiko-7KII (depth capability of 7km) dived to WP-1 (the fourth visit) and recovered the data (Figure 23.4). At this visit, data recording was discontinued. In total, seismic records of 692 days duration (March 2002–Feb. 2004) have been obtained from the WP-1.

The WP-2 observatory was activated in October 2000 using the ROV Kaiko. Due to uncertainties about the performance of SWB, the WP-2 observatory was operated using lithium batteries during the first observation period. The DRU and lithium battery pack were installed in a 65cm diameter titanium sphere. A total lithium battery capacity of 1300 AH was estimated to be required in order to supply power for more than three months. Only one seismometer was activated to minimize power consumption, similar to the case for WP-1. In August 2001, the Kaiko visited the WP-2. Approximately three months of continuous data were recovered. During the second visit, it was confirmed that the SWB system was working properly, and long-term observations using both the SWB and backup lithium batteries commenced, with the lithium pack acting as a backup to the SWB. The backup lithium battery system can supply the power for approximately three months. The Kaiko visited the WP-2 site again in June 2002, and approximately 11 months of continuous data were retrieved for the second observation period. The loss of the Kaiko meant that WP-2 was not visited again until 2005. In June 2005, the Kaiko-7KII made the fourth visit to WP-2, recovered the data (Figure 23.5), and suspended recording. In total, 436 days of data (Oct. 2000–Jan. 2001, Aug. 2001–July 2002) were retrieved. In addition, it was confirmed that the SWB system continued to work for the whole observation period (Shinohara et al., 2009).

23.3 Geophysical records and results from the seafloor borehole observatories

23.3.1 Geodetic records

The JT-1 and JT-2 observatories have broadband seismic sensors, tiltmeters, and volumetric strainmeters, which were grouted at the bottom of ODP holes, where the P-wave velocity is ~2km/s at a depth of ~1100mbsf (Sacks et al., 2000), i.e., in the sedimentary section rather than in the igneous crust. From the JT-1 and JT-2 observatories, we obtained the records from durations of a few months to more than one year. The tiltmeter record is longest, since geodetic data are essential to understanding the dynamics of plate subduction. These tiltmeter data from a seafloor borehole are unique. Although the tiltmeters were fixed in the sedimentary section, the tilt data are of good quality (Figure 23.6). The records from the tiltmeters clearly show earth tides that are comparable with predictions from the gravity theory. Unfortunately, we could not detect events which occurred at the plate boundary such as slow-slip events. The resolution of the tilmeters is such that they were capable of detecting slow-slip events, which have an amplitude comparable to earth tides (e.g., Hirose and Obara, 2005). The tilt data from the borehole observatories convince us that the borehole observatories will play an important role in seafloor geodesy studies.

23.3.2 Seismological records

The seismic data from the seafloor geophysical observatories enable the study of broadband seismic noise in the frequency range from 3 mHz to 50 Hz in the northwestern Pacific. The WP-1 and WP-2 sites, in particular, have more than one year of continuous seismic

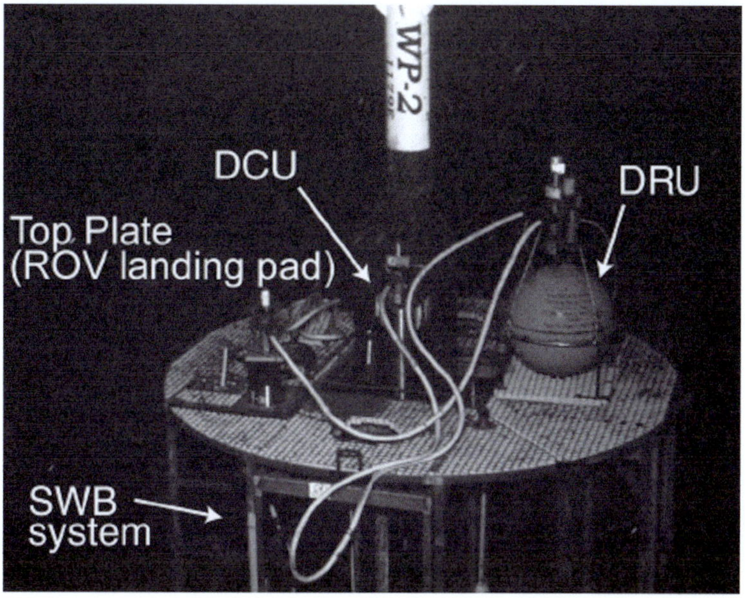

Figure 23.5 Photograph of the WP-2 station taken by the ROV Kaiko 7KII (July 14, 2005). The power is mainly supplied from the Sea Water Battery system. The DRU uses a titanium sphere pressure vessel with a diameter of 65cm. Observations were suspended from 2005.

data. Temporal changes in ambient seismic noise within deep-sea boreholes can be analyzed in detail using these data.

To investigate characteristics of seismic noise, power spectra were calculated using moving windows of one hour duration and grouped by season. Noise levels for the vertical and horizontal seismometer components at the WP-1 and WP-2 observatories are stable for periods of greater than 10s (Shinohara et al., 2006). Noise levels on the vertical component at WP-1 reach −180 dB relative to $1m^2/s^4/Hz$ for periods between 10s and 100s (Figure 23.7). On the other hand, small temporal changes in noise levels at periods around a few seconds are seen. The spectral peaks around a few seconds correspond to microseisms which are believed to originate from ocean waves (Longuet-Higgins, 1950). The seismometer at the WP-1 observatory recorded relatively large noise levels in summer and fall. The noise levels on the horizontal components on the WP-2 seismometer are −160 dB for periods between 10s and 100s (Figure 23.8). The records from the WP-2 observatory reveal high noise levels in a period range of a few seconds during winter. Although the JT-1 has no records in winter, the noise levels on the horizontal components reach −180 dB at periods of 30s (Figure 23.9). The JT-1 observatory has a peak in noise levels near a period of 100s on the horizontal components; however, the WP-1 and WP-2 observatories have no such peak. In general, the peak in the noise spectrum near a period of 100s in

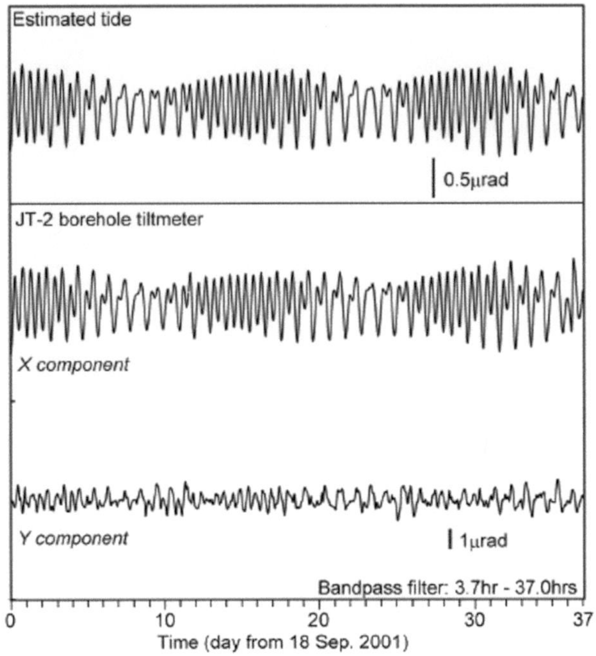

Figure 23.6 Records from the tiltmeter installed in JT-2. The tiltmeter was grouted at the bottom of the hole. The upper trace is a prediction based on the gravity theory; the lower two traces were recorded by the tiltmeter in the JT-2. One component of the tiltmeter record closely follows the calculated tide.

measurements taken at the seafloor or in a shallow borehole is due to deformation of the seafloor by ocean gravity waves (Webb, 1998). The peak on the records at the JT-1 site is explained by ocean gravity waves, and we infer that WP-1 and WP-2 are not sensitive to this noise source due to their installation into the basement deep below the seafloor (Araki et al., 2004). For both the vertical and the horizontal components of the seafloor borehole observatories, the noise levels are close to the Low Noise Model (Peterson, 1993) near 10s period. The horizontal-component records of the WP-1 observatory have much larger noise levels than the vertical-component records for periods longer than 20s. There is a possibility that the horizontal sensors in the WP-1 seismometer have a coupling problem. In addition, the noise level of the vertical component for period longer than 10s in the WP-2 observatory is larger than those of the horizontal components. There is a possibility that the vertical sensor of the WP-2 observatory was damaged during the installation.

It is important to quantitatively assess the performance of the deep-sea borehole seismometers for future study. We compared an earthquake catalog generated from the vertical-component data recorded at WP-2 to the USGS QED catalog. Within the frequency band 20–500s, the deep-sea borehole observatory proved to record all the teleseismic

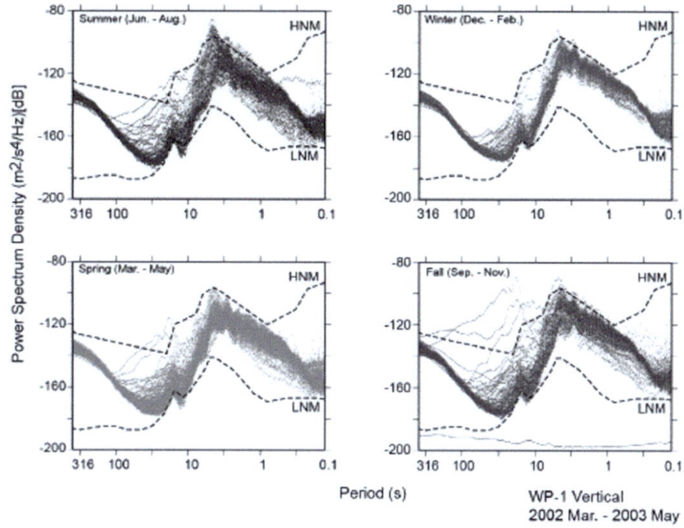

Figure 23.7 Power spectra of vertical-component seismic noise recorded by the borehole seismometer at the WP-1 observatory. Power spectra were estimated using one-hour records from the entire observation period. HNM and LNM indicate the high noise model and the low noise model of Peterson (1993), respectively. The power spectra are grouped by seasons.

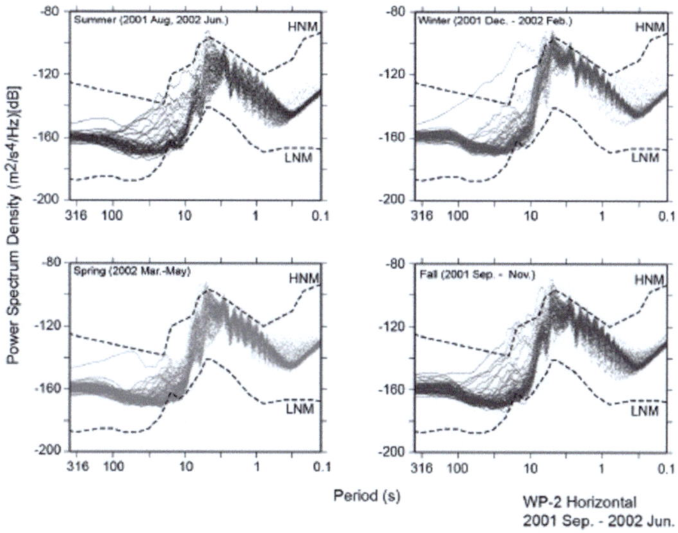

Figure 23.8 Power spectra of horizontal-component seismic noise recorded by the borehole seismometer at the WP-2 observatory. See Figure 23.7 for explanation. The WP-2 observatory has low noise in summer.

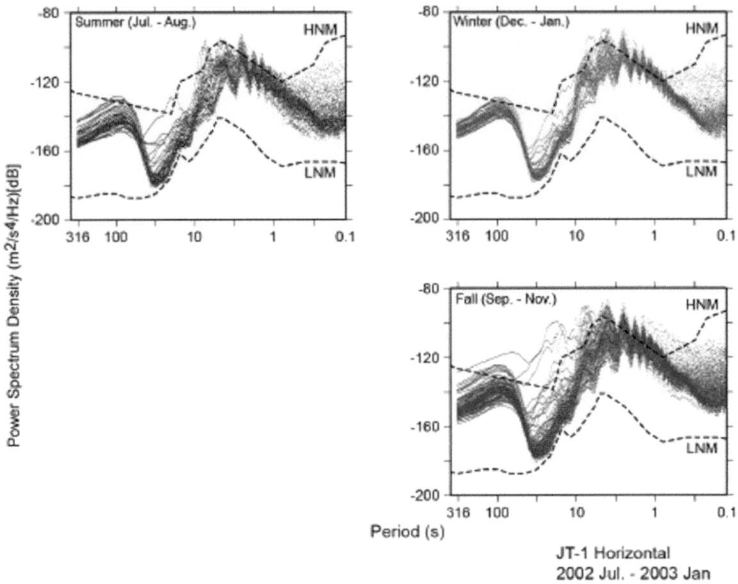

Figure 23.9 Power spectra of horizontal-component seismic noise recorded by the borehole seismometer at the JT-1 observatory. See Figure 23.7 for explanation. The borehole seismometers at the JT-1 and JT-2 observatories, which were installed in the sedimentary layer, have a peak in noise levels around a period of 100s.

events with magnitudes (mb[5] or Ms[6]) greater than 5. Because there is little temporal variation in ambient seismic noise levels, we can confirm more than one seismic event per day. The WP-2 observatory records many events that are not included in the QED catalog. Most of these events were identified by the Japanese seismic network and their magnitudes are less than 4. The epicentral distances of the regional events are less than 2000km. The deep-sea borehole seismic observatories reside in a quiet environment suitable for earthquake observation. Compared with observation on the seafloor, stable ambient seismic noise levels in seafloor boreholes enables us to record more teleseismic events for the same observation duration. Figures 23.10 and 23.11 show vertical component seismograms of events with magnitude (mb or Ms) greater than 6.5 recorded at the WP-1 and WP-2 stations, respectively. The events selected for plotting were determined from the USGS Preliminary Determinations of Epicenters (PDE) catalog. On both record sections, arrivals of various phase are clearly recognized.

5 mb means the body wave magnitude determined by using the amplitude of the initial P-wave.

6 Ms indicates the surface wave magnitude calculated based on Rayleigh surface waves which propagate along the uppermost layers of the Earth.

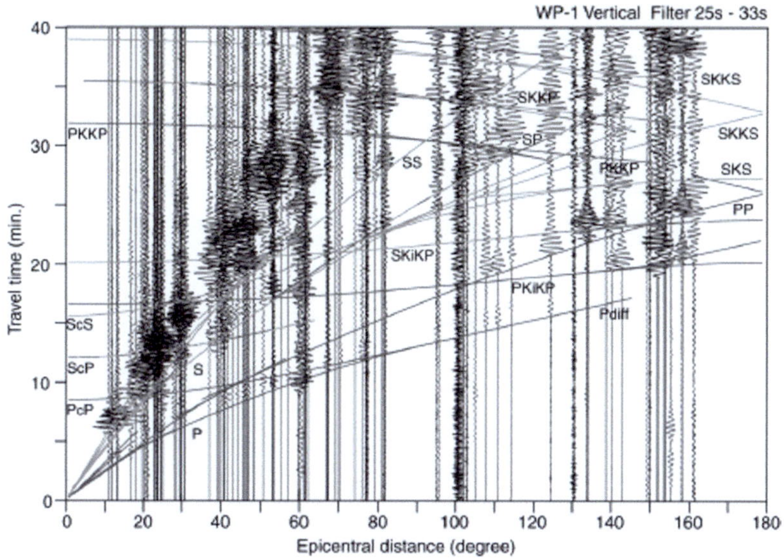

Figure 23.10 Record section of vertical-component data from WP-1. Earthquakes with a magnitude greater than 6.5 are plotted. All traces are plotted from origin times. A bandpass filter (25–33s) was applied and each trace is normalized by its maximum amplitude. Curves indicate travel times expected from the global Earth model: AK-135 (Kennett et al., 1995).

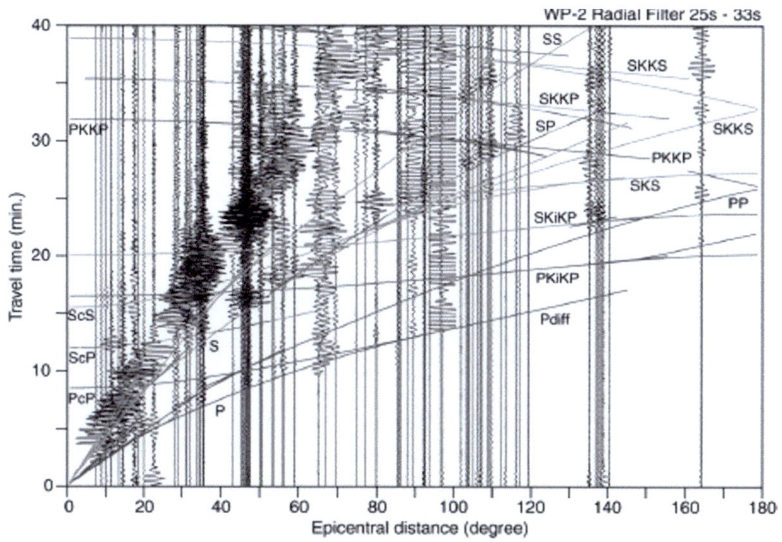

Figure 23.11 Record section of vertical component data from WP-2. See Figure 23.10 for explanation. Earthquakes with a magnitude greater than 6.0 and a good signal-to-noise ratio were selected. Each trace is normalized by its signal-to-noise ratio.

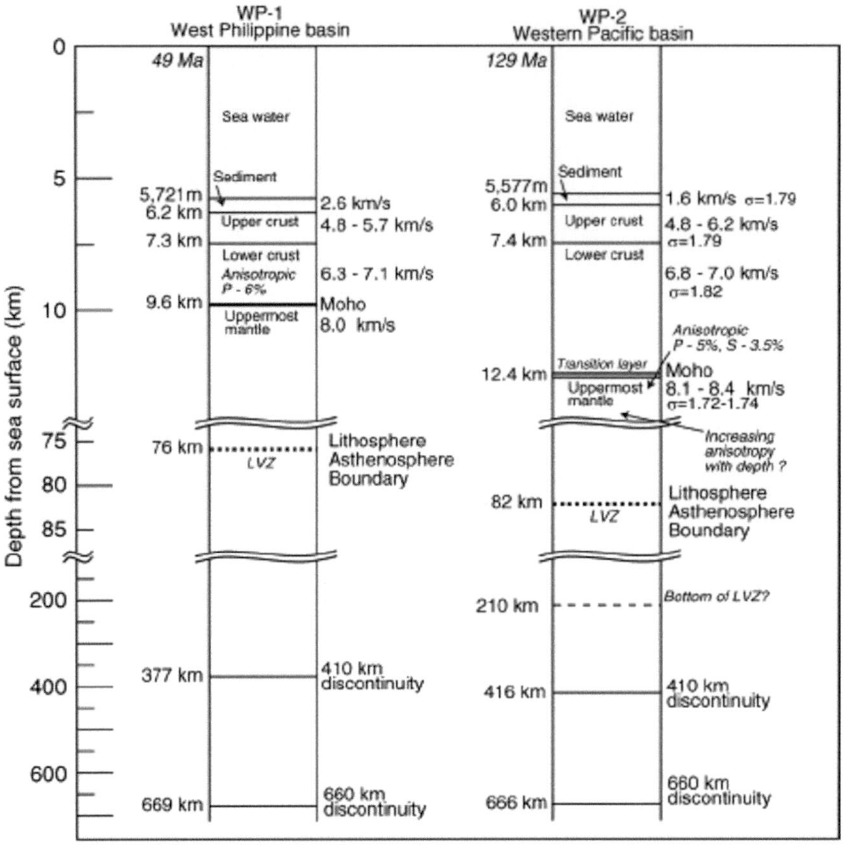

Figure 23.12 Summary of velocity structures beneath the WP-1 and WP-2 observatories. The uppermost mantle and crustal structures were determined by an active seismic sources experiment including the seafloor borehole observatories (Salisbury et al., 2006; Shinohara et al., 2008). The deep structures are estimated using the records of the teleseismic events recorded by the seismometers at WP-1 and WP-2 (Suetsugu et al., 2005; Shinohara et al., 2008; Kawakatsu et al., 2009).

Several deployments of broadband seismometers in seafloor boreholes have been carried out. Ambient seismic noise spectra were obtained in the Atlantic Ocean and the Pacific Ocean off Hawaii (Montagner et al., 1994; Beauduin and Montagner, 1996; Stephen et al., 1999; Collins et al., 2001). We compare the ambient seismic noise spectra from our observatories with those in previous seafloor borehole observations (Shinohara et al., 2006). For both vertical and horizontal components, noise levels at periods smaller than a few seconds are similar; however, there are some differences for periods greater than 10s. The difference of noise levels at periods greater than 10s among the seafloor borehole observations reaches 40dB. Ocean gravity waves are a possible explanation for the large variation in long-period noise. The JT-1 site has a peak in noise levels near a period of 100s that is explained by ocean gravity waves (Araki et al., 2004). A broadband seismometer emplaced within sediments is strongly affected by ocean gravity waves. Another possibility is differences in seismometer coupling to the seafloor. Insufficient coupling should increase seismic noise, especially on the horizontal components. Seismometer damage during installation is another possibility. Generally, the noise levels of our seafloor borehole observatories are comparable with those of previous observations in seafloor boreholes and land seismic stations. We confirm that deep-sea boreholes are the quietest sites for seafloor earthquake studies.

23.3.3 Earth structure from the deep-sea borehole observatories

During every ROV visit, controlled-source seismic experiments were performed to obtain P- and S-wave crustal and upper mantle structure. These experiments utilized temporary deployments of ocean bottom seismometers (OBS) as well as the borehole seismometers (Salisbury et al., 2006, Shinohara et al., 2008). Mantle structure beneath the WP-1 and WP-2 observatories was estimated from teleseismic records (Suetsugu et al., 2005, Shinohara et al., 2008, Kawakatsu et al., 2009) (Figure 23.12).

Site WP-1 is positioned in the middle of the West Philippine Basin behind the Mariana arc and in front of the Ryukyu and Philippine arc-trench systems. A total of three refraction lines were shot to an array laid out in a cross centered at WP-1. One profile, 100km long, was aligned parallel to the paleospreading direction (Hilde and Lee, 1984). The other two lines with a combined length of 120km, were shot normal to the spreading direction and across the extinct Palau-Kyushu Ridge. Twelve OBSs were used in total for the three surveys and air guns with a total size of 100 liters were used as the seismic source. The tau-p inversion method (Shinohara et al., 1994) and two-dimensional ray tracing (Hirata and Shinjo, 1986) were used to estimate the structure. The analysis indicates that basement is covered by ~500m of sediments with an average P-wave velocity of 2.6km/s. The upper crust has thickness of 1–1.2km and seems to be isotropic. P-wave velocities in the upper crust are from ~4.8km/s at the top to 5.7km/s at the bottom. The lower crust is about 2.3km thick with P-wave velocities in the east–west direction increasing with depth from 6.7km/s at the top of the layer to 7.1km/s at the bottom in the east–west direction. On the other hand, lower crustal P-wave velocities in the north–south direction increase with depth from 6.3 to 6.7km/s. Seismic anisotropy in the upper crust of the ocean basins is commonly observed (e.g., Tong *et al.*, 2004). Although anisotropy in the lower oceanic crust is unusual, one possibility of seismic anisotropy in the lower crust is the existence of open cracks normal to spreading as well as the upper oceanic crust. The upper mantle lies 3.3–3.8km below the top of basement and has a P-wave velocity of 8.0km/s in the east–west direction. Mantle anisotropy is not confirmed, because the Pn velocity in the north–south direction could not be obtained due to no deep penetration observed by limitation of the profile length. The total combined thickness of igneous crust in the vicinity of WP-1 is 3.3–3.8km. Although this is approximately half the thickness of normal ocean crust, other refraction results also show that the crust in the Philippine Sea is anomalously thin (Goodman et al., 1989; White et al., 1992).

Kawakatsu et al. (2009) applied sophisticated receiver function analysis to the WP-1 records of teleseismic events and found that there is a velocity boundary below WP-1 at a depth of 76km. The polarity of the receiver functions implies that this interface represents a decrease in velocity with depth, and hence is interpreted to be the lithosphere-asthenosphere boundary.

During the observation period at WP-1, a pop-up type Broadband Ocean Bottom Seismometer (BBOBS) (Kanazawa et al., 2009) was deployed and collected teleseismic records. The broadband waveform data recorded by the WP-1 borehole seismometer and the BBOBS were used to estimate the depths of the mantle discontinuities beneath the west Philippine basin using the receiver function method (Suetsugu et al., 2005). The Ps converted waves from mantle discontinuities were detected on both the WP-1 and BBOBS records. The results are consistent with each other. The depth of the "660km" mantle discontinuity determined using both the WP-1 and BBOBS data was 669km. The estimation error is 9km. Therefore, it is concluded that the depth of the mantle discontinuity around 660km beneath the west Philippine basin is 9km deeper than the global average. A similar analysis shows that the depth of the "410km" discontinuity at WP1 is 377km. The slightly deeper 660km discontinuity can be interpreted to indicate lower temperatures than the global average at a depth of 660km. On the other hand, the elevation of the 410km discontinuity must be due to another cause, for example an anomaly in mantle composition (Salisbury et al., 2006).

The WP-2 observatory is located on 129 Ma lithosphere (Nakanishi et al., 1989). As the case of the WP-1, seismic experiments with OBSs and airguns were performed during every cruise for the ROV visits to the WP-2. To detect seismic anisotropy in the uppermost mantle, the experiments had four profiles with different directions. One profile was perpendicular to the magnetic lineations, and other profile was almost parallel to the lineations. In addition, there are two profiles whose directions are between those. Length of those profile are about 100km. Altogether, airgun data were recorded by 13 OBSs and the WP-2. These data were used for determination of the uppermost mantle and crustal structure. In addition, single-channel-seismic (SCS) reflection records were used. For estimation for seismic structure of crust and uppermost mantle, the tau-p inversion method (Shinohara et al., 1994) and two-dimensional ray tracing (Hirata and Shinjo, 1986) were performed. As a result, it is found that crustal thickness is about 7km beneath the northwestern Pacific basin and the crustal seismic structure around WP-2 is laterally homogeneous. The P- and S-wave structure of the crust and uppermost mantle corresponds to typical oceanic crust. To explain the large-amplitude Pn phase, a crust–mantle transition layer is suggested. Seismic anisotropy in the uppermost mantle is detected beneath the WP-2. The fast direction is almost perpendicular to the magnetic lineations (140°, 320°), and the magnitudes of anisotropy of the uppermost mantle are about 5% for P-waves and about 3.5% for S-waves, respectively (Shinohara et al., 2008). Travel times of earthquakes occurred in Kuril trench region and Izu-Ogasawara trench region recorded by the WP-2, and previous seismological studies (Asada et al., 1984; Shimamura and Asada, 1984), suggest that the lower part of the lithosphere has greater anisotropy than the uppermost mantle. To explain late first arrivals from the earthquakes that occurred in the slow direction with epicentral distances between 1600 and 2200km, a low velocity below a depth of 30km and a rapid increase of velocity at a depth of 210km are inferred. A rapid increase of P-wave velocity at a depth of 210km may correspond to the bottom of the low-velocity zone. Kawakatsu et al. (2009) carried out the same analysis as the WP-1 to the WP-2 records. They concluded that a sharp velocity boundary below the WP-2, which corresponds to the lithosphere-asthenosphere boundary, exists at a depth of 82km. A thickness of approximately 80km for the lithosphere beneath the WP-2, a thickness of the low-velocity zone becomes 130km. Shinohara et al. (2008) performed the receiver function analysis, and reported that the depths of the upper mantle discontinuity beneath the WP-2 are 416 and 666km. Considering estimating errors, the seismic structure of the deep region beneath the WP-2 is considered to be typical beneath the ocean.

23.4 Seafloor cabled observatories

Recently, seafloor-cabled, real-time observation networks have been installed and more are planned. For examples, the JAMSTEC has deployed the Dense Oceanfloor Network system for Earthquakes and Tsunamis (DONET) (Kawaguchi et al., 2009) in the Nankai trough region south of Honshu. The DONET has the science nodes connected to the seafloor-cable permanently, and seafloor instruments deployed after the cable system installation can be connected using underwater-mateable connectors mounted on the science nodes. The use of underwater-mateable connectors makes it possible to replace seafloor instruments.

Kanazawa and Shinohara (2009) describe a new compact Ocean Bottom Cabled Seismometer (OBCS) system that was first deployed in the Japan Sea in 2010. Observation Nodes (ONs) of this system are connected by fiber-optic cable and installed in a continuous S-shaped pattern to distribute the ONs two-dimensionally. The utilization of up-to-date Information and Communication Technologies (ICT) is essential to reduce the size of the ONs and to ensure low-cost deployment. The reliability of the system is maintained by using a redundant system that employs ITC. Each ON in the OBCS has servo-type accelerometers. Pressure gauges could also be connected to the ON. A problem of existing cabled systems is the small number of nodes. However, because the ONs are of low cost, the OBCS system has many nodes. This means that the OBCS can provide spatially dense observations on the seafloor. Because the OBCS uses ICT for data transmission and system control, each ON has Ethernet switching hubs. Therefore, it is easy to add a port to connect an observation system on the seafloor after OBCS deployment. The port uses underwater mateable connectors that supply power and internet connectivity.

Connection of the borehole observatories described here to seafloor cables is feasible because the observatories have underwater mateable connectors and low-power instrumentation. It is important to cooperate with Integrated Ocean Drilling Program (IODP) for establishment of standard technology of construction of a borehole observation system. An advantage of connecting a borehole observation system to a seafloor cable system is a stable power supply from a cable system and real-time data transmission to land via a cable. The combination of a dense seafloor observation network and borehole observatories will be indispensable for researches of plate subduction.

23.5 Conclusions

We deployed and operated successfully four borehole geophysical observatories in the western Pacific area from 1999 to 2001. The four seafloor borehole observatories were installed in cooperation with ODP. The data from those seafloor borehole observatories were retrieved by ROV visits at an interval of about 0.5–1.5 year. The recording periods differ by the sensors, since the observation is sporadic due to limitation of power supply, but continuous seismic data of 700 and 400 days was recovered from WP-1 and WP-2, respectively. From an analysis of the data, we confirm that deep-sea boreholes offer the best environment for seafloor seismic measurements. Controlled seismic experiments were carried out at each observatory site in order to crust and uppermost mantle velocity structure. Analysis of teleseismic data recorded by the borehole seismometers allowed the determination of mantle structure.

It would not be difficult to connect the borehole system to a seafloor cabled system, because the borehole system works with low voltages and uses underwater-mateable connectors connection for the seafloor packages. Power delivery and real-time data transmission via a cable system would allow for long-term observation using boreholes. In order to readily connect seafloor observatories to cable-based networks, is important to establish common interface standards for seafloor instrumentation.

Acknowledgments

The success of the observations was made possible by the contributions of many scientists, engineers, technicians, ship's officers and crew, and ROV operators from various institutions around the world. In particular, we express thanks to Drs William W. Sager, Matthew H. Salisbury, Carlota Escutia, Carl Richter, Mr Mike Storms and many people from ODP who led successful installations of the borehole observatories. For the skillful maneuvers of the ROV by the operation team, JAMSTEC are grateful. We thank Drs Ralph A. Stephen, Hajime Shiobara, Tomoaki Yamada, Kimihiro Mochizuki, Yuka Kaiho and Kazuo Nakahigashi for fruitful discussions and active support on this study. This study was partly supported by the Ocean Hemisphere Project from the Ministry of Education, Culture, Sports, Science and Technology of Japan. Most of the figures were created using GMT (Wessel and Smith, 1998).

References

Araki E., Shinohara M., Sacks S., Linde A., Kanazawa T., Shiobara H., Mikada H. and Suyehiro K. (2004) Improvement of seismic observation in the ocean by use of seafloor boreholes. Bull. Seism. Soc. Am. 94, 678–690.

Asada T., Shimamura H., Asano S., Kobayashi K. and Tomoda Y. (1984) Explosion seismological on long-range profiles in northwestern Pacific and the Mariana sea.

In: T.W.C. Hilde and S. Uyeda (Eds) *AGU Geodynamics Ser., 11*. Am. Geophys. Union,Washington, DC, pp. 105–120.

Beauduin R. and Montagner J.P. (1996) Time evolution of broadband seismic noise during the French pilot experiment OFM/SISMOBS. Geophys. Res. Lett. 23(21), 2995–2998.

Collins J.A., Vernon F.L., Orcutt J.A., Stephen R.A., Peal K.R., Wooding F.B., Spiess F.N. and Hildbrand J.A. (2001) Broadband seismology in the oceans: Lessons from the Ocean Seismic Network Pilot Experiment. Geophys. Res. Lett. 28(1), 49–52.

Goodman D., Bibee L.D. and Dorman, L.M. (1989) Crustal seismic structure beneath the West Philippine Sea, 17°–18°N. Mar. Geophys. Res. 11, 155–168.

Hilde T.W.C. and Lee C.S. (1984) Origin and evolution of the West Philippine Basin: A new interpretation. Tectonophysics 102, 85–104.

Hirata N. and Shinjo N. (1986) SEISOBS—modified version of seis83 for ocean bottom seismometers. *Zisin*. J. Seismol. Soc. Jpn. 2(39), 317–321 (in Japanese).

Hirose H. and Obara K. (2005) Repeating short- and long-term slow slip events with deep tremor activity around the Bungo channel region, southwest Japan. Earth Planet. Space 57, 961–972.

ION/ODP International Workshop (1995) Multidisciplinary observatories on the deep seafloor. France: Marseille.

Ito Y., Obara K., Shiomi K., Sekine S. and Hirose H. (2007) Slow earthquakes coincident with episodic tremors and slow slip events. Science 315, 503–506.

Kanazawa T., Suyehiro K., Hirata N. and Shinohara M. (1992) Performance of the ocean broadband downhole seismometer at Site 794. In: *Proceedings of the Ocean Drilling Program, Scientific Results*, 127/128, pt2, pp. 1157–1171.

Kanazawa T., Sager W.W., Escutia C., et al. (2001) *Proc. ODP, Init. Repts.*, 191 [CD-ROM]. Available from: Ocean Drilling Program, Texas A&M University, College Station TX 777845-9547, USA.

Kanazawa T. and Shinohara M. (2009) A new, compact ocean bottom cabled seismometer system: Development of compact cabled seismometers for seafloor observation and a description of first installation plan. Sea Technology 37–40.

Kanazawa T., Shinohara M. and Shiobara H. (2009) Recent progress in seafloor earthquake observations and instruments in Japan. *Zisin*. J. Seismol. Soc. Jpn. 2, 61, S55–S68 (in Japanese).

Kawaguchi K., Kaneko S., Nishida T. and Komine T. (2009) Cable laying ROV for real-time seafloor observatory network construction. *OCEANS'09 IEEE Bremen*.

Kawakatsu H., Kumar P., Takei Y., Shinohara M., Kanazawa T., Araki E. and Suyehiro K. (2009) Seismic evidence for sharp lithosphere-asthenosphere boundaries of oceanic plates. Science 324(5926), 499–502, doi: 10.1126/science.1169499.

Kennett B.L.N., Engdahl E.R. and Buland R. (1995) Constraints on seismic velocities in the Earth from travel times. Geophys. J. Int. 122, 108–124.

Longuet-Higgins M. S. (1950) A theory of the origin of microseisms. Philos. Trans. R. Soc. London A 243, 1–35.

Miyazaki S. and Larson K.M. (2008) Coseismic and early postseismic slip for the 2003 Tokachi-oki earthquake sequence inferred from GPS data. Geophys. Res. Lett. 35, L04302, doi:10.102.

Montagner J.P., Karczewski J.F., Romanowicz B., Bouaricha S., Lognonne P., Roult G., Stutzmann E., Thirot J.L., Brion J., Dole B., Fouassier D. et al. (1994) The French pilot experiment OFM-SISMOBS: First scientific results on noise level and event detection. Phys. Earth Planet Inter. 84, 321–336.

Nakanishi M., Tamaki K. and Kobayashi K. (1989) Mesozoic magnetic anomaly lineations and seafloor spreading history of the northwestern Pacific. J. Geophys. Res. 94, 15, 437–15, 426.

Obara K. (2002) Nonvolcanic deep tremor associated with subduction in Southwest Japan. Science 296(5573), 1679–1681, doi: 10.1126/science.1070378.

Ozawa S., Miyazaki S., Hatanaka Y., Imakiire T., Kaidzu M. and Murakami M. (2003) Characteristic silent earthquakes in the eastern part of the Boso Peninsula, central Japan. Geophys. Res. Lett. 30, doi:10.1029/2002GL016665.

Ozawa S., Suito H. and Tobita M. (2007) Occurrence of quasi-periodic slow-slip off the east coast of the Boso Peninsula, central Japan. Earth Planet. Space 59, 1241–1245.

Peterson J. (1993) Observations and modeling of seismic background noise. Open-File Report 93-322, U.S. Department of Interior Geological Survey.

Romanowicz B., Stakes D., Montagner J.P., Tarits P., Uhrhammer R., Begnaud M., Stutzmann E., Pasyanos M., Karczewski J.F., Etchemendy S. and Neuhauser D. (1998) MOISE: A pilot experiment towards long term seafloor geophysical observatories. Earth Planets Space 50, 927–937.

Sacks I.S., Suyehiro S., Evertson D.W. and Yamagishi Y. (1971) Sacks-Evertson strainmeter: Its installation in Japan and some preliminary results concerning strainsteps. Pap. Meteorol. Geophys. 22, 195–208.

Sacks I.S., Suyehiro K., Acton G.D. et al. (2000) *Proc. ODP, Init. Repts.* 186 [CD-ROM]. Available from: Ocean Drilling Program, Texas A&M University, College Station TX 777845-9547, USA.

Sakata, S. (1986) Borehole-type tiltmeter and three-component strainmeter. J. Phys. Earth 34.

Salisbury M.H., Shinohara M., Richter C. et al. (2002) *Proc. ODP, Init. Repts.*, 195 [CD-ROM]. Available from: Ocean Drilling Program, Texas A&M University, College Station TX 777845-9547, USA.

Salisbury M.H., Shinohara M., Suetsugu D., Arisaka M., Diekmann B., Januszczak N. and Savov I.P. (2006) Leg 195 synthesis: Site 1201 – a geological and geophysical section in the West Philippine Basin from the 660-km discontinuity to the mudline. In: M. Shinohara, M.H. Salisbury and C. Richter (Eds) *Proc. ODP, Sci. Results,* 195. College Station (Ocean Drilling Program), pp. 1–31.

Shimamura H. and Asada T. (1984) Velocity anisotropy extending over the entire depth of the oceanic lithosphere. In: T.W.C. Hilde and S. Uyeda (Eds) *Geodynamics Ser.,* **11**. Am. Geophys. Union, pp. 121–125.

Shinohara M., Hirata N. and Takahashi N. (1994) High resolution velocity analysis of ocean bottom seismometer data by the τ-p method. Marine Geophys. Res. 16, 185–199.

Shinohara M., Araki E., Kanazawa T., Suyehiro K., Mochizuki M., Yamada T., Nakahigashi K., Kaiho Y. and Fukao Y. (2006) Deep-sea borehole seismological observatories in the western Pacific: Temporal variation of seismic noise level and event detection. Annals of Geophysics 49(2/3), 625–641.

Shinohara M., Fukano T., Kanazawa T., Araki E., Suyehiro K., Mochizuki M., Nakahigashi K., Yamada T. and Mochizuki K. (2008) Upper mantle and crustal seismic structure beneath the Northwestern Pacific Basin using seafloor borehole broadband seismometer and ocean bottom seismometers. Phys. Earth Planet Inter. 170, 95–106, doi: 10.1016/j.pepi.2008.07.039.

Shinohara M., Araki E., Mochizuki M., Kanazawa T. and Suyehiro K. (2009) Practical application of a sea-water battery in deep-sea basin and its performance. J. Power Sources 187, 253–260.

Suetsugu D., Shinohara M., Araki E., Kanazawa T., Suyehiro K., Yamada T., Nakahigashi K., Shiobara H., Sugioka H., Kawai K. and Fukao Y. (2005) Mantle discontinuity depths beneath the West Philippine Basin from receiver function analysis of deep-sea borehole and seafloor broadband waveforms. Bull. Seism. Soc. Am. 95(5), 1947–1956, doi: 10.1785/0120040169.

Suyehiro K., Kanazawa T., Hirata N., Shinohara M. and Kinoshita H. (1992) Broadband downhole digital seismometer experiment at Site 794: A techinical paper. In: *Proceedings of the Ocean Drilling Program, Scientific Results, 127/128*, pt2, pp. 1061–1073.

Suyehiro K., Araki E., Shinohara M. and Kanazawa T. (2002) Deep sea borehole observatories ready and capturing seismic waves in the western Pacific. EOS Trans. AGU 83(53), 621–625.

Suyehiro K., Montagner J.P., Stephen R.A., Araki E., Kanazawa T., Orcutt J., Romanowicz B., Sacks S. and Shinohara M. (2006) Ocean seismic observatories. Oceanography 19(4), 104–109.

Stephen R.A., Collins J.A. and Peal K.R. (1999) Seafloor seismic stations perform well in study. EOS Trans. AGU 80, 592.

Tong C.H., White R.S., Warner M.R. and ARAD Working Group (2004) Effects of tectonism and magmatism on crack structure in oceanic crust: A seismic anisotropy study. Geology 32(1), 25–28. doi: 10.1130/G19962.1.

Webb S.C. (1998) Broadband seismology and noise under the ocean. Reviews of Geophysics 36, 105–142.

Wessel P. and Smith W.H.F. (1998) New improved version of the Generic Mapping Tools released. EOS Trans. AGU 79, 579.

White R.S., McKenzie D. and O'Nions R.K. (1992) Oceanic crustal thickness from seismic measurements and rare earth element inversions. J. Geophys. Res. 97, 19,683–19,715.

24 A first insight into the Marsili volcanic seamount (Tyrrhenian Sea, Italy): Results from ORION-GEOSTAR3 experiment

L. Beranzoli, A. Ciafardini, G. Cianchini, M. De Caro, A. De Santis, P. Favali, F. Frugoni, G. Marinaro, S. Monna, C. Montuori, E. Qamili, T. Sgroi and S. Vitale[1]

24.1 Introduction

The Marsili Seamount is the largest European underwater volcano. It is Plio-Pleistocenic in age, rising up to more than 3000m from the seafloor in the SE Tyrrhenian basin (Central Mediterranean), a back arc basin which began progressively opening 10 Ma ago (Kastens et al., 1988). The seamount lies in a key area for understanding the evolution of the Tyrrhenian region, characterized by high values of heat flow (Della Vedova et al., 2001) and low values of Moho isobaths (Locardi and Nicolich, 1988). In spite of the large dimensions of the Marsili seamount, we still have limited knowledge of its present activity. Ocean exploration is dependent on available technology and infrastructure, which started to develop strongly only after the 1980s. In fact, from its discovery in the 1920s, very little was known of the Marsili Seamount until the late 1990s when new techniques such as multibeam acoustic bathymetry were developed allowed to revealing its morphology.

Dedicated expeditions then produced the first morpho-bathimetric map of the entire Tyrrhenian seafloor, based on multibeam swath-mapping, together with seismic, gravimetric and magnetometric data (e.g., Marani and Gamberi, 2004). Although these have greatly contributed to our understanding, their short duration does not allow the investigation of medium-term geophysical processes.

1 Istituto Nazionale di Geofisica e Vulcanologia, Rome, Italy

Multi-parameter seafloor observatories can provide long-term continuous time-series in deep ocean waters, which are the basis for an original approach to ocean exploration. The observation of phenomena variability over time is a key to understanding many Earth processes, among which we recall hydrothermal activity, active tectonics, and ecosystem life cycles. The development in Europe of multidisciplinary seafloor observatories has been pioneered under the EC Framework Programmes, specifically in the GEOSTAR projects (Beranzoli et al., 1988, 2000). From 2003 to 2005, long-term geophysical and oceanographic monitoring was conducted within the EC ORION-GEOSTAR3 project with two multiparameter observatories deployed on the seafloor 3320m below sea level (b.s.l.) in the vicinity of the Marsili Seamount. The two observatories were equipped with a set of sensors providing long-term continuous time-series of various physical measurements. The acquired time series are the longest continuous data recorded at the Marsili Basin so far. This chapter describes the experiment, and presents results of the data analysis which add valuable information on the poorly-known Marsili.

The next section provides the geological setting necessary to understanding the context of the Marsili Seamount and its basin; Section 24.3 describes the ORION-GEOSTAR3 experiment; some results from analysis of seismic, magnetic and gravimetric data; and finally, in the last section we present our discussion with the main conclusions.

24.2 Geological setting

The development of the Tyrrhenian Sea, the youngest sub-basin of the Mediterranean Sea, is related to the complex large-scale interaction of the converging African and European plates, with contemporaneous collisional and extensional processes which occur along the plate margins.

Opening of the region during the Tortonian led to the formation of the Magnaghi and Vavilov oceanic basins (4.3–2.6 My). During the Plio-Pleistocene the extension changed from E–W to SE–NW, forming the Marsili Basin, a recent (<2 My) oceanic area with back-arc basin characteristics (Malinverno and Ryan, 1986). On the southern and eastern rim of this basin we find the Aeolian volcanic arc, consisting of seven islands and a number of submarine volcanoes west and north-east of the emerged arc (Marani and Gamberi, 2004). The emplacement of basaltic oceanic crust of the Marsili Basin was generated by ESE-directed extension (Kastens and Mascle, 1990), taking place above the presently northwesterly-subducting slab (Montuori et al., 2007 and reference therein). The 100km wide oceanic sector of the Marsili Basin lies at 3500m water depth. The basin is characterized by a Moho depth of 11km (Steinmetz et al., 1983), matched by thinning of the lithosphere to less than 30km (Pontevivo and Panza, 2002), high heat flow with regional values around 120mW/m^2 (Zito et al., 2003), and magnetic anomalies typical of extension basins (Faggioni et al., 1995; Nicolosi et al., 2006). Site 650 of the Ocean Drilling Program (ODP) Leg 107, located on the western rim of the Marsili basin, documents a highly vescicular basalt dated about 1.8–2.0 My, which represents the acoustic basement under 600m of plio-pleistocene sedimentary units (Mascle and Rehault, 1990).

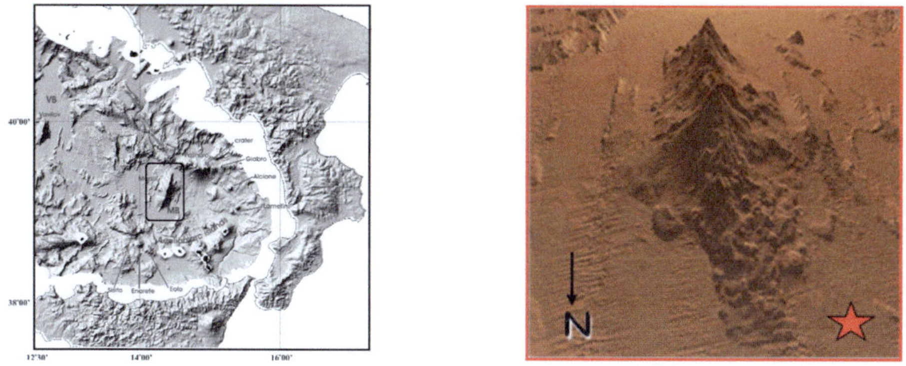

a) b)

Figure 24.1 Bathymetry of (a) the Southern Tyrrhenian Basin and (b) the Marsili Seamount. The volcanic edifice, about 50km long and about 16km wide, is marked by the small square in (a) (Marani and Trua, 2002; Marani and Gamberi, 2004); the star in (b) is the site of the main observatory.

The central part of the basin is occupied by the Marsili Volcanic Seamount, the largest volcano in the Tyrrhenian Sea, rising over 3000m from the abyssal plain to 489m water depth (Figure 24.1). The volcanic edifice is about 50km long and about 16km wide, and is strongly elongated in a NNW–SSE direction. The seamount crest is characterized by a 25km-long narrow belt of volcanic cones along the 1000m isobath. Spherical cones are present on the flanks of the volcano too, particularly on the northwestern side, with diameters ranging between several hundreds and few kilometers. Steep bathymetric gradients separate the crest portion from the deeper volcanic flanks.

Marsili rock samples have been collected through several dredging and coring expeditions (Selli et al., 1977; Savelli and Gasparotto, 1994; Trua et al., 2002). The majority of these samples are basalts and, to a lesser extent, andesites and trachyandesites; their compositions have a calc-alkaline affinity, with medium to high potassium content. The Marsili vertical growth started in the Jaramillo subchrone (1.07 Myr). The volcanic edifice displays a wide positive anomaly of normal magnetization (Faggioni et al., 1995, Cocchi et al., 2009); the youngest volcanic products are high-K andesites lava sampled on the summit of the seamount, with age between 0.1 and 0.2 Myr (Selli et al., 1977), which are characterized by extensive hydrothermal alteration causing demagnetization. Hydrothermal processes on the Marsili Seamount account for a large variety of mineral deposits, implying hydrothermal fluids boiling beneath the seafloor and/or the mixing of the ascending hydrothermal fluids with seawater in fractured rocks (Dekov and Savelli, 2004).

On Marsili, heat flow reaches a value of 250 mW/m^2 (Della Vedova et al, 2001; Mongelli et al., 2004). Using heat flow data collected in various oceanographic surveys, Verzhbitskii (2007) modeled the presence of an oceanic-type lithosphere at shallow depth (up to 5km) under the Seamount. Such a hypothesis is supported by both gravimetric and magnetic modeling (Faggioni et al., 1995; Cella et al., 1998; Caratori et al., 2010) and by the detection of a low apparent electric resistivity under the seamount (Vitale et al., 2009).

The Marsili Seamount shows a remarkable similarity to mid-oceanic ridges and has been interpreted as a super-inflated spreading ridge (Marani and Trua, 2002). The Marsili Basin shows sub-parallel positive-negative magnetic anomalies over the flat-lying seafloor, typical of a linear oceanic spreading, and a central positive anomaly along the Marsili volcano, where vertical accretion acted as the dominant mechanism within the restricted spreading environment of the Southern Tyrrhenian Sea (Nicolosi et al., 2006; Cocchi et al., 2009).

24.3 The ORION experiment

The long-term continuous monitoring of the Marsili Seamount was conducted from 2003 to 2005, in the framework of the ORION-GEOSTAR3 project (EC 6th Framework Programme). The main scientific objective of the experiment was to investigate the Marsili Seamount by means of a multidisciplinary, i.e., geophysical, geochemical, and oceanographic, data analysis over ~1 year. Data were acquired by two GEOSTAR-class deep sea observatories deployed at the base of the seamount at 3320m b.s.l. during the period December 2003–May 2005 (location: 39°29'12" N, 14°19'52" E; red square in Figure 24.1).

The two observatories (hereafter referred to as main and satellite nodes) are single-frame autonomous stations capable of operating down to 4000m (see Beranzoli et al., 2000; Favali et al., 2006).

The GEOSTAR-class observatories are typically formed by a Bottom Station which is managed from the sea-surface by means of a specifically designed vehicle, the MObile Docker for Underwater Sciences (MODUS), through an umbilical cable and winch on board a medium-size research vessel. The umbilical cable has mechanical characteristics suitable for the mobilisation of the Bottom Stations and conductor lines and optic fibers for the communications with both the Bottom Station and MODUS. Some of the GEOSTAR-class observatories are also equipped with an on-board acoustic communication system which can transmit hourly data messages both automatically and upon request. A surface buoy receives the acoustic signals and forwards them to a shore station.

The Bottom Station (Figure 24.2) hosts a wide set of scientific instruments and status sensors according to the specific objectives of the experiment. Sensors are controlled by a central Data Acquisition and Control System (DACS, Gasparoni et al., 2002, 2012) which backs-up data on internal memory. DACS manages a wide set of data streams at quite different sampling rates (from 100 Hz to 1 sample/day) tagging each datum with a common reference time set by a clock with a stability of 10^{-9} needed for processing seismological data. If required, the DACS can also manage software for automatic data preprocessing like the detection of seismic events and water column pressure anomalies.

Table 24.1 lists the sensors hosted on the two ORION observatories (main and satellite node) and reports their main technical characteristics.

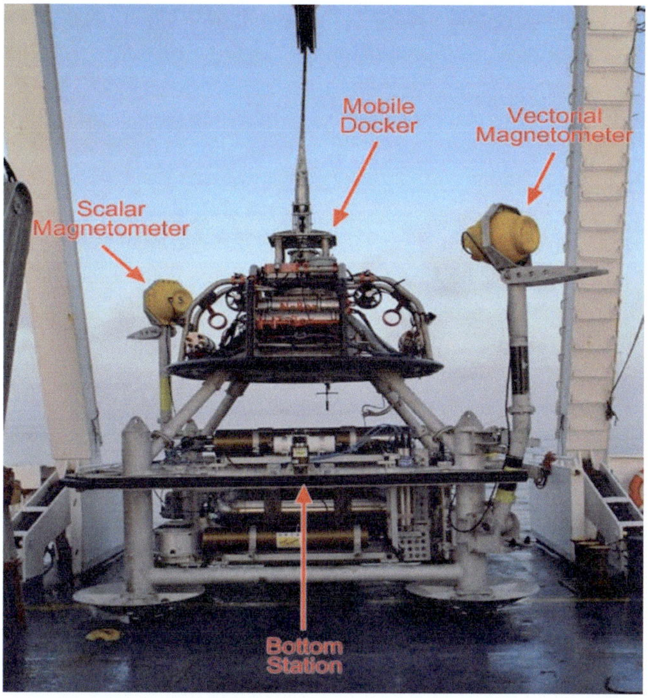

Figure 24.2 ORION-GEOSTAR3 main node on the deck of the R/V Urania before deployment. GEOSTAR-class observatories are usually deployed by MODUS, which is driven by means of a console on board of the ship.

Special care was taken to obtain both optimal deployment and running of the observatories on the seafloor. We tried to minimize observatory frame vibration (induced by water currents) that could affect the seismometer, and observatory electronic noise that could affect the magnetometers. In particular, the seismometer (PMD electro-chemical broadband sensor), installed in an aluminium vessel, is protected by a heavy cylindrical housing. This package, mounted in the central part of the observatory frame and connected by a slack rope, is released after the observatory touch-down. This kind of installation guarantees an optimal coupling of the seismic sensor with the sea bottom as shown by Monna et al. (2005). The magnetometers (up to two) are mounted on the observatory at the end of two booms attached at opposite angles of the frame. These structures keep the sensors as far as possible from electromagnetic noise sources, minimizing their effect on the magnetometric signals. The booms, kept vertical during the descent, are opened upon command once the observatory is safely placed on the seafloor.

As one of the technological objectives of the ORION-GEOSTAR 3 project was to realize and test a prototype of a deep seafloor observatory network, the communications were considerably enhanced in the ORION experiment with respect to previous seafloor

Sensor	**Typical sampling rates**	**Data acquisition (bits)**	**Installation constraints**	**ORION Main node**	**ORION Satellite node**
Three-component broad-band seismometer	100 Hz	24	Positioning (error <100m) ·orientation to the North (known <1°) Good ground coupling Fine leveling (if required)	X	X
Hydrophone	100 Hz	24	Positioning (error <100m)	X	X
Gravimeter	1 Hz	24	Positioning Temperature controlled Fine leveling	X	
Scalar magnetometer	1 sample/min	16	Minimization of possible electro-magnetic interferences	X	
Tri-axial fluxgate magnetometer	1 sample/s	24	Minimization of possible electro-magnetic interferences	X	
Precision tilt meter (X, Y)	10 Hz	24	Northwards orientation	X	X
Tri-axial single-point current meter	2 Hz	16	Avoiding frame interference	X	X
ADCP 300 kHz	1 profile/hour		Avoiding frame interference	X	
Transmissometer	1 sample/hour		Avoiding frame interference	X	
CTD	1 sample/10 min (or 1 sample/hour)			X	
CH_4 sensor	1 Hz	24		X	
H_2S sensor	1 sample/10 min (averaged on 30 samples/s)	24		X	
pH sensor	1 sample/6 hours*		Sampling and self-calibration programmable Self-calibration every 24 samples*	X	
Water sampler			48 bottles, sampling depending on the mission targets	X	

Table 24.1 The sensors hosted on the two ORION observatories.

experiments. In addition to the function of data transfer to the sea surface, the communication system allowed the exchange of data and status parameters between the main node, and the satellite node. A diagram of the general scheme of communications used in ORION-GEOSTAR3 is shown in Gasparoni et al. (2012). Communication among the nodes was achieved through horizontal acoustics based on Multi-Modulation Acoustic Telemetry System (MATS) modems. In this way, the satellite node releases automatic messages to the main node, which is the only one that can communicate to and from the buoy through vertical acoustics. Connection between the buoy and a shore station is ensured by radio and satellite links. Data, specifically pieces of seismic waveforms, can be retrieved on request. The buoy transmission system (DRTS) consists of an electronic unit (MEU) managing the communications and interfacing the acoustic transmission system with two buoy-to-shore data links, VHF radio or IRIDIUM satellite. In case of VHF-link failure, a switch to the satellite transmission is automatically activated. The DACS interface to the communication system was enhanced in order to make data and status parameters available for transmission to the communication system.

24.4 Data analysis

Long-term data acquisition at the base of the Marsili Seamount is an important starting point to investigate all processes with typical time scales from milliseconds (elastic waves propagation) to seasonal (water mass circulation regimes) that may occur in the surrounding area. In this section, we concentrate on the seismic, gravimetric and magnetic data with the aim of testing the present hypothesis about the Marsili activity and possibly to discover unknown features.

24.4.1 Seismometer and gravimeter data

A large variety of clear signals have been recorded by the seismometers installed on the main and satellite nodes. This is particularly due to a good S/N ratio, resulting from the low seismic noise present at the experiment sites.

After the recovery of the observatories we performed data quality analysis on seismic signals. An example of typical seismic noise for the two sites is shown in Figure 24.3: a Power Spectral Density (PSD) of the three seismometer components (E, N and Z) has been performed on eight one-hour long segments of background seismic noise recorded at different periods during the experiment. In particular, the analysis was applied following Welch's method[2] (Welch, 1967) along time-windows of 16384 samples with a window overlap of 50%.

Although the overall shape of the PSD curve of the three components follows the typical trend of the Low and High Noise Models (LNM and HNM; Peterson, 1993), some peculiarities can be pointed out. The maximum of the energy is found within 0.3 and 1 Hz,

2 Welch's method consists of dividing the time series into segments (possibly overlapping), computing the periodogram of each segment and then averaging the periodograms. This average is the power spectral density estimate.

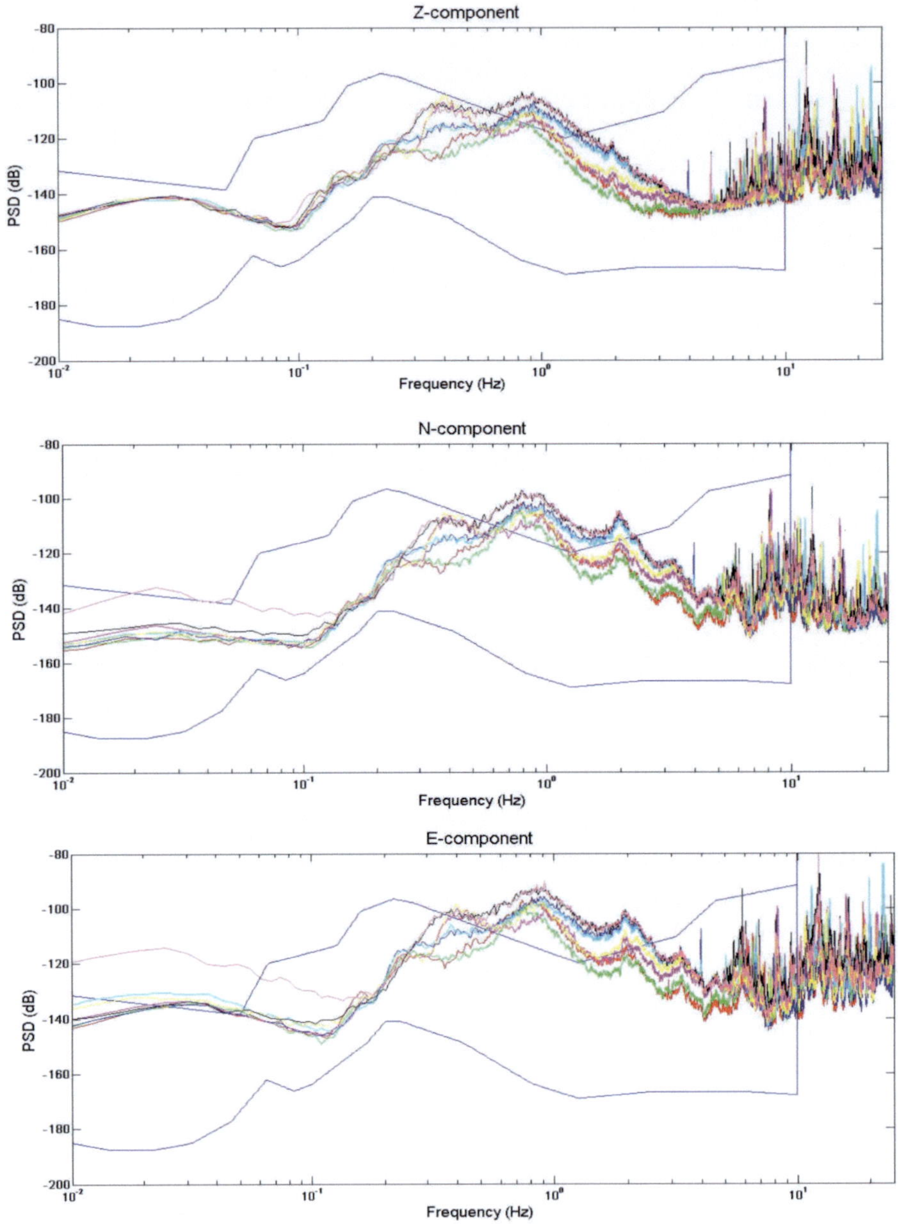

Figure 24.3 Power Spectral Density (PSD) of the three seismic signal components (Vertical: Top; NS: center; EW: bottom) computed on eight one-hour long segments of background seismic noise (one color for each segment) recorded at different times during the experiment. Low and High Noise Models by Peterson (1993) are reported as the upper and lower blue smooth lines in the PSD plot.

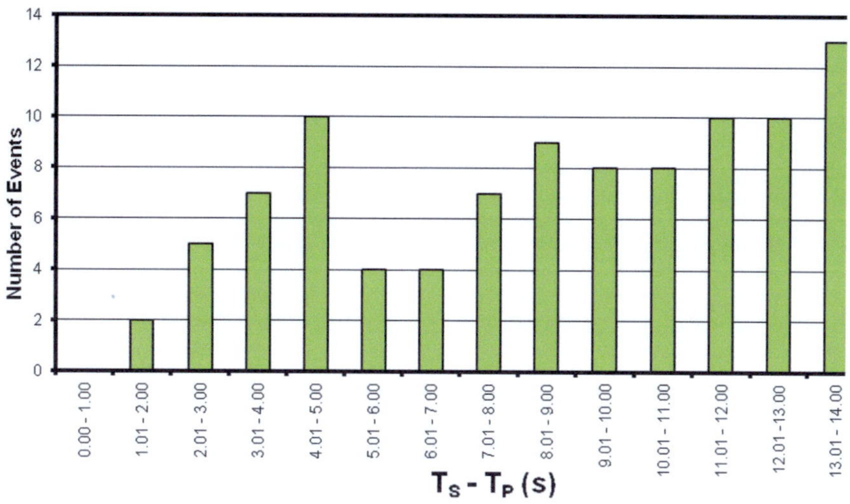

Figure 24.4 Histogram of the number of very local events in the T_S–T_P range 1s–14s. The distribution of number of events is larger for T_S–T_P values 3–5 and 7–14s.

especially for the horizontal components. The peak at 1 Hz is probably due to microseisms induced by local winds (e.g., Monna et al., 2005, for a nearby site in the Ionian Sea). Even moderate winds can produce the 2-s sea waves which cause the microseismic noise at 1 Hz recorded at the seafloor (Webb, 1998). For frequencies larger than 1 Hz, the PSD shows a peak at 2 Hz, generally attributed to volcanic activity; for instance, this peak was also observed on the seismic records acquired off-shore Stromboli (Aeolian Islands) by an OBS module (Sgroi et al., 2009).

The low background seismic noise level on all components allowed the clear detection of teleseismic, regional and local earthquakes.

Interestingly, the ORION seismometers detected some local events not recorded by the closest land stations of the Istituto Nazionale di Geofisica e Vulcanologia (INGV) permanent Seismic Network. These events, called hereafter "very local events", represent well the seismic activity in the Marsili Basin. During the first experiment leg, over 160 very local seismic events were recorded, representing 20% of the total recorded events. All the analyzed time series show evident P and S arrivals and a difference in the S–P arrival time at the Bottom Stations ranging from 1.1s to 20.8s. The histogram of Figure 24.4 shows that the number of events with respect to the value of T_S-T_P (T_S= S-arrival time; T_P= P-arrival time) were in the range 1s–14s. The estimation of rough locations of single station data depends on T_p and T_S, while the distances are derived from travel-time delays assuming an adequate crustal velocity model. In order to solve the problem of the P-wave velocity, we performed the integrated locations of several earthquakes, well recorded both by the ORION observatories Bottom Stations and land stations. We finally estimated an apparent

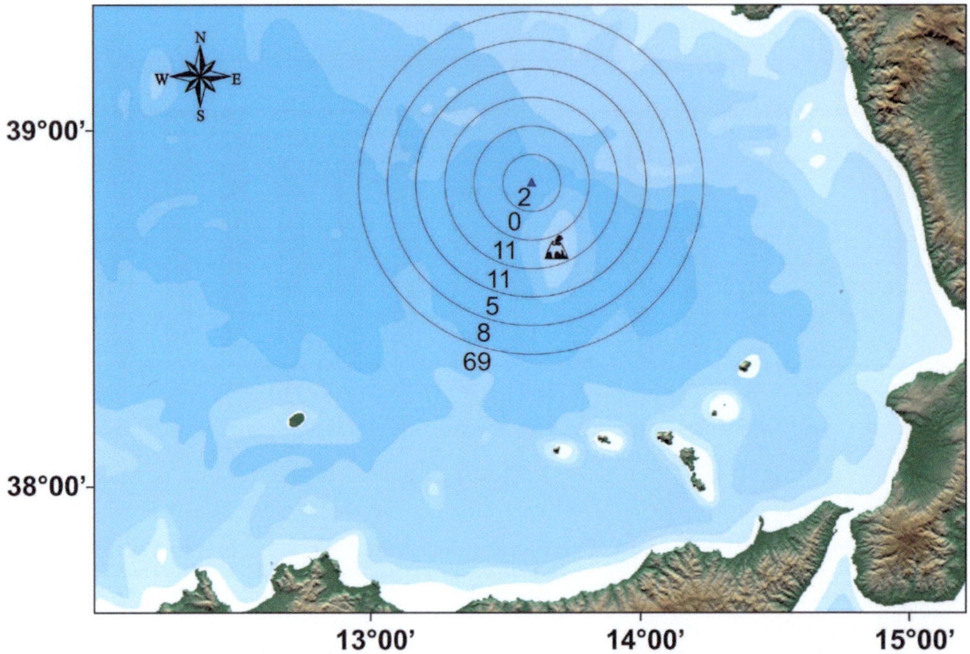

Figure 24.5 Epicentral distance distribution around the GEOSTAR-ORION site. Radii of successive circles increase by 10km. When we consider those seismic events within 60km from the benthic observatory site, the larger seismicity appears at a distance which includes the location of the Marsili Seamount. The seamount top is marked by a stylized volcano.

mean velocity value of 5.7km/s for the P waves through the inversion of the P- and S-arrival times versus the hypocentral distances of these events.

According to the procedure adopted in Sgroi et al. (2007) for the Ionian area, the epicenter location of the very local events was roughly estimated in terms of distance of the epicenters from the seafloor observatory. In Figure 24.5 the different ranges of the epicenter distances are represented with concentric circles respect to the Bottom Stations location, with radii increasing by 10km. The number reported within the single circle indicates the number of very local events whose estimated epicenter locations fall in the distance range comprised between two successive circles. To verify the hypothesis that these events might be associated to the volcanic or tectonic structures of the Marsili area, we would need to expand the dataset and perform a more detailed analysis. Nevertheless, as these events were not recorded by seismic land stations, we may safely hypothesize that they occurred within the Marsili basin area.

The ORION seismometer recorded other interesting short-duration seismic signals at frequencies larger than 10 Hz (Figure 24.6). The rich spectral contribution at these higher

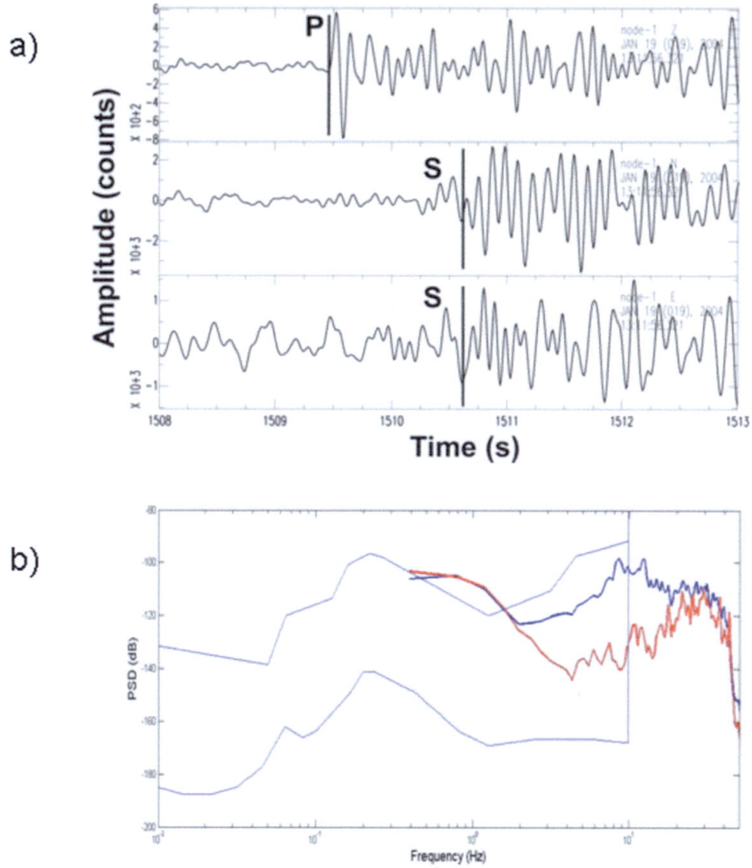

Figure 24.6 An example of a high-frequency signal which occurred at few km from the main node on January 19, 2004. (a) Onset of P and S seismic waves. (b) Power Spectral Density (PSD) of the vertical component of the seismic signal (in blue) compared with the background noise (in red). Low and High Noise Models by Peterson (1993) are also reported.

frequencies could be caused by liquid-supercritical fluid phase transition of seawater, as observed in other cases at shallower depths (e.g., Ohminato, 2006). This kind of phase transition could be explained by the presence of hydrothermal circulation in the Marsili Seamount, as proposed in a recent study on the seismic signals recorded by a broadband ocean bottom seismometer and a hydrophone, deployed from 12 to 21 July, 2006 on the flat top of Marsili (D'Alessandro et al., 2009).

The PSD of the background seismic noise shown in Figure 24.3 points out also on the average noise level below 0.03–0.04 Hz, which becomes higher than the HNM with amplitudes ranging −120 dB to −140 dB. To investigate the possible causes of the noise at lower frequencies, we examine the data of the adjacent gravimeter.

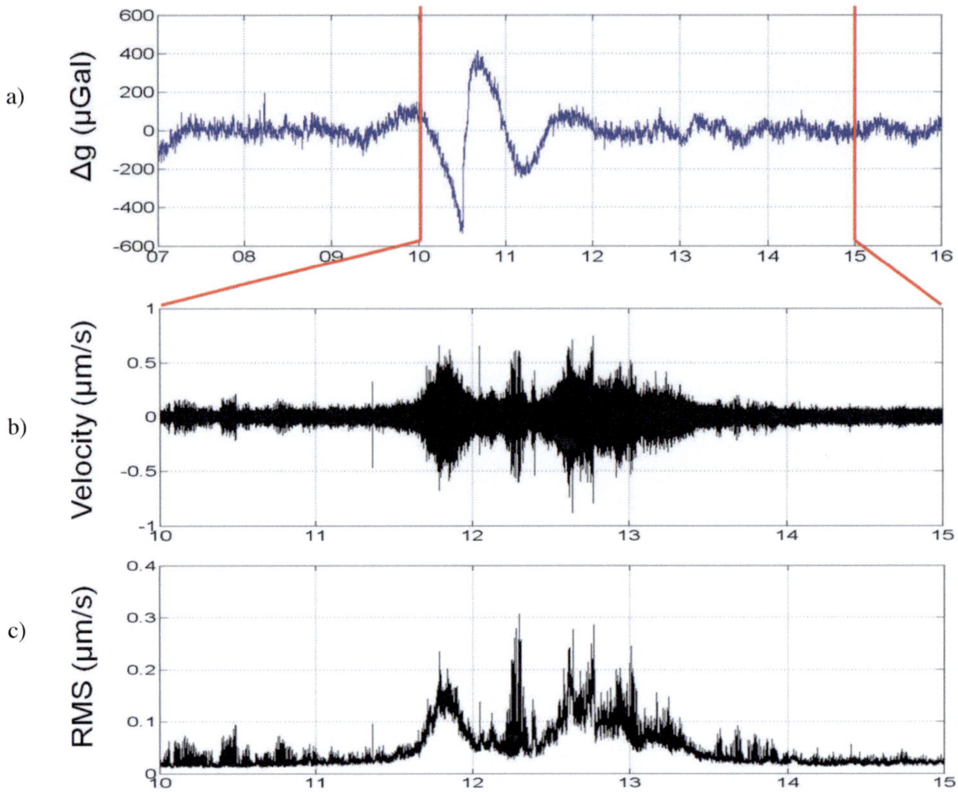

Figure 24.7 (a) 9 hours of gravimeter signal recorded on 12 January, 2004 and band pass filtered within (0.0005,0.1) Hz. (b) 5 hours of seismic record of the vertical velocity component, high pass filtered at 5 Hz. (c) The relative RMS amplitude computed using 1 second sliding time windows without overlap.

The gravimeter on board the seafloor observatory is a very sensitive instrument and it has been specifically designed and built within the laboratories of the Istituto di Fisica dello Spazio Interplanetario (IFSI; Consiglio Nazionale delle Ricerche) (Iafolla et al., 2006). Its mechanical part consists of a proof mass which is connected to an external frame by two torsion arms and acts as a harmonic oscillator with a 15 Hz resonance frequency. The mechanical oscillator is obtained by machining a single plate of aluminum AL 5060. Two external plates are placed on its opposite sides to realize a couple of capacitive detectors working in differential mode. This difference should be zero when the proof mass is under the action of the Earth's gravity.

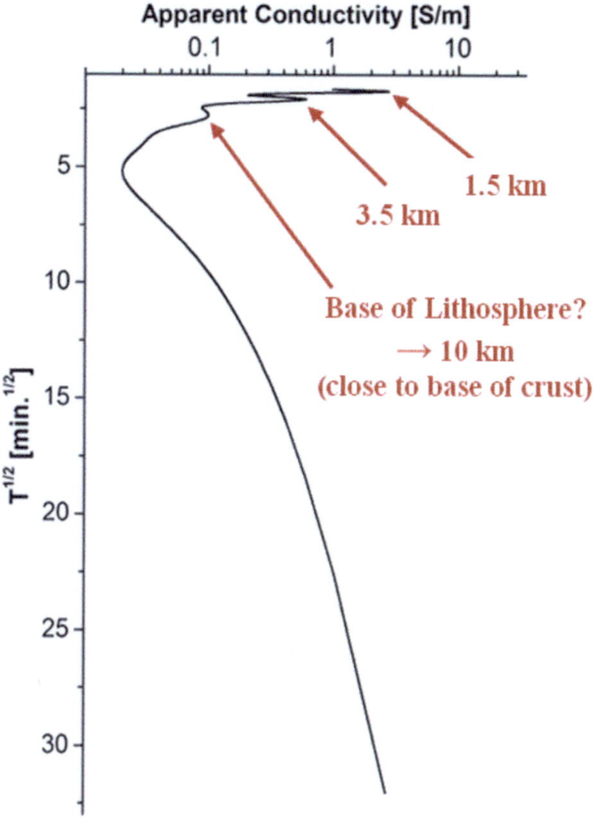

Figure 24.8 From the apparent conductivity curve it is possible to deduce the three main conductive layers (indicated by the arrows) under the Marsili, the deepest of which is probably the magnetic lithosphere (Vitale et al., 2009).

The gravimeter time series has shown changes of the gravity signal from its mean value. After removing mean and linear trends the gravimetric signal has been filtered with a Butterworth band pass filter (0.0005–0.1 Hz), to remove the effect of the earth tides, together with the instrument drift (due to temperature variations) and the longer wavelengths, which are not our interest on this occasion. Figure 24.7(a) shows a microgravity change within 800 microGal (1μGal=10nm/s^2) peak-to-peak observed during the recording of a given day. Interestingly, this fluctuation is simultaneous with the increase of seismic noise at high frequencies (Figure 24.7(b), (c)). This temporal microgravity change with the relative increase of the seismic noise could be associated with the presence of magmatic hydrothermal fluids (Battaglia et al., 2008).

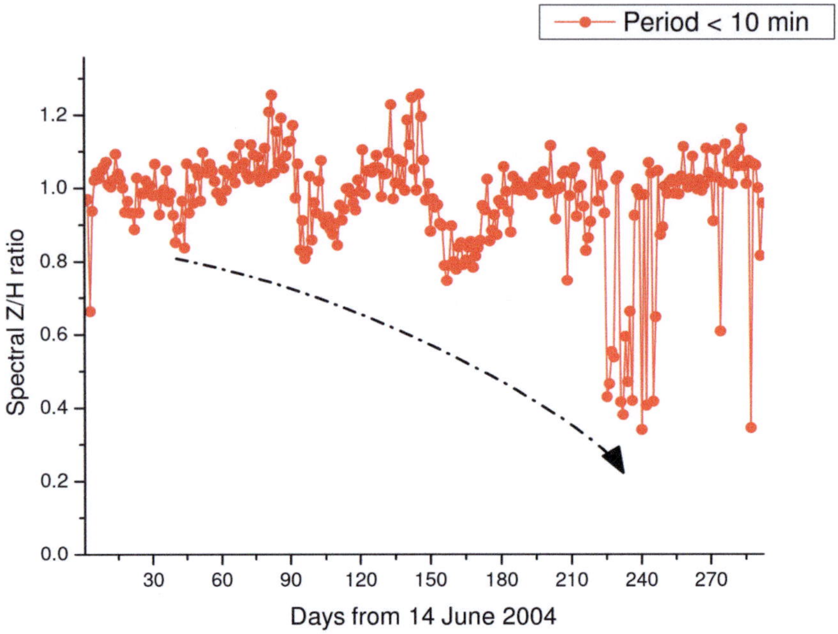

Figure 24.9 Example of the Z/H spectral magnetic ratio for periods of less than 10 minutes; its general pattern of change over time (here indicated by the dashed arrow) could be a symptom of shallow crustal dynamics, likely due to some local Marsili activity.

24.4.2 Magnetic data

The vectorial magnetometer, a sophisticated and accurate three-component magnetic compass, provides precise information about the exact orientation of the benthic station, also useful for the correct orientation of the seismological recordings. Well before the seafloor mission, both magnetometers were checked at the Geomagnetic Observatory of L'Aquila (Central Italy), comparing them with the magnetic instruments normally running in the Observatory. In particular, the magnetometers were installed on the two booms of the benthic station. Both magnetometers had previously been calibrated in order to establish the amount of the eventual contamination by the overall system to the magnetometers signals. When the station was running, the direction of the horizontal magnetic components was also compared with the same components as deduced from land magnetic stations or observatories, in the same manner as was done during the first deep mission close to Ustica Island (Sicily, Italy) in 2000–2001 (see also De Santis et al., 2006a, b).

More specifically, the study of magnetic time variations allowed us to estimate the conductivity structure underneath the area of observation by means of the *skin-depth* δ concept (Vitale et al., 2009; De Santis et al., 2006a, b) which is given by a simple equation:

$$\delta(km)=3.9\left(T/\sigma\right)^{1/2}$$

where T represents the period in minutes of the considered harmonic and σ is the mean seafloor electrical conductivity (in S/m) below the area under investigation. A result is shown in Figure 24.8 where the deepest layer could be interpreted as the lithosphere-asthenosphere discontinuity. This estimation confirmed the low values of both the Moho and lithosphere depths as deduced by seismic data analysis (Locardi and Nicolich, 1988).

Distinct intermittent decreases of the Z/H ratio appear with subsequent larger amplitude (Figure 24.9): since Z/H usually is inversely related to conductivity changes likely due to fluid diffusion, and in turn a decrease of Z/H corresponds to an increase of underneath conductivity, this behaviour would support some dynamics of the Marsili seamount system, in terms of possible deep circulation of thermal mineralized fluids, that sometime increase the conductivity of the seafloor sediments. Of course, further investigation is essential, in terms of both a longer time-series and more sites of geomagnetic observation, to confirm (or not) this aspect.

24.5 Conclusions

The ORION-GEOSTAR 3 deep seafloor experiment was performed from December 2003 to May 2005 in two successive legs, successfully acquiring an enormous amount of unique data. We were able to highlight some unknown features of the mysterious Marsili Seamount. A preliminary data analysis showed interesting variations in the recorded geophysical data which are likely connected with the activity of the Marsili volcanic seamount.

One of the most significant results is the detection of more than 100 very local earthquakes observed in the ORION seismometers time series, and which were not recorded by land stations. A rough estimation of the epicenter of these events shows that about one-third of them could be generated within the Marsili Seamount structure, revealing the presence of weak seismic activity. The interesting increase of seismic signals above 2 Hz could be related to the seamount volcanic activity, as was the case of seismic signals acquired offshore by an OBS module linked to Stromboli activity (Sgroi et al., 2009).

Magnetic data allow us to estimate the conductivity structure at different depths under the Marsili Seamount and gravimetric data show relevant signal patterns at low frequencies. In addition, seismic data show some local activity with recurring seismic signals, both in terms of very local events and in the presence of high frequency content (frequency greater than 10 Hz). Both results would confirm the possible interpretation of the Marsili dynamics deduced from the magnetic ratio Z/H time variations. Significant correlations between recorded time series are probably related to activity and structure of the Marsili volcano, and the high frequency spectral content is an important indicator for the possible existence of a hydrothermal circuit.

All of these indications witness the Marsili volcanic seamount as a complex system, which manifests some important activity. However, all aspects have to be further investigated, in particular through the comparison and eventual correlations of the different signals acquired by new and targeted experiments in the area.

Acknowledgments

In addition to INGV, other institutions contributed to ORION-GEOSTAR 3: Tecnomare S.p.A. and ISMAR-CNR (Italy); TUB, TFH and IFM-GEOMAR (Germany); IFREMER and ORCA (now SERCEL) (France). Thanks are given to the Captains and the crew of the R/V Urania. The project ORION-GEOSTAR 3 has been funded by the European Commission under FP6. We thank the comments kindly provided by Angelo Camerlenghi and Mairi Best.

References

Battaglia M., Gottsmann J., Carbone D. and Fernàndez J. (2008) 4D volcano gravimetry. Geophysics 73(6), WA3–WA18, doi: 10.1190/1.2977792.

Beranzoli L., De Santis A., Etiope G., Favali P., Frugoni F., Smriglio G., Gasparoni F., Marigo A. (1998) GEOSTAR: A Geophysical and Oceanographic Station for Abyssal Research. *Phys. Earth Plan. Inter.* 108, N.2, 175–183.

Beranzoli L., Braun T., Calcara M., Calore D., Campaci R., Coudeville J.M., De Santis A., Etiope G., Favali P., Frugoni F., Fuda J.L., Gamberi F., Gasparoni F., Gerber H., Marani M., Marvaldi J., Millot C., Palangio P., Romeo G. and Smriglio G. (2000) GEOSTAR: The first European long-term seafloor observatory. EOS 81(5), 45–49.

Caratori Tontini F., Cocchi L., Muccini F., Carmisciano C., Marani M., Bonatti E., Ligi M. and Boschi E. (2010) Potential field modeling of collapse prone submarine volcanoes in the southern Tyrrhenian Sea (Italy). Geophysical Research Letters 37, L03305, doi: 10.1029/2009GL041757.

Cella F., de Lorenzo S., Fedi M., Loddo M., Mongelli F., Rapolla A. and Zito G. (2006) Temperature and density of the Tyrrhenian lithosphere and slab and new interpretation of gravity field in the Tyrrhenian Basin. Tectonophysics 412, 27–47.

Cocchi L., Caratori Tontini F., Muccini F., Marani M., Bortoluzzi G. and Carmisciano C. (2009) Chronology of the transition from a spreading ridge to an accretional seamount in the Marsili backarc basin (Tyrrhenian Sea). Terra Nova 21, 369–374.

D'Alessandro A., D'Anna G., Luzio D. and Mangano G. (2009) The INGV's new OBS/H: Analysis of the signals recorded at the Marsili submarine volcano. J. Volcanol. Geotherm. Res. 183, 17–29.

Dekov V.M. and Savelli C. (2004) Hydrothermal activity in the SE Tyrrhenian Sea: An overview of 30 years of research. Mar. Geol. 204, 161–185.

Della Vedova B., Bellini S., Pellis G. and Squarci P. (2001) Deep temperatures and surface heat flow distribution. In: G.B. Vai and I.P Martini (Eds) *Anatomy of an Orogen: The Apennines and Adjacent Mediterranean Basins*, New York: Springer, pp. 65-76.

De Santis A., Di Mauro D., Cafarella L., D'Anna R., Gaya-Piquè L.R., Palangio P., Romeo

G. and Tozzi R. (2006a) Deep seafloor magnetic observations under GEOSTAR project. Ann. Geophys. 49(2/3), 681–693.

De Santis A., Di Mauro D., Cafarella L., Palangio P., Beranzoli L., Favali P. and Vitale S. (2006b) Extending magnetic observations to seafloor: The case of GEOSTAR and ORION missions in the Adriatic and Tyrrhenian Seas. Publs. Inst.Geophys. Pol. Acad. Sc. C-99, 398, 114–122.

Faggioni O., Pinna E., Savelli C. and Schreider A.A. (1995) Geomagnetism and age study of Tyrrhenian seamounts. Geophys. J. Int. 123, 915–930.

Favali P., Beranzoli L., D'Anna G., Gasparoni F., Marvaldi J., Clauss G., Gerber H.W., Nicot M., Marani M.P., Gamberi F., Millot C. and Flueh E.R. (2006) A fleet of multiparameter observatories for geophysical and environmental monitoring at seafloor. Ann. Geophys. 49(2/3), 659–680.

Favali P. and Beranzoli L. (2006) Seafloor observatory science: A review. Ann. Geophys. 49(2/3), 515–567.

Fuda J.-L., Etiope G., Millot C., Favali P., Calcara M., Smriglio G. and Boschi E. (2002) Warming, salting and origin of the Tyrrhenian Deep Water. Geophys. Res. Lett. 29(19), 1898, doi: 10.1029/2001GL014072.

Gasparoni F., Calore D. and Campaci R. (2002) From ABEL to GEOSTAR: Development of the first European deep-sea scientific observatory. In: L. Beranzoli, P. Favali and G. Smriglio (Eds) *Science-technology Synergy for Research in Marine Environment: Challenges for the XXI Century*. Development in Marine Technology Series 12. Amsterdam: Elsevier, pp. 143–159.

Gasparoni et al. This volume, 2014.

Iafolla V., Nozzoli S., Fiorenza E. and Milyukov V. (2006) Deep-sea gravity measurements: GEOSTAR-2 mission results. Ann. Geophys. 49(2/3), 695–704.

Kastens K. and Mascle J. (1990), The geological evolution of the Tyrrhenian Sea: An introduction to the scientific results of ODP Leg 107. In: K.A. Kastens, J. Mascle, et al. (Eds) *Proceedings of the ODP Scientific Results 107*. College Station, TX (Ocean Drilling Program), pp. 3–26

Kastens K.A., Mascle J., Auroux C., Bonatti E., Broglia C., Channell J., Curzi P., Emeis K., Glacon G., Hasegawa S., Hieke W., Mascle G., McCoy F., McKenzie J., Mendelson J., Muller C., Rehault J.-P., Robertson A., Sartori R., Sprovieri R. and Torii M. (1988) ODP Leg 107 in the Tyrrhenian Sea: Insights into passive margin and back-arc basin evolution. Geol. Soc. Amer. Bull. 100, 1140–1156.

Locardi E. and Nicolich R. (1988) Geodinamica del Tirreno e dell'Appennino centro-meridionale: La nuova carta della Moho. Mem. Soc. Geol. It. 41, 1212–1240.

Malinverno A. and Ryan W.B.F. (1986) Extension in the Tyrrhenian Sea and shortening in the Apennines as result of arc migration driven by sinking of the lithosphere. Tectonics 5, 227–245.

Marani M.P. and Gamberi F. (2004). Distribution and nature of submarine volcanic landforms in the Tyrrhenian Sea: the arc vs the backarc. *Mem. Descr. Carta Geol. D'It.*, XLIV,109-126.

Marani M.P. and Trua T. (2002) Thermal constriction and slab tearing at the origin of a super-inflated spreading ridge: The Marsili volcano (Tyrrhenian Sea). J. Geophys. Res. 107, 2188, doi: 10,1029/2001JB000285.

Mascle J. and Rehault J.P. (1990) A revised seismic stratigraphy of the Tyrrhenian Sea: Implications for the Basin Evolution. In: K.A. Kastens, J. Mascle, et al. (Eds) *Proceedings of the ODP Scientific Results 107*, College Station, TX (Ocean Drilling Program), pp. 617–635.

Monna S., Frugoni F., Montuori C., Beranzoli L. and Favali P. (2005) High quality seismological recordings from the SN-1 deep seafloor observatory in the Mt Etna region. Geophys. Res. Lett. 32, L07303, doi: 10.1029/2004GL021975.

Montuori C., Cimini G. and Favali P. (2007) Teleseismic tomography of the southern Tyrrhenian subduction zone: New results from seafloor and land recordings. J. Geophys. Res. 112, B03311, doi: 10.1029/2005JB004114.

Nicolosi J., Speranza F. and Chiappini M. (2006) Ultrafast oceanic spreading of the Marsili Basin, southern Tyrrhenian Sea: Evidence from magnetic anomaly analysis. Geology 34(9), 717–720, doi: 10.1130/G22555.1.

Ohminato T. (2006) Characteristics and source modeling of broadband seismic signals associated with the hydrothermal system at Satsuma–Iwojima volcano, Japan. Jour. Volcan. Geoth. Res. 158, 3–4, 467–490.

Peterson J. (1993) Observations and modelling of seismic background noise. USGS Open File Rep. 93-322, 94.

Pontevivo A. and Panza G.F. (2002) Group velocity tomography and regionalization in Italy and bordering areas. Phys. Earth Plan. Int. 134, 1–15.

Savelli C. and Gasparotto G. (1994) Calc-alkaline magmatism and rifting of the deep-water volcano of Marsili (Aeolian back.arc, Tyrrhenian Sea). Mar. Geol. 119, 137–157.

Selli R., Lucchini F., Rossi P.L., Savelli C. and Del Monte M. (1977) Dati geologici, petrochimici e radiometrici sui vulcani centro-tirrenici. Gior. Geol. 42, 221–246.

Sgroi T., Beranzoli L., Di Grazia G., Ursino A. and Favali P. (2007) New observations of local seismicity by the SN-1 seafloor observatory in the Ionian Sea, off-shore Eastern Sicily (Italy). Geophys. J. Int. 169, 490–501.

Sgroi T., Montuori C., Agrusta R. and Favali P. (2009) Low-frequency seismic signals recorded by OBS at Stromboli volcano (Southern Tyrrhenian Sea). Geophys. Res. Lett. 36, doi: 10.1029/2008GL036477.

Steinmetz L., Ferrucci F., Hirn A., Morelli C. and Nicolich R. (1983) A 550 km long Moho traverse in the Tyrrhenian Sea from O.B.S. recorded Pn waves. Geophys. Res. Lett. 10, 428–431.

Trua, T., Serri G., Renzulli A., Marani M.P. and Gamberi F. (2002) Volcanological and petrological evolution of Marsili seamount (southern Tyrrhenian Sea). J. Volcanol. Geotherm. Res. 114, 441–464.

Verzhbitskii E.V. (2007) Heat flow and matter composition of the lithosphere of the world ocean. Ocean 47(4), 564–570.

Vitale S., De Santis A., Di Mauro D., Cafarella L., Palangio P., Beranzoli L. and Favali P. (2009) GEOSTAR deep seafloor missions: Magnetic data analysis and 1D geoelectric structure underneath the Southern Tyrrhenian Sea. Ann. Geophys. 52(1), 57–63.

Webb S.C. (1998) Broadband seismology and noise under the ocean. Rev. Geophys. 36, 105–142.

Zito G., Mongelli F., de Lorenzo S. and Doglioni C. (2003) Heat flow and geodynamics in the Tyrrhenian Sea. Terra Nova 15(6), 425–432.

25 Development and application of an advanced ocean floor network system for megathrust earthquakes and tsunamis

Y. Kaneda[1,2], K. Kawaguchi[1], E. Araki[1], H. Matsumoto[1], T. Nakamura[1], S. Kamiya[1], K. Ariyoshi[1], T. Hori[1], T. Baba[1] and N. Takahashi[1]

25.1 Introduction

Japan is prone to great earthquakes because of its position near two different subduction zones. The Philippine Sea plate subducts from the southeast, and the Pacific plate subducts from the east. The former was the source of a series of great earthquakes, of which the Tonankai earthquake of 1944 and the Nankaido earthquake of 1946 are the latest events. The latter was the source of the 2011 earthquake off the Pacific coast of Tohoku (Tohoku earthquake) of 11 March 2011 (M9).

The Tonankai and Nankai earthquakes, which are typically considered as a paired event, took place on the Nankai Trough megathrust. Both events had magnitudes of approximately 8, and similar events in these locations have recurrence intervals (e.g., Ando, 1975; Table 25.1, Figure 25.1). These two events induced huge tsunamis on the adjacent coast, and there are many monuments for tsunami victims in the Kochi, Tokushima, Wakayama, and Mie prefectures. The rupture area of the 1944 Tonankai earthquake extended from off the eastern Kii Peninsula to the Enshu-Nada, excluding the Tokai region. Two years later, the 1946 Nankaido earthquake rupture began off the western Kii Peninsula, and then propagated to off the coast of western Shikoku (Hashimoto and Kikuchi, 2002). The previous Nankai Trough megathrust event, the 1854 Ansei earthquake (Ando, 1975) had a rupture pattern similar to the 1944/1946 Tonankai/Nankaido earthquakes. The rupture began off

1 Japan Agency for Marine-Earth Science and Technology, Yokosuka, Japan

2 now at Disaster Mitigation Research Center, Nagoya University, Nagoya, Japan

Year	NANKAIDO	TONANKAI	TOKAI
684	Yes	Possibly	Possibly
887	Yes	None	None
1096	None	Yes	Possibly
1099	Yes	None	None
1361	Yes	None	None
1498	Possibly	Yes	Yes
1605	Yes	Yes	None
1707	Yes	Yes	Yes
1854a	None	Yes	Yes
1854b	Yes	None	None
1944	None	Yes	None
1946	Yes	None	None

Table 25.1 Megathrust earthquake occurrence pattern (after Ando, 1975).

the Tonankai and Tokai areas, then 30 hours later it resumed in the Nankaido area from off the western Kii Peninsula to off Shikoku (Usami, 2003). This pattern, in which the break in the Tonankai area precedes that in the Nankaido area, is perhaps characteristic of the Nankai Trough earthquakes.

This two-step pattern of megathrust seismicity appears to be related to the local plate geometry. Based on the results of structural studies and the seismicity distribution (Special Project for Earthquake Disaster Mitigation in Urban Areas, 2007) (Figure 25.2), the Philippine Sea plate is rather complex. The complex plate interface seemed to affect the spatial and temporal patterns of megathrust earthquakes. By approximating the 3D geometry using variable frictional properties on a planar fault, Hori (2006) numerically simulated the megathrust earthquake patterns and was able to reproduce the pattern of a Tonankai earthquake immediately preceding the Nankaido earthquake. To calculate six earthquake cycles corresponding to 660 years for each model, it took 7.5 hours using 128 nodes (1024 processors) of the Earth Simulator. The processing performance achieved to 3.6 Tflops, which is 44% of the peak speed. In addition, his simulation showed that the recurrence interval of Tonankai/Nankaido earthquake pairs is 90–150 years. While the main features of this simulation are consistent with records of the most recent Tonankai/Nankaido megathrust events, the 2-year gap between the 1944 Tonankai and 1946 Nankaido earthquakes has not yet been reproduced in models. We have no explanation why the delay between the 1944 and 1946 events was so long. Understanding the time delay may be crucial in disaster prevention planning for regional coastal areas.

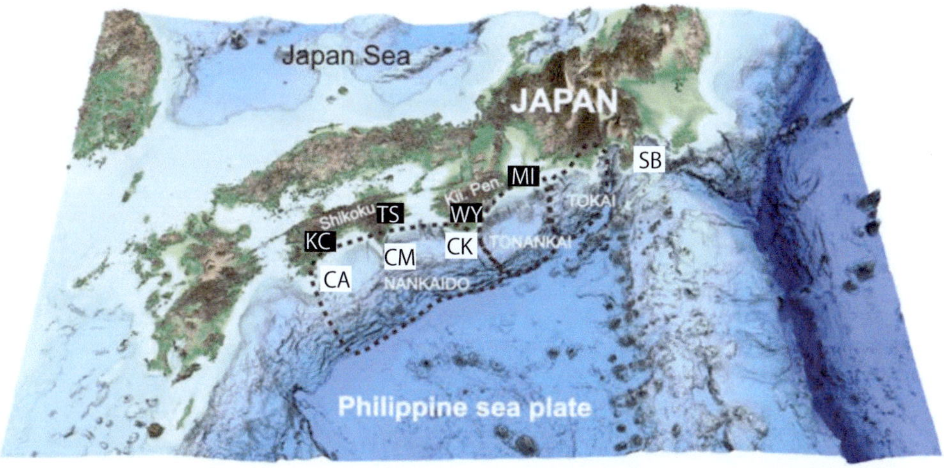

Figure 25.1 Map of the Nankai Trough area. The Philippine Sea plate is subducting beneath the Japan. Three areas surrounded by dotted lines are segments for megathrust earthquakes. CA, CM, CS and SB are Cape Ahizuri, Cape Muroto, Cape Shiono and Sagami Bay, respectively. KC, TS, WY and MI are Kochi, Tokushima, Wakayama and Mie Prefectures.

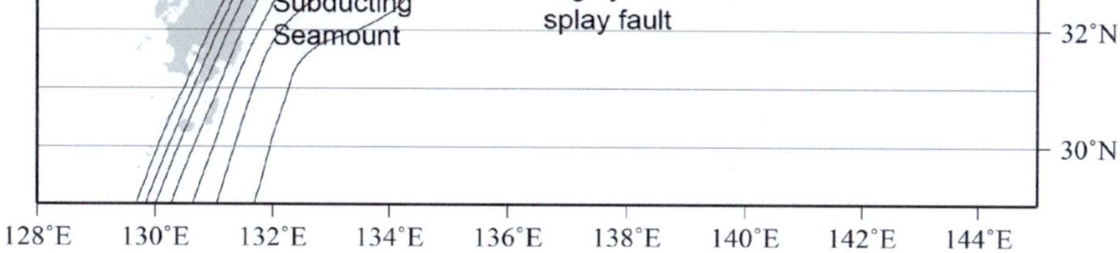

Figure 25.2 Plate geometry of the subducting Philippine Sea plate. Contour lines indicate depth of the plate and its interval is 10km. Numerals are the depth. The ellipses indicate structural characteristics obtained by many seismic surveys.

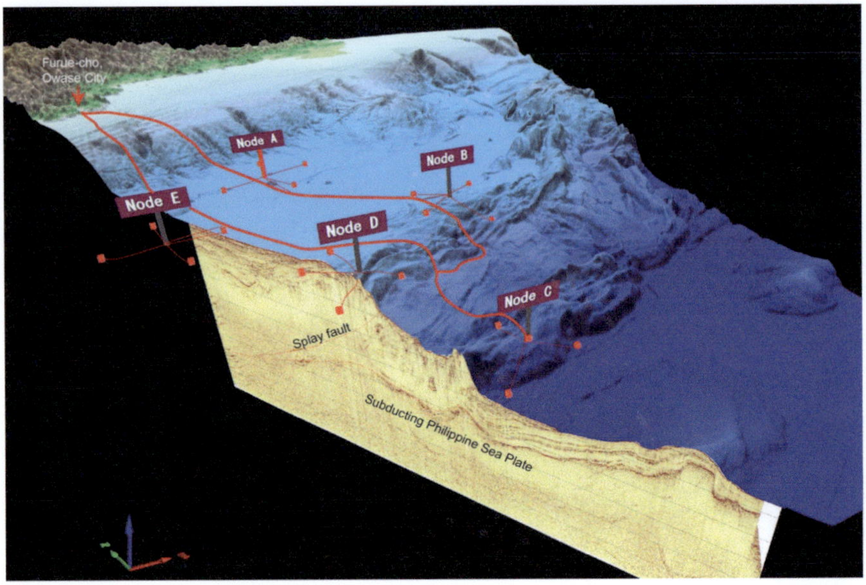

Figure 25.3 Location of DONET. Thick and thin red lines indicate location of backbone cables and extension cables, respectively. Cubes indicate stations. Science nodes are locate at the foot of each signboard. A red arrow indicates the Furue landing station.

In addition, the Nankai Trough region is known for the occurrence of tsunami earthquakes, with little or no strong motion and huge tsunamis. The 1605 Keicho earthquake is considered to be an example. The Tohoku earthquake also generated a huge tsunami as its rupture area extended upward to near the trench (e.g., Ide et al., 2011). In cores drilled in the Nankai Trough megathrust zone, Sakaguchi et al. (2011) found evidence of fast rupture in signs of thermal metamorphism at exits of the plate boundary and splay fault. This suggests that the Nankai Trough region also has a substantial risk of megathrust earthquakes with huge tsunamis like the Tohoku earthquake. Such megathrust earthquake can be identified by old literature and distribution of tsunami sediments; however, the Tohoku earthquake has produced quite a large quantity of victims. To reduce the number of victims, it is essential to monitor in real-time near the source area to detect the large seafloor movement.

The Japan Agency for Marine-Earth Science and Technology (JAMSTEC) has developed several long-term deep seafloor observatories connected by ocean cable networks. The oldest one was constructed off Hatsushima in Sagami Bay (Momma et al., 1994). Similar systems were deployed off Cape Muroto in 1997 (Momma et al., 1997) and off Kushiro, Hokkaido in 1999 (Kawaguchi et al., 2000). These systems, however, have a small number of stations and are not suitable for investigating crustal activity and deformation over large areas. Moreover, they lack redundancy for observation of earthquakes and tsunamis because their inline system incorporates sensors in the main cable, meaning that a single cable break can disable the entire network. At JAMSTEC, we have developed and

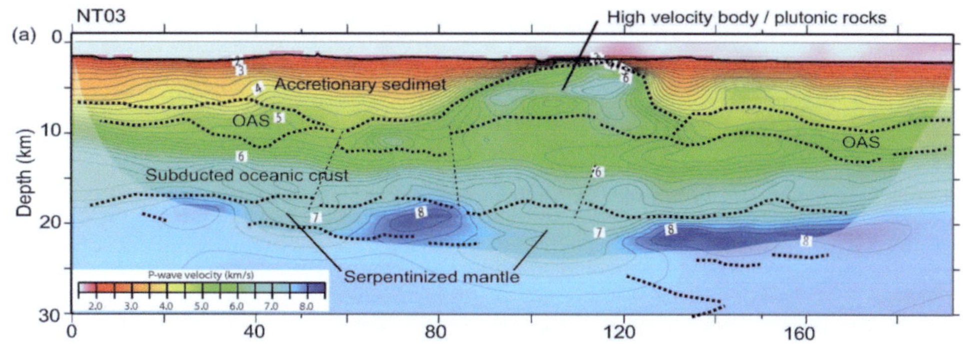

Figure 25.4 Irregular structure located between the Tonankai and Nankai segments (Kodaira et al., 2006). Dotted lines indicate locations of reflectors identified by refraction data. OAS = old accreted sediments.

constructed a new seafloor network system with superior redundancy and ease of replacement, expandability, and maintenance (Kaneda et al., 2009, 2010); it is called the Deep Ocean-floor Network system for Earthquakes and Tsunamis (DONET). The target area is the Tonankai area, where the initial event in megathrust earthquakes occurs (Figure 25.3). DONET has 20 connected stations at present, but it has a total capacity of 40 stations. The stations are distributed from the forearc basin to the deformation front, thus covering both the strong motion area and the tsunami earthquake area. Our main aims for DONET are early detection of earthquakes and tsunamis and development of earthquake models using real-time monitoring data.

25.2 Previous research

The megathrust rupture areas along the Nankai Trough form three seismogenic segments, named Nankaido, Tonankai, and Tokai from west to east (Table 25.1). The rupture zones of the Tonankai and Nankaido earthquakes do not overlap each other (Baba and Cummins, 2005). And the 1946 Tonankai earthquake broke the Tonankai segment and not the adjacent Tokai segment (Ando, 1975). To investigate rupture patterns in the Nankai Trough region, JAMSTEC has carried out seismic surveys using a multichannel reflection system and ocean bottom seismographs since 1997, discovering many heterogeneous structures that may control the rupture patterns. At the segment boundary of the Tonankai/Nankaido earthquake off the Cape Shiono, a dome structure composed of high-velocity materials was detected (Kodaira et al., 2006) (Figure 25.4). In the Tokai region, a series of ridges is being subducted, and the rough topography of the plate boundary appears to control the rupture behavior (Nakanishi et al., 2002; Kodaira et al., 2003). These findings suggest that the

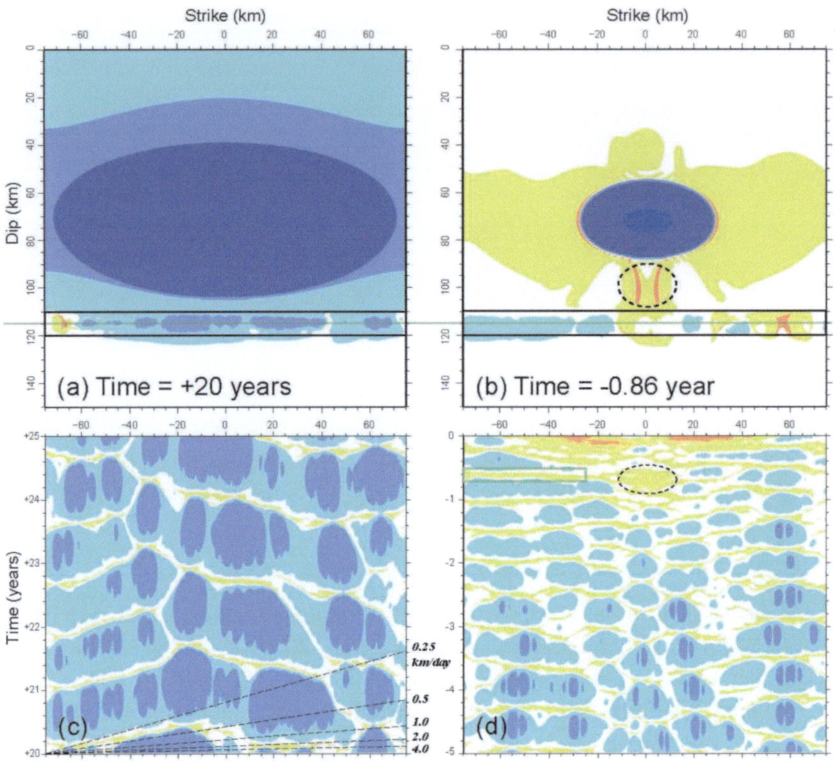

Figure 25.5 Comparison of stick conditions on the plate boundary (Ariyoshi et al., 2010). Blue and red colors indicate struck and broken areas by fast rupture, respectively. (a) Plate boundary condition during struck period. Horizontal axis is parallel to distance along strike of the trough. (b) Plate boundary condition before earthquake occurrence. (c) Spatial and temporal pattern of the slow slip during the struck period. Broken lines indicate indexes of propagation velocity of the slow slip. (d) Spatial and temporal pattern before earthquake occurrence.

irregular structure brings about strong coupling between the plates. In the Nankaido zone, a large seamount is subducting off Cape Muroto (Kodaira et al., 2000) at a location corresponding to the western end of the initial rupture of the 1946 Nankaido earthquake. After stopping there, the rupture resumed propagating to off western Shikoku. Tsunami analysis has also shown that the rupture pattern was strongly heterogeneous (Baba and Cummmins, 2005). There is also irregular structure on the plate boundary off Shikoku (Takahashi et al., 2011) that appears to be closely linked to seismicity patterns. In addition to the topography of the plate boundary, Park et al. (2004) described a splay fault branching from the plate boundary in the Tonankai rupture zone. The results of these studies were incorporated in the plan of the NanTro ocean drilling project in the Nankai Trough. The splay fault was also detected by 3-D structural fine imaging (Moore et al., 2007). Ocean drilling has shown that near-surface traces of the splay fault and plate boundary display evidence of high-velocity slip (Sakaguchi et al., 2011). This is a very significant fact for evaluation of megathrust earthquake models.

Simulations of Nankai Trough megathrust earthquakes showed that the first rupture typically starts in the Tonankai segment and propagates eastward into the Tokai segment and westward into the Nankaido segment (Hori, 2006) In the model, variation of the friction parameter was adopted instead of that of the rough topography on the plate boundary. Understanding that the first rupture favors the Tonankai segment is very important for watching megathrust earthquake occurrence.

Seismicity in the Nankai Trough area is relatively low, but characteristic episodes of low-frequency tremor occur around the plate boundary both updip of the seismogenic zone (including the splay fault) and downdip (Obana and Kodaira, 2007). Ariyoshi et al. (2004) showed that frequencies and propagation velocities of slow events, which include low-frequency tremor and very long period events, increase as the recurrence cycle of megathrust earthquakes progresses (Figure 25.5). This suggests that monitoring of these slow events is useful for evaluation of future megathrust earthquakes. Nakano et al. (2013) showed that seismicity around the DONET area underwent a sustained rise after the Tohoku earthquake. Therefore monitoring of seismicity is also important to understand in situ strain conditions.

Real-time monitoring of the seismogenic zone, in particular the Tonankai segment as the nucleus of the next megathrust earthquake, is very important for understanding in situ strain conditions. Such monitoring should cover the broad area ranging from the coast to the trough axis. And the monitoring has to be useful for detection and observation of many types of signals including crustal deformation, low-frequency events, micro-earthquakes, strong motion events, and tsunamis.

25.3 Configuration of DONET

DONET has been constructed and activated as a real-time monitoring system in the Tonankai segment (Figure 25.3); its objectives are early detection of earthquakes and tsunamis, through ocean floor observation and broadband monitoring, and by development of advanced techniques (Kawaguchi et al., this volume).

Redundancy is an important consideration in the implementation of DONET. During the Tohoku earthquake, some ocean floor network cables were cut by landslides. But DONET can continue observations if the main cable is cut at any location, because its looped design allows electric power supply and data transmission in either direction (Figure 25.6). Moreover, if the sensors, the science nodes linking them to the main cable, or both are damaged or compromised, they are replaceable by using remotely operated and autonomous underwater vehicles. And because each science node has eight ports, it is simple to increase the number of DONET stations as needed.

The other major consideration is accommodating many kinds of sensors to observe a wide range of signals (Figure 25.6). DONET currently has a total of 20 stations (Figure 25.7), consisting of five science nodes each connecting four sensor stations. All stations

Figure 25.6 Sensors used by DONET system. (Upper) Ground motion sensing system. In a pressure-resistant case in the left, strong motion sensor and broadband seismometer attached on the gimbal mechanism. (Lower) Pressure sensing system. On the grating board, pressure gauge, differential pressure gauge, hydrophone and thermometer are attached.

have a ground motion sensing system comprising 3-component strong motion accelerometers and 3-component broad band velocity sensors, and a pressure sensing system including pressure gauges, differential pressure gauges, hydrophone and thermometer. The accelerometers have high-gain and low-gain function to avoid saturation by strong motions. In addition, where possible the ground motion sensing system is buried below the seafloor to reduce noise and improve data quality. The quality of DONET data is better than land stations in general due to the lower-noise environment. Figure 25.8 shows an example of DONET data from an earthquake of M5.7 in northeastern Japan.

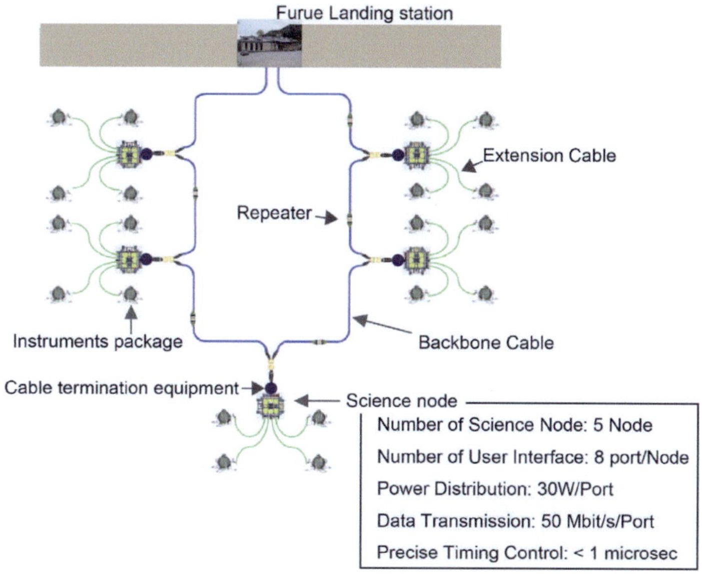

Figure 25.7 Outline of DONET data. From the Furue landing station, the looped backbone cable (blue line) is connected to stations via cable termination equipment and science node. Because each science node has eight ports, we can connect new sensors and stations in response to needs.

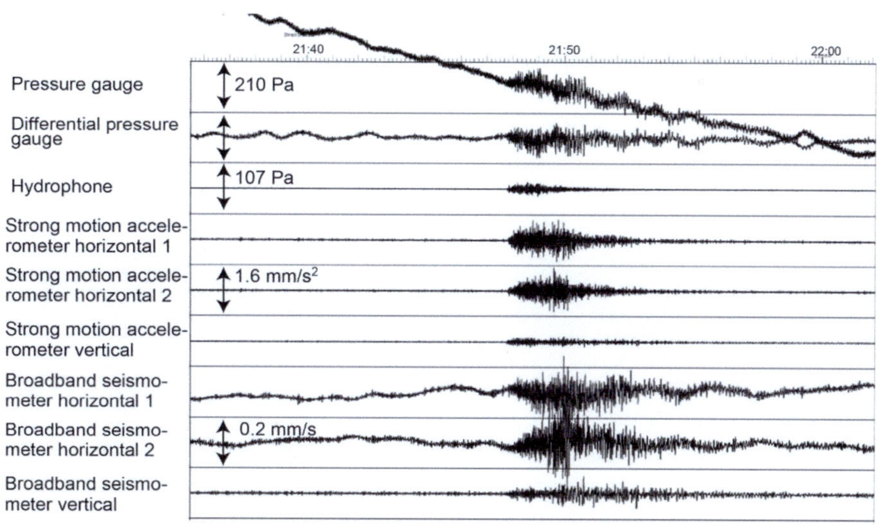

Figure 25.8 Example of seismograms (off Fukushima, M5.7, time: 13 March, 2010) recorded by station KMA3. All traces are filtered by low pass less than 10 Hz.

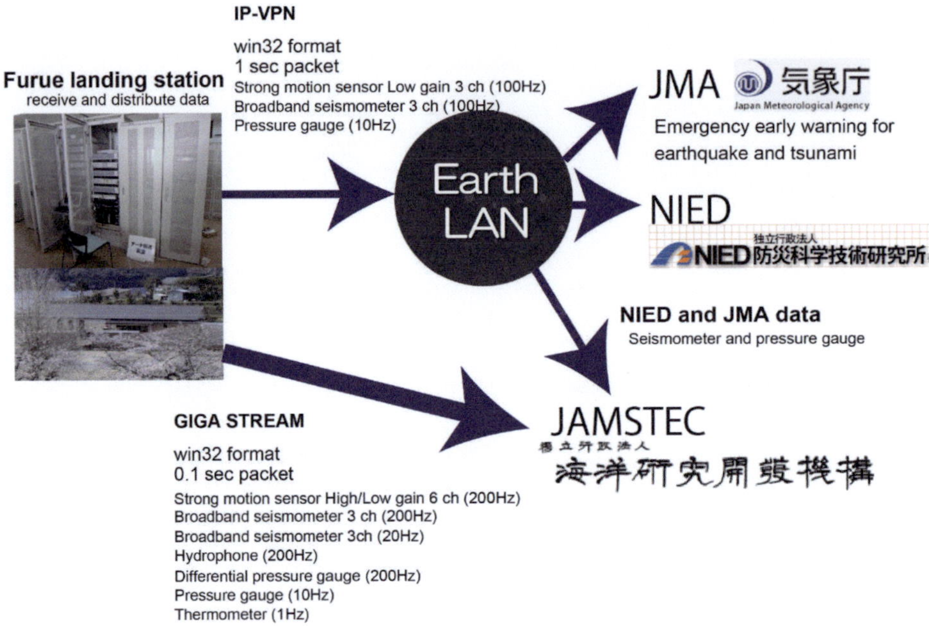

Figure 25.9 DONET data transmission from the Furue landing station. A part of DONET data – strong motion sensor, broadband seismometer and pressure gauge – are sent to JMA and NIED via EarthLAN in real-time.

All data streams are transferred to the Yokohama Institute, JAMSTEC, on a circuit with 10 Mb/s capacities and stored there with a high sampling rate (Table 25.2). Data from strong motion accelerometers and broadband velocity sensors, sampled at 100 Hz, and data from pressure gauges, sampled at 10 Hz, are also transferred to the EarthLAN system, managed by the National Research Institute for Earth Science and Disaster Prevention (NIED) for exchange of earthquake data (Figure 25.9). NIED and the Japan Meteorological Agency (JMA) receive DONET data from EarthLAN in real time in win32 format. A database system is in preparation that will open DONET data to the public.

25.4 Expected results

The Tonankai/Nankaido earthquake rupture area is beneath the ocean, therefore observations from the seafloor offer a great advantage over land-based observations. For early detection of earthquakes and tsunamis for disaster mitigation, the advantage over land-based detection is estimated to be at least a few seconds for earthquakes and about 5–10 min for tsunamis among coastal cities near the Tonankai zone, depending on the location of the focus (Figure 25.10).

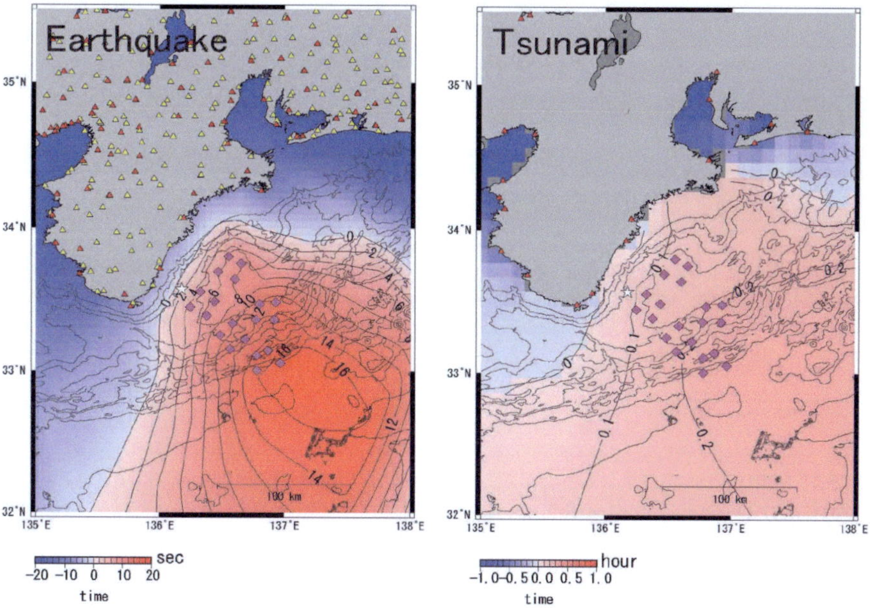

Figure 25.10 Times of early detection for earthquakes and tsunamis compared between events using DONET and without it. When earthquake occurs in the red area, earthquakes and tsunamis can be detected maximum 16 seconds and 12 minutes earlier using DONET data than events using only land stations.

Sensors	Sampling rate (Hz)
Accelerator (high gain)	200
Accelerator (low gain)	200
Broadband velocity sensor	200
Broadband velocity sensor for long term events	20
Pressure gauge	10
Differential pressure gauge	200
Hydrophone	200
Thermometer	1

Table 25.2 Sampling rate of each sensor.

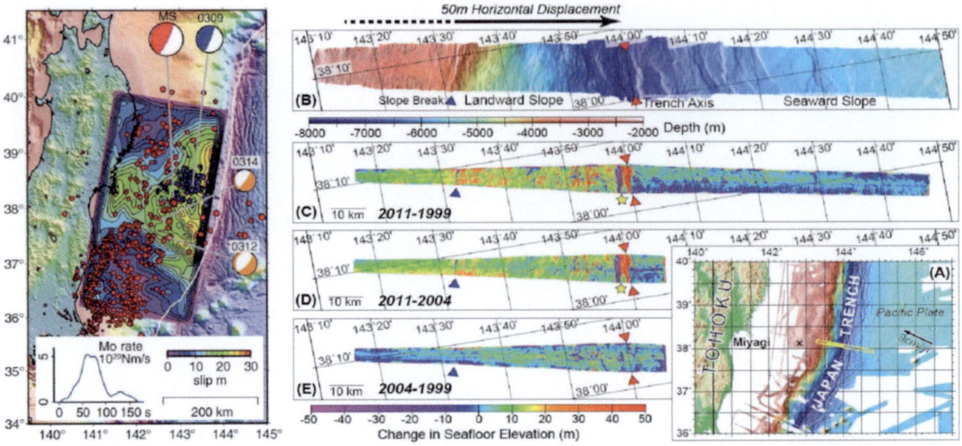

Figure 25.11 Map and seafloor movement of the off Tohoku earthquake (Ide et al., 2011; Fujiwara et al., 2012). (Left) The total slip distribution (Ide et al., 2011). (Right) Seafloor movement (Fujiwara et al., 2012). (A) Line for topographic investigation (yellow line). (B) Current seafloor topography. Red and blue triangles indicate the trench axis and trench slop break. (C) Difference of topography compared with 1999 and 2011. (D) Difference of topography compared with 2004 and 2011. (E) Difference of topography compared with 1999 and 2004.

Here, we introduce an example of the Tohoku earthquake to emphasize the importance of ocean floor observation. As we described above, the Tohoku earthquake generated a huge tsunami. The cause of the tsunami is seafloor movement. By comparing bathymetry before and after the earthquake, Fujiwara et al. (2012) showed that the seafloor near the epicenter area moved eastward with maximum 50m after this earthquake (Figure 25.11). On the other hand, the huge tsunami was recorded by pressure gauges on a cabled off Kamaishi in real time (Figure 25.12). According to these records, there are two types of tsunami height: one is typical tsunami waveform with gradually increasing amplitude, in the other, the amplitude rapidly increases like a pulse on top of the typical tsunami waveform. Probably, the rapid increase of tsunami amplitude is generated by large movement of the ocean floor off Miyagi. Finally, the tsunami's amplitude increased as the waves approached the coast of east Japan, and therefore coastal cities were inundated by a large tsunami and severely damaged. The records of pressure gauges demonstrate the importance of offshore real-time monitoring for the improvement of early warning systems for earthquakes and tsunamis. DONET stations recorded the seismic waves and tsunamis of the Tohoku earthquake (Figures 25.13 and 25.14). Although the tsunami height is amplified at coastal areas, we can estimate the ratio of tsunami amplification from offshore site to onshore site. Normally, the ratio is estimated as approximately 3 to 4 times (Figure 25.15); therefore we can detect tsunamis earlier than a land station using ocean floor network as DONET, and estimate the approximate tsunami amplitude at onshore areas. These advantages are indispensable for disaster reduction of earthquakes and tsunamis.

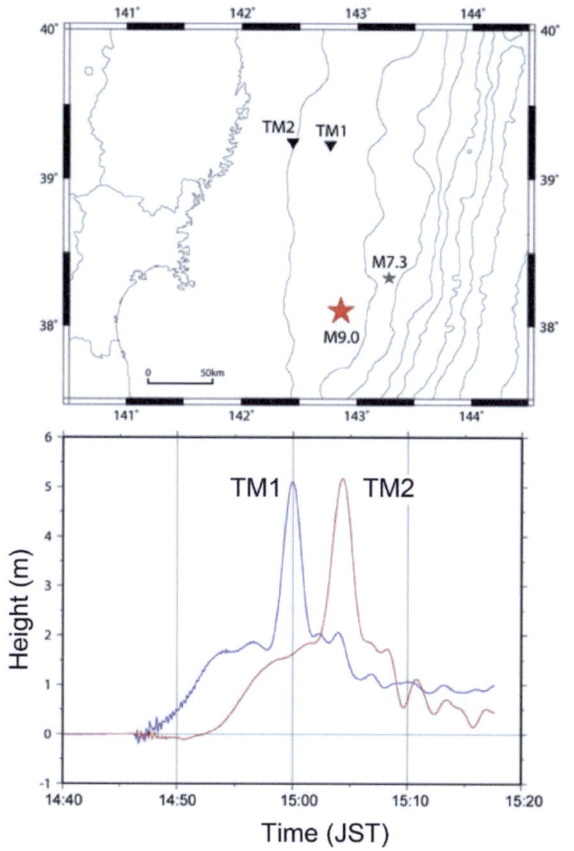

Figure 25.12 Map of the cabled system of off-Kamaishi (upper) and profiles detected by two pressure gauges (lower) (after Homepage of Earthquake Research Institute, Univ. of Tokyo, http://outreach.eri.u-tokyo.ac.jp/eqvolc/201103_tohoku/eng/#sealevel). Blue and red lines indicate profiles of sealevel change detected by TM1 and TM2 stations, respectively.

It is very important to cooperate seafloor observation like DONET and computer simulation technique for actual disaster mitigation (Kaneda et al., 2009). There are two ways to estimate earthquakes and tsunamis for early warning. One is based on comparing simulation results and observed data, and the other is to estimate earthquakes and tsunamis directly using data from many stations. Previous studies have already constructed a detailed structure model around the Tonankai, Nankaido, and Tokai seismogenic zones, and seismic magnitude is estimated from DONET data by simulation of seismic waveforms. For tsunami warnings, it is important to have not only rapid access to first arrivals, but also early estimations of amplitude and duration using amplification of a tsunami as described before.

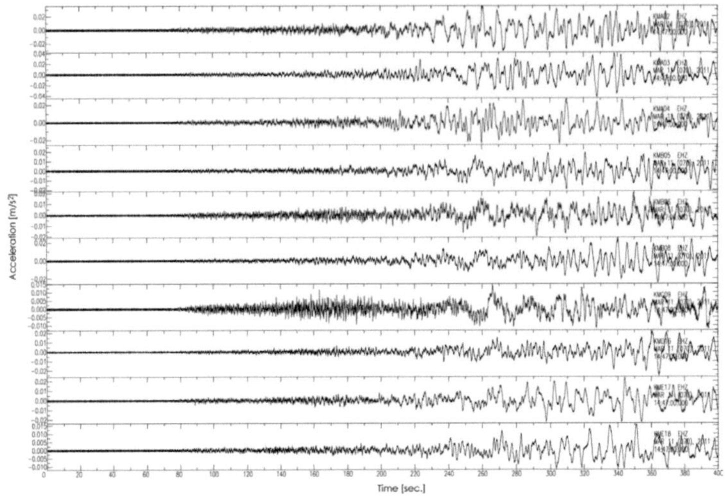

Figure 25.13 Strong motion accelerometers records of DONET stations for the Tohoku earthquake. From top to order, these record are KMA03, KMA04, KMB05, KMB06, KMB08, KMC09, KMD16, KME17, KME17 and KME18.

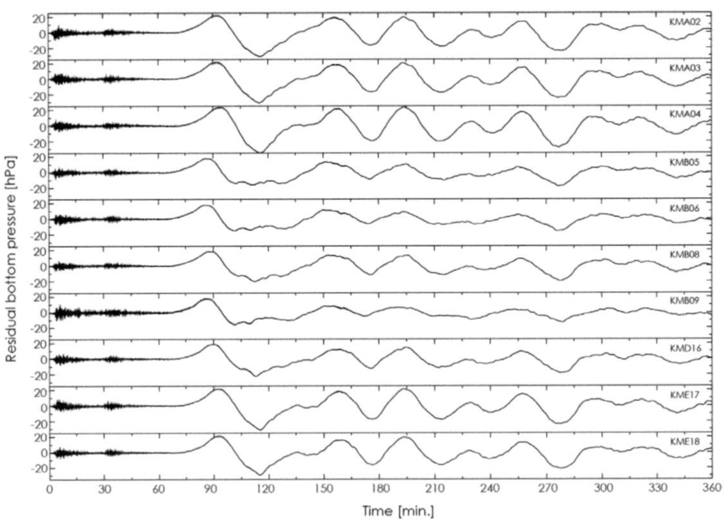

Figure 25.14 Pressure sensor records of DONET stations for the Tohoku earthquake. From the top in order: KMA03, KMA04, KMB05, KMB06, KMB08, KMC09, KMD16, KME17, KME17 and KME18. All records are applied by bandpass filtering with 10,000–100 sec.

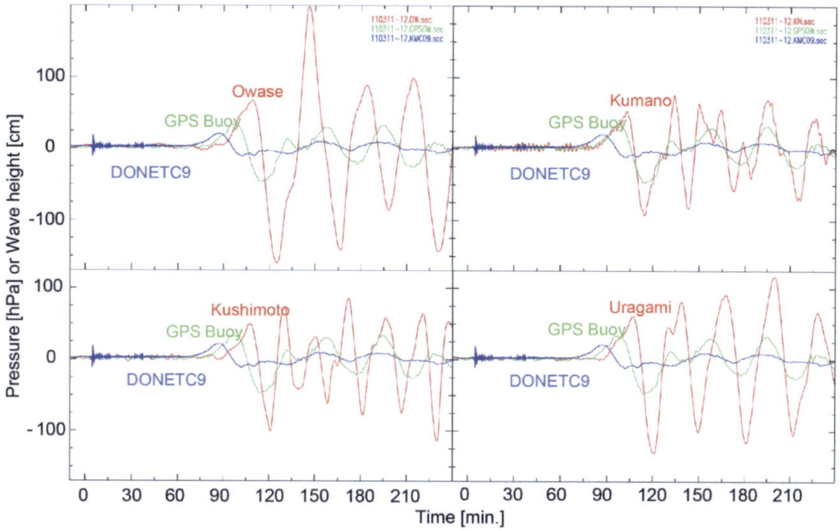

Figure 25.15 Amplification of tsunami compared among DONET, GPS buoy at 20km far from the coastline and tidal gauge of eastern coast of the Kii Peninsula. (Upper left) Owase City. Amplification is approximately 3.5 times. (Upper right) Kumano City. Amplification is approximately 2.75 times. (Lower left) Kushimoto Town. Amplification is approximately 2.5 times. (Lower right) Uragami Town. Amplification is approximately 3.2 times.

Figure 25.16 shows the results of a simulation of the megathrust earthquake recurrence cycle in the Nankai Trough. The complex patterns of the simulation results are qualitatively consistent with the historical earthquakes around the Nankai Trough (Hori, 2006); however, they are not quantitatively consistent, which means that we must improve the simulation model for practical applications. To do this, we first use old documents to study historical earthquakes to understand tsunami magnitudes, crustal deformations, and seismic density. Then we try to semiquantitatively compare the simulation results with this historical information for advanced prediction research. The broadband sensors are very useful in understanding crustal activities. In particular, the pressure data constitute the best sensor for data assimilation representing the vertical component of crustal deformations (Figure 25.17). Data assimilation has two purposes: the determination of initial conditions and the tuning of simulation models. The goal is to achieve these purposes by using a kind of Kalman filter. However, at first we are developing the initial data assimilation method using simple models. In data assimilation, long-term observations are very important for improving the recurrence cycle simulation model around the Nankai Trough. Ocean floor deformation data are quite useful for data assimilation because of their continuity and high quality. Soon we hope to develop advanced data assimilation using both ocean floor deformation data and seismicity for highly reliable simulations of the recurrence cycle of megathrust earthquakes in the Nankai Trough region.

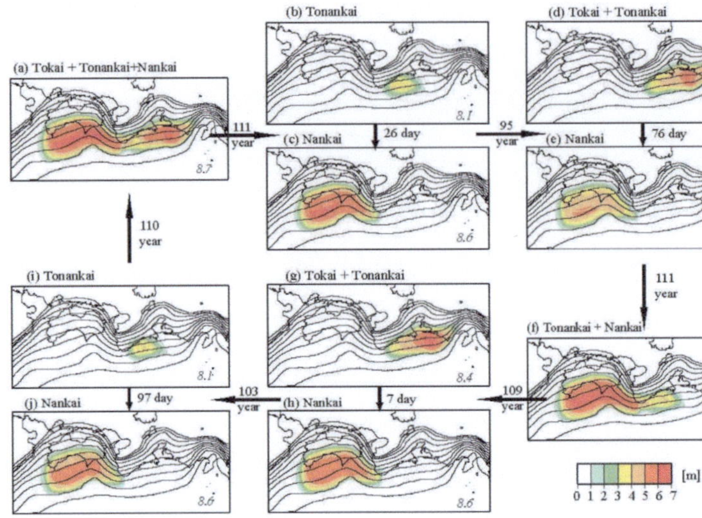

Figure 25.16 Simulated pattern of the megathrust earthquake occurrence from (a) to (j) (Hori, 2006). Italic numerals are calculated moment magnitude. Contour lines indicate depth of the Philippine sea plate with an interval of 10km. Colored area shows distribution of fast rupture, and red shows maximum slip of 6–7m. These figures show patterns of the large earthquake occurrence. After the Tonankai and Nankaido events, whose time interval is significantly short, the time interval becomes longer (from 26 days to 76 days or from 7 days to 97 days) for the following two earthquake cycles. The recurrence interval becomes shorter (from 111 years to 95 years or from 109 years to 103 years) in the two cycles. These cycles are qualitatively consistent with the historical ones from 1707.

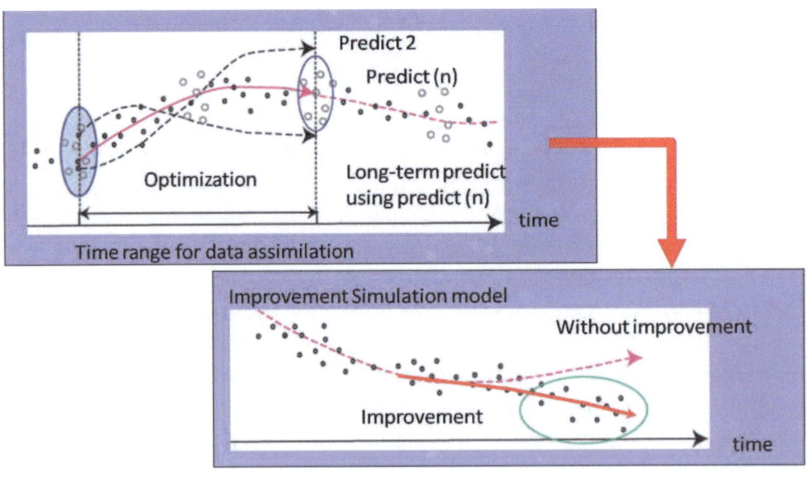

Figure 25.17 Schematic figures of data assimilation. Future condition is predicted using some models and the model are improved by new observed DONET data.

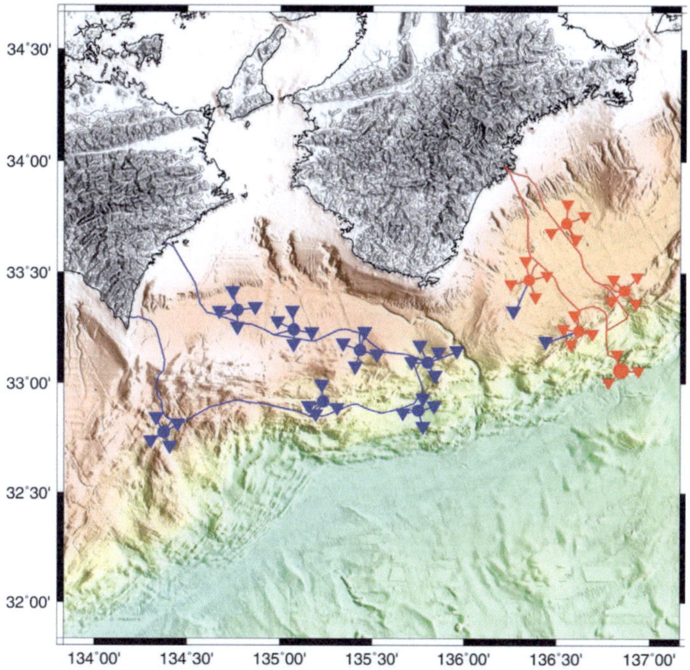

Figure 25.18 Locations of DONET (red) and DONET2 (blue), which is under construction. Circles and reverse triangles indicate locations of science nodes and stations, respectively.

Observation with multiple kinds of sensors may enable us to advance basic science in unexpected ways by using new information from the Earth. For example, it is possible to analyze interactions between the crust and upper mantle by using free oscillation signals from large, distant earthquakes. DONET is a useful tool to research not only earth science but biology. And pressure gauges and hydrophones may provide unique new information on the life of ocean mammals such as whales and dolphins.

25.5 Summary and future plans

We constructed DONET with 20 stations deploying many types of sensors around the Tonankai seismogenic zone, aiming for early detection of earthquakes and tsunamis and improvements in simulations of the repeated occurrence of great earthquakes. We are receiving high-quality data from these stations. In addition to carrying out continuous observations, we are studying numerical simulation techniques, including data assimilation, to realize forecasts of large earthquakes.

To detect small signals, like possible slow slip before the occurrence of megathrust earthquakes, we have to obtain data of high resolution and sensitivity. Therefore, JAMSTEC developed a borehole system composed of accelerometer, broadband seismometer, strain meter and thermometer, which was installed by D/V "Chikyu" in the Nankai Trough area. We have a plan to connect the sensors in a borehole to DONET and realize real-time monitoring of borehole data.

However, the DONET system alone is not enough for monitoring the entire Nankai Trough seismogenic zone from off Tokai to off Kyushu, especially the Nankaido segment. Therefore, we plan to develop and deploy DONET2 off the eastern part of Shikoku. Figure 25.18 shows the concept of DONET2, with its high voltage power system, and the DONET2 array. DONET2 will be even more readily expandable than DONET. In DONET2, each station will have the capacity for over 100 sensors. The project of an off-East Japan cabled network, which is an inline system and different from DONET, is starting to monitor the Japan Trench seismogenic zones. In the more distant future, we foresee deploying ocean floor observatories off Kyushu, because the rupture zone of the 1707 Hoei earthquake, based on recent research such as historical tsunami sediment analyses and tsunami simulations, seems to have extended as far as Kyushu (Furumura et al., 2011). Plans for monitoring the seismogenic zone in southwestern Japan, including the Okinawa Islands, using such as DONET3, DONET4 will be developed in the future.

Plans for an international network are already starting among our DONET, the Taiwanese MACHO system, the NEPTUNE Canada array off western Canada, the Ocean Observatories Initiative (OOI) of the United States, and the European ESONET project. To maximize scientific progress and advances in technology, it is important to foster collaboration among these ocean floor networks by exchanging researchers, engineers, and data.

Acknowledgments

The contract research "Development of Dense Ocean Floor Network System for Earthquakes and Tsunamis" is supported by the Ministry of Education, Culture, Sports, Science and Technology, Japan. We thank the Japan Meteorological Agency and Ministry of Land, Infrastructure, Transport and Tourism for providing us with data of the 2011 Tohoku-oki tsunami from the tide gauges and the GPS buoy.

References

Ando M. (1975) Source mechanisms and tectonic significance of historical earthquakes along the Nankai trough, Japan. Techtonophysics 27, 119–140.

Ariyoshi K., Hori T., Kaneda Y., Ampuero J.-P., Matsuzawa T., Hino R. and Hasegawa A. (2010) Character of deep low-frequency tremors based on chain reaction model. SENAC, 19–24.

Baba T. and Cummins P.R. (2005) Contiguous rupture areas of two Nankai Trough earthquakes revealed by high-resolution tsunami waveform inversion. Geophys. Res. Lett. 32, L08305, doi: 10.1029/2004GL022320.

Furumura T., Imai K. and Maeda T. (2011) A revised tsunami source model for the 1707 Hoei earthquake and simulation of tsunami inundation of Ryujin Lake, Kyushu, Japan. J. Geophys. Res. 116, B02308, doi: 10.1029/2010JB007918.

Hashimoto T. and Kikuchi M. (2002) Source process of the 1946 Nankai earthquake estimated from seismograms. J. Geography 111(1), 118–125.

Hori T. (2006) Mechanisms of separation of rupture area and variation in time interval and size of great earthquakes along the Nankai Trough, southwest Japan. J. Earth Simulator 5, 8–19.

Ide S., Baltay A. and Beroza G.C. (2011) Shallow dynamic overshoot and energetic deep rupture in the 2011 M_w9.0 Tohoku-Oki earthquake. Science 332, doi: 10.1126/science.1207020.

Kaneda Y., Kawaguchi K., Araki E., Matsumoto H., Nakamura T., Kamiya S., Ariyoshi K. and Hori T. (2009) Toward understanding mega thrust earthquakes occurrence system and disaster mitigation around the Nankai trough Southwestern Japan – DONET (Dense Ocean Network for Earthquakes and Tsunamis) and Next DONET. ICES International Symposium, C1.

Kaneda Y. Hirahara K. and Furumura T. (2009) New research project for evaluating seismic linkage around the Nankai Trough – Integration of observation, simulation, and disaster mitigation. Journal of Disaster Research 61–66.

Kaneda Y., Kawaguchi K., Araki E., Matsumoto H., Nakamura T., Kamiya S., Ariyoshi K. and Hori T. (2010) Dense Ocean floor Network System for Mega Thrust Earthquakes & Tsunamis (DONET) – Toward understanding mega-thrust earthquakes, the geohazard and disaster mitigation. SubOptic 2010.

Kawaguchi K., Hirata K., Mikada H., Kaiho Y. and Iwase R. (2000) An expandable deep seafloor monitoring system for earthquake and tsunami observation network. OCEANS 2000 MTS/IEEE Conference 3, pp. 1719–1722.

Kawaguchi K., Kaneko S., Nishida T. and Komine T. (2014) Construction of the DONET real-time seafloor observatory for earthquakes and tsunami sonitoring. (This volume)

Fujiwara T., Kodaira S., No T., Kaiho Y., Takahashi N. and Kaneda Y. (2012) The 2011 Tohoku-Oki earthquake: Displacement reaching the trench axis. Science 334, doi: 10.1126/science.1215141.

Kodaira S., Takahashi N., Nakanishi A., Miura S. and Kaneda Y. (2000) Subducted seamount imaged in the rupture zone of the 1946 Nankaido earthquake. Science 289, 104–106.

Kodaira S., Nakanishi A., Park J.-O., Ito A., Tsuru T. and Kaneda Y. (2003) Cyclic ridge subduction at an inter-plate locked zone off central Japan. Geophys. Res. Lett. 30, doi: 10.1029/2002GL016595.

Kodaira S., Hori T., Ito A., Miura S., Fujie G., Park J., Baba T., Sakaguchi H., Kaneda, Y. (2006) A cause of rupture segmentation and synchronization in the Nankai trough revealed by seismic imaging and numerical simulation, J. Geophys. Res. 111, B09301, doi: 10.1029/2005JB004030.

Momma H., Mitsuzawa K., Kaiho Y. and Hotta H. (1994) Deployment and long observation of long-term deep sea floor observatory off Hatsushima island in Sagami Bay. JAMSTEC J. D.S.R. 10, 363–371 (in Japanese).

Momma H., Fujiwara N., Iwase R., Kawaguchi K., Suzuki S., Kaiho Y. and Kinoshita H. (1997) Long-term deep sea floor observatory off Muroto Cape. JAMSTEC J. D.S.R. 13, 721–731 (in Japanese).

Moore G..F., Bangs N.B., Taira A., Kuramoto S., Pangborn E. and Tobin H. (2007) Three-dimensional splay fault geometry and implications for tsunami generation. Science 318.

Nakanishi A., Shiobara H., Hino R., Kasahara J., Suyehiro K. and Shimamura H. (2002) Crustal structure around the eastern end of coseismic rupture zone of the 1944 Tonankai earthquake. Tectonophysics 354, 257–275.

Nakano M., Nakamura T., Kamiya S., Ohori M. and Y. Kaneda (2013) Intensive seismic activity around the Nankai trough revealed by DONET ocean-floor seismic observations. Earth Planets Space 65, 5-15.

Obana K. and Kodaira S. (2009) Low-frequency tremors associated with reverse faults in a shallow accretionary prism. Earth Planet Sci. Lett. 287, 168–174.

Obara K. (2002) Nonvolcanic deep tremor associated with subduction in southwest Japan. Science 296, 1679–1681.

Park J.-O., Tsuru T., Kodaira S., Cummins P.R. and Kaneda Y. (2002) Splay fault branching along the Nankai subduction zone. Science 297(5584), 1157–1160.

Sakaguchi A., Kimura G., Strasser M., Screaton, E. J., Curewitz D. and Murayama M. (2011) Episodic seafloor mud brecciation due to great subduction zone earthquakes. Geology 39, 919–922, doi: 10.1130/G32043.1.

Special Project for Earthquake Disaster Mitigation in Urban Areas (DAIDAITOKU) (2007) Construction of crustal structural model in offshore region of Japan for prediction of strong motion during subdcution zone earthquakes. Progress Report in 2006, pp.551–558.

Takahashi N., Kodaira S., Nakanishi A., Park J.-O., Miura S., Tsuru T., Kaneda Y., Suyehiro K., Kinoshita H., Hirata N. and Iwasaki T. (2011) Seismic structure of western end of the Nankai trough seismogenic zone. J. Geophys. Res. 107, doi: 10.1029/2000JB000121.

Usami T. (2003) *Compendium of Damage Earthquake in Japan*. University of Tokyo Press (in Japanese).

26 Concluding Remarks: Perspectives and long-term vision

P. Favali[1], L. Beranzoli[1], M.M.R. Best[2,3], J.R. Delaney[4], A. De Santis[1], A.W. Edwards[5] and K. Suyehiro[6,7]

26.1 Vision

As we launch into the 21st century, humankind is faced by unprecedented challenges. As our perspective on our planet approaches the global scale, both scientifically and socially, we realize that in the impact of our behavior we are a geoforming species. At the same time, our understanding of the whole planet is in its infancy. This is particularly the case for the oceans, due to the logistical limitations of studying an environment in which we do not directly live.

In the face of these challenges, this volume brings hope in summarizing the vast volume of work being undertaken to increase our understanding of Earth system processes through the development of interactive ocean networks. These networks can deliver the real-time, multidisciplinary, coordinated data required to understand these complex systems.

1 Istituto Nazionale di Geofisica e Vulcanologia, Italy

2 previously NEPTUNE Canada, Victoria, Canada

3 Consultant, and Adjunct in Department of Earth Sciences, Laurentian University, Sudbury, Canada

4 University of Washington, School of Oceanography, Seattle, USA

5 European Commission, Directorate General for Research and Innovation, Brussels, Belgium

6 previously Integrated Ocean Drilling Program Management International, Inc., Tokyo, Japan

7 Japan Agency for Marine-Earth Science and Technology, Japan

26.2 Visionaries and progress

Visionaries, many included in this volume, saw these challenges and this hope, and thanks to their efforts over many decades a growing network of networks has been ushered in throughout the oceans. Major long-term programs and infrastructure have been, or are being, built around the world, such as in Europe, Canada, the United States, Japan, China, Taiwan and Australia. These facilities, both large and small, provide power and internet to remote environments we otherwise have difficulty studying, allowing continuous multi-variate measurements and their continuous archiving over decades. They include sensors that capture everything from the changing pressure of tsunami waves to close-up video of extreme life at hydrothermal vents. More importantly, they have the capacity to host sensors we have not yet invented. These data allow us to address many of the present fundamental questions on our planet, and also recognize the interconnection between the oceans and the rest of the planet, including humankind.

26.3 Challenges

The challenges of such distributed infrastructures are numerous. The obvious challenge is that of the operation of increasingly complex technical systems in the remote, harsh environments of the marine domain. These challenges are analogous, but different, to those faced in space. Despite technical challenges, these ocean observatories are a reality – up and running and streaming terabytes of data.

While the subsea infrastructure is key, the impact is felt through the resulting data – unprecedented data which is freely and openly available to the world. These terabytes of "Big Ocean Data" pose challenges in archiving and distribution, and particularly in analysis. As the observatory infrastructures form at a local to national scale, there is a need to coordinate the delivery of data such that regional to global pictures can emerge. The capacity to analyze huge amounts of ocean data is in its infancy and it is a challenge that will need the involvement of analysts from around the world. The required capacity is still being built through the development of what can be called Earth/Ocean, or Geo, Informatics. This analysis builds upon present or developing capabilities in other fields, but has the challenge of addressing the range and complexity of what is encompassed by ocean data.

As such, the hardware and software networks cannot bear fruit without the challenges of culture change and capacity building within the human networks of analysts. This Big Data challenge requires a shift from specialization to multidisciplinary and multisectorial science, and a shift towards a new means of data collection and analysis. There are subgroups that provide examples of paths forward, such as seismologists or mid-ocean ridge scientists. However, this challenge of scientific culture change should not be underestimated. It should also be pointed out that the term "Ocean Scientist" covers a very broad group of people, the breadth of which would be analogous to all those disciplines that work on land – it is just our own terrestrial bias that causes us to cluster the various groups together. That said, the level of cooperation and discussion among these very different "Ocean Scientists" is remarkable.

26.4 Public safety

The drive for these observatories has often been, firstly, one of public safety; for example, they often include earthquake and tsunami detection systems. Data from these ocean observatories are therefore immediately relevant to the modeling and early warning systems for such risks. However, as this kind of data again is not typical of that collected in the past, significant work is also involved in integrating the data streams and analyses into public safety systems. The data are also valuable for our long-term understanding and modeling of processes, both acute processes such as earthquakes and tsunamis, but also those that may be slower developing risks such as dead zones and increasing CO_2. In addition, an increasing collaboration between scientists and decision makers will be the basis for a better informed society to mitigate future disasters.

26.5 Paradigm shift

As such, a mark of the 21st century will be this change in our perspective of the planet. This new interactive observation approach is driving a paradigm shift in the way we study the ocean and the ways in which we formulate our questions about the ocean, as we shift from proprietary data, or samples from individual projects, to real-time, multidisciplinary, continuous coordinated suites of shared data from shared infrastructures. This, in turn, drives changes in how we analyze data, requiring a suite of sophisticated quantitative tools to cope with this influx. It requires us to rethink how we develop conceptual models of our growing understanding, how we ask our questions, and how we find solutions. Because of such high resolution continuous sampling, we can now ask questions at a range of time scales, elucidating the various processes contained in these data. We can now pose multivariate questions in space and time, rather than focusing on single data streams. Data and concepts are no longer confined in a small spatio-temporal volume, but extended to the entire Earth's dimension and age. We can now also improve the data input into what has traditionally been the largely satellite-based surficial picture of global observing systems and models. Limited data integration to date indicate that large parameters can only be constrained by volumetric time series ocean data (e.g., carbon cycle and CO_2 sequestration). Even if there is currently a relative paucity of locations, the available data already show the importance to have in situ measurements, especially in deep sea, to focus the phenomena and – overall – their time variability, fixing and better constraining the predictive models. The future is now, and we are sure that more and more discoveries will come.

26.6 Historical significance

We are pushing the boundaries of technology, knowledge, scientific culture and our human sense of our place in our world. The innovation of interactive ocean observing systems will mark this century – a century when we really started to glimpse the full workings of our planet below, at and above the seafloor. Full ocean exploration and understanding will be the real bridge to fill the gap between an inaccessible and unknown planet and a complex, but also wonderful, world still open to new surprises for humankind. It is and will be the duty of ocean scientists to assess both the risks and opportunities to come, in order to preserve and pass on this marvellous heritage to the next generations. Seafloor observatories, together with the technology involved, are the key to allow us to access such a future and the means to make it a reality. A new window is open on the Blue Planet and "A new vision of the Earth from the Abyss" is in our hands.

Acknowledgments

The Editors wish to thank all the referees for their precious and fruitful work that greatly helped to improve the quality and the clarity of this book.

List of the referees in alphabetical order

1 **Tiago Alves**, School of Earth and Ocean Sciences, Cardiff University, UK

2 **Chris R. Barnes**, School of Earth and Ocean Sciences, Victoria University, Victoria, BC, Canada

3 **Mairi M.R. Best**, Department of Earth Sciences, Laurentian University, Sudbury, ON, Canada

4 **Robert Branton**, Ocean Tracking Network; Bedford Institute of Oceanography, Fisheries and Oceans, Dalhousie University, Halifax, NS, Canada

5 **M. Namık Çağatay**, EMCOL, Faculty of Mines, Istanbul Technical University, Istanbul, Turkey

6 **Angelo Camerlenghi**, Istituto Nazionale Oceanografia e Geofisica Sperimentale, Trieste, Italy

7 **Antonio Capone**, Università "Sapienza", Roma, Italy

8 **Arnaud Chulliat**, Institut de Physique du Globe de Paris, Paris, France

9 **Andrew M. Clark**, CSNet International Inc., Melbourne, FL, USA

10 **John A. Collins**, Woods Hole Oceanographic Institution, Woods Hole, MA, USA

11 **Rodolfo Console**, Centro di Geomorfologia Integrata per l'Area del Mediterraneo, Potenza, Italy

12 **Dirk de Beer**, Max Planck Institute for Marine Microbiology, Bremen,, Germany

13 **Holly Given**, Scripps Institution of Oceanography, UC San Diego, La Jolla, CA. USA

14 **Bruce Howe**, School of Ocean and Earth Science and Technology, University of Hawaii at Mãnoa, Honolulu, HI, USA

15 **S. Kim Juniper**, School of Earth and Ocean Sciences, Victoria University; NEPTUNE Canada, Victoria, BC, Canada

16 **Yoshiyuki Kaneda**, Disaster Mitigation Research Center, Nagoya University, Nagoya, Japan

17 **Junzo Kasahara**, Shizuoka University, Shizuoka, Japan

18 **Uli Katz**, Erlangen Centre for Astroparticle Physics, Universität Erlangen-Nürnberg, Erlangen, Germany

19 **William J. Kirkwood**, Monterey Bay Aquarium Research Institute, Moss Landing, CA, USA

20 **Achim J. Kopf**, MARUM - Zentrum für Marine Umweltwissenschaften der Universität Bremen, Bremen, Germany

21 **Frank J. Lowes**, University of Newcastle upon Tyne, UK

22 **Vasilios Lykousis**, Hellenic Centre for Marine Research, Anavyssos, Athens, Greece

23 **Antoni Mànuel Làzaro**, Universitat Politècnica de Catalunya, Barcelona, Spain

24 **Luis M.H.M. Matias**, Lisbon University, Instituto Dom Luiz, Lisbon, Portugal

25 **Hitoshi Mikada**, Department of Civil and Earth Resources Engineering, Kyoto University, Kyoto, Japan

26 **Peter Mikhalevsky**, Science Applications International Corporation, McLean, VA, United States

27 **Jaume Piera Fernández**, Instituto de Ciencias del Mar, Consejo Superior de Investigaciones Científicas, Barcelona, Spain

28 **Scott McLean**, Ocean Networks Canada Innovation Centre, Victoria, BC, Canada

29 **Imants G. Priede**, Institute of Biological & Environmental Sciences, Oceanlab, Aberdeen University, Aberdeen, UK

30 **Giorgio Riccobene**, Laboratori Nazionali del Sud-Istituto Nazionale di Fisica Nucleare, Catania, Italy

31 **Barbara Romanowicz**, Seismological Laboratory, UC Berkeley, Berkeley, CA, United States

32 **Henry A. Ruhl**, National Oceanography Centre, Natural Environment Research Council, Southampton, UK

33 **Mario Sedita**, Laboratori Nazionali del Sud-Istituto Nazionale di Fisica Nucleare, Catania, Italy

34 **Masanao Shinohara**, Earthquake Research Institute, Tokyo University, Tokyo, Japan

35 **Ralph A. Stephen**, Woods Hole Oceanographic Institution, Woods Hole, MA, USA

36 **Christian Tamburini**, Aix Marseille Université, CNRS, Université de Toulon, Marseille, France

37 **Erwan Thébault**, Institut de Physique du Globe de Paris, Paris, France

38 **Christoph Waldmann**, MARUM - Zentrum für Marine Umweltwissenschaften der Universität Bremen, Bremen, Germany

39 **Kelin Wang**, Pacific Geoscience Centre, Geological Survey of Canada, Sidney, BC, Canada

The Editors wish to express their gratitude to **Silvia Nardi** from Istituto Nazionale di Geofisica e Vulcanologia for her constant work as Editorial Assistant.

They also would like to thank **Janet Sterritt** from Springer for her contribution and help in promoting and publicizing the book.

The quality of the work by **David Peduzzi**, **Jim Wilkie** and **Rachael Wilkie** has been particularly appreciated.

Special thanks go to **Philippe Blondel** from the Centre for Space, Atmospheric & Oceanic Science, Department of Physics, University of Bath, Bath, UK for his invaluable comments on all chapters.

The Editors are deeply indebted to **Clive Horwood** from Praxis Publishing, who has enthusiastically encouraged and supported this initiative at every step.

And finally, the Editors gratefully acknowledge all of the authors for their hard and valuable work and for contributing their ideas and expertise to the book.

Index

N

O

P

R

S

T

U

MIX
Papier aus verantwortungsvollen Quellen
Paper from responsible sources
FSC® C105338

If you have any concerns about our products,
you can contact us on
ProductSafety@springernature.com

In case Publisher is established outside the EU,
the EU authorized representative is:
Springer Nature Customer Service Center GmbH
Europaplatz 3, 69115 Heidelberg, Germany

Printed by Libri Plureos GmbH
in Hamburg, Germany